Springer Series in Advanced Microelectronics 30

The Springer Series in Advanced Microelectronics provides systematic information on all the topics relevant for the design, processing, and manufacturing of microelectronic devices. The books, each prepared by leading researchers or engineers in their fields, cover the basic and advanced aspects of topics such as wafer processing, materials, device design, device technologies, circuit design, VLSI implementation, and subsystem technology. The series forms a bridge between physics and engineering and the volumes will appeal to practicing engineers as well as research scientists.



For other titles published in this series, go to
http://www.springer.com/series/4076

Xingcun Colin Tong

# Advanced Materials for Thermal Management of Electronic Packaging

Springer

Xingcun Colin Tong
Laird Technologies
1751 Wilkening Court
Schaumburg, IL 60173
USA
Colin.Tong@hotmail.com

ISSN 1437-0387
ISBN 978-1-4419-7758-8 e-ISBN 978-1-4419-7759-5
DOI 10.1007/978-1-4419-7759-5
Springer New York Heidelberg Dordrecht London



*Cover design:* eStudio Calamar S.L.

Printed on acid-free paper

Springer is part of Springer Science+Business Media (www.springer.com)

*This book is dedicated to my wife Dali, our daughter Lingbo, and our sons William and Alan. Their love fully filled my heart during the long hours of work on this book*

# Preface

The need for advanced thermal management materials in electronic packaging has been widely recognized as thermal challenges became barriers to the electronic industry's ability to provide continued improvements in device and system performance. With increased performance requirements for smaller, more capable, and more efficient electronic power devices, systems ranging from active electronically scanned radar arrays to web servers all require components that can dissipate heat efficiently. This requires that the materials have a high capability for dissipating heat and maintaining compatibility with the die and electronic packaging. In response to these critical needs, revolutionary advances in thermal management materials and technologies for active and passive cooling now promise integrable and cost-effective thermal management solutions. As a result, a large number of papers, articles, and presentations have been published on the development of high-performance materials to solve the vexing problem of device and package-level cooling and thermal management. However, no comprehensive and accessible book has been available on this topic for students, materials scientists, and electronics engineers.

To meet this need, *Advanced Materials for Thermal Management of Electronics Packaging* takes a systems approach ranging from thermal management fundamentals to a balance between cost and performance in materials selection and assessment. Chapter 1 begins with an outline of heat transfer theory and discusses thermal management solutions, materials selection, and component design guidelines. Chapter 2 provides an extensive review of assessment techniques and characterization methodologies for advanced thermal management materials and components. Chapter 3 provides an overview of the state of the art of high-performance advanced electronic packaging materials and their thermal management functions, including properties of key materials, state of maturity, applications, processing, and future directions. Chapters 4 through 8 provide an in-depth introduction to the large and increasing number of advanced thermal management materials, including carbonaceous materials and carbon matrix materials, thermally conductive polymer matrix composites, high thermal conductivity metal matrix composites, ceramic composites, and emerging thermal interface materials. Chapters 9 through 11 discuss advanced materials and design for heat spreaders, air cooling heat sinks, liquid cooling, and thermoelectric cooling devices. Finally, Chapter 12 presents a development roadmap with applications, trends, and perspectives on the future.

It is a great pleasure to acknowledge the help and support I have received from my colleagues who have provided me with various supports and contributed to my understanding of thermal management materials and approaches in electronic packaging. I would like to express my sincere gratitude to my editors, Dr. David Packer and all other editing staff who worked very hard to give the text its final polish.

Schaumburg, IL Xingcun Colin Tong

# Contents

# Abbreviations

| | |
|---|---|
| ABGA | Advanced ball grid array |
| ABS | Acrylonitrile-butadiene-styrene |
| ACPI | Advanced configuration and power interface |
| AES | Auger electron spectroscope |
| AFM | Atomic force microscopy |
| AMC | Aluminum matrix composite, or airborne molecular contamination |
| APG | Annealed pyrolytic graphite |
| ASTM | American Society for Testing and Materials |
| BGA | Ball grid array |
| BIOS | Basic input/output system |
| BIST | Built-in self-test |
| BLT | Bond line thickness |
| BSE | Back scattered electrons |
| BTE | Boltzmann particle transport equation |
| CAD | Computer aided design |
| CBGA | Ceramic ball grid array |
| CCC | Carbon/carbon composite |
| CCD | Charge coupled device |
| CCFL | Cold cathode fluorescent lamp |
| CFCC | Continuous fiber ceramic composite |
| CFD | Computational fluid dynamics |
| CHF | Critical heat flux |
| CMC | Ceramic matrix composite |
| CMOS | Complementary metal oxide semiconductor |
| CNF | Carbon nanofiber |
| CNT | Carbon nanotube |
| COB | Chip on board |
| COP | Coefficient of performance |
| CPU | Central processing unit |
| C-SAM | C-mode scanning acoustic microscope |
| CSF | Combination of solar and fuel cells |
| CTE | Coefficient of thermal expansion |
| CTI | Computer telephony integration |

| | |
|---|---|
| CVD | Chemical vapor deposition |
| CVI | Chemical vapor infiltration |
| CW | Continuous wave |
| dB | Decibel |
| DCB/DAB | Direct copper/aluminum bonded ceramic substrate |
| DF | Density factor |
| DFT | Density functional theory, or discrete Fourier transform |
| DIP | Dual-in-line package |
| DLC | Diamond like carbon |
| DNA | Deoxyribonucleic acid |
| DS | Decision support |
| DSP | Digital signal processing |
| DTM | Dynamic thermal management |
| ECU | Electronics cooling unit |
| EDA | Electronic design automation, or electronic document access |
| EDIFICE | Embedded droplet impingement for integrated cooling of electronics |
| EDM | Electrodischarge machining |
| EDS | Energy dispersive spectroscopy of X-rays |
| EELS | Electron energy loss spectrum |
| EG | Electron gas |
| EIA | Environmental impact assessment |
| EMA | Effective medium approximation |
| EMC | Electromagnetic compatibility |
| EMI | Electromagnetic interference |
| EMS | Electronic music studios |
| EPDM | Ethylenepropylene diene monomer |
| ESD | Electromagnetic discharge |
| ESPI | Electronic speckle pattern interferometry |
| EV | Electric vehicle |
| FCBGA | Flip chip ball grid array |
| FEA | Finite element analysis |
| FET | Field effect transistors |
| FIT | Failure in time |
| FLG | Few layer graphene |
| FMEA | Failure mode and effect analysis |
| FPBGA | Flip chip plastic ball grid array |
| GAEC | Gas-assisted evaporative cooling |
| GGI | Gold–gold interconnection |
| GLP | Graphite loaded polymer |
| HAST | Highly accelerated stress test |
| HEV | Hybrid electric vehicle |
| HFE | Hydroflouroether |
| HIP | Hot isostatic press |
| HOPG | Highly oriented pyrolytic graphite |

| | |
|---|---|
| HPHT | High pressure and high temperature |
| HRTEM | High resolution transmission electron microscopy |
| HVAC | Heating, ventilating, and air conditioning |
| HVOS | High-velocity oxyfuel spraying |
| Hz | Hertz |
| IACS | International annealed copper standard |
| IC | Integrated circuit |
| ICA | Isotropic conductive adhesive |
| ID | Inside diameter |
| IGBT | Insulated-gate bipolar transistor |
| IMS | Insulated metal substrate, metal core board |
| IPEM | Integrated power electronics module |
| IR | Infrared |
| JCPDS | Joint committee on powder diffraction standards |
| JTRS | Joint tactical system |
| LCD | Liquid crystal display |
| LCP | Liquid crystal polymer |
| LD | Laser diode |
| LDA | Laser diode array |
| LED | Light-emitting diode |
| LEM | Lunar excursion module |
| LMA | Low melting alloy |
| LRU | Line replaceable unit |
| MBLT | Minimum bond line thickness |
| MCAD | Mechanical computer-aided design |
| MEMS | microelectromechanical system |
| MH | Metal hydride |
| MIL | Military |
| MLC | Multilayer ceramic |
| MMC | Metal matrix composite |
| MPU | Memory protection unit |
| MRU | Modular refrigeration unit |
| MSRS | Miniature scale refrigeration system |
| MTBF | Mean time between failure |
| MTTF | Mean time to failure |
| MWNT | Multi-walled carbon nanotube |
| NEMS | Nanoelectromechanical systems |
| NMP | Net material product, or *N*-methylpyrrolidone |
| OD | Outside diameter |
| OEE | Overall equipment effectiveness |
| OEM | Original equipment manufacturer |
| OPC | Object linking and embedding for process control |
| OTD | On time delivery |
| PAA | Polyacrylic acid |
| PAI | Polyamideimide |

| | |
|---|---|
| PAN | Polyacrylonitrile |
| PBGA | Plastic ball grid array |
| PC | Polycarbonate, or personal computer |
| PCB | Printed-circuit board |
| PCD | Polycrystalline diamond |
| PCF | Phase change film |
| PCM | Phase change material |
| PCMA | Phase change metal alloy |
| PDE | Partial differential equation |
| PDP | Program data processor |
| PE | Polyethylene, or postal explorer |
| PEDOT:PSS | Poly(3,4-ethylenedioxythiophene) poly(styrenesulfonate) |
| PEEK | Polyaryletherketones |
| PEI | Polyetherimides |
| PEK | Polyaryletherketones |
| PEM | Proton exchange membrane |
| PES | Polyethersulfone |
| PET | Polyethylene terephthalate, or positron emission tomography |
| PFC | Perfluorinated chemicals or perfluorochemicals, or power factor correction |
| PG | Pyrolytic graphite |
| PGA | Pin grid array |
| PI | Polyimides |
| PIP | Polymer infiltration and pyrolysis |
| PLD | Pulsed laser deposition |
| PM | Preventive maintenance |
| PMC | Polymer matrix composite |
| PMMA | Poly(methyl methacrylate) |
| POM | Polyoxymethylene |
| PPO | Polyphenylene |
| PPS | Polyphenylene sulfide |
| PSA | Pressure sensitive adhesive |
| PSH | Polymer solder hybrid |
| PSS | Plasma sound source or polystyrene sulfonic acid |
| PSul | Polysulfone |
| PTC | Positive temperature coefficient |
| PTFE | Polytetrafluoroethylene |
| PV | Photovoltaic |
| PVA | Polyvinyl alcohol |
| PWB | Printed-wire board |
| PyC | Pyrocarbon |
| QDSL | Quantum dot superlattice |
| QFP | Quad flat package |
| RIM | Reaction injection molding |

| | |
|---|---|
| RMI | Reactive melt infiltration |
| RoHS | Restriction of hazardous substances |
| RTM | Resin transfer molding |
| SAM | Scanning acoustic microscopy |
| SEM | Scanning electron microscope |
| SFF | Small form factor |
| SiC | Silicon carbide |
| SiP | System in package |
| SIP | Slurry infiltration process |
| SLS | Reaction-bonded SiC with laser sintered process |
| SMA | Shape memory alloy |
| SMT | Surface mount technology |
| SOC | System on a chip |
| SOI | Silicon on insulator |
| SPM | Scanning probe microscopy |
| STD | Standard |
| SUB | Single vacuum bag |
| SWNT | Single walled carbon nanotube |
| TAB | Tape-automated bonding |
| TBGA | Tape ball grid array |
| TE | Thermoelectric |
| TEC | Thermoelectric cooler |
| TEM | Transmission electron microscopy |
| TGF | Thermal gap filler |
| TIM | Thermal interface material |
| TML | Total mass loss |
| TPE | Thermoplastic elastomer |
| TPS | Thermal protection system |
| TPU | Thermoplastic polyurethane |
| TQFP | Tape-bonded quad flat package |
| TTR | Transient thermal reflectance |
| ULSI | Ultra large scale integration |
| UV | Ultraviolet light |
| VCC | Vapor compression cooling |
| VCM | Volatile condensable material |
| VGCF | Vapor grown carbon fiber |
| VGCNF | Vapor grown carbon nanofiber |
| VPE | Vapor phase epitaxy |
| VR | Virtual reality |
| VRLA | Valve regulated lead acid |
| WBG | Wide band gap |
| WEEE | Waste electrical and electronic equipment |
| WIP | Work in progress |
| *ZT* | Dimensionless thermoelectric figure of merit |

# Chapter 1
# Thermal Management Fundamentals and Design Guides in Electronic Packaging

**Abstract** With increased performance requirements for smaller, more capable, and more efficient electronic systems, thermal challenges have become critical issues in electronic packaging design. Breakthroughs are needed in advanced cooling solutions and pragmatic design at chip, board, and system package levels. In response to these critical needs, advanced materials and process improvements in packaging and cooling technology are required to provide high thermal transfer efficiency, well-controlled thermal transient behavior, environmental compatibility, low weight, and acceptable material and fabrication cost. The lowest device temperature and expected reliability will be achieved when the lowest package thermal resistance is achieved. This chapter provides an outline of thermal challenges in advanced electronic packaging, heat transfer theory, and its application in thermal design, development of thermal management solutions, advanced thermal management materials selection and component design guideline, as well as environmental compatibility requirements in thermal management of electronic packaging.

## Rationale of Thermal Management

Electronic packaging of microelectronics can be identified as the art of enclosing, interconnecting, powering, cooling, chip packaging, and protecting chips from the ambient environment as well as protecting the ambient environment from the chips (Rasmussen 2003). It provides signal and power transmission, thermal dissipation, electromagnetic interference (EMI) shielding as well as physical and environmental protection, and act as the bridge that interconnects the integrated circuit (IC) chips and other components into an electronic system to form electronic products. Electronic packaging of a typical electronic system can be divided into several packaging levels from bare chip, packaged chip, printed circuit board assembly, electronic subassembly, and electronic assembly to system or final electronic product. The trend in electronic packaging is to simplify and/or reduce the number of packaging levers.

From a thermal management perspective, electronic packaging is usually classified as chip-level packaging, board-level packaging, and system-level packaging. Each level of the packaging has distinctive interconnection devices associated

X.C. Tong, *Advanced Materials for Thermal Management of Electronic Packaging*, Springer Series in Advanced Microelectronics 30,
DOI 10.1007/978-1-4419-7759-5_1, © Springer Science+Business Media, LLC 2011

with it. Chip level package usually refers to packaged electronic functional devices (e.g., active, passive, or electromechanical devices, etc.) or packaging of silicon chips into dual-in-line packages, small outline ICs, chip carriers, multichip packages, and the chip level interconnects that join the chip to the lead frames, or utilizing tape-automated bonding or chip on board to assemble the chip. The board-level packaging is the assembly of the chip or multichips together with additional components, such as capacitors, resistors, inductors, switches, etc. on a printed circuit board (PCB). Printed conductor paths connect the leads of components to PCBs and to the electrical edge connectors for off-the-board interconnection. System-level packaging is normally the outer enclosure or shell of the electronics, such as the casing of a hearing aid, mobile phone, electronic organizer, etc., with PCB-to-PCB interconnections or card-to-motherboard interconnections, or it can be a rack or frame hold several shelves of subassemblies that are connected together to make up a complete system or a final electronic product. In the past years, increasing frequency and power density coupled with lower product costs have been driving new packaging technologies. Some examples of this are migration from wirebond to flipchip interconnect, higher levels of integration in semiconductors and increased usage of hybrids and multichip modules, such as multicore processors and system-in-package. This trend in microprocessor architecture results in increased heat densities which mandate that thermal management be given a high priority in electronic packaging design so as to maintain system performance and reliability.

With continuous demands of high performance, low cost, and miniaturized microelectronic devices, electronic packaging will continue to undergo evolution at every new generation of microelectronics and product technologies. Theses changes have been introduced and will continue creating a new set of changes that require advanced packaging technology including innovative thermal management solutions. In addition to the technical challenges, market forces such as declining product prices, increased user experience through miniaturized devices, wireless connectivity, and longer battery life would make these challenges even more complex (Mallik et al. 2005). The challenges in thermal management can be viewed in terms of three different but nonseparable problems (Hannemann 2003). (1) The chip temperature must be maintained at a relatively low level despite high local heat density. (2) High heat loads must be handled at the assembly or module level. (3) The thermal environment of an electronic system, such as the computer machine room, office space, or telecommunications central office must be controlled and the overall rack heat load dealt with. As a result, thermal management would be a serious concern behind any new electronic product designs. Two major objectives should be achieved through thermal management: Prevention of catastrophic thermal failure, and extension of the useful lifetime of the electronic system. Catastrophic thermal failure is usually the result of semiconductor material failure as a result of overheating, or thermal fracture of a mechanical element in an electronic package, such as case or substrate. The failure rate increases exponentially with operating temperature.

## *Heat Sources and Thermal Effects on Integrated Circuit Operation*

The insatiable demand for higher performance processors has led to a steady escalation in power consumption across all the market segments, such as mobile and performance desktops as well as servers and workstations (Viswanath et al. 2000). Increasing power density and current levels in the microprocessors have been main heat sources and cause concerns to do the thermal management of the on-chip hotspots as well as the package and interconnection Joule heating. The result in temperature increasing will affect the operation of the active and passive devices in the integrated circuits. If the temperature increase is high enough, the active or passive devices being heated may either permanently degrade or even totally fail. Such failures include thermal runaway, junction failure, metallization failure, corrosion, resistor drift, and electromigration diffusion (Krum 2004). Therefore, it is crucial to minimize any temperature increase in an electronic package.

### Power Density

In general, for microprocessors such as complementary metal oxide semiconductor (CMOS), power dissipation is proportional to the capacitance of the logic elements, the square of the operating voltage swing, and the operating frequency (Hannemann 2003):

$$P \approx NCV^2f, \tag{1.1}$$

where $P$ is power dissipation of CMOS in Watt; $N$ is the number of devices per chip; $C$ is the capacitance of the logic elements in Farad; $V$ is the operating voltage in Volt; and $f$ is the operating frequency in Hz. While the logic element capacitance declines with feature size, and while operating voltages have been significantly reduced, the increase in the number of devices per chip and the operating frequency have driven power levels for the next generation of microprocessors to very high levels (70–200 W). As the frequency scales higher over time, so does the power dissipation of the microprocessors. The improvements in process, such as introduction of multicore processors have been able to hold the power increase to reasonable levels, but it is definitely trended higher. A similar trend is reflected in the average heat flux (power dissipated per unit die area) on the processor, indicating a linear increase over time. This is due to the fact that the power reduction obtained from architecture and process modifications is not commensurate with the scaling in die size, voltage, and frequency to support a cap in power consumption. In addition, the wider range of power and frequency offerings will enable performance and cost trade offs to be made between the various market segments. The need for higher performance and an increased level of functional integration as well as die size optimization on the microprocessor leads to preferential clustering of higher power units on the processor. In turn, this leads to a higher heat-flux concentration in certain areas of the die and lower heat fluxes in other regions on the die, which manifests itself as large temperature gradients on the die (Viswanath et al. 2000). This thermal no-uniformity typically refers to hotspots, where power densities of 300 W/cm$^2$ or over are possible. To help quantify the nonuniform power effects, a density factor (DF) has been introduced, which is the ratio of the actual package thermal resistance at the hottest spot to the die-area-normalized uniform power resistance or thermal impedance, and has the unit of inverse area ($A^{-1}$). DF can be used to quantify the impact of nonuniform die heating on thermal management by the following equation (Mallik et al. 2005):

$$\psi_{jc} = R_{jc} \times \text{DF}, \tag{1.2}$$

where $\psi_{jc}$ is the junction-to-case thermal resistance of the package; and the $R_{jc}$ is the thermal interface material resistance. It can be seen that for the same $R_{jc}$, the power maps with a higher DF will result in a higher package thermal resistance, which in turn requires more effective thermal management solutions. The future trend is that the DF is increasing for emerging generation of microprocessor architectures due to the increasing local power density at the hotspots, therefore, the thermal management solutions have to meet more stringent heat-flux requirements that are significant multiples of the average heat flux at some areas of the chip-package interface.

### Joule Heating

Joule heating is generated when current passes through a resistor, following Joule's law: $P = I^2R$, where $P$ is power (W), $I$ is the current (A), and $R$ is resistance (Ω). As current levels continue to rise, Joule heating along the power delivery path becomes significant. Because it is not uncommon for a high-end microprocessor to draw currents in excess of 100 A, for instance, even a resistance of 0.5 mΩ along the path will result in a power dissipation of 5 W. In an electronic package, higher current combined with the need to reduce the package size, i.e., thinner and narrower conductors and finer pitch power delivery interconnects, would lead to a high amount of heat generated within the package and electronic interconnection system. This would require thermal management of the entire interconnection system which may include the flip chip joints, the substrate, the socket, and the attaching solder balls (Mallik et al. 2005). For example, one challenging key area is the increasing temperature of the flip-chip die bumps. Without proper attention to thermal design, the bump temperature could be significantly higher than the transistor junction temperature causing bump electromigration problems. The joule heating of interconnects on the package can be effectively managed through careful design, which involves minimizing current concentration and spreading heat through thermal design and thermal management materials selection.

## *Thermal Failure Induced by Different Coefficient of Thermal Expansions*

An electronic package is comprised of various conducting and insulating materials which have different coefficients of thermal expansions (CTE). Figure 1.1 illustrates a cross-sectional view of a ceramic ball grid array package assembly, where a silicon chip is mounted on a multilayer ceramic substrate module through solder joint embedded in epoxy underfill, and the module is attached to a printed circuit board (PCB) through solder ball interconnections to form a final second level assembly. In addition, a metal heat spreader can be attached to the silicon chip through thermal interface material, or a heat sink is attached to the module to dissipate the excessive heat. When the chip is powered up so that the package is subjected to a temperature change, each material with different CTE deforms at a different rate. This nonuniform CTE distribution will produce thermally induced mechanical stresses within the package assembly.

When the assembly is cooled from an assembly temperature, for instance the PCB contracts more than the module. This uneven contraction produces a global bending of the whole assembly as well as relative horizontal displacements between the top and bottom of solder balls. When it is cooled to room temperature, the free thermal contraction of the solder joint at the interfaces is constrained by adjacent materials which have a lower CTE. In general, if the global effect reinforces the local effect at a point in the package, the concentrated strain will be accumulated during thermal cycles, which would result in the premature failure of the device during operation, such as heat sink bending, the thermal interface or solder joint failing, the ceramic substrate warping, or completely failing and cracking. To reduce thermal stress or eliminate thermal failure requires both a selection of the proper materials and a minimization of the temperature changes through thermal management.

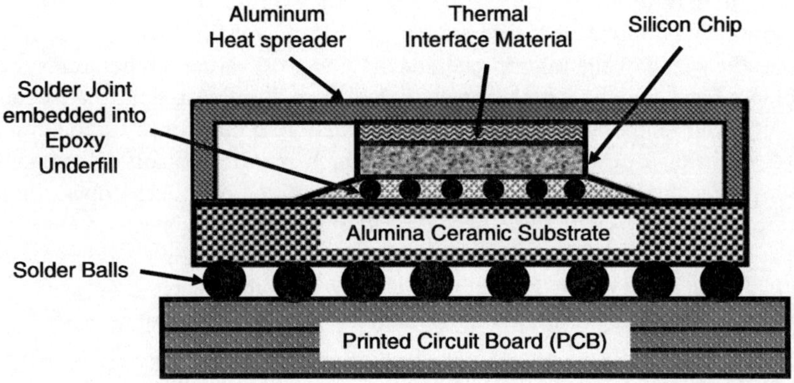


**Fig. 1.1** Schematic illustration of a typical ceramic ball grid array package assembly

### *Thermal Failure Rates*

The failure rate of an electronic component can be estimated by the Arrhenius equation (Krum 2004):

$$f = A\varepsilon^{-e_A/kT}, \tag{1.3}$$

where $A$ is Arrhenius constant; $f$ is failure rate, $e_A$ is activation energy in electron volts (eV); $k$ is Boltzmann's constant ($8.63 \times 10^{-5}$ eV/K); and $T$ is junction temperature in $K$. The activation energies vary for different failure mechanisms, for example, $e_A = 1.3$ eV for charge injection; 1.0 eV for contamination; 0.53–0.70 eV for corrosion; 0.68–0.95 eV for electromigration; 0.3–0.7 eV for gate oxide; 0.55–1.0 eV for gold–aluminum film joint; 0.68–0.95 for hillock; 0.73 eV for aluminum wire–gold film; and 0.69–1.0 eV for gold–aluminum wire couples (Krum 2004). For an activation energy of 0.65, the failure rate would increase by a factor of 2.0 for a temperature increase from 50°C to 60°C. This 0.65 eV value of activation energy is usually used as a rule of thumb, i.e., for every 10°C increase in temperature, the failure rate doubles.

### *Thermal Management Challenges and Common Concerns*

The current trend in microprocessor architecture is to increase the level of integration (higher power), shrink processor size (smaller die), and increase clock speeds (higher frequency). This results in an increase in both the raw power as well as the power density on silicon. The drive to manage yield and reliability is resulting in the need for lower operating temperatures. There are two major reasons to maintain the operating temperature of a device at a certain level. (1) The reliability of circuits (transistors) is exponentially dependent on the operating temperature of the junction. As such, small differences in operating temperature (order of 10–15°C) can result in approximately two times the difference in the lifespan of the devices. (2) The other factor is the speed of the microprocessor. At lower operating temperatures, microprocessors can operate at higher speeds due to reduced gate delay. A secondary effect of lower temperatures is related to a reduction in idle power dissipation (also known as leakage power) of the devices, which manifests itself as reduction in overall power dissipation. These two factors combined dictate the operating temperature of devices as a function of the speed of the device (Viswanath et al. 2000).

This in turn translates to a shrinkage in temperature budgets for thermal design. Careful management of the thermal design space from the silicon to the system level is therefore critical to ensure a viable solution space for succeeding generations of processors (Viswanath et al. 2000). On another hand, the continuing increase in electronic packaging density has resulted in demands for advanced packaging materials with low weight, low cost, high-thermal conductivities, and matched CTEs with those of ceramic substrates and semiconductors to minimize thermal stresses that can cause component or interconnection failure. Various more comprehensive and accurate modeling, and analysis of the electronic packaging systems are needed to understand the thermal behavior and the behavior for cost efficiency, and optimize the thermal management, which requires identification of critical issues such as shock, vibration, and usage scenarios, considering potential coupling effects, and developing integrated, interdisciplinary solutions. Systematic thermal solutions also need universal design regulations and thermal testing standards.

These challenges have involved all electronic market segments. For instance, the mobile processor market segment is typically constrained by battery life (~2–3 h), and the form-factor must be small and light to allow for portability. If the desktop market is more cost sensitive, the mobile market is more space and weight sensitive. These sensitivities place boundariess on the effective power removal capabilities of the electronic package. The long-term solution would be: (1) The design and architecture of the microprocessor must be such that it optimizes performance and power consumption; (2) New cost-effective technologies in microprocessor and system packaging including effective thermal management solutions must be developed.

As a result, the common concerns of the thermal management for electronic packaging can be summarized as follows.

1. For the current generation of microelectronic products, the two main areas of concern are (a) the thermal control of high-power microprocessors and/or similar ultra-large scale integration components, given the very high local heat density; and (b) the thermal management of the high overall heat loads generated by equipment in racks and machine rooms (Hannemann 2003). For example, telecommunications equipment rooms, whether central offices or smaller remote huts, have reached their limit in terms of power density. Computer facilities have lower overall heat loads, but concentrations of power within these rooms are a serious issue. Office areas (typified by the "workstations" data) have rising heat loads that within an open office environment pose a very significant HVAC (heating, ventilating, and air conditioning) problem. These problems are beginning to be addressed with liquid-cooled equipment frames (allowing heat to be transported to a remote location) and spot cooling units (Hannemann 2003). Furthermore, telecommunications central offices are already stretching power availability and power handling limits. Photonic components used in these systems (as well as data networking systems) have unique and serious microcooling and materials challenges. For example, semiconductor lasers have chip-level heat fluxes on the order of 2,000 W/cm$^2$. These devices also have very narrow thermal control limits, because performance is significantly impacted by operating temperature (e.g., wavelength). Historically, this segment has been a major applications space for thermoelectric cooling–and conventional, cost-effective thermoelectric coolers are reaching their limits in newer applications (Hannemann 2003).
2. For these systems, such as mainframe computer systems, storage arrays, server farms, and network offices, localized thermal control for very high-power chips

is a major issue as chip powers increase beyond 100 W. Ducted forced air cooling and active heat sinks (heat sinks with small, dedicated air movers) will continue to be used, but miniaturized liquid cooling loops and in some cases microelectromechanical systems-based coolers such as embedded droplet impingement for integrated cooling of electronics require development. These devices, however, must meet stringent cost and reliability requirements–no more than $100 or so as a target cost, and perhaps 1,000–5,000 failures in time (one failure per $10^9$ h) for reliability performance. Also, rack-level cooling via liquid cooled frames and compact rack-mountable air conditioners (both meeting size, cost and reliability targets) are of interest. While these cooling approaches may seem straightforward, components such as compressors and extended surface arrays meeting the size, cost, and lifetime constraints are not yet available. Furthermore, equipment rooms also pose HVAC design problems. Improved modeling tools and sensing and control devices are needed (Hannemann 2003).

3. Office systems include high performance workstations, printers, desktop computers, and office- and equipment-closet installed networking and telecommunications products. Cost/performance microprocessors and networking chips will reach 80–100 W. The cooling of these chips will pose unique challenges because some options open to large system designers will be unavailable. Cooling solutions must be extremely compact, quiet, and cost effective. Proven-reliability microheat pipes, breakthrough microsized liquid loops, and especially optimized active heat sinking systems may provide thermal management solutions. Acoustic performance will remain very important in this segment (Hannemann 2003).
4. Other important electronic systems must operate in harsh environments, calling for ruggedized cooling solutions with some unique requirements (such as transient high-heat loads) but no or less pressure on cost and reliability. These applications include automotive, telecommunication, space, and military systems (Hannemann 2003).
5. Another challenge is that the increasing emphasis placed on thermal design at the early design stages has inadvertently increased electromagnetic compliance (EMC) problems. Thermal design often conflicts with EMC design, so fixes that are implemented to address thermal concerns often exacerbate or create EMC problems. The most obvious example is that thermal design requires large holes to enable adequate air flow while EMC design requires small holes to reduce emissions. Another example is that thermal connections (metal studs) are often added to conduct heat away from a hot component. For example, in a small module where forced cooling is not available, it may be necessary to conduct the heat to the enclosure. Switching currents can couple to the stud and cause it to radiate like an antenna. In thermal design, a large surface area is often used to increase convective heat flow. The same large surface areas may also increase capacitive coupling to the enclosure, which enables displacement current to flow from components such as heat sinks to the chassis and then onto cables connected to the chassis (Manning and Johns 2007). Therefore, the concurrent thermal and EMC design process should be implemented.

Consequently, the importance of thermal management has waxed and waned through the technology and product generations of the past decades. Current integrated circuit and photonic technologies as well as the ubiquity of electronic system applications are once again providing a serious challenge to thermal management in terms of basic theory, tools, components, and innovative design. Breakthroughs are needed in advanced cooling and pragmatic design at all levels (Hannemann 2003). The thermal engineering for electronics is on the cusp of a renaissance.

## Overall Picture of Thermal Management in Different Packaging Levels

With increasing power densities, the thermal management becomes more and more a central issue in electronic packaging. An overall picture of thermal management in different packaging levels including chip level, board level and system level is shown in Figure 1.2. Chip-level thermal management mainly relates to dissipating and heat spreading within the semiconductor device and chip package. The thermal management technologies include conventional copper, copper-tungsten, or copper-moly heat spreaders as well as advanced thermal design, on-chip thermoelectric cooling, microscale heat pipes, and microchannel cooling, etc. Board-level thermal management indicates heat transfer or spreading through PCB and from the chip

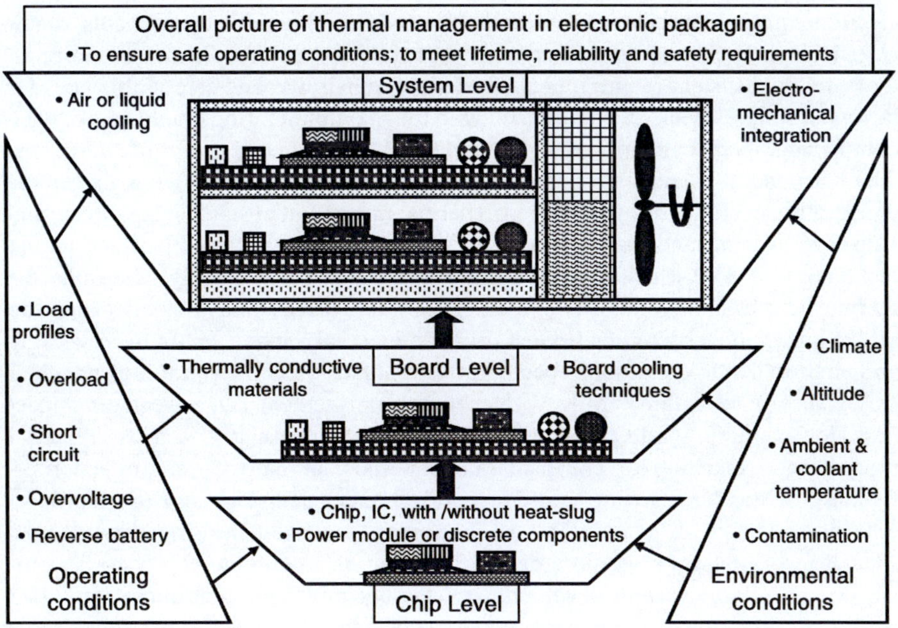


**Fig. 1.2** Overall picture of thermal management in different electronic package levels

package to the chassis or electronic system. Conventional technologies include copper and aluminum spreaders. System-level thermal management refers to heat dissipation through system heat exchanger and heat transport from the chassis to the system heat exchanger. Conventional technologies include air-cooled heat sink and radiators, forced air flow, and pumped liquid loops. For some high-power semiconductor devices, a cooling mechanism may directly connect the chip package to the system heat exchanger by passing board level. This reduces the number of thermal interfaces and improves the cooling efficiency, but may be at a price of added costs.

Although the thermal design of the chip, board, and system levels shares the common places, each level has its unique cooling requirements. For example, chip-level cooling designs are dominated by high-heat flux and miniaturization issues. Board level requires transport of large amounts of heat with minimal temperature gradients. System-level heat exchanger designs are required to sink the most heat with minimal volume and mass. In some cases where the temperature of the environment varies greatly, thermal or temperature control mechanisms are needed to protect the electronics by maintaining a relatively constant temperature. The temperature control may be implemented at different levels. For example, variable speed air blowers or liquid bypass can be added to board and system levels to adjust the cooling effectiveness to realize temperature control. With development of the innovative electronic packaging accompanied by advanced thermal management technologies, however, the picture of thermal management in different package levels would undoubtedly change.

Whatever change or modification may occur, the basic target of thermal management is always to ensure safe and reliable thermal rating conditions for all electronic packaging components, including active and passive components, die or chip, PCB, interconnections, subsystem, and system structures. In consideration of the individual system requirements on lifetime, reliability, and safety, this goes far beyond a simple check on compliance with the maximum rating conditions of each component. Special attention must be paid to lifetime data and useful life deratings, which in general are not only dependent on the absolute temperature but also on the temperature cycling during operation. Thermal management starts at the device and chip level, for instance, with or without a heat-slug for an SMT (surface mount technology) device. Such a small technical decision can basically determine the thermal behavior and reliability of the whole system just as many other decisions do on the board and system level. Each decision must take into account the operating and environmental conditions specified for the final system. Even things which at first glance have nothing to do with thermal management can have their impact here, for example, waste or recycling regulations which aim at a better material separability, reduction of encapsulation materials, or ban of certain materials. Because of the far-reaching consequences of many decisions and the high-cost risks of a redesign, it is important that the thermal management is not considered a separate work-package within a design flow, but as a continuous process accompanying the whole system development. This becomes more and more important with the continuously increasing power densities of modern power electronic systems (März 2003).

## *Chip Level Packaging Thermal Management*

The emergence of nanoelectronics has led to localized areas of high-heat flux that are dominating the performance of electronics at the chip level. While the traditional thermal management techniques mainly provided system-level cooling sufficient to meet past electronics cooling requirements, they are facing challenges to meet future chip-level cooling requirements in most of their current forms. Therefore, chip-level thermal management has been explored and focused on advanced chip-level thermal design, and miniaturization of thermal management techniques to the microscale, which is receiving increased attention as a solution to the future thermal management technology shortfall. Microscale thermal management offers several enticing opportunities (Moran 2001): (1) The ability to "spot cool" high-heat flux regions with unparalleled resolution to bring down critical junction temperatures; (2) Potential for macrolevel performance leaps by optimizing microlevel heat transfer; (3) Improved integration of thermal management at the chip level using compatible semiconductor materials and fabrication techniques; (4) Enabling of board-level and system-level miniaturization to support the pervasive trend toward increased capabilities in smaller devices.

There has been a wide range of new technology developments in chip-level microscale thermal management, such as microscale channel cooling, wicking structure, enhanced thermal conduction, and refrigerators or coolers. For instance, microchannels have been utilized to provide increased surface area for improved heat transfer. These devices are generally limited to laminar flow regimens because the increased surface area also increases pressure drop and the subsequent pumping power required. The channels can be micromachined from bulk semiconductor material, or fabricated by other means with a variety of materials. Microtubes, for example, have been developed using copper, silver, platinum, glasses, polymers, alloys, and other materials with diameters ranging from 0.5 to 410 μm (Moran 2001). Used in conjunction with evaporator and condenser regions, microscale wicking structures have been developed and become a critical component of microheat pipes. The wicking structure can be fabricated with a coherent pattern of 2.5 μm diameter pores (channels) which length is equal to 260 μm, utilizing surface tension forces of the working fluid to induce flow without active pumping. The wick is part of a micromachined silicon loop heat pipe concept for removing heat at the chip level with a predicted heat flux cooling capacity greater than 100 W/$cm^2$ (Moran 2001). Nanomaterials have been explored to offer enhanced microscale thermal conduction. For example, nanocrystalline diamond film from fullerenes resulting in freestanding diamond structures as thin as 0.3 μm has particular interest in high-performance electronics cooling due to their unusual combination of superior thermal conductivity (~2,000 W/m K), yet low electrical conductivity. Carbon nanotubes offer the potential for ultrahigh strength structures with outstanding thermal conduction characteristics (e.g., >5,000 W/m K). In addition, microrefrigerators and coolers, such as thin-film thermoelectric cooler and miniaturized Stirling cycle refrigerator, have a unique characteristic for chip-level cooling that differentiates them from all other potential thermal management technologies: the ability to generate cooling temperatures well below the ambient temperature. This key advantage potentially allows junction temperatures to be driven much lower resulting in improved reliability, faster performance, and operation in higher temperature environments (Moran 2001).

On the other hand, optimal thermal design and proper materials selection in chip-level packaging plays an important role in thermal management. In most of advanced electronic devices, the bulk of the heat is concentrated to the active transistors on a semiconductor chip, assembled in some kind of package. In general, this package is mounted on a PCB and attached to electric conductors. The transfer of the heat that is originated on the transistor junction goes to the lead frame and further on to connecting pins of the packaging to some extent via thin bonding wires, but mainly by conduction paths as the adhesive between chip and other parts of the packaging, As the connecting pins are usually soldered to electric conductors of copper, a metallic bridge is formed for the continued heat transfer and the distribution of the heat to the PCB, some of the connecting pins can also have only a heat conducting function. As the dielectric material of the PCB is not a good heat conductor, it is important to include as much metal as possible in the stack-up of the PCB; in this way the heat is spread effectively and hotspots are flattened. Preferably this metal is in most cases copper, as copper is an excellent conductor of both current and heat. As in this interpretation of the heat transport, the temperature gradient is directed from the junction toward the PCB, the PCB must be the colder part. This means the PCB must be cooled by a proper cooling system, which can effectively dissipate the heat through conduction, convention, radiation or combined approaches (Siebert 2005).

## Board Level Packaging Thermal Management

Board-level packaging thermal management has been having more and more attention paid to it, starting from the early stage of an IC design. Often, when designing a new board for an electronic device, a thermal simulation can initially be done to identify hotspots. Problems identified at this stage can often be addressed by layout changes that can be made nearly without cost at this stage of the process.

In a typical board-level thermal simulation process flow, the systems architect will develop the initial concept design by creating a functional block diagram. The hardware design engineer then derives the first physical layout directly from the block diagram. At an early stage in the design process, long before the mechanical engineer gets involved, the electrical engineer can use board-level simulation to evaluate the new board design in an existing system. For instance, a three-dimensional (3-D) computational fluid dynamics solver can be used to predict air flow and temperature for both sides of the board, as shown in Figure 1.3 (Petrosjanc 2009). Often the designer will identify hotspots, and cooling management can thus be considered from the earliest stages of the design process. Changes made to the

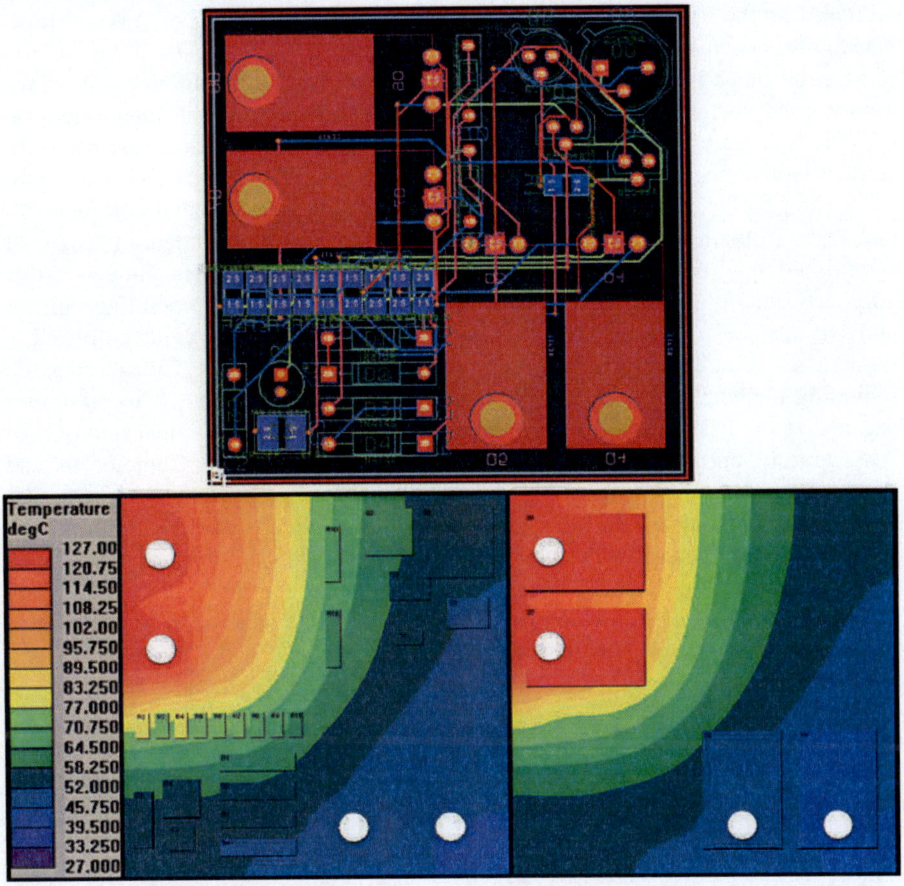


**Fig. 1.3** Example of thermal simulation for an audio powder amplifier printed circuit board (Petrosjanc 2009)

functional block diagram are instantly reflected in the physical layout and thermal representation. At this stage, many more alternatives exist to deal with thermal problems. Rather than being limited to expensive additions, engineers can consider a wide range of processes such as changing the board layout, adding copper inserts, or changing the package style. The board-level model can also be imported into a system-level thermal model, such as the one that may have been created when the original system was designed. This saves the mechanical engineer time in updating the system-level model, if necessary, while reducing the chance of errors caused by miscommunications. The results from the system-level analysis can also be exported to the board-level simulation, making it possible for the electrical engineer to apply the system-level air flow and temperatures to the board being designed. This approach keeps all team members in sync and enables them to contribute to concept development in real time (Bornoff 2005).

Based on the board-level thermal simulation and thermal design, a board-level packaging system with effective thermal management can be built up for the integration of power semiconductors, sensors, and control electronics. The commonly used techniques include monolithic integration, hybrid integration on ceramic substrates, DCB/DAB (direct copper/aluminum bonded ceramic substrates) technique, lead frame technique (e.g., in combination with ceramic substrates and molded packages), IMS (insulated metal substrate, metal core boards), and PCB. In the low-voltage, low-power range the monolithic integration is well established, e.g., in form of the widespread Smart-Power switches. However, this approach quickly comes to its economic limits with increasing operating voltage and currents because of the large difference in complexity and costs per chip area of modern semiconductor processes for power and logic devices. Considering the mounting techniques for discrete components, the hybrid and DCB technologies can realize the best thermal performance with substrate materials such as $Al_2O_3$ or AlN ceramic, but also show the highest costs per substrate area. Common to lead frame, DCB and IMS technology is generally the single-layer copper structure which makes it very difficult to route more complex circuits and greatly limits the package density. The comparatively coarse trace widths in DCB and partially in the lead frame technology also limit the integration of more complex signal electronics. On the other hand the thick copper traces ($> 0.3$ mm), which can be realized in DCB technology easily, allow control of high currents in the range of several hundred amperes. Comparably, the printed circuit board technique is based on the worst heat-conducting substrate material, nevertheless, has the charm to open up the great potential of a modern, very innovative, and very high production volume technology, which by far outperforms the competitive techniques with respect to integration density, flexibility, and costs per substrate area (Moran 2001). The base material of a PCB is typically an epoxy laminated cotton paper or woven fiberglass. The active and passive devices are usually directly soldered on the PCB with copper interconnects. The effective thermal management on a PCB level is typically achieved or enhanced through (1) replacing nonavailable on-board copper area with cooling pins; (2) contacting board to a heat spreader or a heat sink or both; (3) using thermal vias (through-hole, blind or buried) where possible; (4) using maximum copper filing in all available layers acted as the heat spreader; (5) providing a good thermal coupling to components with high thermal conductivity and large surface; (6) using any available forced or convective air flow; (7) introducing emerging thermal management techniques.

In many power electronic systems the integration density in the signal and control electronics parts has reached a level that requires multilayer boards. The integration of the power electronics parts on the same board does not only cause the cancellation of interconnection elements, but also allows new dimensions of multifunctional board integration. This results in many benefits of a multilayer PCB integration of power electronics, such as an SMT-compatible assembly process, integrable heat path with electrical insulation from heat sink, integrable EMI shields and low-noise ground planes, high package density due to multilayer routing, integrable windings for planar magnetics, and cancellation of interconnects between power and signal electronics. Therefore, advanced board-level packaging can be obtained with optimal cooling of all active and passive power components; very low dead volumes, high power density, very low circuit parasitics, good EMI behavior, and high mechanical stability and functional reliability (März 2003).

However, besides all thermal optimizations on a board level, in many cases money spent on better components or more intelligent circuit topologies is paid back by less expensive heat sinks, improved reliability, and–in the long term–less energy costs (März 2003).

## *System-Level Packaging Thermal Management*

System-level thermal management has been widely used with various techniques which can be conveniently grouped into passive and active methods. Passive methods, which do not require any input power, tend to be very reliable and relatively easy to implement. However, they are also performance limited and therefore inadequate for many high power applications. Typical passive methods include (Moran 2001): (1) natural convection (finned heat sinks, ventilation slots, board/component placement, etc.); (2) radiation (paints, coatings, mechanical surface treatments, component layout, etc.); (3) conduction (heat spreaders, thermal conduction structures, vias, elastomers, pastes, adhesives, pads, chip and board packages, etc.); and (4) emerging thermal management technologies. Strictly speaking, for example, capillary and phase change methods (e.g., heat pipes, wicks, melting wax, boiling, etc.) are also passive methods. However, they generally provide higher performance capabilities and are not as easily implemented as the other passive methods. In addition, phase change devices are limited in terms of their operating temperatures, which are determined by the saturation temperature of the working fluid (or melting temperature of the solid).

Contrastingly, active thermal management techniques require input power to provide increased performance/capacity, but generally at the price of lower reliability and added complexity. Active methods usually include external forced convection (fans, nozzles, etc.), pumped loops (heat exchangers, cold plates, etc.), and refrigerators and coolers (thermoelectric/Peltier, vapor-compression, vortex, gas cycles, etc.). It should be noted that heating (e.g., resistance, induction, etc.) is also an active thermal management tool for applications where a lower temperature boundary must be avoided. Aerospace applications requiring heating are common due to the effective thermal radiation sink temperature of space being a few degrees above absolute zero. On earth, cold region applications may also require heating. In addition, measurement electronics or other systems where a fixed temperature band must be maintained often require an integrated cooling and heating system with temperature feedback control (Moran 2001).

There are quite a few emerging technologies that are under development and evaluation. One area is to enhance convection cooling by improving the heat transfer coefficient through extended heat sink surface and high air flow fans with a built-in feature for acoustic noise cancellation. Attention has also been paid to the development of heat spreader materials and components, such as carbon fiber, graphite, thermally conductive composites, vapor chambers, heat pipes, and nanomaterials. The integrated solutions also being explored, for instance, are a close-loop liquid cooling system can implement cold plates or microchannels with either single-phase or two-phase liquid cooling. Furthermore, refrigeration to achieve negative thermal resistance has been developed with the focus on reducing the size and cost of the compressor and the heat exchanger. Solid-state refrigeration or thermoelectric cooling is another example for hotspot cooling of devices with highly nonuniform power dissipation or full-chip cooling in conjunction with vapor chamber heat sinks. In addition, emerging nanomaterials hold promise of providing highly conductive thermal interface materials and reducing interconnect Joule heating (Mallik et al. 2005).

Another research area is the system-level dynamic thermal management (DTM), which uses both hardware and software support in a synergistic fashion and hence leads to a significant execution time overhead. For example, a system-level framework for doing fine-grained coordinated thermal management uses a hybrid of hardware techniques (like clock gating) and software techniques (like thermal-aware process scheduling), leveraging the advantages of both approaches in a synergistic fashion. While hardware techniques can be used reactively to manage the overall temperature in case of thermal emergencies, proactive use of software techniques can build on top of it to balance the overall thermal profile with minimal overhead using the operating system support (Kumar et al. 2008). This would build up a new routine for future system-level thermal management of electronic packaging.

## Thermal Management Solutions

Effective thermal management of an electronic system requires identifying critical issues such as shock, vibration, and usage scenarios, considering potential coupling effects, and developing an integrated, interdisciplinary solution (Madrid 2000). This usually can be achieved based on understanding the thermal behavior of a system and optimizing that behavior for cost efficiency through modeling and analyzing the system, including for instance (1) system-level cooling and active control; (2) board-level and chip-level thermal design and thermal management; (3) device or die electrothermal modeling and optimal thermal design; and (4) micro- or nanoscale thermal design and engineering or processing. As all different component reliability is inversely proportional to temperature gradients and sustained exposure at the elevated temperatures, power dissipation is a key function of any thermal solution. General thermal management solutions, therefore, typically include hardware assembly, software control, and optimal thermal design as well as the combination of these approaches. System hardware solutions are based on internally distributing and externally dissipating the thermal energy to the ambient environment. System software solutions typically regulate the dissipative power of the device based on active system feedback controls. Optimal thermal design is required for all level cooling and thermal control, particularly effective for device-level micro- or nanoscale thermal engineering and processing. A well-designed hardware thermal solution combined with software thermal management preserves a system's functionality and extends the life cycle of an electronic device (Madrid 2000).

## *Hardware Solutions*

Various hardware thermal management devices are able to dissipate a range of heat fluxes over a range of temperature differences. As the package level goes from the chip die to its packaging, to a board, to a motherboard, to a system, the area over which the heat is distributed increases so the heat flux decreases correspondingly. This allows heat removal at one or more packaging levels, where the heat flux and temperature differences are consistent with the available cooling method (Couvillion 2006). As shown in Figure 1.4, such hardware typically consists of the use of natural convection, forced convection, fluid-phase change, thermionic and liquid cooling devices, as well as interface materials, mounting assembly, allocated system, and component and motherboard placement.

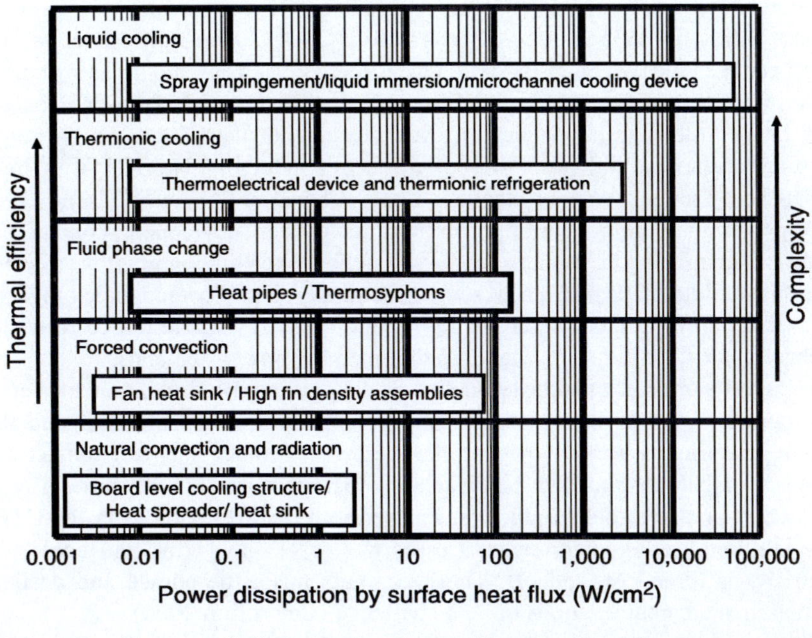


**Fig. 1.4** Achievable power dissipation by heat fluxes with various thermal management solutions

Typical natural convection heat sinks are passive in nature and manufactured from copper or aluminum sheet, extruded aluminum, machined, or cast alloys. The heat sink cools a device by expanding the surface area of the part to which it is attached, increasing the amount of heat that can be dissipated by the ambient air.

Forced convection requires forced air velocity generated through the incorporation of either a dedicated or system-level fan in order to increase thermal efficiency. Fan heat sinks, high fin density assemblies, as well as board-level coolers are manufactured and configured for either impingement or cross-flow environments. Fan solutions include passive-active and active approaches. With a passive-active approach, fan heat sink solutions provide air flow and require little or no system air flow, while active fan heat sink solutions incorporate a fan that is attached to the solution. The air flow that a fan produces blows parallel to the fan's blade axis (Madrid 2000).

Fluid-phase change is a recirculating process, typically employs closed loop heat pipes which allow the rapid exchange of heat transfer through evaporation and condensation. Heat pipes are integrated into other heat sink technologies to further increase the thermal efficiency when greater density is required or physical size restrictions exist. Considerations in choosing heat pipes include: single-slot height, power consumption, noise elimination/reduction, low maintenance, sealed enclosure cooling, and extended ambient temperatures.

Thermionic cooling solutions, such as thermoelectrical and thermionic refrigeration, utilize solid-state conversion of low thermal conductivity but good electrical conductivity materials to achieve refrigeration. The electrons in the n-type material and the holes in the p-type material all carry heat away from the top metal-semiconductor junction, which leads to cooling of the junction, by means of a process called the Peltier effect. Conversely if a temperature difference is maintained between the two ends, higher thermal energy electrons and holes will diffuse to the cold side, creating a potential difference that can be used to power an external load. Due to the high dependence of the operation of thermionic cooling on the electronic behavior and properties of the semiconductor materials, it is highly influenced by the presence of magnetic fields (Chen and Shakouri 2002).

A great many liquid cooling solutions have been developed. One of them comprises channeled cold plates along with a heat exchanger and pump system in order to circulate fluids past a heat source. Generally, liquid cooled technologies are reserved for applications containing high heat flux density where forced convection or phase change systems are unable to dissipate the power demands.

Table 1.1 summarizes some typical hardware cooling solutions and their benefits and drawbacks. All of these thermal management solutions must be designed along with the electrical system. However, the thermal resistance between contact interfaces is usually high, and electrical insulating materials used in the electrical system are usually poor thermal conductors. Furthermore, the mismatches in the CTE of materials bonded together create stresses as a package warms and cools when electricity is turned on and off. These conflicts make the choice and design of thermal management methods quite a challenge (Couvillion 2006).

**Table 1.1** Typical conventional and emerging hardware cooling solutions

| Cooling types | Hardware device | Character description | Advantages | Disadvantages |
|---|---|---|---|---|
| Air cooling | Heat sink | The simplest of all the technologies and use natural or forced convections as the cooling method. Heat dissipation is directly proportional to available surface area | Extremely simple option; requires no pumping mechanisms; minimal maintenance; systems are inexpensive | Very low heat removal capacity; high thermal resistance; requires cleaning and dehumidification environments; added size and weight; acoustic noise due to fans |
| | Thermotunneling | Quantum electron tunneling provides cooling rates of about 3–5 W/cm$^2$, but has potential to achieve around 100 W/cm$^2$. Electrode work function needed at around 1–1.5 eV to improve tunneling effect | Can be packaged in arrays to cool virtually any size heat load; operates at 70–80% of the maximum theoretical efficiency for cooling; low cost | Very low heat removal capacity; tunneling achieved less than 10% of the total area; tension is introduced during the electroplating process; high work function of electrodes is needed |
| | Ionization | Ionization sticks to air cooling while providing cooling capacity comparable to liquid cooling at 40 W/cm$^2$. Management of nanosand, charge insulation and leftover voltage during use is needed | No moving parts, high reliability; best heat flux removal (~40 W/cm$^2$) achieved in air cooling techniques; small size, possible usage as a layer onto the chip | Possible problems of insulation due to the charge; leftover voltage will be converted to heat which hampers the cooling; possible problems of nanosand (dust, oils) that can inhibit heat-flow |
| Refrigeration | Vapor compressor | Vapor compressor is more efficient as compared to simple air cooling. The miniaturization of the vapor compression system is the critical success factor | Can generate cooling temperature below the ambient temperature; more efficient as compared to air cooling and cold plates | Difficulty in miniaturizing the system; additional components condenser, compressor and evaporator makes the system bulky |
| | Stirling cooler | Higher reliability is possible by using diaphragms instead of pistons and displacers. Scalability to micron level at acceptable cost is a key success factor | Higher efficiencies are possible, as compared with vapor compressor; low weight; possible size reduction to 1 cm | Diaphragms or pistons are moving parts, thus reducing reliability; cost is high |

(*continued*)

**Table 1.1** (continued)

| Cooling types | Hardware device | Character description | Advantages | Disadvantages |
|---|---|---|---|---|
| | Thermoelectric cooler | Thermoelectric Coolers are based on the Peltier Effect and can be scaled to micron level for on-chip cooling. Fabrication at the micron level is a key success factor | Solid state devices with no moving parts; can be made very small; can generate cooling temperature below the ambient temperature | Low efficiencies resulting in high power requirements and added waste heat; low cooling power density as compared to liquid cooling; high cost especially as size decreases |
| | Hetero-structure cooler | Two stage thermionic/thermoelectric micro-refrigerators may provide large cooling power densities (600 W/cm$^2$). Cost is a key success factor for commercial applications | Very small size and standard thin film fabrication method; possible to cool hotspots directly; high cooling power density and better transient response time as compared to thermoelectric cooler | Cost is high with packaging issues; high thermal conductivity of materials used; high contact resistance between metal and cap layer |
| | Superlattice structures | Improved performance over thermoelectric cooler in terms of speed, cooling power density and size | Extremely small size and thin film to provide site-specific microcooling of hotspots; provide best in class thermoelectric materials figure of merit | Lead (Pb) component in the thermoelectric superlattice |
| Liquid Cooling | Cold plates | Inexpensive as compared to other liquid based technologies and better performance than heat sinks. Power dissipation to weight ratio is a key success factor | Compared to heat sinks, water cold plates have very low thermal resistance and high thermal conductivity; can be effectively used with microchannels and heat pipes to increase the cooling capacity; inexpansive | Low heat removal capacity as compared to other liquid cooling technology; added size and weight; insulation to prevent leakage |
| | Jet impingement | Higher heat flux in two phase mode, although with increased complexity. Optimized device geometry and flow rates handle high pressure drops is critical | Very high heat fluxes can be achieved; critically designed nozzles not required; flow can be localized if only a single jet is used; can operate in single phase or two phase mode | High pressure drops, which may require larger pumps and optimized device geometry and flow rates; maximum chip heat flux removal is limited by the critical heat flux (2 phase); vapor droplets adherence to chips may hamper cooling |

| | | | |
|---|---|---|---|
| Spray cooling | Provides customized (temperature specific) and localized cooling. Geometry of nozzles and flow rates is a key success factor | Can provide higher heat flux as compared to jet impingement; gives good coverage due to atomizing action; extreme variations in flow rate magnitudes are possible; spray pattern can be flexible allowing for tracking of power density variations with time; localized cooling possible | Requires special spray nozzles; nozzles are sensitive to manufacturing tolerances and quality; nozzle action can change in time due to erosion, corrosion build up, and contaminants; spray velocity and momentum are critical |
| Two phase cooling device | Two phase liquid cooling provides high flux removal capacity and can cool hotspots directly | High cooling capacity and high heat transfer coefficients; can provide localized on spot cooling and easily be incorporated with microchannels; reduced sensitivity to gravity and acceleration as compared with spray cooling | The system tend to more complex with fabrication and miniaturization of no-moving-part pumps; extra insulation and sealing required; use of dielectrics or liquid metal alloys may lead to compatibility problems and environmental problems |

## *Software Solutions and Software-Based Dynamic Thermal Management*

Apart from the hardware cooling solutions, temperature control on a board was also achieved traditionally by using the system BIOS (basic input/output system) and passive cooling functionality. Passive cooling reduces the speed or disables some on-board devices in order to decrease power consumption, thereby reducing overall system temperature. Active cooling as discussed earlier, on the other hand, increases power consumption by activating fans for instance, that increase air flow, thereby reducing the system temperature.

In a typical software-based approach, the system will display central processing unit (CPU) die temperature in real time, allowing the user to read the normal temperature of the system and set accurate values for resume and overheat. With an advanced configuration and power interface (ACPI) solution, temperature control is moved from the BIOS to the operating system. In keeping with more sophisticated temperature control features, the inclusion of the ACPI in operating systems lets the user make more intelligent decisions with a better follow up on the CPU load and applications control. The ACPI thermal design is based on regions called thermal zones. Some systems may have more than one thermal sensor to allow subdividing the system in many more thermal zones. A benefit of ACPI is that it standardizes thermal control methods. In contrast, BIOS control methods are proprietary and do not let applications use them transparently. ACPI control also provides the choice of passive cooling, active cooling, or a mix of both (Madrid 2000).

In fact, both hardware- and software-controlled solutions usually exist for power and thermal management of processor-based systems. Software-based solutions are primarily utilized in connection with mobile platforms. The software-controlled techniques involve an interrupt generated when a processor temperature setting is exceeded. The processor may be throttled after detecting an above temperature condition by polling processor temperature. Generally, the software-controlled solutions have a slower response time than the hardware-controlled solutions. In addition, it tends to overshoot and undershoot problems with software-controlled solutions. The sensors utilized in software-controlled solutions are relatively slow and inaccurate. The on-die sensor (which is normally a diode) is not located on the hottest part of the processor die. The software-controlled solution is based on the premise that the platform exposes a variety of trip points to the operating system. A trip point is a temperature for a particular thermal region when some action should be taken. As the temperature goes above or below any trip point, the platform is responsible for notifying the operating system of this event and the operating system then takes an appropriate action. When a temperature crosses a passive trip point, the operating system is responsible for implementing an algorithm to reduce the processor's temperature. It may do so by generating a periodic event at a variable frequency. The operating system then monitors the current temperature as well as the last temperature and applies an algorithm to make performance changes in order to keep the processor at the target temperature. Software-controlled throttling is exposed to the operating system, allowing the operating system to know the processor performance at all times. This becomes especially important with future operating systems that guarantee some quality of service based upon the processor performance to the executing applications. This concept is known as guaranteed bandwidth allocation and is based on the processor's current performance level (Cooper 2006).

Another typical software-based solution is DTM. This refers to a range of possible hardware and software strategies which work dynamically, at run-time, to control a chip's operating temperature. In contrast, the packaging and fans for a CPU or computer system were traditionally designed to be able to maintain a safe operating temperature even when the chip was dissipating the maximum power possible for a sustained period of time, and therefore generating the highest amount of thermal energy. This worst-case thermal scenario is highly unlikely, however, and thus such worst-case packaging is often expensive overkill. DTM allows packaging engineers to design systems for a target sustained thermal value that is much closer to average-case for real benchmarks. If a particular workload operates above this point for sustained periods, a DTM response will work to reduce chip temperature. In essence, DTM allows designers to focus on average, rather than worst-case, thermal conditions in their designs. Until now, techniques developed to reduce average CPU power have garnered only moderate interest among the designers of high-end CPUs because thermal considerations, rather than battery life, were their primary concern. Therefore, in addition to reducing packaging costs, DTM improves the leverage of techniques such as clock gating designed to reduce average power (Brooks and Martonosi 2001).

The key goal of DTM is to provide inexpensive hardware or software responses that reliably reduce power, while impacting performance as little as possible. Implementing an effective DTM system involves (1) Selecting simple and effective triggers; (2) Identifying useful response mechanisms; and (3) Developing policies for when to turn responses on and off. For trigger selection, DTM allows arbitrary tradeoffs between performance and savings in cooling hardware. Conservative target selections can still lead to significant cost improvements with essentially zero performance impact, because the trigger point is rarely reached for many applications. In addition, DTM makes other techniques targeting average power more interesting to the designers of high-end CPUs. Effective DTM makes average power the metric of interest even for high-end CPU designers, since packages need no longer be designed for worst-case power. With DTM, lowering average CPU power will reduce the trigger value needed for a particular level of performance, and thus will reduce packaging costs. Not unexpectedly, the triggering delay is a key factor in the performance overhead of DTM. More lightweight, fine-grained policies, such as the microarchitectural techniques often allow the temperature to stay close to the target level with a small performance penalty. Furthermore, the fine-grained policies are less affected by rapid fluctuations in the temperature. Because of these growing opportunities for microarchitectural DTM techniques, a methodology for evaluating new DTM approaches has been explored. This mechanism correlates power and performance, and looks for "low-hanging fruit;" that is, techniques that can cut power by significantly more than they hurt performance are needed. Identifying these sorts of wasted work, particularly on an application-specific basis, appears to be a promising way of discovering new microarchitectural DTM techniques in the future (Brooks and Martonosi 2001).

In fact, several DTM techniques have been developed. One proposed method is transient thermal management, which involves using heat storage devices to store the heat produced by the processor during power intensive computations and dissipate it gradually over time. This method is combined with a dynamic thermal management technique to lower the power consumption (Cao et al. 1996; Wirth 2004). A system has been developed by monitoring the temperature of the processor and as it reached the set maximum it stalled any power intensive instructions. This system allows the processor to still run normally for low power instruction, which benefited the user over a system which slowed the processor down for all applications (Wirth 2004).

Another DTM technique is to design a system that slows down the entire processor (Brooks and Martonosi 2001). While most of the DTM systems that have been simulated measure the power and temperature and then cause the system to change when the processor is consuming too much power or the temperature is too high, a predictive voltage scaling method uses an algorithm that predicts the power to be used and then sets the voltage and frequency accordingly. This method may have better response time than reactive systems (Srinivasan and Adve 2003). A different idea involving voltage and frequency scaling has been developed with the design that involves separating the chip into different sections in which the voltage and frequency can be scaled independently of the other sections. This design would minimize the performance loss on those tasks that are not power intensive. A similar method uses a performance manager to analyze the power requirements of the next task and then set the voltage and frequency based on values given in a look up table. A performance manager is especially viable in embedded processors because the workload is typically limited to a small range of instructions for which the entire thermal management system can be optimized. In addition, chip architectures have been developed, which have several built-in features to run at lower power during times when the device is not being used and allow for both dynamic voltage and frequency scaling (Clark et al. 2002). Taking the thermal considerations as a part of the design process rather than an afterthought, combining DTM systems to create even more efficient designs for future processors would be a future direction. Still, many of these DTM techniques need to be tested on actual devices to measure their true performance and feasibility. Chip designers will work on developing lower power chips and hardware that has DTM techniques built into the architecture. Meanwhile, better ways of judging performance by developing good, workable thermal requirements have been explored. Whatever direction the future of the industry leads, the problem of heat dissipation has no magic solution. As the processor density continues to increase, so does the need for better methods of heat dissipation (Wirth 2004).

## *Optimal Thermal Design of a Package*

The optimal thermal design is a fundamental approach for future micro- and nanoscale electrothermal modeling, engineering, and processing as well as present packaging thermal management, based on the minimization of thermal resistance. The thermal resistance at device level is associated with conduction of heat from the die to its substrate to the surface of the package. There is often significant thermal resistance in the attachment of the die to the substrate, such as contact or interface resistance. From the chip-level package surface, heat can be removed by a coolant and/or conducted to a board. At the board level, heat is removed by the cooling system or moved to a higher packaging level (Couvillion 2006). The overall thermal resistance of a package is almost entirely determined by (NEC Electronics 2003) package structure, package size, chip dimensions, airflow rate, as well as hardware cooling assemblies.

Different package structures have different thermal resistance characteristics. Packages such as ABGAs (advanced ball grid arrays) and FCBGAs (flip chip ball grid arrays), which feature a copper lid to which the chip is directly attached with thermally conductive paste, offer excellent thermal resistance characteristics. In the case of PBGAs (plastic ball grid arrays), thermal resistance can be lowered by using a four-layer substrate instead of a two-layer substrate, and it can be further lowered by placing solder balls directly underneath the thermal via holes.

In general, the larger the size of the package, the lower its thermal resistance is. This is particularly true for ABGAs and TBGAs (tape ball grid arrays), which have a copper lid offering excellent thermal conductivity characteristics. In the case of packages with lower thermal conductivity such as FPBGAs (flip chip plastic ball grid arrays), there is a weaker correlation between thermal resistance and package size, and there is little variation in thermal resistance among different size packages if the chip dimensions are the same.

Large chip size benefits to the low thermal resistance. The thermal conductivity of silicon (the material chips are made of) is about 100 times higher than that of mold resin and about 10 times higher than that of package substrates; therefore, the surface area of the chip itself greatly contributes to heat dissipation.

Airflow rate and hardware cooling assembly are effective to reduce the thermal resistance. The airflow rate is not directly related to the thermal conductivity of the package itself, but forced air cooling, such as through the use of a fan, efficiently transfers heat from the package surface or printed wiring board to the surrounding atmosphere and thus reduces thermal resistance.

There have been many approaches for optimal thermal design to reduce the thermal resistance and maximize the intrinsic performance and lifespan of semiconductor devices, helping meet market needs for higher performance as well as saving costs by curtailing the need for heat dissipation measures. For example, the thermal resistance of QFPs (quad flat packages) can be reduced through appropriate material selection and lead frame design (NEC Electronics 2003): (1) Employment of thermally enhanced resin. Because mold resin is the lowest thermally conductive material among package materials, thermal enhancement of the mold resin can significantly lower the thermal resistance of packages. One way to do this is to replace the existing filler with high thermal conductivity filler. Furthermore, the thinner the resin on the chip, the shorter the distance over which heat is transmitted, thus, mounting a heat sink on the surface of the package enhances the heat dissipation effect to a greater extent in the case of tape-bonded quad flat packages, which have a thinner resin thickness than QFPs. (2) Lowering thermal resistance through lead frame design. Thermal resistance can be lowed by changing the material of the lead frame from alloy 42 (Fe–Ni) to a copper alloy; designing the die pad on which the chip is to be mounted as large as possible; and attaching a heat spreader or other cooling system to the die pad so that heat is transmitted almost throughout the entire package.

Similarly, the thermal resistance of BGAs (ball grip arrays) can be lowered by optimal material selection, package structure placement, and substrate design. As in the case of QFPs, the thermal resistance of BGAs can be decreased by using material with a higher thermal conductivity for the filler. For package structure and substrate design, the following measures may be taken (NEC Electronics 2003): (1) Use of thermal balls as a low-cost solution. In terms of package structure, a heat dissipation path is secured from the rear side of the chip to the solder balls immediately beneath the chip by providing a large number of solder balls on the rear side of the chip, and thermally connecting these solder balls to the die pad via through-holes (thermal vias). The balls at the center of a package are electrically grounded and commonly called "thermal balls" as they play a thermal dissipation role by conducting heat to the printed wiring board. This is the cheapest way to dissipate heat. The thermal balls also serve as ground pins and neighboring balls can be assigned as signal pins, meaning that the actual number of pins can be increased. (2) Use of two inner layers of package substrate as ground layers. Generally, a printed wiring board having four or more layers, including power and ground layers, is used as the package substrate for BGAs to ensure satisfactory electrical characteristics. However, the use of a four-layer structure to lower the thermal resistance, rather than improve electrical characteristics, has been increasing. In this case, heat from the chip is transmitted to the inner layers via the die pad's through-holes, which also serve as grounds, and two out of the four inner layers are used as grounds to secure a heat dissipation path. To further decrease thermal resistance, a substrate with a thick embedded metal core layer has been used. (3) Use of an embedded heat spreader. If the combination of thermal balls and a four-layer package substrate still fails to satisfy the thermal resistance requirement, a heat spreader can be embedded in the package. Such a heat spreader serves to diffuse the heat transmitted through the mold resin to the surface of the package. However, the reduction in thermal resistance that can be achieved this way is limited because the heat spreader is not in direct contact with the die pad and the chip. (4) Modification of printed wiring board design. The thermal resistance of a BGA can also be lowered by modifying the design of the printed wiring board. The thermal resistance changes according to the number of thermal holes, the number of layers of the package substrate, and the presence or absence of a heat dissipation path. Reassessment of the thermal design of the entire package, including the printed wiring board, can result in a low-cost package that meets thermal resistance requirements. (5) Selection of cavity-down type PBGA. Cavity-down type BGAs in which the chip is flipped and attached to the heat spreader that is exposed at the surface, is the most effective solution for lowering thermal resistance. This also holds true for QFPs. Packages of this type include TBGAs, ABGAs, and FCBGAs. Heat is directly conducted from the chip to the copper plate on the package surface, thereby achieving low thermal resistance.

In addition, measuring the temperature of the chip after it has been installed in a system is very important and can be used to verify or optimize the thermal design. Such measurement data are also useful for estimating the power consumption of devices to be developed in the future. Furthermore, if the power consumption of the system is known to be considerably above the value estimated at the design phase, it is very important to know the junction temperature of the device under actual use in the system.

## Fundamentals of Heat Transfer and Thermal Calculation in Electronic Packaging

The objective of thermal management in electronic packaging is to efficiently remove heat from the semiconductor junction to the ambient environment. This process includes: (1) heat transfer within the semiconductor component package; (2) heat transfer from the package to a heat dissipater, such as heat spreader or the initial heat sink; (3) heat transfer from the heat dissipater to the ambient environment through the ultimate heat sink, or other cooling systems. Achieving an efficient thermal connection through these paths requires a thorough understanding of heat transfer fundamentals as well as knowledge of available interface materials and how their key physical properties affect the heat transfer process (Chomerics 1999).

Heat transfer is the movement of heat flow as a result of a temperature difference. Temperature represents the amount of thermal energy available, whereas heat flow represents the movement of thermal energy from place to place. On a microscopic scale, thermal energy is related to the kinetic energy of molecules. The higher a material's temperature, the greater the thermal agitation of its constituent molecules is, which is manifested both in linear motion and vibrational modes. It is a natural process for regions containing greater molecular kinetic energy to pass this energy to regions with less kinetic energy. Several material properties serve to modulate the heat transferred between two regions at differing temperatures, such as thermal conductivities, specific heats, material densities, fluid velocities, fluid viscosities, surface emissivities, and more. Taken together, these properties serve to make the solution of many heat transfer problems an involved process.

The mechanisms of heat transfer depend on the media involved and usually divided into conduction, convention, radiation, and multimode, which is a combination of one or more of the above. When a temperature gradient exists within a continuous, nonmoving medium, solid or stationary fluid, heat is transferred through the medium via the conduction mode. Convection heat transfer occurs when a surface is in contact with a moving fluid, liquid, or gas at a different temperature. Heat is exchanged between the fluid and the surface. Radiation heat transfer occurs when two surfaces at different temperatures exchange energy in the form of electromagnetic energy emitted by the surfaces. Radiation can occur in a vacuum because no medium between the two surfaces is required as in conduction and convection (Couvillion 2006).

### *Conduction*

Thermal conduction is a process in which heat flows through a solid, liquid, or gas or between two media that are in intimate contact. Conduction is the dominant mechanism for heat transfer within solids, involving the transfer of kinetic thermal energy from one electron to another to cause no visible motion of the particles of the body. Conduction through dielectric solids is almost entirely due to lattice vibrations, while conduction through metallic solids has added energy transport by free electrons. Liquids also conduct thermal energy, but to a significantly lesser extent than solids. When a material changes phase from a solid to a liquid, there is a lessening of its intermolecular bonds and a deterioration of the ordered state of the solid, therefore, more freedom for thermal motion of the molecules makes its thermal conductivity lower. When a liquid changes to a gas, there is a further loosening of molecular bonds that allows random motion of the gas molecules with only restraints of random collisions. As a result, the thermal conductivity of gases is quite low (Krum 2004). Based on Fourier's law of heat conduction, the rate at which heat is conducted through a material is proportional to the area normal to the heat flow, the temperature gradient along the heat flow path, and the thermal conductivity of the material. For a one dimensional, steady state heat flow, as shown in Figure 1.5a, the rate is expressed by Fourier's equation:

$$q = kA\frac{\Delta T}{L}, \tag{1.4}$$

or

$$q'' = \frac{q}{A} = -k\frac{\mathrm{d}T}{\mathrm{d}x}, \tag{1.5}$$

where $q$ is rate of heat transfer (W); $q''$ is heat flux (W/m$^2$); $k$ is thermal conductivity (W/m K); $A$ is cross sectional transfer area (m$^2$); $\Delta T$ is temperature difference (°C); $L$ is conduction path length. Thermal conductivity, $k$, is an intrinsic property of a homogeneous material which characterizes the material's ability to conduct heat. This property is independent of material size, shape, or orientation. For nonhomogeneous materials, however, for instance those having glass mesh or polymer film reinforcement, the term of relative thermal conductivity is usually used because the thermal conductivity of these materials depends on the relative thickness of the layers and their orientation with respect to heat flow (Chomerics 1999).

Another inherent thermal property of a material is its thermal resistance, $R$, which can be derived from (1.4) and (1.5), and is defined as:

For a slab,

$$R = \frac{\Delta T}{q} = \frac{L}{kA}. \tag{1.6}$$

For a cylindrical shell:

$$R = \frac{1}{2\pi kH} \ln\left(\frac{r_2}{r_1}\right), \tag{1.7}$$

where $H$ is the length of the cylinder; $r_1$ is the inside radius of the cylinder shell; and $r_2$ is the outside radius of the cylinder shell.

For a spherical shell:

$$R = \frac{1}{4\pi k} \ln\left(\frac{1}{r_1} - \frac{1}{r_2}\right), \tag{1.8}$$

where $r_1$ is the inside radius of the spherical shell; and $r_2$ is the outside radius of the spherical shell.

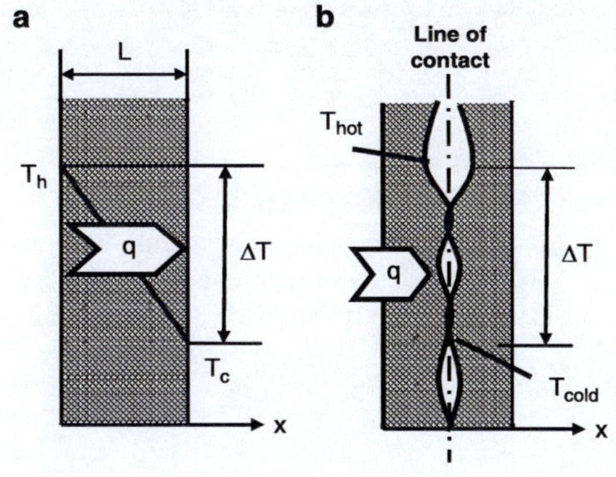


**Fig. 1.5** Schematic illustration of (**a**) one-directional conduction; (**b**) and contact interface

For a conical frustum:

$$R = \frac{L}{\pi k r_1 r_2}, \tag{1.9}$$

where $L$ is the length of the conical frustum; $r_1$ and $r_2$ is the radius of the top and bottom of the conical frustum, respectively.

The thermal resistance is a measure of how a material of a specific thickness resists the flow of heat. For homogeneous materials, thermal resistance is directly proportional to the thickness. For nonhomogeneous materials, the resistance generally increases with thickness but the relationship may not be linear. Thermal conductivity and thermal resistance describe heat transfer within a material once heat has entered the material. When different surfaces contact each other, the contact interface between two surfaces can also produce a resistance to the flow of heat, as shown in Figure 1.5b. Actual contact occurs at the high points, leaving air-filled voids where the valleys align. Air voids resist the flow of heat and force more of the heat to flow through the contact points. This constriction resistance is referred to as surface contact resistance and can be a factor at all contacting surfaces. The total impedance ($\theta$) of a structure is defined as the sum of its material thermal resistance and any contact resistance between the contacting surfaces. Surface flatness, surface roughness, clamping pressure, material thickness, and compressive modulus have a major impact on contact resistance. Because these surface conditions can vary from application to application, thermal impedance of a structure will also be application dependent (Chomerics 1999).

When several materials are stacked in series, such as a die attached with epoxy to a substrate that is soldered to a package base, the total thermal resistance becomes the sum of the individual thermal resistances. For $N$ thermal resistances in series, the total thermal resistance $\theta$ is

$$\theta = \theta_1 + \theta_2 + \theta_3 + \cdots + \theta_N. \tag{1.10}$$

The temperature at a particular interface of an electronic package may be calculated as follows (Krum 2004):

$$T_{j,j-1} = T_{\text{hs}} + q \sum \theta_{j-\text{hs}}, \tag{1.11}$$

where $T_{j,\,j-1}$ is the temperature at interface of layers $j$ and $j-1$; $T_{\text{hs}}$ is the contact surface temperature of the heat sink; $\sum \theta_{j-\text{hs}}$ is the sum of thermal resistances from interface of $j$ and $j-1$ to the heat sink.

When there is more than one heat path, for instance $N$ thermal paths, from the dissipating element to ambient, the total thermal resistance is calculated as

$$\frac{1}{\theta} = \frac{1}{\theta_1} + \frac{1}{\theta_2} + \frac{1}{\theta_3} \cdots + \frac{1}{\theta_N}. \tag{1.12}$$

## *Convection*

Convection is the thermal energy transfer between two surfaces as a consequence of a relative velocity between them. It occurs only in fluids wherein the transfer mechanism is the mixing of the fluids. Although each of the surfaces may be a fluid, the most practical application is a solid surface and the other is fluid (Krum 2004). When heat conducts into a static fluid it leads to a local volumetric expansion. As a result of gravity-induced pressure gradients, the expanded fluid parcel becomes buoyant and displaces, thereby transporting heat by fluid motion, i.e., convection, in addition to conduction. Such heat-induced fluid motion in initially static fluids is known as free convection, as shown in Figure 1.6a. For cases where the fluid is already in motion, heat conducted into the fluid will be transported away chiefly by fluid convection. These cases, known as forced convection as shown in Fig. 1.6b, require a pressure gradient to drive the fluid motion, as opposed to a gravity gradient to induce motion through buoyancy. The heat exchange between solid surface and circulating fluid can be described by Newton's law:

$$q = hA\Delta T, \tag{1.13}$$

or

$$q'' = \frac{q}{A} = h\Delta T, \tag{1.14}$$

where $q$ is heat transfer rate (W); $A$ is surface area ($m^2$); $q''$ is heat flux ($W/m^2$); $\Delta T = T_s - T_\infty$, $T_s$ is surface temperature (°C), and $T_\infty$ is fluid temperature (°C); $h$ is convention heat transfercoefficient ($W/m^2$ K). For electronic packaging, the thermal resistance for convection, $R$ (K/W), can be expressed as

$$R = \frac{\Delta T}{q} = \frac{1}{hA}. \tag{1.15}$$

The convention heat transfer coefficient $h$ mainly depends on nature of the fluid motion, fluid properties, and surface geometry, and is usually obtained experimentally. For example, $h = 1.4\ (\Delta T/A^{1/2})^{1/4}$ for the nonlinear natural air convection; $h = 4.0\ (V_f/L)^{1/2}$ for linear forced air convection, where $V_f$ is air flow velocity, $L$ is the character length in flow direction.

In the electronic packaging thermal management, natural convection can be used for passive cooling, while forced convection mainly for active cooling. Natural convection is caused entirely by differences in density within the fluids resulting from different temperatures and does not use externally forced air movement. Heat flows by conduction or contact from the solid surface to the fluid particles in intimate contact with the surface, and there is a resulting boundary layer of hot

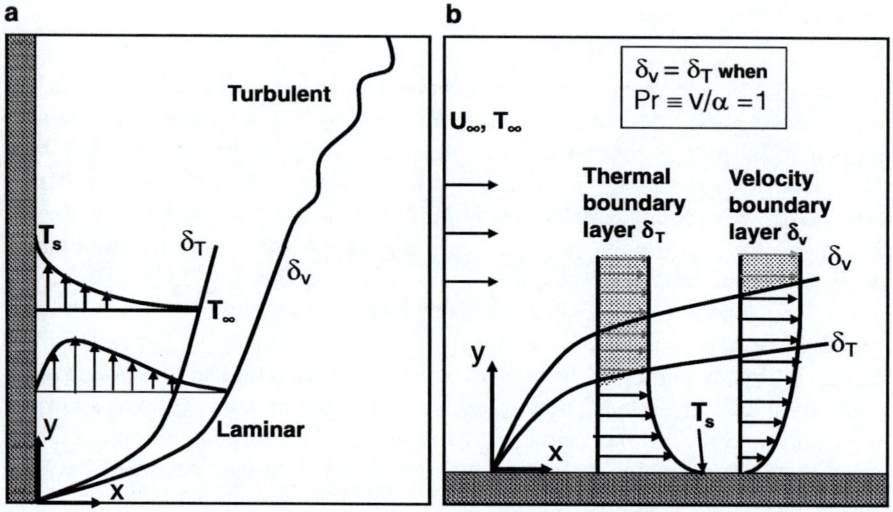


**Fig. 1.6** Schematic illustration of **(a)** natural convention; and **(b)** forced convention

air immediately adjacent to the surface. In forced convection, thermal energy is transferred from the solid to the adjacent fluid particles in the same manner as in natural convection. However, the subsequent fluid action occurs through artificially induced fluid motion generated by fans, pumps, or blowers. In air cooling, for instance, three types of air moving devices are usually used: centrifugal, propeller, and axial flow. Centrifugal fans are designed to move small volumes of air at high velocities and are capable working against a high resistance. Propeller types are designed to move large volumes of air at low velocities. Axial flow fans are an intermediate type of air mover between the centrifugal and propeller types. In addition, forced convection can be divided into laminar flow and turbulent flow. For air, the transition from laminar to turbulent flow usually occurs at a velocity of 180 linear feet per minute (54.864 m/min). Turbulent flow, characterized by the irregular motion of fluid particles, has eddies in the fluid in which the particles are continuously mixed and rearranged. The heat is transferred from eddies back and forth across the streamlines. Therefore, the greater heat transfer occurs for turbulent flow (Krum 2004). Whatever laminar flow or turbulent flow, the convection cooling method is wholly dependent on the movement of the fluid surrounding the heat dissipating element.

## *Radiation*

All objects with a temperature above 0 K emit thermal radiation. Radiation cooling is the transfer of heat by electromagnetic emission, and is maximized in a complete vacuum. Radiation from solid objects may be considered to be a totally surface-related phenomenon, and the radiators can be classified as black bodies, gray bodies, and selective radiators. A black body is defined as a surface that absorbs the entire thermal radiation incident upon it, neither reflecting nor transmitting any of the incident radiation. Good absorbing materials are also good emitting materials. The black body, at any given temperature, radiates more energy, both in the total spectrum and for each wavelength interval, than any other temperature radiator and more than any gray body or selective radiator at the same temperature. A gray body is defined as a radiator that has the same spectral emissivity for all wavelengths. A selective radiator is one in which the emissivity varies with wavelength (Krum 2004). Materials used for electronic packaging are usually gray bodies. All materials radiate thermal energy in amounts determined by their temperature, where the energy is carried by photons of light in the infrared and visible portions of the electromagnetic spectrum. When temperatures are uniform, the radiative flux between objects is in equilibrium and no net thermal energy is exchanged. The balance is upset when temperatures are not uniform, and thermal energy is transported from surfaces of higher to surfaces of lower temperature. The transfer of energy by electromagnetic waves can be expressed by Stefan–Boltzmann's Law, as shown in Figure 1.7, when two surface, gray diffuse enclosure presents with radiatively nonparticipating media,

$$q_1 = -q_2 = \frac{\sigma(T_1^4 - T_2^4)}{\frac{1-\varepsilon_1}{\varepsilon_1 A_1} + \frac{1}{A_1 F_{12}} + \frac{1-\varepsilon_2}{\varepsilon_2 A_2}}. \tag{1.16}$$

For small surface in large surrounds,

$$q_1 = \varepsilon_1 \sigma A (T_{s1}^4 - T_{s2}^4). \tag{1.17}$$

For Infinite parallel plates,

$$q_1 = \frac{\sigma A (T_1^4 - T_2^4)}{\frac{1}{\varepsilon_1} + \frac{1}{\varepsilon_2} - 1}. \tag{1.18}$$

For a three surface gray diffuse enclosure,

$$q_1 + q_2 + q_3 = 0, \tag{1.19}$$

where $\varepsilon$ is surface emissivity ($0 < \varepsilon < 1$), for enclosure with black surfaces, $\varepsilon_i = 1$; $\sigma$ is Stefan–Boltzmann constant, $5.67 \times 10^{-8}$ W/m$^2$ K$^4$; $T_i$ is surface temperature (K) of the emission material. The thermal resistance can be expressed as

$$R = \frac{\Delta T}{q} = \frac{1}{h_r A}, \tag{1.20}$$

For electronic devices, $h_r \approx 2\text{–}6\ \mathrm{W/m^2\,K}$. Radiation cooling is dependent on the temperature difference between objects, their emissivity, and surface area. From a materials standpoint, the thermal design can be optimized with the emissivity parameters.

Space cooling is a particular example of radiation cooling application. In the vacuum of space, radiation plays a major role in the cooling of the electronic systems. Internal to the spacecraft, conduction plays a major role where material interfaces have increased thermal resistance due to the lack of air or any other gas to conduct the heat. In addition, the thermal resistance in high vacuum is further increased by the reduction in joint thermal conductance. For instance, the interface of dry aluminum on aluminum in a vacuum has a thermal conductance approximately half of that seen in air. This degradation varies with each application (Krum 2004).

## *Multimode Heat Transfer in Electronic Packaging*

A major property of packages is how they dissipate the heat generated when a current flows through a resistor in an electric circuit. An electronic system may emit smoke or catch fire if its device generates more heat than anticipated. Excessive heat may also degrade the performance of the device by lowering its operating

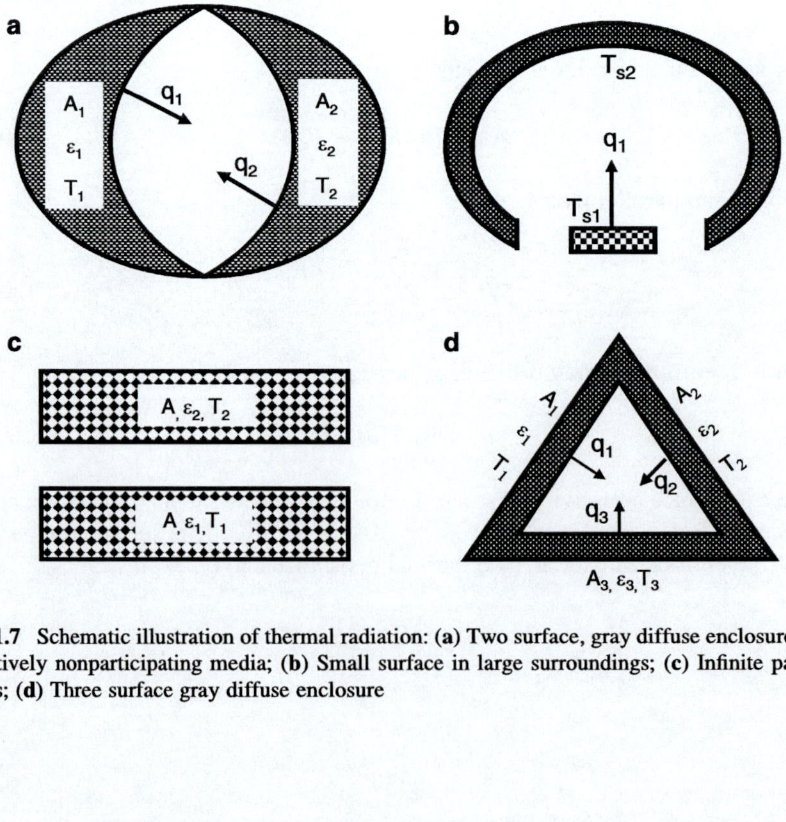


**Fig. 1.7** Schematic illustration of thermal radiation: (**a**) Two surface, gray diffuse enclosure with radiatively nonparticipating media; (**b**) Small surface in large surroundings; (**c**) Infinite parallel plates; (**d**) Three surface gray diffuse enclosure

speed, and in the worst case, damage the device, rendering it inoperable. Even if the worst case can be avoided, reliability is adversely affected through device malfunctions and a shorter system life. Heat transfer in electronic packages is via multiple, coupled modes simultaneously. Figure 1.8 shows heat dissipation paths and causes of thermal resistance, which are influenced by (1) chip area, heat generation (hotspots), and power consumption; (2) Package materials, structure, dimensions and heat spreader or heat sink; and (3) operating environment, such as cooling conditions, structure of mounting printed wiring board, mounting density, and ambient temperature. Heat dissipation has been designed to be done mostly through the printed wiring board. Because heat radiation effectively occurs only when the surface area of the package is extremely large, heat is actually dissipated from the package via (1) from the surface of the package into the atmosphere; (2) from the external cooling system to the printed wiring board and then into the atmosphere; (3) from the heat source to the sides of the package. Of these three paths, the heat dissipation path via the printed wiring board is the most effective and accounts for 80% of total heat dissipation according to some calculations.

The problem to be solved involves multiple media. Appropriate governing equations must be written based on conservation principles for each medium, along with boundary, interface and initial (if transient) conditions. Traditionally, a single technique is applied at one level of the packaging hierarchy at a time. Multiscale approaches are being developed to analyze across the packaging hierarchy, i.e., modeling performed at the chip-carrier, board, and system levels. The common techniques used include (1) resistor network approach; (2) analytical approach; and (3) numerical solution of governing equations.

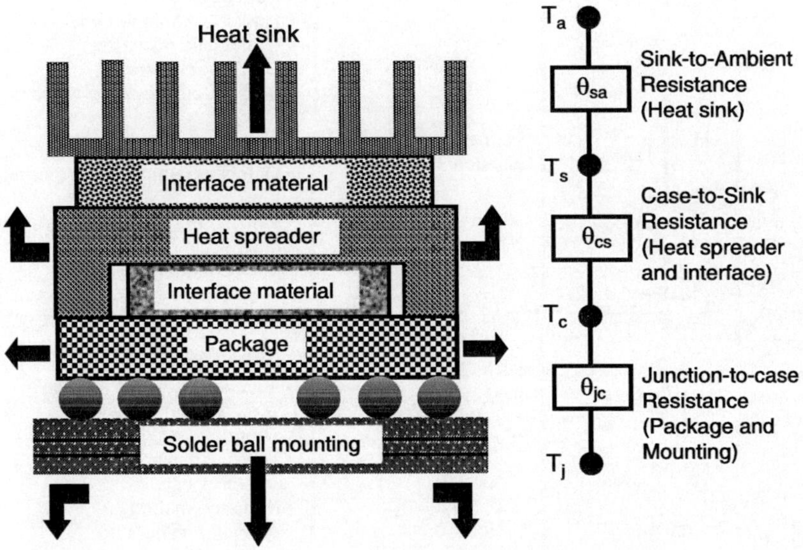


**Fig. 1.8** Schematic illustration of heat flow paths in an electronic package

## *Microscale Heat Transfer*

Advances in microfabrication processes of microelectronic devices, such as field effect transistors that contain semiconductor (e.g., silicon), insulator (e.g., silicon dioxide), and metallic (e.g., copper interconnects) layers only a few nanometers thick, have posed a challenge on accurate thermal modeling and design of microelectronic devices and thin film structures at micro- and nanoscales. Microscale heat transfer analysis is unconventional and often challenging because the Fourier heat conduction equation or continuum assumption fails as the characteristic dimension of the structures becomes comparable with the mean free path of energy carriers in phonons of semiconductors and electrons of metals. A phonon refers to the quanta of energy of lattice vibrations in semiconductors such as Si and Ge. Phonons can be treated as particles despite the fact that they are propagating wave packets, which carry energy across the lattice (Asheghi and Liu 2007).

Figure 1.9 gives a schematic hierarchy of microscale heat transfer in electronic packaging. This graph also provides a general guideline for the appropriate treatment of phonon transport in nanostructures. Phonon transport can be predicted using the

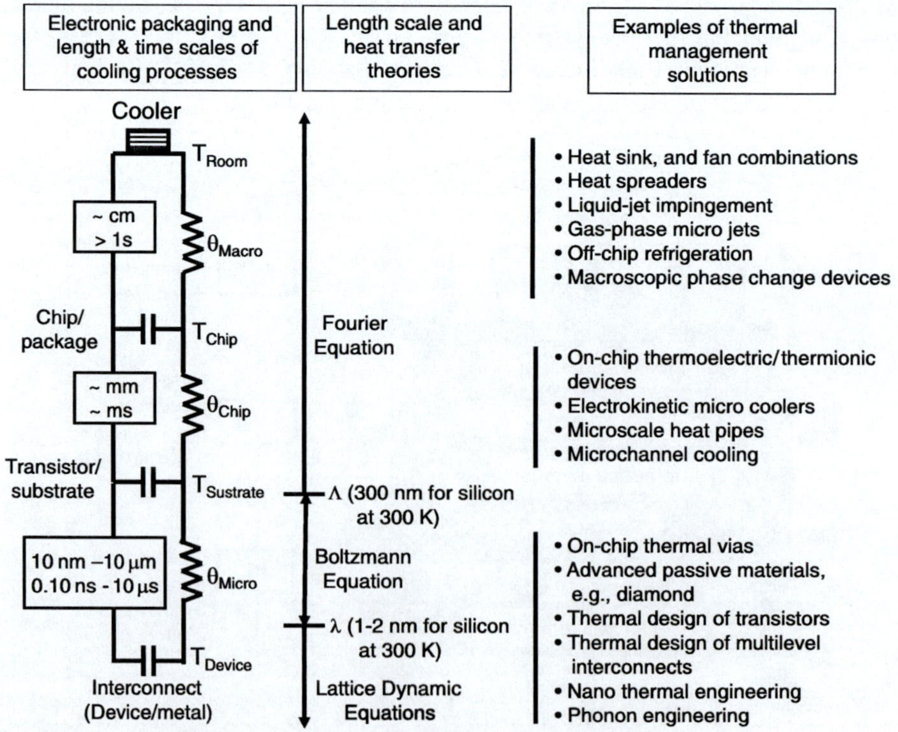


**Fig. 1.9** Schematic hierarchy of microscale heat transfer in an electronic package

Boltzmann particle transport equation (BTE), which is required only when the scattering rates of electrons or phonons vary significantly within a distance comparable to their respective mean free paths (Asheghi and Liu 2007). BTE is a general transport equation, and Fourier equation is a special case of simplified BTE. BTE describes particle-like behavior for phonons, photons, and electrons using statistical particle assembles, which consists of diffusion term, acceleration term, and scattering term (Narumanchi et al. 2006).

$$\frac{\partial f}{\partial t}+v\cdot\nabla f+F\cdot\frac{\partial f}{\partial p}=\left(\frac{\partial f}{\partial t}\right)_{\text{scat}}, \tag{1.21}$$

$$\left(\frac{\partial f}{\partial t}\right)_{\text{scat}}=\frac{f_0-f}{\tau}+q_{\text{electron-phonon}}, \tag{1.22}$$

$$\frac{1}{\tau}=\frac{1}{\tau_{\text{i}}}+\frac{1}{\tau_{\text{u}}}+\frac{1}{\tau_{\text{b}}}, \tag{1.23}$$

where $f$ is statistical distribution of particles per unit volume and solid angle, depending on time, location and momentum, such as phonon frequency; $v$ is phonon velocity vector, anisotropic and frequency dependent, typically modeled as isotropic for simplicity and lack of comprehensive data; $F$ is external acceleration field and negligible for phonons. Equation (1.22) shows that the scattering term is proportional to difference between phonon equilibrium distribution ($f_0$) and actual distribution ($f$) over time to reach equilibrium, i.e., relaxation time $\tau$, which can be obtained from (1.23), where $\tau_i$ *is due to imperfections;* $\tau_u$ *is due to phonon–phonon scattering; and* $\tau_b$ *is due to phonon-boundary scattering.*

The BTE simply deals with the bin counting of the energy carrier particles of a given velocity and momentum, scattering in and out of a control volume at a point in space and time. Analysis of the heat transfer in microelectronic devices, interconnects, and nanostructures using the BTE is very cumbersome and complicated, even for simple geometries, and have been the topic of research and development in the field of micro- and nanoscale heat transfer for the past two decades (Asheghi and Liu 2007).

Lattice dynamic equation can be used to simulate heat conduction across broad length scales with continuum and subcontinuum effects, and solve the BTE by calculating relevant variables on discrete nodal points for both equilibrium and nonequilibrium phenomena. Starting with (1.21) and (1.22), multiplying the equation with $h\omega_{\text{p}}D_{\text{p}}(\omega)$, integrating over frequency, and the energy formulation can be obtained (Narumanchi et al. 2006):

$$\frac{\partial e}{\partial t}+v\cdot\nabla e=\frac{e_0-e}{\tau} \tag{1.24}$$

Or for one-dimensional space

$$\frac{\partial e}{\partial t} + v_x \cdot \frac{\partial e}{\partial x} = \frac{e_0 - e}{\tau} \tag{1.25}$$

This method can be extended to include electron transport and phonon-election coupled problems.

For example, the direct simulation of the Boltzmann equation of phonon is being developed and applied to the heat conduction analysis of thin film. However, when dealing with liquid or interphase phenomenon, which is inevitable for phase-change heat transfer, the most powerful tool for the investigation of the microscopic phenomena in heat transfer is the molecular dynamics method. In principal, the molecular dynamics method can be applied to all phases of gas, liquid, and solid and to interfaces of these three phases (Maruyama 2000).

Based on the microscale heat transfer simulation, intelligent electrothermal design along with careful floor planning and microscale thermal management at the device level can largely reduce the temperature rise within a device. This will help to prevent the problem at the early stages and at the device level, rather than to pass the problem to the package-level thermal design (Asheghi and Liu 2007).

## Design for Advanced Thermal Management of Electronic Packaging

The thermal design of electronic packaging has become the most important aspect of product design that ensures reliability of the electronic device and thus its ability to compete in a demanding market. The primary objective of the thermal design is to plan the thermal balance of the product so that it operates reliably within the specified environmental conditions throughout its service life (Hienonen et al. 1997). This is generally accomplished by comprehensively considering all factors during thermal design, such as (1) optimizing the heat remove paths from the integrated circuits, i.e., backside of the chip or the substrate for instance; (2) choosing the optimal cooling methods; (3) using highly conductive materials or thermal vias as possible to consume a large of the substrate, reduce interconnect capacity and thermal resistances; (4) minimizing stresses induced in the chips and substrate due to mismatches in thermal coefficients of expansion; and (5) assuring that the set of design goals, the modeling type, the design accuracy, and the test and measurement methods work together so that unintentional over- or underdesign is avoided and that excessively expensive modeling and/or lengthy testing are avoided. All of these considerations will impact performance, cost, and reliability of the product. Furthermore, gathering information for the most critical failure mechanisms of the intended components and materials is necessary to ensure that the main activities related to the thermal design focus on details where possible failures primarily affect the usability and reliability of the product (Hienonen et al. 1997).

## *Thermal Design Guidelines*

Thermal design is a crucial part of the overall research and development process because thermal control solutions decisively influence the technologies that can or must be used in the electronic product. It is important that thermal, electrical, mechanical, ergonomic, electromagnetic design characteristics are dealt with concurrently with other aspects of the product design. The influence of thermal design on the product is shown in Figure 1.10. Almost all the properties of components and materials change with the temperature and humidity, therefore, controlling these variables is an important part of product design and must be taken into account in thermal design. The thermal control solutions used in the product design invariably limits the scope of design in the areas of mechanics, ergonomics, electromagnetic shielding, and electronic circuits. On the other hand, thermal design solutions can greatly improve the product's functionality, reliability, and corrosion resistance if they are included in the design objectives.

The first task during the thermal design (as shown in Figure 1.10) is to identify the concepts and objectives of the thermal design. The most common objectives of the thermal design in electronics are to (1) keep the junction temperature of the semiconductor devices at or below a given temperature such as 125°C, and keep the temperature of all components inside the product within the limits set by their specifications; (2) level internal temperature differences; and (3) lead excessive heat waste away from the product in a technically and economically sensible way. Prevention of excessive cooling of internal parts of some products in outdoor conditions is also a part of the thermal design. Thermal design is based on the understanding of heat transfer processes from component to unit level through paths of conduction, convection, and radiation. Three general ways of analyzing the design situation include analytical analysis, mathematical modeling, and experiments.

Once the objectives have been set, and when the problem at hand requires the creation of a mathematical model or analytical software, all the boundary conditions influencing the thermal design must be defined, such as (Hienonen et al. 1997): (1) Mechanical constrains, including package or product size and dimensions, and surface treatments of the chosen materials. (2) Electrical constraints, including the power of the unit, the component layout, the PCB mounting layout, electromagnetic compliance requirements for the packaging. (3) Thermal characteristics of components and materials, including maximum allowable operating temperatures of components, reliability target of the unit, temperature dependence of the thermal and other properties of the materials. (4) Environment constraints, including temperature limits of the operating environment, sources of heat in the environment, thermal sinks (mounting platform, ventilation, rain, and wind), and requirements set by the corrosion protection, requirements set by the

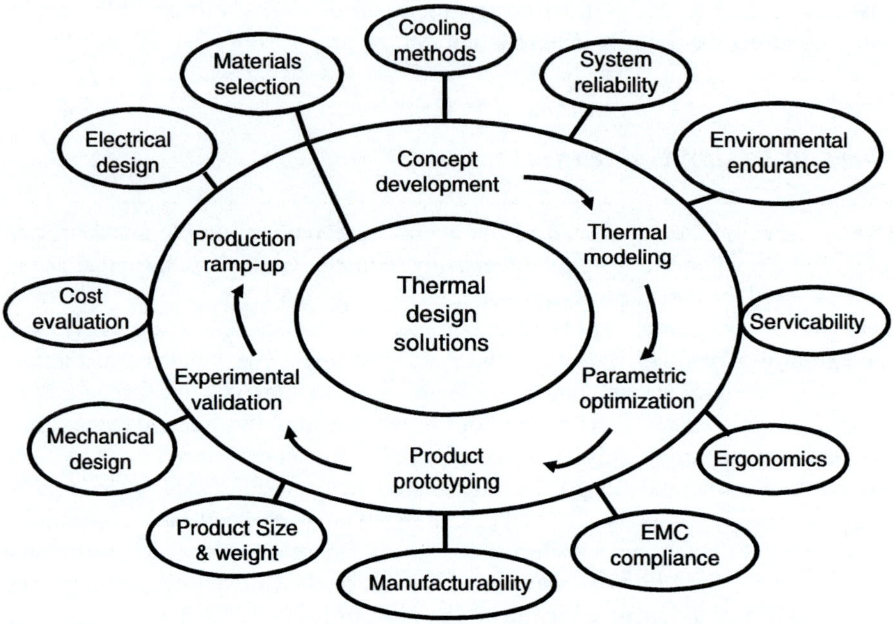


**Fig. 1.10** Schematic thermal design guidelines and its influence on the product of electronic packaging

different individual installations. (5) Special requirements by customer or marketing department, such as no fans, no heaters, ergonomics, acoustic noise and appearance.

Once all boundary conditions have been defined based on the product specification, the analysis cases and results sought must be clearly defined. The analysis cases typically include the hottest possible case (such as highest power level, largest heat input from the environment, and highest ambient temperature) and coldest possible case. A case that strongly affects the reliability of the unit is one that includes large power level changes or wide fluctuations of environmental temperature, leading to cyclic temperature changes that stress all interface and materials. Easing these circumstances by thermal design may be the basis of the thermal design process (Hienonen et al. 1997).

Thermal modeling can be performed with analytical analysis or mathematical modeling. For example, one-dimensional or two-dimensional heat equations can be used to evaluate the thermal performance of some electronic components. In most cases, however, numerical modeling is needed to perform thermal analysis for real thermal products, for instance, by solving the flow field and the heat transfer either simultaneously or iteratively. This is because the thermal phenomena in most electronic components and systems are complex and 3-D. In addition, the thermal design in electronic systems not only involves heat transfer but also fluid flow. Furthermore, when mathematical modeling is used as a tool for thermal analysis, experimental validation is always required to validate the accuracy of the mathematical model. The experimental validation involves conducting experiments that the numerical results can be compared with. With the validated thermal models, various studies can be conducted to study the influence of the design parameters on the thermal performance of the power electronic components and systems (Pang 2005). By identifying the critical design parameters during the parametric studies, the critical design parameter can then be optimized to achieve the best result. Optimization of the design generally concentrates on selectively choosing the best nominal values of design parameters that optimize performance reliability at lowest cost. Then, the best design parameter quantity will be chosen as the output of the design for prototype preparation. Finally, the prototype can be built using the optimized design parameters for further performance evaluation by experimental validation and verification. The verification focuses on the final design and the results of the prototype test and on decisions about possible corrective action if needed before production ramp-up.

With the increase functionality in the power electronics, however, thermal design has become more complex. The need for having a concurrent engineering design process in which electrical, packaging, and thermal designs are performed in parallel is essential in order to achieve a good product design (Pang 2005). In the concurrent engineering design process, the optimized parameters are checked with electrical performance limitations and package manufacturing constraints for optimal performance and manufacturability (Kromann et al. 1998). It involves analyzing all the possible solutions before deciding on the final solution. During the thermal design process, a design guideline had better build up by a standard or principle by which to make a judgment or determine a policy or course of action. Design guidelines can be in the form of theoretic, empirics, or practices that relate the thermal behavior of the power electronics package to the thermal design requirements. By understanding the thermal behavior of the component, the relevant materials, geometries, and cooling strategies can be chosen to optimize the thermal performance of the component. The role of the design guidelines is to provide some insight to the design parameters during the design process. Although the designer may be faced with the available choices for the design, the main point in using the design guidelines is to recognize the available information as well as the known constraints. The design guideline is used to realize common thermal design goals in the electronic components, as mentioned earlier, which typically relate to the thermal performance, reliability, and cost, as well as weight and size reduction. Package design constraints can constitute information on the package manufacturing, assembly, electrical performance, and environment limitations (Pang 2005).

## *Thermal Modeling and Simulation*

The implementation of thermal modeling in the thermal design process can help the designer to understand the thermal behavior of the electronics components and their interrelationship between the electronics components as well as the thermal dynamics of the system, and thereby to ensure good thermal management for the electronics components in the electronics systems.

In a traditional standard industry semiconductor package, heat is generally conducted from the chip through the wire bonds and solder to the bottom of the package base, and then through an interface material into a heat spreader and heat sink for either air-cooled or water-cooled applications. Heat is also conducted through some thick-compound material or electrical insulated layer to the top of the package. However, almost all of the heat is conducted through the package base rather than to the top of the package due to the higher thermal resistance of the compound material or the insulation layer. In many cases, mathematical models can be used to analyze the thermal behavior of the kind of packages.

On the other hand, the heat transfer in some advanced electronic packages, such as planar multilayer structures, is usually more complicated than in the traditional electronic packages because it is generally 3-D with spatially distributed thermal resistances and capacitances. Therefore, analytical approaches and numerical models are usually required to analyze and simulate the thermal behavior of the planar multilayer structures and to predict the temperature distribution of these structures. In fact, numerical simulation methods for the thermal modeling of electronic packages have routinely been widely used in most design processes. From early spreadsheet-type approaches, the sophisticated finite element analysis or computational fluid dynamics tools are more and more popular, and the mechanical computer aided-design data are directly interfaced into their analysis. Modeling is taken as a necessary step, especially at the early stages of thermal design, during which feasibility studies can narrow the spectrum of possible design choices. Modeling is also used at later stages of the design primarily for verification and design optimization.

## Principles of Thermal Modeling

The basic goal of thermal modeling is for cooling electronic packages to improve their reliability. As heat is the major cause of failure of electronic products, thermal modeling is used to find possible means that allow the thermal behavior of the electronic product to stay within the specified ranges at extreme limits of the environmental parameters without the building and testing of expensive prototypes. Also, the thermal model is planned and created to find and screen solutions to thermal problems related to those packaging structures and electronic components that are most critical to the design and of most interest to the designers of the product (Hienonen et al. 1997). For instance, if the operation of the product requires the surface temperature of a microprocessor to be below a certain limit, the model will be constructed so that the best accuracy can be achieved in this particular area and those modes of heat transfer that have the greatest influence on the temperature of this component are modeled accurately. At the same time, the adequate accuracy at the system level should be assured to make the whole model reliable (Hienonen et al. 1997). Therefore, different design targets and objectives will result in different directions of the model modification.

On the other hand, the same thermal model can be used for analysis at different extreme conditions by changing the power levels, environmental boundary conditions, and the thermal characteristics of the packaging structure and materials. One of the most important design criteria for a thermally good structure or system is its insensitivity to changes of individual thermal parameters. As a good practice, sensitivity analysis should always be performed for the thermal model by changing various parameters one by one and recording the subsequent changes of temperature at points of interest; thereby the stability of the system and the inaccuracy caused by incorrect thermal model parameters can be determined, and the model reliability can be increased. For example, the inaccuracy of a good model can be decreased from ±10ºC up to ±3ºC (Hienonen et al. 1997).

In addition, if necessary, thermal testing should be used when designing the thermal model. For instance, an accurate description of the behavior of the thermal interface of structures can only be obtained by thermal tests, such as how heat flows and what the real values of the heat transfer coefficient are over the contact surfaces. Using these test results, final fine tuning of the thermal model can be done to reach the real condition. To avoid mistakes and design iterations, however, the test plan must be drawn up and recorded thoroughly, then documented along with the measurement data. The measurement points chosen in the planning of the tests must be simply relatable to the thermal model (Hienonen et al. 1997).

## General Approaches

The basic approaches of thermal modeling are generally divided into mathematical model and numerical simulation. Dealing with average thermal parameters such as air temperature, pressure, flow velocity, and overall power, the mathematical model is usually used to simulate the thermal behavior of electronic product in specific situations where the internal and external thermal conditions for the unit vary. Numerical models are usually used to simulate very complicated packages and are full of details ranging from modeled components, printed circuit boards, and board edge guides to the system level. Mathematical or theoretical, numerical and accompanied with experimental validation are equally important and indispensable to the design of a viable, reliable, and cost-effective product. Mathematical modeling of a simplified structure of interest can be especially useful for the selection and mastering of the preprocessing numerical model, while the experimental validation approach enables one to quantitatively assess the role of various uncertainties in the materials properties, geometrical characteristics, and loading conditions.

Predictive modeling is an effective tool for the prediction and prevention of mechanical and functional failure of microelectronic systems, subjected to thermal loading. A very detailed model can predict, for instance, the natural convection from the outer surfaces of the electronic product cabinet, and the radiative heat transfer between component boards and even the conduction of heat in a transistor or a microprocessor along its leads, as well as the effect of the temperature on each component.

Whenever using mathematical model or numerical model, one of the most challenging aspects of thermal modeling is to identify potential problem areas in all kinds of environmental and operational situations, and to understand the methods of heat transfer in such situations in order to increase the geometric accuracy and accuracy of the calculated heat transfer in the right places and in the right way. Often, the places where increased model accuracy is required are not initially known but emerge during the first analysis runs if the modeling work is done carefully. The accuracy of the important parts of the model can be improved by modifying the model iteratively. This entails not only better geometric accuracy but also takes into account all modes of heat transfer. Radiative heat transfer between the internal surfaces of the package or enclosure is often omitted under the impression that only convection is meaningful. In modeling convection, it must be taken into account that the air flow pattern can change drastically over a short distance. Laminar flow can change abruptly to become strongly turbulent, with dramatic impact on the convective heat transfer coefficient of adjacent areas. Such local variations in heat transfer coefficients make modeling work difficult. The accuracy and level of detail of the thermal model determines the quality and usefulness of the results. A simple model-checking routine can easily increase the reliability of the analysis results, including: (1) heat transfer paths must be continuous; (2) thermal balance: in thermal equilibrium, the sum of the power generated in the system and the power absorbed by the system or otherwise received by it must equal the power expelled by the system into its environment (Hienonen et al. 1997).

The complexity of the electronic package being modeled and the purpose of the modeling to a large extent prescribe the selection of modeling tools. For the simplest cases, mathematical calculation and modeling are adequate. As the problems and geometry get more complex, the analysis and numerical tools become more elaborate and powerful. A series of software types have been developed with various numerical programs that solve flow field and radiation simultaneously for time conduction. Whatever tool is used, the prerequisite for successful thermal design is a thorough understanding of heat transfer phenomena and ways to change the relative magnitudes of heat transfer modes in each case, thus guiding the thermal behavior of the product in the desired direction (Hienonen et al. 1997).

### Example Methods for Thermal Modeling of Electronic Packaging

The generation of a thermal mathematical model requires a large amount of input data, such as mechanical structure of the analysis object; thermal properties of materials and components; the surface treatments; environmental effects; system overall power level and its distribution; and effects caused by the aging of materials.

## Chip Modeling

For a chip with packaging, heat sinks, and cooling systems, the system can be modeled with two parts. First, the temperature distribution in a chip including substrate and interconnects is governed by the heat conduction equation (Wang and Chen 2004):

$$\rho C_p \frac{\partial T(\vec{r},t)}{\partial t} = \nabla \cdot [\kappa(\vec{r},T)\nabla T(\vec{r},t)] + g(\vec{r},t), \tag{1.26}$$

where $T$ is the time dependent temperature, $\rho$ is the density of the material, $C_\mathrm{p}$ is the specific heat, $\kappa$ is the thermal conductivity, and $g$ is the heat energy generation rate. The physical meaning of (1.20) can be described from the law of energy conservation. For a control volume, the rate of energy stored causing the temperature increase is $\mathrm{d}E/\mathrm{d}t = \int \rho C_\mathrm{p}(\partial T/\partial t)\mathrm{d}V$, the rate of heat conduction through surface $\mathrm{d}\vec{A}$ is $(\kappa\nabla T \cdot \mathrm{d}\vec{A})$ and all surface of the control volume is $Q = \int k\nabla T \cdot \mathrm{d}\vec{A} = \int \nabla \cdot [\kappa\nabla T]\mathrm{d}V$, and the powered is $Q_\mathrm{p} = \int g\mathrm{d}V$.

Second, packaging, heat sinks, and cooling systems are modeled as one-dimensional equivalent thermal resistance network. Suppose that the package surfaces are held isothermal. If the package surfaces are not isothermal, a 3-D model is needed for better accuracy to include the contribution due to heat spreading within the package. The effective heat transfer coefficient in the direction of heat flow, $\vec{i}$, is modeled as $h_i^\mathrm{e} = 1/A^i R_\theta^i$, where $A^i$ is the effective area normal to $\vec{i}$ and $R_\theta^i$ are the equivalent thermal resistance. The equivalent convection boundary conditions are

$$\kappa(\vec{r},T)\frac{\partial T(\vec{r},t)}{\partial n_i} = h_i^\mathrm{e}(T_\mathrm{a} - T(\vec{r},t)), \tag{1.27}$$

where $T_\mathrm{a}$ is the ambient temperature and $\partial/\partial n_i$ is the differentiation along the outward direction normal to the boundary surface.

## Substrate Modeling

The term $\nabla \cdot [\kappa(\vec{r},T)\nabla T(\vec{r},t)]$ in (1.26) can be replaced by $\kappa(T)\nabla^2 T(\vec{r},t)$ for homogeneous materials, and then a second-order parabolic partial differential equation can be obtained

$$\rho C_p \frac{\partial T(\vec{r},t)}{\partial t} = k\left[\frac{\partial^2 T(\vec{r},t)}{\partial x^2} + \frac{\partial^2 T(\vec{r},t)}{\partial y^2} + \frac{\partial^2 T(\vec{r},t)}{\partial z^2}\right] + g(\vec{r},t). \tag{1.28}$$

If the substrate is descretized with size $\Delta x$, $\Delta y$, and $\Delta z$ in $x$, $y$, and $z$ directions, respectively. Then the temperature $T(x; y; z; t)$ at node $(i; j; k)$ can be replaced by

$T(i\Delta x; j\Delta y; k\Delta z; t)$, which is denoted as $T_{i,j,k}$. To have second order accuracy in space $O[(\Delta x)^2; (\Delta y)^2; (\Delta z)^2]$, the central-difference discretization is applied on (1.22). Therefore, the difference equation at node $(i, j, k)$ can be expressed as

$$\begin{aligned}\rho C_p \Delta V \frac{dT_{i,j,k}}{dt} = & -k\frac{A_x}{\Delta x}(T_{i,j,k} - T_{i-1,j,k}) - k\frac{A_x}{\Delta x}(T_{i,j,k} - T_{i+1,j,k}) \\ & -k\frac{A_y}{\Delta y}(T_{i,j,k} - T_{i-1,j,k}) - k\frac{A_y}{\Delta y}(T_{i,j,k} - T_{i+1,j,k}) \\ & -k\frac{A_z}{\Delta z}(T_{i,j,k} - T_{i-1,j,k}) - k\frac{A_z}{\Delta z}(T_{i,j,k} - T_{i+1,j,k}),\end{aligned} \tag{1.29}$$

where $\Delta V = \Delta x \Delta y \Delta z$ is the control volume of node $(i, j, k)$, $A_x = \Delta y \Delta z$, $A_y = \Delta x \Delta z$, and $A_z = \Delta x \Delta y$.

Interfaces

The efficiency and quality of heat transfer depend very strongly on the characteristics of the thermal interfaces. One of the most important parts of the modeling is to simulate the heat transfer process at each interface and their proportional shares in the transfer of total waste heat. Thermal interface can be divided into internal and external. External interfaces transfer heat to the environment, and vice versa. Therefore, the characteristics and means of heat transfer at those surfaces determine the average temperature difference between the unit and its environment. Internal interfaces define the heat transfer paths inside the enclosure. Large internal temperature differences can be a sign of poor design of internal heat transfer paths (Hienonen et al. 1997).

### *Experimental Verification*

Experimental verification is used to test and measure product prototypes at various stages of thermal design and product development, complying with the thermal design objectives. Testing includes both functional operation tests and conditioning to different internal operational loads and external environmental loads using calibrated measurement and testing equipment documented and regulated in all quality assurance systems. Although modeling and simulations can give fairly useful information about thermal characteristics when looking for viable options, experimental verification of the performance of the electronic device and cooling system is still imperative in many cases. This is because thermal parameters are not known well enough to verify the system by analytical means only, due to the complexity of materials, components, and constructions used.

Tests and measurements of experimental verification involve various stages of the thermal design and the product development, including (Hienonen et al. 1997): (1) Thermal testing of preliminary structural models when searching for correct thermal structural solutions. (2) Investigation of functional prototypes to determine the final characteristics before production. (3) Acquisition of field data from production units intended for users by monitoring devices, such as remote measurement technologies, service data from the warranty period. (4) Two technically essential areas: (a) verification of thermal characteristics, such as temperatures, powers, flows, with tests that simulate the operational conditions of a device or a system; (b) verification of the performance and reliability of a device and detection of potential failure mechanisms related to temperature and its variations under the operational conditions of the device.

These fundamental objectives must be taken into consideration when planning verification tests and measurements, which must be designed in close cooperation with the mechanical, EMC, and electrical design as well as other disciplines. Tests must be attached to each device prototype and selected appropriately to each prototype's characteristics. The precursors of the product characteristic during product development, such as cardboard, structure, and thermal models should be tested as appropriate to guide the thermal design and minimize the invariably complicated and time-consuming prototype manufacture (Hienonen et al. 1997).

## Materials Selection for Advanced Thermal Management

Advanced electronic systems require improved thermal management to sustain customer expectations of reliability levels. Customer expectations must be satisfied in an environmentally friendly manner and meet the volume, weight, cooling requirements, manufacturability, and reparability of electronic systems. Materials for thermal management in electronics can basically be classified into two main groups which include interface and bulk materials. Interface materials are formulated to provide a low thermal resistance path between the heat-producing device and the second group of materials which move the thermal energy of the device over a larger area and deposit it into a thermal sink. The interface group is usually relatively flexible and is required to overcome the surface irregularities between the device and the heat-sink surfaces for the lowest thermal resistance. In addition to good thermal transfer, these materials may also serve to perform mechanical attachment and offer compliance to provide stress/strain relief, therefore, this group is also considered to include coating and bonding techniques. The materials into which the device thermal energy passes require high thermal conductivity to move the heat power effectively. Other factors which may be significant for these materials include fast thermal response (high thermal diffusivity) to control thermal transient behavior, low weight, and acceptable material and fabrication cost. The lowest device temperature and hence expected reliability will be achieved when the lowest device-to-sink thermal resistance is achieved (Young et al. 2006).

The device-to-sink thermal resistance conventionally consists of a thermal resistance from the active side to the backside of the die, a thermal resistance across the interface between the die and heat spreader, conductive and spreading thermal resistances from heat spreader, another interface between the heat spreader and heat sink, a thermal resistance associated with the heat sink itself. In many cases, the majority of the device's thermal budget is not taken up by a heat sink, but rather by the interface materials and heat spreaders. Proper selection of thermal interface materials, bulk thermal spreading and dissipating materials, and increasingly, ensuring that these materials interact optimally, is critical to thermal management.

### *Interface Joining Materials*

The interface joining materials are chosen to build up mechanical tolerances in the thermal path with some compliance or flexibility, but at the same time to provide lower thermal interface resistances. Therefore, these jointing materials are required that prevent the formation of voids, often a cause of increased thermal resistance. The materials used in assemblies must have similar coefficients of thermal expansion to avoid thermal stress failures. The joining techniques must also take into account the wider range of materials being used and be able to form a low-resistance path between dissimilar materials. There is also pressure to reduce thickness of interface layers in order to reduce thermal resistance. Common materials used for device-to-heat sink interfacing are flexible polymeric films or greases loaded with thermally conductive particles. The filler materials may be metallic or thermally conductive ceramic particle, such as alumina or boron nitride if electrical insulation is required (Young et al. 2006).

In fact, several classes of interface materials have been developed. For most high-power microprocessors, very thin bondlines are desired and grease or grease-like materials are usually used. Phase change materials and thermal gels are thermal grease replacements. Phase change materials have the advantage of being preapplied to a component. Thermal greases, however, remain the standard for thermal resistance performance.

Materials that will form a relatively rigid bond, such as adhesives and solders, also may be used to form a thin bondline. However, these materials are usually used in conjunction with small die, or heat spreaders with a low CTE, to minimize die and bond stress that could lead to die cracking or delamination. The thermal interface materials dominate the total thermal resistance of high power microelectronics.

Tapes typically are used for lower power applications, as are gap-filling pads. The latter also may be used when compliance to variations in component height are required, as the thermal resistance sensitivity to changes in thickness is not high (Dean 2003).

Consequently, interfaces have a significant impact on the thermal impedance of electronic systems and in practice they can be the dominant factor in achieving effective thermal transfer. Interface materials and processes are the methods used to join an electronic device to the thermal transfer medium such as substrate, heat pipe, and heat sink including coatings and bonding techniques. In this respect they may need to perform the tasks of attachment, stress/strain relief, and thermal transfer over a wide range of temperatures (Young et al. 2006).

## *Bulk Materials for Heat Spreading and Dissipating*

Electronic systems ranging from active electronically scanned radar arrays to web servers all require components made out of bulk materials capable of dissipating heat and maintaining compatibility with the package and die. In response to the needs, many thermal management materials have been developed, such as low-CTE, low-density materials with thermal conductivities ranging between 400 and 1,700 W/m K, and many others with somewhat lower conductivities. Some are low cost; others have the potential to be low cost in high volume.

For instance, the high thermal conductivity materials have been used to make heat spreaders that are attached to the die with the thermal interface material between. The heat spreader has the primary function of spreading the thermal energy from the small footprint of the die to a larger area in order to make more efficient use of convection from the heat sink or package surface.

The surfaces of heat spreaders are often finished to prohibit corrosion or oxidation, improve cosmetics, or allow for marking of some sort. Typical surface finishes are anodizing (aluminum), sputtering, plating, and oxide growth such as black oxide on copper. The effect of the surface finish on thermal properties should be considered. Another key aspect of a heat spreader is the surface roughness and flatness. Both these requirements will have an impact on the average bondline of the thermal interface material and, hence, can significantly impact package thermal performance (Dean 2003).

Another common component is heat sinks usually made of aluminum. Aluminum has the advantage of low cost, easy machining and forming and a corrosion resistant surface which can be further enhanced by anodizing. Its thermal conductivity at around 180 W/m K is low compared with copper (~379 W/m K) but aluminum is often preferred on cost and weight factors unless the thermal load is very high.

Other examples are carbon-based materials and active cooling techniques using a liquid. The carbon materials, usually in the form of graphite or diamond, may be combined with metals to give materials that are easier to process into manufactured items. The graphitic materials are lower density than copper or aluminum and can offer higher thermal conductivities. The carbonaceous materials also can be processed to form low-density thermal insulators to protect electronics from heat sources, or to shield parts of the assembly from excessive temperature rise, i.e., in laptop computers and telephone handsets. Phase change is often used in active cooling systems to remove heat from electronics. Almost all of the active cooling systems require a subsequent heat exchanger to transfer the thermal energy to the external environment.

Passive cooling components and materials are generally preferred to active cooling for reasons of cost, complexity, and reliability. The available performance of passive systems is still extending. This is enabled both by improvements in the materials engineering and also by the opportunity for improved design which comes from system-level modeling. However, developments in active cooling are addressing many of the issues which have limited its scope in the past (Young et al. 2006).

In addition, advanced innovative bulk materials must offer substantial improvement relative to mainstream materials in order to have a realistic possibility of being specified. Even if materials do have sufficient technical merit there is a massive investment and lead time required in order to move toward market readiness with appropriate qualification approvals aligning with matched production capacity and downstream integration.

## *Materials and Components Integration*

As requirements on the materials increase, it will no longer be sufficient to choose a good performing material for each component of the heat dissipation path, how these materials interact with each other must be considered. A material that spreads heat very well, but is not wet by an interface material may not perform as well as a lower conductivity material that does wet. Phonon scattering at the interfaces and at plating/surface finish interfaces may need to be optimized in a thermal design. Four approaches have been taken to reduce total thermal resistance (Dean 2003): (1) increase thermal conductivity of the interface materials and bulk heat spreading and dissipating materials; (2) increase wetting or bonding to decrease contact resistance at the surface; (3) increase flatness of the component like spreader to decrease the thickness of the interface to reduce heat transfer path; and (4) eliminate the number of the interfaces in the thermal management package.

### Higher Conductivity Materials

While progress has been made, the bulk conductivity of most thermal interface materials is relatively low. Many thermal interface materials now have bulk conductivity in the range of 2–3 W/m K. Improving material conductivity by a factor of 3 to 5 (to ~10 W/m K) would result in a reduction in bulk and thus total thermal resistance of approximately 0.10°C cm$^2$/W. This would drive the bulk resistance down to 0.03–0.04°C cm$^2$/W. Further improvements in thermal conductivity would have diminishing returns unless the contact resistance was also improved. Improving the conductivity of the heat spreading and dissipating materials through the use of engineered composite materials or heat pipe/vapor chambers will also be required to drive total device thermal resistance down. Because the spreading resistance is typically higher than the through-thickness thermal resistance, the use of anisotropic materials is increasing (Dean 2003).

### Increasing Wetting or Bonding Forces

Increasing the wetting or bonding to a surface will decrease the contact resistance. For most grease or grease-like interface materials, contact resistance has historically been a fairly low fraction of the overall thermal resistance. However, as the material thermal conductivity has increased and bondline thickness decreased, the contact resistance has begun to become significant. Contact resistance can be reduced through improving the bonding or wetting of the interface material to each surface at the interface. Materials should no longer be chosen individually, but rather the effect of the materials working together will be evaluated. It will be important for suppliers and users of interface materials to work with suppliers and users of heat spreaders to ensure optimal performance. Synergy between metal finishes and interface materials should allow contact resistance to be decreased by 50% (Dean 2003).

### Decreasing Interface Thickness

With everything else held constant, a thinner bondline between components will produce a lower interface thermal resistance. This will require flatter components. The tightening of flatness tolerances and shift to die referenced cooling has dropped interface thicknesses by a factor of 2 or more (from ~50 to 25 μm or less). Further improvements in spreader or lid flatness will enable thinner interfaces and better package thermal performance (Dean 2003).

### Elimination of the Number of Interfaces

Most high-performance devices use a thermal lid or heat spreader. This creates at least two thermal interfaces in the heat removal path from the die backside to the ambient: the die to spreader and spreader to heat sink interfaces. With current performance levels, eliminating one of the interfaces will eliminate a significant fraction of the overall junction to ambient thermal resistance. Thermal lids are used to protect the die and spread the heat from the concentrated die footprint to a larger area so that the heat sink is more efficient in dissipating heat to the environment. Heat sinks with thicker bases or high performance bases, will help address the latter concern, minimizing the need for a package heat spreader. When a bare die package is used, the interface material between the die and heat sink is softer, or more compliant, to prevent damage during assembly or operation. Improvements in interface materials combined with designs to protect bare die will enable the wider spread use of bare die packaging and its inherent elimination of one of the two interfaces in the heat transfer path (Dean 2003).

As package and heat sink design have improved significantly, the materials used largely determine the thermal performance of electronic packages. Thermal resistance across physical interfaces has progressed from an almost negligible portion of the total junction to ambient thermal resistance to the dominant factor in total thermal resistance. Heat spreading and reduction of hotspots within a die are creating a need for higher conductivity materials. Further increases in device power dissipation will require significant improvements from materials. Increased thermal conductivity ($k \geq 10$ W/m K), improved ability to work synergistically with other packaging components and package designs to allow thinner bondlines, and increased use of bare die packaging will be needed to meet the further challenges (Dean 2003).

With the continuing trend in electronic systems toward higher power and increased packing density, advanced materials for efficient thermal management has become a crucial need. Typical electronic devices and their packages consist of a variety of different types of materials, including metals, semiconductors, ceramics, composites, and plastics. The most important physical properties in the use of materials in thermal management are thermal conductivity and the CTE. It is also quite evident that in general there is an enormous disparity between the CTE of high conductivity metals used for heat sinks (aluminum and copper) and insulators used for electronic substrates (alumina, BeO, AlN, etc.). One of the many challenges of electronic packaging is bridging this thermal expansion gap in a manner that does not compromise the thermal efficiency of the package. To solve this problem, composite materials have been developed and utilized. The printed circuit board is an example of a polymer matrix composite. Metal matrix composites (MMCs) are fabricated using a high thermal conductivity metal matrix such as aluminum or copper, with a low CTE material added to reduce the overall CTE of the composite. By proper adjustment of the relative composition of the composite, the CTE can approach that of silicon and insulating materials while maintaining high thermal conductivity. For instance, the kinetic spray to fabricate MMCs has been developed. Further improvement of the material properties of MMCs for thermal management, and advanced manufacturing techniques that will allow cost-effective MMC fabrication on a production scale.

## Environmental Compliance of Thermal Management Materials

Environmental compliance must be evaluated during thermal design of electronic packaging because it has become a key advantage of competitive electronic products. The requirements are regulated by the Restriction of the Use of Certain Hazardous Substances in Electrical and Electronic Equipment (RoHS) and Waste Electrical and Electronic Equipment (WEEE) directives promulgated by the European Union (EU). The directives make it critical for selections of thermal management materials. These environmental compliance regulations provide forceful guidelines for designing electronic products with environmental compliance.

## *RoHS*

RoHS Directive 2002/95/EC, together with WEEE Directive 2002/96/EC became European Law in February 2003, setting collection, recycling and recovery targets for all types of electrical goods. Producers must comply with all provisions of the WEEE directive after August 13, 2005, while the RoHS directive requires that producers prohibit any of the banned substances on the market after July 1, 2006. Producers failing to comply with the RoHS and WEEE directives' requirements face legal penalties and potential restriction from selling products in the EU.

Table 1.2 lists materials that are currently restricted by the RoHS directive (Tong 2009). These substances are restricted to the ppm (parts per million) threshold level in all applications. All homogeneous material in purchased articles (i.e., materials, components, subassemblies, or products) must be free of the substances or cannot contain higher concentrations than the defined ppm (parts per million) threshold levels as listed in the table. Table 1.3 lists the banned substances in electronic packaging, while the substances must be reported when concentration exceeds the indicated ppm threshold level as listed in Table 1.4 (Tong 2009). Exemptions to the maximum allowed concentrations of restricted materials are identified for cases where technology does not yet allow for substitutions, or where alternatives may have a worse impact on human health and the environment. Some exemptions include mercury in several kinds of fluorescent lamps; lead in steel, copper, and

**Table 1.2** Restriction of hazardous substances restricted substances

| Material/substance category | Maximum concentration in weight |
|---|---|
| Asbestos | Not detectable |
| Certain Azo colorants and Azo dyes | Not detectable |
| Cadmium/cadmium compounds | 75 ppm |
| Hexavalent chromium/hexavalent chromium compounds | 1,000 ppm |
| Lead/lead compounds | 1,000 ppm 300 ppm (PVC cables only) |
| Mercury/mercury compounds | 1,000 ppm |
| Ozone depleting substances (continuous fiber ceramics, Hydro chlorofluorocarbons, HBFCs (Hydrobromofluorocarbons), carbon tetrachloride, etc.) | Not detectable |
| Polybrominated biphenyls | 1,000 ppm |
| Polybrominated diphenylethers | 1,000 ppm |
| Polychlorinated biphenyls | Not detectable |
| Polychlorinated naphthalenes (more than three chlorine atoms) | Not detectable |
| Radioactive substances | Not detectable |
| Certain short chain chlorinated paraffins | Not detectable |
| Tributyl tin and triphenyl tin | Not detectable |
| Tributyl tin oxide | Not detectable |

**Table 1.3** Banned substances in electronic packaging

| Material/substance category | Maximum concentration in weight |
|---|---|
| Cadmium, mercury, lead, chromium VI | Total amount 100 ppm |
| Chlorofluorocarbons | Not detectable |
| Hydro chlorofluorocarbons | Not detectable |

**Table 1.4** Substances that must be reported when concentration exceeds the threshold level

| Material/substance category | Maximum concentration in weight (ppm) |
|---|---|
| Antimony/antimony compounds | 1,000 |
| Arsenic/arsenic compounds | 1,000 |
| Beryllium/beryllium compounds | 1,000 |
| Bismuth/bismuth compounds | 1,000 |
| Brominated flame retardants (other than PBBs (Polybrominated biphenyl) or PBDEs (Polybrominated diphenylethers)) | 1,000 |
| Nickel (external applications only) | 1,000 |
| Certain phthalates | 1,000 |
| Selenium/selenium compounds | 1,000 |
| Polyvinyl chloride (disclosure is limited to "is present'/'is not present" in amounts that exceed threshold) | 1,000 |

aluminum alloys; lead in some types of solder; and military applications. RoHS Article 3(a) states that RoHS covers electrical and electronic equipment "which is dependent on electric currents or electromagnetic fields in order to work properly and equipment for the generation, transfer, and measurement of such currents and fields falling under the categories set out in Annex IA to Directive 2002/96/EC (WEEE) and designed for use with a voltage rating not exceeding 1,000 V for alternating current and 1,500 V for direct current." With that said, a microwave oven would not be covered by RoHS because it cannot perform its intended function with the power in the off position. On the other hand, a talking doll can still be used as a doll even when the batteries are removed, therefore, it isn't covered by RoHS. RoHS does allow for noncompliant components after the July 1, 2006, but only as spare parts for equipment on the market before July 1, 2006.

## *WEEE*

The WEEE directive imposes the responsibility for the disposal of WEEE on the manufacturers of such equipment. Those companies should establish an infrastructure for collecting WEEE in such a way that "Users of electrical and electronic equipment from private households should have the possibility of returning WEEE at least free of charge." Also, the companies are compelled to use the collected waste in an ecologically friendly manner, either by ecological disposal or by reuse/refurbishment of the collected WEEE. The WEEE identifies producers as any company that sells electronic and electrical equipment directly or indirectly under its own brand name. The intent of the WEEE directive is to require producers to design products and manufacturing processes which prevent the creation of WEEE and barring that, reuse, recycle, dispose of, or incinerate WEEE. The WEEE directive calls for set percentages of IT (information technology) and telecommunications equipment (Category 3 includes personal computers, wireless devices, and similar devices) to be recovered and reused, or recycled (minimum 65%), incinerated (maximum 10%), or safely disposed of (maximum 25%).

Therefore, the goal of the thermal design and material selection should be compliant with RoHS and WEEE directives in a minimized cost manner. Meanwhile, designing for environmental compliance must assure that time-to-market is minimized to take the product's competitive advantages.

## Summary

Escalation of heat flux and power dissipation of electronic chips resulted from the demand for high-performance microprocessors. Meanwhile, the desire for smaller form-factor system and lower semiconductor operating temperatures is compounding the thermal challenge. Thermal design for a microprocessor can no longer be treated in isolation. Power and performance tradeoffs and smart circuit design techniques are required to conserve power consumption. Advanced materials and process improvements in packaging and cooling technology are required to minimize thermal resistance. Therefore, viable thermal design and cooling solutions are critical for development of high performance microprocessors and cost-effective electronic packaging.

In fact, the importance of thermal management has waxed and waned through the technology and product generations of the past decades. Advanced integrated circuit and photonic technologies as well as the ubiquity of electronic system applications are providing a serious challenge to heat transfer design and product development in terms of basic theory, tools, components, and innovative design. Breakthroughs are needed in advanced cooling and pragmatic design at all package levels.

The mechanisms of heat transfer depend on the media involved and are usually divided into conduction, convention, radiation, and multimode, which is a combination of one or more of the above. When a temperature gradient exists within a continuous and nonmoving medium such as solid or stationary fluid, heat is transferred through the medium via the conduction mode. Convection heat transfer occurs when a surface is in contact with a moving fluid, liquid, or gas at a different temperature. Heat is exchanged between the fluid and the surface. Radiation heat transfer occurs when two surfaces at different temperatures exchange energy in the form of electromagnetic energy emitted by the surfaces. Radiation can occur in a vacuum because no medium between the two surfaces is required as in conduction and convection. The multimode problems present in most electronic packages with various mediums. Appropriate governing equations must be written based on conservation principles for each medium, along with boundary, interface, and initial conditions. Traditionally, a single technique is applied at one level of the packaging hierarchy at a time. Multiscale approaches are being developed to analyze across the packaging hierarchy, i.e., modeling performed at the chip, board, and system levels. The common techniques used include (1) a resistor network approach; (2) an analytical approach; and (3) a numerical solution of governing equations.

General thermal management solutions typically include hardware assembly, software control, and optimal thermal design as well as the combination of these approaches. System hardware solutions are based on internally distributing and externally dissipating the thermal energy to the ambient environment. System software solutions typically regulate the dissipative power of the device based on active system feedback controls. Optimal thermal design is required for cooling and thermal control of all level packages, particularly effective for device-level or chip-level micro/nanoscale thermal engineering and processing.

The thermal design of electronic packaging has become the most important aspect of product design that ensures reliability of the electronic device and thus its ability to compete on a demanding market. The primary objective of the thermal design is to plan the thermal balance of the product so that it operates reliably within the specified environmental conditions throughout its service life. This is generally accomplished by comprehensively considering all factors during thermal design, such as (1) optimizing the heat removal paths from the integrated circuits; (2) choosing the optimal cooling methods; (3) using highly conductive materials or thermal vias as possible to consume a large amount of the substrate and reduce interconnect capacity and thermal resistances; (4) minimizing stresses induced in the chips and substrate due to mismatches in thermal coefficients of expansion; and (5) assuring that the set of design goals, the modeling type, the design accuracy, and the test and measurement methods work together so that unintentional over- or underdesign is avoided and that excessively expensive modeling and/or lengthy testing is avoided. All of these considerations will impact performance, cost, and reliability of the product. Furthermore, gathering information on the most critical failure mechanisms of the intended components and materials is necessary to ensure that the main activities related to the thermal design focus on details where possible failures primarily affect the usability and reliability of the product.

Advanced materials are becoming critical for thermal management of microelectronic systems. The systems ranging from active electronically scanned radar arrays to web servers all require materials capable of dissipating heat and maintaining compatibility with the package and die. In response to these needs, many thermal management materials have been developed, including low-CTE, low-density materials with thermal conductivities ranging between 400 and 1,700 W/m K. These materials have been used for fabrication of servers, laptops, PCBs, PCB cold plates/heat spreaders, cellular telephone base stations, hybrid electric vehicles, power modules, phased array antennas, thermal interface materials (TIMs), optoelectronic telecommunication packages, laser diode and light-emitting diode packages, and plasma displays.

In addition, environmental compliance must be evaluated during thermal design of electronic packaging because it has become a key advantage of competitive electronic products. The requirements are regulated by the RoHS and WEEE directives promulgated by the EU. The directives make selections of thermal materials critical. These environmental compliance regulations provide forceful guidelines for designing electronic products with environmental compliance.

## References

Asheghi M, Liu W (2007) Microscale heat transfer. Electronics Cooling. http://www.electronics-cooling.com/articles/2007/feb/a2. Accessed 16 February 2010.

Bornoff R (2005) Get on board: solving thermal problems at board level. Printed Circuit Design & Manufacture 22:38–41.

Brooks D, Martonosi M (2001) Dynamic thermal management for high-performance microprocessors. Proceedings of the 7th International Symposium on High-Performance Computer Architecture, Monterrey, Mexico. pp. 171–182. IEEE, Piscataway.

Cao L et al (1996) Transient thermal management of portable electronics using heat storage and dynamic power dissipation control. Electronic Components and Technology Conference, Orlando. pp. 205–211. IEEE, Piscataway.

Chen G, Shakouri A (2002) Heat transfer in nanostructures for solid state energy conversion. Journal of Heat Transfer 124:242–252.

Chomerics X X (1999) Thermal management products. http://www.chomerics.com/products/documents/thermcat/THERMALMP.pdf. Accessed 06 February 2010.

Clark L et al (2002) An embedded 32-b microprocessor core for low-power and high- performance applications. IEEE Journal for Solid-State Circuits, 56:1599–1608.

Cooper B (2006) Operating system coordinated thermal management. US patent 7054720.

Couvillion R J (2006) Thermal considerations. In Ulrich R K, Brown W D (eds) Advanced Electronic Packaging, 2nd edn. John Wiley & Sons, Hoboken.

Dean N (2003) Materials for thermal management. Advanced Packaging. http://ap.pennnet.com/display_article/169366/36/ARTCL/none/none/1//Materials-for-Thermal-Management/. Accessed 26 February 2010.

Hannemann R (2003) Thermal control of electronics: perspectives and prospects. http://web.mit.edu/hmtl/www/papers/HANNEMANN.pdf. Accessed 30 January 2010.

Hienonen R, Karjalainen M, Lankinen R (1997) Verification of the thermal design of electronic equipment. VTT Automation. http://www.vtt.fi/inf/pdf/publications/1997/p.320.pdf. Accessed 23 February 2010.

Kromann G, Pimont V, Addison S (1998) CFD and EDA tools the interoperability of flotherm® and board station®/AutoTherm®: Concurrent design of a motorola powerPC™ RISC microprocessor-based microcomputer. Freescale Semiconductor, Inc., http://www.freescale.com/files/archives/doc/white-paper/603604THERMAL.pdf. Accessed 26 February 2010.

Krum A (2004) Thermal management. In Harper C A (ed) Electronic packaging and interconnection handbook, 4th edn. McGraw-Hill Professional, New York.

Kumar A, Li S, Peh L-S, Jha N K (2008) System-level dynamic thermal management for high-performance microprocessors. IEEE Transactions on Computer-Aided Design of Integrated Circuits and Systems 27:96–108.

Madrid A (2000) Thermal management for applied computing. Intel Developer Update Magazine (April). http://www3.intel.com/technology/magzine/computing/ac04002.pdf. Accessed 16 January 2010.

Mallik D et al (2005) Advanced package technologies for high-performance systems. Intel technology Journal 9:259–271.

Manning A, Johns D (2007) Thermal/EMC simulation addresses new mechanical design challenges. http://www.conformity.com/artman/publish/printer_228.shtml. Accessed 03 February 2010.

Maruyama S (2000) Molecular dynamics method for microscale heat transfer. In Minkowycz W J and Sparrow E M (Eds), Advances in Numerical Heat Transfer 2:189–226, Taylor & Francis, New York.

März M (2003) Thermal management in high-density power converters. IEEE International Conference on Industrial Technology 2:1196–1201.

Moran M (2001) Micro-scale avionics thermal management. NASA/TM-2001-211095. http://gltrs.grc.nasa.gov/GLTRS. Accessed 09 February 2010.

Narumanchi S V J, Murthy J Y, Amon C H (2006) Boltzmann transport equation-based thermal modeling approaches for hotspots in microelectronics. Heat and Mass Transfer 42:478–491.

NEC Electronics (2003) Thermal design of a package. http://www.necel.com/pkg/en/characteristic/heat01_03.html. Accessed 21 February 2010.

Pang Y F (2005) Assessment of thermal behavior and development of thermal design guidelines for integrated power electronics modules. Ph.D. Dissertation. Verginia Polytechnic Institute and State University, Blackaburg.

Petrosjanc K O (2009) Thermal design system for chip- and board-level electronic components. http://ewdtest.com/conf/EWDTS09/57.pdf. Accessed 12 February 2010.

Rasmussen F E (2003) Electrical interconnections through CMOS wafers. PhD Thesis. Technical University of Denmark, Hellerup.

Siebert W P (2005) Alternative electronic packaging concepts for high frequency electronics. PhD Thesis. Royal Institute of Technology, Tellinge (Sweden).

Srinivasan J, Adve S (2003) Predictive dynamic thermal management for multimedia applications. 17th Annual International Conference of Supercomputing, San Francisco. pp. 109–120. ACM, New York.

Tong X C (2009) Advanced materials for electromagnetic interference shielding. CRC Press, Boca Raton.

Viswanath R, Wakharkar V, Watwe A, Lebonheur L (2000) Thermal performance challenges from silicon to systems. Intel Technology Journal Q3. ftp://download.intel.com/technology/itj/q32000/pdf/thermal_perf.pdf. Accessed 26 January 2010.

Wang T, Chen C (2004) SPICE- Compatible thermal simulation with lumped circuit modeling for thermal reliability analysis based on modeling order reduction. http://cc.ee.ntu.edu.tw/~cchen/research/ISQED2004_Wang.pdf. Accessed 26 February 2010.

Wirth E (2004) Thermal management in embedded systems. BSc Thesis, University of Virginia, Charlottesville.

Young R et al (2006) Developments and trends in thermal management technologies – a mission to the USA. www.lboro.ac.uk/research/iemrc/documents/. . ./CB2007.pdf. Accessed 26 February 2010.

# Chapter 2
# Characterization Methodologies of Thermal Management Materials

**Abstract** The materials selection for thermal management of electronic packaging is influenced by thermal, electrical, physical and thermomechanical requirements of the device and its surrounding electrical system and by the environment to which the device will be exposed. The reliability of the finished device and electric system will depend not only on the characteristics of the individual materials but also on the interaction of package and thermal management materials at interfaces during exposure to such stresses as thermal gradients, temperature cycling, moisture, and contamination. This chapter will introduce general characterization methodologies of thermal management materials for assessing performance and reliability of electronic packaging, including thermal properties, electrical properties, thermomechanical analysis, as well as material microstructure and interface characterization, surface finish and contact interface compatibility, and reliability analysis and environmental evaluation.

## Thermal Properties and Measurement Techniques

The major thermal properties involved in the characterization of the thermal management materials for electronic packaging include thermal conductivity and diffusivity, specific heat capacity, coefficient of thermal expansion (CTE), and thermal shock resistance. A variety of measurement techniques have been developed to evaluate these properties for thermal management materials, here mainly focusing on some classic testing methods.

### *Thermal Conductivity and Diffusivity*

The accurate measurement and characterization of the thermal conductivity of bulk materials can pose many challenges. For instance, loss terms of the heat input intended to flow through the sample usually exist and can be most difficult to quantify. Many testing methods exist with an accuracy of within 5%, such as the

X.C. Tong, *Advanced Materials for Thermal Management of Electronic Packaging*, Springer Series in Advanced Microelectronics 30,
DOI 10.1007/978-1-4419-7759-5_2, 

steady-state technique, the 3$\omega$ technique, and the thermal diffusivity measurement (Uher 2005). Each of these techniques has its own advantages as well as its inherent limitations, with some techniques more appropriate to specific sample geometry, such as the 3$\omega$ technique for thin films (this method will be discussed in the section "Thermal Characterization of Micro/Nanomaterials"). This section will focus on the measurement techniques that are more appropriate for bulk-like solid-state materials.

Thermal conductivity, $k$, is the property of a material that indicates its ability to conduct heat. It can be defined by first order Fourier–Biot equation as the quantity of heat, $\Delta Q$, transmitted during time $\Delta t$ through a thickness $\Delta x$, in a direction normal to a surface of area $A$, due to a temperature difference $\Delta T$, under steady state conditions and when the heat transfer is dependent only on the temperature gradient. That is

$$k = \frac{\Delta Q}{\Delta t} \times \frac{1}{A} \times \frac{\Delta x}{\Delta T}. \tag{2.1}$$

Thermal diffusivity, $\alpha$, is the rate at which a temperature disturbance on one side of the body travels to another part of the body, defined by second order Fourier–Biot equation as

$$\frac{\partial T(r,t)}{\partial t} = \alpha \nabla^2 T(r,t). \tag{2.2}$$

Thermal conductivity and thermal diffusivity are related by

$$k = \alpha \rho C_p, \tag{2.3}$$

where $k$ is thermal conductivity in W/m K; $\alpha$ is thermal diffusivity in $m^2/s$; $\rho$ is material density in $kg/m^3$; and $C_p$ is specific heat capacity in J/kg K. At high temperatures, or when the material is available only in small sizes, $\alpha$, $\rho$, and $C_p$ are often measured to determine thermal conductivity $k$. This kind of measurement method is classified as transient method, as discussed later in this chapter.

Thermal conductivity approximately tracks electrical conductivity for metals according to the Wiedemann–Franz law, as freely moving valence electrons transfer not only electric current but also heat energy. However, the general correlation between electrical and thermal conductance does not hold for other nonmetal materials due to the increased importance of phonon carriers for heat in nonmetals. For example, highly electrically conductive silver is less thermally conductive than diamond, which is an electrical insulator. Additionally, thermal conductivity depends on many properties of a material, notably its structure and temperature.

As shown in Figure 2.1 and Table 2.1, a number of techniques have been developed to measure thermal conductivity and thermal diffusivity, each of them suitable for a limited range of materials, depending on the thermal properties and the medium temperature (Maglić et al. 1984, 1992). The exist measurement

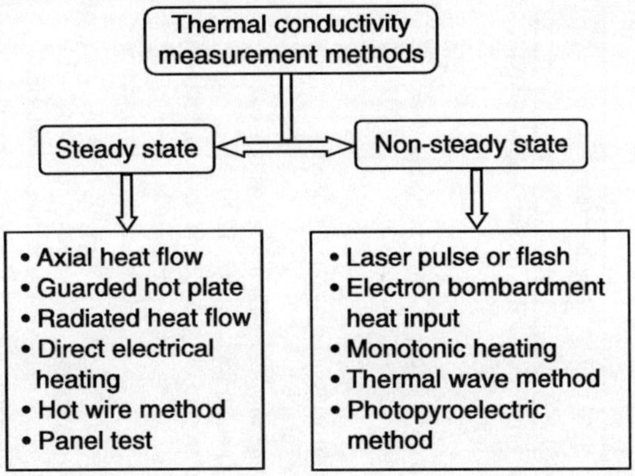


**Fig. 2.1** Illustration of typical thermal conductivity measurement techniques

methods can be divided into two classes: steady-state and non-steady-state or transient methods. A distinction can be made between steady-state and transient techniques. The steady-state techniques generally perform a measurement when the temperature of the material that is measured does not change with time. This makes the signal analysis straightforward. The disadvantage is that it usually takes a well-engineered experimental setup. Comparably, the transient techniques perform a measurement during the process of heating up. The advantage is that measurements can be made relatively quickly because there is no need to wait for a steady-state situation. The disadvantage is that the mathematical analysis of the data is in general more difficult.

For conductivity measurement, steady-state methods (up to 1,200 K) and pulse methods (in particular over 1,500 K) are usually used. For instance, if the material under test is a conductor, the specimen can be self heated by passing of an electric current. Transient methods are usually preferred for diffusivity measurement. Measuring diffusivity requires an accurate recording of the time dependence of temperature following a transient or periodic temperature perturbation at a specimen boundary. As a consequence of the wide ranges of thermal property, there is no single method of measurement that can be used for measurement of either property, in particular thermal conductivity. To obtain acceptable values for the measured property, the material type and its range of property value over its operational temperature range will particularly influence the type of method used and the size and conjunction of the test specimen and apparatus.

Therefore, thermal conductivity is measured by steady-state techniques and thermal diffusivity by transient techniques. It is possible to use some of the latter in a modified way to also measure thermal conductivity, or converse the diffusivity data to the conductivity data. However, a measurement method has to be selected depending on the following criteria: (1) appropriate sample size and shape; (2) applicable temperature range, which is limited for individual techniques; (3) suitable thermal conductivity range, because low conductivity materials such as insulating materials or foams need different methods than for high conductivity materials such as metals.

**Table 2.1** Thermal conductivity and diffusivity measurement methods

| Measurement method | Specimen material type | Temperature range (K) | Conductivity range (W/m K) | Uncertainty (%) | Advantages | Disadvantages |
|---|---|---|---|---|---|---|
| Axial heat flow | Metals and metallic alloys; cylindrical shape specimen | 90–1,300 | 10–500 | 0.5–2.0 | High accuracy; utilize electrical resistivity for heating source | Heat losses above ~500 K |
| Direct electrical heating | Wires, rods, tubes of electrical conductors | 400–3,000 | 10–200 | 2.0–5.0 | Wide temperature coverage. Fast and use electric properties | Complex equipment |
| Radial heat flow | Solids and powders in cylindrical form | 298–2,600 | 0.01–200 | 3.0–15 | Good accuracy with wide temperature coverage | Large specimens |
| Guarded hot plate | Solid, opaque, insulators | 80–1,500 | 0.0001–1.0 | 2.0–5.0 | Apply to wide range of materials. High accuracy | Complex costly slow (3–12 h) |
| Hot wire method | Refractory materials | 298–1,800 | 0.02–2 | 5.0–15 | Small size probe; drop in specimen | Apply to low conductivity materials |
| Panel test | Refractory materials | 600–1,600 | 0.05–15 | 15 | Simplicity | High temperature gradients; slow measuring |
| Laser flash | Solids, liquid metals; polymers; ceramics. Disk specimen 6–16 mm in diameter | 100–3,300 | 0.1–1,500 | 1.5–5 | Wide temperature coverage. Simple, rapid measurement for thermal diffusivity | Not convenient for translucent materials. Complex error analysis |
| Thermal wave | Solids, liquid metals, gases. Specimen shape: rods, cylinders | 60–1,300 | 0.5–500 | 1.0–9 | Apply to a wide range of materials. Multiproperty measurement | Complex math analysis. Complex error analysis |
| Electron bombardment heat input | Solid and liquid metals. Nonmetals | 330–3,200 | 50–400 | 2.0–10 | High temperature coverage. Small specimen. AC techniques applicable | High vacuum; complex experimental apparatus |
| Monotonic heating | Ceramics, plastics, composites | 4.2–3,000 | 50–400 | 2.0–12 | Simple apparatus. Simple measurement. Wide temperature coverage | Inappropriate for good thermal conductors. Low precision |
| Photothermal methods | Small specimens of most solid material types | 200–800 | 0.1–200 | 1.0–10 | Test thermal diffusivity or thermal conductivity in appropriate models including NDT mode | Complex math analysis. Complex error analysis |

Steady-state techniques typically include axial flow, radial flow, and guarded hot plate and hot-wire methods; while the most commonly used transient method is the laser flash method.

## Axial Flow Methods

Axial flow methods have been long established with key measurement issues centered mainly on reduction of radial heat losses in the axial heat flow developed through the specimen from the electrical heater mounted at one end, in which the power dissipation of this heater is used in calculating column heat flux. These losses are minimal at low temperatures. As the specimen temperature moves above room temperature, control of heat losses becomes more and more difficult. Therefore, a great deal of attention centers on important experimental parameters such as the ratio of effective specimen conductance to lateral insulation conductance (the higher the better) and to the quality of guarding, which is the match of the axial gradient in the specimen to that of the surrounding insulation. In practice, cylindrical symmetry heat transfer is used. The subcategories can be divided into the following categories listed below (Anter 2007).

### Absolute Axial Heat Flow

This method is mostly used in subambient environments. The system requires very precise knowledge of the electrical power feeding the heater, and the losses from the hot heater surfaces also play a major role.

### Comparative Cut Bar (ASTM E1225 Test Method)

This is the most widely used method for axial thermal conductivity testing. The principle of the measurement is to have the heat flux pass through a known sample and an unknown sample and compare the respective thermal gradients, which will be inversely proportional to their thermal conductivities. Most commonly, the unknown sample is sandwiched between two known samples, the references, to further account for minor heat losses that are very difficult to eliminate, as shown in Figure 2.2. $K_R$ is the thermal conductivity of the references. From this, the thermal conductivity of the unknown sample, $K_s$ can be calculated as:

$$\frac{Q}{A} = K_s \frac{\Delta T_s}{L} = K_R \frac{\Delta T_1 + \Delta T_2}{2L}, \tag{2.4}$$

or

$$K_s = K_R \frac{\Delta T_1 + \Delta T_2}{2\Delta T_s}. \tag{2.5}$$

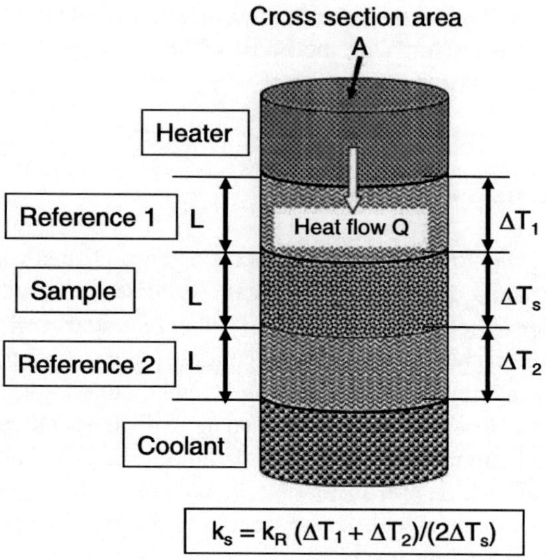


**Fig. 2.2** Schematic of a comparative-guarded-axial heat flow system. The sample and two reference bars have the same height *L* and the same cross-section area *A*

Guarded or Unguarded Heat Flow Meter Method (ASTM C518, E1530 Test Methods)

A flux gauge is used, which is very similar in its purpose, to the references in the comparative cut bar method. In practice, the reference material has a very low thermal conductivity and, therefore, it can be made very thin. Usually, a large number of thermocouple pairs are located on both sides of the reference plate, connected differentially to yield directly an electrical signal proportional to the differential temperature across it. The assembly is cast into a protective coating for durability. This type of flux gauge is mostly used with instruments testing very low thermal conductivity samples, such as building insulations. In a similar fashion, flux gauges can be constructed from just about any material, thick or thin, depending on the material's thermal conductivity. Common requirements for all flux gauges are that the material used for the measuring section be stable, not affected by the thermal cycling, and the gauge be calibrated by some method independently. A very large variety of testing instruments use this method.

**Guarded Hot Plate Method (ASTM C 177 Test Method)**

The guarded hot plate is a widely used and versatile method for measuring the thermal conductivity of insulations. The specimens are often rather large. A flat, electrically heated metering section surrounded on all lateral sides by a guard heater section controlled through differential thermocouples, supplies the planar heat

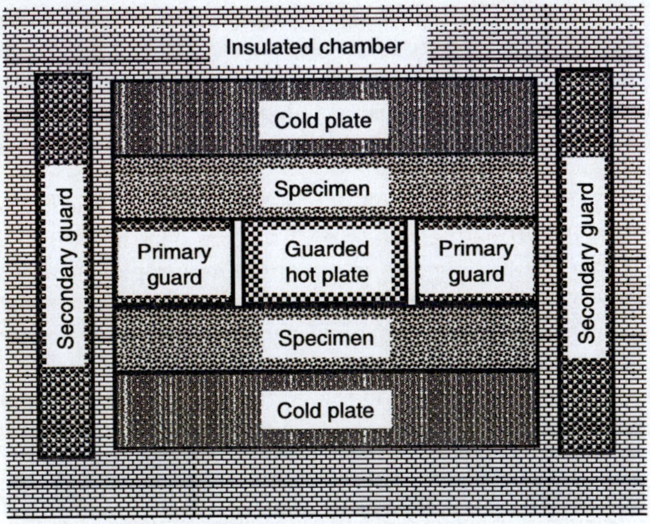


**Fig. 2.3** Schematic of guarded hot plate method

source introduced over the hot face of the specimens. The most common measurement configuration is the conventional, symmetrically arranged guarded hot plate where the heater assembly is sandwiched between two specimens, as shown in Figure 2.3. In the single-sided configuration, the heat flow is passing through one specimen and the back of the main heater acts as a guard plane creating an adiabatic environment.

This is an absolute method of measurement and its applicability requires: (1) the establishment of steady-state conditions and (2) the measurement of the unidirectional heat flux in the metered region, the temperatures of the hot and cold surfaces, the thickness of the specimens and other parameters which may affect the unidirectional heat flux through the metered area of the specimen. Three different categories of measurement systems can be distinguished by an apparatus working around room temperatures, an apparatus working below room temperatures (down to about −180°C), and an apparatus working at high temperature (600°C or above). A given apparatus is most often best adopted for measurement in one of these temperature ranges.

## Hot Wire Method (ASTM C1113 Test Method)

The hot wire method is a standard transient dynamic technique based on the measurement of the temperature rise in a defined distance from a linear heat source (hot wire) embedded in the test material. If the heat source is assumed to have a constant and uniform output along the length of test sample, the thermal conductivity can be derived directly from the resulting change in the temperature over a known time interval. Hot wire methods are most commonly used to measure the thermal conductivity of refractory materials, such as insulating bricks and powder or fibrous materials. Because it is basically a transient radial flow technique, isotropic specimens are required. The technique has been used in a more limited way to measure properties of liquids and plastics materials of relatively low thermal conductivity.

The modification of this long-established technique is the probe method. The hot wire probe method utilizes the principle of the transient hot wire method. The heating wire as well as the temperature sensor (thermocouple) is encapsulated in a probe that electrically insulates the hot wire and the temperature sensor from the test material. This configuration is particularly practical where the specimen conductivity is determined from the response of a "hypodermic needle" probe inserted in the test specimen. Thus the method is conveniently applied to low-conductivity materials in powder or other semirigid form. A probe device can be used to measure the thermal properties of soils in situ, but most commonly a closely controlled furnace is used to contain the sample and produce the base temperatures for the tests. The probe contains a heater and a thermocouple attached to it. When a certain amount of current is passed through the heater for a short period of time, the temperature history of the heater's surface will take on a characteristic form. In the initial phase, the temperature will rapidly rise, and as the heat begins to soak in, the rate of rise becomes constant. When the thermal front reaches the outer boundary of the sample, the rise will slow down or stop altogether due to losses into the environment. From the straight portion of the rate curve (temperature vs. time) the thermal conductivity can be calculated.

### Laser Flash Method

The laser flash method is the most common transient method for thermal diffusivity and conductivity measurement, which was developed by Parker, Butler, Jenkins, and Abbott of the U.S. Navy Radiological Defense Laboratory in 1961 (Parker et al. 1961). This method is based on depositing a very short but intense laser energy pulse on one surface of a disk-shaped sample, while monitoring the temperature excursion of the opposite face. From the characteristic time dependence of the temperature rise, called thermogram, thermal diffusivity can be calculated using Parker's formula (Parker et al. 1961; Gaal et al. 2004)

$$\alpha = \frac{138d^2}{t_{1/2}}, \tag{2.6}$$

where $d$ is the sample's thickness and $t_{1/2}$ is the time necessary for the signal to reach 50% of its maximum value. The testing apparatus has been developed not only the basic technology concerning the homogenization of laser beam by the optical fiber system, the measurement of the transient temperature on the rear surface of the specimen by high-speed infrared thermometer with a temperature scale, contact measurement at steady state temperature and the calculation of thermal diffusivity by the curve fitting method, but also the new technology of the differential laser flash calorimetry consists of a specimen holder which can hold two specimens (reference and measured) up to the high temperatures for simultaneous measurements of thermal diffusivity and specific heat capacity and a high-speed infrared thermometer which can simultaneously measure the temperature of the rear face of two specimens (Shinzato and Baba 2001).

Figure 2.4 shows a typical differential laser flash apparatus. The apparatus consists of a pulse laser, an optical fiber system with mode mixer as a pulse beam introduction unit, chamber with a sample support, a high-speed infrared thermometer (InSb elements, cooled by liquid nitrogen), differential amplifiers, and a transient memory. In addition, this apparatus is equipped with a beam profile monitor to provide the visual checking of the uniformity of a pulsed laser beam, and a vacuum pump, a circulating water-cooling unit and a personal computer for the analysis of measured data. An Nd-YAG laser (wavelength: 1.06 μm) is used as a pulsed laser and the half width of the laser pulse is selectable to either 300 or 500 μs mode with a switch. The output power of the laser can be changed up to 2 J for 300 μs mode and 5 J for 500 μs mode. The pulsed laser beam is focused into the optical fiber and homogenized in beam profile by the mode mixer and irradiated onto the specimens. The charging time required for one pulse shot is 30 s. The chamber can be operated in the range from room temperature to 1,800 K and the vacuum level of 0.05 Pa or lower to prevent oxidation and dirt proof and to suppress the heat loss caused by convection and can also be operated in an inert gas

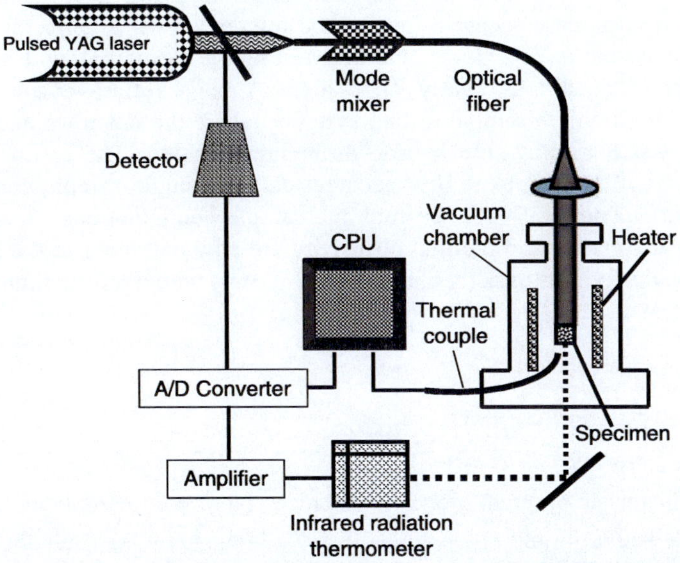


**Fig. 2.4** Schematic of laser flash apparatus

atmosphere. The specimen is opaque and is a disk of 10 or 5 mm in diameter and from 0.5 to 3 mm in thickness, and black coating to absorb the laser beam is usually required for the specimen which is transparent, translucent, or low emissivity. The specimen with the diameter of 5 mm is used for the simultaneous measurements of thermal diffusivity and specific heat capacity. A high-speed infrared radiation thermometer equipped with five sensing elements has been developed and calibrated with a black body for the temperature scale. The precision of temperature calibration is ± 0.2 K. The temperature response of the infrared radiation thermometer is fast enough to respond the emitting signals, up to 10 kHz of infrared-emitting diode, without decrease of amplification factor or delay of phase (Shinzato and Baba 2001).

The advantages of the method originally was not only its speed (the actual test takes less than 1–2 s for most solids), but the ability to use very small, 5–12 mm diameter and 1–3 mm thick samples. This advantage became a problem for some composite materials, when such small samples are considered nonrepresentative of a larger body. However, the development of a system for large samples (30 mm diameter and up to 5 mm thickness) has alleviated this limitation to a great extent. Nevertheless, the need for specific heat capacity data remained the main limitation for this method in generating thermal conductivity data. While it was found that with careful experimental techniques, one could actually calibrate the response of the detector that generates the signal for the curve, the procedure itself imposed new limitations of its own. The process, simple in principle, involves testing a sample of known heat capacity first and then the unknown, and comparing the magnitude of the two resultant curves. Obviously, for such a differential computation the experimental variables (environment, electronics, energy pulse, etc.) must stay constant for both the sample and the reference. In reality this was and still is a serious problem for any instrument that can test only one sample at a time. It is easy to see that even the best case scenario (testing the reference over a temperature range, waiting for system to cool down, and then testing the unknown), will make close duplication of conditions nearly impossible. A major improvement has been achieved with multiple sample testing systems, where the unknown and the reference are tested side-by-side in true differential fashion. The results typically improved from the ±7 to ±10% accurate data for single sample conventional systems, to ±2 to ±3% for the multiple sample configurations. Because both specific heat capacity and thermal diffusivity are now measured in the same test, thermal conductivity is directly obtained in real time, with a priority knowledge of density (Gaal et al. 2004).

### *Coefficient of Thermal Expansion*

The CTE is a basic physical property which can be of considerable importance in thermal managing design applications of a material. Materials usually expand as increase in temperature leads to greater thermal vibration of the atoms in the material, and hence to an increase in the average separation distance of adjacent atoms. The linear CTE, $\alpha$, designates how much a material will expand for each degree of temperature increase, as given by:

$$\alpha = \frac{\Delta L}{L_0} \Delta T, \tag{2.7}$$

where $\Delta L$ is the change in length of material in the direction being measured; $L$ is original overall length of material in the direction being measured; $\Delta T$ is the change in temperature over which $\Delta L$ is measured. The magnitude of the CTE depends on the intermolecular or atomic bonding structure of the material. For example, strongly bonded structures such as ceramics have relatively low CTEs compared with metals, and loosely bonded structures such as polymers generally have high CTEs, especially those that are thermoplastics or elatomers.

Common methods for measuring CTE are shown in Table 2.2, including mechanical dilatometer, interferometry techniques, and thermomechanical analysis (Maglić et al. 1984). With typical CTEs in the range of 5–50 ppm/K, the changes in dimensions are extremely small and difficult to measure. Among these measuring methods, interferometry provides high accuracy, which looks at the changes in the interference pattern of monochromatic light, usually from a laser. With this technique it is possible to plot strain against temperature throughout a heating or cooling cycle. The slope of the strain/temperature curve at a given temperature is the instantaneous CTE. Furthermore, to give comparisons with earlier figures, the average slope over a temperature range can also be derived from the data.

## *Specific Heat Capacity*

The specific heat capacity is another common physical property used to characterize the ability of the thermal managing materials to absorb heat. It is the amount of heat per unit mass required to raise the temperature by 1°C. The relationship between heat and temperature change is usually expressed by:

$$C = \frac{\Delta Q}{m\Delta T}, \tag{2.8}$$

**Table 2.2** Thermal expansion measurement methods

| Measurement method | Temperature range (°C) | Minimum coefficient of thermal expansion ($1.0e^{-6}$/K) | Accuracy (%) | Characteristics |
|---|---|---|---|---|
| Mechanical dilatometer | −150 to 2,000 | Above 5 | 5.0–7.0 | Medium precision |
| Interferometry techniques | −150 to 700 | Below 5 | 1.0–3.0 | High accuracy |
| Thermomechanical analysis | −120 to 600 | Above 5 | 3.0–12 | Inconsistent precision |

**Table 2.3** Specific heat measurement methods

| Measurement method | Specimen materials | Temperature range (K) | Uncertainty (%) | Advantages | Disadvantages |
|---|---|---|---|---|---|
| Adiabatic | All | 4–1,300 | 1.0–3.0 | Very versatile. High sensitivity. Solid and liquid specimen applicable | Specimen in container. Limitation in high temperatures. Based on enthalpy data |
| Drop | All | 300–2,000 | 1.0–3.0 | Solid and liquid specimen | Specimen in container. Time consuming. Based on enthalpy data |
| Levitation | Electrical conductor (sphere) | 1,000–2,500 | 2.0–5.0 | Solid and liquid specimen | Specimen: electric conductor. Based on enthalpy data |
| Modulation | All | 80–3,000 | 2.0–5.0 | Multiproperty measurement | Limited to solid specimen |
| Pulse | Electrical conductor (wire, rod, tube) | 1,000–7,000 | 2.0–10 | Solid and liquid specimen applicable. Wide temperature coverage. Multiproperty measurement | Specimen: electric conductor. Complex instrumentation |
| Differential scanning | All | 100–1,000 | 1.0–10 | Fast and economical | Specimen in container. Limitation in high temperature |

where $C$ is specific heat in J/kg K, $\Delta Q$ is the heat energy put into or taken out of the material (J/mol), $m$ is the mass of the material (kg), $\Delta T$ is the temperature differential (K). The relationship does not apply if a phase change is encountered, because the heat added or removed during a phase change does not change the temperature. There have been several methods developed, as shown in Table 2.3 (Maglić et al. 1984).

### *Thermal Shock Resistance*

Thermal shock will cause cracking as a result of rapid temperature change. It usually occurs when a thermal gradient causes different parts of an electronic package to expand by different amounts. This differential expansion can be understood in terms of stress or strain, equivalently. At some point, this stress overcomes the strength of the material, causing a crack to form. If nothing stops this crack from propagating through the material, it will cause the electronic package to fail. The thermal shock resistance of a material can be expressed as (Krupke et al. 1986):

$$R = \frac{k\sigma(1 - v)}{\alpha E}, \tag{2.9}$$

where $k$ is the thermal conductivity, $\sigma$ is maximal tension the material can resist, $\alpha$ is the thermal expansion coefficient, $E$ is the Young's modulus, and $v$ is the Poisson ratio. From (2.9), thermal shock can be prevented by: (1) reducing the thermal gradient of the electronic package by (a) changing its temperature more slowly and (b) increasing the material's thermal conductivity; (2) reducing the material's CTE; (3) increasing its strength; (4) increasing its toughness by (a) crack tip blunting, i.e., plasticity or phase transformation and (b) crack deflection.

The typical measuring methods are shown in Table 2.4. Among these methods, the impulse excitation technique has proved to be a useful tool. It can be used to measure Young's modulus, shear modulus, Poisson's ratio, and damping coefficient in a nondestructive way. The same test piece can be measured after different thermoshock cycles and in this way the deterioration in physical properties can be mapped out. The impulse excitation technique uses natural frequency, dimensions, and mass of a test piece to determine Young's modulus, Shear modulus, Poisson's ratio, and damping coefficient and then calculate the thermal shock resistance using (2.9). Dimensions and mass of a test piece can be easily measured.

**Table 2.4** Thermal shock assessment methods

| Heating method | Materials applicable | Advantages | Disadvantages |
|---|---|---|---|
| Induction (5–40 kW capacity) | Al, Cu, steels, Ni-superalloys | Rapid heating; complex sample geometries | Coil design experience; electric noise; high cost |
| Quartz lamp (radiation) | Ni, Co alloys; metallic composites | Inexpensive; uniform temperature; screening materials | Slow cooling rates; enforced cooling needed |
| Fluidized bed | Ni-based superalloys | Good for screening temperature fluidization resistance | Surface oxidation; need calculation of $\sigma - \varepsilon$ temperature transients |
| Flame burner heating | Ni-superalloys; steels | Screening of temperature fluidization resistance; surface corrosion representative of service | Oxidation; temperature transients |
| Dynamometer (friction heating) | 0.5–0.7% C steels | Very high temperature on surface reached; representative of service | Oxides are wedged into cracks; friction changes with time |
| Impulse excitation | Materials that have enough stiffness to vibrate | Nondestructive; multiproperty measurement | Create vibration in a well prepared sample, and need a reference piece in some cases |
| Thermal cycling | Electronic packaging and thermal managing components and materials | Nondestructive; applicable a wide range of components and materials | Time consuming for some components and materials |

Natural frequency is determined by gently tapping the test piece and analyzing the vibration. Tapping can be done using a small hammer or an automated tapping device. There are different ways to detect vibrations in the test piece: piezoelectric sensor, microphone, laser-vibrometer, accelerometer, etc. To optimize the results a microphone or a laser-vibrometer can be used because there is no contact between the test piece and the sensor. Laser-vibrometers are preferred to measure signals with extreme frequencies. These kinds of signals are usually only induced in very thin test pieces. Accuracy is determined by the quality of support of the test piece and the correctness of measurement of the dimensions and mass of the test piece. It is possible to do measurements within an accuracy rate of 0–1%. The best way to support a test piece is to use nylon fibers, but most systems use polyurethane foam strips as support. This makes it very difficult to measure small test pieces because the damping effect of the support is too big.

The most common method used for thermal shock testing is thermal cycling, which performed to determine the resistance of the part to sudden changes in temperature (JEDEC 2004). The parts undergo a specified number of cycles, which start at ambient temperature. The parts are then exposed to an extremely low (or high) temperature and, within a short period of time, exposed to an extremely high (or low) temperature, before going back to ambient temperature. Failure acceleration due to temperature cycling mainly depends on: (1) the difference between the high and low temperatures used; (2) the transfer time between the two temperatures; and (3) the dwell times at the extreme temperatures.

After the final cycle, external visual examination of the components or electronic packages should be performed under microscope. Electrical testing of the samples to device specifications also needs to be performed to detect electrical failures accelerated by the temperature cycle. For instance, for reliability testing or qualification of a electronic device or package, 1,000 temperature cycles are usually performed, with interim visual inspection and electrical test read points under microscope at a magnification of 200 and 500 times. Failure mechanisms accelerated by thermal shock include die cracking, package cracking, neck/heel/wire breaks, and bond lifting, etc.

## *Thermal Characterization of Micro/Nanomaterials*

With developments of two-dimensional (2-D) (thin film) and 1-D micro- or nanostructured materials, many limits of bulk (3-D) material properties are being exceeded, i.e., the thermal conductivity of materials can be significantly improved or reduced depending on the system requirements. For instance, 1-D carbon nanotube (CNT) structures have very high thermal conductivity values. Conversely, thermoelectric thin films exhibit thermal conductivities which are lower than their bulk materials. This reduction in thermal conductivity is created by increased phonon interference and scattering caused by the introduction of low-dimensional structures. Typically 2-D structures are fabricated by depositing a repeating series of thin films onto a thick bulk substrate. Determining the thermal conductivity of these thin films can be a challenge because of the small thickness of the films. The deposited film often cannot be removed from the substrate because the substrate provides mechanical support and interface adhesion strength for the film. Therefore, in experimental measurements, the substrate and the film must be measured together and then the properties of the substrate must be subtracted from the properties of the film/substrate composite. For measurements of thermal conductivity, if a steady state conduction method is used, errors can be introduced by the parasitic heat flow through the substrate and by radiation losses from the test sample. If the thermal conductivity of the substrate is relatively high, then the total heat flow through the substrate will be large compared with the heat flow through the film and subtraction of the substrate's contribution to heat conduction will significantly magnify measurement errors. If the thermal conductivity of the substrate is low, then radiation losses from the surface will dominate the heat flow through the test sample and the uncertainties of the radiative behavior will complicate the measurement of the thermal conductivity of the film when measurements are made at room temperature or above (Aller 2007).

Therefore, measuring the thermal conductivity of these micro- or nanostructured materials can prove challenging because of their small size and potentially large aspect ratios. The classic methods used for bulk material measurement would exceed their limitation. Hence, the $3\omega$ method and thermoreflectance approach have been developed for measuring the thermal conductivity and specific heat of thin films and nanomaterials such as vertically grown CNTs, which would have increasing demand in thermal management of advanced electronic packaging.

## The $3\omega$ Method

The $3\omega$ method is a transient method of accurately measuring the specific heat and thermal conductivity of a material. It uses a simple heater geometry, whose behavior can be solved from the heat diffusion equation. The heater is excited at a frequency of $\omega$, this excitation produces temperature fluctuations in the material at a frequency of $2\omega$, and measurements of thermal conductivity and specific heat can be taken from the $3\omega$ response. The heater is used simultaneously as a heater and a thermometer, therefore, no additional probes are required to carry out the method. The effect of black-body radiation produces less than 10% error in the measurement at 1,000 K and that this error was proportional to $T^3$, quickly becoming insignificant at lower temperatures (Aller 2007).

### 2-D Thin Films

The schematic illustration of a four pad measurement system of the $3\omega$ method is shown in Figure 2.5. A thin metal band of width $2b$ ($10 < 2b < 100$ μm), and of length $L$ ($L \approx 4$ mm), with four contacts, is prepared on the surface of the thin film

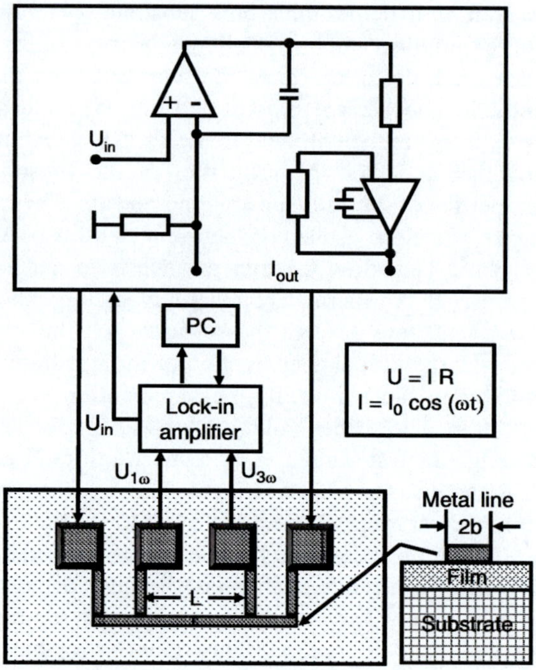


**Fig. 2.5** Schematic illustration of four pad measurement system of the $3\omega$ method

by photolithography. The resistance of the heater band is measured in function of the temperature by the four-wire method, so as to know the resistance between the two inside connections. The results are fitted with a linear model, in order to calculate the temperature coefficient of resistivity (Bourlon 2005)

$$\alpha_R = \frac{1}{R}\frac{dR}{dT}. \tag{2.10}$$

The metal used for the heater is selected to have a temperature coefficient $\alpha_R$ as high as possible to increase the signal to noise ratio (*S/N*), such as gold or nickel. An alternating current

$$I(t) = I_0 \cos(\omega t), \tag{2.11}$$

where $\omega$ is the angular modulation frequency of the electric current, heats the metal band and the sample at an angular frequency $2\omega$ by Joule effect. Because of the increase of the temperature of the heater band

$$\Delta T = \Delta T_0 \cos(2\omega t + \varphi), \tag{2.12}$$

the resistance $R$ oscillates at the angular frequency $2\omega$ and can be approximated as (Bourlon 2005)

$$R = R_0(1 + \alpha_R \Delta T) = R_0[1 + \alpha_R \Delta T_0 \cos(2\omega t + \varphi)]. \tag{2.13}$$

The amplitude $\Delta T_0$ and the phase $\varphi$ are directly linked to the thermal conductivity of the sample and to the angular frequency $\omega$. The voltage $V$ across the heater is obtained by Ohm's law (Bourlon 2005):

$$\begin{aligned} V(t) &= R(t)I(t) \\ &= R_0 I_0 \cos(\omega t) + \frac{1}{2} R_0 I_0 \alpha_R \Delta T_0 \cos(\omega t + \varphi) \\ &+ \frac{1}{2} R_0 I_0 \alpha_R \Delta T_0 \cos(3\omega t + \varphi). \end{aligned} \tag{2.14}$$

The first term, which is just the normal AC voltage at the drive frequency, does not contain any information on the amplitude $\Delta T_0$ and the phase $\varphi$. The two other terms come from mixing the current at frequency $\omega$ with the resistance oscillations at $2\omega$. And as $R_\alpha \approx 3$ m K$^{-1}$ for the gold heater, and $\Delta T_0 \approx 1$ K, they are relatively small. The first of these terms is at the same frequency as the driving term. However, the last term appears at the third harmonic of the leading frequency, and for that reason it can be extracted from the total signal thank to a lock-in amplifier. The amplitude of this term is about 1% (Bourlon 2005)

$$V_{3\omega} = \frac{1}{2} R_0 I_0 \alpha_R \Delta T_0. \tag{2.15}$$

and its phase is $\varphi$.

For the 2-D heat conduction across a multilayer film-on-substrate system, the complex heater temperature rise can be expressed as (Bourlon 2005)

$$\Delta \tilde{T} = \Delta T_0 \mathrm{e}^{\mathrm{i}\varphi} \approx \frac{P_1}{\pi k_{y1}} \int_0^\infty \frac{1}{A_1 B_1} \frac{\sin^2(\lambda b)}{\lambda^2 b^2} \, \mathrm{d}\lambda, \tag{2.16}$$

where

$$A_{i-1} = \frac{A_i \frac{k_{yi} B_i}{k_{y_{i-1}} B_{i-1}} - \tanh(\varphi_{i-1})}{1 - A_i \frac{k_{yi} B_i}{k_{y_{i-1}} B_{i-1}} - \tanh(\varphi_{i-1})}, \quad i = 2, \ldots, n, \tag{2.17}$$

$$\varphi_i = B_i d_i, \tag{2.18}$$

$$k_{xy} = \frac{k_x}{k_y}. \tag{2.19}$$

In the above expression, $P_1$ is the power per unit length (W/m), $n$ is the total number of layers including the substrate; subscript $i$ corresponds to the $i$th layer starting from the top, subscript $y$ corresponds to the direction perpendicular to the film/substrate interface (cross plane), $b$ is the heater half width, $k$ is the thermal conductivity of the layer, $\omega$ is the angular modulation frequency of the electric current, $d$ is the layer thickness, and $\alpha$ is the thermal diffusivity. The effect of the thermal conductivity anisotropy is introduced through the term $k_{xy}$, which is the ratio of the in plane to cross plane thermal conductivity of the layer. For the substrate layer $i = n$, and if the substrate is semiinfinite $A_n = -1$. When the substrate has a finite thickness, the value $A_n$ depends on the boundary condition at the bottom surface of the substrate: $A_n = -\tanh(B_n d_n)$ for an adiabatic boundary condition or $A_n = -1/\tanh(B_n d_n)$ if the isothermal boundary condition is considered.

For 2-D heat conduction across a thin film on semiinfinite substrate, if the heater band is deposed on an electrical insulating material, and if the heat flux is homogenous on all the width of the heat band, the complex heater temperature rise could be calculated by (Bourlon 2005)

$$\Delta\tilde{T} = \Delta T_s \approx \frac{P_l}{\pi k}\int_0^{\infty} \frac{\sin^2(\lambda b)}{\lambda^2 b^2 (\lambda^2 + q^2)^{1/2}} \mathrm{d}\lambda, \tag{2.20}$$

where $|1/q| = \sqrt{\alpha_s/2\omega}$ is the penetration depth of the thermal wave, $\alpha_s$ is the thermal diffusivity of the substrate.

For 2-D heat conduction across a thin film on a semiinfinite substrate, the temperature rise of the heater $\Delta T_0$ is composed of two parts: the first, $\Delta T_s$, is due to the substrate and the second, $\Delta T_F$, is due to the thermal resistance of the thin film (Bourlon 2005).

$$\Delta T_0 = \Delta T_s + \Delta T_F. \tag{2.21}$$

If the width of heater is large compared with the thickness of the film, $b \gg d_F$, then the direction of the thermal flux is perpendicular to the film/heater interface; if the thermal conductivity of the thin film is smaller, then the thermal conductivity of the substrate, $k_F \ll k_s$; if the contact thermal resistances between the heater and the film and between the film and the substrate are negligible; and if the thermal diffusivity of the thin film, $\alpha_F$, is important; then the temperature rise of the heater due to the thermal resistance of the thin film is (Bourlon 2005):

$$\Delta T_F = \frac{P_l d_F}{2bk_F}. \tag{2.22}$$

The precision of this measurement will depend on the precision of the measurements of the width of the heater band, $2b$, and of the thickness of the thin film, $D_F$, but also on the quality of the photolithography necessary to structure the heater. The measurement will be the more precise the bigger the difference of the thermal conductivity of the thin film and of the substrate.

### 1-D Nanomaterials

Taking a uniform rod- or filament-like specimen with four-probe configuration as for electrical resistance measurement, the two outside probes are used for feeding an electrical current, and the two inside ones for measuring the voltage across the specimen with the following conditions: (1) the specimen is suspended in a point between the two voltage probes to allow the temperature fluctuation; (2) all the probes are highly thermal-conductive, so the heat generated in the specimen can be dissipated into the substrate through the probes; (3) the specimen is maintained in a high vacuum and the whole setup is heat-shielded to the substrate temperature, so the radial heat loss through gas convection and radiation will be minimized. When an alternative electrical current of the form $I_0 \sin \omega$ passes through the specimen, the heat generation and transfer along the specimen will follow the partial differential equation (PDE) (Lu et al. 2001):

$$\rho C_p \frac{\partial}{\partial t} T(x,t) - \kappa \frac{\partial^2}{\partial x^2} T(x,t) = \frac{I_0^2 \sin^2 \omega t}{LS} [R + R'(T(x,t) - T_0)] \tag{2.23}$$

with boundary conditions $T(0,t) = T_0$, $T(L,t) = T_0$, $T(x,-\infty) = T_0$. Where $C_p$, $k$, $R$ and $\rho$ are the specific heat, thermal conductivity, electric resistance and mass density of the specimen at the substrate temperature $T_0$, respectively. $R' = (\mathrm{d}R/\mathrm{d}T)_{T_0}$. $L$ is the length of the specimen between voltage contacts, and $S$ is the cross section of the specimen. Through Fourier analysis, the $3\omega$ voltage approximation can be derived as (Lu et al. 2001):

$$V_{3\omega,\mathrm{rms}}(t) \approx \frac{4 I_{\mathrm{rms}}^3 L R R'}{\pi^4 k S \sqrt{1 + (2\omega\gamma)^2}}, \tag{2.24}$$

where $\gamma = (L^2/\pi^2\alpha)$ is the specimen's characteristic thermal time constant in axial direction; $\lambda = \sqrt{\alpha/2\omega}$ is thermal wavelength; and $\alpha = (k/\rho C_p)$ is thermal diffusivity.

For a CNT specimen, when $L = 0.9$ μm, $S = 1.25\mathrm{e}^{-16}$, $I_0 = 1.41$ μA, $R = 3$ kΩ, $R' = 140$ Ω/K, $k =$ 1–1,000 W/m K, $\rho = 1$ kg/m$^3$, $C_p =$ 1–1,000 J/kg K, the operational limits would be $(I_0^2 R' L / n^2 \pi^2 k S) \ll 1$ ($2.0\mathrm{e}^{-4}$ for $k_{\max}$; and 0.2 for $k_{\min}$); $\alpha = (k/\rho C_p) = 1{,}000\,\mathrm{m^2/s}$(max) or $1\mathrm{e}^{-3}$ m$^2$/s; and $\lambda(1\,\mathrm{kHz}) = \sqrt{\alpha/2\omega} = 0.28\,\mathrm{m}$ (max) or $2.8\,\mathrm{e}^{-4}$m (min). As $\lambda \gg L$, (2.24) can be expressed as

$$V_{3\omega,\mathrm{rms}} \approx \frac{4 I_{\mathrm{rms}}^3 L R R'}{\pi^4 k S}. \tag{2.25}$$

### Thermoreflectance Approach

The basic principle of the transient thermal reflectance (TTR) method is to heat a sample by laser irradiation and probe the changes in the surface reflectivity of the heated material. The source of energy in the TTR method is normally provided by a pulsed laser with short pulse duration. During each pulse, a given volume below the sample surface heats up due to the absorbed laser light energy. The depth of the volumetric heating is determined by the optical penetration depth, which is a function of laser wavelength and surface material properties. After each laser pulse, the sample cools down to the initial ambient temperature. During this process, a probing CW (continuous wave) laser light reflected from the sample surface at the heating spot center is collected on a photodetector (1 ns maximum rise time) that reads the instantaneous surface reflectivity (Kim 2007).

The TTR method is typically used for the measurement of thin film through-plane thermal conductivity or interface thermal resistivity. The TTR system uses the fact that the relative change in the temperature of the surface material is linearly proportional with the relative change in the reflectivity within a wide but finite temperature range (Kim 2007):

$$\frac{\Delta T}{T} = \frac{\Delta R}{R}, \tag{2.26}$$

where $T$ is the temperature and $R$ is the reflectivity of the materials. The changes in surface reflectivity, are then recorded by an oscilloscope (at rates of up to 5 Giga-samples per second). The result of the experiment is a transient normalized temperature response, which represents the overall heat transfer behavior of the layers of materials including the unknown material under test. To extract the thermal conductivity from the recorded temperature response, an identical mathematical representation of the corresponding physical measurement problem is solved numerically with guessed thermal properties with the intention of matching the experimental and numerical transient normalized temperature responses. A mathematical optimization technique makes it possible to systematically vary the desired unknown properties and compare each resulting numerically obtained response to the reference experimental data until the error between them is minimized in the root-mean-square sense. The final numerical solution, therefore, yields the desired unknown parameters, which represent the best fit to the actual thermal properties of the physical sample. By using a two-parameter optimization technique, the method can yield the thermal conductivity of the material under test and the interface resistance between this material and the absorption layer on top of it. The transient heat transfer in the TTR method can be described by the use of the heat equation, as follows (Kim 2007):

$$\rho C_p \left( \frac{\partial T}{\partial t} \right) = \nabla(K \nabla T) + \dot{Q}_{\mathrm{ab}}, \tag{2.27}$$

where $\rho$ is the mass density of the material, $C_p$ is its specific heat, $t$ is the time, $K$ is its thermal conductivity, and $\dot{Q}_{\mathrm{ab}}$ is the heat source created by absorption of the laser light energy. Moreover, the heat transfer inside the sample under test is governed by the 2-D heat equation in polar coordinates and can be written as follows:

$$\frac{\partial T}{\partial t} = \alpha(T)\left(\frac{\partial^2 T}{\partial r^2} + \frac{1}{r}\frac{\partial T}{\partial r} + \frac{\partial^2 T}{\partial z^2}\right) + \frac{1}{\rho C_p}\dot{Q}(r,z,t), \tag{2.28}$$

where $r$ and $z$ are dimensionless coordinates and $\alpha$ is the thermal diffusivity of the material, which is $\alpha = K/(\rho C_p)$. The heat source $\dot{Q}(r,z,t)$, can be expressed as (Kim 2007):

$$\dot{Q}(r,z,t) = I(t)(1-R)\gamma e^{-\gamma z}, \tag{2.29}$$

where $\gamma$ is the absorption coefficient of the top layer, and $I(t)$ is a Gaussian temporal distribution as follows (Kim 2007):

$$I(t) = \frac{2F}{\tau\sqrt{\pi}} e^{-4\left(\frac{t-t_0}{\tau}\right)^2}, \tag{2.30}$$

where $F$ is the fluence of laser irradiation, $\tau$ is the duration of the heating laser pulse which is 6.1 ns, and $t_0 = 7$ ns is the time at which the intensity reaches its maximum value. When the dimension of the probing spot is close to two orders of magnitude smaller than the heating laser spot, the heated sample under test can be treated as a semiinfinite solid for the 1-D problem. The diameter of the heated spot (YAG Laser) is 185 μm, for example, while the probing spot is around 2.5 μm, which is small enough to make $(\partial T/\partial r) \approx 0$ in the probing spot area. Thus the 1-D heat equation can be induced from 2-D equation. Thus, the 1-D heat equation is induced from the 2-D (2.28) as follows (Kim 2007):

$$\frac{\partial T}{\partial t} = \alpha\left(\frac{\partial^2 T}{\partial z^2}\right) + \frac{1}{\rho C_p}\dot{Q}_{ab}(r,z,t). \tag{2.31}$$

The first boundary condition is an adiabatic boundary condition at the top of the sample as $\partial T/\partial t = 0$, at $z = 0$ because the time scale of the TTR is in the nanosecond range, natural convection and conduction from the sample to the surrounding medium (air) can be neglected. The second boundary condition is an isothermal boundary condition at the bottom of the sample as $T = T_{chuck}$, at $z = \infty$ because the sample is located on a thermochuck that keeps the temperature constant at $T_{chuck}$. Initially, because the materials are at ambient temperature, the initial condition is written as $T = T_{ambient}$, at $t = 0$. During the heating and cooling process, the instantaneous surface reflectivity is acquired by the probing CW laser light reflected from the sample surface at the center of the heated spot, and then the thermal diffusivity of the material is extracted by solving the 1-D inverse heat (2.31) based on (2.26). This would be accomplished by first numerically simulating the transient heating caused by the laser pulse and then fitting the experimental results with the computed data in the TTR system (Kim 2007).

## Electrical Properties and Measurement Techniques

### *Electrical Conductivity and Resistivity*

Electrical conductivity or resistivity is key physical property of all thermal managing materials. Electrical conductivity, $k$, is a measure of how well a material accommodates the movement of an electric charge. It is the ratio of the current density, J (A/m$^2$), to the electric field strength, E (V/m).

$$k = \mathbf{J}/\mathbf{E}. \tag{2.32}$$

Its SI (International System of Units) derived unit is the S/m (Siemens per meter), but conductivity values are often reported as percent IACS (International Annealed Copper Standard). The conductivity of the annealed copper ($5.8108 \times 10^7$ S/m) is defined to be 100% IACS at 20°C, adopted since 1913. Electrical resistivity, $\rho$ ($\Omega$ m), is the reciprocal of conductivity. It is the opposition of a body or substance to the flow of electrical current through it, resulting in a change of electrical energy into heat, light, or other forms of energy. The amount of resistance depends on the type of material and the applied magnetic field.

The resistivity of any material varies with temperature. For instance, the resistivity of metals usually increases as temperature increases, while the resistivity of semiconductors usually decreases as temperature increases. For temperature range that is not too great, this variation can be represented approximately as a linear relation:

$$R_{\mathrm{T}} = R_0[1 + \alpha(T - T_0)], \tag{2.33}$$

where $R_{\mathrm{T}}$ and $R_0$ are the values of the resistance at temperature $T$ and $T_0$, respectively. $T_0$ is often taken to be either room temperature or 0°C. $\alpha$ is the temperature coefficient of resistivity. Pure metals have a small, positive value of $\alpha$, which means that their resistance increases with increasing temperature. There are materials in which resistance decreases with increasing temperature. A thermistor is an example of such a material. It is made of semiconductors, such as oxides of manganese, nickel and cobalt mixed in the desired proportion with a binder and pressed into shape. Thermistors are very sensitive to even small changes of temperature, therefore they are often used as thermometers. The change of resistance of a thermistor caused by temperature change is a nonlinear function and can be approximated by the following formula:

$$R_{\mathrm{T}} = R_0 \exp[\beta(1/T - 1/T_0)], \tag{2.34}$$

where $R_{\mathrm{T}}$ and $R_0$ are the resistance values at absolute temperatures $T$ and $T_0$ (K), $\beta$ (K) is a constant over a limited temperature range and characterizes a property of material. Some materials have very complicated temperature dependencies of resistance. For example, nichrome wire, used as a heating element in most space heaters, practically does not change its resistance in the temperature range between 0 and 100°C. For other materials, such as carbon resistors, the resistance may be constant for a narrow temperature range and show a large effect beyond that range.

The measuring methods for electrical conductivity mainly include eddy current testing, two-point technique, four-point technique, and van der Pauw technique.

### Eddy Current Testing

Eddy current testing uses electromagnetic induction for conducting examinations. When alternating current is applied to the conductor, a testing probe, or sensor, a magnetic field develops in and around the conductor. This magnetic field expands as the alternating current rises to maximum and collapses as the current is reduced to zero. If another electrical conductor, testing sample, is brought into the close proximity to this changing magnetic field, current will be induced in this second conductor, which is measured and correlated with the conductivity of the testing sample.

The testing for electrical conductivity is usually to null an absolute probe in air and place it in contact with the sample surface. And this technique is generally limited to nonmagnetic materials, as the change in impedance of the coil probe can be correlated directly to the conductivity of the material. Because conductivity changes with temperature, the measurements should be made at a constant temperature and adjustments made for temperature variations when necessary. The thickness of the specimen should generally be greater than three standard depths of penetration. This is so the eddy currents at the back surface of the sample are sufficiently weaker than the variations in the specimen thickness that are not seen in the measurements.

Except electrical conductivity measurement, eddy current testing as a nondestructive test tool can also be used for crack detection, material thickness measurements, coating thickness measurements, as well as conductivity measurements for material identification; heat damage detection, case depth determination, and heat treatment monitoring. The major advantage of this method is that it can be used for measuring complex shapes and wide size ranges of conductive materials with various defects. The main limitations of eddy current testing method include: (1) only conductive materials can be inspected; (2) sample surface must be accessible to the probe, and surface finish or roughness may interfere testing result; (3) depth of signal penetration is limited; and (4) flaws such as delaminations that lie parallel to the probe coil winding and probe scan direction are undetectable.

### Two-Point Technique

The resistivity of a material can be obtained by measuring the resistance and physical dimensions of a bar of material. The material is cut into the shape of a rectangular bar of length $l$, height $h$, and width $w$. Copper wires are attached to both ends of the bar. A voltage source applies a voltage $V$ across the bar, causing a current $I$ to flow through the bar. The amount of current $I$ that flows through the bar is measured by the ammeter, which is connected in series with the bar and voltage source. The voltage drop across the ammeter should be negligible. The resistance $R$ of the bar is given by $R = V/I$, where $R$ = resistance in Ω; $V$ = voltage in volts, $I$ = current in amps. The two-point resistivity of the material can be obtained by (Heaney 1999):

$$\rho = (Rwh)/l, \tag{2.35}$$

where $\rho$ is the resistivity in Ω m, $R$ is the measured resistance in Ω, and $w$, $h$, and $l$ are the measured physical dimensions of the sample bar in meters. In practice, measuring resistivity with a two-point technique is often not reliable, because of some resistance between the contact wires and the material, or in the measuring equipment itself. These additional resistances make the resistivity of the material measure higher than it really is.

### Four-Point Technique

Four-point measurement technique on a bar of material uses four wires attached to the sample bar. A current source forces a constant current through the ends of the sample bar. A separate ammeter measures the amount of current $I$ passing through the bar. A voltmeter simultaneously measures the voltage $V$ produced across the inner part of the bar. The four-point resistivity of the material can then be expressed as (Heaney 1999):

$$\rho = (Vwh)/(Is), \tag{2.36}$$

where $\rho$ = resistivity in Ω m, $V$ = voltage measured by the voltmeter in volts, $w$ = width of the sample bar measured in meters, $h$ = height of the sample bar measured in meters, $I$ = current the ammeter measures flowing through the sample in amperes, $s$ = distance between the two points where the voltmeter wires make contact to the bar, measured in meters.

### Van der Pauw Technique

The four-point measurement technique described earlier has assumed the material sample has the shape of a rectangular thin film or a bar. There is a more general four-point resistivity measurement technique that allows measurements on samples of arbitrary shape, with no need to measure all the physical dimensions of the sample. This is the van der Pauw technique. There are four conditions that must be satisfied to use this technique (Heaney 1999): (1) the sample must have a flat shape of uniform thickness; (2) the sample must not have any isolated holes; (3) the sample must be homogeneous and isotropic; (4) all four contacts must be located at the edges of the sample.

In addition to these four conditions, the area of contact of any individual contact should be at least an order of magnitude smaller than the area of the entire sample. For small samples, this might not be possible or practical. If sufficiently small contacts are not achievable, it is still possible to do accurate van der Pauw resistivity measurements, using geometric correction factors to account for the finite size of the contacts.

## *Permittivity and Its Characterization*

Permittivity, $\varepsilon$, also called electric permittivity or dielectric constant, is a constant of proportionality that exists between electric displacement (D) and electric field intensity (E): $D = \varepsilon E$. The vacuum permittivity $\varepsilon_0 = 1/(c^2 \mu_0) \approx 8.85 \times 10^{-12}$ farad per meter (F/m), where $c$ is the speed of light; $\mu_0$ is the permeability of vacuum. The linear permittivity of a homogeneous material is usually given relative to that of vacuum, as a relative permittivity $\varepsilon_r = \varepsilon/\varepsilon_0$. When $\varepsilon_r$ is greater than 1, these substances are generally called dielectric materials, or dielectrics, such as various ceramics, plastics, and certain metal oxides. A high permittivity tends to reduce any electric field present. For instance, the capacitance of a capacitor can be increased by increasing the permittivity of the dielectric material. Common methods used for the measurement of permittivity include loaded resonant waveguide cavity, open ended coaxial line, and loaded coaxial transmission line. Loaded resonant waveguide cavity methods offer the good measurement accuracy, particularly for measurement of imaginary part of the permittivity for determining the energy losses in a material. The sample material is mounted inside a resonant cavity to fit well with certain dimensions and thereby loading the cavity. This method only covers a narrow frequency band to maintain the resonance status of a cavity. However, this can be improved to some extend by including higher order resonant modes.

The open ended coaxial line is a nondestructive test method. The coaxial probe is pressed against a test sample to measure the reflection coefficient. And then the permittivity can be determined from the reflection coefficient. Because this method is sensitive to air gaps that disturb the electric field, the probe and the sample, its measurement accuracy is not as good as the other two methods (Tong 2009).

In the loaded coaxial transmission line method the material under test is placed to fill the volume between the inner and outer conductor in a section along the transmission line. The dimensions of the sample piece are critical to ensure a precise fit. The material may load the line and cause a change of characteristic impedance. Both reflection and transmission through the fixture is used when calculating the test material data. The measurement accuracy of this method is not as good as that of loaded cavity resonance measurement methods. An advantage of the loaded coaxial transmission line method is that it offers the possibility to perform measurements in a broad frequency band (Tong 2009).

The selection of the measurement method to use depends on restrictions on sample preparation and desired frequency range and accuracy of the data. The loaded cavity resonance measurement method gives high accuracy data for narrow-band measurements at frequencies of over 100 GHz. The open ended coaxial line offers a method where only a flat surface of sample material is needed. The loaded coaxial transmission line method makes it possible to measure over wide frequency range with better accuracy than what the open ended coaxial line method offers (Tong 2009).

## Thermomechanical Characterization

To produce low cost and compact electronics goods, high input/output surface mounted chip carriers on printed circuit boards (PCBs) or printed wire boards, such as multilayer PCBs, flip chip, flip-chip with underfill, chip scale package, and ball grid array packages, have been widely used in electronic packaging assemblies. One critical reliability concern about this type of electronic packaging is the thermomechanical stress and strains induced by temperature change. For example, the in-plane strain affects the thermal fatigue life of surface mount solder interconnections, while the out-of-plane strain affects the mechanical integrity of the plated-through holes of the PCBs. The primary causes for this strain are thermal mismatch in properties among constituent materials of the package as well as mechanically induced flexure. A variety of techniques have been used to characterize the stress and strain in the electronic packaging. For instance, a micromechanical test apparatus was developed for use with a compact four-beam moiré interferometer. The interferometer was used to evaluate whole field displacements of plastic ball grid array packages over a range of temperatures and under four-point bending.

In fact, thermomechanical failure has become one of major concerns for electronic packaging design, especially because the high degree of integration, speed, and power level and the high cost of production have imposed more stringent reliability requirements on individual components. Therefore, a fully understanding of thermomechanics of component materials is very important as they apply to electronic packaging. Thermal stress effects can be evaluated by experimental validation, matched with simulation by coupling a heat transfer analysis (steady-state or transient) and a structural analysis (static stress with linear or nonlinear material models or mechanical event simulation).

This section will introduce basic concepts of thermally induced stress, strain, and materials behavior characterization methods, as well as the finite-element method and its application to thermomechanical analysis, and various failure theories, including fatigue and fracture with analytical solutions, which provide insight into the behavior of layered assemblies of electronic packaging and thermal cooling system.

## *Characterization Techniques of Thermally Induced Stress and Strain*

As shown in Figure 2.6, stress is defined as force per unit area, which can be interpreted as internal tractions that act on a defined internal datum plane within a solid. Depending on its action direction, it can usually be divided into normal stress and shear stress. Strain is defined as the amount of deformation an object experiences compared with its original size and shape under the action of stress. Stress components acting perpendicular to the surfaces are called normal stresses; stress components acting along planes are referred as shear stresses. In the static case, when all forces acting on three orthogonal faces of a body are in equilibrium, the matrix of nine stress components in a Cartesian coordinate system is called stress tensor.

The strain tensor, $\varepsilon$, is a symmetric tensor used to quantify the strain of an object undergoing a small 3-D deformation: the diagonal coefficients $\varepsilon_{ii}$ are the relative change in length in the direction of the $i$ direction (along the $x_i$-axis); the other terms $\varepsilon_{ij} = 1/2\gamma_{ij}$ $(i \neq j)$ are the shear strains, i.e., half the variation of the right angle (assuming a small cube of matter before deformation). The deformation of an object is defined by a tensor field, i.e., this strain tensor is defined for every point of

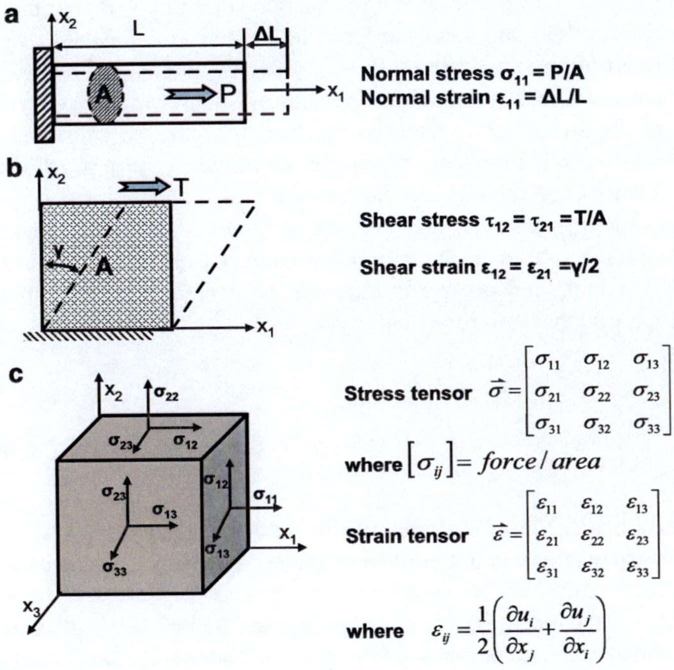


$$\vec{\sigma} = \begin{bmatrix} \sigma_{11} & \sigma_{12} & \sigma_{13} \\ \sigma_{21} & \sigma_{22} & \sigma_{23} \\ \sigma_{31} & \sigma_{32} & \sigma_{33} \end{bmatrix}$$

$$\vec{\varepsilon} = \begin{bmatrix} \varepsilon_{11} & \varepsilon_{12} & \varepsilon_{13} \\ \varepsilon_{21} & \varepsilon_{22} & \varepsilon_{23} \\ \varepsilon_{31} & \varepsilon_{32} & \varepsilon_{33} \end{bmatrix}$$

$$\varepsilon_{ij} = \frac{1}{2}\left(\frac{\partial u_i}{\partial x_j} + \frac{\partial u_j}{\partial x_i}\right)$$

**Fig. 2.6** Schematic of concept of stress and strain. **(a)** Normal stress and strain in 1-D system; **(b)** shear stress and strain in 1-D system; **(c)** stress and strain tensors in 3-D system the object. This field is linked to the field of the stress tensor by the generalized Hooke's law.

In case of small deformations, the strain tensor is the linearized Green strain tensor or Cauchy's infinitesimal strain tensor, defined by the equation:

$$\varepsilon_{ij} = \frac{1}{2}\left(\frac{\partial u_i}{\partial x_j} + \frac{\partial u_j}{\partial x_i}\right),$$

where $u$ represents the displacement field of the object's configuration (i.e., the difference between the object's configuration and its natural state). This is the symmetric part of the Jacobian matrix. The antisymmetric part is called the small rotation tensor (Symon 1971).

In the electronic packaging, changes in temperatures causes thermal effects on materials. Some of these thermal effects include thermal stress, strain, and deformation. Thermal deformation means that as the thermal energy (and temperature) of a material increases, so does the vibration of its atoms/molecules; and this increased vibration results in what can be considered a stretching of the molecular bonds–which causes the material to expand. Of course, if the thermal energy (and temperature) of a material decreases, the material will shrink or contract. The thermal strain, which results from unrestrained expansion due to temperature change, $\Delta T$, is given by $\varepsilon_1 = \varepsilon_2 = \varepsilon_3 = \alpha(T)\Delta T$, where $\alpha(T)$ is the CTE, $T$ is the temperature, and $\Delta T$ is the temperature change. For many materials, $\alpha$ is not a strong function of the temperature and can be treated as a constant over a reasonably large variation in temperature.

In electronic packaging including the thermal management system, the main objectives of the stress–strain analysis are to determine the causes of failure, to conduct design optimizations and predict reliabilities. Because most of the current prediction theories and models use thermomechanical stresses and strains as parameters for estimating damages and failures, accurate determinations of strains and stresses are indispensable in the reliability analysis and predictions. In order to obtain critical strains and stresses in the package structures, several methods have been developed, such as micromechanical tester, moiré interferometry, and speckle techniques.

#### Micromechanical Tester

The micromechanical tester or diamond indenter is usually used to evaluate the mechanical response of a small volume of material near the surface of a sample by probing the surface with a diamond indenter of known geometry. A piezoelectric transducer drives the indenter into the sample surface while load, depth of penetration, and displacement data are continuously collected. There are several diamond indenters, such as Vickers (square pyramid), Knoop (elongated square pyramid), Berkovich (triangular pyramid), and right circular cones with tip radii of 1, 5, 10, and 50 μm. A

PC (personal computer) interface is used to control the machine and collect data. The micromechanical tester is generally equipped with a hot stage: fixed temperatures from ambient to 100°C; and a tangential load cell for measuring lateral forces.

The micromechanical tester can be used to characterize the mechanical properties and adhesion strength of films. Indentation, microscratch, load relaxation, indentation fatigue, indentation creep, and dynamic mechanical experiments can be performed on coatings to obtain information on hardness, elastic modulus, storage and loss moduli, friction coefficients, film adhesion, and fracture toughness. Three- and four-point bending, tensile, compression, crush, peel, and fiber pushout tests can also be performed.

Scratch tests are performed by drawing the indenter across the surface as it penetrates. Delamination of films allows qualitative and, in some cases, quantitative evaluation of interfacial properties.

Virtually any test that requires pushing or pulling under carefully controlled conditions within the physical limits of the micromechanical tester can be performed. The micromechanical tester can be used for testing materials and coatings of all kinds from highly viscous polymers to diamond-like carbon. Maximum sample size: specimens 2 cm high and 5 cm across. Larger samples can be accommodated with some geometrical constraints. A smooth surface produces best results. Load resolution: 50 μN; depth resolution: 5 nm.

## Moiré Interferometry

Moiré interferometry is a whole-field optical interference technique with high resolution and high sensitivity for measuring the in-plane displacement and strain fields. This method has been successfully used to measure the thermomechanical deformations of electronic packages with the objective to study package reliability.

Moiré interferometry relies upon the interference of laser light to produce maps of displacement. Interference occurs through the superposition of the electric field vectors of rays of light, with the same frequency, meeting at a point in space. Depending on both the amplitudes and difference in phase between superimposed rays of light, the resulting intensity for two interfering light rays is described by (Stout 1994):

$$I = a_1^2 + a_2^2 + 2a_1a_2 \cos \Phi, \tag{2.37}$$

where $a_1 =$ the electric field amplitude of beam 1, $a_2 =$ the electric field amplitude of beam 2, $\Phi =$ the phase difference between beams 1 and 2. When two interfering light rays have equal amplitudes, (2.34) simplifies to

$$I = 4a^2 \cos^2\left(\frac{\Phi}{2}\right), \tag{2.38}$$

where $a = a_1 = a_2 =$ the electric field amplitude of both beams. If both beams are out of phase by $\pi$ radians, completely destructive interference occurs.

During measurement in an interferometer, a high frequency diffraction grating is transferred on to the testing area of a specimen. The optical arrangement in the moiré interferometer produce a virtual reference grating of spatial frequency $f$ by means of the interference of two coherent beams (beam 1 and beam 2) from a laser light source. The virtual reference grating is superimposed on the specimen grating to form a fringe pattern, which is a displacement contour map. Each fringe passes through points with the same displacement in the electrical field corresponding to the $x$ and $y$ direction. Fringes are numbered consecutively, with fringe order $N_x$ and $N_y$. The formulas for determining the displacements at each point in the field from the fringe orders can be written as (Post and Han 2009):

$$U = \frac{1}{f} N_x; \quad V = \frac{1}{f} N_y, \tag{2.39}$$

where $U$ and $V$ are displacements in the $x$- and $y$-directions, respectively. For the strains at each point

$$\varepsilon_x = \frac{\partial U}{\partial x} = \frac{1}{f}\left(\frac{\partial N_x}{\partial x}\right), \tag{2.40}$$

$$\varepsilon_y = \frac{\partial V}{\partial y} = \frac{1}{f}\left(\frac{\partial N_y}{\partial y}\right), \tag{2.41}$$

$$\gamma_{xy} = \frac{\partial U}{\partial y} + \frac{\partial V}{\partial x} = \frac{1}{f}\left(\frac{\partial N_x}{\partial y} + \frac{\partial N_y}{\partial x}\right), \tag{2.42}$$

where $\varepsilon$ and $\gamma$ are normal and shear strains, respectively. Thus, the strains are determined by the rate of change of fringe orders in the patterns, or the fringe gradient surrounding each point. Stresses are determined from the strains, using the stress–strain relationships (or the constitutive equations) for the specimen material.

Moiré measurements are performed routinely in the interferometric domain with fringes representing subwavelength displacements per contour. Since moiré responds only to geometric changes, it is equally effective for elastic, viscoelastic, and plastic deformations, for isotropic, orthotropic, and anisotropic materials, and for mechanical, thermal, and dynamic loadings (Post and Han 2009). Moiré interferometry can be used to contour surfaces at any wavelength longer than 10 μm (with difficulty) or 100 μm with reduced environmental requirements and no intermediate photographic recording setup. For nondestructive testing, holographic interferometry has a precision of a small fraction of a micrometer and is useful over a deformation amplitude of a few micrometers, whereas moiré interferometry has a precision ranging from 10 to 100 μm to mm, with a correspondingly increased useful range of deformation amplitude. A widely used moiré interferometer in electronic package analysis is the Portable Engineering Moiré Interferometer originated from IBM. It uses a grating frequency of 1,200 lines/mm, which yields a spacing of the interference fringe corresponding to 417 nm of in-plane displacement. The sensitivity is adequate for measuring the overall thermal deformation of electronic packages. However, it is not sufficient for measuring thermal deformation in high-density electronic packages, particularly for small features, such as solder balls. Such measurement can be performed using phase-shifting moiré interferometry or speckle techniques.

## Speckle Techniques

The speckle pattern is formed by the self-interference of a large number of random coherent waves scattered from a rough object surface or propagated through a medium of random refractive index fluctuations. The remarkably simple and effective way to use the speckle effect in the measurement of displacement and deformation is speckle photography and speckle interferometry. They can give point-by-point or whole-field data, and the level of sensitivity can be adjusted. Speckle interferometry is based on the coherent addition of scattered light from an object surface with a reference beam that may be a smooth wave front or a scattered field from a reference object or from the sample. Compared with the speckle photography method, speckle interferometry is more sensitive (An 2002).

Laser speckles is a phenomenon that is experienced if an optically rough surface is illuminated by a highly coherent laser source. They are formed by the interference of dephased but coherent wavelets, emanating from different microscopic elements of the specimen's surface and cause a randomly looking pattern. This speckle pattern is characteristic for a particular surface element just like a fingerprint and if that surface element is undergoing a pure displacement $\delta_{surf}$, the associated speckles, as recorded by a camera at a fixed location, are displaced likewise by $\delta_{image}$ image. If the specimen, in addition to pure displacement, is also strained, the associated structural changes of the speckles will for most applications cause a negligible additional speckle-displacement. The usual digital image-processing algorithm, is used to track speckles, based on pattern matching by some kind of correlation algorithm, which usually gives displacement values that are integer multiples of a quant (usually the size of the pixel pitch of the camera; Zimprich and Zagar 2007).

The electronic image acquisition and computer image processing has been applied to the speckle metrology and lead to a versatile method called electronic speckle pattern interferometry (ESPI), television holography, or electrooptic holography. The advances in the development of CCD (charge-coupled device) cameras and computer-based image processing have propelled ESPI to the forefront of thermomechanical analysis of electronic packaging. The attractiveness and versatility of speckle metrology lies in its ability to measure (An 2002): (1) deformation or displacement with variable sensitivity to in-plane and out-of-plane direction; (2) 3-D object shape surface roughness; (3) vibration.

ESPI has been applied for noncontact, real-time evaluation of thermal deformation in a flip-chip package. The spatial resolution of ESPI was increased to submicron scale by magnifying the areas studied in order to measure the deformation of such small-scale components as the solder in the flip-chip package. Thermal deformation in the horizontal and vertical directions around the solder joints, for instance, was measured as 2-D mappings during heating from 25 to 125°C. ESPI was successful in obtaining information on the complicated deformation field around the solder joints. Furthermore, the shear strain could also be calculated using the measured thermal deformation around each solder joint. The applicability of ESPI to flip-chip packages could be verified by comparing the ESPI results with those of finite-element analysis (FEA).

### *Fundamental Equations of Deformable Bodies*

Deformation analysis of electronic system is usually based on continuum mechanics using mathematical techniques and physical laws of statics, kinematics, and dynamics. Statics deals with the equilibrium of forces acting on a body at rest or moving with constant velocity. Dynamics deals with bodies in motion, which is divided into kinematics and kinetics. Kinematics is concerned only with the geometry of motion. It is the study of the positions, angles, velocities, and accelerations of body segments and joints during the motion. Kinetics is concerned with the force analysis of bodies in motion (Szostak-Chrzanowski et al. 2006). The basic conditions and relations of continuum mechanics involve equations of equilibrium, strain displacement equations, constitutive equations, and boundary conditions.

The conditions of equilibrium are fundamental equations of continuum mechanics. The acting forces on the object must be in equilibrium. The forces acting on the object are divided into external forces acting on boundaries and body forces. An acting force $F$ per unit area $A$ is defined as a stress $\sigma$. If $\sigma_{ij}$ denotes stress tensor and the stress tensor is symmetric, that is, $\sigma_{ij} = \sigma_{ji}$, the equation of equilibrium is (Szostak-Chrzanowski et al. 2006)

$$\sigma_{ij}\mathrm{d}A_i = \mathrm{d}P_j. \tag{2.43}$$

Eigenvalues of the stress tensor are known as principal stresses. The principal stress directions are the eigenvectors of the stress tensor. In case of a body in motion, the motion is described by dynamic equations. For small motions, the velocity is a partial derivative of displacement function $U$ and is given as (Szostak-Chrzanowski et al. 2006):

$$U_i' = \frac{\partial U_i}{\partial t}. \tag{2.44}$$

Negative product of acceleration and mass density $\rho$, known as d'Alambert inertia force per unit volume, is added to the acting force $F_i$. This gives the dynamic equation using Newton's law in the form (Szostak-Chrzanowski et al. 2006):

$$\sigma_{ij}\mathrm{d}A_i = -\mathrm{d}F_i + \mathrm{d}(\rho U'_i). \tag{2.45}$$

The kinematic relations are between strain and displacement and relate to the geometry of the motion which leads from undeformed to deformed position. The kinematic relations are independent of acting forces and type of material behavior (elastic or inelastic). The strain tensor is denoted as $\varepsilon_{ji}$ and has nine components in the 3-D space. In case when displacements $U$ are small it means, $U_{i,j} = \partial U_i/\partial x_j \ll 1$, the strain tensor components are (Szostak-Chrzanowski et al. 2006)

$$\varepsilon_{i,j} = \frac{1}{2}(U_{i,j} + U_{j,i}). \tag{2.46}$$

Deformation of a body is fully described if nine components (in 3-D space) of deformation tensor can be determined at any point of the body. The deformation tensor may be decomposed into strain tensor $\varepsilon$ and rotation (deviatoric) tensor $\omega$ (Szostak-Chrzanowski et al. 2006):

$$\frac{\partial U_i}{\partial x_j} = \varepsilon_{ij} + \omega_{ij}, \tag{2.47}$$

where $\omega_{ij} = (1/2)((\partial U_i/\partial x_j) - (\partial U_j/\partial x_i))$ and $\omega_{ij}$ is skew symmetric: $\omega_{ij} = -\,\omega_{ji}$.

The stress–strain relations are mathematical descriptions of mechanical properties of the material using constitutive matrix. Linear elastic constitutive laws may model almost all materials subjected to sufficiently small loads. For linear elastic material, Hooke's law for a general anisotropic solid is given as: $\sigma_{ij} = E_{ijlm}\varepsilon_{lm}$, where $E_{ijlm}$ represents components of the elasticity tensor and $i$, $j$, $l$, $m$ have values 1, 2, 3. This is the so-called first constitutive equation (constitutive law). In real materials, complex transport equations must be solved to determine the time and spatial response of charges, for example, the Boltzmann equation or the Fokker–Planck equation or the Navier–Stokes equations with a general form:

$$F(\sigma_{ij}, \varepsilon_{ij}\, \varepsilon'_{ij}, T, t) = 0. \tag{2.48}$$

An entire physical apparatus for dealing with these matters has developed, such as linear response theory, Green–Kubo relations and Green's function (many-body theory). These complex theories provide detailed formulas for material parameters such as permittivities, permeabilities, conductivities, and so forth.

## *Constitutive Behavior*

While there is rich diversity of structure at various levels in any piece of electronic packaging material, a fundamental assumption of continuum theory is to assume materials consist of aggregates of representative volume elements (RVE). The deformation response of the material is then the result of the collective response of the RVEs. This is what is known as the constitutive behavior of the material. Such volume elements are large enough that material inhomogeneities can be ignored and average material property values can be used, but also small enough that it is entirely legitimate to apply the methods of mathematical analysis. In the case of electronic packaging and thermal managing materials, the constitutive behavior involves thermoelastic (silicon carbide die, encapsulant, ceramic and FR4 substrate, dielectric films, thermal interface materials [TIMs], etc.), elastic-plastic (metal films, traces, bonding wire, thermal managing materials), and viscoplastic/creep (solder, TIMs) behaviors.

The constitutive equation of thermoelastic behavior for anisotropic materials can be derived from (2.48):

$$\sigma_{ij} = C_{ijkl}(\varepsilon_{kl} - \alpha_{kl}\Delta T), \tag{2.49}$$

where $C$ is a fourth-order tensor of elastic moduli, whose components are formed relative to the crystal lattice. $C_{ijkl}$ is symmetric in indices $i, j$ and $k, l$ and generally has the following symmetry: $C_{ijkl} = C_{klij}$, assuming the existence of an elastic strain energy function. The number of independent components of $C$ depends on the symmetry property of a particular crystal. If the crystal is isotropic, then there are only two independent elastic constants. For isotropic materials, $C_{ijkl} = \lambda\delta_{ij}\delta_{kl} + \mu(\delta_{ik}\delta_{jl} + \delta_{il}\delta_{jk})$, $\alpha_{ij} = \alpha\delta_{ij}$, and the following equations can be derived:

$$\sigma_{11} = \frac{E}{1+v}\left[\varepsilon_{11} + \frac{v}{1-2v}(\varepsilon_{11} + \varepsilon_{22} + \varepsilon_{33})\right] - \frac{\alpha E}{1-2v}\Delta T, \tag{2.50}$$

$$\sigma_{22} = \frac{E}{1+v}\left[\varepsilon_{22} + \frac{v}{1-2v}(\varepsilon_{11} + \varepsilon_{22} + \varepsilon_{33})\right] - \frac{\alpha E}{1-2v}\Delta T, \tag{2.51}$$

$$\sigma_{33} = \frac{E}{1+v}\left[\varepsilon_{33} + \frac{v}{1-2v}(\varepsilon_{11} + \varepsilon_{22} + \varepsilon_{33})\right] - \frac{\alpha E}{1-2v}\Delta T, \tag{2.52}$$

$$\sigma_{12} = \frac{E}{1+v}\varepsilon_{12}, \tag{2.53}$$

$$\sigma_{13} = \frac{E}{1+v}\varepsilon_{13}, \tag{2.54}$$

$$\sigma_{23} = \frac{E}{1+v}\varepsilon_{23}. \tag{2.55}$$

For elastic-plastic behavior, (2.48) transforms to $\sigma_{ij} = F(\varepsilon_{kl}, T)$. The stress–strain relationship is nonlinear and depends on temperature $T$. Loading and unloading follow different stress–strain curves.

Creep is the plastic deformation of a material at a high homologous temperature below a material's yield stress. For creep behavior, (2.48) transforms to $\varepsilon_{ij} = F(\sigma_{kl},T,t)$. Creep can be subdivided into three categories primary, tertiary, and steady state creep. The qualitative behavior of the strain vs. time is shown in Figure 2.7. A material deforming by creep spends most of its time in the steady state region which is by far the dominant region when considering the effects of creep. The primary creep region is characterized by the following equation (Rhoads 2008):

$$\varepsilon_{\mathrm{pc}} = \beta t^{1/3}, \tag{2.56}$$

where $\beta$ is a constant. Primary creep strain is usually less than 1% of the sum of the elastic, steady state, and primary strains. The mechanism in the primary region is the climb of dislocations that are not pinned in the matrix. The primary region is strongly dependent on the history of the material. If the material had been heavily worked before the creep test, there would have been many more dislocations present and the characteristics of the primary creep region would have been much different. When the amount of strain is high creep fracture or rupture will occur. In the tertiary region the high strains will start to cause necking in the material just as in the uniaxial tensile test. This necking will cause an increase in the local stress of the component which further accelerates the strain. Eventually the material will pull apart in a ductile fracture around defects in the solid. These defects could be precipitated at high temperatures or grain boundaries at lower temperatures. In any case the importance of the tertiary region to normal operation and creep design criteria is minimal. In Figure 2.7, the time scale of the tertiary region is greatly expanded for the purpose of clarity. Considering the small amount of time in addition to the fact that the tertiary region develops a plastic instability similar to necking, operating in the tertiary region is not feasible. Therefore, it is a conservative estimate to approximate the end of serviceable life of any component to

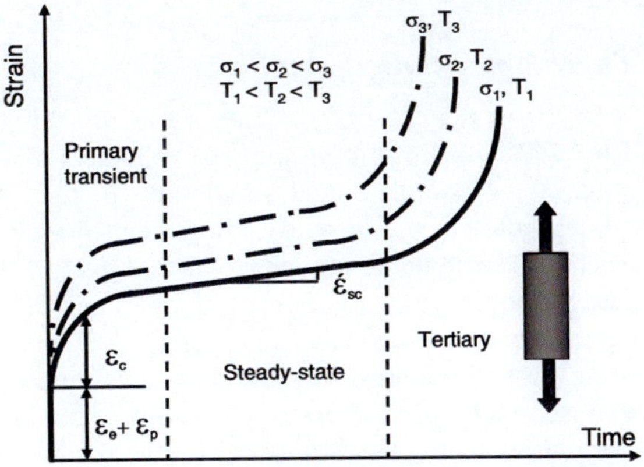


Fig. 2.7 Illustration of typical creep behavior

coincide with the end of the steady state creep region. The extra time in the tertiary region may be useful to consider only under accidental conditions.

The steady state region is so-called because the strain rate is constant. In this region the rate of strain hardening by dislocations is balanced by the rate of recovery. When the homologous temperature of a creep sample is between 0.3 and 0.7 of its melting temperature, the mechanism of creep is dislocation climb. This is the climb of dislocation jogs that impede the normal motion of the dislocation. This dislocation climb mechanism can be described by the following equation (Rhoads 2008):

$$\frac{\dot{\varepsilon}_{ss}}{\dot{\varepsilon}_0} = A\left(\frac{\sigma}{\sigma_0}\right)^m e^{-\frac{H}{kT}}, \tag{2.57}$$

where $H$ is the activation energy for diffusion. $A$ is a constant, and $\sigma$ are the applied stresses. When higher temperatures are applied and the temperature is above 0.7 of the melting temperature, a different mechanism takes place. The creep is analogous to viscous flow in this region, and the mechanism of the creep is diffusion of atoms from one place to another. This type of creep can be described by (Rhoads 2008):

$$\frac{\dot{\varepsilon}_{ss}}{\dot{\varepsilon}_0} = \frac{1}{d_g^2}\frac{\sigma}{\sigma_0} e^{-\frac{H'}{kT}}, \tag{2.58}$$

where $d_g$ is the diameter of the grains in the material, and $H'$ is the diffusion activation energy. As grain size increases the creep rate decreases. In case for high temperature applications large grains are necessary to reduce the creep rate because they reduce the number of grain boundaries and increase the bulk diffusion distances.

## *Thermomechanical Analysis*

The traditional use of prototypes to validate a new product design is rapidly being replaced by computational simulation. In fact, simulation driven design using virtual prototypes is currently being adopted by best in class manufacturers as the most consistent method to reduce time to market and hit their revenue, cost, launch date, and quality targets. For metal structures, simulation is now well established, but multimaterial structures incorporating ceramics and joints, like electronic packaging and its cooling system, pose a particular challenge. Thermomechanical analysis is one part of the computational simulation of the electronic packaging thermal products, which is based on the mathematical calculation and numerical simulation.

## Plain Stress and Plain Strain

Many problems in thermomechanical analysis of electronic packaging and cooling systems can be treated by 2-D, or plane theory of elasticity. Plane stress and plane strain are two general types of problems involved in this plane analysis. Take a thin chip as an example, which is loaded by thermally induced forces applied along the boundary edges parallel to the plane of the chip. If the $x$–$y$ plane is parallel to the plane of the chip, then the stress components $\sigma_z$, $\tau_{xy}$, and $\tau_{yz}$ are zero on both plane faces and are assumed to be zero within the chip. The state of stress is then given by $\sigma_x$, $\sigma_y$ and $\tau_{xy}$ and is so-called plane stress. Moreover, plane strain is more useful for packing thermal system. Consider a simple package assembly of a chip bonded to a heat spreader, the stresses that develop in the bond between the chip and the heat spreader when there is a temperature change applied to the assembly. For plane analysis, the assumptions are that any external forces acting do not vary with $z$ and the strains $\varepsilon_z$, $\gamma_{yz}$ and $\gamma_{xz}$ on a section near the middle of the assembly are zero. In this case, Hook's law can be expressed as (Schmidt 2006):

$$\sigma_z = -\nu(\sigma_x + \sigma_y) - E\alpha\Delta T. \tag{2.59}$$

For $\sigma_z$, where thermal strain has been included, and the plane strain can be expressed (Schmidt 2006):

$$\varepsilon_x = \frac{1}{E}\left[\sigma_x(1-\nu^2) - \nu(1+\nu)\sigma_y\right] + (1+\nu)\alpha\Delta T, \tag{2.60}$$

$$\varepsilon_y = \frac{1}{E}\left[\sigma_y(1-\nu^2) - \nu(1+\nu)\sigma_x\right] + (1+\nu)\alpha\Delta T. \tag{2.61}$$

## Beams and Laminate Assemblies

In some cases, electronic packaging and cooling system can be simplified as beams. The beam is a 3-D body, however, its length must be significantly larger than the depth (three or more times the depth). Internally, beams experience compressive, tensile and shear stresses as a result of the loads applied to them. As shown in Figure 2.8, one simple prismatic beam of length $L$ will be considered, with an axial force $F_x$, transverse distributed load $p(x)$, and shear $\tau_0$ as external force acting on the beam. For this beam, the stress $\sigma_y$ and $\sigma_z$ are zero and the normal stress $\sigma_x$ and axial strain $\varepsilon_x$, based on the Hook's law (Schmidt 2006):

$$\sigma_x = \frac{F_x}{A} + \frac{My}{I}, \tag{2.62}$$

$$\varepsilon_x = \frac{F_x}{AE} + \frac{My}{EI} + \alpha\Delta T, \tag{2.63}$$

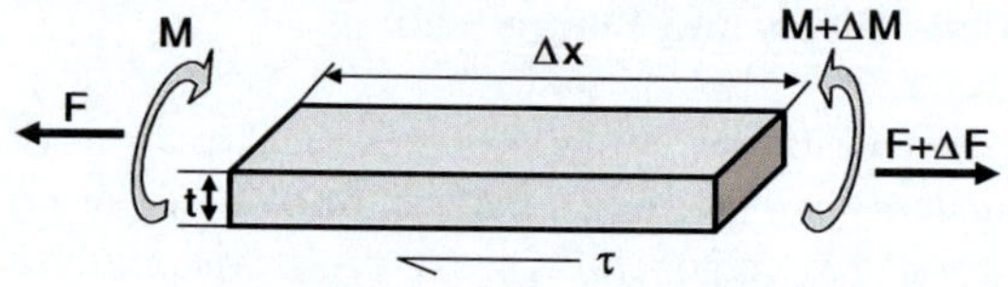


**Fig. 2.8** Illustration of a free-body beam

where $A$ is the cross-sectional area, $y$ is the distance from the centroid of the cross section to any point on the cross section, $M$ is moment, and $I$ is the area moment of inertia about the $z$ axis at the centroid of the cross section. When a strip is cut from a wide beam is under consideration, then a form of plane strain conditions exist and, for the strip, $\sigma_y = 0$ and $\sigma_z = \nu\,\sigma_x$, the strain becomes

$$\varepsilon_x = \frac{(1-\nu)F_x}{AE} + \frac{(1-\nu^2)My}{EI} + \alpha\Delta T. \tag{2.64}$$

### Bimaterial Assembly

The typical bimaterial problems and solutions to simulate the electronic bonding system include bimaterial assembly axial effects, bimaterial assembly bending effects, and peeling stress.

The problem of bimaterial-axial effects is used to analyze two materials simply bonded together at their interface, as shown in Figure 2.9. Denote the top material layer by 1 and the lower material layer by 2, with the thickness, moduli, and coefficients of linear thermal expansion, designated by $t_1$, $t_2$, $E_1$, $E_2$, $\alpha_1$ and $\alpha_2$, respectively. A uniform temperature change, $\Delta T$, in two materials is assumed. When $t_0 \to 0$, the bond property between two materials can be neglected, or does not influence the behavior of the system. Assuming the external force on the system is zero, therefore, $F = F_1 = F_2$. The radius of curvature $R$ after the system experiences a temperature change can be related to the force $F$. The sum of the moments about $z$ axis at $y = 0$ can be expressed as (Schmidt 2006):

$$M_R = M_1 + M_2 - \frac{t_1 + t_2}{2}F = 0. \tag{2.65}$$

For a beam of unit width cut from the assembly, the moment curvature relation gives

$$M_i = \frac{E_i I_i}{(1-\nu_i^2)R} = \frac{E_i t_i^3}{12(1-\nu_i^2)R} = \frac{D_i}{R}, \quad i = 1, 2. \tag{2.66}$$

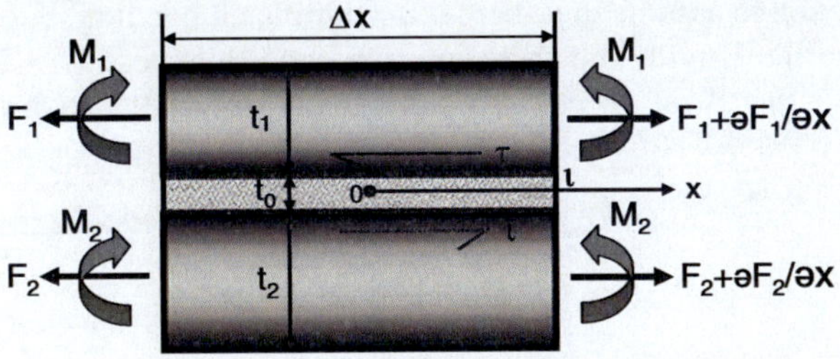


**Fig. 2.9** Illustration of a bimaterials assembly (when bond thickness $t_0 \to 0$) or a bond assembly ($t_0 \ll t_1$ or $t_2$, and the bond is compliable)

where $D_{\mathrm{i}} = ((E_i t_i^3)/12(1 - v_i^2))$, the radius of curvature, $R$, for each member is assumed equal. This is valid if the strips are thin. Substituting (2.63) into (2.62), the relationship between $R$ and material properties, thickness, and the force $F$ can be obtained:

$$R = \frac{2(D_1 + D_2)}{(t_1 + t_2)F}. \tag{2.67}$$

On the other hand, the strains at the interface can be arisen due to the axial force, the bending moment, and the temperature change. If the strain at the lower surface of the upper member is equal to the strain at the upper surface of the low member when the system deforms, the strain compatibility condition will be meet (Schmidt 2006):

$$\frac{F}{E_1 t_1} + \frac{t_1/2}{R} + \alpha_1 \Delta T = -\frac{F}{E_2 t_2} - \frac{t_2/2}{R} + \alpha_2 \Delta T. \tag{2.68}$$

And a direct solution for the force $F$ can be obtained by combining (2.64) and (2.65):

$$F = \frac{(\alpha_2 - \alpha_1)\Delta T}{\lambda}, \tag{2.69}$$

where

$$\lambda = \lambda_1 + \lambda_2 + \lambda_{12},\ \lambda_i = \frac{1 - v_i}{E_i t_i},\ \text{and}\ \lambda_{12} = \frac{(t_1 + t_2)^2}{4(D_1 + D_2)}.$$

This equation is the simplest form of solution for modeling electronic packaging and cooling system, but it is only valid away from edges of the strip because the force $F$ was considered constant along the length. As the ends of the strip are approached, the force $F$ must go to zero.

For the situation to assume that there is no significant bending, but the axial force does vary along the length, and therefore, the bond thickness $t_0 > 0$, as shown in Figure 2.9, and the shear force, $\tau$, in the bond material, can be expressed as (Schmidt 2006):

$$\frac{\partial^2 \tau}{\partial x^2} - \frac{1}{K}(\lambda_1 + \lambda_2)\tau = 0, \tag{2.70}$$

where $K = t_0/G$. Let $k^2 = 1/K(\lambda_1 + \lambda_2)$, then (2.70) becomes

$$\tau = A \sinh kx + B \cosh kx, \tag{2.71}$$

Here, the coordinate system has its origin at the center of the assembly, and the assembly is symmetric about the origin. $A$ and $B$ can be solved by $\tau = 0$, at $x = 0$; and $K(\partial\tau/\partial x) = (\alpha_1 - \alpha_2)\Delta T$, at $x = l$. Therefore, (2.71) becomes (Schmidt 2006):

$$\tau = \frac{(\alpha_1 - \alpha_2)\Delta T}{Kk \cosh kl} \sinh kx, \tag{2.72}$$

when $x = l$, the maximum shear stress can be obtained:

$$\tau_{\max} = \frac{(\alpha_1 - \alpha_2)\Delta T}{Kk} \tanh kx. \tag{2.73}$$

In many situations, there is not a continuous bond between the chip and heat spreader, but only an edge bond. If the bond region begins at $x = c$, and ends at $x = l$, the shear stress can be expressed as (Schmidt 2006):

$$\tau = \frac{(\alpha_1 - \alpha_2)\Delta T[\sinh k(x - c) + kc \cosh k(x - c)]}{Kk[kc \sinh k(l - c)]}. \tag{2.74}$$

When the bending effect is included, the shear stress, $\tau$, and the peeling stress (the stress normal to the interface), $p(x)$, can be derived (Schmidt 2006):

$$\tau = \frac{k(\alpha_2 - \alpha_1)\Delta T \sinh(kx)}{\lambda \cosh(kl)}, \tag{2.75}$$

$$p(x) = \frac{-a(\alpha_1 - \alpha_2)\Delta T \cosh(kx)}{K \cosh(kl)}, \tag{2.76}$$

where $k^2 = (\lambda_1 + \lambda_2 + \lambda_{12})/(K_1 + K_2 + K); K_i = (2(1 + v_i)t_i)/3E_i = t_i/3G_i\ i = 1, 2;$ $D = D_1 + D_2; a = (D_1t_2 - D_2t_1)/2D$.

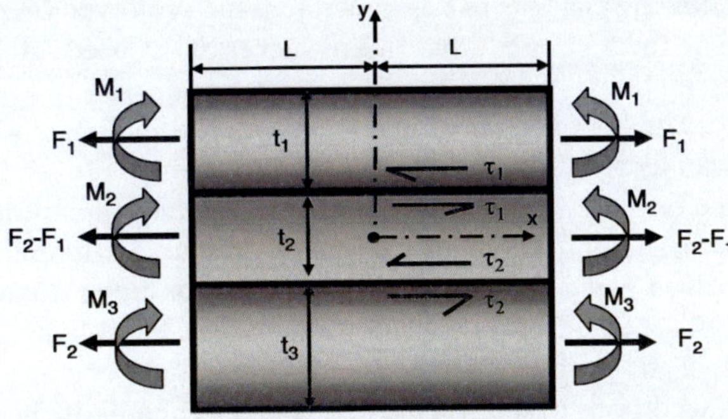


**Fig. 2.10** Illustration of a tri-materials assembly

## Trimaterial Assembly

The trimaterial assembly analysis can be used to simulate the three-layer electronic structure, such as a heat spreader, an adhesive or solder, and a chip. Figure 2.10 shows the model with three layers, designated as 1, 2, and 3, and a free-body diagram for a cut at some arbitrary $x$ location. The approach is essentially the same as in the bimaterial assembly analysis, including (1) equilibrium is used to relate the radius of curvature to the axial forces, and then (2) the compatibility conditions at the interfaces are used to eliminate all unknowns except for the shear stress. This procedure is general and can easily be applied to any number of layers. The shear stress can be derived as (Schmidt 2006):

$$\tau_1 = k\Delta T[(\alpha_2 - \alpha_1)\beta_1 + (\alpha_3 - \alpha_2)\beta_2]\frac{\sinh(kx)}{\cosh(kl)}, \tag{2.77}$$

$$\tau_2 = k\Delta T[(\alpha_2 - \alpha_1)\beta_2 + (\alpha_3 - \alpha_2)\beta_1]\frac{\sinh(kx)}{\cosh(kl)}, \tag{2.78}$$

where $k^2 = (r \pm (r^2 - 4K_{12}K_{23}s)^{1/2})/2K_{12}K_{23}, r = \lambda_{12}K_{23} + \lambda_{23} + K_{12}, K_{ij} = K_i + K_j, \lambda_{20} = \lambda_2 - ((t_1 + t_2)(t_2 + t_3)/4)D, \lambda_{ij} = \lambda_i + \lambda_j + ((t_i + t_j)^2/4D), D = D_1 + D_2 + D_3$, and $s = \lambda_{12}\lambda_{23} - \lambda_{20}^2; \beta_1 = \lambda_{23}/(\lambda_{12}\lambda_{23} - \lambda_{20}); \beta_2 = \lambda_{20}/(\lambda_{12}\lambda_{23} - \lambda_{20}); \beta_3 = \lambda_{12}/(\lambda_{12}\lambda_{23} - \lambda_{20})$.

## Numerical Method

The analytical approaches discussed previously are useful for assessing trends in the behavior of electronic components and provide useful insight into the behavior of an assembly. Once a preliminary design is developed for an assembly, extensive experimental prototyping and evaluation or numerical simulation should be done to make sure the device have the required reliability. Compared with experimental evaluation, numerical modeling and computer simulation have the advantage of allowing changes in design without the prototype fabrication expense. As a numerical method, FEA has been successfully used in the simulation of electronic assemblies.

The FEA is a numerical technique for finding approximate solutions of PDEs as well as of integral equations. The solution approach is based either on eliminating the differential equation completely (steady-state problems), or rendering the PDE into an approximating system of ordinary differential equations, which are then solved using standard techniques such as Euler's method, Runge–Kutta, etc. The FEA method originated from the needs for solving complex elasticity, structural analysis problems in civil and aeronautical engineering. Its development can be traced back to the work by Alexander Hrennikoff in 1941 and Richard Courant in 1942. While the approaches used by those pioneers are dramatically different, they share one essential characteristic: mesh discretization of a continuous domain into a set of discrete subdomains, usually called elements. Hrennikoff's work discretizes the domain by using a lattice analogy while Courant's approach divides the domain into finite triangular subregions for solution of second order elliptic PDEs that arise from the problem of torsion of a cylinder. Courant's contribution was evolutionary, drawing on a large body of earlier results for PDEs developed by Rayleigh, Ritz, and Galerkin. Development of the finite element method began in earliest in the middle to late 1950s for airframe and structural analysis and gathered momentum at the University of Stuttgart through the work of John Argyris and at Berkeley through the work of Ray W. Clough in the 1960s for use in civil engineering. By late 1950s, the key concepts of stiffness matrix and element assembly existed essentially in the form used today and NASA issued request for proposals for the development of the finite element software NASTRAN in 1965. The method was provided with a rigorous mathematical foundation in 1973 with the publication of Strang and Fix's *An Analysis of The Finite Element Method*, and has since been generalized into a branch of applied mathematics for numerical modeling of physical systems in a wide variety of engineering disciplines, e.g., electromagnetism and fluid dynamics (Wikipedia 2009).

There have been a number of commercial FEA software packages available for application to mechanical and thermal design in electronic packaging, such as ABAQUS, ADINA, ANSYS, MSC/NASTRAN, and MARC. However, the drawback of the numerical analysis such as FEA is that there is no way to be sure the results are accurate. Therefore, caution must be used to always compare the compute results to estimates of the solution. Whenever possible, comparison to analytical results or experimental results is the best check.

### *Thermomechanical Failures*

Thermomechanical failure is generally defined as any thermally induced change in the shape, material properties, or integrity of a part, which makes it incapable of performing its intended function. The failure of an electronic device can be the result of one or any combination of responses of the device to loads and the environment. Failure mode seeks to predict failure conditions and correlate a calculated level of stress in the device with limits for the material obtained from tests of the material for each component.

### Static Failure

The major static failure modes in the design of microelectronic components are ductile rupture and brittle rupture. Ductile rupture occurs when the plastic deformation in a material results in a progressive local reduction in cross-sectional area. This type of failure can occur in solder joints. Brittle failure occurs when the deformation results in a breaking apart of the interatomic bonds. Silicon and ceramics often fail by brittle fracture as a result of the thermal expansion mismatch between these materials and other bonded layers. The criteria for the ductile failure can be expressed as (Schmidt 2006):

$$S = \frac{\sqrt{2}}{2}\left[(\sigma_1 - \sigma_2)^2 + (\sigma_2 - \sigma_3)^2 + (\sigma_3 - \sigma_1)^2\right]^{1/2} \leq \sigma_{yP}. \tag{2.79}$$

When $S$ is effective stress, $\sigma_{yP}$ is the material yielding strength. If the value of $S$ is less than $\sigma_{yP}$, then there is no yielding and the design can be considered adequate, otherwise, the design must be modified. The criteria for the brittle failure can be expressed as (Schmidt 2006):

$$|\sigma_1| < \sigma_u \quad \text{or} \quad |\sigma_3| < \sigma_u \quad \text{when}\ \sigma_1 < \sigma_2 < \sigma_3. \tag{2.80}$$

Brittle failure occurs when the materials no yielding and the stress–strain curve continues smoothly to fracture at ultimate stress $\sigma_u$. The failure can occurs rapidly with little or no indication of the impeding failure.

### Failure of Fracture Mechanics

The theory of fracture mechanics is used to predict under what circumstances an existing imperfection, usually a crack, will propagate and eventually result in total failure. It uses methods of analytical solid mechanics to calculate the driving force on a crack and those of experimental solid mechanics to characterize the material's resistance to fracture. The term stress intensity $K_{\mathrm{I}}$ and the material fracture toughness are defined as (Lawn 1993):

$$K_{\mathrm{I}} = \sigma\sqrt{\pi a}, \tag{2.81}$$

and

$$K_c = \sqrt{EG_c} \quad \text{(for plane stress)}, \tag{2.82}$$

$$K_c = \sqrt{\frac{EG_c}{1-\nu^2}} \quad \text{(for plane strain)}, \tag{2.83}$$

$$G_c = \frac{\pi\sigma_f^2 a}{E}, \tag{2.84}$$

where $K_I$ is the stress intensity, $K_c$ the fracture toughness, ν is Poisson's ratio, $\sigma_f$ is the stress at fracture, $\sigma$ is the applied stress, $a$ is half the crack length, and $E$ is the Young's modulus. If $K_I < K_c$, the crack will not propagate; otherwise, the fracture will occur.

### Fatigue Failure

Fatigue is the progressive and localized structural damage that occurs when a material is subjected to cyclic loading. The maximum stress values are less than the ultimate tensile stress limit, and may be below the yield stress limit of the material. Electronic components are frequently subjected to repeated or cyclic loads. Fatigue damage of the electronic component is usually nucleated at persistent slip band or preexisting defects on or near its free surface. A small crack may then form and grow into a major crack under cyclic loading. There major approaches have been developed and used to analyze and design against fatigue failure, including the stress-based approach, the strain-based approach, and the fracture mechanics-based approach.

Electronic components in a package or cooling system may be subjected to complex loadings so that the stress at a given point in the material often occurs in more than one direction. If sufficiently severe, such combined stresses can act together to cause the material to fail. A failure criterion that predicts failure when a certain combination of stresses reaches a limit, can be expressed as (Lawn 1993):

$$\bar{\sigma} = f(\sigma_1, \sigma_2, \sigma_3) < \sigma_c, \tag{2.85}$$

where $\bar{\sigma}$ is the effective stress, $f(\sigma_1, \sigma_2, \sigma_3)$ is a given function of the principle stress $\sigma_1, \sigma_2, \sigma_3$, and $\sigma_c$ is a material property parameter. Historically, most attention has focused on situations that require more than $10^4$ cycles to failure where stress is low and deformation primarily elastic.

In high-cycle fatigue situations, materials performance is commonly characterized by an *S–N* curve, also known as a Wöhler curve. This is a graph of the magnitude of a cyclical stress (*S*) against the logarithmic scale of cycles to failure (*N*). *S–N* curves are derived from tests on samples of the material to be characterized where a regular sinusoidal stress is applied by a testing machine which also counts the number of cycles to failure. This process is sometimes known as coupon testing. Each coupon test generates a point on the plot though in some cases there is a runout where the time to failure exceeds that available for the test. Analysis of fatigue data requires techniques from statistics, especially survival analysis and linear regression.

In 1945, M.A. Miner popularized a rule that had first been proposed by A. Palmgren in 1924. The rule, variously called Miner's rule or the Palmgren–Miner linear damage hypothesis, states that where there are $k$ different stress magnitudes in a spectrum, $S_i(1 \leq i \leq k)$, each contributing $n_i(S_i)$ cycles, then if $N_i(S_i)$ is the number of cycles to failure of a constant stress reversal $S_i$, failure occurs when (Schijve 2009):

$$\sum_{i=1}^{k} \frac{n_i}{N_i} = C. \tag{2.86}$$

$C$ is experimentally found to be between 0.7 and 2.2. Usually for design purposes, $C$ is assumed to be 1. This can be thought of as assessing what proportion of life is consumed by stress reversal at each magnitude then forming a linear combination of their aggregate.

Though Miner's rule is a useful approximation in many circumstances, however, it has two major limitations: (1) It fails to recognize the probabilistic nature of fatigue and there is no simple way to relate life predicted by the rule with the characteristics of a probability distribution. (2) There is sometimes an effect in the order in which the reversals occur. In some circumstances, cycles of low stress followed by high stress cause more damage than would be predicted by the rule. It does not consider the effect of overload or high stress which may result in a compressive residual stress. High stress followed by low stress may have less damage due to the presence of compressive residual stress.

Anderson, Gomez, and Paris derived relationships for crack growth with cycles $N$, in terms of the cyclical component $\Delta K$ of the stress intensity factor $K$ (Schijve 2009):

$$\frac{\mathrm{d}a}{\mathrm{d}N} = C(\Delta K)^m, \tag{2.87}$$

where $a$ is the crack length and $m$ is typically in the range 3–5 (for metals).This relationship was later modified by Forman in 1967 to make better allowance for the mean stress, by introducing a factor depending on $(1 - R)$ where $R =$ min. stress/max stress, in the denominator.

Where the stress is high enough for plastic deformation to occur, the account in terms of stress is less useful and the strain in the material offers a simpler description. Low-cycle fatigue is usually characterized by the Coffin–Manson relation (Schijve 2009):

$$\Delta\varepsilon_\mathrm{p} = 2\varepsilon_\mathrm{f}(2N)^c, \tag{2.88}$$

where $\Delta\varepsilon_p$ is the plastic strain amplitude; $\varepsilon_f$ is an empirical constant known as the fatigue ductility coefficient, the failure strain for a single reversal; $2N$ is the number of reversals to failure ($N$ cycles); $c$ is an empirical constant known as the fatigue ductility exponent, commonly ranging from −0.5 to −0.7 for metals.

Based on these theories, the fatigue failure in electronic packaging and cooling system have been explored, however, majority of studies in this area are in its infancy and still experimental. Thus, the thermal fatigue limits of materials and structures are currently determined primarily based on the results of thermal cycling tests.

## Analytical Techniques for Materials Characterization

Some analytical technical techniques, as summarized in Table 2.5, have been widely used to characterize materials employed in fabrication of electronic packaging and thermal management system. It is necessary to observe and obtain materials characteristics, such as composition, structure, morphology, and interface joining, in the synthesis, design, and development of electronic, photonic, and thermal modules, as well as in the analysis of their failure.

### *Optical Microscopy*

Optical microscopy, or light microscopy, refers to the sample inspection with a type of microscope which uses visible light and a system of lenses to magnify images of small samples. A basic optical microscope has the following parts (Siliconfareast 2010): (1) a lamp to illuminate the specimen; (2) a nose piece to hold four to five objectives used in changing the viewing magnification; (3) an aperture diaphragm to adjust the resolution and contrast; (4) a field diaphragm to adjust the field of view; (5) an eye piece to magnify the objective image (usually by 10×); and (6) a stage for manipulating the specimen. During optical microscope inspection, the specimen is positioned perpendicularly to the axis of the objective lens. Light is then shown on the sample, which reflects some light back to the lens. The image seen in the microscope depends not only on how the specimen is illuminated and positioned, but on the characteristics of the specimen as well.

Optical microscopes are commonly classified as either low-power or high-power microscopes. Low-power microscopes are those which typically magnify the specimen at 5× to 60×, although some can magnify up to 100×. High-power microscopes, on the other hand, typically magnify the specimen at 100× to 1,000×.

There are three modes by which optical microscopy is commonly conducted, namely, brightfield illumination, darkfield illumination, and interference contrast. Brightfield illumination is the normal mode of viewing with an optical microscope. This mode provides the most uniform illumination of the sample. Under this

**Table 2.5** Analytical techniques for material characterization

| Technique | Incident probe | Analyzed phenomenon | Depth of analysis | Spatial resolution and detection limit | Applications |
|---|---|---|---|---|---|
| Optical microscopy | Visible light | Light reflection from surface microstructure | – | >2 μm<br>– | Analyze samples in air or water, the images are in natural color with magnifications up to 1,000 times |
| X-ray diffraction | X-rays | Angles of X-rays relative to incident beam and sample that are diffracted by sample | 100 Å to 1,000s Å | >10 μm<br>1% | Crystal structure, texture, orientation, phase analysis, element composition |
| Scanning electron microscopy | Electron beam | Electrons | Several nm to several μm | 1–50 nm<br>0.1% weight | Morphology, topography, composition, and phases |
| Transmission electron microscopy | Electron beam | Electron beam passes through and interacts with the specimen | Several hundred nm | 0.2 nm<br>0.1% weight | Morphology, topography, composition, and phases, microstructure and nanostructure |
| Scanning acoustic microscopy | Sound | Transducer and receiver | Several mm | Several μm | Cracks, delamination, voids |
| Atomic force microscopy | Atomic force | Atomic force | – | Fractions of nm | Nanoscale images |

mode, a full cone of light is focused by the objective on the sample. The image observed results from the various levels of reflectivities exhibited by the compositional and topographical differences on the surface of the sample. Under darkfield illumination, the inner circle area of the light cone is blocked, such that the sample is only illuminated by light that impinges on its surface at a glancing angle. This scattered reflected light usually comes from feature edges, particulates, and other irregularities on the sample surface. Darkfield illumination is therefore effective in detecting surface scratches and contamination. Interference contrast makes use of polarized light that is divided by a Wollaston prism into two orthogonal light packets. These slightly displaced light packets hit the specimen at two different points and return to the prism through different paths. The differences in the routes of the reflected packets will produce interference contrasts in the image when the packets are recombined by the prism upon their return. Surface defects or features such as etch pits and cracks that are difficult to see under brightfield illumination can stand out clearly under Nomarski mode (Siliconfareast 2010).

The stereo microscope is designed differently from the compound optical microscopes, and serves a different purpose. It uses two separate optical paths with two objectives and two eyepieces to provide slightly different viewing angles to the left and right eyes, which produces a 3-D visualization of the sample being examined. Unlike compound microscopes, illumination in a stereo microscope most often uses reflected (episcopic) illumination rather than transmitted (diascopic) illumination, that is, light reflected from the surface of an object rather than light transmitted through an object. Use of reflected light from the object allows examination of specimens that would be too thick or otherwise opaque for compound microscopy. However, stereo microscopes are also capable of transmitted light illumination as well, typically by having a bulb or mirror beneath a transparent stage underneath the object, although unlike a compound microscope, transmitted illumination is not focused through a condenser in most systems. Stereoscopes with specially equipped illuminators can be used for dark field microscopy, using either reflected or transmitted light (Lipson et al. 1995).

Various video dual CCD camera pickups have been fitted to stereo microscopes, allowing the images to be displayed on a high-resolution liquid crystal display monitor. Software converts the two images to an integrated Anachrome 3-D image, for viewing with plastic red/cyan glasses, or to the cross converged process for clear glasses and somewhat better color accuracy.

### *X-Ray Diffraction*

X-ray diffraction is a versatile, nondestructive analytical technique for identification and quantitative determination of the crystallographic structure, chemical composition, and physical properties of solid materials, powders, and thin films. The technique is based on observing the scattered intensity of an X-ray beam hitting a sample as a function of incident and scattered angle, polarization, and wavelength or energy. Modern computer-controlled diffractometer systems use automatic routines to measure, record and interpret the unique diffractograms produced by individual constituents in even highly complex mixtures.

X-rays are only a small part of the electromagnetic spectrum with wavelengths ($\lambda$) ranging from 0.02 to 100 Å (Å = Angstroms = $10^{-8}$ m). X-rays used to study crystals have $\lambda$ on the order of 1–2 Å (i.e., Copper $K\alpha = 1.5418$ Å). Visible light has much larger $\lambda$s (4,000–7,200 Å) and thus, X-rays are much more energetic (i.e., can penetrate deeper into a material). This can easily be seen by inspection of the Einstein equation ($E = h\nu = hc/\lambda$; $E$ is energy, $\nu$ frequency, $c$ speed of light which is constant for electromagnetic radiation, $\lambda$ wavelength, $h$ Plank's constant).

Diffraction of an X-ray beam striking a crystal occurs because the $\lambda$ of the X-ray beam is similar to the spacing of atoms in materials (1–10 Å). When an X-ray beam encounters the regular, 3-D arrangement of atoms in a crystal most of the X-rays will destructively interfere with each other and cancel each other out, but in some specific directions they constructively interfere and reinforce one another. It is these reinforced (diffracted) X-rays that produce the characteristic X-ray diffraction patterns that used for material identification.

W.L. Bragg (early 1900s) showed that diffracted X-rays act as if they were reflected from a family of planes within crystals. Bragg's planes are the rows of atoms that make up the crystal structure. These reflections were shown to only occur under certain conditions which satisfy the equation: $n\lambda = 2d \sin \theta$ (Bragg equation), where $n$ is an integer $(1, 2, 3, \ldots, n)$, $\lambda$ is the wavelength, $d$ is the distance between atomic planes, and $\theta$ is the angle of incidence of the X-ray beam and the atomic planes. $2d \sin \theta$ is the path length difference between two incident X-ray beams where one X-ray beam takes a longer (but parallel) path because it reflects off an adjacent atomic plane. This path length difference must equal an integer value of the $\lambda$ of the incident X-ray beams for constructive interference to occur such that a reinforced diffracted beam is produced. For a given $\lambda$ of incident X-rays and interplanar spacing ($d$) in a material specimen, only specific $\theta$ angles will satisfy the Bragg equation (Gorring 1998).

Photographic plates were traditionally used to record the intensity and position of diffracted X-rays. Modern systems use diffractometers which are electronic X-ray counters (detectors) that can measure intensities much more accurately. Computers are used to process data and make necessary complex calculations. There are two main techniques to be discussed as below.

### Single-Crystal Method

This method refers to X-ray beam is focused on a single crystal. Primary application is to determine atomic structure (symmetry, unit cell dimensions, space group, etc.). Older methods (Laue method) used a stationary crystal with white X-ray beam (X-rays of variable $\lambda$) such that Bragg's equation would be satisfied by numerous atomic planes. The diffracted X-rays exiting the crystal all have different $\theta$ and thus produce spots on a photographic plate. The diffraction spots show the symmetry of the crystal.

Modern methods (rotation, Weissenberg, precession, four-circle) utilize various combination of rotating-crystal and camera setup to overcome limitations of the stationary methods (mainly the number of diffractions observed). These methods use monochromatic X-rays, but vary $\theta$ by moving the crystal mounted on a rotating stage. Usually employ diffractometers and computers for data collection and processing (Gorring 1998).

### Powder Method

The powder method indicates an X-ray beam focused on a powder pellet or powder smeared on a glass slide, which is essential for materials that do not form large crystals (i.e., clays) and eliminates the problem of precise orientation necessary in single-crystal methods. This method is primarily used for material identification, and also can be used to determine material compositions (if *d*-spacing is a function of material chemistry) and to determine relative proportions of materials in a mixture. Monochromatic X-rays are focused on pellet or slide mounted on rotating stage. Because the sample is powder, all possible diffractions are recorded simultaneously from hypothetical randomly oriented grains. Mount is then rotated to ensure all diffractions are obtained (Gorring 1998).

X-ray powder diffractometry uses monochromatic X-rays on powder that is mounted on a glass slide that is attached to a stage which systematically rotates into the path of the X-ray beam through $\theta = 0$–$90°$. The diffracted X-rays are detected electronically and recorded on an inked strip chart. The detector rotates simultaneously with the stage, but rotates through angles $2\theta$. The strip chart also moves simultaneously with the stage and detector at a constant speed. The strip chart records the intensity of X-rays as the detector rotates through $2\theta$. Thus, the angle $2\theta$ at which diffractions occur and the relative intensities can be read directly from the position and heights of the peaks on the strip chart. Then use the Bragg equation to solve for the interplanar spacing ($d$) for all the major peaks and look up a match with JCPDS cards (JCPDS refers to Joint Committee on Powder Diffraction Standards).

Powder diffraction is commonly used to identify unknown substances, by comparing diffraction data against a database maintained by the International Centre for Diffraction Data. It may also be used to characterize heterogeneous solid mixtures to determine relative abundance of crystalline compounds and, when coupled with lattice refinement techniques, such as Rietveld refinement, can provide structural information on unknown materials. Powder diffraction is also a common method for determining strains in crystalline materials. Thin film diffraction and grazing incidence X-ray diffraction may be used to characterize the crystallographic structure and preferred orientation of substrate-anchored thin films. High-resolution X-ray diffraction is used to characterize thickness, crystallographic structure, and strain in thin epitaxial films. It employs parallel-beam optics. X-ray pole figure analysis enables one to analyze and determine the distribution of crystalline orientations within a crystalline thin-film sample. X-ray rocking curve analysis is used to quantify grain size and mosaic spread in crystalline materials.

In a summary, the angle of reflection of X-rays from a sample is related to the crystal structure and composition of the material. Typical X-ray analyses include lattice parameter measurements, crystallite size and distribution, texture analysis-preferred orientation of crystallites in a material, internal and residual stress measurements, CTE measurement, phase and composition identification, and film thickness measurements. The X-ray can penetrate into the sample with a depth of 100–1,000 Å, spatial resolution is more than 10 μm, and detection limit is 1%.

## *Scanning Electron Microscopy*

The scanning electron microscope (SEM) is used to image the sample surface with much higher magnification than with an optical microscope by scanning it with a high-energy beam of electrons in a raster scan pattern. The electron beam interacts with the atoms that make up the sample producing signals that contain information about the sample's surface topography, composition and other properties such as electrical conductivity.

The SEM is not only an isolated instrument, but also represents a platform. When combined with scanning probe microscopy, the electron microscope can be used to further control manipulation of nanostructures or select an area for observation with high precision. In situ phase transitions can be seen when cryogenic or heating stages are installed in the chamber. The combination with a focused ion beam is used for specimen preparation in transmission electron microscopy (TEM). The types of signals produced by a beam of highly energetic (0.1–50 keV) electrons in SEM include the emission of secondary electrons, backscattered electrons (BSE), characteristic X-rays, photons, specimen current and transmitted electrons. Because the bombarding electron beam is scanned in the *X–Y* plane, an image for each of these different processes can be mapped with a suitable detector. A detector for secondary electrons, standard to all basic SEMs, records topography of the surface under observation with resolution on the order of 1–2 nm and magnification range from 10× to 500,000×. Due to the way these images are created, SEM micrographs have a very large depth of field yielding a characteristic 3-D appearance useful for understanding the surface structure of a sample. In addition, information on composition, phase, electrical, optical, thermal, and other properties can be mapped with excellent resolution with appropriate detectors. BSEs are beam electrons that are reflected from the sample by elastic scattering. BSEs are often used in analytical SEM along with the spectra made from the characteristic X-rays. Because the intensity of the BSE signal is strongly related to the atomic number of the specimen, BSE images can provide information about the distribution of different elements in the sample. For the same reason, BSE imaging can image colloidal gold immuno-labels of 5 or 10 nm diameter. Characteristic X-rays are emitted when the electron beam removes an inner shell electron from the sample, causing a higher energy electron to fill the shell and release energy. These characteristic X-rays are used to identify the composition and measure the abundance of elements in the sample (Schatten and Pawley 2007): (1) SEM-secondary and BSEs are detected from a sample with a focused primary electron beam, resulting in a surface image with a great depth of field; (2) EDX (Energy-Dispersive X-Ray Analysis)–The energy of X-rays emitted from a sample that is irradiated with the primary electron beam lead to elemental analysis and mapping; (3) high-resolution SEM imaging (1–50 nm), high-speed acquisition (30–60 s) with depth of analysis: few nanometers to few micrometers; (4) live SEM observation of the specimen in five to six orders of magnification (10× to 500,000×); (5) lateral resolution of EDX is ~0.5 μm, limit of detection ~0.1% weight, sample depth 0.02 μm; (6) nondestructive specimens; (7) vacuum compatibility required. Vacuum chamber accommodates specimens up to 100 mm in diameter; (8) versatility: multiple modes of operation available; and (9) readily accessible cross-sectional measurements.

## *Transmission Electron Microscopy*

TEM is an imaging technique whereby a beam of electrons is transmitted through an ultrathin specimen, interacting with the specimen to produce a high-resolution image, magnified and focused by an objective lens and appears on an imaging screen, a fluorescent screen in most TEMs, plus a monitor, or on a layer of photographic film, or to be detected by a sensor such as a CCD camera. TEM is the most powerful electron microscope, with a resolving power 10 times better than that of a SEM and 1,000 times better than that of an optical microscope. A fine electron beam passes through the specimen, which must therefore be sliced extremely thinly–typically to about 100 nm. The TEM can resolve objects 0.001 μm (0.04 millionth of an inch) apart, a gap that is 100,000 times smaller than the unaided eye can see.

A TEM consists of a tall, evacuated column at the top of which is a heated tungsten filament that emits electrons. The beam of accelerated electrons are moved at uniform high velocity down the column by a high voltage (around 100,000 V) and pass through the slice of specimen at a point roughly half way down. Because the density of the specimen varies, the "shadow" of the beam falls on a fluorescent screen near the bottom of the column and forms an image. A camera is mounted beneath the screen to record the image. The electron beam is controlled by magnetic fields produced by electric coils, called electron lenses. One electron lens, called the condenser, controls the beam size and brightness before it strikes the specimen. Another electron lens, called the objective, focuses the beam on the specimen and magnifies the image about 50×. Other electron lenses below the specimen then further magnify the image. On the screen a high-resolution image of the specimen can be viewed. The illuminated areas of the image correspond to those that have been hit by electrons which have passed through those areas of the specimen at which the electrons did not strike any solid material. The dark areas of the image correspond to those areas of the specimen, through which electrons could not pass through at all. At such points on the specimen, electrons have struck atoms of the specimen material and have been scattered.

The ultrathin layer of the specimen (some hundreds of nanometers thick), is first prepared properly, before being mounted on the microscope platform. Non-conductive specimen preparation entails drying it, embedding the specimen in plastic, and then coating it with gold or carbon, to facilitate conduction of electrons. The specimen is also stabilized. Further, the entire microscope assembly unit, inclusive of the specimen, is placed under vacuum to prevent the focused electron beam from getting scattered by air molecules. Cryogenic specimen preparation helps in better output. The whole microscope unit should be placed on the ground level to avoid influence of mechanical vibration on the image output (Raju 2010).

Typical applications of the TEM include: (1) observation of microstructure and nanostructure: size and morphology; (2) phase identification down to the nanoscale; (3) transmission electron diffraction; (4) double-tilt sample holders permit crystallographic studies, including defect analysis (dislocations, stacking faults, etc.); (5) chemical information discerned from emission of characteristic X-rays (energy dispersive spectroscopy of X-rays, or energy dispersive spectroscopy of X-rays [EDS]); (6) chemical information discerned from ionization edges in the electron energy loss spectrum (EELS); (7) high resolution imaging (~0.2 nm resolution); (8) formation of small electron probe allows diffraction, EDS, or EELS at the nanoscale; and (9) cross-section analysis (layer thickness, interface quality).

The specimens must be prepared as a thin foil, or etched so some portion of the specimen is thin enough for the beam to penetrate. Preparation techniques to obtain an electron transparent region include ion beam milling and wedge polishing. Materials that have dimensions small enough to be electron transparent, such as powders or nanotubes, can be quickly produced by the deposition of a dilute sample containing the specimen onto support grids. The suspension is normally a volatile solvent, such as ethanol, ensuring that the solvent rapidly evaporates allowing a sample that can be rapidly analyzed.

The imaging techniques are particularly important. Faults in crystals affect both the mechanical and the electronic properties of materials, so understanding how they behave gives a powerful insight. By carefully selecting the orientation of the sample, it is possible not just to determine the position of defects but also to determine the type of defect present. If the sample is orientated so that one particular plane is only slightly tilted away from the strongest diffracting angle (known as the Bragg angle), any distortion of the crystal plane that locally tilts the plane to the Bragg angle will produce particularly strong contrast variations. However, defects that produce only displacement of atoms that do not tilt the crystal to the Bragg angle (i.e., displacements parallel to the crystal plane) will not produce strong contrast.

Furthermore, the high-resolution transmission electron miscroscopy (HRTEM) technique allows the direct observation of crystal structure and therefore has an advantage over other methods in that there is no displacement between the location of a defect and the contrast variation caused in the image. However, it is not always possible to interpret the lattice images directly in terms of sample structure or composition. This is because the image is sensitive to a number of factors (specimen thickness and orientation, objective lens defocus, spherical and chromatic aberration), and although quantitative interpretation of the contrast shown in lattice images is possible, it is inherently complicated and may require extensive simulation of the images. Computer modeling of these images has added a new layer of understanding to the study of crystalline materials.

There are a number of drawbacks to the TEM technique. Many materials require extensive sample preparation to produce a sample thin enough to be electron transparent, which makes TEM analysis a relatively time consuming process with a low throughput of samples. Graphene, a carbon nanomaterial, relatively transparent, very hard and just one atom thick, is currently being used as a platform on which the materials to be examined are placed. Being almost transparent to electrons, a graphene substrate has been able to show single hydrogen atom and hydrocarbons. The structure of the sample may also be changed during the preparation process. Also the field of view is relatively small, raising the possibility that the region analyzed may not be characteristic of the whole sample. There is potential that the sample may be damaged by the electron beam, particularly in the case of plastic or biological materials.

### *Scanning Acoustic Microscopy*

Scanning acoustic microscopy (SAM) is a nondestructive technique that can be used to image the internal features of a specimen or an integral structure. SAM can be used in microelectronic packaging and thermal management system for detecting popcorn cracking/delamination, die attach voiding, evaluating flip chip underfill integrity, and lid seal integrity in hermetically sealed packages, and detecting delaminations of submicron thickness, which are difficult to detect using X-ray radiography. Both delamination/cracking and die attach voiding are assembly related defects that can increase the susceptibility of components to failure in storage or use, although they may not constitute failures by themselves. Delamination and cracking can result in sheared or lifted wirebonds, passivation cracking, metallization shifting, intermittent electrical failures, and metallization/bond pad corrosion. Die attach voiding can lead to die cracking, die attach fracture, or thermal runaway due to poor heat dissipation through the die attach.

SAM works by directing focused sound from a transducer at a small point on a target object. Sound hitting the object is either scattered, absorbed, reflected (scattered at 180°) or transmitted (scattered at 0°). Typically, either the reflected or transmitted sound is gathered and measured. Based on the measurement, a value is assigned to the location investigated. The transducer is moved slightly and then insonified again. This process is repeated in a systematic pattern until the entire region of interest has been investigated. Often the values for each point are assembled into an image of the object. The contrast seen in the image is based either on the object's geometry or material composition. The resolution of the image is limited either by the physical scanning resolution or the width of the sound beam, which in turn is determined by the frequency of the sound. A typical scanning acoustic microscope, for instance, may employ either pulse echo or through transmission inspection to scan for disbonds or delaminations. Pulse echo inspection consists of interpreting echos sent back by the package while through transmission inspection consists of interpreting the sound wave at the other end of the package, after it has passed through the latter. The ultrasonic wave frequency used usually ranges from 5 to 150 MHz. The sound wave may be generated by a piezoelectric crystal, or transducer, that has been cut to provide a specific frequency. It is activated by a high voltage pulse from a transmitter, which is also known as the pulser. The activation would cause the transducer to vibrate at the specified frequency, which transmits an ultrasonic wave through the package. This wave travels to the specimen through a medium or coupling, which is usually deionized water because sound waves could not travel through air at the frequencies used. The wave travels through the specimen's material at the material's velocity, with a portion of it being reflected back every time it hits an interface within the material. In the pulse echo method, the same transducer is used as sender and receiver of the sound waves. Pulses are repeated using repetition rates at which the echoes from one pulse will not interfere with those of another, for example, 10–20 KHz. The echoes received by the transducer are converted to voltages, amplified, digitized, and presented to the user as an image. In the through transmission technique, separate transducers are used to send and receive sound waves, both of which are on opposite sides of the specimen. The absence and presence of signals mean bad and good bonding, respectively (Siliconfareast 2010).

The resolution of a microscope can be defined as the minimum size of a feature that can be determined. The resolution of the light microscope is determined by the equation (Connor et al. 1998):

$$\omega = \frac{0.61\lambda_0}{N}, \tag{2.89}$$

which is based on the Rayleigh criterion. The resolution of the reflection acoustic microscope is somewhat better than the resolution based on the Rayleigh criterion and is given by the equation:

$$\omega = \frac{0.51\lambda_0}{N}, \tag{2.90}$$

where $N$ is the numerical aperture and is equal to sin $\theta_0$, which is the semiangle subtended. The wavelength, $\lambda_0$, is given by $v_0/f$, where $v_0$ is the velocity of sound in the fluid, and $f$ is the frequency. A conservative estimate of the resolution of the acoustic scans can be between 90 and 180 μm.

As the frequency increases, the resolution improves but the depth of penetration decreases; thus, a tradeoff exists between improved resolution and depth of penetration. The best obtainable resolution is determined by the depth of penetration desired. The desired depth of penetration coupled with a minimum pulse length enables determination of the highest usable frequency. These factors are thus used to determine the best resolution (Connor et al. 1998).

## *Atomic Force Microscopy*

The atomic force microscope (AFM) is a very high-resolution type of scanning probe microscope, with demonstrated resolution of fractions of a nanometer, more than 1,000 times better than the optical diffraction limit. The AFM is one of the foremost tools for imaging, measuring and manipulating matter at the nanoscale. The term "microscope" in the name is actually a misnomer because it implies looking, while in fact the information is gathered by "feeling" the surface with a mechanical probe. Piezoelectric elements that facilitate tiny but accurate and precise movements on electronic command enable the very precise scanning. The AFM was developed to overcome a basic drawback with STM (Scanning Tunneling Microscope), which can only image conducting or semiconducting surfaces. The AFM, however, has the advantage of imaging almost any type of surface, including polymers, ceramics, composites, glass, and biological samples.

The AFM consists of a microscale cantilever with a sharp tip probe at its end that is used to scan the specimen surface. The cantilever is typically silicon or silicon nitride ($Si_3N_4$) with a tip radius of curvature on the order of nanometers. When the tip is brought into proximity of a sample surface, forces between the tip and the sample lead to a deflection of the cantilever according to Hooke's law: $F = -kz$, where $F$ is the force, $k$ is the stiffness of the lever, and $z$ is the distance the lever is bent. The force is not measured directly, but calculated by measuring the deflection of the lever, and knowing the stiffness of the cantilever. Depending on the situation, forces that are measured in AFM include mechanical contact force, van der Waals forces, capillary forces, chemical bonding, electrostatic forces, magnetic forces, Casimir forces, solvation forces, etc. Typically, the deflection is measured using a laser spot reflected from the top of the cantilever into an array of photodiodes. Other methods that are used include optical interferometry, capacitive sensing, or piezoresistive AFM cantilevers. These cantilevers are fabricated with piezoresistive elements that act as a strain gauge. Using a Wheatstone bridge, strain in the AFM cantilever due to deflection can be measured, but this method is not as sensitive as laser deflection or interferometry (Search 2010).

If the tip were scanned at a constant height, there would be a risk that the tip would collide with the surface, causing damage. Hence, in most cases a feedback mechanism is employed to adjust the tip-to-sample distance to maintain a constant force between the tip and the sample. Traditionally, the sample is mounted on a piezoelectric tube that can move the sample in the $z$ direction for maintaining a constant force, and the $x$ and $y$ directions for scanning the sample. Alternatively a "tripod" configuration of three piezo crystals may be employed, with each responsible for scanning in the $x$, $y$ and $z$ directions. This eliminates some of the distortion effects seen with a tube scanner. The resulting map of the area $s = f(x, y)$ represents the topography of the sample.

The AFM can be operated in a number of modes, depending on the application. In general, possible imaging modes are divided into static, also called contact modes and a variety of dynamic or noncontact modes (Wikipedia 2009):

### Contact Mode

In constant force mode, the tip is constantly adjusted to maintain a constant deflection, and therefore constant height above the surface. It is this adjustment that is displayed as data. However, the ability to track the surface in this manner is limited by the feedback circuit. Sometimes the tip is allowed to scan without this adjustment, and one measures only the deflection. This is useful for small, high-speed atomic resolution scans, and is known as variable-deflection mode. Because the tip is in hard contact with the surface, the stiffness of the lever needs to be less than the effective spring constant holding atoms together, which is on the order of 1–10 nN/nm. Most contact mode levers have a spring constant of $<1$ N/m.

### Lateral Force Microscopy

Lateral force microscopy measures frictional forces on a surface. By measuring the "twist" of the cantilever, rather than merely its deflection, one can qualitatively determine areas of higher and lower friction.

### Noncontact Mode

Noncontact mode belongs to a family of AC modes, which refers to the use of an oscillating cantilever. A stiff cantilever is oscillated in the attractive regime, meaning that the tip is quite close to the sample, but not touching it. The forces between the tip and sample are quite low, on the order of pN ($10^{-12}$ N). The detection scheme is based on measuring changes to the resonant frequency or amplitude of the cantilever.

### Dynamic Force/Intermittant-Contact/"Tapping Mode" AFM

In this mode, a stiff cantilever is oscillated closer to the sample than in noncontact mode. Part of the oscillation extends into the repulsive regime, so the tip intermittently touches or "taps" the surface. Very stiff cantilevers are typically used, as tips can get "stuck" in the water contamination layer. The advantage of tapping the surface is improved lateral resolution on soft samples. Lateral forces such as drag, common in contact mode, are virtually eliminated. For poorly adsorbed specimens on a substrate surface the advantage is clearly seen.

### Force Modulation

Force modulation refers to a method used to probe properties of materials through sample/tip interactions. The tip (or sample) is oscillated at a high frequency and pushed into the repulsive regime. The slope of the force–distance curve is measured which is correlated to the sample's elasticity. The data can be acquired along with topography, which allows comparison of both height and material properties.

### Phase Imaging

In phase-mode imaging, the phase shift of the oscillating cantilever relative to the driving signal is measured. This phase shift can be correlated with specific material properties that effect the tip/sample interaction. The phase shift can be used to differentiate areas on a sample with such differing properties as friction, adhesion, and viscoelasticity. The technique is used simultaneously with dynamic force mode so that topography can be measured as well.

The AFM has several advantages over the SEM. Unlike the electron microscope which provides a 2-D projection or a 2-D image of a sample, the AFM provides a true 3-D surface profile. Additionally, samples viewed by AFM do not require any special treatments, such as metal/carbon coatings, which would irreversibly change or damage the sample. While an electron microscope needs an expensive vacuum environment for proper operation, most AFM modes can work perfectly well in ambient air or even a liquid environment. In principle, AFM can provide higher resolution than SEM. It has been shown to give true atomic resolution in ultra-high vacuum. A disadvantage of AFM compared with the SEM is the image size. The SEM can image an area on the order of millimeters by millimeters with a depth of field on the order of millimeters. The AFM can only image a maximum height on the order of micrometers and a maximum scanning area of around 150 × 150 μm.

Another inconvenience is that an incorrect choice of tip for the required resolution can lead to image artifacts. Traditionally the AFM could not scan images as fast as an SEM, requiring several minutes for a typical scan, while an SEM is capable of scanning at near real-time (although at relatively low quality) after the chamber is evacuated. The relatively slow rate of scanning during AFM imaging often leads to thermal drift in the image, making the AFM microscope less suited for measuring accurate distances between artifacts on the image. However, several fast-acting designs were suggested to increase microscope scanning productivity including what is being termed video AFM (reasonable quality images are being obtained with videoAFM at video rate–faster than the average SEM). To eliminate image distortions induced by thermodrift, several methods were also proposed (Search 2010).

AFM images can also be affected by hysteresis of the piezoelectric material and cross talk between the (*x*, *y*, *z*) axes that may require software enhancement and filtering. Such filtering could "flatten" out real topographical features. However, newer AFM uses real-time correction software (for example, feature-oriented scanning) or closed-loop scanners which practically eliminate these problems. Some AFM also uses separated orthogonal scanners (as opposed to a single tube) which also serve to eliminate cross-talk problems (Lipson et al. 1995).

Due to the nature of AFM probes, they cannot normally measure steep walls or overhangs. However, specially made cantilevers can be modulated sideways as well as up and down (as with dynamic contact and noncontact modes) to measure sidewalls, at the cost of more expensive cantilevers and additional artifacts.

## Surface Finish Requirement and Contact Interface Compatibility

In the thermal management of electronic packaging, a proper material surface finish is usually required for the following reasons: (1) corrosion and oxidation protection for the base material of a component; (2) establishing a thermally or electrically conductive interface between two mating parts; (3) improving surface solderability; (4) maintaining galvanic compatibility between different contact materials. The surface finish is usually applied by electroplating; other processes such as painting, cladding or spraying also have been chosen for some specific applications.

### *Corrosion and Oxidation Protection*

In many cases, thermal management materials are subject to corrosion in typical operating environments through oxidation, sulfidation or galvanic incompatibility, etc. Application of surface finish will seals off the thermal management components from the environment and therefore prevents corrosion. For this reason, the contact finish itself must be corrosion resistant, usually provided by noble metal finishes or by nonnoble metal finishes through forming passivating surface via tin or nickel plating for instance.

#### Noble Finish Selection

Noble finishes such as Au, Pd, and some alloys of these metals are intrinsically corrosion or oxidation resistant. When using this kind of finish, however, the extrinsic factors, such as contamination, base metal diffusion or intermetallic compound formation, and contact wear, would degrade their functions. Nickel underplating is generally applied in providing such protection. Typical surface finish platings are 0.4–0.8 μm thick for the Au, or in some cases Au over Pd alloy, and 1.25–2.5 μm for the nickel underplating. These finishes vary in their degree of nobility and will be considered separately (Tong 2009).

Gold (Au)

Au provides an ideal contact surface finish due to its excellent electrical and thermal characteristics, as well as corrosion and oxidation resistance in virtually all environments. Pure gold, however, is relatively soft, therefore, alloyed gold plating is usually applied, with cobalt being the most common, at levels of a few tenths of a percent to increase the hardness and wear resistance (Tong 2009). Cost reduction objectives have led to the use of (1) selective plating practices such as reductions of plating area and thickness; (2) alternative noble metals, most commonly gold flashed palladium or palladium alloy.

Palladium (Pd)

Pd is not as good as gold in corrosion resistance or electrical/thermal conductivity. It is, however, significantly harder than gold, which improves durability performance. Palladium can catalyze the polymerization of organic deposits and result in contact resistance increase under fretting motions. Therefore, palladium is not as noble as gold although the effect of these factors on contact performance depends on the operating environment. In most applications, palladium is used with a gold flash (around 0.1 μm thick) to provide a gold contact interface (Tong 2009).

Noble Metal Alloys

Noble metal alloys mainly include gold alloys and palladium alloys. The major gold alloy used is 69 wt%Au–25 wt%Ag–6 wt%Pt. There are two palladium alloys in use, 80 wt%Pd–20 wt%Ni and 60 wt%Pd–40 wt%Ag. The Pd–Ni alloy is electroplated and the Pd–Ag finish is primarily an inlay. In general these finishes include a gold flash to counter the lower corrosion resistance of these alloys compared with gold. When applying these noble metal finishes onto thermal management materials, Ni underplating is usually used to provide benefits such as (Tong 2009): (1) Ni, through the formation of a passive oxide surface, seals off the base of porosities and surface scratches sometimes presented in the noble metal finished; (2) Ni provides an effective barrier against the diffusion of base metal constituents to the surface plating where the intermetallic compounds could be formed; (3) Ni provides a hard supporting layer under the noble surface which improves contact durability; (4) Ni provides a barrier against the migration of base metal to the surface finish, reducing the potential of the contact interface contamination due to the base metal element migration. These benefits allow for equivalent or improved performance at reduced noble metal thickness. The effects of discontinuities are moderated, and the durability is improved. Ni underplating also allows a reduction in the size of the contact area which must be covered with the noble metal finish. All these functions serve to maintain the nobility of the contact surface finish at a reduced cost.

## Nonnoble Finishes

Nonnoble contact finishes differ from noble finishes in that they always have a surface oxidation film, which can be disrupted during contact surface mating and the potential for recurrence of the films during the application life time of the contact thermal management components. Sn, Ag, and Ni are typical nonnoble finishes usually used. Sn is the most commonly used nonnoble finish; Ag offers advantages for high conductive contacts; and Ni is used in high temperature applications.

### Tin (Sn)

The utilization of Sn surface finish derives from the fact that the oxide film is easily disrupted on mating, and metallic contact areas are readily established. Generation of a tin contact interface for tin finishes result from cracks in the oxide under an applied load. The load transfers to the soft ductile tin which flows easily, opening the cracks in the oxide, and tin then extrudes through the cracks to form the desired metallic contact regions. Here normal force alone may be sufficient; however, the wiping action that occurs on mating of contact spring gaskets for instance, virtually ensures oxide disruption and creation of a metallic interface. The potential problem with tin is the tendency for reoxidation of the tin at the contact interface if it is disturbed. This process is called fretting corrosion. The tin finish surface would be re-oxidized continuously when it is repeatedly exposed. The end of the result is a buildup of oxide debris at the contact interface leading to an increase in contact resistance. Driving force for fretting motions include mechanical (vibration and disturbances/shock) and thermal expansion mismatch stresses. Two approaches to mitigating fretting corrosion in tin and tin alloy finishes are high normal force (to reduce the potential for motion) and contact lubricants (to prevent oxidation). Each has been used successfully, and each has its limitations. High normal forces limit the durability capability of the finish, which is already low due to tin being very soft, and result in increased mating forces. Contact lubricants require secondary operations for application, have limited temperature capability, and may also result in dust retention (Tong 2009).

Another potential issue for tin plating finish is tin whisker growth. Tin whisker is an electrically conductive single crystal structure which spontaneously grows from the tin plating. Over time the whisker may grow to be several millimeters long. Whisker formation is usually identified as one of two mechanisms. (1) Spontaneous whisker formation: the metal atoms forming the whisker diffuse through the deposit layer to the location from which the whisker then grows. In this case, they do not diffuse through any other phase layer. (2) Squeeze whisker formation is due to externally applied compressive stress. Whereas the growth rate of spontaneous whiskers can be up to 1 cm/year under the most favorable conditions, squeeze whiskers can grow at up to 1 m/year (Tong 2009). Tin whiskers are capable of causing electrical failures ranging from parametric deviations to sustained plasma arcing that can result in catastrophic short circuits. A great many attempts have been made to mitigate the effects of tin whisker growth. At this time, the only sure way of avoiding tin whiskers is not to use parts plated with pure tin for some critical applications. Compared with bright tin, matte tin is less prone to forming tin whisker. Even with matte tin plating, the whisker formation is still a majority concern especially for some critical applications. In general, 2–5 μm is the most dangerous tin plating thickness range for tin whisker growth. Below 0.5 μm and above 20 μm are the relatively safe ranges. For thicknesses below 1 μm and over 5 μm, tin whiskers are more resistant to grow. Examples of typical approaches to mitigate tin whisker formation include (Tong 2009): (1) solder dipping pure tin plated components using a leaded solder. The effectiveness of this approach at covering all pure tin plated surfaces can be variable. (2) Application of conformal coat material to pure tin plated surfaces. Conformal coat appears to reduce the growth rate of tin whiskers, but whiskers are still capable of growing through some conformal coat materials such as polyurethanes. (3) Plate or replate the component surfaces using finishes such as thicker matte tin (usually not less than 5 μm), tin/lead, nickel/tin or nickel which are substantially less prone to whisker formation.

Silver (Ag)

Silver can react with sulfur and chlorine to form silver sulfide and silver chloride films on its surface. Silver sulfide films tend to be soft and readily disrupted, and do not result in fretting corrosion. Silver chloride films, however, are harder, more adherent, and more likely to have detrimental effects on contact performance. In addition, silver is susceptible to electromigration, which can be a problem in some applications (Tong 2009). Therefore, silver has limited use in contact surface of thermal management components. Because of its high electrical and thermal conductivity and resistance to welding, however, silver finish still is a candidate for high conductive contacts. The thickness of silver finish is generally in the range of 2–4 μm.

Nickel (Ni)

Nickel plating forms a passivating oxide film that reduces its susceptibility to further corrosion. This passive film also has a significant effect on the contact resistance of nickel. Because both nickel oxide and base nickel are hard, nickel finishes usually require higher contact normal forces to ensure the oxide film disruption. The self limiting oxide on nickel, however, makes it a candidate for high temperature applications. With the same mechanism as tin finishes, nickel contact finishes are also susceptible to fretting corrosion. When Ni underplating is covered with thin noble metal finishes for instance, wear-through of the finish due to mating cycles or fretting action, which exposes the nickel underplating, can result in fretting corrosion (Tong 2009).

## *Solderability of Surface Finishes*

Solderability of some thermal management component surface finishes is required when the component needs to be soldered with other components. Solder bonds depend on whether the solder is either soluble in or capable of forming a metallic bond (a thin layer of intermetallic layer) with the material being soldered. Solderability requires that the metal surface finish be clean and remain clean and wettable by the solder with or without the aid of flux. Different metals have different affinity to a particular solder. Generally, tin plating is a most common surface finish that has an excellent solderability. However, the solderability can degrade over time. Loss of solderability in tin and tin alloy plating for instance, usually stems from three conditions: (1) excessive thickness of the intermetallic layer; (2) excessive amounts of oxides and other surface contaminants on the plating; (3) high levels of codeposited carbon from the organic brighteners in bright tin coatings. Each of these conditions can become worse over time: storage conditions, especially temperature can influence the rate of degradation and therefore shelf life of the plating. Therefore, the solderability of the surface finish needs to be inspected before soldering. Two most common methods used to predict and measure solderability of a surface finish are "dip and look" method and the wetting beam balance test. The dip and look test is performed according to JEDEC J-STD002A and a similar test using lead free solder. Parts are dipped in flux for 5 s and then immersed in the liquid solder for 5 s. The part is examined for the extent of solder wetting on the surface. If the wetting area is greater than 95% coverage of solder, the solderability of the surface finish is deemed acceptable. This is the test most commonly performed in the manufacturing environment to ensure on-going product solderability. The wetting beam balance test is another method used for solderability evaluation. In this test, the part is first fluxed and then partially submerged into a solder pot. During the submersion process, the force required to submerge the part is measured using a delicate balance. The insertion process produces a characteristic diagram of force vs. displacement with several key features which related to solderability. During insertion, the surface tension of the solder initially resists insertion by the terminal which appears as a downward dip in the curve. At some point, the solder begins to wet the surface of the terminal, drawing the terminal into the solder pot. The wetting force produces an upswing in the force curve. The first key metric is the zero cross time. This indicates the time required for the wetting force to overcome the surface tension of the solder. Acceptability conditions vary, but generally times of less than 1 s are desired (Tong 2009).

### *Effects of Mating Cycles and Operating Environments on Contact Surface Finishes*

The contact durability of some spring thermal management components depends on contact surface finish, the normal force, contact geometry and material's stress relaxation resistance. With respect to contact surface finish, the dominant factors of its durability performance include the hardness of the finish, contact geometry, normal force, surface roughness, and state of lubrication of the interface. Gold-flashed palladium or palladium alloy finishes provide the highest durability capability. Silver and tin finishes are severely limited in the number of mating cycles they can support. Contact lubricants are usually used to improve the mating cycle capability of noble finishes. An appropriate contact lubricant can improve the durability capability by an order of magnitude under favorable conditions. In addition, surface finish thickness largely affects the durability, with a roughly linear relationship between mating cycles and thickness for a given contact force/geometry configuration (Tong 2009).

Operating environments including temperature and corrosion severity will impact finish durability. The corrosion resistance of noble metal finishes decreases in the following order: gold, palladium, and palladium–nickel (80 wt% Pd–20 wt% Ni) alloy. The corrosion of the noble metal finishes are mainly caused by chlorine and sulfur. Nickel underplating is generally used to reduce the corrosion susceptibility of the noble metal coated parts. Tin finishes usually exhibit good stability with respect to corrosion, although fretting corrosion must always be taken into consideration for tin finishes. Temperature limitations for noble metal finishes exhibit a similar pattern to corrosion resistance. In both cases, the order is determined by the presence of the nonnoble constituent in the palladium alloys, because it is alloying elements that are susceptible to corrosion. Hard gold finishes, even though the alloy element is of the order of tenths of a percent, are also subject to oxidation as operating temperature increase. In general, soft, or pure, gold finishes are recommended for temperature above 125°C. Tin has a temperature limitation due to an increasing rate of intermetallic compound formation, a reduction in the already low mechanical strength and an enhanced oxidation rate. The interaction of these factors results in a recommendation that tin not be used above 100°C in conventional contact surfaces. Thermal cycling is another important environmental consideration and is arguably the major driving force for fretting corrosion of tin system. In addition, thermal cycling accelerates the effects of humidity on contact performance degradation (Tong 2009).

### *Galvanic Corrosion and Contact Interface Compatibility*

Thermal management component design requires an understanding of galvanic corrosion and contact interface compatibility. Galvanic corrosion is the process by which the materials corrode in contact with each other. Three conditions must exist for galvanic corrosion to occur (Tong 2009): (1) two electrochemically dissimilar metals are in contact with each other; (2) an electrically conductive path presents between the two metals; and (3) a conductive path like electrolyte must be present for the metal ions to move from the more anodic metal to the more cathodic metal. If any one of these three conditions does not exist, galvanic corrosion will not occur. Usually when design requires that dissimilar metals come in contact, the galvanic compatibility is managed by plating or coating with other finishes.

## Reliability Analysis and Environmental Performance Evaluation

The reliability of an electronic packaging and thermal cooling system indicates the probability that the system will be operational within acceptable limits for a given period of time. Advanced electronic packages are very complicated systems containing many thin layers, narrow conducting wires, and tiny solder joints. The dimensions of these structures are in the micrometer scale and getting smaller. Because of the fine features and large number of parts involved in each device, the probability of system failure is high unless high reliability of each device is ensured. In addition, more and more electronic packages are being used in very harsh environments, for instance, under-the-hood applications in automobiles, missiles stored in desert sand, airplanes flying in high attitude where temperatures can drop below negative 60°C, space stations exposed to strong radiation as well as extreme temperatures, etc. These harsh environments have imposed stringent requirements on the reliability of microelectronic systems. The packages must be designed to sustain high and low temperature extremes, to survive humid and corrosive surroundings, and to be protected from ultraviolet radiation. The mean time to failure for modern electronic packaging and thermal cooling system may range from several days to several decades at room temperature. Reliability tests cannot be performed for such long durations, but can simulate the real situation based on well-designed, well-understood, and thoroughly implemented accelerated testing. Reliability testing and simulating can be used to determine which failure modes apply to a given part, how probable it is that these failure modes will occur while the part is in service, and how they might be prevented during the design and manufacture of this part. Therefore, identifying and understanding the mechanisms that cause component failure is the key to make a reliable electronic packaging and thermal management system. In thermal reliability analysis, the focus is on the failures caused by thermomechanical conditions that the electronic packaging and thermal cooling system experiences during manufacturing and service.

## *Failure Modes and Mechanisms*

### Failure Theories

Reliability is defined as the probability that a component functions as designed, whereas failure indicates the probability that a component does not perform the function any more. The fraction of a group of original devices that have failed at time $t$ is called as the cumulative failure function, $F(t)$, while the fraction of a group of original devices surviving at time $t$ is called the reliability function $R(t)$. Therefore, $R(t) = 1 - F(t)$. The failure rate, or the failure occurrences during a given time interval, $f(t)$ is defined as (Jha and Gupta 2003)

$$f(t) = \frac{dF(t)}{dt} = -\frac{dR(t)}{dt}, \tag{2.91}$$

$$F(t) = \int_0^t f(s)ds. \tag{2.92}$$

The failure density $f(t)$ usually follows the bathtub curve. There is an initial high rate of failure due to manufacturing defect escapes; this stage is called infant mortality. After the initial failures, the majority of the remaining eligible products stay functional for its designed life. Towards the end, wearout starts to occur and there is a resurgence of high rate of failure indicating the end of useful life. The mean time to failure, $\mathrm{MTTF} = \int_0^\infty tf(t)dt$. Weibull distribution is commonly used to characterize the failure density (Jha and Gupta 2003):

$$f(t) = \frac{\beta}{\lambda}\left(\frac{t}{\lambda}\right)^{\beta-1} \exp\left[-\left(\frac{t}{\lambda}\right)^{\beta}\right], \tag{2.93}$$

where $\beta$ is the shape parameter, a dimensionless number between 0.5 and 2.0. It is a measure of how the failure frequency is distributed around the average lifetime. $\lambda$ is life time parameter, a measure of the average time-to-failure. It has unit of time and equals to the time at which 62.3% of the samples failed.

### Chemical Failure Mechanisms

In an electronic package, corrosion of metallization and bonding region areas is the most common chemical failure mode. Corrosion is the result of two electrochemical reactions that occur when metal comes onto contact with an aqueous phase containing dissolved ions. The anodic reaction process begins with the oxidation of atoms of a solid metal, changing their valence from zero to a positive number. This produces water-soluble metal ions that move into the aqueous solution, resulting in loss of the solid metal. The electrons liberated from the anodic reaction are then carried out of the metal phase by the cathodic reaction that uses these free electrons to reduce some species in the aqueous phase. For example, a region of the Cu surface as the anode, where oxidation occurs: Cu (s) $\rightarrow$ $Cu^{+2}$(aqueous solution) + $2e^-$. The electron given up by Cu reduces atmospheric oxygen to water at the cathode, which is another region of the same Cu surface: $O_2$ (g) + $4H^{+1}$ (aqueous solution) + $4e^-$ $\rightarrow$ $2H_2O$ (l). The $Cu^{+2}$ ions are further oxidized by oxygen: $4Cu^{+2}$(aqueous solution) + $O_2$ $\rightarrow$ $2Cu_2O$ (s).

In addition, galvanic corrosion of metallization joints, moisture-induced corrosion, and stress corrosion of metallic constituents are the major chemical failure mechanisms in the electronic packaging and thermal cooling system.

### Physical Failure Mechanisms

With increasing use of polymer materials in electronic packaging and thermal cooling system, physical aging of polymers has been one of the major failure modes. As physical aging is a gradual continuation of the glass transformation, many physical and mechanical properties of the polymer are affected in the same direction as during cooling through the $T_g$ range. After aging, the polymer tends to become denser, stiffer, and more brittle; its yield strength may greatly be reduced; and its damping decreases and so does its creep rate.

Electromigration, a direct transport of atoms in a metal produced by an electric current, can be another failure mode especially when the electric current density is on the order of $10^4$–$10^5$ A/cm$^2$ or higher. Electromigration-induced damage usually appears in the forms of voids and hillocks in thin-film conductors due to the massive atomic transport. Voids can grow and link together to cause electrical or thermal discontinuity in conductor lines or thermal dissipation systems. Hillocks can also grow and extrude out materials to cause short-circuit failure between adjacent conductor lines in integrated circuits.

### Thermomechanical Failure Mechanisms

Electronic packaging and thermal cooling system are complicated material integrates operating under electrical, thermal and mechanical loading conditions. Many of these materials are organic materials that have highly nonlinear properties and that are very process and scale sensitive. These complex mechanical systems usually lead to complicated thermomechanical failure mechanisms, such as die or other component cracking, adhesive delamination, solder fatigue failure, interconnection failure, and material fracture and interface voids.

## *Reliability Qualifications*

To ensure product reliability, extensive reliability tests need to be performed during product development and before a new product can be shipped. Commonly accepted accelerated tests include thermal cycling and thermal shock, steady state thermal soaking, mechanical vibration, voltage extremes and power cycling, high humidity and high pressure, and combination of the above.

### Thermal Cycling and Thermal Shock

The single, dual, and triple chamber can be used for temperature cycling. In single-chamber cycling, the load is placed in a stationary chamber, and is heated or cooled by introducing hot or cold air into the chamber. In dual-chamber cycling, the load is placed on a moving platform that shuttles between stationary chambers maintained at fixed temperatures. In triple-chamber temperature cycling, there are three chambers and the load is moved between them. This test is conducted to determine the ability of electronic packaging and thermal cooling system, including components and solder interconnects, to withstand mechanical stresses induced by alternating high and low temperature extremes. Permanent changes in electrical and/or physical characteristics can result from these mechanical stresses (JESD22-A104C). Typical testing conditions can be found from JESD22-A104C.

The thermal shock test is conducted to determine the resistance of a part to sudden exposure to extreme changes in temperature and to the effect of alternate exposures to these extremes. The conditions and recommended fluids can be found from JESD22-A106B.

Thermal cycling is an excellent test method for first order reliability assessment. However, it is not good for products with closely matched thermal coefficients of expansion because second-order effects, such as out-of-plane warpage, are not duplicated in this test method.

Comparability, power cycling, or powered functional cycling is similar to thermal cycling but more closely matches the actual service condition. It consists of a power-on transient in which the components go from room temperature to some steady-state temperature in a matter of minutes. This steady-state condition then lasts for several hours or longer before a power-off (cooling) transient occurs. The power-off condition then lasts for several hours. The transient conditions are difficult to determine either experimentally or analytically. For this test, the same support structure should be modeled as closely as possible. The same environment conditions, such as cooling air flow, should be duplicated. With such care this approach will yield highly accurate reliability information. However, this method is more difficult, expensive, and time consuming.

### Steady-State Temperature/Humidity Bias Life Test

The steady-state temperature humidity/bias life test is performed for the purpose of evaluating the reliability of nonhermetic packaged electronic devices in humid environments. It employs conditions of temperature, humidity, and bias which accelerate the penetration of moisture through the external protective material (encapsulant or seal) or along the interface between the external protective material and the metallic conductors which pass through it. The test requires a temperature–humidity test chamber capable of maintaining a specified temperature and relative humidity continuously, while providing electrical connections to the devices under test in a specified biasing configuration. Test conditions consist of a temperature, relative humidity, and duration used in conjunction with an electrical bias configuration specific to the device. The temperature of 85 ± 2°C (dry bulb) and relative humidity of 85 ± 5% should be applied to the entire useable test area. The wet bulb temperature is 81.0°C, and vapor pressure is 49.1 kPa, and the duration of 1,000 (−24, +168) hours applied continuously except during any interim readouts. The detail testing conditions can be found from EIA/JESD22-A101-B.

### Mechanical Vibration

The vibration test is usually conducted in accordance with MIL-STD-167B. The vibration frequency is swept from 4 to 22 Hz. MIL-STD-167B requires an exploratory vibration test (10-min resonance survey sweep), a variable frequency test (5-min dwell at each frequency), and a 2-h endurance test at the resonant frequency.

Vibrating tables or shake tables are usually used to test electronic packaging and thermal cooling systems to determine or demonstrate their ability to withstand vibration. These machines are capable of producing three different types of vibration profile: sine sweep, random vibration, and synthesized shock. The part being tested will typically be instrumented with one or more accelerometers to measure how the component responds to the vibration input. A sine sweep vibration profile, for instance, typically starts vibrating at low frequency and increases in frequency at a set rate (measured in Hertz). The amplitude (measured in Gs) of the sine wave may increase or decrease as well. A sine sweep is intended to look for resonant frequencies in the part. A random vibration profile will excite different frequencies along a spectrum at different times. Significant calculation goes into making sure that all frequencies get excited to within an acceptable tolerance band. A random vibration test can be anything as short at 30 s up to several hours. It is intended to synthesize the effect of, for example, a car driving over rough terrain or a rocket taking off. A synthesized shock pulse is a short duration high level vibration calculated as a sum of many half-sine waves covering a range of frequencies. It is intended to simulate the effects of an impact or explosion. A shock pulse test typically lasts less than a second.

## References

Aller J (2007) Challenges of measuring thermal conductivity on deposited thin films and advantages of the 3-Omega method. http://courses.ucsd.edu/rherz/mae221a/reports/Aller_221A_F07.pdf. Accessed 03 March 2010.

An W (2002) Industrial applications of speckle techniques – measurement of deformation and shape. Ph.D. thesis. Royal Institute of Technology. Stockholm, Sweden. http://www.diva-portal.org/diva/getDocument?urn_nbn_se_kth_diva-3342-2__fulltext.pdf. Accessed 09 March 2010.

Anter (2007) Principal methods of thermal conductivity measurements. Technical Note #67. Anter Corporation. http://www.anter.com/TN67.htm. Accessed 01 March 2010.

Bourlon A J G (2005) Thermal conductivity measurement by the 3$\omega$ method. Technical Note PR-TN 2005/01035. Philips Electronics, Koninklijke.

Connor Z M, Fine M E, Achenbach J D, Seniw M E (1998) Using scanning acoustic microscopy to study subsurface defects and crack propagation in materials. JOM-e, http://www.tms.org/pubs/journals/JOM/9811/Connor/Connor-9811.html. Accessed 16 March 2010.

Gaal P S, Thermitus M-A, Stroe D E (2004) Thermal conductivity measurements using the flash method. J Therm Anal Calorim **78**:185–189.

Gorring M L (1998) X-ray diffraction. http://www.csam.montclair.edu/earth/eesweb/gorring/geos443/443notes/xray.html. Accessed 12 March 2010.

Heaney M B (1999) Electrical conductivity and resistivity. CRC Press LLC, Boca Raton, FL. http://www.autex.spb.ru/download/wavelet/books/sensor/CH43.PDF. Accessed 03 March 2010.

JEDEC (2004) Thermal shock. JESD22-A106B. JEDEC Solid State Technology Association, Arlington.

Jha N K, Gupta S (2003) Testing of digital systems. Cambridge University Press, Cambridge.

Kim I C (2007) Experimental investigation of size effect on thermal conductivity for ultra-thin amorphous poly methyl methacrylate (PMMA) films. M.S. thesis. Texas A&M University.

Krupke W F, Shinn M D, Marion J E, Caird J A, Stokowski S E (1986) Spectroscopic, optical, and thermomechanical properties of neodymium-and chromium-doped gadolinium scandium gallium garnet. JOSAB **3**(1):102–114.

Lawn B R (1993) Fracture of brittle solids, 2nd edn. Cambridge Solid State Science Series, Cambridge.

Lipson A, Lipson S G, Lipson H (1995) Optical physics. 3rd edn. Cambridge University Press, Cambridge.

Lu L, Yi W, Zhang D L (2001) 3$\omega$ Method for specific heat and thermal conductivity measurements. Rev Sci Instrum **72**:2996–3004. DOI:10.1063/1.1378340.

Lundgren U (2004) Characterization of components and materials for EMC barriers. PhD dissertation Lulea University of Technology, Sweden.

Maglić K D, Cezairliyan A, Peletsky V E (eds) (1984) Compendium of thermophysical property measurement methods. Vol. 1. Survey of measurement techniques. Plenum Press, New York.

Maglić K D, Cezairliyan A, Peletsky V E (eds) (1992) Compendium of thermophysical property measurement methods. Vol. 2. Recommended measurement techniques and practices. Plenum Press, New York.

Parker W J, Jenkins W J, Butler G P, Abbott G L (1961) Flash method of determining thermal diffusivity, heat capacity and thermal conductivity. J Appl Phys **32**:1679–1684

Post D, Han B (2009) Moiré interferometry. In: Sharpe Jr W N (ed) Springer handbook of experimental solid mechanics. http://www.springer.com/978-0-387-26883-5. Accessed 06 March 2010.

Raad P E, Komarov P L, Burzo M G (2005) An integrated experimental and computational system for the thermal characterization of complex three-dimensional submicron electronic devices. 11th THERMINIC, Belgirate, Italy, September 2005.

Raju A (2010) Transmission electron microscopy. http://www.sfermion.com/a253165-transmission-electron-microscopy.cfm. Accessed 12 March 2010.

Rhoads J L (2008) Basic explanation of creep processes. http://www.nuc.berkeley.edu/thyd/ne161/jlrhoads/creep.html. Accessed 10 March 2010.

Schatten H, Pawley J B (2007) Biological low-voltage scanning electron microscopy. Springer, New York.

Schijve J (2009) Fatigue of structures and materials, 2nd edn. Springer, New York.

Schmidt W F (2006) Mechanical design considerations. In: Ulrich R K and Brown W D (eds) Advanced electronic packaging, 2nd edn. Wiley, Hoboken.

http://www.search.com/reference/Atomic_force_microscope

Shinzato K, Baba T (2001) A laser flash apparatus for thermal diffusivity and special heat capacity measurements. J Therm Anal Calorim **64**:413–422.

http://www.siliconfareast.com/optical.html

http://sottosgroup.beckman.uiuc.edu/papers/stout.pdf

Symon K (1971) Mechanics. Addison-Wesley, Reading, MA.

http://www.fig.net/commission6/baden_2006/pdf/mod1/szostak-chrzanowski.pdf

Tong X C (2009) Advanced materials for electromagnetic interference shielding. CRC Press, Boca Raton.

Uher C (2005) Thermal conductivity of metals. In: Tritt T M (ed) Thermal conductivity: theory, properties, and applications. Springer, Berlin.

http://en.wikipedia.org/wiki/Finite_element_method

Zimprich P, Zagar B G (2007) Advanced laser speckle techniques characterize the complex thermomechanical properties of thin multilayered structures. Proc Estonian Acad Sci Eng **13**(4):394–408.

# Chapter 3
# Electronic Packaging Materials and Their Functions in Thermal Managements

**Abstract** Advanced electronic packaging materials play a key role in the proper functioning and useful life of the packaged electronic assembly. These functions mainly include electrical conduction, electrical insulation, mechanical support and structural profiles, environmental protection, as well as thermal conduction and dissipation. Therefore, electronic packaging materials shall possess required electrical, mechanical, thermal, chemical, and physical properties to provide the electronic system with excellent performance and reliable functions. For example, metals provide the means for conducting signals throughout the system via thin-film conductors, wires, interconnects, vias, etc.; insulating materials are used to prevent loss of signal currents by confining them to the metal path; structural or multifunctional materials are used to provide physical functions and mechanical support; and there are materials in which the primary function is for thermal dissipation or to prevent the system from the environment. This chapter will provide an overall review for the state of the art of high-performance electronic packaging materials and their thermal management functions, including properties of key materials, state of maturity, applications, processing, and future directions. These materials mainly include metallic materials, ceramics and semiconductors, electronic glasses, polymers, multimaterial laminates, printed circuit board (PCB) materials, thermal interconnection materials, low thermal conductivity materials, and advanced thermally conductive materials.

## Materials Selection for Electronic Packaging

Materials selection for an electronic packaging application involves a variety of considerations in order for there to be a balance between function, performance, manufacturability, reliability, and cost. The procedure of a proper material selection includes (Fackler 2002): (1) Determining the primary technical performance requirement for the design application; (2) defining any overriding considerations that narrow the field of choice; (3) reviewing typical application requirements regarding materials selection; (4) listing the most likely set of candidate materials; (5) comparing the properties of the selected materials and select one or more

X.C. Tong, *Advanced Materials for Thermal Management of Electronic Packaging*, Springer Series in Advanced Microelectronics 30,
DOI 10.1007/978-1-4419-7759-5_3, 

materials that appear to offer the most likely solution; and (6) validating or testing selected material performance against the original design criteria to very suitability.

The primary technical performance offered by the selected material is to achieve the technical objectives of the design. Dominant technical performance criteria include electrical conductivity, thermal conductivity, thermal emissivity, thermal expansion, chemical inertness, corrosion resistance, thermal stability or applied temperature range, strength, density, electromagnetic and electrostatic shielding, magnetic shielding, fatigue resistance, hardness, ductility, wear resistance, sublimation, combustibility, creep resistance, and moisture absorption. For example, when choosing electrical conductors, a minimum conductivity or maximum electrical resistance requirement must be defined because electrical resistance creates voltage drop and heat generation, either of which may be desirable or an undesirable consequence. Furthermore, it is necessary to make sure that the material electrical resistivity is satisfactory over the temperature range of interest because electrical resistivity is temperature sensitive. Electromagnetic shielding and electrostatic shielding may need to be implemented for achieving electromagnetic compliance during electronic packaging design.

The thermal conductivity of polymer materials is orders of magnitude lower than that of metals, with ceramics in between. The low thermal conductivity of polymers is problematic to maintaining acceptable temperatures in high-power systems. In addition, polymers generally show much higher coefficient of thermal expansions (CTEs) than metals and ceramics. It is usually desirable to operate well below the glass transition temperature of polymers to prevent the thermal stresses that can result from large CTEs. At the glass transition temperature, the long strands of polymer molecules are vibrating so strongly that they break the weak bonds between the individual strands, resulting in a sudden increase in CTE of the polymer (Brown and Ulrich 2006). Chemical inertness of materials to resist chemical attack from acids, salts, gases and water, or chemical corrosion due to the environment is of primary importance. Galvanic corrosion can occur when two dissimilar materials are coupled in a corrosive electrolyte. In a galvanic couple, one of the metals in the couple becomes the anode and the other metal becomes the cathode. The less noble material becomes the anode. The anodic metal corrodes faster than it would all by itself. The cathodic metal corrodes slower than it would all by itself.

The strength and stiffness of packaging materials are primarily important for providing the mechanical support and physical protection for components in the system. Ceramic materials tend to have very high stiffness, but their ductility tends to be low, compared with metals and plastics, so that they are prone to cracking if deformed by such forces as thermal mismatch or impact. Metals have high strength compared with polymers but generally also have higher density. Material density establishes the inertia and resonant frequency of mechanical components and thus the degree to which they are highly affected by mechanical vibration and mechanical shock loading. Polymers have less ultimate strength than metals and ceramics but better ductility, enabling them to tolerate considerably more deformation before failing. However, ductility in plastic materials is strongly affected by temperature and deformation rate. Material ductility varies widely among even a given general class of materials and each individual material must be reviewed for ductility. Furthermore, the harder and smoother the material surface is, the greater its resistance to wear. Wear life is also a function of the magnitude of load-induced localized compressive stresses and the methods used to reduce friction such as lubrication. Creep is not only a characteristic of some metals, but is also present in several inorganic or organic materials as well. Creep failure shall be considered although it is often very slow, and it may occur only during certain conditions, such as mechanical joints or seals that utilize solders, adhesives, elastomers, and other nonmetallic materials.

Sublimation is the loss of material when the material is exposed to vacuum or elevated temperatures. Sublimation conversion may occur uniformly to the surface of a uniform and pure material, or composite materials in which one or more constituent materials may sublimate leaving a residual material possessing poor mechanical or electrical properties. Combustibility is the property of a material, under a given set of conditions, to rapidly oxidize or burn. It is most often associated with plastic and organic materials; however, some metals will burn in the presence of oxidizing gases or chemicals. Combustion may lead to the generation of noxious gasses and destructive deterioration of the packaging components in the vicinity of the combustion. Moisture absorption is an important factor to reliability performance of electronic packaging. For metals, moisture absorption is not a significant consideration; however, the presence of moisture will hasten galvanic corrosion when different metals are in intimate contact. For electrical insulators, such as porous sintered metal structures and ceramics, moisture is a significant consideration because the insulation capability is reduced and the insulator may become electrically conductive to various degrees when moist. In some cases, particularly for organic materials, moisture absorption leads to a loss of strength and physical deterioration. Furthermore, in conditions of high humidity and moisture in the presence of inorganic salts, conditions are created that encourage the growth of fungus. Fungus growth can lead to material degradation in appearance and loss of strength and development of undesirable electrical shorts (Fackler 2002).

Though the primary function of an electronic package is to provide power and signal input and output to the chip, it must also provide mechanical integrity and several other critical functions as well. Among these are dissipating excess heat from the chip and protecting it from corrosive environments and rough handling and vibration. All of these functions are further compounded in the multichip modules where in addition to all the previously mentioned requirements, the package must further provide efficient wiring channels among an array of single-chip devices. The most common used materials are metals, polymers, elastomers, ceramics, glasses, and composites. Successful design and material selection exploits and brings out the true potential of materials chose to meet a certain profile of properties and performances.

## Metallic Materials

### *Monolithic Metals*

A monolithic metal is an element that can readily lose electrons to form positive ions, called cations, forming metallic bonds with other metal atoms, and forming ionic bonds with non-metals. Metals are opaque, lustrous, and good conductors of heat and electricity. Metals are usually inclined to form cations through electron loss, reacting with oxygen in the air or electrolytes to form oxides over changing timescales. The transition metals, such as iron, copper, zinc, and nickel, take much longer to oxidize. Others, like palladium, platinum, and gold, do not react with the atmosphere in general. Some metals form a barrier layer of oxide on their surface which cannot be penetrated by further oxygen molecules and thus retain their shiny appearance and good conductivity for a long period of time, such as aluminum, galvanized or stainless steels and titanium. Painting, anodizing, or plating are usually used on the active metals for protecting them from corrosion. In electronic packaging, both ferrous and nonferrous metals and their alloys are used. Ferrous metals and their alloys are those that have iron as their primary element, and nonferrous metals and their alloys have other elements. Of the ferrous metals, steels are most commonly used, while many of the nonferrous metals have special uses in electronics. Table 3.1 shows typical properties of some selected metals and alloys used in electronic packaging. According to the chemical compositions, standard steels can be classified into three major groups: carbon steels, alloy steels, and stainless steels. Carbon steels consist of alloying elements that do not exceed these limits: 1% carbon, 0.6% copper, 1.65% manganese, 0.4% phosphorus, 0.6% silicon, and 0.05% sulfur. Alloy steels contain alloying elements that exceed the element limits for carbon steels, also include steels that contain elements not found in carbon steels such as nickel, chromium (up to 3.99%), cobalt, etc. Stainless steel contains at least 10% chromium, with or without other elements. Based on the structures, stainless steels can be grouped into three grades. The first being austenitic stainless steels that typically contain 18% chromium and 8% nickel, and are nonmagnetic in annealed condition, which can only be hardened by cold working. The second, ferritic stainless steels containing very little nickel and either 17% chromium or 12% chromium with other elements such as aluminum or titanium. They are always magnetic, which can be hardened only by cold working. Finally, martensitic stainless steels typically contain 12% chromium and no nickel, which is magnetic, and can be hardened by heat treatment. In addition, tool steels typically have excess carbides (carbon alloys) which make them hard and wear-resistant. Most tool steels are used in a heat-treated state, generally hardened and tempered.

Aluminum is a metal that is silverfish-white in color that has a strong resistance to corrosion and, like gold, is rather malleable. It is a relatively light metal compared with metals such as steel, nickel, brass, and copper with a specific gravity of 2.7. Aluminum is easily machinable and can have a wide variety of surface finishes.

**Table 3.1** Typical properties of some selected metals and alloys used in electronic packaging

| Metals and alloys | Thermal conductivity at 0–100°C (W/m K) | Electrical resistivity at 20°C (Ω m × $10^{-8}$) | Thermal expansion (1/K × $10^6$) | Density (20°C) (g/cm$^3$) | Melting point (°C) |
|---|---|---|---|---|---|
| Aluminum (Al) | 238 | 2.7 | 23.5 | 2.7 | 660 |
| Al 2024-T4 | 121 | 5.82 | 23.2 | 2.78 | 502–638 |
| Al 5052-H32 | 138 | 4.99 | 23.8 | 2.68 | 607.2–649 |
| Al 6061-T4 | 154 | 4.32 | 23.6 | 2.71 | 582–651.7 |
| Al 7075-T6 | 130 | 5.15 | 23.6 | 2.81 | 477–635 |
| Antimony (Sb) | 17.6 | 42.0 | 8.11 | 6.7 | 630 |
| Beryllium (Be) | 167 | 5.0 | 12.0 | 1.8 | 1,284 |
| BeCu (C17200) | 105 | 7.68 | 17.5 | 8.36 | 871–982 |
| BeCu (C17510) | 245 | 3.79 | 17.6 | 8.82 | 1,029–1,068 |
| Bismuth (Bi) | 7.9 | 116 | 13.4 | 9.8 | 271 |
| Cadmium (Cd) | 92.0 | 7.4 | 31 | 8.6 | 321 |
| Chromium (Cr) | 69.0 | 13.0 | 6.5 | 7.2 | 1,875 |
| Copper (Cu) | 393 | 1.7 | 17.0 | 8.9 | 1,491 |
| Red brass (15%Zn) | 151 | 4.66 | 18.7 | 8.8 | 990–1,025 |
| Yellow brass (35%Zn) | 119 | 6.38 | 20.3 | 8.8 | 905–930 |
| Phosphor bronze (C5210) | 63 | 13.0 | 18.0 | 8.8 | 880–1,030 |
| Cu–Ni–Sn (C72700) | 50 | 14.0 | 17.1 | 8.88 | 1,020–1,110 |
| Germanium (Ge) | 59.0 | >$10^6$ | 5.8 | 5.3 | 937 |
| Gold (Au) | 263 | 2.3 | 5.8 | 19.3 | 1,064 |
| Indium (In) | 81.9 | 9.1 | 25.0 | 7.3 | 156 |
| Iron (Fe) | 71.0 | 9.7 | 6.8 | 7.9 | 1,525 |
| Invar (64%Fe–35%Ni) | 13.8 | 82 | 1.3 | 8.13 | 1,427 |
| Kovar (54%Fe–29% Ni–17%Co) | 16.3 | 49 | 5.1–5.5 | 7.85 | 1,450 |
| Steel 1015 | 51.9 | 15.9 | 11.9 | 7.87 | |
| Stainless Steel 301 | 16.2 | 72 | 16.6 | 8.03 | 1,399–1,421 |
| Lead (Pb) | 34.3 | 21.0 | 20.0 | 11.7 | 328 |
| Lithium (Li) | 71.1 | 9.4 | 55.0 | 0.5 | 179.0 |
| Magnesium (Mg) | 167 | 4.0 | 26.0 | 1.7 | 650 |
| Mg–Al (AZ31) | 96 | 9.2 | 26.0 | 1.77 | 605–630 |
| Molybdenum (Mo) | 142 | 5.6 | 5.2 | 10.2 | 2,595 |
| Nickel (Ni) | 88.0 | 6.8 | 13.4 | 8.9 | 1,254 |
| Nichrome (80%Ni–20%Cr) | 13.4 | 108 | 14.0 | 8.4 | 1,400 |
| Ni–Cu (N04400) | 21.8 | 54.7 | 13.9 | 8.8 | 1,300–1,350 |
| Palladium (Pd) | 71.0 | 10.9 | 10.9 | 12.0 | 1,550 |
| Platinum (Pt) | 71.0 | 10.7 | 9.1 | 21.5 | 1,770 |
| Rhodium (Rh) | 84.0 | 4.7 | 8.4 | 12.4 | 1,959 |
| Silicon (Si) | 83.0 | >$10^{10}$ | 7.5 | 2.3 | 1,414 |
| Silver (Ag) | 418.0 | 1.6 | 19.0 | 10.5 | 961 |
| Tantalum (Ta) | 54.0 | 14.0 | 6.5 | 17.0 | 2,985 |
| Tin (Sn) | 65.0 | 12.9 | 12.0 | 7.3 | 232 |
| Titanium (Ti) | 17.0 | 55.0 | 8.9 | 4.5 | 1,670 |
| Ti–6Al–4V | 6.7 | 178.0 | 8.6 | 4.4 | 1,604–1,660 |
| Tungsten (W) | 165 | 6.0 | 4.5 | 19.4 | 3,380 |
| Zinc (Zn) | 111 | 26.0 | 30.0 | 7.1 | 420 |
| Zirconium (Zr) | 20.0 | 45.0 | 5.9 | 6.5 | 1,860 |

It also has good electrical and thermal conductivities and is highly reflective to heat and light. At elevated high temperatures (200–250°C), aluminum alloys tend to lose some of their strength. However, at subzero temperatures, their strength increases while retaining their ductility, making aluminum an extremely useful low-temperature alloy.

Beryllium copper is a ductile, weldable, machinable, high strength and highly conductive alloy. It can be heat-treated to improve its strength, durability, and electrical conductivity. Because beryllium is toxic, there are some safety concerns for handling its alloys. In solid form and as finished parts, beryllium copper presents no particular health hazard. However, breathing its dust, as formed when machining or welding, may cause serious lung damage. Beryllium compounds are known human carcinogens when inhaled. As a result, beryllium copper is sometimes replaced by safer copper alloys such as Cu–Ni–Sn alloy, Cu–Ti alloy or bronze.

Other nonferrous metals can be used in electronic packaging, are elementary or alloyed cobalt, copper, lead, magnesium, nickel, zinc, and two groups of precious metals (gold, palladium, platinum, rhodium, silver) and refractory metals (tungsten, tantalum, titanium, molybdenum, niobium, or columbium, zirconium).

## *Metallic Composites*

As discussed earlier, metals such as copper and aluminum are commonly used when materials of high thermal conductivity are required. However, these metals usually have high value of CTE and will not meet the requirement when the adjoining component has a low CTE. Traditionally, some metallic composites, as shown in Table 3.2, have been selected to obtain tailored CTEs, meanwhile maintain a relatively high thermal conductivity.

### Tungsten/Copper

The high thermal conductivity of copper and the low CTE of tungsten make tungsten/copper composites attractive for electronic packaging thermal management applications. These kinds of composite materials contain large amounts of tungsten distributed in a continuous matrix phase, and hence own the excellent performances such as heat-resistant, ablate resistant, high-intensity, high density, low CTE, and good thermal and electrical conductivity. Tungsten/copper composites is usually fabricated through molten copper infiltration into porous, presintered tungsten compact, or through liquid-phase sintering of compacts pressed from mixed powders. Liquid-phase sintering is a common process and offers the advantages of low processing costs, established technology, and generally attractive mechanical properties. However, liquid-phase sintering is restricted to certain geometry and a limited range of tungsten and alloying compositions. For high tungsten contents, it is very hard to achieve full densification composites due to

**Table 3.2** Typical properties of some selected metallic composites

| Metallic composites | Thermal conductivity at 0–100°C (W/m K) | Thermal expansion at 20–100°C (1/K × $10^6$) | Density at 20°C (g/cm$^3$) |
|---|---|---|---|
| W–10wt%Cu | 180–190 | 5.6–6.3 | 16.6–17.0 |
| W–15wt%Cu | 190–200 | 6.3–7.0 | 16.2–16.6 |
| W–20wt%Cu | 200–220 | 7.8–8.5 | 15.4–15.8 |
| W–25wt%Cu | 220–240 | 9.5–10.2 | 14.8–15.2 |
| Mo–30wt%Cu | 190–200 | 7.8–8.4 | 9.6–9.8 |
| Mo–40wt%Cu | 200–220 | 9.0–9.6 | 9.5–9.7 |
| Mo–50wt%Cu | 220–250 | 10.1–10.7 | 9.3–9.5 |
| Al–62wt%Be | 212 | 13.9 | 2.1 |
| Al–40wt%Be | 210 | 16.0 | 2.28 |
| Be–20vol%BeO | 210 | 8.7 | 2.045 |
| Be–40vol%BeO | 220 | 7.5 | 2.277 |
| Be–60vol%BeO | 230 | 6.1 | 2.513 |
| Al–65vol%Si | 124 | 8.27 | 2.35 |
| Al–55vol%Si | 140 | 11 | 2.50 |

the insolubility of tungsten in liquid copper, which severely limits densification during the solution-reprecipitation stage of liquid-phase sintering. Therefore, some high performance tungsten/copper composites have been processed by spray conversion process, such as liquid-phase sintered nanocrystalline composites. Solid-state processes at sufficiently low temperatures have also been used to avoid undesired reactions between the tungsten and the matrix phase, such as hot isostatic pressing (HIPing), hot extrusion, and dynamic compaction. The HIPing and hot extrusion processes can also be used to improve mechanical properties of conventional liquid-phase sintered materials.

The tungsten/copper composite components can be made with varying metal matrix density to achieve optimum combination of CTE and thermal conductivity values; excellent dimensional stability unaffected by high temperature cycling; being machinable to exceptional flatness and surface finish; metallization capability includes electrolytic and electroless plating and thin film deposition. However, relatively high cost and high density may limit applications of the composite materials.

### Molybdenum/Copper

Molybdenum/copper composites have similar performances to tungsten/copper composites. However, their densities are lower than tungsten/copper composites; therefore, they are more suitable for space flight and navigation industries. Molybdenum/copper composites can provide a controllable expansion composite matrix, yielding superior material stability and uniform thermal expansion characteristics. However, high cost, high density, and difficulty to machine are major limitations to their wider applications.

## Silicon/Aluminum, Beryllium/Aluminum and Beryllium/ Beryllium Oxide

Silicon/aluminum (Si/Al), beryllium/aluminum (Be/Al) and beryllium/beryllium oxide (Be/BeO) composites provide advantageous materials for a number of applications, including uses in aerospace and electronics. The low weight and high stiffness of these composites allow use of these materials in these applications. These composites are characterized by their high fatigue strength, high heat capacitance, moderately low CTE, high thermal conductivity, high stiffness, and low weight.

One current problem in the use of beryllium–aluminum alloys (Be/Al alloys) is the difficulty in joining parts made of this alloy. Be/Al alloys have a surface oxide layer, which makes it difficult to join or coat this surface. To deal with this problem, prior attempts to join Be/Al alloy parts have used mechanical joining, adhesive bonding, or dip brazing. Mechanical joining (e.g., by threaded attachments, nuts and bolts) may localize the strength of attachment, add significant weight to the structure and provide a more uneven thermal profile. Adhesive bonding of Be/Al parts does not provide sufficient mechanical strength for many applications, does not provide a uniform media for thermal transfer, and does not provide uniform material for thermal expansion. Dip brazing of Be/Al parts requires that these parts be precisely constructed to form narrow faying surfaces and small joint gaps at the joined surfaces to retain the brazing alloy in the joint gaps during the brazing process. To form a moderate to complex assembly, it is difficult to manufacture parts that would allow adequate braze fillet formation. Braze fillets are advantageous in Be/Al part assembly where narrow faying surfaces exist because joint stresses are carried by these fillets, rather than by the narrow faying surfaces. In addition, dip brazing subjects the Be/Al parts to the conditions of the molten dip bath. This process can result in distortion of the part shape, or alternatively can produce substantial deposits of entrapped flux, further adding to part weight.

Dispersed beryllium oxide in the Be/BeO composite is present as small, individual particles with single crystal structures ranging in size from about 1 μm to about 50 μm at loadings of from about 10% to about 70% by volume. Higher volume fractions of beryllium oxide result in lower thermal expansion coefficients and higher thermal conductivities. It should also be stressed that processing becomes more difficult with volume fractions of greater than about 60%. Therefore, preferred volume loadings are in the range from about 20% to about 60% by volume, more preferably in the range of about 40–60% by volume. The Be/BeO composite material is fabricated by blending to form a homogeneous composite powder. The composite powder is then formed into a desired shape and densified. Densification is accomplished by the HIP process, with the resulting fillet being further processed into the desired shape with required dimensions (Grensing 1992).

## Metal Foam and Metallic Cellular Materials

The metal foam materials consist of interconnected cells in a random orientation. Two mechanisms have been found to be important to the heat transfer enhancement associated with the use of metal foam materials: interactions between the solid foam material and a through-flowing fluid; and the importance of achieving a quality metal-to-foam bond (Haack et al. 2001). Applications in electronics cooling and compact heat exchangers have been investigated, revealing promising advances in the rate of heat removal or transfer under experimental conditions. Some applied research has been performed, applying metal foam in unique designs to radiators and advanced reactors (Haack et al. 2001).

The typical metal foam structure consists of ligaments forming a network of inter-connected dodecahedral-like cells. The cells are randomly oriented and mostly homogeneous in size and shape. Pore size may be varied from approximately 0.4 to 3 mm, and the net density from 3 to 15% of a solid of the same material. Heat transfer enhancement using porous metal foams depends on both the cellular structure of the foam material and the thermal properties of the metal foam. Metal foam thermal conductivity is dependent upon the overall density of the piece and the metal from which the foam is made. Conductive pathways through the porous material are limited to the ligaments of the material. Experimental measurements have determined a functional relationship between the foam thermal conductivity and density as (Ashby et al. 2000; Haack et al. 2001):

$$K_s \rho_r^{1.8} < K_f < K_s \rho_{xr}^{1.65}, \tag{3.1}$$

where $\rho_r$ is the foam relative density, $K_f$ is the foam conductivity, and $K_s$ is the solid conductivity. Higher material conductivity is associated with higher density materials, and significant increase in thermal conductivity results from an increase in material density. On the other hand, heat transfer by metal foams as a result of thermal dispersion effects is proportional to cell size (Hunt and Tien 1998).

Metal foams have been manufactured for many years using a variety of techniques. Metallic sintering, metal deposition through evaporation, electrodeposition or chemical vapor decomposition (CVD), and investment casting have created open cell foams. In foam creation through metal sintering, metallic particles are suspended in slurry and coated over a polymeric foam substrate. The foam skeleton vaporizes during heat treatment and the metallic particles sinter together to create the product. This method is believed to be the most cost effective and the most amenable to mass production. The CVD method utilizes chemical decomposition of a reactive gas species in a vacuum chamber to deposit material onto a heated substrate, such as polymer or carbon/graphite, depending upon the temperature of the deposition process. In this method, production rates are limited by the rate at which material is deposited on the substrate. High quality highly refractory metals and ceramics may be created with this method.

Molten metal infiltration has been utilized to make aluminum and copper foam materials (Haack et al. 2001). With this method, a foam precursor is coated with a ceramic casing and packed into casting sand. The casting assembly is heated to decompose the precursor and harden the casting matrix. Molten metal is then pressure infiltrated into the casting, filling the voids of the original matrix. After solidification, the material is broken free from the mold. This method has the advantage of being capable of producing pieces in widely used metals and alloys with solid struts. However, the process requires several steps and specialized equipment and does not lend itself to rapid production processes.

Of the methods most suitable to produce metal foam materials, the metal sintering method offers the most promise for mass production. Necessary production equipment is easily automated and yields high-quality, low-cost metal foam materials for use in a variety of applications (Haack et al. 2001).

## Ceramics and Semiconductors

Ceramics are inorganic, nonmetallic solids, or metal-nonmetal compounds, and some of them have been used for electronic packaging applications due to their unique electrical, chemical, mechanical, and physical properties. Ceramics used in electronic packaging can usually be classified into three distinct material categories: (1) oxides: alumina, zirconia; (2) nonoxides: carbides, borides, nitrides, silicides; and (3) composites: particulate reinforced, combinations of oxides and nonoxides. Each one of these classes can develop unique material properties. Oxide ceramics are oxidation resistant, chemically inert, and electrically insulating. They generally have low thermal conductivity, slightly complex manufacturing and low cost for alumina, more complex manufacturing and higher cost for zirconia. Non-oxide ceramics have low oxidation resistance, extremely high hardness, chemically inertness, high thermal conductivity, and electrically conducting, difficult energy dependent manufacturing and high cost. Ceramic-based composites are of toughness, low and high oxidation resistance (type related), variable thermal and electrical conductivity, complex manufacturing processes, and high cost (Taylor 2001).

High purity ceramics used in electronic packaging are made with precise methods to ensure that the desired properties of these advanced materials are achieved in the final product. Oxide ceramics are prepared with high purity starting materials (powders) using mineral processing techniques to produce a concentrate followed by further processing (typically wet chemistry) to remove unwanted impurities and to add other compounds to create the desired starting composition. This is a very important stage in the preparation of high-performance oxide ceramics. As these are generally high purity systems, minor impurities can have a dynamic effect, for example, small amounts of MgO can have a marked effect upon the sintering behavior of alumina. Various heat treatment procedures are utilized to create carefully controlled crystal structures. These powders are generally ground to an extremely fine or ultimate crystal size to assist ceramic reactivity. Plasticizers and binders are blended with these powders to suit the preferred method of forming (pressing, extrusion, slip or tape casting, etc.) to produce the raw material. Both high- and low-pressure forming techniques are used. The raw material is formed into the required green shape or precursor (machined or turned to shape if required) and fired to high temperatures in air or a slightly reducing atmosphere to produce a dense product (Taylor 2001).

The production of nonoxide ceramics is usually a three-stage process involving first, the preparation of precursors or starting powders; second, the mixing of these precursors to create the desired compounds (Ti + 2B, Si + C, etc.), and third, the forming and sintering of the final component. The formation of starting materials and firing for this group require carefully controlled furnace or kiln conditions to ensure the absence of oxygen during heating because these materials will readily oxidize during firing. This group of materials generally requires temperatures that are quite high to effect sintering. Similar to oxide ceramics, carefully controlled purities and crystalline characteristics are needed to achieve the desired final ceramic properties (Taylor 2001).

Ceramic-based composites can be composed of a combination of oxide ceramics/nonoxide ceramics (granular, platy, whiskers, etc.), oxide/oxide ceramics, nonoxide/nonoxide ceramics, ceramics/polymers, etc. An almost infinite number of combinations are possible. The object is to improve either the toughness or hardness to be more suited to a particular application. This is a somewhat new area of development and compositions can also include metals in particulate or matrix form (Taylor 2001).

Typical electronic ceramic materials include electrical porcelains which have varied properties based on the amount of alumina incorporated into the material. Alumina is mainly used for improving mechanical strength, thermal conductivity, and flexure strength. However, the use of alumina substantially increases the material CTEs and elevation of the material dielectric constant (permittivity). Ceramics are often bonded to metal substrates to perform useful electronic functions. Other oxides such as beryllia may be used to achieve high thermal conductivity. Aluminium nitride also has high thermal conductivity with thermal expansion nearly that of silicon, while boron nitride combines high thermal conductivity with machinability in a softer yet durable material (Fackler 2002).

Advanced ceramics have been used in the packaging of discrete semiconductor devices, single- and multiple-integrated circuits, hybrid circuits, and circuit boards for high performance packaging. They can be a constituent part or the only material used in substrates, circuit components and conductors, sealing compounds, and packaging covers or lids. One of their more attractive properties is that packages made entirely of ceramic, or ceramic in combination with metals, can be made hermetic. The particular application of a ceramic material in electronic packaging and thermal management system, as is the case for other types of materials, dictates which of its properties are important. In general, there is no material that possesses all the requirements for a given application. Therefore, trade-off decisions must be made with regard to the various material properties to achieve optimization with respect to application requirements (Brown and Ulrich 2006).

For instance, ceramic superconductors allow the flow of electricity with little or no resistance or heat loss. This gives the potential for supercomputers that are the size of desktop calculators to perform calculations thousands of times faster than any computer today. Other ceramic applications in electronics include sensors, actuators, electro-optic materials, semiconductors, and multilayer capacitors. Ceramic coatings are also used to protect or lubricate materials such as metals. The ceramic coatings can prevent electronic shutdowns, component failure and excessive wear and tear in computers and other electronic devices. Ceramics used in electronics are chemically inert, able to withstand high temperatures, and less prone to corrosion. Because of these characteristics, ceramics can provide a stable and safe environment for electronic circuitry.

Semiconductor materials typically have electrical conductivity in between a conductor and an insulator; the conductance can vary over that wide range either permanently or dynamically, depending on the current or voltage applied to a control electrode, or on the intensity of irradiation by infrared, visible light, UV, or X-rays. In semiconductors at room temperature, just as in insulators, very few electrons gain enough thermal energy to leap the energy band gap from the valence band to the conduction band, which is necessary for electrons to be available for electric current conduction. For this reason, pure semiconductors and insulators in the absence of applied electric fields, have roughly similar resistance. The smaller band gaps of semiconductors, however, allow for other means besides temperature to control their electrical properties. Semiconductors' intrinsic electrical properties are often permanently modified by introducing impurities by a doping process. Upon the addition of a sufficiently large proportion of impurity dopants, semiconductors will conduct electricity nearly as well as metals. Depending on the kind of impurity, a doped region of semiconductor can have more electrons or holes, and is named n-type or p-type semiconductor material, respectively. That is, an n-type semiconductor carries current mainly in the form of negatively charged electrons, in a manner similar to the conduction of current in a metal wire. A p-type semiconductor carries current predominantly as electron deficiencies (holes). A hole has a positive electric charge, equal and opposite to the charge on an electron. In a semiconductor material, the flow of holes occurs in a direction opposite to the flow of electrons. Junctions between regions of n- and p-type semiconductors create electric fields, which cause electrons and holes to be available to move away from them, and this effect is critical to semiconductor device operation. Also, a density difference in the amount of impurities produces a small electric field in the region which is used to accelerate nonequilibrium electrons or holes.

In addition to permanent modification through doping, the resistance of semiconductors is normally modified dynamically by applying electric fields. The ability to control resistance/conductivity in regions of semiconductor material dynamically through the application of electric fields is the feature that makes semiconductors useful. It has led to the development of a broad range of semiconductor devices such as transistors and diodes. Semiconductor devices that have dynamically controllable conductivity, such as transistors, are the building blocks of integrated circuits devices. These transistors are combined with passive components implemented from semiconductor material such as capacitors and resistors, to produce complete electronic circuits.

In most semiconductors, when electrons lose enough energy to fall from the conduction band to the valence band (the energy levels above and below the band gap), they often emit light. This photoemission process underlies the light-emitting diode (LED) and the semiconductor laser. Conversely, semiconductor absorption of light in photodetectors excites electrons to move from the valence band to the higher energy conduction band, thus facilitating detection of light and indicating its intensity. This is useful for fiber optic communications, and provides the basis for energy from solar cells.

Elemental semiconductors include antimony, arsenic, boron, carbon, germanium, selenium, silicon, sulfur, and tellurium. Silicon is the most common of these, forming the basis of most integrated circuits. Common semiconductor compounds include gallium arsenide, indium antimonide, indium phosphide, or alloys such as silicon germanium, aluminium gallium arsenide, and the oxides of most metals. Of these, gallium arsenide (GaAs) is widely used in low-noise, high-gain, weak-signal amplifying devices.

Some organic materials also have semiconductor properties, including single molecules, short chain (oligomers) and long chain (polymers) organic semiconductors, such as pentacene, anthracene and rubrene; poly(3-hexylthiophene), poly (*p*-phenylene vinylene), F8BT, as well as polyacetylene and its derivatives. Like inorganic semiconductors, organic semiconductors can be doped. Highly doped organic semiconductors, for example Polyaniline (Ormecon) and PEDOT: polystyrene sulfonic acid are also known as organic metals. Several kinds of carriers mediate conductivity in organic semiconductors. These include π-electrons and unpaired electrons. Almost all organic solids are insulators. But when their constituent molecules have π-conjugate systems, electrons can move via π-electron cloud overlaps. Polycyclic aromatic hydrocarbons and phthalocyanine salt crystals are examples of this type of organic semiconductor. In charge transfer complexes, even unpaired electrons can stay stable for a long time, and are the carriers. This type of semiconductor is also obtained by pairing an electron donor molecule and an electron acceptor molecule.

## Electronic Glasses

Due to their insulating property, surface smoothness, controllable thermal deformability, ability to serve as a bonding phase, glasses have been used as substrate, delay lines, passivation layers, capacitors, resistor and conductor bonding phases, package sealants, and insulating bushings. Glasses are amorphous noncystalline materials that perform like liquids which have been sufficiently cooled to become substantially stiff and rigid. The inclusions of additives can substantially alter glass melting point, thermal expansion, and electrical properties. Glasses may be clear, fused with other materials to make components such as resistors and capacitors, and are vulnerable to impact mechanical loads and thermal shock loads that cause glass to exhibit brittle fracture (Fackler 2002).

In an electronic packaging and thermal management system, glasses should be selected to provide optimum compatibility with the device requirements and usage, as well as the application technique used to apply it. The most critical characteristics of a glass include type of glass, CTE at the glass set point, glass transformation temperature ($T_g$), particle size distribution, glass flow and wettability at the processing temperature, compatibility of the glass composition with the application. Glasses are generally of two types, vitreous and devitrifying (crystallizing). Vitreous glasses are thermoplastic and flow at the same temperature each time they are fired. Devitrifying glasses are thermosetting and crystallize during firing to form glass-ceramic, which have different properties than the original glass. The actual crystallization and resultant glass-ceramic can be modified by changing firing conditions and particle-size distribution. They typically have greater strength and allow higher device operating temperatures than the vitreous form. Glasses may also be a composite of several different glasses, or glass and ceramic fillers. This is typically done to meet specific requirements, such as thermal expansion and firing temperature, which are not possible with a single glass. The CTE at a set point should be as close as possible to the substrate material to prevent stress within the glass and/or substrate that can lead to cracking and failure. The largest recommended difference between glass and a substrate is $\pm 5 \times 10^{-7}/°C$; this is generally thought of as a matched-fit. In some special configurations, a compression-fit can allow the joining of two significantly different thermal expansion materials by an intermediate glass. Each glass has a maximum operating temperature, or the temperature a device can be operated without degrading the glass which can lead to device failure. This is typically somewhat lower than the $T_g$ of the glass. A devitrifying glass, after firing, will have a maximum operating temperature dependent on the crystalline phase formed and is higher, sometimes significantly, than the original glass. The powder type should only be as small as necessary to achieve the desired fired thickness and line definition using the application technique chosen. Typically, the smaller a powder type is, the more difficult to process efficiently. In the case of devitrifying glasses (crystallization during firing), changes in the particle size can affect fired CTE, flow and wetting, and densification and strength. Glass powders are applied to the substrates and parts in a variety of techniques, e.g., screen printing, doctor blading, electrophoresis, spin-coating, performing, etc. This typically involves the mixing of the glass powder in a liquid vehicle to form slurry. The vehicle that is used to apply the glass is typically composed of a binder which promotes substrate adhesion and film green strength, and a solvent. The higher the resin content of the vehicle, the greater the green strength and the higher the viscosity of the vehicle; and subsequently, the viscosity of the glass slurry. Resins should be selected to allow decomposition, or burn-out, below the glass $T_g$. Each of the techniques indicated above have optimum slurry viscosities dictated by the equipment you are using and the glass application characteristics you are attempting to achieve. Glass slurry viscosity is controlled by the ratio of the glass powder to the vehicle ratio and by the particle size distribution of the glass. The finer the glass powder is, the higher the slurry viscosity for the same ratio; and also the more difficult to remove. After applying the slurry, the solvent is removed during a drying stage by heating in air at 125–150°C for 10–20 min. The binder is then removed in either a separate binder burn-out stage, or in conjunction with the final firing stage. The binder decomposition is most efficiently done in air, at a temperature below the $T_g$ of the glass. Insufficient binder burn-out can lead to excessive porosity and possible reduction of any reducible heavy metal components, such as PbO, present in the glass. The time and temperature required is dependent upon the binder selected, thickness of the glass layer, and the particle size distribution of the glass.

The appropriate firing temperature and time should be optimized for specific firing equipment and application to ensure sufficient flow and wetting of the substrate. Most devitrifying glasses require extended firing times to achieve the appropriate level of crystallization. Variations in time and temperature for these types of glasses can significantly change properties, particularly CTE, and performance. Final firing should typically be done in air or other oxidizing atmospheres. If it is necessary to prevent oxidation of metal parts, a neutral atmosphere should be utilized, such as nitrogen. Reducing atmospheres can cause reduction of heavy metal glass components. Heating rates are generally determined by part size, configuration, and thermal conductivity and should assure that the substrate/parts and the glass are in equilibrium. Devitrifying glasses are particularly sensitive to heating rates (Ferro 2008).

Solda-lime glasses are usually used to seal hermetic packages and to form insulator bushing for feed-through devices. Borosilicate glasses have excellent resistance to chemicals, high electrical resistance, and low dielectric constants and are used as bonders for compounds associated with component construction. Lead alkali borosilicate glasses have lower melting point and are used for adhesive and sealing applications, including semiconductor processing (Fackler 2002).

Glass-ceramics are a subgroup of ceramics and usually formed by adding glass to materials such as alumina, in slurry that then can be used in forming green sheet, or shaped into a variety of useful configurations. They have a crystalline structure and are machinable materials. With a dielectric constant of around 5, glass-ceramics are one of the best high-performance dielectric materials. Their properties actually approach those of ideal substrate materials for electronic packaging except for their low thermal conductivity. In addition to their low dielectric constant, they have a good thermal match with silicon and are very compatible with gold and copper metallization (Brown and Ulrich 2006).

## Polymers

Polymers, or plastics, are used extensively in electronic packaging and thermal managing systems such as packaging molding, encapsulation, passivation or stress buffer coatings, interlayer dielectric die-attach adhesives, PCB substrates, and thermal interface materials (TIMs). Although polymers do not have the high hermeticity of ceramics or the strength of metals, they play an important role in packaging and thermal management as a result of their ease of processing, flexibility, insulating properties, and low cost. Polymers are usually classified as thermoplastics, thermosets, and elastomers.

## *Thermoplastics*

Thermoplastic polymers or resins are high-molecular-weight polymers that can be melted and reformed repeatedly because they have no covalent cross links between the separate polymer molecules. For temperatures below the glass transition temperature, they are stiff because the molecules are held in place with one another by secondary forces that act as weak cross links. As the temperature is raised, a point is reached ($T > T_g$) where the individual molecules gain enough kinetic energy that they overcome some of the intermolecular secondary forces, causing them to wiggle independent of one another. This will result in the material becoming much softer and the CTE increasing. As the temperature is raised further, a point is reached where the energy of the system is sufficient to overcome the remaining secondary bonds and the polymer melts into liquid. Because the viscosity of the resulting liquid is very high due to molecular entanglement, molding requires enough pressure energy to force the material into a mold (Brown and Ulrich 2006). Thus, they can be formed into useful shapes while in the melted or viscous phase. In most manufacturing processes, thermoplastics are heated, then formed by injection molding, foaming, stamping, extrusion, or thermoforming and finally cooled so the end product retains its shape. Many types of thermoplastics are available with a wide range of useful properties. They can be made as flexible as rubber, as rigid as metal, and concrete or as clear as glass, for use in a wide range of packaging and thermal management components. Some can withstand temperatures of up to 600°F (315°C). Others retain their properties down to −100°F (−73°C). They do not oxidize, are highly corrosion resistant, and are excellent thermal and electrical insulators. Thermoplastics are used as pure polymers or enhanced with filters and reinforcements to form composite materials (Fackler 2002; Furness 2008).

Acrylonitrile-butadiene-styrene (ABS) polymers are amorphous polymers readily synthesized by adding styrene and acrylonitrile to a polybutadiene latex. In the resulting polymer, some styrene acrylonitrile (SAN) is grafted onto the polybutadiene backbone chain while the remainder of the SAN forms a continuous matrix. It is the SAN that is grafted onto the polymer chains that makes the two phases compatible. This essentially gives the ABS its strength and toughness. A wide range of ABS materials can be formulated from different combinations of the above three components. ABS polymers tend to creep, particularly at elevated temperatures; ABS materials also have poor solvent resistance, but they have high impact strength and can be processed by injection molding (Furness 2008). ABS thermoplastics are usually used to make enclosures, insulators, appearance sheet materials, and a variety of molded components for electronic packaging, mostly in consumer goods and other low-cost applications.

Fluoropolymers are essentially those in which some or all of the hydrogen atoms in polyethylene have been replaced by fluorine, with polytetrafluoroethylene (PTFE) perhaps the best known. PTFE is a tough, flexible, crystalline polymer that retains ductility down to −150°C. Its solvent and chemical resistance is the best of all the thermoplastics and it has the lowest coefficient of friction of any known solid (0.02) (Furness 2008). They have high heat resistance, low dielectric losses, zero moisture absorption, high arc resistance and creep resistance. However, it must be molded by a powder sintering technique, although it can be extruded very slowly, and it is very expensive with low strength and stiffness. Applications of PTFE are therefore limited to those that make use of its special properties. Teflon is an example of a fluoropolymer used for wire insulation, electrical insulation, low load bearings, and dielectric purposes (Fackler 2002). Other fluoropolymers include: polyvinyl fluoride, polyvinylidene fluoride, perfluoroalkoxy tetrafluoroethylene, and polychloro trifluoroethylene.

Polyamides (PAs–nylons) have good strength and toughness with excellent fatigue resistance. However, they are prone to absorb moisture to equilibrium with the surrounding humidity, ranging from 8–10% for PA6 and PA66 to 2–3% for PA11 and PA12 at saturation. Its dielectric strength and resistivity varies with the humidity. Mechanical properties are affected by moisture with toughness improving with the absorption of moisture whereas modulus is reduced. PAs are resistant to hydrocarbons, esters and glycols, but swell and dissolve in alcohols. They are also attacked by acids but generally stable to alkalis. PA6 and PA66 are mainly used in textiles, but they also find application where toughness is a requirement, for example, zip fastener teeth, gears, wheels and fan blades. PA11 is more flexible than PA66 and is typically used for petrol and hydraulic hose as well as powder coatings for metals. Strength and rigidity of these materials can be dramatically enhanced by the addition of glass or carbon fiber reinforcement; the level of saturation water absorption is also reduced. Nylons can be processed with all the typical thermoplastic processing methods. Molded nylons are used as coil bobbins, wire ties, connectors, and wire jackets (Sampson and Mattox 1997; Furness 2008).

Polyarylates (PAryls) are a family of aromatic polyesters that are considered tough, resistant to UVradiation, resistant to heat and, in amorphous form, transparent. The basic mechanical properties of amorphous PAryls are similar to polycarbonates. The elastic rebound of PAryls is exceptional which makes it a logical candidate for snap fit applications. They are susceptible to environmental stress cracking in the presence of aliphatic or aromatic hydrocarbons and their properties deteriorate rapidly in water. The materials are characterized by stable electrical properties over the range of temperatures and frequencies, clarity, high arc resistance, and low dielectric constant. They can be shaped with all normal molding processes. The melt viscosity is very low, permitting to fabricate thin-wall components. Polyarylates have been used to mold three-dimensional circuit board, and applied to solar panels, lighting, fire helmets, electrical connectors, insulating film, terminal blocks, and fuse holds (Fackler 2002; Furness 2008).

Polyethylene terephthalate (PET) and polybutylene terephthalate (PBT) are the most common thermoplastic polyesters. They are similar to PA6 and PA66 in many respects but with much lower water absorption. However, they are prone to hydrolysis, and prolonged contact with water at temperatures as low as 50°C has a detrimental effect on properties. PET is used in the manufacture of biaxially oriented film and bottles, the latter suitable for carbonated drinks. The purpose of the orientation is to enhance rigidity, strength and toughness and also to improve barrier properties, which allows thinner bottles to be made. PBT displays a good combination of stiffness and toughness and can withstand continuous service at 120°C (Furness 2008). The most important grades are those reinforced with glass. Applications for PBT include electrical connectors, pump components, and gears as well as underbonnet and exterior parts for cars.

Polycarbonate (PC) is unusually tough and exhibit high impact strength, due to the nature of its chemical bonding. It is also transparent and almost self extinguishing, with a relatively high continuous use temperature of around 115°C. Chemical resistance is not outstanding and it needs the addition of light stabilizers for any UV resistance. Glass fibers enhance the stiffness but reduce toughness, as might be anticipated. Polycarbonate is a versatile blending material, with blends of PC/PET and PC/ABS available commercially (Furness 2008). Applications of polycarbonate include: glazing panels, windows and other functional and appearance-related components, light fittings, safety helmets and medical components (Fackler 2002; Furness 2008).

Polyimides (PI) are used for radomes, antenna housings, films, print circuit boards, and connectors. They have high radiation resistance, low CTE, high electrical resistance, low dielectric loss, are wear resistant, and can retain their mechanical properties to 250°C. They exhibit low flammability and smoke emission characteristics and offer the lowest minimum service temperature of thermoplastics. They are relatively expensive and can be difficult to process (Furness 2008). Thermoplastic PI requires high temperatures and pressures and is usually processed by powder metallurgy, and autoclave or compression molding. They are chemically resistant but are degraded by hot caustic solutions and highly polar solvents.

Polyamideimide (PAI) was initially developed as high temperature enamel but was later modified for processing by injection and compression techniques. No other commercially available unreinforced thermoplastic is as strong as PAI over its operating range. Applications include valves, bearings, gears, electrical connectors, and jet engine parts. If wear resistance is important, grades filled with graphite and PTFE are available (Furness 2008).

Polyetherimides (PEI) are amorphous, high-performance thermoplastics with a continuous use temperature of around 170°C. PEI resin has low dielectric losses over a wide temperature range, good mechanical strength, and is easily molded into complex shapes. The strength, creep, and impact properties of PEIs make them ideal for underbonnet or similar components (Furness 2008). They are also used in high temperature switchgear and electrical connectors, circuit-breaker housings, chip carriers, bobbins, fuse blocks, and circuit board insulators which can be vapor-phase soldered and electroless plated. A number of medical equipment components are manufactured using PEIs, taking advantage of their excellent resistance to repeated sterilization using steam, autoclave, gamma radiation or ethylene oxide. Microwave cookware is another application (Sampson and Mattox 1997; Fackler 2002).

Polyoxymethylene (POM acetal) polymers are highly crystalline thermoplastics that are commercially available as homopolymers or copolymers. This crystallinity is responsible for their excellent solvent resistance, fatigue resistance, surface finish and predictable mechanical properties over a wide temperature range. POMs are superior to PAs in stiffness, creep resistance, fatigue strength, and water absorption but have inferior impact and abrasion resistance. In general, the copolymers are thermally more stable, with similar mechanical properties to the homopolymer, albeit with slightly reduced tensile properties. The greater stiffness and strength of the homopolymer have promoted its use in cams, gears, and exterior car door handles. The copolymer is preferred where chemical resistance and resistance to hydrolysis in particular are important. Examples of the applications of POM copolymer are electric kettles, wash basins and shower heads, as well as snap fit components and toys (Fackler 2002; Furness 2008).

Polyphenylene (PPO) is a high-strength, tough, and heat-resistant polymer, but in the unmodified state it is extremely difficult to process, and also, it is relatively expensive. Fortunately, it is miscible with polystyrene, and the resulting amorphous blends are easily processed and less expensive than PPO, with little loss in mechanical properties. Stiffness and strength are approximately 50% higher than high impact ABS, with similar creep behavior. Modified PPO grades are also self extinguishing when ignited (Furness 2008). Resistance to solvents is poor, a characteristic of styrene-based polymers. As well as glass fiber reinforced grades, these materials are available in structural foam grades. Modified PPOs are used for electrical fittings, car fascia panels, TV components, and computer housings. Foamed modified PPO is particularly suited to the last example (Fackler 2002).

Polyaryletherketones semicrystalline polymers have excellent mechanical properties, good thermal stability and good chemical resistance. Despite a $T_g$ of 145°C, the continuous service rating of PEEK is 250°C. PEEK is inherently fire retardant. It is easier to burn a hole through an aluminum sheet than through one made from PEEK. However, these materials are very expensive and difficult to process (Furness 2008). They have been used for high temperature wire covering and PCBs. Fiber reinforced grades have been used in applications that include valves, pumps, and missile nose cones (Fackler 2002).

Polysulfone (PSul) is an amorphous, transparent polymer with good heat resistance and stiffness. It can be processed by conventional thermoplastic techniques despite the fact that it has a continuous use temperature of 150°C. Resistance to ionizing radiation is good but is poor to UV radiation. PSul is susceptible to stress cracking in certain solvents (Furness 2008).

Polyethersulfone (PES) has a higher continuous use temperature (180°C) and a modified PES has been shown to operate for tens of thousands of hours at 200°C without significant loss in properties. This high temperature stability is not matched by weathering or UV resistance which is poor. The polysulfones are used in electrical and electronic applications, medical components requiring repeated sterilization, microwave cookware, and underbonnet and aerospace components (Furness 2008).

Polyphenylene sulfide (PPS) is a crystalline material, usually reinforced with glass fibers or glass fibers and mineral fillers. The chemical and ionizing radiation resistance of PPS is excellent and the maximum service temperature for PPS is about 200°C, although it will withstand 350°C for short periods of time. While PPS will burn and char in the presence of a flame, it is self extinguishing and any smoke that does form is lower in toxicity compared with that given off by many polymers. There are some similarities between PPS and polysulfones, with PPS usually the less expensive option. Uses of PPS include chemically resistant coatings, chemical pumps, and electrical components (Furness 2008).

Liquid crystal polymers (LCPs) comprise a diverse family although most are based on polyesters and polyamides, whose molecules have a tendency to align themselves and remain in that alignment. In their molecular structure, LCPs do not fit into the conventional polymer categories of amorphous and semicrystalline, displaying a high degree of crystallinity in the melt phase, hence called liquid crystal. LCPs are essentially composed of long, rod-like molecules that align themselves in the direction of material flow. This alignment is maintained as solidification takes place; thus they are referred to as self reinforcing, however, this leads to anisotropic properties. Despite offering the best high temperature and fire resistance properties of all the thermoplastics, with certain grades able to operate at temperatures around 300°C, LCPs are relatively easy to process, although the higher the temperature resistance the more difficult the processing. The crystalline nature imparts excellent resistance to solvents, industrial chemicals, and UV and ionizing radiations (Furness 2008). The low melt viscosity allows them to be molded into intricate shapes with thin cross sections. Warpage control and low mold shrinkage are exceptional. LCPs have been used as chip carriers, sockets, connectors, bobbins, and relay cases (Sampson and Mattox 1997).

### *Thermosets*

Thermosets and thermoset materials are cross-linked polymers that are cured or set using heat or heat and pressure. That is, the covalent cross links are formed during curing, and the cross-linking reactions are used to form covalent bonds between different molecules resulting in a more rigid, chemically resistant and higher $T_g$ materials than a thermoplastic. The material is supplied to the molder in non-cross-linked form so it can be melted and formed into the desired shape. Covalent cross links are usually formed in response to either an added cross-linking chemical or by heat or both. Once this conversion occurs, the further addition of the heat does not cause the material to melt but can soften or degrade the material. When the dissociation energy of the primary covalent bonds is reached, the bonds of both the chains and cross links fail, and the polymer degrades into a variety of organic gases.

For most polymers, this occurs at around 250–350°C (Brown and Ulrich 2006), which indicates that thermosets generally have a higher resistance to heat than thermoplastics.

Thermoset materials are usually supplied in the form of resin and hardener, and the final polymerization occurs during curing, hardening, vulcanizing, or cross linking with or without the help of heating. Vulcanization is a thermosetting reaction that uses a cross-linked compound or catalyst. In rubber-like materials, vulcanization results in greatly increased strength, stability, and elasticity. Traditionally, sulfur is used as the vulcanizing agent for natural rubber. The curing process involves the generation of heat by exothermic reaction within the material, and this heat must be shunted away to prevent cracking and other deterioration. The material also shrinks up to 10% by volume during cure, thus having an impact on the configuration of the mold. The addition of fillers, pigments, and lubrications may also be used to control shrinkage (Fackler 2002). Thermosets and thermoset materials may contain filler materials such as powders or fibers to provide improved strength and/or stiffness. Fibers can be either chopped or wound, and commonly include glass, fiberglass, or cloth. Some products contain solid lubricant fillers such as graphite or molybdenum disulfide. Others contain aramid fibers, metal powders, or inorganic fillers with ceramics and silicates.

Many types of thermosets and thermoset materials are commercially available. Filled or reinforced products consist of resins and modifiers such as pigments, plasticizers, or chopped fibers. By contrast, unfilled resins, base polymers, and raw materials do not contain additives. Typically, raw materials are available as pellets, powders, granules, or liquids. Stock shapes such as bars, sheets, and film are also available. Molding compounds are designed for processes such as reaction injection molding and resin transfer molding. Composite materials consist of a matrix and a dispersed, fibrous or continuous second phase. Casting resins include a catalyst or hardener. Electrical resins and electronic-grade polymers and elastomers are used in potting or encapsulating compounds, conductive adhesives, and dielectric sealants. Specialized elastomers and rubber materials are used in optical and photonics applications and in mold and tooling fabrication. Monomers and intermediates are commonly available. Various properties of thermoset materials can be controlled by formulating the resin and the hardener/catalyst packages to include retarders, inhibitors, accelerators, adhesion promoters, and plasticizers that may adjust the viscosity, the time available between end-user mixing and initial curing, the working time available, the cure temperature, the toughness, corrosion resistance, etc. Many resin formulations are combustible and require fire retardant additives in the formulation if the finished product is to be used in locations where materials must be flame resistant.

The allyl based polymers have good dimensional stability, electrical insulation property, thermal resistance, and moisture resistance. Glass fiber filled diallyl phthalate can be laminated and most commonly used to make inserts in connectors, electronic components, housings, and to form insulating standoff posts, radomes, and ducts.

Epoxy materials cover a wide range of complex cross-linked polymers with a corresponding range of mechanical properties and of applications. The room-temperature curing varieties (as opposed to oven-temperature curing varieties) can be post-cured at higher temperature, have low shrinkage (about 2–3%), good chemical resistance, excellent adhesive and hand-laminating properties, provide a better matrix in terms of strength for fiber reinforced materials. The drawback is the higher cost, a need for an accurate resin/hardener ratio (epoxies use a hardener not a catalyst and the ratio will vary) when mixing and the need for skin protection when using the resins. In electronic packaging and thermal managing system, epoxies are used to encapsulate circuits, fabricate PCBs with glass fabrics and phenolic laminates, and form TIM as adhesives with thermally conductive fillers.

Phenolics are the reaction product of phenol and formaldehyde, and used for high temperature, relatively high mechanical strength, and low-cost applications. They have poor arc resistance, and are not elastic and may fracture upon bending or impact loading. Phenolics are used to fabricate knobs, handles, connectors, chip carriers, bobbins, and in laminate form are used to build small enclosures.

Polyesters are useful for their low-pressure processing conditions, low cost, good electrical properties, and high arc resistance. They are formed by the reaction of an organic acid with alcohol and then further reacted with a vinyl monomer and organic peroxide (Sampson and Mattox 1997). Polyesters provide moderate strength, are easily mixed and cure at room temperatures. They are the most common resins used in industrial applications but a high shrinkage (perhaps 8% by volume) during cure must be allowed for and although most of the shrinkage occurs during the initial cure period of 24 h or so some additional shrinkage will continue during the full cure period of perhaps 60–90 days which has a detrimental effect on the surface finish; in laminations the fabric weave shows through. Polyester moldings are usually in bobbins, terminal boards, housings, and connectors. Films are used as wire insulation, coil insulation, and protective covers. Pultrusions are used as bus supports and spaces. Radomes are often filament wound or compression molded (Sampson and Mattox 1997).

PIs exhibit low outgassing, good radiation resistance, and have excellent high and low temperature performance. However, they are sensitive to moisture and are degraded by organic acids and alkalis (Fackler 2002). PIs are extensively used as multilayer circuit board coatings, chip carriers, laminates, film insulation, flexible cable, flexible circuits, tape, tape wire wrap, sleeving, moldings, coil bobbins, wire enamel, fiber reinforcements, and laminates.

Polyurethanes are formed by reacting diidocyanate with glycol. They exhibit elastomeric properties and have tear resistance, abrasion resistance, and unusual toughness. However, polyurethanes are sensitive to chlorinated solvents, ketones, and strong acids and bases. The operating temperature is limited and should not exceed 120°C. They are usually used as paints, wire enamels, conformal coatings, embedding compounds and foam dielectric and thermal insulation. Some materials may be poured and used as encapsulants to protect circuits from the environment (Sampson and Mattox 1997; Fackler 2002).

Silicones are becoming more popular for their use as sealants, elastomers, adhesives, paints, conformal coatings, encapsulants and interface materials due to their thermal stability and low modulus which can protect the components within the microelectronic package. Silicones have an inherent large free volume within the amorphous structure due to the large bond lengths and angles between the repeating silicon and oxygen atoms that make up the majority of the polymeric structures. This, along with unusually weak intermolecular forces, gives silicone many unique properties such as thermal stability in extreme temperatures, $T_g$ typically less than $-115°C$, low modulus, low shrinkage (<1%), dielectric strength about 500 V/mil (0.001 in.) or 20 kV/mm, versatile usage configurations (optically clear, thermally conductive, electrically conductive, silica filled). They have a broad operation temperature range (−40 to 250°C), excellent arc resistance, are non-burning and water resistant when fully cured (Fackler 2002).

Molding compounds are made by combining a silicon resin with fillers, such as fused silica, carbon black, or titanium oxide. Fused silica enhances mechanical and rheological properties, while carbon black and titanium oxide are used as radiation screens. Silicon has relatively large CTE, which is about double that of most polymers, but the low modulus usually more than makes up for it (Brown and Ulrich 2006).

### *Elastomers*

Elastomers are usually vulcanized thermosets but may also be thermoplastic. The long polymer chains cross link during curing. The elasticity is derived from the ability of the long chains to reconfigure themselves to distribute an applied stress. The covalent cross linkages ensure that the elastomer will return to its original configuration when the stress is removed. As a result of this extreme flexibility, elastomers can reversibly extend from 5 to 700%, depending on the specific material. Without the cross linkages or with short, uneasily reconfigured chains, the applied stress would result in a permanent deformation. Temperature effects are also present in the demonstrated elasticity of a polymer. Elastomers that have cooled to a glassy or crystalline phase will have less mobile chains, and consequentially less elasticity, than those manipulated at temperatures higher than the glass transition temperature of the polymer. It is also possible for a polymer to exhibit elasticity that is not due to covalent cross-links, but instead for thermodynamic reasons. The properties of elastomers are employed as sealing gaskets and drive belts and other uses that involve bending and recovery of shape.

## Multimaterial Laminates

Multimaterial laminates usually indicate the lay up of fully isotropic laminate that exhibits a concomitant stiffness isotropy or quasihomogeneous anisotropic laminate that has multifunctional property or identical stiffness anisotropy in extension, shear, bending, and twisting. A laminate is a flat plate or curved shell consisting of two or more plies stacked and bonded as an integral component for multifunctional or structural applications. Each ply is a uniform-thickness layer of material. The layup of a laminate is generally tailored to match the functional or stiffness and strength requirements for loadings from various directions. The constituent material of a uniform-thickness ply may be homogeneous or heterogeneous (including porous material); isotropic or nonisotropic; honeycomb-like or otherwise mechanically formed; or of certain combination of the above. However, the multimaterial not only refers to the variation of the effective ply material property between plies, also include the variation of the constituent material property within a ply in some applications. Laminated plates and shells have found a wide range of applications in electronic packaging or aerospace structures where, multifunctional properties, high strength-to-weight and high stiffness-to-weight ratios are desired. Fiber-reinforced composite laminates such as graphite/epoxy and Kevlar/epoxy are used to combine with or to replace the conventional aluminum-, titanium-alloy structural components for weight reduction and other improvements (Wu and Wu 1996). This section will focus on the application of multimaterial laminates to electronic packaging and thermal managing system, which would provide special performance, such as tailored coefficient of expansion, relatively high heat conductivity, and proper specific heat, dielectric constant, and heat dissipation factor.

### *Multilayer Materials*

Multilayer materials can be designed from the atomic level to micro-scale and macro-scale. For instance, multilayers can be fabricated in layers atom-by-atom ranging in scale from atomic to microscopic, which exist as fully dense, fine grain, high interface concentration solids and that multilayer laminate component layers can be as thin as two atoms with a strength that nearly approaches theoretical limits. The potential applications for multilayer laminates include new manufacturing strategies, short-wavelength optics, high-performance capacitors for energy storage, industrial capacitors, integrated-circuit interconnects, tribological coatings, ultra-high strength materials, and coatings for high temperature applications such as gas turbine engines.

Metal multilayers are a unique way to manufacture reliable conductive members for microelectromechanical systems devices. These members are particularly suited for electrostatic switches. Unlike elemental Al which experiences curvature problems during processing, the residual stress in these beams can be controlled and calibrated during stress measurements to have repeatable shapes and stresses upon release.

There are many applications of materials where retention of a thin film of aqueous fluid is desirable. For example, the retention of an aqueous fluid layer is beneficial for lubrication of catheters, the retention of an aqueous fluid layer can reduce protein fouling on the surface of pacemakers and artificial vascular grafts, or the retention of an aqueous fluid layer can prevent the colonization of a surface by bacteria as they are unable to attach properly. In another aspect, the facile movement of an eyelid over a contact lens is important for the comfort of the wearer; this sliding motion is facilitated by the presence of a continuous layer of tear fluid on the contact lens, a layer which lubricates the tissue/lens interface.

Multilayer material processing technology has been used for fabrication of multilayer electronic circuit or packaging. The circuit has three or more conductive layers superposed with at least a portion of each adjacent set of layers overlapping, such that the layers are mechanically connected, and such that interconnect pads located on the various conductive layers are electrically connected, to transfer signals between the various layers. Adjacent layers are electrically insulated from one another, apart from the electrical connections between the interconnect pads (Casson et al. 1997). Conventional multilayers have been made by imaging, exposing, and etching various interlayers of circuitry and then bringing all layers together in alignment. The layers are then laminated together in plies using a suitable insulating adhesive. The multilayer product is then usually pressed (e.g., by means of a platen). Subsequent to pressing, a cure of the adhesive is often required. Through holes are then drilled to form barrels which are plated to create electrical contact between layers. Modifications have been made for a multilayer circuit board having a plurality of conductive layers, which has reliable interconnections between layers that are capable of withstanding thermal cycling. Furthermore, a need exists for a method of making such multilayer circuits which is less complex and which produces reliable interconnections between many conductive layers. For example, a conductive layer is a layer of conductive material generally disposed in a plane having a conductive pattern and usually having a plurality of contact pads. Such a layer is typically made of copper or another conductive metal and is capable of transmitting electronic signals between components and other conductive layers electrically connected to the conductive layer. Furthermor, an interconnecting adhesive layer is a layer of conductive adhesive material having a nonconductive adhesive with a plurality of deformable metallic particles dispersed substantially uniformly throughout (Casson et al. 1997).

Another example is multilayer ceramic (MLC) substrate, which used for interconnecting a semiconductor device or similar heat generating device to the next level of packaging such as a board. The MLC substrates provide wiring circuitry which provides for the transmission of signals and power to and from the semiconductor device. In addition, the typical MLC substrate may provide some level of cooling of the semiconductor device via transfer of heat from the semiconductor device through the MLC substrate. It is more usual to provide some external apparatus, such as a cap and heat sink, to remove heat from the semiconductor device. This external apparatus often has internal channels for the flow of a cooling fluid. Numerous proposals have been made for removing the heat generated by semiconductor devices by such external apparatus (Bezama et al. 1999).

Typical adhesive materials used for fabricating the mutimaterial laminates are titanate, zirconate, stannate, and oxide additives designed for use in ceramic capacitor dielectric formulations; organic binder materials and systems designed for coating and casting dielectric tapes and other film products; dielectric formulations for ceramic capacitors includes products compatible with precious metal and base metal electrode systems; electrode pastes for use in multilayer components as well as silver inductor pastes for high-frequency applications; and pastes designed especially for compatibility with the leading equipment available to the industry as well as with a variety of dielectric systems.

## *Metallic Laminate Materials*

Table 3.3 shows typical metallic laminate materials, such as copper–Invar–copper (Invar indicates nickel steel FeNi36) or copper–moly–copper laminate, commonly included in electronic packages. A copper–Invar–copper laminate may be present within a dielectric substrate. Exterior surfaces of the substrate may be circuitized such that a plated-through-hole passes through the copper–Invar–copper laminate and electrically couples the surface circuitizations. After a through hole is formed in the substrate, such as by laser drilling, and prior to a plating of the through hole to complete formation of the plated-through-hole, the exposed copper and Invar surfaces may be precleaned by immersing the substrate in an acid solution. Such acid precleaning removes oxides previously formed on the copper surfaces and cleans the Invar surface. A suitable acid for this purpose is, inter alia, a mixture of sulfuric acid and phosphoric acid. The copper–Invar–copper laminate in an acid solution results in formation of a galvanic cell that etches the Invar but does not etch the copper, thus forming pocket voids within the Invar. The pocket voids, if deep enough, may prevent the Invar from being copper plated in subsequent copper-plating steps. Even if the pocket voids do not impair subsequent copper plating, the structural integrity of the copper plating within the pocket voids will be weak and subject to rupture from thermal or mechanical stresses (Galasco et al. 2001).

After the acid precleaning, a substrate surface and the through hole are both plated with copper. The copper plating may be accomplished by first forming a thin

**Table 3.3** Typical properties of metallic laminates

| Metallic laminates | | Thermal conductivity at 0–100°C (W/m K) | | Longitudinal thermal expansion at 20–100°C (1/K × $10^6$) | Electrical conductivity (% IACS) | Density at 20°C (g/cm$^3$) |
|---|---|---|---|---|---|---|
| | | *X–Y* direction | *Z* direction | | | |
| Copper–Invar–copper | | | | | | |
| Volume | 5–90–5 | 52 | 15 | 1.5–3.5 | 10 | 8.18 |
| ratio | 12.5–75–12.5 | 110 | 19 | 2.4–5.6 | 25 | 8.33 |
| (%) | 20–60–20 | 167 | 20 | 3.6–6.8 | 40 | 8.43 |
| Copper–molybdenum–copper | | | | | | |
| Volume | 5–90–5 | 166 | 151 | 5.0–5.2 | 40 | 10.2 |
| ratio | 13–75–13 | 208 | 170 | 5.5–6.5 | 51 | 9.82 |
| (%) | 20–60–20 | 243 | 194 | 5.7–6.7 | 60.4 | 9.66 |

porous copper layer by electroless plating, or by coating with another type of conductive material such as conductive graphite or a conductive polymer, followed by forming a relatively thicker layer of copper by electroplating. The electroless plating of copper may be accomplished by any known process such as by first depositing on the surfaces to be plated an adsorbing material, such as a cationic surfactant, which serves as an attractant for the next material to be deposited, namely a seed material such as palladium. Following the palladium seeding, the substrate is electroless plated with copper.

After the electroless plating of copper and before the electroplating of copper, a cleaning step removes oxides from exposed copper surfaces by immersing the substrate in an acid solution such as a sulfuric acid solution. Inasmuch as the thin copper plating in a through hole is porous, acid may migrate through the pores of the thin copper plating and become trapped between the thin copper plating and the Invar surface covered by the thin copper plating. Subsequent electroplating of copper seals the trapped acid. As with the acid precleaning described previously, the Invar in contact with the sealed acid is subject to pocket void formation due to galvanic action. The pocket voids resulting from the cleaning step are typically larger and deeper than are the pocket voids resulting from the precleaning step, because the sealed acid will continuously contact the Invar for an indefinite period of time. Because of the pocket voids, the structural integrity of the copper plating that covers the pocket voids will be weak and subject to rupture from thermal or mechanical stresses (Galasco et al. 2001).

Copper–Invar–copper and copper–moly–copper are commonly used for heat sinks. The CTEs of theses laminates can be tailored by adjusting the relative thickness of the copper and constraining layers. However, their major limitations are high densities and their thermal conductivities may not high enough for some applications.

## Printed Circuit Board Materials

A variety of different materials have been used to fabricate PCBs. These materials can also be assembled in a variety of different ways potentially using multiple laminates, graded or composite materials and different plating finishes, such as Au, Ni, Sn, and Pb, and plated through via structures. Hybrid structures combining different materials have been explored and utilized, allowing a combination of properties to be used for different circuit areas.

PCB materials generally consist of an insulating support and outer layers of highly conductive material on one or both exterior surfaces. The insulating support is usually made with glass or ceramic fiber and resin laminates, and adhered with one or more layers of electrically resistive materials and then attached with highly conductive materials. The resistive materials can be chromium–antimony, chromium– manganese, chromium–phosphorus, chromium–selenium, chromium–tellurium, cobalt–antimony, cobalt–boron, cobalt–germanium cobalt–indium, cobalt–molybdenum, cobalt–phosphorus, cobalt–rhenium, cobalt–ruthenium, cobalt–tungsten, cobalt–vanadium, iron–vanadium, nickel–antimony, nickel–boron, nickel–chromium, nickel–germanium, nickel–indium, nickel–molybdenum, nickel–phosphorus, nickel–rhenium, nickel–vanadium, and palladium–molybdenum. The resistive materials usually have a maximum bulk resistivity greater than 100 Ω cm, as being plateable from aqueous solution to reproducibly yield adherent deposits capable of withstanding bonding to the insulating support without loss of physical integrity, as being nonradioactive, as having melting point and crystallographic phase transitions, if any, at temperatures greater than 400°C, as having a temperature coefficient of resistivity less than ±300 ppm from −65 to +125°C when properly deposited, as having a diffusion coefficient into alpha phase copper less than $2.89 \times 10^{-22}$ moles/cm$^2$/s, as having current versus voltage characteristics typical of presently available resistors, and, as having sufficient chemical resistance to withstand normal use conditions when properly protected by passivation, anodization, overplating, or coating with an organic or inorganic layer (Castonguay 1974).

The resistive material can deposited from a solution bath onto a conductive layer such as copper foil. The double layer foil is then laminated, resistive side at the interface, with one or more plies of fiberglass fabric preimpregnated with an appropriate formulation of curable organic resins. A layer of highly thermally conductive materials such as Al and Cu foils are usually used in the laminate construction to provide thermal dissipation for the moderation of electrical heating effects of resistors which will be formed on the laminate surface. The thermally conductive layer may be laminated to the side opposite the resistive cladding or within the several plies of preimpregnated reinforcement. By a photoetching process, the panel with the developed image is formed and then rinsed in water and dried. Finally, the conductive and resistive patterns are individually defined in appropriate electrical contact with each other (Castonguay 1974).

Other substrate material for PCBs can be a composite material composed of plastic, and ceramic and metal wires fixed in the composite material at given pitches, wherein the metal wires ensure electrical connection between the two surfaces of the substrate material. The substrate material possesses a good moldability and excellent insulation, tailored thermal expansion and abrasion resistance, by varying the kinds and compounding ratio of the plastic and the ceramic. In other words, the amounts of the plastic and the ceramic in the composite material can be determined appropriately depending upon the properties (e.g., insulation, thermal expansion, and abrasion resistance) and application purpose of the composite material. However, it is preferred that the ceramic particles, or the ceramic fiber is present in the composite material in an amount of 40–90% by volume because the resulting composite material can show low thermal expansion and, during curing, a small volume contraction. The ceramics used in the composite material can be alumina, zirconia, silicon nitride, or glass (e.g., silica glass), which are in the form of particles or fiber. As for the plastic, it can be a thermoplastic resin or a thermosetting resin–For the thermoplastic resin, it can be one of any resins, such as a polyvinyl chloride, a polyethylene or a polypropylene. These resins may be used in combination of two or more kinds. Similarly, the thermosetting resin can be a phenolic resin, an epoxy resin or a urea resin. These resins may also be used in combination of two or more kinds. When the composite material is made with a mixture of a plastic (e.g., an epoxy resin) and a ceramic (e.g., glass fiber or beads), no anisotropy in thermal expansion, excellent insulation, low thermal expansion, abrasion resistance, and high strength are expected. As to the kind of the metal wires used in the composite material, there is no particular restriction as long as the metal is conductive. However, the metal with high electrical conductivity is preferred because low electrical resistance is required for the fine metal wires. Because gold is too expensive, in practical application, any metal selected from copper, copper alloys, aluminum, or aluminum alloys is preferred. The most preferred alloy is beryllium copper because it has good electrical conductivity and has high Young's modulus and stiffness (this enables application of an appropriate tension when beryllium copper wires are stretched in a mold). Desirably, the plastic and the ceramic are each subjected to a coupling treatment and the composite material and the metal wires are bonded to each other by a coupling agent. This would make the resulting composite material have higher stability, higher bonding strength, and higher peeling resistance. An effective coupling agent used for the composite material can be one of silane based materials, such as vinyl, epoxy, methacryloxy, amino, chloropropyl, or mercapto. The primer made with a solution of such a silane material and water or an organic solvent can also be an effective coupling agent. In addition. titanium or aluminum based coupling agents have also been developed.(Suzuki et al. 2002).

For aluminum and aluminum alloy substrates, a low-temperature thick-film system has been developed to adapt it to ferritic and austenitic stainless steel, and glass substrates, which, unlike the common alumina substrate material, cannot usually be exposed to the standard high temperature of 850°C thick-film firing cycle. Such substrate materials are useful for several important applications: high-response piezoresistive thick-film sensors (steel), displays (glass), high power electronics (aluminum based materials), and heaters (steel and aluminum). Thick-film technology applied to piezoresistive force or pressure sensing typically uses alumina as a substrate material. However, alumina is not optimal for piezoresistive sensing applications as its elastic modulus is high and its strength rather low. Additionally, alumina is brittle and therefore ill-suited to harsh environments and its heat dissipation capabilities are limited. Aluminum alloys and stainless steels offer advantages in applications such as high-power electronics or high-range load cells due to their excellent thermal dissipation, mechanical sturdiness, easy packaging and, in the case of the Al–Si system, adjustable CTE. Metallic materials also offer advantages such as robustness and ease of fabrication. However, the high temperatures associated with commercial thick-film processing (850°C) are not compatible on the one hand with aluminum owing to its low melting point and on the other hand with high-strength steel, owing to degradation of mechanical properties of steel due to annealing or dimensional changes associated with martensitic transformation (which tend to destroy the thick-film layers). Additionally, standard thick-film materials are thermally matched to alumina, which has a rather low CTE of 7 ppm/K, whereas steels and aluminium alloys range from 11 to 24 ppm/K. Appropriate low-temperature thick-film systems (dielectrics, resistors and conductors) are have been developed and utilized, such as silver- and gold-based thick film materials systems for single and multilayer circuits.

## Interface Materials

Interfaces between electronic packaging materials or components have a significant impact on the thermal impedance of electronic systems and in practice they can be the dominant factor in achieving effective thermal transfer. TIMs are used to join an electronic device to the thermal transfer medium such as substrate, heat pipe, and heat sink, or join the thermal management components to each other. In some cases they may need to perform the tasks of attachment, stress/strain relief and thermal transfer (Young et al. 2006). For example, in layered components, one goal appears to be decreasing the number of the layers while at the same time increasing the functionality and durability of the remaining layers and surface/support materials. At same time, TIMs have been used to dissipate the increased heat flux. TIMs are critical to protect active semiconductor devices, such as microprocessors, from exceeding the operational temperature limit. They enable thermal bonding of the heat generating device (e.g., a silicon semiconductor) to a heat sink or a heat spreader (e.g., copper and/or aluminum components) without presenting an excessive thermal barrier. Different TIMs may also be used in the assembly of other components of the heat sink or the heat spreader stack that comprise the overall thermal impedance path. Formation of a small thermal barrier is an important property of a TIM. The thermal barrier can be described in terms of effective thermal conductivity through the TIM and is preferably as high as possible. The effective thermal conductivity of the TIM is primarily due to the interfacial heat transfer coefficient as well as the intrinsic bulk thermal conductivity of the TIM. A variety of other properties are also important for a TIM depending on the particular application, for example, an ability to accommodate or avoid thermal expansion stresses when joining two materials, an ability to form a mechanically sound joint that is stable during thermal cycling, a lack of sensitivity to moisture and temperature changes, manufacturing feasibility, and cost.

Several classes of materials are currently being used as TIMs, for example, thermal greases, thermal gels, adhesives, elastomers, thermal pads, and phase change materials. Thermal greases or phase change materials have lower thermal resistance than elastomer tape because of the ability to be spread in very thin layers and provide intimate thermal contact between adjacent surfaces. Typical thermal impedance values range between 0.05 and 1.6°C cm$^2$/W. However, a serious drawback of thermal grease is that thermal performance deteriorates significantly after thermal cycling, such as from −65 to 150°C, or after power cycling when used in VLSI (Very-large-scale integration) chips. The most common thermal greases use silicone oils as the carrier. It has also been found that the performance of these materials deteriorates when large deviations from surface planarity cause gaps to form between the mating surfaces in the electronic devices or when large gaps between mating surfaces are present for other reasons, such as manufacturing tolerances, etc. When the heat transferability of these materials breaks down, the performance of the electronic device in which they are used is adversely affected.

Although the foregoing TIMs are adequate for many current semiconductor devices, the increased performance of semiconductor devices in the near future will render the presently known TIMs inadequate. Specifically, the thermal conductivity of current nonmetallic TIMs generally does not exceed about 5 W/m K and is typically less than about 1 W/m K. However, TIMs that form thermal interfaces with effective thermal conductivities of about 50 W/m K or greater will be needed in the future. These needs include: (1) design and produce thermal interconnects and TIMs, layered materials, components, and products that meet customer specifications while minimizing the size of the device and number of layers; (2) produce more efficient and better designed materials, products and/or components with respect to the compatibility requirements of the material, component or finished product; (3) produce materials and layers that are more compatible with other layers, surfaces, and support materials at the interface of those materials; (4) develop reliable methods of producing desired thermal interconnect materials, TIMs and layered materials and components/products comprising contemplated thermal interface and layered materials; (5) develop materials that possess a high thermal conductivity and a high mechanical compliance; and (6) effectively reduce the number of production steps necessary for a package assembly, which in turn results in a lower cost of ownership over other conventional layered materials and processes.

One alternative to the foregoing nonmetallic TIMs is a solid metal sheet or preform made of a typical solder alloy. The metal TIMs ensure high thermal conductivity value (e.g., about 80 W/m K for an indium sheet). Metal TIMs may also exhibit a favorable solder or wetting behavior upon reflow which facilitates a low thermal interfacial resistance. During reflow, the solder and substrate are heated, the solder melts and wets by surface tension and/or local surface alloying. The interfaces consist of intermetallics or interdiffused metals with thermal properties that are frequently less desirable than those of the bulk TIM metal but much better than polymer based TIMs. In most cases, however, metallic TIMs must be subjected to reflow in order to form reliable thermal interfaces. Metallic TIMs, however, can fail in certain applications due to the relatively large differences of CTEs between the TIM and the semiconductor and/or heat sink components and the lack of compliance. For example, the TIM for bonding components of electronic devices, may comprise a solder component comprising a bonding component selected from the group consisting of In, In–Sn alloy, Au–Sn alloy, Bi alloy, or mixtures of them; and an additive component selected from among (1) a CTE modifying component having a CTE that is less than about 10 ppm/°C, (2) a thermal conductivity enhancement component having a thermal conductivity that is at least about 100 W/m K, and (3) the combinations above, and an intrinsic oxygen getter selected from the group consisting of rare earth metals, alkali metals, alkaline-earth metals, refractory metals, zinc or its alloys, and mixtures of them. Furthermore, some active solders can be used to wet metallic and nonmetallic surfaces without extrinsic fluxing. These active solders comprise a bonding component selected from the group consisting of In, Cu, Au, Sn, Pb, Sb, Ag, Bi, or their alloys, and mixtures of them, and an intrinsic oxygen getter selected from the group consisting of rare earth metals, alkali metals, alkaline-earth metals, refractory metals, zinc or its alloys, and mixtures of them. In addition, a multilayer solder preform for bonding components of electronic devices can be formed with a first solder preform layer comprising a solder metal bonding component and an additive selected from among thermal conductivity enhancement components, CTE modifying components, and mixtures of them. There is a second solder metal preform layer applied to the bottom surface of the first solder preform layer; and a third solder metal preform layer applied to the top surface of the first solder preform layer (Lewis et al. 2007).

One commercialized active solder series is developed by S-Bond, which enables wetting directly to metal and ceramic surfaces without the need for flux. This is achieved through the modification of conventional solders by the addition of titanium (0–5%) and/or rare earth elements. These active elements migrate to the interface and react with the opposing material surface to remove oxides and nitrides and transport them into the bulk of the solder as an inert material. This process can be conducted in an air environment but requires the application of a low-level mechanical shearing action to break a thin oxide (Sn–Ti) layer on the molten solder and initiate the reaction with the component interfaces. The level of shear is claimed to be small and can be delivered by brushing or scraping the surface, sliding the joining surfaces relative to one another, or by the application of high frequency vibration to the parts to be joined. Once the oxide layer has been disrupted, the bulk solder reacts very rapidly with the substrate surfaces to form either a metallurgical or atomic attraction/van der Waals bond. The S-Bond active solder is potentially an attractive product for electronic thermal management systems with two key features: (1) The ability to directly join combinations of high thermal conductivity materials with a reduced number of interface (thermal impedance) layers; (2) Clean (no-flux) processing. The latter is becoming increasingly important as package density increases, systems become more complex, and environments more severe. The process is relatively immature and consequently new applications will need to be fully tested to establish the performance of the joint and its impact on the application and product reliability. The need to precoat materials prior to soldering may limit some high-volume applications unless alternative metallization techniques (e.g., sputtering) can be developed (Young et al. 2006).

Consequently, use of a metallic solder as the TIM layer can result in improved thermal conductivity. However, in most cases, metallic TIM layers and processes for their application to the component surfaces have many drawbacks, including some of the aforementioned limitations. For example, achieving a durable intermetallic bond in the soldering process may require the reflow of the TIM, which may include heating a solid TIM preform located between two components. Heating at temperatures that potentially damage heat sensitive components may be required, and components having sufficiently different coefficients of thermal expansion may produce an unreliable intermetallic bond. Another impediment to achieve a reliable bond in the soldering process is that metal surfaces of the components are readily subject to oxidation. Accordingly, a chemical solder flux may be used in the soldering process. Use of a solder flux in soldering the TIM layer may result in permanent bonding of surfaces of each of the coupled components. A permanent attachment at both surfaces, however, makes subsequent disassembly of a final product difficult, and may be even more undesirable when using a TIM layer to thermally couple a heat dissipating device and a test vehicle (Hurley et al. 2005).

In addition, nanomaterials such as carbon nanotubes (CNTs) are increasingly finding applications in thermal management materials and are also being considered as potential interface and attachment techniques. The details will be discussed in the following chapters.

## Low Thermal Conductivity Materials

Low thermal conductivity materials can generally be classified into thermal insulation materials, relatively low thermal conductivity materials, and thermoelectric materials with low ratio of thermal conductivity over electrical conductivity.

Thermal insulation materials come in a range of forms; primarily they are low-density foams or fibrous materials that restrict conductive and convective heat transfer. Depending on the thermal environment, appropriate materials may be polymers, glasses, ceramics, or carbons. A range of these materials is used in aerospace applications to shield and safeguard electronic components; for protection of control equipment that tend to be bespoke designs. With the introduction of more electronics into demanding environments, for example, all-electric aircraft and jet engines, there will be requirements for even higher performance thermal insulation which may require new materials and approaches. For instance, as a space application the reflective insulation/heat radiator combination is a key approach to ensure that the electronics within satellites are maintained within their operating temperature range. For the Mars Pathfinder Rover, NASA used solid silica aerogel to insulate the electronics and restrict heat loss. Silica aerogel is the lowest density solid material and is a remarkable insulator because silica is a poor conductor and the materials structure restricts convective transfer; typical conductivity is 0.02 W/m K. Aerogels can be manufactured from a number of materials, and in the early 1990s carbon aerogels were developed. A major factor in preventing the more extensive use of aerogels has been their cost of production. However, take-up of the materials has been significant in thermal insulation applications such as building and pipelines, and more recently they have been used for automotive "under-the-hood" thermal management. With the ongoing cost reductions and the development of new materials and composites this class of super-insulator should become very attractive for electronics thermal shielding. Coatings may have a role, particularly in hybrid constructions. An oxide nanocomposites material may be applied as a liquid coating, exhibiting a total conductivity of 0.017 W/m K because the nano-sized structure presents tortuous thermal paths and consequently poor thermal transfer. The creation of a high interface density within a nanolaminate material has produced a very effective barrier to heat transfer and a material with a thermal conductivity three times smaller than conventional insulator. The heat flow is limited at the interface as the differences in lattice vibration between the materials restrict the transfer of energy across the interface. In addition, a new insulation material with the lowest thermal conductivity for a fully dense solid has been created by using a novel approach to produce thin films of tungsten diselenide. The unusual structure is crystalline in two directions but with a rotational disorder in the direction of low heat conduction. Though someway it is away from immediate application, understanding of the principles may point the way toward new methods for producing improved insulation (Young et al. 2006).

Relatively low thermal conductivity materials, such as carbon foams, can be processed to have high or low thermal conductivity. These materials have been developed as heat shields but are also finding applications in composites and lightweight structures. Variant carbon foams are supplied having thermal conductivity values in the range 0.3–0.06 W/m K with solid fractions in the range 39–1.4% respectively. These foams are produced using a low-cost technology so that these materials may realize higher levels of market penetration, albeit in applications other than mainstream electronics (Young et al. 2006).

The thermoelectric materials with low ratio of thermal conductivity over electrical conductivity have long been developed and will be explored to find more and more new materials. The details will be discussed in Chapter 11.

## Advanced Thermally Conductive Materials

Advanced materials have played an important role in electronic packaging and thermal managing systems. In response to critical needs, there have been revolutionary advances in thermal management materials, such as tailored CTE, low-density materials with thermal conductivities ranging between 400 and 1,700 W/m K. Some are low cost; others have the potential to be low cost in high-volume. Production applications include servers, laptops, PCBs, PCB cold plates/heat spreaders, cellular telephone base stations, hybrid electric vehicles, power modules, phased array antennas, TIMs, optoelectronic telecommunication packages, laser diode and LED packages, and plasma displays (Zweben 1998). As shown in Table 3.4, among the advanced materials, high performance composite materials provide great advantages over conventional materials for electronic packaging thermal control, including extremely high thermal conductivities (more than twice that of copper); low tailorable CTEs; weight savings up to 80%; extremely high strength and stiffness; low-cost, net-shape fabrication processes; and cost reductions as high as 65%. In addition, composites will provide even greater benefits with continuous development. Novel composite materials have been developed that provide unique combinations of properties that make them outstanding candidates for packaging and thermal management applications. These include polymer matrix composites (PMCs), metal matrix composites (MMCs), ceramic-matrix composites, and carbon/carbon composites (CCCs). PMCs, MMCs, and CCCs are particularly the key composites of interest for thermal management (Zweben 1998).

In the polymer matrix composites, polymers reinforced with continuous vapor grown carbon fibers (VGCFs) nanocarbon fiber offer the highest thermal conductivity of all commercial PMCs. The in-plane thermal conductivity and elastic modulus of this composite are much higher than the corresponding properties of aluminum, and its density is much lower. The in-plane CTE is negative due to that

**Table 3.4** Physical properties of advanced thermally conductive materials

| Base material (matrix) | Reinforcement | | Thermal conductivity (W/m K) | CTE $10^{-6}$/K | Electrical resistivity (Ω cm) | Density (g/cm$^3$) |
|---|---|---|---|---|---|---|
| | Type | Vol% | | | | |
| Diamond | – | 0 | 900–2,320 | 1.0–1.5 | $10^{13}$–$10^{16}$ | 3.52 |
| Carbon fiber | – | 0 | 100–500 | 4–5(L); −1(T) | 0.00155 | 1.81 |
| Carbon nanotubes | – | 0 | 6,600 | −1.0 | $10^{-4}$ | 2.0 |
| Vapor grown carbon fibers | – | 0 | 1,200–1,950 | −1.0 | $10^{-4}$ | 1.8–2.0 |
| Carbon | VGCF | 70 | 910 | – | – | 1.8–2 |
| Epoxy | Carbon fiber | 60 | 310 (*X*, *Y*) | −1.1 | – | 1.85 |
| Epoxy | VGCF | 73 | 660 | – | – | 1.87 |
| SiC | VGCF | 20 | 310 | – | – | 2.9 |
| SiC | Diamond | | 500 | – | – | 3.1–3.3 |
| Copper | Diamond | 50 | 420 | 5.5 | – | 5.35 |
| Copper | Carbon fiber | 28 | 290 (*X*, *Y*) | 6.5 | – | 7.2 |
| Copper | CNTs | 50 | 1,024 | – | – | 5.45 |
| Copper | VGCNF | 50 | 840 | 5.5 | – | 5.45 |
| Copper | Carbon particles | 50 | 150–300 | 8.0–9.0 | – | 5.9 |
| Aluminum | VGCF | 37 | 642 | 5.0 | – | 2.44 |
| Aluminum | Carbon whisker | 40 | 230 | 4.0 | – | 2.40 |
| Aluminum | SiC particles | 60 | 180 | 9.0–10 | – | 2.9 |
| Aluminum | Graphite particles | 50–70 | 190–270 | 6.5–9.5 | – | 2.2–2.0 |

carbon fibers generally have negative CTEs. Although the CTE of this composite does not fall in the desired range of 4–7 ppm/K, it is much closer than those of aluminum and copper. Although continuous fibers offer the best properties, PMCs reinforced with discontinuous fibers can be used with relatively inexpensive net-shape processes such as injection molding (Zweben 1998).

MMCs reinforced with nanofibers such as VGCFs lead to tailorable CTE, low density, and high thermal conductivity. This makes these composites an optimal material for heat sinks in high power electronic devices working at high temperatures and high power densities, such as those based on new GaN semiconductors. Traditionally, there are two methods for the production of these VGCFs named as "fix bed" and "flying seed". The former is based on the growth from catalytic particles and is used to obtain high quality of VGCFs. The main disadvantage is that leads to low production yield. The latter is based on the continuous feed of catalytic seeds in a furnace and subsequently collection of the fibers at the bottom furnace exit. It is being used for mass production of novel carbon nanofibers (VGCNFs), typically about 10 μm in length and diameter in the 30–150 nm range. Although CNTs shows considerably higher conductivity (higher than 6,000 W/m K at room temperature is reported based on theoretical calculations), VGCNFs provide much better availability and performance-cost ratio for industrial production of electronic packaging components. For instance, the VGCNFs can be coated with copper and then consolidated into dense product by semisolid state hot pressing (Barcena et al. 2002). Other MMCs with high thermal conductivities are diamond reinforced composites and the composites made by infiltrating porous carbon/carbon composites with aluminum or copper. For materials with highly oriented fibers, such as ThermalGraph, axial thermal conductivities are as high as 890 W/m K (Zweben 1998).

CCCs are also very attractive thermal management materials for use in weight-sensitive applications due to their low specific gravity and relative high thermal conductivity. For example, CCCs made from P120 carbon fibers, which have a nominal axial thermal conductivity of about 600 W/m K, are reported to have an in-plane isotropic thermal conductivity of 250 W/m K, with a specific gravity of only 1.85.

Some promising reinforcements include SiC particles with much higher thermal conductivities (over 400 W/m K has been measured for high-purity single crystals), and diamond, aluminum nitride, and cubic boron nitride particles. Commercial pitch-based and VGCFs can approach a thermal conductivity of 2,000 W/m K. Commercial diamond and diamond-coated carbon fibers can also provide the thermal conductivities approaching 2,000 W/m K. Reinforcement coatings to promote adhesion, reduce reinforcement matrix reactions, and improve thermal conductivity. With these reinforcements, continuous development of polymer-, metal-, and carbon-matrix material systems would be expected to tailor for use in thermal management applications. The development efforts are likely to result in improved processes for the production of net-shape components. This will eliminate machining costs and make composites more competitive with traditional materials (Zweben 1998).

## References

Ashby MF, Evans A, Fleck NA, Gibson LJ, Hutchinson JW, Wadley HNG (2000) Metal foams: a design guide. Butterworth-Heinemann, Boston.

Barcena J, Maudes J, Coleto J, Obieta I (2002) Novel copper/carbon nanofibers composites for high thermal conductivity electronic packaging. http://escies.org/GetFile?rsrcid=1691. Accessed 9 March 2010.

Bezama RJ, Casey JA, Pavelka JB, Pomerantz GA (1999) Method of forming a multilayer electronic packaging substrate with integral cooling channels. US Patent 5870823.

Brown WD, Ulrich R (2006) Materials for microelectronic packaging. In: Ulrich RK, Brown WD (eds) Advanced electronic packaging, 2nd edn. Wiley, Hoboken.

Casson KL et al (1997) Multilayer electronic circuit having a conductive adhesive. US Patent 5688584.

Castonguay RN (1974) Printed circuit board material incorporating binary alloys. US Patent 3857683.

Fackler WC (2002) Materials in electronic packaging. In: Kutz M (ed) Hand book of materials selection. Wiley, Hoboken.

Ferro (2008) Electronic material systems. http://www.ferro.com/non-cms/EMS/EPM/GLASSES/Product-Selection-and-Application.pdf. Accessed 3 March 2010.

Furness J (2008) Thermoplastics-material glossary. http://www.azom.com/details.asp?ArticleID=506. Accessed 6 March 2010.

Galasco RT, Lehman LP, Magnuson RH, Topa RD (2001) Biased acid cleaning of a copper--invar–copper laminate. US Patent 6179990.

Grensing FC (1992) Method of making beryllium-beryllium oxide composites. US Patent 5124119.

Haack DP, Butcher KR, Kim T, Lu TJ (2001) Novel lightweight metal foam heat exchangers. http://www.porvairadvancedmaterials.com/papers/Mech%20Engineer%20Congress%20Paper.pdf. Accessed 8 March 2010.

Hunt ML, Tien CL (1988) Effects of thermal dispersion on forced convection in fibrous media. Int J Heat Mass Transf **31**: 301–309.

Hurley E, Rumer C, Christner R, Renfro T (2005) Metallic solder thermal interface material layer and application of the same. US Patent 0211752.

Lewis BG et al (2007) Thermal interface material and solder perform. US Patent 7187083.

Sampson RN, Mattox DM (1997) Materials for electronic packaging. In: Harper CA (ed) Electronics packaging & interconnection handbook, 2nd edn. McGraw-Hill, New York.

Suzuki T, Otagiri T, Kawai S, Ishikawa S (2002) Printed circuit board material and method of manufacturing board material and intermediate block body for board material. US Patent 6379781.

Taylor DA (2001) Advanced ceramics-the evolution, classification, properties, production, firing, finishing and design of advanced ceramics. Mater Aust 33(1): 20–22.

Wu K-M, Wu TJ (1996) Multimaterial fully isotropic laminates and multimaterial quasi-homogeneous anisotropic laminates. US Patent 5532040.

Young R et al (2006) Developments and trends in thermal management technologies – a mission to the USA. www.lboro.ac.uk/research/iemrc/documents/.../CB2007.pdf. Accessed 26 February 2010.

Zweben C (1998) Advances in composite materials for thermal management in electronic packaging. JOM. http://findarticles.com/p/articles/mi_qa5348/is_199806/ai_n21422924. Accessed 16 March 2010.

# Chapter 4
# Monolithic Carbonaceous Materials and Carbon Matrix Composites

**Abstract** The carbon materials used for electronic packaging and its thermal management system are usually in the form of monolithic carbonaceous materials, such as graphite or diamond, as well as carbon matrix composites, which may be combined with metals or other materials to give the complex materials that are easier to process into manufactured components. The graphitic materials, for instance, are lower density than metallic materials such as copper or aluminum, and can offer higher thermal conductivities. Furthermore, the carbonaceous materials can also be processed to form low-density thermal insulators to protect electronics from heat sources, or to shield parts of the assembly from excessive temperature rising, for example, in laptop computers and telephone handsets. Carbon matrix composites have combined properties including high thermal conductivity, high stiffness, and low coefficient of thermal expansion (CTE). This chapter will give a brief review about carbon materials for electronic packaging thermal management applications, including natural and industrial graphite, pyrolytic graphite, graphite/carbon foams, carbon fibers, diamond, carbon nanotube and nanofiber, graphene and buckypaper, as well as carbon matrix composites.

## Introduction

Carbon is one of the most abundant elements on Earth, and plays a dominant role in the chemistry of life. Carbon has four electrons in its outer valence shell. Because this energy shell can hold eight electrons, each carbon atom can share electrons with up to four different atoms. Carbon atoms can attach, bond, or combine with other elements as well as with itself in many different ways to create different materials with varying properties.

As shown in Figure 4.1, carbon alone forms the differing arrangements of atoms in the different forms of carbon: diamond, graphite, amorphous carbon, fullerene, and nanotube, creating materials with very different properties. Diamond, the hardest known natural material, has a very strong tetrahedron structure; each atom is tightly bonded to four other carbon atoms. In contrast, each carbon atom in graphite is strongly bonded to three carbon atoms in the same plane and only

X.C. Tong, *Advanced Materials for Thermal Management of Electronic Packaging*, Springer Series in Advanced Microelectronics 30,
DOI 10.1007/978-1-4419-7759-5_4, 

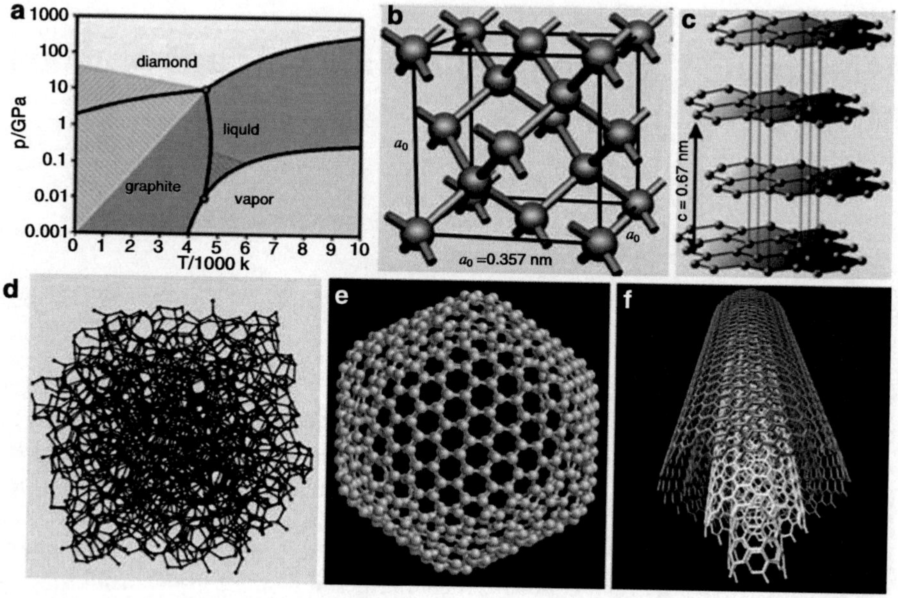


**Fig. 4.1** Phase diagram (**a**) and allotropes of carbon: diamond (**b**), graphite (**c**), amorphous carbon (**d**), fullerene (**e**), and carbon nanotube (**f**)

weakly bonded to carbon atoms in other planes. The carbon atoms in the plane are arranged in the shape of hexagons, called graphene sheets, while the weak bonding out of plane allows the sheets to slide easily past each other. Fullerenes are nanometer-sized soccer ball-shaped molecules of carbon atoms, which consist of hexagons and pentagons that form a spherical shape. Carbon nanotube looks like a single graphene sheet rolled up into a cylinder. If the sheet is rolled up in different ways, different patterns can be obtained along the cross section of the tube, such as armchair, zig–zag, chiral, and in-plane sheet or bukypaper. Carbon nanotube can also be described as single-walled carbon nanotube (SWNT), resembling a single graphene sheet rolled up, or multi-walled carbon nanotube (MWNT), like several SWNTs nested inside each other.

The bonds in diamond, within the plane of graphite and in fullerence molecules, nanotubes and buckypapers, are the strongest covalent bonds that have been found, which leads to some exceptional intrinsic properties, such as (1) the highest room temperature thermal conductivity in diamond and within the covalent bond plane of graphite, fullerence, nanotube, and buckypaper; (2) exceptional thermal stability of the structure in graphite; and (3) the greatest elastic modulus in diamond, within the graphite plane, and in SWNTs. Typical physical properties of some carbonaceous materials are shown in Table 4.1.

Graphite can be made with a very wide range of properties and degree of anisotropy depending on the selection of raw materials and processing. For example, pyrolytic graphite, formed from the gas phase pyrolysis of hydrocarbons can

**Table 4.1** Typical physical properties of some carbonaceous materials

| Carbon materials | Thermal conductivity (W/m K) | | Coefficient of thermal expansion (CTE) $10^{-6}$/K | Electrical resistivity (Ohm cm) | Density (g/cm$^3$) |
|---|---|---|---|---|---|
| | Through thickness | In plane | | | |
| Single graphite crystal | 10 | ~2,000 | 27 (L); −0.5 (T) | 0.004 (L); 0.06 (T) | 2.25 |
| Natural graphite sheet | 3.0–10 | 140–500 | 50 (L); −0.5 (T) | – | 1.1–1.7 |
| Natural graphite/epoxy laminate | 6.5 | 370 | 50 (L); −0.5 (T) | – | 1.94 |
| Pyrolytic graphite (PG) | 3.5 | 300–400 | 6.5 (L); −0.5 (T) | 0.0005 (L); 0.5 (T) | 2.18–2.22 |
| Annealed pyrolytic graphite | 8 | 1,700 | 20(L); −1 (T) | 0.000476 (L); 0.2 (T) | 2.24–2.25 |
| Natural diamond | – | 900–2,320 | 1.0–1.5 | $10^{13}$ – $10^{16}$ | 3.52 |
| Sintered polycrystalline diamond | – | 500 | – | – | 4.1 |
| Chemical vapor deposition (CVD) diamond plate/film | >2,000 | 600–1,600 | 1.0–2.0 | – | 3.51 |
| Polyacrylonitrile (PAN)-based continuous carbon fiber | – | 160 | 4–5 (L); −1(T) | 0.00155 | 1.81 |
| Pitch based continuous cabon fiber | – | 1100 | – | – | – |
| Pitch based discontinous cabon fiber | – | 800 | – | – | – |
| 2D Graphene | – | 4,840–5,300 | – | $10^{-6}$ | 2.0 |
| Single-walled carbon nanotubes (SWCNTs) | – | 6,600 | −1.0 | $10^{-4}$ | 2.0 |
| Graphite nanoplatelet flake | – | 2,000 | – | – | – |
| Vapor grown carbon fibers (VGCFs) | – | 1,200–1,950 | −1.0 | $10^{-4}$ | 1.8–2.0 |
| 55 vol% K1100 carbon–carbon Fiber ratio = warp($x$)/fill ($y$) = 6:1 | 50($z$) | 700($x$); 55($y$) | – | – | 1.8 |
| 55 vol% K1100 carbon–carbon Fiber ratio = warp($x$)/fill ($y$) = 1:1 | 50($z$) | 450($x$); 450($y$) | – | – | 1.8 |
| 50 vol% K321 carbon–carbon Fiber ratio = warp($x$)/fill ($y$) = 4:1 | 45($z$) | 368($x$); 97($y$) | – | – | 1.75 |
| 50 vol% K321 carbon–carbon Fiber ratio = warp($x$)/fill ($y$) = 1:1 | 32($z$) | 201($x$); 200($y$) | – | – | 1.75 |

exhibit thermal conductivity in the basal plane of up to 1,700 W/m K, and 10 W/m K in perpendicular to the basal plane. Pitch-based graphite fibers can achieve thermal conductivity parallel to the fiber direction of ~1,000 W/m K. At the other extreme, polycrystalline graphite materials can be manufactured with almost isotropic properties, the thermal conductivity in this case being a rather uninteresting 100 W/m K. Graphite can be spun or extruded to form graphite filaments or fibers can be extruded or spun to form roving mat or prepreg for secondary processing or composite material fabrication. Carbon/graphite foam derived from a pitch precursor, on the other hand, can be considered an interconnected network of graphitic ligaments. Because of the continuous graphitic network, the foam-reinforced composites will display higher isotropic thermal conductivities than carbon fiber reinforced composites. Furthermore, the lack of interlaminar regions, which develop in traditional prepregged carbon fiber reinforced composites, results in enhanced mechanical properties such as shear strength and fracture toughness. The use of carbon fibers has evolved from structural reinforcement to a thermal management material, with the emphasis in applications such as high-density electronic modules, communication satellites, and automotive systems. The relatively low-cost alternative of carbon fibers is carbon foam. The carbon foam and its composites have exhibited relatively high thermal conductivity, low weight, and low CTE, which are attractive for thermal management of electronic packaging.

Unlike most electrical insulators, diamond is a good thermal conductor because of the strong covalent bonding within the crystal. In contrast to metals, where heat is conducted by electrons, lattice vibrations are responsible for diamond's high thermal conductivity. Monocrystalline synthetic diamond enriched in $^{12}C$ isotope (99.9%) has the highest thermal conductivity of any known solid at room temperature: >2,000 W/m K, five times more than copper. Because diamond has such high thermal conductance, it is already used in semiconductor manufacturing to prevent silicon and other semiconducting materials from overheating. At lower temperatures conductivity becomes even better as its Fermi electrons can match the phononic normal transport mode near the Debye point, and transport heat more swiftly, to reach ~80,000 W/m K at 100 K ($^{12}C$ enriched diamond).

The nanotube's thermal conductivity is predicted to be 10 times higher than silver. Unlike metals, which conduct heat by moving electrons, carbon nanotubes (CNTs) conduct heat by wiggling the bonds between the carbon atoms themselves. The thermal conductivity of CNTs, however, is dependent on the temperature and the large phonon mean free paths. The thermal conductivity for a single rope nanotube at room temperature could vary between 1,800 and 6,000 W/m K.

In addition, high thermal conductivity of carbon–carbon (C–C) composites has made them good candidates for thermal management of electronic packaging. Additionally, C–C composites have very low thermal expansion coefficients, making them dimensionally stable at a wide range of temperatures. C–C composites retain mechanical properties even at temperatures (in nonoxidizing atmospheres) above 2,000°C. They are also highly resistant to thermal shock or fracture due to rapid and extreme changes in temperature. The material properties of a C–C composite vary depending on the fiber fraction, fiber type selected, textile weave type and similar factors, and the individual properties of the fibers and matrix material. Fiber properties depend on precursor material, production process, degree of graphitization and orientation, etc. The tensioning step in fiber formation is critical in making a fiber (and therefore a composite) with any useful strength at all. Matrix precursor material and manufacturing method have a significant impact on composite strength. Sufficient and uniform densification is necessary for a strong composite. Furthermore, the elastic modulus is very high, from 15 to 20 GPa for composites made with a 3-dimensional (3-D) fiber felt to 150–200 GPa for those made with unidirectional fiber sheet. Other properties include low weight, high abrasion resistance, high electrical conductivity, low hygroscopicity, non-brittle failure, and resistance to biological rejection and chemical corrosion. C–C composites are very workable, and can be formed into complex shapes.

In noncomposite form, up to now, only chemical vapor deposition (CVD) diamond grown in thin sheets and highly oriented pyrolytic graphite (HOPG) have been offered practical possibilities for passive thermal management although extensive carbon materials are developed.

## Natural and Industrial Graphite

Natural graphite is a mineral consisting of graphitic carbon. It varies considerably in crystallinity, and is subdivided into three types of material: amorphous, flake, and high crystalline. Amorphous graphite is still crystalline, but the least graphitic of the natural graphite with the graphite content from 25 to 85%. Amorphous graphite is found as minute particles in beds of mesomorphic rocks such as coal, slate, or shale deposits. The graphite content ranges from 25 to 85% dependent on the geological conditions. Flake graphite is found in metamorphic rocks uniformly distributed through the body of the ore or in concentrated lens-shaped pockets with carbon concentrations of 5 and 40%. Crystalline vein graphite is believed to originate from crude oil deposits that through time, temperature, and pressure have converted to graphite.

Natural graphite consists of up to 98% crystalline graphite, is very soft, and is an electrical and thermal conductor. Natural graphite does not dissolve in acids or bases and is nonmagnetic under normal conditions. The pronounced layered structure of graphite leads to a clear anisotropy of its physical characteristics. Electrical and thermal conductivity and thermal coefficient of expansion longitudinally or vertically relative to the direction of the layers differ in size. Natural graphite can be directly manufactured into versatile products for thermal management of electronic packaging, such as thermal interface sheet materials, heat spreader, and heat sink components can be produced from the same raw materials. These materials retain the thermal anisotropy features of single crystal graphite, with control of thermal anisotropy being possible by manipulation of the processing. The material is attractive as a thermal interface material because of its combination of good surface conformability and high thermal conductivity. The high in-plane thermal conductivity results in spreading and evening out of the hot spots resulting from uneven loading, surface distortion, and uneven heat distribution on the die surface. Meanwhile, application of insulating plastic films on the surface of the graphite and edge protection schemes, such as natural graphite/epoxy laminates, can be used to eliminate any concerns of flaking of electrically conductive graphite particles (Smalc et al. 2005).

Heat spreaders can be made from natural graphite sheet materials. These spreaders take advantage of the anisotropic thermal properties of natural graphite. Natural graphite exhibits a high thermal conductivity in the plane of the sheet combined with a much lower thermal conductivity through the thickness of the sheet. For example, natural graphite material exhibits a thermal conductivity ranging from 140 to 500 W/m K in in-plane (parallel to the layer planes) and from 3 to 10 W/m K through thickness (perpendicular to the planes). As a result, a natural graphite sheet can function as both a heat spreader and an insulator and can be used to eliminate localized hot spots in electronic components. By comparison, conventional isotropic materials used for spreading heat, such as AA1100 series aluminum and C11000 series copper, have a thermal conductivity of 200 and 387 W/m K in all three directions. Natural graphite has a density ranging from 1.1 to 1.7 g/cm$^3$ compared with 2.7 and 8.89 g/cm$^3$ for aluminum and copper respectively. Also, because of their excellent flexibility, natural graphite materials are able to conform well to surfaces under low contact pressures. This combination of properties makes natural graphite a potential substitute for aluminum and copper materials as heat spreaders. In some cases, a natural graphite heat spreader can also replace a conventional thermal management system consisting of a heat sink and cooling fan (Smalc et al. 2005).

Graphites may be characterized as laminated structures of carbon, that is, structures consisting of superposed layers or laminae of carbon atoms joined together by weak van der Waals forces. In considering the graphite structure, two axes or directions are usually noted, to wit, the "*c*" axis or direction and the "*a*" axes or directions. For simplicity, the "*c*" axis or direction may be considered as the direction perpendicular to the carbon layers. The "*a*" axes or directions may be considered as the directions parallel to the carbon layers or the directions perpendicular to the "*c*" direction. The graphites suitable for manufacturing flexible graphite sheets possess a very high degree of orientation.

The bonding forces holding the parallel layers of carbon atoms together are only weak van der Waals forces. Natural graphites can be chemically treated so that the spacing between the superposed carbon layers or laminae can be appreciably opened up so as to provide a marked expansion in the direction perpendicular to the layers, that is, in the "*c*" direction, and thus form an expanded or intumesced graphite structure in which the laminar character of the carbon layers is substantially retained.

Graphite flake which has been chemically or thermally expanded, and more particularly expanded so as to have a final thickness or "*c*" direction dimension which is as much as about 80 or more times the original "*c*" direction dimension, can be formed without the use of a binder into cohesive or integrated sheets of expanded graphite, e.g., webs, papers, strips, tapes, typically referred to as flexible graphite. The formation of graphite particles which have been expanded to have a final thickness or "c" dimension which is as much as about 80 times or more the original "c" direction dimension into integrated flexible sheets by compression, without the use of any binding material, is believed to be possible due to the mechanical interlocking, or cohesion, which is achieved between the voluminously expanded graphite particles.

In addition to flexibility, the sheet material has also been found to possess a high degree of anisotropy to thermal and electrical conductivity and fluid diffusion, somewhat less, but comparable with the natural graphite starting material due to orientation of the expanded graphite particles substantially parallel to the opposed faces of the sheet resulting from very high compression, e.g., roll processing. Sheet material thus produced has excellent flexibility, good strength, and a very high degree or orientation. There is a need for processing that more fully takes advantage of these properties.

Briefly, the process of producing flexible, binderless anisotropic graphite sheet material, e.g., web, paper, strip, tape, foil, mat, or the like, comprises compressing or compacting under a predetermined load and in the absence of a binder, expanded graphite particles which have a "c" direction dimension which is as much as about 80 or more times that of the original particles so as to form a substantially flat, flexible, integrated graphite sheet. The expanded graphite particles that generally are worm-like or vermiform in appearance will, once compressed, maintain the compression set and alignment with the opposed major surfaces of the sheet. Properties of the sheets may be altered by coatings and/or the addition of binders or additives prior to the compression step. The density and thickness of the sheet material can be varied by controlling the degree of compression.

Lower densities are advantageous where surface detail requires embossing or molding, and lower densities aid in achieving good detail. However, higher in-plane strength, thermal conductivity and electrical conductivity are generally favored by more dense sheets. Typically, the density of the sheet material will be within the range of from about 0.04 to about 1.4 g/cm$^3$.

Flexible graphite sheet material typically exhibits an appreciable degree of anisotropy due to the alignment of graphite particles parallel to the major opposed, parallel surfaces of the sheet, with the degree of anisotropy increasing upon roll pressing of the sheet material to increased density. In roll-pressed anisotropic sheet material, the thickness, i.e., the direction perpendicular to the opposed, parallel sheet surfaces comprises the "c" direction and the directions ranging along the length and width, i.e., along or parallel to the opposed, major surfaces comprises the "a" directions and the thermal, electrical and fluid diffusion properties of the sheet are very different, by orders of magnitude typically, for the "c" and "a" directions.

Graphite/epoxy laminate has been used for fabrication of heat sink. As shown in Table 4.1, the graphite/epoxy laminate exhibits markedly different properties in-plane compared with out-of-plane (through-thickness). In the in-plane direction, the properties are dominated primarily by the in-plane properties of the graphite; while in the through-thickness direction, the properties are dominated by the epoxy and out-of-plane properties of the graphite. The in-plane thermal conductivity of the material is ~370 W/m K, which is 77% higher than aluminum and comparable with copper. The through-thickness value of thermal conductivity is low (~6.5 W/m K) for the graphite/epoxy material making it uniquely thermally anisotropic. The material anisotropy is also reflected in the difference in thermal expansion coefficient in the in-plane and through-thickness directions. In the in-plane direction, the CTE is slightly negative, while in the through-thickness direction it is ~50 × $10^{-6}$ m/m/°C, measured over the temperature range of 30.0–100.0°C. The material is an electrical conductor, although its electrical conductivity, unlike its thermal conductivity, is two orders of magnitude lower than aluminum and copper alloys. The material is not as strong as aluminum and copper alloys. The material is also relatively soft and will dent if dropped on an edge. The high in-plane thermal conductivity and low density of the graphite/epoxy material make it an ideal material for the fins of a heat sink where low through-thickness thermal conductivity is of little consequence. Graphite/epoxy fins can provide the same thermal performance as copper at only ~22% of the weight. In the production of a traditional heat sink, however, thought must be given to the manufacture of not only the fins but also the base. The base usually needs to spread the heat in all three directions, depending on the dimensions of the heat source relative to the heat sink footprint. The low through-thickness thermal conductivity of the graphite/epoxy material is a disadvantage in the manufacture of a heat sink base. Despite this disadvantage, effective heat sinks can still be manufactured from graphite/epoxy materials by appropriate selection of the high thermal conductivity orientations in the base. This is possible to make because the fins and base can be made independently from each other and joined at some later stage of the process (Norley and Chen 2002).

Graphite based thermal management components include heat sinks, heat pipes and heat spreaders. All offer thermal conductivity equivalent to or better than copper or aluminum, but are a fraction of the weight of those materials, and provide significantly greater design flexibility. Graphite-based thermal management products take advantage of the highly directional properties of graphite to move heat away from sensitive components. Compared with typical aluminum alloys used for heat management, graphite components can exhibit up to 300% higher thermal conductivity, with values comparable with copper (~400 watts per meter degree Kelvin, i.e., W/m K) or greater attainable. Furthermore, aluminum and copper are isotropic, making it impossible to channel the heat in a preferred direction.

Lamination, molding, and embossing methods can be utilized to produce a variety of different component forms of the graphite products. In addition, due to its high specific surface area (e.g., 15 $m^2$/g), the natural graphite product with proper geometry can be made with very high electromagnetic interference shielding effectiveness (up to 130 dB) over the frequency range of 1–2 GHz.

## Pyrolytic Graphite

Pyrolytic graphite (PG) is a unique form of graphite manufactured at high temperatures by CVD from hydrocarbon gases on a mandrel surface. The CVD reactions are based on the thermal decomposition or pyrolysis of hydrocarbons. The most common precursor is methane ($CH_4$), which is generally pyrolyzed at 1,100°C or above, under a pressure ranging from 0.1 to 100 MPa (0.001–1 atm). The reaction can be formulized as follows (Pierson 2008):

$$CH_4 \rightarrow C + 2H_2. \tag{4.1}$$

Other common precursors are ethylene ($C_2H_6$), acetylene ($C_2H_2$), and propylene ($C_3H_6$). Acetylene can also be decomposed at lower temperature (300–750°C) and at pressure up to 1 atm, in the presence of a nickel catalyst. The propylene decomposes in the 1,000–1,400°C temperature range at around 13 MPa (Pierson 2008).

The result can be an ultrapure coating or oriented graphite sheet which is near theoretical density and extremely anisotropic. This anisotropy results from the highly oriented layers of hexagonal structure. Therefore, PG exhibits thermal conductivity consistent with the best conductors in the horizontal plane and lower than alumina plate in the vertical plane (Table 4.1). Mechanical, thermal, and electrical properties are generally far superior to conventional graphite. PG is a nonporous and self-lubricating, light-weight material, showing excellent thermal shock resistance. It can be easily machined with normal metalworking tools. The engineered flexible PG sheets can be used for thermal interface material, thermal vias, and other thermal managing components of electronic packaging.

After annealing under pressure at temperatures about 3,000°C, the annealed pyrolytic graphite transforms into the HOPG. Figure 4.2 shows the major difference between the as-deposited PG and HOPG. The highly aligned crystalline graphite HOPG has an in-plane thermal conductivity of 1,700 W/m K, and relatively high strength parallel to the hexagonal layer lattice or basal plane. This makes HOPGs

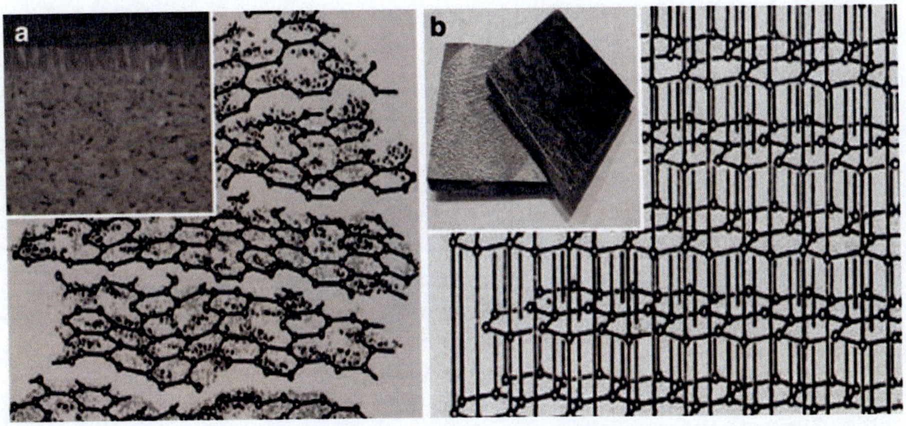


**Fig. 4.2** Pyrolytic graphite (**a**) as deposited: hexagonally ordered 2-D crystal structure, third dimension is disordered, typical density 2.3 g/cm$^3$, thermal conductivity 200–400 W/m K; (**b**) oriented with heat treatment: ordered 3-D crystal structure with additional ordering of 2-D structure, typical density 2.26 g/cm$^3$, thermal conductivity >1,500 W/m K

the most sophisticated large-area thermal management materials in premium passive heat transfer applications. For instance, HOPG can offer more than five times the power dissipation capability of aluminum. However, a great deal of materials engineering know-how needs involved in realizing this, given the inherent fragility and extreme anisotropy. The benefits of using high conductivity materials are determined by the other thermal resistances in the system. High-quality thermal interfaces are needed to exploit the potential of the material. In addition, the overall mechanical properties of the HOPG for some thermal managing components may not be good enough because of the weak van der Waals forces that bond the lattice between each basal plane. However, encapsulating HOPG within a structural shell can overcome this structural limitation. The HOPG encapsulation scheme combines the beneficial properties of two materials in a configuration that allows the cost effective optimization of the assembly. The encapsulated HOPG composite can be an ideal thermal management material because of its high thermal conductivity (up to 1,400 W/m K), low mass density (as low as 1.9 g/cm$^3$), high stiffness (up to 50 Msi), and the ability to have an engineered CTE (Montesano 2006).

Applications of HOPG have been those where cost is not a driving factor and have included avionic, telecom, power electronics, high-end servers, and high-brightness light-emitting diodes.Other methods of integrating the material with lower expansion advanced materials are also being developed, for example as an embedded insert in Al-SiC flip-chip lids which offer thermal conductivity of 1,350 W/m K in the *XY* plane. The poor Z-plane conductivity of the material can reduce the thermal coupling efficiency in dealing with hot spots in some configurations. This problem can be attacked using the technique of introducing diamond pin inserts, for example, used in one radar module application (Young et al. 2006).

## Carbon–Graphite Foams

Carbon foams were first developed in the late 1960s as a reticulated vitreous (glassy) carbon foam, produced by carbonizing thermosetting organic polymer foams through heat treatment. Reticulated carbon foams have been used as the template for many of the metal and ceramic foams currently used in industry. From 1970s to early 1990s, carbon foams were produced on a trial basis from cork, coal, and other alternative precursors in an attempt to modify properties and reduce cost. The majority of these carbon foams were used for thermal insulation and structural applications. Mesophase-derived graphitic foams and PG-coated reticulated vitreous carbon foams were specifically for replacing expensive 3-D woven fiber performs in polymer composites and as replacements for honeycomb materials with the high specific strength and high-modulus ligaments. High thermal conductivity graphitic foams with bulk thermal conductivities of 40–180 W/m K have been developed since 1997. By combining an open cellular structure with a thermal conductivity to weight ratio of greater than 200 (compared with 45 for copper), this material presents a unique opportunity to radically change the approach to solving

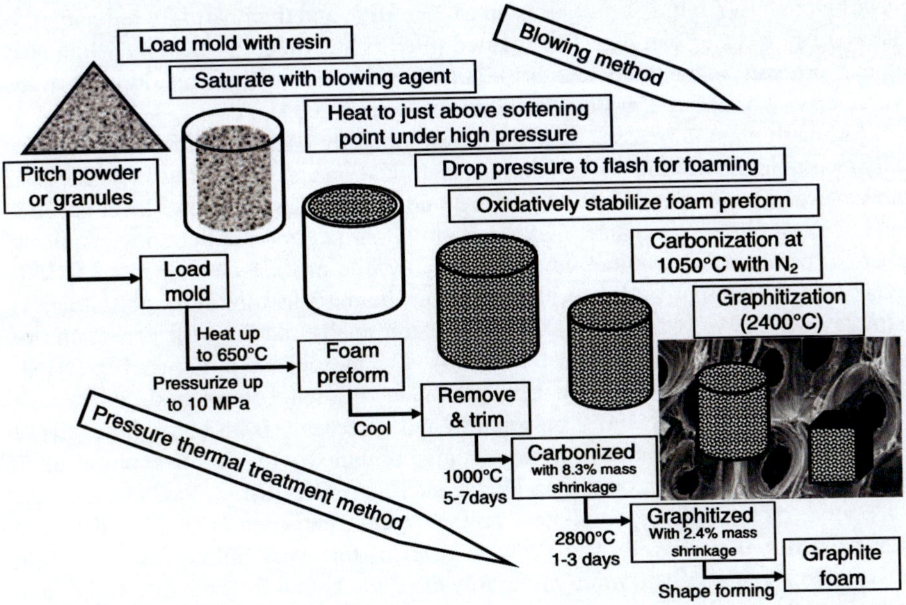


**Fig. 4.3** Illustration of fabrication process of high thermal conductivity graphite foams

many heat transfer problems. This graphite material has been examined for the core of heat transfer devices such as radiators and heat sinks, evaporative cooling, and phase change devices. Furthermore, the ability of the graphite foam to intercalate lithium and absorb acoustic energy makes them candidates for several applications beyond thermal management (Klett et al. 2003).

## *Fabrication Process*

Conventionally, carbon foams have been basically prepared by two general routes: blowing methods and pressure thermal treatment methods, as shown in Figure 4.1. Highly graphitizable foams have been produced by thermal treatment of mesophase pitches under high pressure. These foams tend to have high thermal and electrical conductivities (Klett 2000; Stansberry and Lewis 2008). Mesophase pitch is heated to approximately 50 to about 100°C above the softening point of the pitch to coalesce the pitch. The pitch mixture is then pressurized to 7–10 MPa or higher with an inert gas, followed by heating of the pitch. Gases are evolved from the pitch that is sufficient to foam the pitch at 7–10 MPa to produce open-cell foam containing interconnected pores with a size range of 90–200 μm. After heat treatment to 2,800°C, the solid portion of the foam develops into a highly crystalline graphitic structure with an interlayer spacing of 0.366 nm. The foam is asserted to have compressive strengths of 3.4 MPa for a density of 0.53 g/cm$^3$. Another process was used to produce carbon foams with densities ranging from 0.678 to 1.5 g/cm$^3$ by heating pitch in a mold at pressures up to 206 MPa, and then partially reducing the pressure to induce foaming. The foamed pitch is stabilized by heating while still under pressure. The foam can be highly graphitized, with a resulting thermal conductivity of about 250 W/m K (Hardcastle et al. 2004).

Although many processes have been developed for carbon foams, which generally available require the use of complex stabilization and thermosetting processes, and many polymer-based carbon foams do not exhibit desirable interlayer spacing and a crystalline size value making them ill-suited for applications requiring thermal or electrical conductivity. Comparably, the process developed by ORNL (Oak Range National Laboratory) process for the manufacture of graphite foam is simple and free of oxidative stabilization traditionally required for processing of pitches and mesophases (Klett et al. 2003). The mesophase pitch used in ORNL process is derived from either coal tar, petroleum, or model compounds, containing from about 60 to about 100% mesophase and preferably from about 90 to about 100% mesophase. The cross-linking agents combined with the mesophase pitch include sulfur, quinone, oxidizing agents such as $(NH_4)_2S_2O_3$, $NaClO_3$, oxygen, chlorates, dichromates, persulfates, and inorganic carbenes such as $SnCl_2$ and $PbCl_2$; sulfur is preferred. The addition level of the cross-linking agent is from about 1 to about 10% and more preferably of from about 4 to about 6% of the total weight of the cross-linking agent/mesophase pitch mixture. The cross-linking agent reacts with the pitch by heating under pressure and provides for cross-linking of the aromatic rings by removal of pendant hydrogen on each ring and bonding with an element from the cross-linking agent to form a functional byproduct. In the case of sulfur, $H_2S$ is generated. This functional byproduct acts to generate foaming, and furthermore, also acts as a thermal setting agent for the pitch so that complex stabilization and thermosetting processes are not needed.

During processing, a mesophase pitch precursor is heated in an oxygen-free environment to about 50°C above its softening point. Once the pitch has melted, the furnace pressure is elevated and the temperature rises at a controlled rate. While the pitch is molten, it begins to evolve low molecular weight species. These volatile gases form bubbles at nucleation sites on the bottom and sides of the crucible and rise to the top, beginning to orient the mesophase crystals in the vertical direction. With time, a significant amount of the mesophase crystals are oriented vertically. At high temperatures the mesophase begins to pyrolyze (polymerize) and create additional volatile species. This pyrolysis weight loss can be very rapid and is dependent on the precursor. The process is accompanied by an increase in the molecular weight of the precursor, which, in turn, increases the melt viscosity of the liquid mesophase. As the rapid evolution of gases progresses, the increase in viscosity tends to capture the bubbles in place, forcing the material to foam in the unrestrained direction, denoted as the $z$-direction. As the temperature of the furnace is further increased, the foamed mesophase continues to pyrolyze, further increasing the viscosity of the material until it has sufficiently cross linked and is rendered infusible (cannot be melted).

The formed foam typically exhibits uniformly shaped bubbles with a normal distribution. The average pore size, orientation, and distribution is determined primarily by the pitch viscosity and processing pressure during foaming. Additionally, the mesophase foam will have a preferred orientation of crystals in the *z*-direction with an accompanying anisotropy of properties in the *z*-direction compared with those in the *x*–*y* plane even though the visual physical structure (bubble-shaped) may not be anisotropic (Klett et al. 2003).

There is a delicate relationship between the viscosity temperature behavior, melting temperature, and pyrolysis temperature of the mesophase pitch. Initially the pyrolysis gases develop at a temperature such that the viscosity is sufficient to result in stable foam. Premature pyrolysis gas evolution causes the pitch to froth (like boiling milk) resulting in foam with a significant density gradient (this may be desirable in some applications). However, if the gases are evolved too late when the pitch viscosity is high, the bubbles may not be uniform, and cracking can occur as a result of thermal stresses. If the pyrolysis gases are evolved very slowly, as for certain high melting point Conoco pitches, the pores will tend to be smaller overall. Bubble formation is closely related to the autoclave operating pressure as well as temperature. Typically, the higher the autoclave gas pressure, the higher the temperature that gas evolution occurs, and the smaller the resulting pores. However, depending on the unique rheological properties of the starting pitch, the cell walls have different thicknesses, the bubble sizes can be dramatically different, and the mechanical and thermal properties can be affected. Unlike other foaming techniques, such as lost foam processing or foam blowing, the resultant properties of the graphitic foam, (such as bubble size, ligament size, relative density, thermal and mechanical properties) are not independent properties. They are all dependent on the starting precursor's melt viscosity, pyrolysis temperature, and other pitch rheological properties. The foamed mesophase is carbonized by heating to between 600 and 1,000°C to yield relatively pure carbon foam. The foam is an excellent thermal insulator in this state, with a bulk thermal conductivity of about 1.2 W/m K for foam with a density of 0.5 g/cm$^3$. Because the carbonized foam was formed with a mesophase that was not oxidatively stabilized during the pyrolysis/carbonization stages, the mesophase crystals are not inhibited and can grow to very large sizes. Consequently, when the carbon foam is converted to a highly graphitic foam by heat treatment to more than 2,800°C under an argon purge, the resultant graphite crystals are highly aligned and significantly larger than those found in mesophase-derived carbon fibers. Hence, the ligaments of the graphite foam produced with this method are more thermally conductive than even the best mesophase pitch-based graphite fibers (Klett et al. 2003).

## *Thermal Conductivity and Heat Transformation*

As shown in Table 4.1, mesophase pitch-derived graphite foam made with the ORNL process exhibit high bulk thermal conductivities, 175 W/m K in average, at densities up to 0.6 g/cm$^3$. The foam conductivity is very dependent on the process used to make the foam through changes in density of the foam, while the final heat treatment parameter affects the final thermal conductivity by affecting the structure of the ligaments. In general, the thermal conductivity in the $z$-direction is significantly higher than that in the $x$ or $y$ directions. The thermal conductivity in the $x$–$y$ directions is independent of the graphitization rate, while the thermal conductivity increases in the $z$ direction with increasing graphitization rate (Klett et al. 2003).

Typically, manufactured graphite exhibits a peak in thermal conductivity around room temperature. As the order and perfection of the graphitic crystals improve and approach those of perfect graphite, this peak in conductivity shifts to lower temperatures around 80 K. Below this maximum temperature the thermal conductivity increases with temperature due to increasing specific heat. The reduction in thermal conductivity above the maximum temperature is caused by phonons scattering from intrinsic lattice defects and self scattering. The exact location of the peak is largely controlled by the concentration of lattice defects. Defect-free graphite usually exhibits a peak at very low temperatures, whereas manufactured graphite typically exhibits a peak close to room temperature (Klett et al. 2003).

The mechanism of heat transfer in the graphene lattice can be described by vibrational modes represented as phonons. The heat transfer down the graphite lattice is extremely fast due to the very stiff nature of the covalent bonds. However, when a phonon reaches a defect in the structure, the vibration of the atoms is interrupted and the phonon is considered "scattered." In addition, the position and vibration of atoms in neighboring planes may impede the vibration of the atoms in the plane of interest. The crystal perfection controls thermal conductivity. In addition, the crystal lattice must be oriented in such a manner that the vibration of atoms in neighboring planes does not interfere with the phonon transport in the plane of interest. This requires perfectly aligned, defect-free graphene sheets with true 3-D crystallinity. The hexagonal graphite structure requires an ABA (Atom Stacking Order) stacking sequence, which requires that the graphene sheets be perfectly flat. If there is any curvature, the ABA stacking sequence will be disrupted over a given distance (depending on the radius of curvature), and the result will be scattering of phonons. As the radius of curvature decreases, there will be more scattering of phonons and decreased heat transfer. Furthermore, bending of a graphene plane is most likely due to defects (or vacancies) in the graphene lattice which forces the structure to curve, such as in fullerenes. Even where the graphene lattice is perfect, it must be oriented in the appropriate direction. If there is any rotation of the planes, the atoms will not line up in the proper ABA stacking sequence, which will result in a larger interplanar distance due to the interaction of van der Walls forces and a decreased heat transfer through phonon scattering. A rotation of only 4° can result in disruption of the lattice. Phonon transport is dominated by vibrational modes of the crystal lattice. The more perfect the lattice, the less the vibrations interact destructively to reduce heat transfer. However, there are many ways to disrupt and interfere with the phonon transfer. These include: (1) defects in the crystal lattice (vacancies, extra atoms, imperfect bonding, etc.); (2) folds in the graphite structure (lattice boundaries); and (3) interlayer orientation and preferred orientation. The phonon mean free path is effectively the average distance along the graphene plane a phonon travels before it is scattered, either by a defect or by another phonon (destructive vibrational interference). Typically for most synthetic graphite, this length is small. However, in highly ordered graphite (like HOPG) the phonon mean free path can be quite large (on the order of 1,000 nm). Therefore, to achieve high thermal conductivity in the graphite crystal, the structure must be comprised of aligned, straight graphene planes, resulting in the optimum contributions to phonon transport. Furthermore, the structure must have very large mean free paths for the phonons, requiring a relatively defect free structure to minimize phonon scattering along the plane. At low temperatures, the phonon self scattering is insignificant. As the temperature rises, these interactions become more significant than lattice boundaries. For all graphite, however, the thermal conductivity decreases with increasing temperature above the maximum temperature (Klett et al. 2003).

Unlike metallic or glassy foams, foams made from graphite contain material that is locally anisotropic and orientation of the graphite basal planes parallel to the heat flow is critical. The theoretical thermal conductivity of perfect graphite along crystallographic basal planes is more than 2,000 W/m K at room temperature, but less than 10 W/m K perpendicular to the basal planes. Therefore, in regions with poor crystallographic alignment, such as at the junctions, the local thermal conductivity will be lower than that in the ligaments. Such regions will significantly affect the thermal conductivity of the bulk foam. Therefore, bulk thermal conductivity of the graphite foam can be estimated by (Klett et al. 2003):

$$\lambda_{\text{bulk}} = \alpha \left( \frac{\rho_{\text{bulk}}}{\rho_{\text{solid}}} \right)^m \lambda_{\text{ligament}}, \tag{4.2}$$

where $\lambda$ is the factor incorporating the effect of pore shape on the inverse in heat flow length path; m is the factor incorporating effect of density on volume ratio of ligament to junctions and the result on increased heat flow path; $\rho_{\text{bulk}}$ is bulk density, g/cm$^3$; $\rho_{\text{solid}}$ is density of ligaments, g/cm$^3$; $\lambda_{\text{ligament}}$ is average thermal conductivity of ligaments. The constants $\alpha$ and $m$ are functions of the pore structure and can be determined from measured thermal properties of the foams by regression techniques. For the graphite foams made by ORNL, $\alpha = 0.728$, $m = 1.427$, $\lambda = 1{,}311$ W/m K (Klett et al. 2003).

Consequently, modifying the junctions is likely crucial to improve the thermal properties of the material. Therefore, process changes that only affect the structure slightly, such as heat treatments, will not necessarily have a large impact on the thermal properties. The change of process that modifies the pitch precursor to produce a junction which is less cracked and more ordered offers great potential to improve the properties of the foams (Klett et al. 2003).

## Thermally Conductive Carbon Fibers

The application of carbon fiber can be dated back to 1879, when Thomas Edison used carbon fibers as filaments for early light bulbs. These carbon fibers were made out of cellulose-based materials such as cotton or bamboo. Carbonization took place with pyrolysis process by baking bamboo filaments at high temperatures in a controlled atmosphere. The resulting carbonized bamboo filaments were fire-resistant and capable of enduring the intense heat needed for incandescence.

Later in the1950s the high tensile strength carbon fibers were produced from a rayon precursor, and it was replaced by more effective materials such as polyacrylonitrile (PAN) in 1962, isotropic pitch precursor in 1969, and mesophase pitch precursor in 1975 (Arai 1993). High-performance carbon fibers with light-weight, high-strength, and high-elasticity modulus are produced from PAN and mesophase pitch precursors and used mainly as base materials for advanced composites. In particular, mesophase pitch-based carbon fibers are more easily produced with high elasticity modulus grades and much more superior thermal conductivity than PAN-based carbon fibers, therefore, they have been very attractive as materials of components for high-temperature surface as in space and aircraft applications as well as for thermal management of advanced electronic packaging.

The fabrication process of thermally conductive carbon fibers is shown in Figure 4.4. The precursor fibers are prepared, converted to thermosetting fibers by infusibilization, and then cabonized and graphitized at a high temperature in an inert atmosphere. The difference in the precursor fibers governs the basic structure and final performance of the finished carbon fibers (Arai 1993). Pitch-based carbon fibers can be derived from optically isotropic pitch anisotropic (liquid crystalline or mesophase) pitch. These two pitches can be developed with good spinnability by reforming coal tar or heavy oil from petroleum refining or petrochemical production. General purpose carbon fibers with a tensile strength of 0.6–1 GPa and tensile

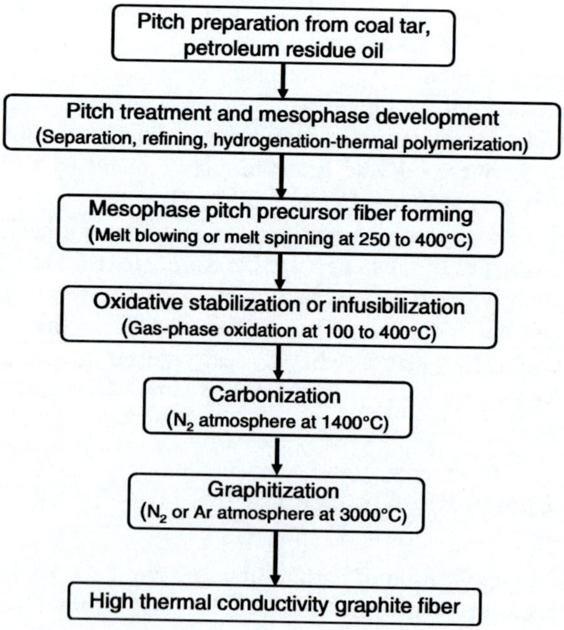


**Fig. 4.4** Illustration of fabrication process of mesophase-pitch-derived graphite fibers

elasticity modulus of about 40 GPa are obtained from the optically isotropic pitch, while high thermal conductivity carbon fibers with a tensile strength of 2–4 GPa and a tensile elastic modulus of 150–900 GPa are prepared from the optically anisotropic mesophase pitch. As shown in Figure 4.4, the starting heavy oil is separated, refined, hydrogenated, or otherwise chemically treated, and then thermally polymerized into pitch for spinning to form mesophase pitch fibers with a diameter of several to 10 μm. In fact, the melt spinning is the preferred technique for producing mesophase precursor fibers. During the process, the mesophase pitch is melted and forced through a die containing a number of extrusion capillaries. As the precursor emerges from the capillaries, it is stretched with shearing force and solidified to form fibers. These thermoplastic pitch fibers are oxidized and cross linked by gas-phase oxidation into infusible thermosetting fibers. The subsequent steps of carbonization and graphitization remove almost all noncarbon elements from the fibers and grow graphite crystals (Arai 1993). The final graphite fibers for composites are usually activated by surface treatment to well adhered with matrix materials. Therefore, pitch-based carbon fibers are produced through a systematic combination of various techniques, such as chemical reaction, melt spinning and high-temperature firing, to excellent quality including superior thermal conductivity.

The mesophase pitch precursor for high thermal conductivity carbon fibers is a kind of nematic liquid crystal in which molecules predominantly composed of condensed polycyclic aromatic polymers with a planar structure are stacked. As the molten mesophase pitch flows through the extrusion capillary, it is deformed by shear. The planar molecules are arranged in the direction of shear deformation, developing an extremely high degree of molecular orientation. Therefore, a pitch fiber with the planar structure molecules of mesophase pitch is highly oriented in the melt spinning step. Because the component molecules of metosphase pitch tend to assume a planar structure, the pitch fibers also have regularity in the cross-sectional direction normal to the fiber axis with oriented transverse structures, such as radial, onion, and random structures. If relaxation does not occur during infusibilization or cross-linking stabilization, the orientation of these macrostructures has no appreciable changes or merely perfects in the subsequent steps of infusibilization, carbonization, and graphitization. On the other hand, microstructures of the fibers significantly change during carbonization and graphitization. The planar molecules of mesophase pitch develop a six-member ring structure at about 1,300°C, stretch the folded portions with the expansion of the planar structure, and exhibit a graphite crystal structure with a wide layer structure during heat treatment at 2,000°C or above (Arai 1993). The expansion direction of graphite layer planes and the inclination or orientation degree along the fiber axis, and graphitization parameters, such as interlayer spacing, thickness and expansion, govern the thermal conductivity, tensile strength and elastic modulus of the final carbon fibers. In other words, the final carbon fibers should preferably have graphite crystals well oriented in the direction of the fiber axis and proper degree of graphitization to obtain superior thermal conductivity, and required strength and elastic modulus. Therefore, this highly oriented graphite structure in the direction of the fiber axis provides excellent thermal conductivity and chemical stability for thermal management of electronic packaging.

Another kind of thermally conductive carbon fiber that is attractive to thermal management of electronic packaging would be vapor grown carbon fiber (VGCF). The density of the VGCF is 1/4 or less than that of copper and the room temperature thermal conductivity of a graphitized VGCF is up to 1,950 W/m K, which is approximately five times higher than that of copper. Additionally, the electrical resistance of the VGCF is extreme low, i.e., about 0.6 μΩm, which is similar to that of titanium. The VGCF also has good tensile stress and stretching coefficient; therefore, the VGCF is excellent for developing composite materials having high conductivity.

The traditional VGCF are short fibers having a length of at most hundreds micrometers, so the composite materials fabricated by these short fibers have a lot of discontinuity, and thereby the conductivity of these composite materials is merely equal to or lower than that of aluminum. In another words, although the VGCF has a thermal conductivity of as high as 1,950 W/m K, but the composite materials fabricated by discontinuous short fibers do not have the expected heat dissipation effect. In order to make the composite materials have the advantageous of high conductivity from VGCFs, it is necessary to increase the length of them. Thus, efforts have been made for developing methods to produce a VGCF having a semicontinuous structure and a length to be increased dramatically. The reaction apparatus for continuously producing a carbon fiber by the vapor-growth process consists of several parts: CO, methane, acetylene, ethylene, benzene, and toluene can be used as the carbon source which is the raw material of the carbon fiber. A carbon source which is a gas at an ordinary temperature and pressure is supplied in the gaseous state by being mixed with a carrier gas. In the case of a liquid, the carbon source is vaporized in a vaporizer, mixed with a carrier gas and then supplied, or is sprayed in a liquid state into the heating zone. For example, the carrier gas used is nitrogen gas, which is an inert gas, or hydrogen gas, which is reducing gas. In some cases, the carbon source is supplied into a depressurized apparatus. As for a catalyst of a process for producing a VGCF, a supported catalyst in which a metal is supported on a support such as alumina, or an organic metal compound such as ferrocene, is used. In the case of using a supported catalyst, the supported catalyst is subjected to a necessary pretreatment by placing and heating it in a heating zone, and then a carbon source is supplied and reacted thereon. Alternatively, the reaction is performed by continuously or pulsedly supplying a supported catalyst which has been pretreated, from outside of the system. It is also possible that an organic metal compound readily dissolvable in a carbon source, such as ferrocene, is employed as a catalyst precursor, and continuously or pulsedly fed to the heating zone together with a carbon source, and thereby a carbon fiber is produced by utilizing metal particles generated by thermal decomposition of the catalyst precursor as a catalyst. The production process of a carbon fiber by a vapor-phase method is roughly classified into the following three types according to the method of supplying a catalyst or a precursor compound of the catalyst (Higashi et al. 2007): (1) a substrate or boat comprising an alumina or graphite supporting a catalyst or a precursor compound thereof is placed in a heating zone, and contacted with a gas of a carbon source supplied in a vapor phase; (2) a particulate catalyst or a precursor compound thereof is dispersed in a liquid-state carbon source or the like, and continuously or pulsedly supplied to a heating zone from outside of the system, and thereby contacted with a carbon source at a high temperature; and (3) a metallocene, carbonyl compound and like dissolvable in a liquid-state carbon source is used as a catalyst precursor compound, and a carbon source comprising this catalyst precursor compound dissolved therein is supplied to a heating zone, whereby a catalyst and the carbon source, which is a hydrocarbon or the like, are contacted at a high temperature.

Figure 4.5 shows typical photos of VGCF: the thermal conductivity of as-grown VGCF typically 20 W/m K; heat treated VGCF under 3,000°C could reach 1,950 W/m K (Young et al. 2006). (1) Fixed catalyst growth on substrate; (2) formed semialigned (several centimeters in length) or semicontinuous fibers; and (3) prototype of a heat spreader made with VGCF reinforced carbon composite (C-VGCF) composite, which are as promising as next-generation thermal management materials of heat spreader or other components for commercial and industrial electronics.

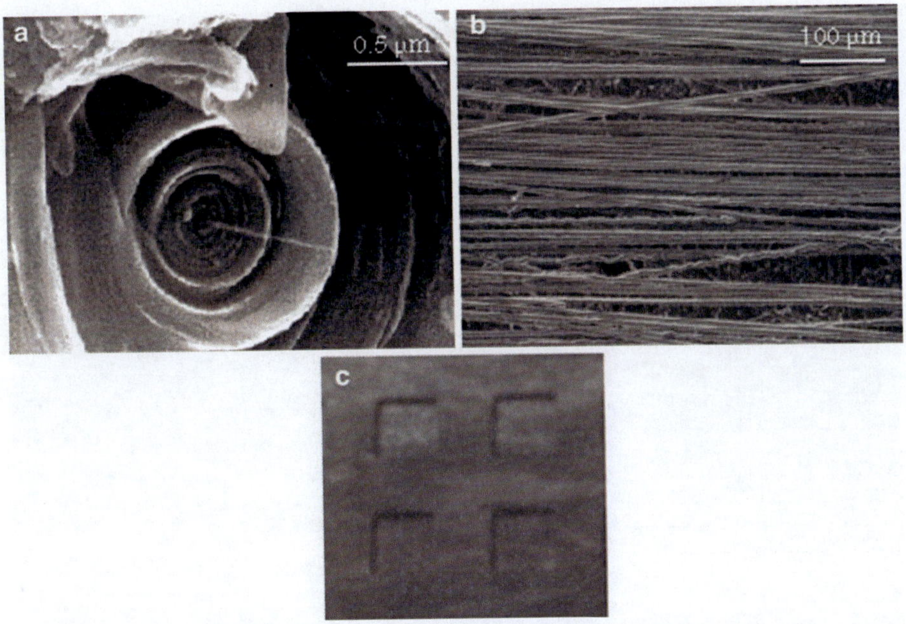


**Fig. 4.5** Vapor grow carbon fiber (VGCF): the thermal conductivity of as-grown VGCF typically 20 W/m K; heat treated VGCF under 3,000°C could reach 1,950 W/m K (Young et al. 2006). **(a)** Fixed catalyst growth on substrate; **(b)** formed semialigned (several centimeters in length) or semicontinuous fibers; **(c)** prototype of a heat spreader made with C-VGCF composite

## Diamond

Diamond exhibits a rare combination of physical, optical, and thermal properties that make it an attractive material for many applications. It offers ultrahigh thermal conductivity, transparency across a variety of wavelengths, and is wear resistant, electrically insulating and chemically inert. For thermal management of electronic packaging, CVD diamond has mainly been used in small-scale chip cooling applications, while particulate diamond is being used in composite structures.

Since the 1950s, synthetic single crystal and sintered polycrystalline diamonds have been produced by high temperature and pressure methods with the presence of a metal solvent or catalyst to accelerate the transformation of graphite to diamond during growth or sintering. These high temperatures and pressures require expensive equipment and controls and make the production of coated surfaces, curved shapes, and large pieces very difficult. The sintered products contain less than 100% diamond. The residual sintering aid and catalytic materials eliminate sintered high-pressure diamonds from most electronic and optical applications (Bigelow 1993). However, the synthetic diamond particles can be used for thermally conductive reinforcement in composite materials and structures. Furthermore, interest in diamond composites has been growing as a result of the falling prices of both synthetic and natural diamond, largely driven by sources in China. Figure 4.6

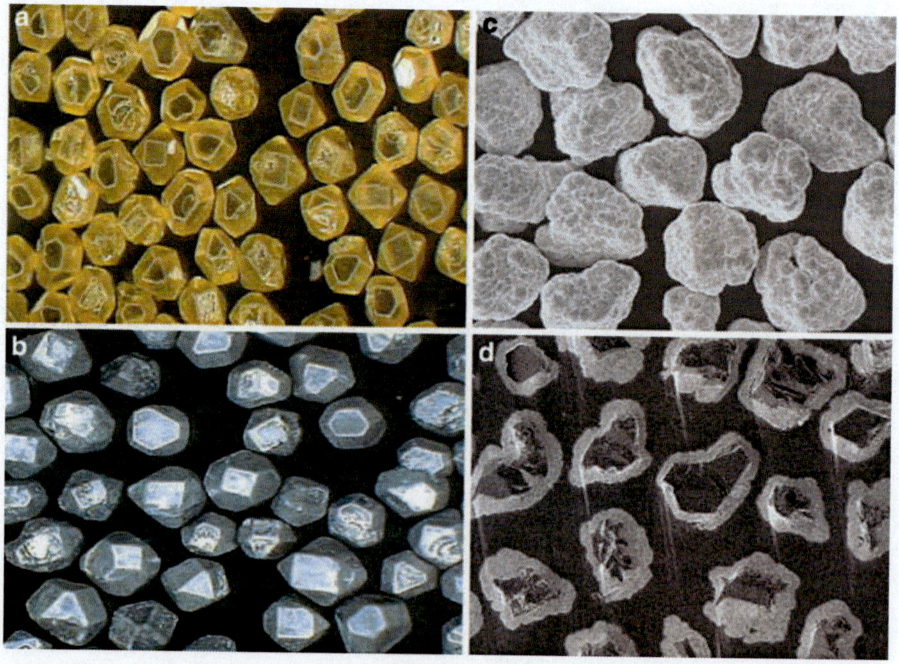

**Fig. 4.6** Synthetic diamonds: (**a**) as-synthesized; (**b**) coated with molybdenum or SiC conversion; (**c**) coated with copper, and (**d**) cross-section of diamond particles coated with copper

shows typical synthetic diamond particle morphology, including (1) as-synthesized; (2) coated with molybdenum or SiC conversion; (3) coated with copper; and (4) cross section of diamond particles coated with copper. The diamond surface coating or SiC conversion is for improve the wettability between the diamond particles and the composite matrix. The common isotropic high thermal conductivity diamond composites are diamond–copper or diamond–aluminum composites. Diamond loadings are 50% volume fraction and this could be increased to 75–80% if necessary. Thermal conductivity of 600–800 W/m K (isotropic) can be produced for diamond–copper composites with CTE in the range 5–8 ppm/K. Unlike many other materials, the thermal conductivity of diamond–aluminum composites rises with increasing temperature: values are in the range of 350 (ambient) to ~520 W/m K at 150°C and thermal expansion rises from 7.5 to 10 ppm/K over the same range. Properties obviously scale with diamond loading: another data set for diamond aluminum rates it at thermal conductivity 650 W/m K, CTE ~7.5 ppm/K, density 3.2 g/cm$^3$.

However, the major disadvantage of the diamond composites is the difficulty of machining the composite. This problem has been addressed by net-shape processing routes using high-pressure squeeze casting or melt infiltration process. Exploitation is hampered by surface roughness (10 μm and 1.3–5 Ra for electrodischarge machining and diamond wheel cutting respectively). Applications of these composites have been explored for heat spreader or heat sinks (Young et al. 2006).

On the other hand, CVD processes for polycrystalline diamond growth, which overcome many of the limitations of the high-pressure synthetic process, has been under intense development since the late1980s. These techniques typically include microwave plasma assisted CVD, hot filament deposition, plasma jet deposition, and combustion flame (laminar or turbulent oxy-acetylene) deposition (Choudhary and Bellare 2000). In these CVD processes, diamond is deposited from a gas phase, and can therefore coat large areas and curved surfaces, or be grown thick enough to become free-standing parts which can be cut and finished to the size and shape desired. For instance, the CVD processes typically include the maintenance of plasma containing atomic hydrogen over the growth surface, and require a source of carbon such as methane ($CH_4$). Hydrogen is the key element which stabilizes the diamond surface, and permits the growth of diamond while suppressing the growth of graphitic carbon. The growth process typically requires temperatures of 800–1,000°C, but can occur even at temperatures as low as 200°C at much lower rates (Bigelow 1993). Essentially, all the deposition techniques rely on the ability to set up a dynamic nonequilibrium system, in which only sp$^3$ carbon bonding can survive. This is achieved by the presence of hydrocarbon radicals, and more importantly, by large quantities of atomic hydrogen in the deposition gas. The hydrocarbon radicals provide the vapor source of carbon for diamond deposition. The atomic hydrogen is believed to play two important roles: it eliminates sp$^2$ bonded carbon (graphite), while stabilizing the dangling bonds of the tetrahedrally bonded carbon sp$^3$. By creating a plasma system where atomic hydrogen predominates, any graphitic bonding is etched away, leaving only high-quality diamond.

The CVD diamond can offer similar properties of single-crystal diamond, but it is a family of materials with process-dependent properties. The material is somewhat anisotropic, with through-thickness thermal conductivities somewhat higher than in-plane values. Some forms of CVD diamond are optically transparent, making it an attractive material for applications requiring both this property and high thermal conductivity. Furthermore, the growth rate of polycrystalline diamond has a strong influence on the final quality of the material. Relatively high growth rates are possible, but at the expense of sample-phase purity. This has resulted in the development of different grades of CVD diamond to cover the wide diversity of its applications. There are two main types. High quality, low growth rate white transparent diamond contains virtually no graphite, and is used in electrical and optical applications. High growth rate color diamond, which contains small amounts of graphite, especially at the grain boundaries, is used for thermal management applications such as heat spreaders. Polycrystalline diamond via a CVD process can grow on substrates, typically refractory metal or Si, although other substrate materials can be used. This CVD diamond can be produced as flat plates or thin-walled 3-D structures, such as hemispheres and parabaloids. The diamond growth can be tailored to the specific application, which results in a range of products from tool-grade material with high fracture toughness to high clarity optical window material. Adjustments in growth conditions produce parts with thickness varying from about a half of a micron to over a millimeter. For example, diamond heat spreader parts typically range in thickness from 100 to 500 μm, which can be matched to the thermal needs with thermal conductivities ranging from 500 to 2,100 W/m K through plane (Choudhary and Bellare 2000). This property, coupled with its high electrical resistivity, makes it an ideal material for use as heat spreaders for high-powered electrical devices.

Thermal management applications of diamond require a close-contact attachment of the diamond heat spreader onto the device to be cooled. Various mounting techniques such as mechanical clamping, solder bonding, brazing, or adhesive bonding can be used. Diamond surfaces at atmospheric conditions are often covered by chemisorbed species such as oxygen, hydrogen, or hydroxyl groups, or by physisorbed layers of water or organic materials. Adhesion of solid materials in the presence of such layers results in weak, van der Waals type bonding with poor wetting and adhesion. For strong bonding of solid materials to diamond surface, chemical or covalent type carbon–metal bonding interactions at the interface is desirable. Strong carbide forming elements such as Ti, Zr, V, Nb, Ta, Cr, or Si generally provide reliable metal–diamond bonding. These elements also exhibit a strong affinity to oxygen, thus they are capable of removing chemisorbed or physisorbed oxygen and other species from the diamond surfaces. However, oxidized surfaces of these metals can induce poor wetting and reduced bond strength during subsequent bonding processing such as soldering or brazing. To overcome this problem, a common practice is to cover the reactive metal surface with a protective inert metal such as Au, Ag, Pt, Cu, or Sn. An alternative approach is to alloy the soldering or brazing material with a small amount (typically less than several percent) of the carbide forming elements, which enhances the adhesion at the diamond-metal interface. Because the use of minimal amount of bond materials is desirable to minimize thermal impedance, the interface layers or bonding materials are often prepared by thin film deposition methods, e.g., by sputtering. An exemplary metallization for solder bonding of CVD diamond for thermal management in laser diode devices uses a three-layer configuration of Ti/Pt/Au. The titanium layer provides bonding to diamond, the platinum layer is for diffusion barrier, and the gold overcoat serves as a protective layer to minimize Pt solder reactions and also as a bond layer to the Au-Sn eutectic solder. The thicknesses of sputter-deposited metallization layers on diamond are, for example, 100 nm thick Ti/200 nm thick Pt/500 nm thick Au, attached with 2.5 μm total of 80Au-20Sn eutectic solder layer (either as multiple alternating layers of Sn and Au or as a single-alloy layer) to bond with laser diode. The solder bonding of the laser diode onto the diamond heat spreader is carried out by rapid thermal annealing, e.g., at 300–350°C for several seconds (Jin and Mavoori 1998). In addition, when bonding the diamond to other materials, a smooth surface is required. While the unpolished roughness of the growth side of the diamond begins at around 10% of the diamond thickness, the peak-to-valley polished roughness can be reduced to about 10 nm with a polishing step. Diamond parts can be laser cut to the size and shape needs of the package of the end user. Engineered laser systems and -ontrolled motion allow specialized cutting of complex geometries.

In addition to heat dissipation, CVD diamond can provide tailorable CTE to reduce thermal stresses arising from differences in CTE. The CTE of CVD diamond, 1.0–1.5 ppm/K, is lower than desired for some applications. This can be overcome by adding layers of copper on top and bottom, creating a laminate with a tailorable CTE. The same principal is used for low-CTE laminated metal heat spreaders, such as copper-Invar-copper and copper-molybdenum-copper. Size has a significant effect on thermal stresses. In general, thermal stresses increase linearly with planar dimensions (length and width). Consequently, CTE differences may not be significant when device dimensions are small. Additionally, the density of CVD diamond is much lower than that of most packaging materials, which is important in weight-critical applications, such as aircraft and spacecraft electronic systems, notebook computers, and other mobile devices.

Because diamond growth and postprocessing conditions can be altered to create parts with different thermal properties and cost, thermal modeling is an important step in the design stage to maximize the cost-value relationship of the CVD diamond part. In general, cost increases with plate planar dimensions, thickness, and thermal conductivity. Thermal modeling often shows that the heat spreader required in diamond material is often thinner and has smaller in-plane dimensions than heat spreaders made from traditional materials, which have much lower thermal conductivities. Optimizing the diamond heat spreader for each application has enabled more frequent use of diamond in high-power packages. The reason that cost increases with thermal conductivity is that there is an inverse relationship between growth rate and thermal conductivity. That is, it takes longer to make a heat spreader with high thermal conductivity, increasing manufacturing cost. As the cost of CVD diamond decreases, its use will certainly increase and diversify further. However, as with any new material, two areas must be addressed if it is to become a commercially viable product. First, production costs must be reduced. This will be achieved not only by economy of scale as demand increases, but by the development of a new generation of deposition systems which will enable high-quality diamond to be grown over ever-larger areas. Second, new doping, processing, and joining technologies need to be developed. This is not trivial, because the properties that make diamond attractive for so many applications (i.e., hardness, chemical inertness, etc.) also make it difficult to dope and process.

Consequently, in view of thermal management applications, what is needed most is better adherence to functional substrates; higher uniformity and homogeneity of the coatings over large areas; increased deposition rates; higher purity; higher nucleation density; more randomly oriented structures; lower surface roughness; more efficient polishing methods; improved electrical contacting of the diamond layers; epitaxial layers; 3-D growth of large single crystal; and whiskers and/or fibrous growth of diamond. In fact, the diamond fibers have been made by CVD onto metallic wires or nonmetallic fibers. Two types of hollow fibers can be made by (1) etching out metallic or ceramic cores, leading a smooth diamond core surface; (2) depositing diamond on to helical tungsten wire coils, the core surface is rougher than for the etched core. Hollow cores can lead to a reduction in composite density and may provide channels for gases or liquids for convection cooling (Partrigge et al. 1995). The achievement of high thermal conductivity may be limited in practical composites, because of the thermal conductivity gradient inherent in CVD diamond films, a need for large numbers of small-diameter fibers, and a limit on fiber volume fraction. However, the use of hollow fibers to combine conductive and convective cooling will increase the options in the design of thermal management systems.

## Carbon Nanotubes

The thermal management in electronic packaging especially nanosize devices has increasingly become important as the size of the device reduces. Therefore, the thermal conduction of nanometer materials may play a fundamentally critical role that controls the performance and stability of nano/micro devices. Among various potential candidates for future MEMS/NEMS applications, CNTs stand on a unique position for thermal management of electronic packaging due to their remarkable properties, such as superior thermal conductivity, great strength, light weight, special electronic structures, and high stability. More commonly, CNTs have been explored for use of thermal conductivity enhancements in thermal conducting materials or composites. A limiting factor in the use of CNTs is the transfer of heat flux from one nanotube to another in an efficient manner. Butt joints give poor thermal resistance, and research into devising a low thermal resistance joint is taking place. CNTs can be woven into mats to produce a low-density, high-thermal conductivity material. This can be put into a metal composite by pressure or squeeze casting, or epoxy added as a filler to give rigid mats.

CNTs are long, slender fullerenes where the walls of the tubes are hexagonal carbon and are capped at both ends with C60 fullerene hemispheres. CNT can be visualized as a graphene sheet that has been rolled into a tube. CNTs are either SWNTs or MWNTs, where an SWNT is a tube with only one wall, and MWNTs have many concentric tubes where the walls of the tubes are held together by weak van der Walls forces (Banda 2004). CNT is made virtually from the fullerene structure that is planar closed polyhedra formed by $sp^2$ bonded hybridization, in which the curvature and cylindrical symmetry cause very important modifications compared with planar graphite. The properties of CNTs depend on chirality, diameter, and length of the tubes. The chirality and the tube diameter introduces significant changes in the electronic density of states, and hence, provide unique electronic character for the CNTs. The other factor is topology, or the closed geometry of the individual layers in each tube, which has a profound effect on the physical properties; in contrast to planar graphite where the individual graphene layers are stacked on top of each other with weak van der Walls forces of attraction. The combination of size, structure, and topology endows nanotubes with important and unique mechanical (stability, strength, stiffness, and elastic deformability), thermal, and surface properties. By rolling a graphene sheet to form a CNT, SWNT, or MWNT, the total energy of the nanotube is increased by the strain energy associated with the curvature of the nanotube, the strain energy increases with decreasing nanotube diameter. SWNTs have a tensile modulus that ranges from 270 GPa to 1 TPa and a tensile strength ranging from 11 to 200 GPa which is 10–100 times higher than the strongest steel at a fraction of the weight. SWNTs have a very low density of $\sim$1.3 g/cm$^3$ because of the hollow structure of SWNTs. SWNTs also exhibit exceptional thermal and electrical properties. SWNTs are thermally stable up to 2,800°C in vacuum, their thermal conductivity is about twice as high as that of diamond; these high values are due to the large phonon mean free paths present in SWNTs, the fact that their electric-current carrying capacity is 1,000 times higher than that of copper wires (Banda 2004).

Synthesis and production methods for CNTs have been developed, typically including the arc-discharge, laser ablation, CVD from hydrocarbons, and gas phase catalytic growth from carbon monoxide. The electric arc discharge technique generally involves the use of two high purity graphite rods as the anode and cathode, the rods are brought together under helium atmosphere and a voltage is applied until a stable arc is achieved. The material then deposits on the cathode to form a build up consisting of an outside shell of focused material and a softer fibrous core containing nanotubes and other carbon particles, and the CNTs are obtained by purification. In laser ablation technique, a pulsed or continuous laser is used to vaporize a graphite target held in a controlled atmosphere oven at temperatures near 1,200°C. The condensed material is then collected on a water-cooled target. Laser vaporization can yield more SWNTs with narrower size distribution than those produced in the arc discharge process. MWNTs can also be produced with pure graphite using this process. However, both arc discharge and laser ablation techniques produce small quantities of CNTs with many impurities. Gas phase techniques are therefore developed to overcome these limitations. CVD from hydrocarbons is a gas-phase technique that utilizes hydrocarbon gases as the carbon source for production of CNTs. The CNTs are formed by the decomposition of a carbon containing gas, the hydrocarbons pyrolize readily on surfaces when they are heated above 600–700°C. This is a continuous process as the carbon source is continuously replaced by flowing gas. High-pressure carbon monoxide is another gas-phase technique to exclusively produce SWNTs with carbon monoxide as the carbon source. SWNTs are produced by flowing carbon monoxide with a small amount of $Fe(CO)_5$ through a heated reactor that is maintained between 800 and 1,200°C. The products of Fe $(CO)_5$ thermal decomposition react to produce iron clusters in gas phase. These iron clusters act as nuclei upon which SWNTs nucleate and grow. This is an effective method to continuously produce high purity SWNTs. In addition to the SWNTs, the raw product material contains amorphous carbon and iron (Fe) particles coated by thin carbon layers. The unwanted products can be selectively removed by a multi-stage procedure which is thermal oxidation in a wet $Ar/O_2$ environment and washing with concentrated hydrochloric acid (Banda 2004).

Both SWNTs and MWNTs can be reinforcements for composite structures, but in the case of MWNTs, very weak van der Walls forces exist to link the individual graphene shells together, allowing the inner shells to rotate and slide freely; therefore, using MWNTs as intrinsic reinforcements for composite structures may not allow the maximum strength or thermal conductivity to be achieved. For this reason, SWNTs are most desired in CNT-reinforced composites. CNTs can be embedded into matrices like polymers to make light-weight composites with improved conductivity, mechanical and dielectric properties. The practical applications of CNT composites have so far been largely limited by their poor processibility. The intrinsic van der Walls attraction among the tubes in combination with their high surface area and high aspect ratio often leads to significant agglomeration of CNTs. The surface of CNTs is also nonreactive which makes it difficult in achieving efficient dispersion, as they mix and blend with the host matrix. To effectively use CNTs as intrinsic reinforcements in composite structures and ensure a good thermal transfer between CNTs and the polymer matrix, uniform dispersion within the polymer matrix and improved nanotube/matrix wetting and adhesion are critical. Different methods have been used to efficiently disperse the CNTs in polymer and other matrices. CNTs are ultrasonicated in a solvent to disperse them before (or during) they are added to the polymer matrix or before another dispersion technique is applied. Appropriate parameter control for the ultrasonic process may be the key here to get the best dispersion result. Microscale twin screw extruder melt mixing and coagulation are some mechanical methods that minimize the aggregate formation by applying appropriate shear to CNT polymer composite. Dispersion of CNTs is also done by using dispersants as processing aids in CNT polymer composites. CNTs can also be chemically modified and functionalized to disperse them better in the polymer matrix. The functionalization chemistry of the open ends, the exterior walls, and the interior cavity of the CNTs play a vital role in tailoring the properties of CNTs. CNT sidewall surface is chemically modified by fluorinating and defluorinating them, CNTs then serve as a staging point for a wide variety of sidewall chemical functionalizations that results in better dispersion. Another method to disperse CNTs is by in situ polymerization in which CNTs are added while the polymerization is in process (Banda 2004).

On the other hand, CNTs can be aligned before or after being embedded into the composite matrix to take advantage of their anisotropic structure and to have improved properties in the direction of the alignment. By aligning the CNTs in the polymer matrix, the strength, stiffness, and electrical and thermal properties of the composite can be controlled. Depending on the application of CNTs, they can be aligned by arc discharge technique and CVD by applying an electric field while fabricating the CNTs or they can be aligned after they are dispersed in a solvent. The advantage of aligning CNTs while producing them by the CVD process is that it allowed assembling individual CNTs into desired architectures by placing them at specific locations with controlled orientations. The disadvantage to this method is that the CNTs are not perfectly aligned when they are embedded into composites as they lose their alignment as a result of dispersion. An electric field can be applied to a CNT suspension to assemble, rearrange, and orient the CNTs. Mechanical methods such as the microscale twin screw extruder, melt processing, and mechanical stretching of CNT composites are also used to align the CNTs in polymer composites. Furthermore, strong magnetic field can be applied to magnetically align CNTs in polymer composites. Aligning CNTs in polymer composites mechanically and by applying an electric field has rendered better alignment compared with aligning CNTs by a magnetic field. CNTs are not easily influenced by magnetic fields; in addition, very high magnetic fields are usually required (Banda 2004). As an example of aligned CNT application for thermal management of electronic packaging, the thin film of CNTs and carbon have been developed for thermal interface materials to replace thermal grease or phase-change materials.

Consequently, while individual CNT has demonstrated exceptional thermal-conductive properties, when used as filler in composite materials, their aggregate properties fall far short of what should be expected. This is due partly to the presence of defects in the CNTs, the tendency to bundle so they do not fully disperse in the composite, and interference effects between individual isolated CNTs. These limitations can be overcome by joining adjacent CNTs at several points, which provides multiple pathways around defects, allowing a continuous path for mechanical and thermal transfer. This improvement can be further enhanced by rotating the orientation of aligned alternating layers of CNTs. The effect is similar to that which gives Plywood its great strength by alternating wood grain orientation in successive layers. One typical approach is to first functionalize, or add chemical groups to the surface of the CNTs, to make them soluble in a mildly polar organic solvent. This allows them to be separated into individual nanotubes, although they are still randomly aligned. Another process called Langmuir-Blodgett, can be used to produce a single layer (monolayer) of CNTs, or forming buckypaper with help of magnetic field. Rearrangements of the carbon atoms produce a new structure which connects the CNTs with bonds that are as strong as the CNTs themselves. Planner CNTs have been directly grown to form thin film for potential 2-D applications.

Unfortunately, when CNTs are added to various composite materials, the improvement in thermal conductivity is modest. Even in bundles of CNTs, the aggregate thermal conductivity is still only comparable with that of silver. Difficulties arise from a combination of factors including poor thermal transfer between disjoint CNTs, the presence of defects, and interference effects between isolated CNTs. Cross linking addresses all of these shortcomings by causing the assemblage of CNTs to act as a whole, rather than a collection of individuals. Thermal transfer between CNTs is readily affected through cross link bridges, alternate path around defects is also provided, and interference effects are suppressed by collective contributions of local groups of interconnected CNTs. The exceptionally high thermal conductivity of CNTs means that only a very thin film will be needed, offsetting the initial cost of this material. Also, CNTs, being relatively chemically inert, are compatible with semiconductor processes. MWNTs are much less expensive than SWNTs and are suitable for this application, further enhancing mass production economics.

In addition to high-thermal conductivity, the heat spreader must also match the CTE of the component being cooled. Otherwise, as the temperature rises, they will expand at differing rates, producing stresses that can crack the device or create reliability problems over several repeated temperature cycles. This is why a diamond particle reinforced composite is usually used for high-end heat spreaders, rather than a thin film of pure diamond. The very low CTE of diamond produces unacceptable stresses. The ability of CNTs to withstand repeated bending and kinking without damage provides a built-in stress releasing mechanism to absorb CTE mismatch. In this regard, CNTs have also been proposed as shock absorbers.

## Graphene

Graphene is a one-atom-thick planar sheet of $sp^2$-bonded carbon atoms that are densely packed in a honeycomb crystal lattice. The C–C bond length in graphene is about 0.142 nm. Graphene is the basic structural element of some carbon allotropes including graphite, CNTs and fullerenes. It can also be considered as an infinitely large aromatic molecule, the limiting case of the family of flat polycyclic aromatic hydrocarbons called graphenes. The graphene exhibits extremely high intrinsic thermal conductivity (3,080–5,300 W/m K), high electron mobility ($\sim 2 \times 105$ $cm^2/V$ s), and low resistivity ($\sim 10^{-8}$ Ωm). The unique electrical and thermal properties of graphene make it a promising material for the low-noise transistors, electrical interconnects, and thermal management (Balandin et al. 2008; Ghosh 2008). The advanced electronic devices have been explored to provide improved heat removal capabilities from electronic, optoelectronic, and photonic devices and integrated circuits via incorporation of high thermally conducting channels made of graphene. The embodiments allow for better thermal management of the electronic and optoelectronic devices and circuits and reduced power consumption.

## Carbon–Carbon Composites

C–C composites with pitch fiber reinforcements would be very attractive for thermal management of high-power electronic packaging because of their light weight, high strength and stiffness, low CTE, excellent thermal conductivity, high thermal shock resistance, and low recession in high-pressure ablation environments. They can be manufactured by different techniques such as solid pyrolysis of the preform of carbon fiber/thermosetting resins, liquid pitch infiltration and carbonization, and gaseous CVD. The matrix is either a graphitizable carbon or nongraphitizable carbon, and the carbonaceous reinforcement is fibrous in form. The composite may also contain other components in particulate or fibrous forms. C–C composites may be manufactured with different orientation of the reinforcing carbon fibers: unidirectional structure, bidirectional structure (cloth made of multiple carbon fiber yarns), and multidirectional structure.

As shown in Figure 4.7, the fabrication process of C–C composites involves preparation of a three-directional or multidirectional porous carbon fiber preforms having the desired shape or a porous C–C skeleton with carbon fibers and a carbonaceous matrix followed by their densification. The densification can be achieved through vapor phase infiltration within the pressurized hydrocarbon gases such as methane, propane, propylene, acetylene, and benzene infiltrating into porous fibrous structure heated to a temperature of 1,000–1,400°C in the absence of oxygen and are made to crack therein. Commercially, isothermally

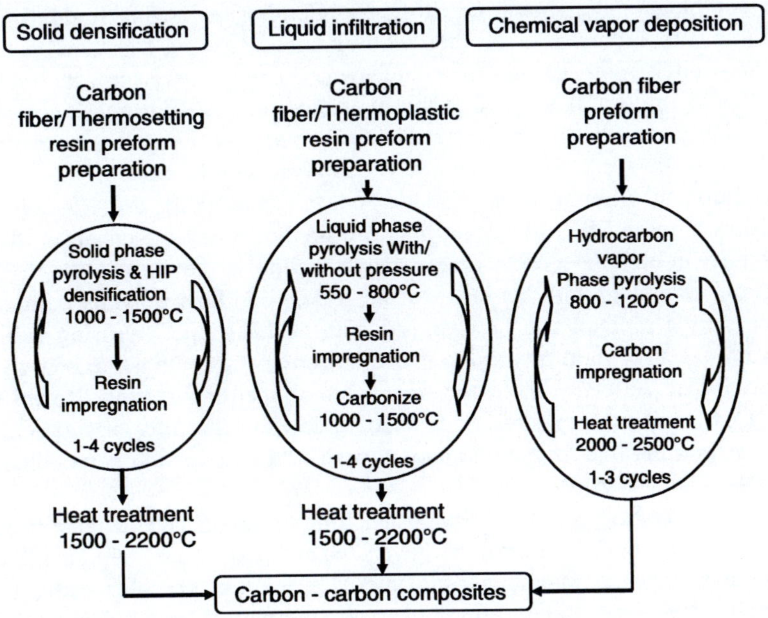


**Fig. 4.7** Illustration of fabrication process of high-thermal conductivity C–C composites

heated stacks of components are impregnated simultaneously in a large-size furnace. Principally, the gases should infiltrate into the pores and interfilamentary spaces and then crack. But these gases have a tendency to crack at the outer surface itself which blocks the passages and causes closed pores. In order to have dense composites, the surfaces are ground and the components are reinfiltrated. The process duration is determined by the thickness of the preform, through which the gas diffuses. CVD is a very slow process and it usually takes months to get dense C–C composites. In the following graphitization heat treatment, the deposited amorphous carbon is transformed into crystalline graphite. Porous structures can also be impregnated with liquid-phase pitch or phenolic resin or coal tar followed by carbonization and high-temperature heat treatment (Manocha 2003). Infiltration-pyrolysis cycle is repeated several (up to four or higher) times until the desired density is achieved. At the graphitization heat treatment stage amorphous carbon is transformed into crystalline graphite. The temperature of the treatment may vary within the range of 2,700–5,400°F (1,500–3,000°C). Typical graphitization temperature is 4,530°F (2,500°C). Graphitization of carbon–carbon composites results in increase of thermal conductivity, modulus of elasticity and strength of the composite. Pyrolysis of the pitch fiber/resin matrix can be done under normal pressure or under high pressure (Manocha et al. 2003).

As C–C composites are of heterogeneous structure consisting of fibers, matrix and pores, estimation of their thermal transport properties becomes complex. However, C–C composites with tailored thermal conductivities can be fabricated by proper choice of constituents, their configuration, and processing conditions. A comparison of thermal conductivities of C–C composites with different fiber/matrix combinations is shown in Table 4.1. Composites having highly oriented graphitic fibers or matrices or their combination, like VGCFs and matrix or mesophase pitch-based carbon fiber and matrix, exhibit very high thermal conductivities of up to 700 W/m K in the fiber direction. Though these composites exhibit highly anisotropic character and low conductivities in transverse directions, there is always scope of improvement by varying the fiber architecture and their addition in different forms and ways (Manocha 2003). For example, C–C composites based on K1100 carbon fiber are predicted to have enhanced thermal conductivity in both the $x$ and $y$ directions, as well as in the through-the-thickness $z$ direction, compared with laminates of K1100 carbon fiber polymer composites. A space shuttle radiator can be designed with no heat pipes by using laminates of K1100 (6:1, fiber ratio of warp to fill) C–C composites with a maximal thermal conductivity of 700 W/m K in the fiber-warp-aligned $x$ direction (Rawal et al. 1994). CTE of the composites is dictated largely by the fiber orientation. It is 0–1 ppm/°C in the fiber direction and 6–8 ppm/°C in a direction perpendicular to the fibers (Manocha 2003).

In spite of advantages of C–C composites for thermal management applications, the increasing use of C–C composites have been affected by the prohibitive product expense and long-lead manufacturing times due to slow, complex carbonization and densification steps. Hence, affordable, more thermally conductive materials are needed to meet future thermal management requirements.

## Summary

Elemental carbon in its chemical allotropes of graphite, diamond, and nanotuber occurs in a great variety of species and has been developed to a large number of highly specialized applications as structural and functional materials, in particular for thermal management of electronic packaging. These applications have been augmented by the creation of mesoscopic composites on the basis of carbon-reinforced carbon materials. Carbon plays a major role in nanosciences and the application to electronic packaging. Fullerenes, nanotubes, whiskers, and carbon onions are typical representatives of nanostructured carbons to which all variables also apply in addition to the small scale of each object. Diamond as CVD diamond, as diamond-like carbon, and as homoepitactic diamond plays a strategic role in the electronics industry, including thermal management components, protective layers on chips and hard disks, and high-power transistors. Highly oriented graphitic carbon and graphite foam can be used as heat shields, heat sink, and other heat dissipating components in electronic packaging. Carbon fibers are large-scale products used as reinforcement structures in polymers, metals, and carbon materials. Electronic applications as thermal vias, cables, heaters and super capacitors are under development or existing in niche applications.

With carbon fibers reinforcements in different forms and directions and thermosetting resins or thermoplastic pitches or hydrocarbon gases as matrix precursors, C–C composites can attain densities between 1.6 and 1.98 g/cm$^3$. Reinforcing fibers are more anisotropic in structure and properties than the carbon matrix. Therefore, the thermal conductivity and other properties of C–C composites are dominated by fiber volume content and fiber orientation. Unlike polymer matrices, carbon matrices contribute significantly to the ultimate properties of the composites, especially in case of pitch and CVD-derived carbon matrices. C–C composites are a family of materials with choice of variation in fiber and matrix architecture, structure, microstructure, mechanical, thermal, and physical properties. Hence, the C–C composites have a great potential to make high performance components in thermal management of electronic packaging.

The use of cost-effective carbon materials in consumer electronics is accelerating. Reduced cost synthetic diamond and carbon fibers make composite materials more attractive. Materials engineering of the interfaces within composite materials is key in determining bulk thermal properties. Nanomaterials and nanocomposites are important for next-generation high-conductivity laminates and anisotropic materials.

## References

Arai Y (1993) Structure and properties of pitch-based carbon fibers. Nippon Steel Technical Report No. 59 October 1993: 65–70.

Balandin AA, et al (2008) Superior thermal conductivity of single-layer graphene, Nano Letters 8: 902–907.

Banda S (2004) Characterization of aligned carbon nanotube/polymer. Master thesis. Virginia Commonwealth University, Richmond.

Bigelow LK (1993) Diamond coatings. Journal De Physique IV(3): 897–902.

Choudhary D, Bellare J (2000) Manufacture of gem quality diamonds: a review. Ceramics International 26: 73–85.

Ghosh S (2008) Extremely high thermal conductivity of graphene: Prospects for thermal management applications in nanoelectronics, Applied Physics Letters 92: 151–911.

Hardcastle L, Sheppard RG, Dingus DF (2004) Process for making porous graphite and articles produced thereform. US Patent 6776936.

Higashi T, Kambara E, Tsuji K, Aoki T (2007) Vapor-grown carbon fiber and production process thereof. IPC PCT/JP2006/305473.

Jin S, Mavoori H (1998) Processing and properties of CVD diamond for thermal management. Journal of Electronic Materials 27(11): 1148–1153.

Klett J, McMillan A, Gallego N, Walls C (2003) The role of structure on the thermal properties of graphite foams. http://www.ms.ornl.gov/researchgroups/CMT/FOAM/klett-jmatsci-2003.pdf. Accepted 26 March 2010.

Klett JW (2000) Process for making carbon foam. US Patent 6033506.

Montesano MJ (2006) Annealed pyrolytic graphite. Advanced Materials & Processes. http://www.asminternational.org/pdf/amp_articles/AMP200606_Graphite.pdf.

Manocha et al (2003) Microstructure of carbon/carbon composites reinforced with pitch-based ribbon-shape carbon fibers. Carbon 41: 1425–1436.

Norley J, Chen G (2002) High performance, lightweight graphite heat sinks/spreaders. Proceedings of PCIM 2002: Power electronics – thermal management and packaging. May 14–16. Nurnberg, Germany.

Partrigge PG, Lu G, May P, Steeds JW (1995) Potential high-strength high thermal conductivity metal-matrix composites based on diamond fibers. Diamond and Related Materials 4:848–851.

Pierson HO (2008) Handbook of carbon, graphite, diamond and fullerenes. Elsevier, New York.

Rawal SP, et al (1994) Thermal conductance maps of an advanced composite radiator. Proc. 39th Int. SAMPE Symposium, Anaheim, Calif., April 11–14, 1994, pp. 2168–2182.

Smalc M, et al (2005) Thermal performance of natural graphite heat spreaders. Proceedings of IPACK, ASME InterPACK, July 17–22. San Francisco, California, USA.

Stansberry PG, Lewis IC (2008) Process for the production of highly graphitizable carbon foam. US Patent 0286191.

Young R, et al (2006) Developments and trends in thermal management technologies – a mission to the USA. www.lboro.ac.uk/research/iemrc/documents/. . ./CB2007.pdf. Accessed 26 February 2010.

# Chapter 5
# Thermally Conductive Polymer Matrix Composites

**Abstract** Thermally conductive polymer matrix composites are used increasingly for thermal management of electronic packaging system. The polymer matrix includes both thermosetting and thermoplastic types. Different kinds of fillers or reinforcements have been developed to process composite materials with desired thermal, mechanical, and electrical properties. Reinforcement fillers have an important role to play in maximizing polymer performance and production efficiency. Cost reduction, density control, optical effects, thermal conductivity, magnetic properties, flame retardancy, and improved hardness and tear resistance have been increased the demand for high-performance fillers. Several types of reinforcements, especially nanoparticulate fillers, have been used in polymer matrix composites: vapor grown carbon fiber (VGCF), carbon foam, carbon nanotube (CNT), and other thermal conductive particles, such as ceramic, carbon, metal or metal-coated particles, as well as metal or carbon foams. Nanoparticles of carbides, nitrides, and carbonitrides can be used to reinforce polymer matrix nanocomposites with desirable thermal conductivity, mechanical strength, hardness, corrosion, and wear resistance. To achieve these of desirable properties, however, polymer matrix and layout or distribution of reinforcements including nanoparticles need to be optimized. This chapter will give a brief review on the advanced thermally conductive polymer matrix composites, including polymer matrix types, reinforcement selection and its effect on thermal conductivity, fabrication and manufacturing process, and typical applications for thermal management of electronic packaging system.

## Introduction

Thermally conductive polymer matrix composites have been increasingly used for thermal management of electronic packaging system. Composites with continuous reinforcements such as fibers, woven or cloths, are usually used as substrates, heat sinks, and enclosures. Composites with discontinuous reinforcements such as particles or whiskers, are generally used for die attach, electrically/thermally conductive adhesives, encapsulations, thermal interface materials (TIMs), and electrical interconnections (Chung 2003). The composite matrix can be thermal

X.C. Tong, *Advanced Materials for Thermal Management of Electronic Packaging*, Springer Series in Advanced Microelectronics 30,
DOI 10.1007/978-1-4419-7759-5_5, © Springer Science+Business Media, LLC 2011

plastic or thermosetting polymer. Thermoplastic composites can be reworked by heating for the purpose of repair, whereas thermosetting composites do not allow reworking, but are attractive for their thermal stability and dielectric properties. The polymer matrices for thermally conductive composites shall preferably exhibit low dielectric constant, low dissipations factor, low coefficient of thermal expansion (CTE), and being compliable for electronic packaging applications. The composites can be electrically conducting or electrically insulating, or both electrically and thermally conducting, depending on the reinforcement materials used. With proper reinforcement display, the electrical or thermal performance of the composites can be isotropical or anisotropical (Chung 2003).

Typical composites for electronic packaging components exhibit tailored CTE by varying the architecture of reinforcements, such as carbon fibers, silicon carbide (SiC) particles, boron nitride particles, titanium nitride particles, and diamond particles. The resulting composite materials with these reinforcements have not shown substantial improvements in thermal conductivity. VGCFs that are grown through the pyrolysis of hydrocarbon gas in the presence of a metal catalyst, have a higher thermal conductivity than any other carbon fiber and may be produced at a lower cost. VGCFs also exhibit the highest degree of graphitic perfection of any known carbon fiber. Polymer matrix composites formed from the interwoven VGCF mats, exhibit high thermal conductivity (>400 W/m K), low density, and variable CTE (Ting 2002).

On the other hand, the through-thickness thermal conductivity in the polymer matrix composite becomes particularly important in applications such as composite space electronics enclosures where the heat dissipation is entirely dependent on thermal conduction to a heat sink. The spreading of heat at the composite surface and subsequent localized conduction in the through-thickness direction down to high thermal conductivity fiber is thought to be the key to designing a light weight, thermally efficient enclosure. Consequently, there exists a need for lightweight, thermally conductive through-thickness composite materials. For instance, low conductivity in the through-thickness direction of unidirectional fiber reinforced composites poses problems in any design that requires a thermal path to a high thermal conductivity fiber, particularly where large amounts of heat are input over small areas. As one approach, introducing metal particles into the matrix will increase the through-thickness conductivity throughout the composite; however, the weight density of the metallic reinforcements over a polymer matrix composite is greatly increased. Locally increasing the through-thickness thermal conductivity and then allowing the high thermal conductivity fibers to spread and orient the heat flow to a heat sink appears to be an optimal solution (Roberts 1999).

## Polymer Matrix Types

The polymer matrices in advanced thermally conductive composites include both thermosetting and thermoplastic types. Table 5.1 gives general properties of some typical polymer matrices. The most widely used thermosetting resins are polyester,

**Table 5.1** General properties of typical polymer matrices of polymer matrix composites (Alvino 1994)

| Properties | Polyesters | Epoxies | Phenolics | Polyethylene | Polypropylene | Acrylics | Silicones |
|---|---|---|---|---|---|---|---|
| Thermal conductivity (W/m K) | 0.15–0.4 | 0.14–0.87 | 0.87 | 0.33 | 0.17–0.25 | 0.21–0.23 | 0.19–0.22 |
| Coefficient of thermal expansion ($10^{-6}$/K) | 18–80 | 18–63 | 14.4–40 | 31.7–67.3 | 15.1–23.4 | 11.9–23.8 | 178.2 |
| Specific heat (J/K kg) | 840–2,090 | 1,674 | 1,172–1,674 | 1,925–2,301 | 1,883–2,008 | 1,464 | – |
| Glass transition temperature (°C) | 70–90 | – | – | −25 | −1 to −13 | −24 to −115 | – |
| Melt temperature (°C) | 220–285 | – | – | 83–135 | 160–168 | <100 to >200 | – |
| Volume resistivity (Ω cm) | $10^5$ | $10^9$–$10^{16}$ | $10^9$–$10^{12}$ | $10^{15}$–$10^{19}$ | $10^{17}$ | $10^{14}$–$10^{16}$ | $10^{14}$–$10^{15}$ |
| Dielectric strength (kV/mm) | 14–23 | 11.0–17.0 | 7.9–18.7 | 18.9 | 17.7–27.6 | 15.8–20.9 | 4.9–21.7 |
| Dielectric constant | | | | | | | |
| 60 Hz | 3.1–4.3 | 3.3–5.7 | 5.0–16 | 2.3 | 2.2 | 3.4–4.5 | 2.7–5 |
| 1 MHz | 2.9–3.9 | 2.7–4.8 | 4.0–7.0 | – | 2.2 | 2.5–3.2 | 2.6–4 |
| Tensile strength (MPa) | 55–172 | 28–186 | 21–117 | 11.0–62 | 20.7–96.5 | 34.5–75.8 | 2.4–10.3 |
| Tensile modulus (GPa) | 1.65–24 | 0.35–3.45 | 5.52–23 | 0.028–6.21 | 0.90–1.66 | 1.93–3.10 | 0.062 |
| Elongation (%) | 1–300 | 1.0–60 | 0.4–1.0 | 80–1,000 | 3–200 | 2.0–30 | 20–800 |
| Density (g/cm$^3$) | 1.2–1.4 | 0.75–1.15 | 1.35–1.75 | 0.91–0.97 | 0.905 | 1.18 | 1.05–1.19 |
| Flammability | Self extinguishing | Self extinguishing | V0, V1, HB | HB | HB | – | |
| Water absorption 24 (%) | 0.3 | 0.2–1.0 | 0.05–0.7 | <0.01 | 0.01–0.03 | 0.3 | 0.12 |

which provide a combination of low cost, versatility in processing, and relatively good property and performance. Polyester is generally made by reacting dibasic acids such as maleic anhydride or phthalic anhydride with dihydric alcohols such as ethylene glycol (antifreeze) in approximately equal amounts. The resulting polymer is a short chain polymer with a molecular weight of about 5,000 (about 35–40 repeat units or mers) and is a stable liquid. The liquid resin is set into an amorphous solid by cross linking the polyester chains to each other. The cross linking occurs by the addition of a small monomer molecule such as styrene. The monomer, like the active sites on the polyester chain has an unsaturated C=C bond and it is this that provides the bridge between the polyester chains (Pilling 2005).

Epoxy resins are of particular interest to structural and electronic applications because they offer a unique balance of chemical and mechanical properties combined with extreme processing versatility. The performance of epoxy resins is highly dependent on the formulation, including the base resin, curatives, and modifies (Nutt 2001). Epoxy resins are much more expensive than polyester resins because of the high cost of the precursor chemicals, most notably epichlorohydrin. However, the increased complexity of the epoxy polymer chain and the potential for a greater degree of control of the cross linking process gives a much improved matrix in terms of strength and ductility. Most epoxies require the resin and hardener to be mixed in equal proportions and for full strength require heating to complete the curing process. This can be advantageous as the resin can be applied directly to the fibers and curing need only take place at the time of manufacture, such as prepreg or preimpregnated fiber. Epoxy polymers are made by reacting epichlorohydrin with bisphenol-A in an alkaline solution which absorbs the HCl released during the condensation polymerization reaction. Each chain has a molecular weight between 900 and 3,000 (about 3–10 mers) with an epoxide grouping at each end of the chain but none within the polymer chain. The epoxy is cured by adding a hardener in equal amounts and being heated to about 120°C. The hardeners are usually short chain diamines such as ethylene diamine. Heat is usually required because the cross linking involves the condensation of water which must be removed in the vapor phase (Pilling 2005).

Silicones are used mainly for TIMs in electronic packaging. Silicone polymers composed of alternating atoms of silicon and oxygen with organic substituents attached to the silicon atoms, which may exist as liquids, greases, resins, or rubbers. They have good resistance to water and oxidation, stability at high and low temperatures, and lubricity. Silicones are obtained by the condensation of hydroxy organosilicon compounds formed by the hydrolysis of organosilicon halides. The first products are usually low in molecular weight ($n = 2$–7), and usually consist of a mixture of linear and cyclic species, especially the tetramer. Fluids having a wide range of viscosity are prepared by polymerizing further, using a monofunctional trichlorosilane to limit molecular weights to the value desired. Elastomers are made by polymerization of the purified tetramer using an alkaline catalyst at 100–150°C (212–302°F). Properties can be varied by partial replacement of some of the methyl groups by other substituents and by the use of reinforcing fillers. The wide range of structural variations makes it possible to tailor compositions for many kinds of applications. Low molecular weight silanes containing amino or other functional groups are used as treating or coupling agents for glass fiber and other reinforcements in order to cause unsaturated polyesters and other resins to adhere better. The liquids, generally dimethyl silicones of relatively low molecular weight, have low surface tension, great wetting power and lubricity for metals, and very small change in viscosity with temperature. They are used as hydraulic fluids, as antifoaming agents, as treating and waterproofing agents for leather, textiles, and masonry, and in cosmetic preparations. The greases are particularly desired for applications requiring effective lubrication at very high and at very low temperatures. Silicone resins are used for coating applications in which thermal stability in the range 300–500°C (570–930°F) is required. The dielectric properties of the polymers make them suitable for many electrical applications, particularly in electrical insulation that is exposed to high temperatures and as encapsulating materials for electronic devices. Silicone rubbers are compositions containing high molecular weight dimethyl silicone linear polymer, finely divided silicon dioxide as the filler, and a peroxidic curing agent. The silicone rubbers have the remarkable ability of remaining flexible at very low temperatures and stable at high temperatures.

There are also many other polymers that can be chosen as the matrix. Basically, the matrix should provide (Trostyanskaya 1995): (1) good wettability with reinforcements to couple over their surfaces as a result of chemical and polar interactions preserving a clear phase boundary; (2) controllable or limited chemical reaction with reinforcements to be able to modify bulk properties of the reinforcement; (3) monolithization and configuration forming or shaping in regimes preventing thermodegradation and mechanical fracture or destruction of reinforcements and disturbing their distribution; (4) continuity of an uniform matrix distribution over the whole interstitial space at a required filling volume of the reinforcement.

One typical example is to use polymer as the matrix of the conductive adhesives that have been used for die mounting and terminal bonding of components in some types of hybrid circuits, especially for TIM applications. Polymers are long-chain molecules, such as epoxies, acrylics, and urethanes that are widely used to produce structural products such as films, coatings, and adhesives. Although polymers occur naturally, most are now synthesized. Their properties can be tailored to meet thousands of different applications. Polymer-based adhesives are used in virtually every industry because of this capability to customize performance. Polymers have excellent dielectric properties and, for this reason, are used extensively as electrical insulators. Although a narrow class of conductive polymers, called intrinsically conductive polymers, does exist, their other properties do not lend themselves for use as conductive adhesives. Therefore, adding conductive fillers to nonconductive polymer binders makes virtually all-conductive adhesives. The most common conductive adhesives are silver-filled thermosetting epoxies that are typically provided as thixotropic pastes. They are used to electrically interconnect and mechanically bond components to circuits. Heat is most often used to activate a catalyst or coreactant hardener that converts the paste to a strong, conductive solid. The products, which conduct equally in all directions, are referred to as isotropic conductive adhesives. These metal-filled thermosetting conductive adhesives have been used as die attach materials for many decades and are still the most popular products for bonding integrated circuits to lead frames. More recently, metal-filled thermoses have been formulated as component assembly materials. Novel polymer-based materials are now being used to replace metallurgical solders, especially for surface mount assembly (Nordic 2010).

Two-type polymers are usually used for the conductive adhesives: thermoplastics and thermosetting polymers. Thermoplastic binders, those polymers that are already polymerized, have not found widespread use for component assembly, as have the thermoses isotropic type conductive adhesives. One problem is that useful thermoplastics are solids that must either be melted or dissolved in solvent to be used. Very limited use of isotropic thermoplastics has occurred in calculators, but the difficulty in application has severely limited them. Anisotropy conductive adhesives do use thermoplastic binders because they do not require selective application. Most anisotropy products are provided in film form. The material is used by first applying it to the circuit (and sometimes to the component). Because electrical conductivity only occurs in the *z*-axis where opposing conductors are forced together, the film can be applied to the entire circuit. Assembly is accomplished by forcing components against the adhesive-coated circuit conductors while adding heat. A thermoplastic binder will soften and bond to the adherents. Thermoplastic materials can also be mixed with thermosetting polymers to allow lower temperature assembly by higher temperature performance. It should be emphasized that thermoplastics can be remelted. They are not altered during the assembly heating process like thermoses. They also have excellent storage characteristics and do not require refrigeration like the one-part thermoses (Nordic 2010).

Most thermosetting polymers, especially the pastes used for isotropically conductive adhesives, are polymer precursors (ingredients that will polymerize). Epoxies typically consist of a low weight liquid with reactive epoxy groups and a coreacting hardener. The addition of heat causes the two ingredients to chemically react forming very high weight, cross-linked polymers. Cross links, or chemical bonds between adjacent chains, produce the thermoses characteristic of shape retention on heating. The thermoplastics are made up of polymer chains that are independent (not linked). Heat allows the individual chains to move past one another and to be reshaped. The reapplication of heat again softens the thermoplastic making them somewhat analogous to solder. The three-dimensional (3-D) network of cross links in the thermoses prevents chain movement. Thermoses adhesives are provided as both single component (one-part) and as unmixed two-part is stable almost indefinitely. Prior to use, the ingredients must be mixed, however. This often introduces air, a serious problem, and the user may not have the equipment to produce good mixing. Both users and manufacturers prefer one-part systems. The one-part system's major disadvantage is that the mixed system has a limited storage life unless refrigerated. Recalling that a chemical reaction rate increases for every 10°C, the converse is also true. Refrigeration slows the polymerization rate to a level where storage can be extended to 6 months or more. While it is possible to make up systems that are stable at room temperature, higher temperatures over extended times are required for curing. However, improvements in catalysts and hardeners continue and pot life, now up to 6 days at room temperature for some of the products that cure quickly (3–6 min) even as low as 130–150°C, can be expected to continue to improve (Nordic 2010).

## Reinforcements of Conductive Polymer Composites

Different kinds of fillers or reinforcements have been developed to process composite materials with desired thermal, mechanical, and electrical properties. Fillers may be in the form of fibers or in the form of particles uniformly distributed in the polymer matrix material. The properties of the polymer composite materials are strongly dependent on the filler properties as well as on microstructural parameters such as filler diameter, length, distribution, volume fraction, and the alignment and packing arrangement of fillers. It is evident that thermophysical properties of fiber-filled composites are anisotropic, except for the very short, randomly distributed fibers, whereas, thermophysical properties of particle-filled polymers are isotropic (Tavman 2004).

For instance, silver is by far the most popular conductive filler for conductive polymer adhesives, although gold, nickel, copper, and carbon are also used. Metal-plated con-conductors are also used, especially for anisotropy adhesives. Silver has high thermal and electrical conductivity. The metal forms a silver oxide layer at the surface. This layer, unlike many other oxide layers, has good electrical conductivity. Due to this special characteristic the aging of the silver flakes will not result in significantly reduced conductivity in a composition including flakes. A disadvantage of silver is its tendency to migrate. Furthermore, it is relatively expensive. Even so, silver is unique among all of the cost-effective metals by nature of its conductive oxide. Oxides of most common metals are good electrical insulators and copper powder, for example, becomes a poor conductor after aging. Nickel and copper-based adhesives do not have good stability. Copper has good conductivity and is not very expensive but in the presence of oxygen it will form a continually growing oxide layer, which reduces the conductivity of the composition. This means that copper cannot offer stability with regard to thermal and electrical conductivity. Nickel forms well-defined oxides layers. Both for nickel and copper the relatively low thermal and electrical conductivity limits the use of this material to low-cost applications that do not require high thermal and electrical conductivity. Aluminum has the same tendency of forming an oxide layer resulting in reduced electrical conductivity as with copper, but it is usually used as low-cost thermally conductive filler for TIMs. Even with antioxidants, copper-based materials will show an increase in volume resistivity on aging, especially under high humidity conditions. Silver-plated copper has found commercial application in conductive inks and this type of filler should work in adhesives as well. While composites made with pure silver particles often show improved conductivity when heat aged, exposed to heat and humidity or thermal cycled, this is not always the case for silver-plated metals, such as copper flake. Presumably, the application of heat and mechanical energy allows the particles to make more intimate contact, but the silver-plated copper may have coating discontinuities that allow oxidation of the copper. Therefore, silver is found to be the most frequently used filler particle material in conductive adhesives. Gold has lower conductivity than silver and is much more expensive but it has one great advantage compared with silver, and that is a very low ability to migrate. As silver, gold forms a thin and relatively conductive oxide layer. Gold is often chosen for use in military and space electronics (hybrid circuits) to avoid the risk of silver migration. In addition to the pure metals, attempts have been made to develop composites for use as filler material. Examples are copper and glass plated with silver, and coated reformative (Nordic 2010).

Fillers for anisotropy conductive adhesives are often very different from those for isotropic adhesives. There is usually only one layer of conductive particles between the two adherents in anisotropy configurations. The conductive particle makes a mechanical contact between the conductors with the polymer binder supplying the tensional force. Although both hard and soft conductors are being used, most systems have moved toward resilient particles that deform and act like small springs. The most common type has a plastic core that is overcoated with a good conductor like gold or silver. The most popular conductive particle is a polymer sphere that has been first plated with nickel and then pure gold (Nordic 2010).

It is not only the choice of filler material but also its size and shapes that influence the properties of the final adhesive. Some of the properties that can be influenced by particle size and shape are electrical conductivity, thermal conductivity, tensile strength, viscosity, weight loss, and rheology. The production of silver flakes are often based on silver powder, which is mechanically worked using fluid energy milling or ball milling. The latter is the most frequently used. Fluid energy milling is a method in which a mixture of silver powder and surfactant is accelerated in a stream of compressed air or superheated steam. When the particles collide under the turbulent flow, the densification takes place (particle size decreases but density increases). In ball milling the silver powder mixed with solvent and surfactant is tumbled in a rotary mill loaded with balls. When collision occurs the powder is flaked. The chosen solvent must have a high flash point and low vapor pressure because of heat generation. The surfactant used must be soluble in the solvent to ensure even distribution. Often fatty acids are used as surfactant. A surfactant is used either to prevent agglomeration (fluid energy milling) or to prevent cold welding (ball milling) in both milling methods. A part of this surfactant is absorbed at the surface of the flakes produced resulting in a coating that can affect the properties of the final flake. It is very important for the resulting conductivity of the adhesive that the surfactant shows compatibility with the resin. The size of the resulting flakes, after cleaning and drying is reported to be 2–30 μm. Three other manufacturing processes for silver filler particles are chemical reduction of an alkaline silver nitrate solution–the resulting flake size being 0.5–10 μm; electrochemical cathodic precipitation in nitrate or sulphate solution–the resulting flake size being 1.2–50 μm; atomization-melted silver is induced into a high speed water- or gas-flow resulting in flakes that are 5–100 μm. The electrical conductivity of an isotropic adhesive is dependent on the amount of contact points between the filler particles. This means that the conductivity will increase with increased amounts of filler particles. However, as the amount of filler increases the amount of polymer must obviously decrease which means that there is a limit for the amount of filler that can be used without too much decrease in the properties of the polymer matrix (adhesion and tensile strength). Eighty weight percent of silver is found to be the limit in epoxy adhesives. The achieved particle size influences the amount of filler material that provides the optimum balance between electrical conductivity and tensile strength. It is found that the larger the flake particles, the greater the number of contacts. This means that larger particles, when formed as flakes, should be preferred to smaller ones (tap density, packing ability of flakes, increase with increased particle size). With regard to viscosity, as the flake particles become larger, the increase in frictional forces between particle surface results in higher viscosities. With regard to filler particle shape some influence on rheology, conductivity, and tensile strength is also observed. The use of spherical filler particles decreases the viscosity while flakes, needles, and fibers offer ideal overlapping conditions for isotropic adhesives. The manufacturer therefore can optimize the adhesive to a specific use by mixing filler particles with different shapes. In addition, the tensile strength of the final adhesive is found to be influenced by the properties of the surface of the filler particles/coatings. It is found that when nickel is used as filler, the strength increases until 40 wt% is added, whereas it only increases up to 10 wt% for silver. This is explained by the rough surface of nickel compared with a smoother surface of silver. This must only be taken as an example of the relationship between surface conditions and strength and not as a general rule in comparing nickel with silver (Nordic 2010).

Consequently, reinforcement fillers have an important role to play in maximizing polymer performance and production efficiency. Cost reduction, density control, optical effects, thermal conductivity, magnetic properties, flame retardancy and improved hardness and tear resistance have increased the demand for high performance fillers. Several types of reinforcements, especially nanoparticulate fillers, have been used in polymer matrix composites: VCGF, carbon foam, CNT, and other thermal conductive particles, such as ceramic, carbon, metal or metal-coated particles, as well as metal or carbon foams. Nanoparticles of carbides, nitrides, and carbonitrides can be used to reinforce polymer matrix nanocomposites with desirable thermal conductivity, mechanical strength, hardness, corrosion, and wear resistance. To achieve these of desirable properties polymer matrix and layout or distribution of nanoparticles need to be optimized. For example, TiC nanoparticles as reinforcement were produced by the sol-gel method. The nanocomposites were obtained in situ, in the reaction mixture and synthesizing copolymers (Meija 2007).

## Design and Modeling of Conductive Polymer Composites

Different types of fillers with high thermal conductivities, such as graphite, metals and ceramics, have been added to polymers to create composite materials with improved mechanical or electrical properties. A great many modeling efforts have been made and found that it is relatively difficult to predict the effective thermal conductivity of a polymer composite material when incorporated with large volume content of filler. For small volume content of filler, the effective thermal conductivity of the composite material generally increases linearly with an increasing volumetric fraction. The Maxwell-Eucken model can predict the effective thermal conductivity of this composite material. However, for large volume content of filler, the thermal conductivity behaves nonlinearly with the volumetric fraction. Many analytical and numerical models have been developed for dispersed phase, at small volume content, the filler is likely evenly dispersed in the homogeneous phase of the matrix. Hence, they are hardly touching each other. At high volume content, it is likely that some filler is touching others and forming conductive chains of filler. These conductive chains, if aligned along the direction of the temperature gradient, will create a preferable path for heat transfer. These kinds of conductive chains may behave differently with percolation of electrically conductive connection, however, this arrangement of thermally conductive filler, in the attached phase, subsequently leads to a drastic increase in the effective thermal conductivity of the composite material.

Many experiments have been conducted to determine the impact of incorporating filler to enhance the thermal conductivity of thermally conductive elastomers. Among them, Agari is most representative. Based on the thermal conductivity measurement of three composites: polyethylene with carbon black, polyethylene with graphite filler, and polyvinyl chloride with graphite filler (maximum volume content of graphite was about 30%), Agari stated that the data deviated from the Maxwell-Eucken equation at about 10% volumetric fraction (Agari and Uno 1986). At large volume content of filler, the measured value was much higher than the prediction using existing correlations. The thermal conductivity was measured for polyethylene and polystyrene composites filled with high volume content of quartz or alumina particles. At above 30% volume content, these fillers start to interact with each other and form an “attached” system. Semiempirical correlation has been developed to fit the data obtained from experiments. Using the experimental data, many analytical and numerical models have been developed and work well for the disperse phase; these model predictions are reliable as long as the filler is homogeneously dispersed. However, if the filler is in the attached phase, only a few models show promising results. Maxwell and Rayleigh developed the effective medium approximation model. This model assumes an isotropic distribution of fillers. Since then, many modifications have been added to this model, making it increasingly sophisticated. Recent versions include shape, size effects, and interfacial resistance. However, these models are unable to predict the effective thermal conductivity accurately if touching between filler particles exists. Agari and Uno (1986) used a generalization of parallel and series conduction models to estimate the thermal conductivities of filled polymers. His modified model considered the fillers being randomly dispersed and predicted an isotropic thermal conductivity with moderate success at low volume content. An assessment was also performed on model predictions and their comparison with experimental data at high volume fraction. Discrepancy existed, especially at high volume content of filler.

The general observation was that empirical correlations did not capture the structural complexity of the composite materials well. Hatta and Taya (1986) solved the heat conduction equation of a coated-fiber embedded in an infinite matrix subjected to a far-field heat flux using an analytical model. Analytical solutions for both single and multiple fibers were developed. No clear closed-form solution could be generated for high volume fraction with touching fibers. Davis and Artz (1995) used finite element analysis to investigate heat conduction of composite with bimodal distributions of particle size. Prediction is reliable as long as filler particles are not touching others. Contact resistance between matrix and particles (Kapitza resistance) was investigated by Hasselman and Johnson (1987). They developed formulas for spherical, cylindrical, and flat-plate geometry to predict an effective thermal conductivity of composites with interfacial thermal barrier resistance. The model worked well for a much-diluted system. If the interfacial resistance was zero or infinite, particle size effect on the effective thermal conductivity was minimal. For a finite resistance, decreasing the particle size usually decreased the effective thermal conductivity for a given volume content. Hence, just incorporating nanoparticles of the same size in a composite system might not work, but adding nanoparticles only in the depletion zone of a composite with filler of microns in diameter should be analyzed (Rightley et al. 2007).

### *Theoretical Modeling*

Many theoretical, numerical, and experimental models have been proposed about reinforcement selection and their layout design for achieving high thermal conductivity of polymer composites. Figure 5.1 shows several kinds of filler layouts. The upper or lower bounds of effective thermal conductivity are given when materials are arranged in either parallel or series with respect to heat flow. For the parallel conduction model (Tavman 2004):

$$k_\mathrm{c} = \varphi k_\mathrm{f} + (1 - \varphi) k_\mathrm{m}, \tag{5.1}$$

and for series conduction model:

$$\frac{1}{k_\mathrm{c}} = \frac{\varphi}{k_\mathrm{f}} + \frac{1 - \varphi}{k_\mathrm{m}}, \tag{5.2}$$

where, $k_\mathrm{c}$, $k_\mathrm{m}$ and $k_\mathrm{f}$ are thermal conductivities of composite, matrix and filler, respectively, and $\varphi$ is the volume fraction of filler. In the case of geometric mean model, the effective thermal conductivity is given by:

$$k_\mathrm{c} = k_\mathrm{f}^{\varphi} k_\mathrm{m}^{(1-\varphi)}. \tag{5.3}$$

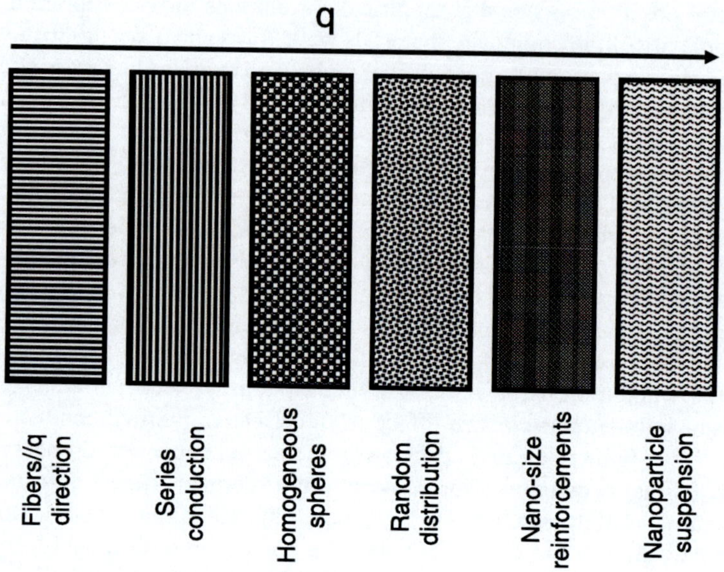


**Fig. 5.1** Typical layout of reinforcements

Using potential theory, Maxwell obtained a simple relationship for the conductivity of randomly distributed and noninteracting homogeneous spheres in a homogeneous medium (Maxwell 1954):

$$\frac{k_c}{k_f} = 1 + \frac{3\varphi}{\left(\frac{k_m + 2k_f}{k_m - k_f}\right) - \varphi}. \tag{5.4}$$

Rayleigh models heterogeneous medium for large volume fraction $\varphi$ as homogeneous spheres embedded in the intersections of a cubic lattice (Tavman 2004):

$$\frac{k_c}{k_f} = 1 + \frac{3\varphi}{\left(\frac{k_m + 2k_f}{k_m - k_f}\right) - \varphi + 1.569\left(\frac{k_m - k_f}{3k_m - 4k_f}\right)\varphi^{10/3}}. \tag{5.5}$$

The interaction between spheres is found to be small even when the Rayleigh derivation is used. Thus, the simpler derivation from Maxwell is generally used to simplify the calculations.

For nonspherical fillers in a continuous matrix, when square arrays of long cylinders parallel to $z$ axis, Raleigh's derivation is described as (Tavman 2004):

$$\frac{k_{c,zz}}{k_f} = 1 + \left(\frac{k_m - k_f}{k_f}\right)\varphi. \tag{5.6}$$

When the composite is anisotropic (the thermal conductivity is not the same in all directions) (Tavman 2004):

$$\frac{k_{c,zz}}{k_f} = \frac{k_{c,yy}}{k_f} = 1 + \frac{2\varphi}{\left(\frac{k_m + k_f}{k_m - k_f}\right) - \varphi + \left(\frac{k_m - k_f}{k_m + k_f}\right)(0.30584\,\varphi^4 + 0.013363\,\varphi^8 + \cdots)}. \tag{5.7}$$

Complex nonspherical approximation models simple unconsolidated granular beds as complex nonspherical fillers in a continuous solid matrix (Tavman 2004):

$$\frac{k_c}{k_f} = \frac{(1-\varphi) + \alpha\varphi(k_m/k_f)}{(1-\varphi) + \alpha\varphi}, \tag{5.8}$$

where $\alpha = 1/3 \sum_{k=1}^{3} \left[1 + ((k_m/k_f) - 1)g_k\right]^{-1}$, $g_k$ are "shape factors" for granules.

When the solid matrix contains gas pockets, assumes thermal radiation important and parallel planar fissures perpendicular to the direction of heat conduction (Tavman 2004):

$$\frac{k_c}{k_f} = \frac{1}{1 - \varphi + \left(\frac{k_m}{k_f\varphi} + \frac{4\sigma T^3 L}{k_f}\right)^{-1}}, \tag{5.9}$$

where $\sigma$ is the Stefan–Boltzmann constant, and $L$ is the total thickness of the material in direction of heat conduction.

For a parabolic distribution of the discontinuous phase reinforcement in a solid matrix, assuming that the constants of the parabolic distribution is a function of the discontinuous phase volume fraction, the effective thermal conductivity of the composite is given for the case $k_f > k_m$ (Cheng and Vachon 1969; Tavman 2004):

$$\frac{1}{k_c} = \frac{1}{\sqrt{C(k_f - k_m)\,(k_m + B(k_f - k_m))}} \times \ln\frac{\sqrt{k_m + B(k_f - k_m)} + \frac{B}{2}\sqrt{C(k_f - k_m)}}{\sqrt{k_m + B(k_f - k_m)} - \frac{B}{2}\sqrt{C(k_f - k_m)}} + \frac{1-B}{k_m}, \tag{5.10}$$

where $B = \sqrt{3\varphi/2}$, $C = -4\sqrt{2/3\varphi}$.

Considering the effect of the shape of the particles and the orientation or type of packing for a two-phase system or single phase reinforcement composite, the effective thermal conductivity can be expressed as (Lewis and Nielsen 1970):

$$k_c = k_m \frac{1 + A\beta\varphi}{1 - \beta\varphi\psi}, \tag{5.11}$$

where $\beta = ((k_f/k_m) - 1)/((k_f/k_m) + A)$, and $\psi = 1 + ((1 - \varphi_m)/{\varphi_m}^2)\varphi$, $A$ is constant and depends upon the shape and orientation of the dispersed particles.

$\varphi_m$ is the maximum packing fraction of the dispersed particles. For randomly packed spherical particles $A = 1.5$ and $\varphi_m = 0.637$, whereas for randomly packed aggregates of spheres or for randomly packed, irregularly shaped particles $A = 3$ and $\varphi_m = 0.637$.

Based on the generalization of models for parallel and series conduction in composites as (5.1) and (5.2), the effective thermal conductivity of the polymer composites is given (Agari and Uno 1986):

$$\log k_c = (1 - \varphi)\log(C_1 k_m) + \varphi \log(C_2 k_f), \tag{5.12}$$

where, $C_1$, $C_2$ are experimentally determined constants of order unity. $C_1$ is a measure of the effect of the particles on the secondary structure of the polymer, like crystallinity. $C_2$ measures the ease of the particles to form conductive chains, the more easily particles are gathered to form conductive chains, $C_2$ becomes closer to 1. However, experimental data are needed for each type of composite in order to determine the necessary constants (Tavman 2004).

For a specific situation, such as the polymer composite reinforced with pitch-based carbon fibers, the in-plane thermal conductivity for a ply of carbon fiber is determined by the rule of mixtures (Bootle 2001):

$$k_x = k_f \varphi \sin^2\alpha, \tag{5.13}$$

$$k_y = k_f \varphi \cos^2\alpha, \tag{5.14}$$

where $\varphi$ is fiber volume percentage, $k_f$ is the longitudinal fiber thermal conductivity, and $\alpha$ is the angle of fibers in a particular ply. The thermal conductivity through the thickness of the laminate $k_z$, typically in the region of 1.5 W/m K, is dominated by the resin thermal conductivity and radial thermal conductivity of the fiber, $k_f$. However, the $k_f$ usually various depending on the plane through the fiber, therefore, a statistical term must be included for $k_f$ when calculating it. For practical purposes, it is generally better to work from measured data. For instance, it is possible to increase $k_z$ up to about 4–6 W/m K by adding boron nitride powder to the laminate.

Thus far the effective thermal conductivity of composites and mixtures as discussed above is derived from continuum-level phenomenological formulations that typically incorporate only the particle shape and volume fraction as variables and assume diffusive heat transport in both phases, providing a good description of systems with micrometer or larger-size particles. However, this approach fails to describe thermal transport in nanocomposites, as interfaces between materials become increasingly important on small length scales. For sufficiently small particles, the properties of the polymer/nanoparticle interface also control thermal transport in the composite. Interface effects in thermal transport can be captured by effective medium models if the introduction of particle fillers does not significantly alter the thermal conductivity of the matrix material. Because local vibrations of the atoms and molecules dominate heat transport in an amorphous polymer, this assumption should be well satisfied in a polymer matrix composite. In the case of a low volume fraction $\varphi$ of high thermal conductivity spherical particles, the thermal conductivity of the composite $k_c$ can be expressed as (Tavman 2004):

$$k_c = k_m(1 + 3\varphi)\frac{\gamma - 1}{\gamma - 2}, \tag{5.15}$$

where, $k_m$ is the thermal conductivity of the matrix and $\gamma = rG/k_m$; $r$ is the radius of the particle and $G$ is the thermal conductance per unit area of the particle/matrix interface. A critical particle radius is defined as $r_c = k_m/G$; the thermal conductivity of the composite is increased by particles of size $r > r_c$ and decreased for $r < r_c$. The changes in conductivity are a relatively weak function of ($rG$), but the thermal conductivity is sensitive to $G$ over a wide range of particle sizes $0.3 < r/r_c < 8$ (Tavman 2004). Classical molecular dynamics simulations are emerging as a powerful tool for calculations of thermal conductance and phonon scattering, and may provide for a lively interplay of experiment and theory to understand the conductivity of nanocomposite materials in the near future.

## *Computational Modeling*

A set of computational tools have been applied to model the heat conduction across the conductive polymer composites, such as TIMs and calculate the effective thermal conductivity of the composite for different volume contents of filler. The analysis procedure involves (Rightley et al. 2007): (1) generating a computational geometric model representing the structure of the composite; (2) simulating the heat transfer across the composite for a set of boundary conditions using a finite element thermal analysis tool; and (3) calculating and evaluating the results.

### CUBIT: A Computational Mesh Generator

CUBIT is a 2-D or 3-D finite element mesh generation toolkit for solid models (Rightley et al. 2007). Usually a geometry file can be ported into CUBIT to generate a computation mesh system for analysis. It can produce quadrilateral and hexahedral meshes, as well as triangle, tetrahedral, and hex-dominant meshes. CUBIT follows a toolkit approach, offering a variety of meshing techniques: 2-D and 3-D mapping, multiple sweeping, paving, and the latest meshing techniques, such as plastering, whisker weaving, and grafting. It runs on both UNIX and PC platforms. In addition to mesh generation, CUBIT has some basic features to construct a 3-D solid model. Other complex solid modeling capabilities include advanced geometry creation and modification, Boolean operations, web-cutting, healing, autodecomposition, defeaturing, and nonmanifold topology (Rightley et al. 2007).

### Calore: A Finite Element Thermal Analysis Program

Calore is a computational program for transient 3-D heat transfer analysis. It is built upon the SIERRA finite element framework to run on both desktop and parallel computers (Rightley et al. 2007). Advanced thermal analysis capabilities include anisotropic conduction, enclosure radiation, thermal contact, and chemical reaction. The governing energy conservation equation in Calore is expressed as (Rightley et al. 2007):

$$\frac{\partial}{\partial t}(\rho \cdot c \cdot T) + \sum_{i=x,y,z} \frac{\partial}{\partial x_i}(\rho \cdot c \cdot u_i \cdot T) = \sum_{i=x,y,z} \frac{\partial}{\partial x_i}(q_i) + S, \tag{5.16}$$

where $q_i = \sum_{i=x,y,z} k_{ij}(\partial T/\partial x_i)$, $\rho$ is density, $c$ is heat capacity, $T$ is temperature, $u_i$ is convective velocity, $q$ is heat flux, and $S$ is volumetric heat generation rate. Boundary conditions available include specified temperature, heat flux, force or free convection, and surface radiation.

Calore also has many state-of-the-art computational features, such as element death, automatic selection of the time step size, mesh adaptivity, and dynamic load balancing for massively parallel computing (Rightley et al. 2007).

### Code Validation and Modeling Strategy

For a given volume content of filler, the measured thermal conductivity of the composite is reasonably repeatable, even though the filler is randomly distributed in the adhesive polymers. These measured values generally follow a normal distribution with a very small standard deviation, provided that the samples are produced from a well-controlled and consistent manufacturing process. The computational model can be assessed by comparing its predictions against measurement. The computational model in the initial analysis has the filler randomly distributed. For a low volume content of filler, the numerical algorithm that searches and places filler particle into an open space within a defined volume works very well. However, when the volume content of filler gets too large, the "search-and-place" algorithm becomes inefficient and time consuming. To overcome this deficiency, a new numerical algorithm was developed that has filler initially distributed in a crystallographic structure, such as face-centered cubic (fcc) or body-centered cubic (bcc) crystal structure. Then for the next 100 iterations, it allows the filler to move randomly. This approach works well to build a computational model and calculate an effective thermal conductivity of a composite. Nevertheless, randomly distributed filler particles have a drawback. It generates significant variations in the calculated thermal conductivity as a result of the randomness of filler distribution. A large number of simulations are needed to suppress the noises and calculate an average value. Obtaining an average is important to recognize the pattern of thermal conductivity behavior as a function of volume content of filler (Rightley et al. 2007).

### Modeling Filler in the Attached Phase

When a group of filler particles are in contact with one another, forming a chain of attached filler particles in the direction of temperature gradient, the effective thermal conductivity can drastically increase. This effect is predicted especially for those cases when the volume content of filler is very high. For the composite with filler in the bcc lattice structure, the drastic increase of conductivity occurs at the volumetric fraction of about 60%. Similarly, for the composite with filler in the fcc lattice structure, the drastic increase occurs at the volumetric fraction of about 65%. When chains of attached filler are created in the direction of temperature gradient, a preferable path for heat conduction will be generated, leading to a significant reduction of temperature at the hotspot. Hence, the effective thermal conductivity of the composite is much higher (Rightley et al. 2007).

### Effect of Contact Resistance

For the modeling results presented thus far, the contact resistance between adhesive polymer and filler particle is assumed to be zero. The effect of nonzero contact resistance between two dissimilar materials has been analyzed; it would be interesting to evaluate the contact resistance between adhesive polymer and filler on the overall effective thermal conductivity of the composite. The cases analyzed have the following conditions: volume content = 40%; contact conductance between adhesive polymer and filler particles, $h = 40\ \mathrm{W/cm^2\ K}$. For those cases in which the filler particles are in contact with each other to form a string of particles, the effective thermal conductivity of the composite does not change drastically if the contact resistance between polymer and filler increases significantly. For example, for a string of 12 contact filler particles, the calculated effective thermal conductivity of the composite decreases by less 4%, from 0.0203 to 0.0195 W/cm K if the contact conductance between polymer and filler decreases to 40 $\mathrm{W/cm^2}$ K. However, for those cases in which the filler particles are dispersed evenly without any contact with any other particle, the effective thermal conductivity of the composite does change drastically (Rightley et al. 2007).

## *Percolation Theory*

Consider a large square lattice in which each site is either occupied with a probability $p$ or empty with a probability $1-p$ to simulate the structure of the polymer matrix composite reinforced with thermally conductive reinforcements. An occupied site, i.e., reinforcement, is assigned a conductivity $k_s$ and an unoccupied site, i.e., polymer matrix, a conductivity $k_f$.

The fundamental premise of the percolation theory is contained in the idea of a sharp increase in the effective conductivity of the disordered media, polymer matrix composite, at a critical volume fraction of the reinforcement known as the percolation threshold $\varphi_c$ at which long-range connectivity of the system appears.

When $k_s \neq 0$, $k_f = 0$, and $\varphi_s < \varphi_c$, no macroscopic conducting pathway exists and the composite remains in the insulating phase. When $\varphi_s \geq \varphi_c$, however, the system becomes conducting as a cluster of bonds of conductivity $k_s$ almost certainly forms a connected bridge between the two boundaries of the disordered media across which the potential is applied. In the vicinity of the transition volume fraction $\varphi_c$, one has (Stauffer 1992; Karayacoubian 2006):

$$k_e \approx 0, \quad \text{when}\, \varphi_s < \varphi_c \tag{5.17}$$

$$k_e \approx k_s(\varphi_s - \varphi_c)^t, \quad \text{when}\, \varphi_s \geq \varphi_c \tag{5.18}$$

where critical exponent $t$ has a universal value of $t = 2.0$ in 3-D and $t = 1.3$ in 2-D problems.

The existence of a critical percolation threshold for electrical conductivity has since been demonstrated for a wide variety of fillers, all at concentrations below the maximum packing fraction. The percolation threshold in an actual granular aggregate is in general a function of the lattice structure of the phases, and ranges from $\varphi_c = 0.2$ for a fcc arrangement to $\varphi_c = 0.7$ for a honeycomb arrangement and can be calculated exactly for certain simple lattices. The critical point can be approached in different ways. For example, when $k_s \rightarrow \infty$, and $k_f \neq 0$, the effective conductivity diverges as $\varphi_c$ is approached from below and is written as (Karayacoubian 2006):

$$k_e \approx k_f(\varphi_c - \varphi_s)^{-t}, \quad \text{when}\, \varphi_s < \varphi_c \tag{5.19}$$

$$k_e \approx \infty, \quad \text{when}\, \varphi_s \geq \varphi_c \tag{5.20}$$

and when both $k_s$ and $k_f \neq 0$ and $k_f/k_s \rightarrow 0$, one has (Karayacoubian 2006):

$$k_e \approx k_s^{1-t} k_f^t, \quad \text{when}\, \varphi_s = \varphi_c \tag{5.21}$$

Because of the complexity of implementing percolation theory and the eventual necessity of numerical work, the method has not been very popular. In addition, it has been suggested that in most cases, the presence of a network of filler particles does not change the basic mechanism of thermal transport in composite systems. A thermal transport network may develop only at the maximum packing fraction. Moreover, all the previous discussions on the percolation model are for site percolation. There is another type of percolation known as bond percolation, which is applicable to fiber-type reinforcements. The bond percolation model has been used to a TIM made from CNT and Ni particles, showing the existence of a percolation threshold at a very small volume fraction of CNT. This indicates the potential of using CNTs in TIMs (Prasher 2006).

## General Fabrication and Manufacturing Processes of Polymer Matrix Composites

The process to fabricate and manufacture polymer matrix composites (PMCs) have been pursued and ranged from hand lay-up with labor- and cost-intensive autoclave processing to the use of automated processes such as sheet molding, bulk molding, injection molding, extrusion, and pultrusion. Transport processes during fabrication and manufacturing of PMCs encompass the physics of mass, momentum, and energy transfer on all scales. There are simultaneous transfers of heat, mass, and momentum at micro-, meso-, and macroscales, which often occurs with chemical reactions in a multiphase system with time-dependent material properties and boundary conditions. Based on the dominant flow process, PMCs manufacturing processes can be categorized as short-fiber or particulate suspension methods; squeeze flow methods; and porous media methods (Advani 2001).

Short-fiber or particulate suspension manufacturing consists of manufacturing processes that involves the transport of fibers, whiskers or particulates, and resin as a suspension into a mold or through a die to form the composite. In such processes, the reinforcements in the molten deforming resin can travel large distances and are usually free to rotate and undergo breakage. The reinforcement distribution in the final product is linked with the processing method and the flow of the suspension in the mold. Typical processing methods include injection molding, compression molding, and extrusion processes, which are most commonly used for producing highly conductive PMCs reinforced with discontinuous fibers, whiskers, particulates, or nanoparticles. Squeeze flow manufacturing methods usually involve continuous or long, aligned, discontinuous fibers either partially or fully preimpregnated with thermoplastic resin, in which the fibers and the resin deform together to form the composite shape. These processes typically include thermoplastic sheet forming, thermoplastic pultrusion, and fiber tape-laying methods. Porous media manufacturing methods usually involve continuous and nearly stationary fiber networks into which the resin will impregnate and displace the air to form the composite product in an open or a closed mold. Low viscosity thermoset is almost always used in such processes, typically including liquid composite molding, thermoset pultration, filament winding, and autoclave processing.

The common feature to all polymeric composite processes is the combining of a resin, a curing agent, some type of reinforcing fiber, and in some cases, a solvent. Typically, heat and pressure are used to shape and cure the mixture into a finished product. Formulation is the process where the resin, curing agent, and any other component required are mixed together. This process may involve adding the components manually into a small mixing vessel or, in the case of larger processes, the components may be pumped into a mixing vessel. Prepregging is the process where the resin and curing agent mixture are impregnated into the reinforcing fiber. Somme common used processes are discussed below.

Sheet molding is usually used to make composite that contains unsaturated polyester resin, thermoplastic resin, styrene monomer, initiator, catalyst, thickener (earth oxide or earth hydroxide), filler, and a mold release agent. This resin paste is combined with fiber reinforcement via a continuous line process to form a prepreg, which is stored between thin plastic sheets at low temperature for later use in molding finished parts. During molding, the cure of the paste is advanced to produce the final rigid product. Bulk molding also uses precompounded reinforcements intended for fast processing by compression, transfer, and injection molding.

Most parts made by hand lay-up or automated tape lay-up must be cured by a combination of heat, pressure, vacuum, and inert atmosphere. To achieve proper cure, the part is placed into a plastic bag inside an autoclave. A vacuum is applied to the bag to remove air and volatile products. Heat and pressure are applied for curing. Usually an inert atmosphere is provided inside the autoclave through the introduction of nitrogen or carbon dioxide, as shown in Figure 5.2. In many circumstances, an autoclave provides enhanced processing flexibility compared with other common processing techniques such as ovens and presses. However, composite fabrication by autoclave is often costly in terms of labor consumption as well as capital investment. Furthermore, autoclave fabrication techniques typically limit the size of the parts which can be produced. One technique utilized to overcome disadvantages of autoclave fabrication is single-vacuum-bag (SVB) processing in an oven utilizing vacuum bag pressure. To date, this is believed to be one of the most -ffective out-of-autoclave fabrication techniques for fiber-reinforced resin matrix composites. However, this process and technique is often ineffective when a reactive resin matrix or solvent containing prepreg is present. A reactive resin (e.g., poly(amide acid)/N-methyl pyrrolidone [NMP]) typically generates reaction by-products (e.g., water) during curing at elevated temperatures. In order to produce a void-free quality laminate, it is often imperative to deplete these volatiles and solvents before commencing forced consolidation. The traditional SVB assembly inherently hinders and/or retards the volatiles

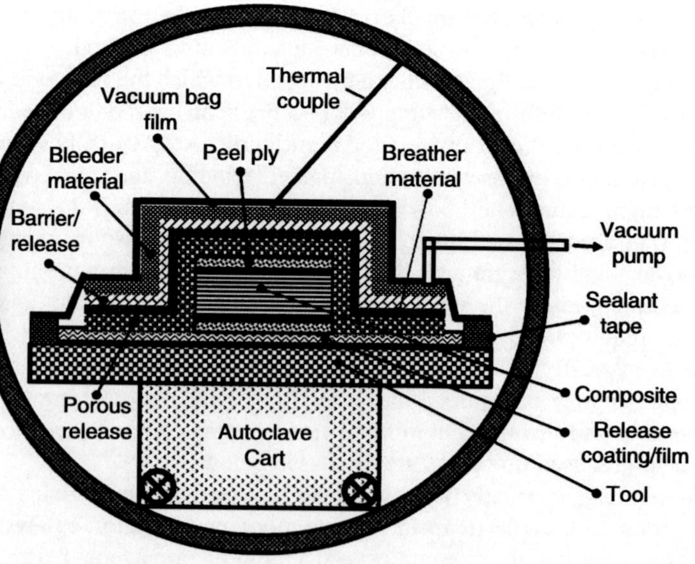


**Fig. 5.2** Schematic illustration of autoclave processing

depletion mechanisms during composite fabrication because a vacuum-generated compaction force is applied to the laminate during volatile depletion. In addition, the one atmospheric pressure associated with the SVB processing tends to create excessive resin flash out of the composite during the B-stage period. As a result, resin content and net shape of the consolidated laminate become difficult to control. Polymeric prepreg material is commonly impregnated with a solution of resin to provide tack and drape for handle ability. The SVB assembly and process are simply too primitive, too time consuming (i.e., costly), and ineffective in removing solvent and reaction by-product during composite fabrication (Hou and Jensen 2007).

Open molding processes are those where the part being manufactured is exposed to the atmosphere. The worker typically handles the part manually, and there is a higher potential for exposure. The resin mixture may be a liquid being formed onto a reinforcing material or it may be in the form of a prepreg material being formed for final cure. Closed molding processes are those in which all or part of the manufacture takes place in a closed vessel or chamber. The liquid resin mixture or prepreg material may be handled or formed manually into the container for the curing step. In the case of liquid resin mixtures, these may be pumped into the container, usually a mold of some type, for the curing step. Sequential or batch processes involve manufacture of a single part at a time, in sequence. This type of process is usually required where the part being made is small and complex in shape, when the curing phase is critical, when finishing work must be minimized, or where a small number of parts are involved. Continuous processes are typically automated to some degree and are used to produce larger numbers of identical parts relatively quickly. These processes are typified by pumping of the resin mixture into the mold, followed by closed curing. One of the older plastics processes, injection molding is also the most closed process, as shown in Figure 5.3. It is not normally used in PMC processes due to fiber damage in the plasticating barrel. Thermoplastic granules are fed via a hopper into a screw-like plasticating barrel where melting occurs. The melted plastic is injected into a heated mold where the part is formed. This process is often fully automated. As shown in Figure 5.4, resin transfer molding is used when parts with two smooth surfaces are required or when a low-pressure molding process is advantageous. Fiber reinforcement fabric or mat is laid by hand into a mold, and resin mixture is poured or injected into the mold cavity. The part is then cured under heat and pressure.

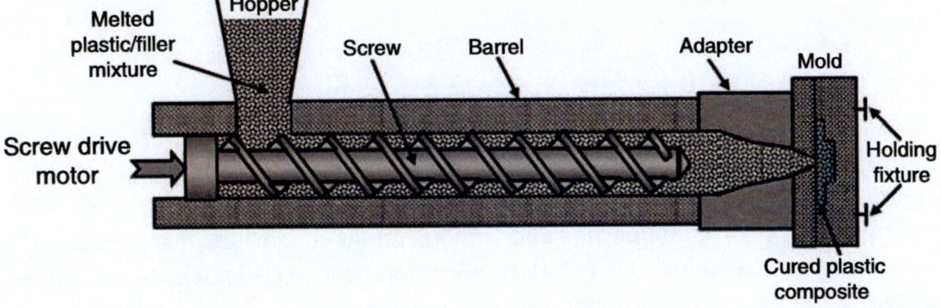


**Fig. 5.3** Schematic injection molding/extrusion process

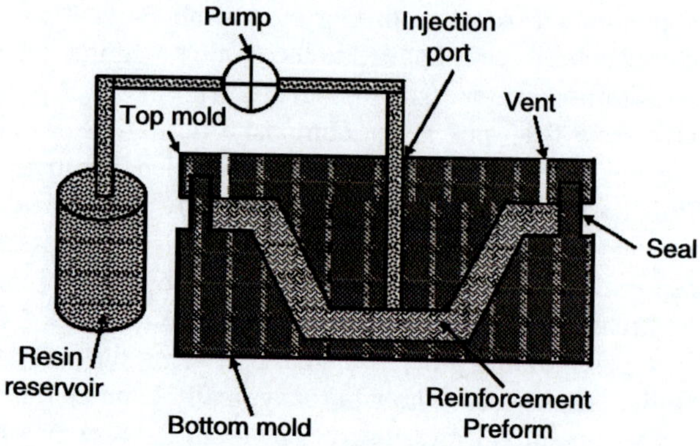


**Fig. 5.4** Schematic resin transfer molding process

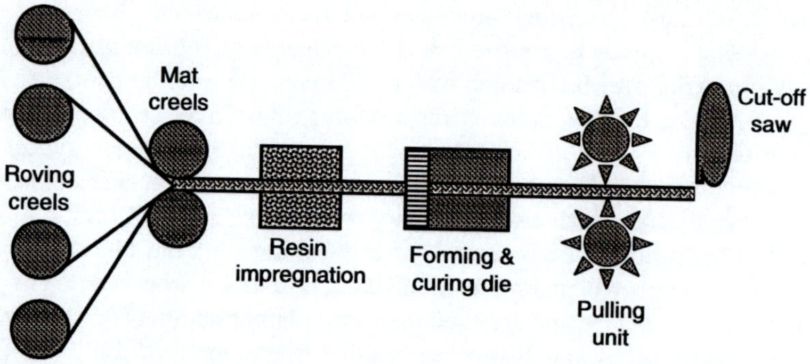


**Fig. 5.5** Schematic pultrusion process

In the pultrusion process, as shown in Figure 5.5, continuous roving strands are pulled from a creel through a strand-tensioning device into a resin bath. The coated strands are then passed through a heated die where curing occurs. The continuous cured part, usually a rod or similar shape, is then cut to the desired length.

Most of the parts made in PMC processes require some machining and/or finishing work. This traditionally involves drilling, sanding, grinding, or other manual touch-up work. These processes vary widely, depending on the size of the finished part and the amount of finishing work required, and often requires complex drill and trims fixtures (OSHA 2010).

## Typical Applications for Thermal Management

The development of carbon fibers, whiskers, and foams with extremely high thermal conductivities has resulted in thermally conductive polymer matrix composites, metal matrix composites and carbon/carbon composites that increasingly are being used in commercial and industrial equipment and microelectronic, optoelectronic and microelectromechanical system packaging.

## *Polymer–Carbon Composites*

Carbon fibers derived from polyacrylonitrile (PAN) precursors have been the dominant reinforcements in high strength PMCs since their commercialization in the late 1960s, however they are limited to use in applications for which high thermal conductivity is not a requirement due to their relatively low thermal conductivities. The development of pitch-based carbon fibers, VGCBs, carbon foam, and CNTs having high thermal conductivities opens up significant new markets and potential applications. Thermally conductive carbon fibers which are made from petroleum pitch, coal tar pitch, and gaseous hydrocarbon precursor materials are the most highly developed and of greatest commercial interest.

An important consideration in evaluating competing thermal management materials is that convection often is the limiting factor, rather than material thermal conductivity in some applications. That is, in many cases, heat dissipation is not improved by increasing thermal conductivity beyond a certain value. This makes it possible, for example, to replace aluminum, and even copper, with carbon-fiber-reinforced polymers having much lower thermal conductivities in some designs.

Table 5.2 shows thermal conductivity of some typical polymer–carbon composites, sparked in their potential to improve the materials' thermal, electrical, and mechanical performances. The thermal conductivities of carbon fiber-in-oil suspensions as carbon fiber-in-polymer composites could also be significantly enhanced by up to, for example, 150% with just 1% volume fraction loading of fibers.

Mesophase pitch-derived graphitic foam can be considered an interconnected network of graphitic ligaments and, exhibit isotropic material properties. More importantly, such a foam will exhibit extremely high thermal conductivities along the ligaments of the foam (up to five times better than copper) and, therefore, will exhibit high bulk thermal conductivities. The graphite foam reinforced composite structure can be made using equipment consisting of a flexible bag for holding a foam preform. A resin inlet valve and resin source are attached to one end of the bag. A vacuum port and vacuum source are attached to the other end of the bag. The bag containing the preform is thermally coupled to a heat source. A vacuum is

**Table 5.2** Thermal properties of some typical carbon reinforced polymer composites

| Reinforcement | Matrix | Inplane thermal conductivity (W/m K) | Through-thickness thermal conductivity (W/m K) | Coefficient of thermal expansion (ppm/K) | Density (g/cm$^3$) |
|---|---|---|---|---|---|
| Continuous carbon fibers | Polymer | 330 | 3.0–10 | −1 | 1.8 |
| Discontinuous carbon fibers | Polymer | 10.0–100 | 3.0–10 | 4.0–7.0 | 1.7 |
| Graphite foam (88%) | Polycyanate | 76 | 129 | – | 1.37 |
| Graphite foam (84%) | Epoxy | 36 | 42 | – | 1.31 |
| Graphite foam (85%) | Polyester | 34 | 40 | – | 1.33 |

created within the bag. The preform is then heated. Resin is introduced into the bag through the resin inlet valve. After the resin is amply impregnated into the preform, the inlet valve is closed and the preform is cured. Once the preform is cured, the resulting composite structure is removed from the bag. As the self-rigidizing carbon or graphite foam reinforcement is both isotropic and continuous, the resulting composites are more uniformly strong in all directions and have greater resistance to shear forces than composites formed from fibrous preforms.

The use of polymer carbon composites is growing considerably. They have been used in thermal applications as both heat conductors or insulators because of their thermal conductivity and thermal expansivity that can be varied and even oriented (as in the case of fibers) in a preferred direction. They can also be suitable for electronic applications especially because specific values of their electric conductivity can be attained, as well as provide new suitable materials in the battery industry. However, structural applications are by far the most studied and applied for this type of composites; the importance of understanding and prediction of their mechanical properties of polymer carbon composites are reflected in the many studies for a variety of structures of the carbon dispersed phase, from macrolaminates to the CNTs (single and multiwalled). Coupled to a carbon fiber reinforcement which can be used as a sensor or local heating source, some PMCs can be self-healing of cracks without significant opening. Polymer–carbon composites can also be the dispersed phase in other composites as in the case of concrete matrix composites.

## *Polymer–Metal Composites*

Polymer composites with metal fillers possess intrinsic advantages including low cost and ease of processing, high electric conductivity, paramagnetism, high thermal conductivity, as well as good mechanical properties. In other words, the interest in polymer–metal composites arises from the fact that the electric and magnetic characteristics of such materials are close to those of metals; there can also be significant improvement in the thermal properties of the pure polymers and a mechanical reinforcement effect might also be achieved, whereas the processability is the same as for the neat polymer, a great advantage for speed of production and processing costs.

The thermal and/or electric conductivity of PMCs is one of the main properties of interest in electronic packaging. For example, poly(ethylene oxide) with added aluminum particles have found that only 1–2 wt% is needed in order to increase the conductivity of the pure polymer by over one order of magnitude, reaching a value of $5 \times 10^{-6}$ Scm$^{-1}$. There is a big advantage of using aluminum instead of a ceramic as the filler: in order to get the maximum conductivity with the ceramic, a concentration of around 40 wt% is needed; this of course is reflected in the cost of production of the material. For copper and nickel powders as fillers in an epoxy resin and in poly(vinyl chloride), the concentration dependence of the electric and thermal conductivity of the resulting composites: less than 1 vol% was sufficient in order to reach the percolation concentration and achieve a big jump in electric conductivity (around 12 orders of magnitude). On the other hand, the thermal conductivity increases linearly with the metal powders concentration. PMC with a soft Al–Fe–Si magnetic powder to a polymeric matrix to produce magnetic films to be used for shielding of electromagnetic waves was found that the level of shielding is mostly dependent of the film thickness and density and related to an increase of magnetic permeability and electric conductivity. In addition, metallic particles are also added to polymers to improve their mechanical properties. However, polypropylene composites with silver powder (several tens of microns) as filler have been found that the tensile modulus, strength, and elongation at break decrease initially when adding the Ag particles and then start increasing when adding more powder. Impact strength also decreased with Ag powder content. This negative effect on the composites properties is attributed to a discontinuity in the structure as well as poor interaction between the two phases. Nevertheless, flexural modulus and strength increase with filler content due to an increase in rigidity. A different way of improving mechanical properties of PMCs is irradiating the material. This improvement is attributed to a cross-linking effect of the polymer chains due to the irradiation making the material stronger. Interest in PMCs is increasing because of their potential to be used as sensors and actuators, especially for biomedical and biological applications. When a voltage is applied to the composite in a circuit, there is a mechanical response depending of the voltage value. This kind of materials can be used potentially as sensors for active noise damping and biosensors. Their large actuation strain and flexibility make them suitable for artificial muscles and other bioinspired devices, meanwhile keep a tailored thermal properties including appropriate thermal conductivity and thermal expansion coefficient (Mejia 2007).

## *Polymer–Ceramic Composites*

Polymer–ceramic composites consist of an organic matrix, for instance polysiloxanes, and inorganic fillers such as oxides or silicates. The properties of the composites are mainly influenced by chemical cross linking between the organic components and between the organic components and functional groups on the surfaces of the filler particles. Polymer–ceramic composites combine typical properties of ceramics, including high thermal stability, thermal conductivity, and dimensional stability with typical properties of plastics, namely plastic formability and functional variability. The composite materials can be processed by plastic forming techniques, in particular injection molding. Polymer–ceramic composite materials have been used for power tool and automotive industries where the materials must endure high thermal loads. Examples include the encapsulation of high-performance resistors or metallic contact bars in heating registers as well as the use of polymer–ceramic composites for thermally loaded components such as brush holders in electric motors. Other applications include an adhesive for attaching electrical components to printed circuit boards, high power components such as transistors to heat sinks, laminates, hybrid substrates for electrical components, and encapsulating compositions for use in integrated circuits where dissipation of heat is a critical requirement of the encapsulant.

One major application of the polymer–ceramic composites is used for TIMs. There has been a heat-reducing grease of the type which uses silicone oil as a base material and a zinc oxide or alumina powder, as well as aluminum nitride as a thickener to achieve improved thermal conductivity. The silicone grease is prepared by mixing a silicone oil with a thickener usually having a low affinity for the silicone oil, so that they have a problem of separating the oil from the composition (in terms of the degree of oil separation) upon long-term standing at a high temperature or by a long-range repetition of cooling and heating cycles, and so on. This problem frequently arises in cases where the thickener used has a relatively large particle size and excellent thermal conductivity. To solve this problem, one silicone grease composite was made, which comprises 10–50 parts by weight of an organopolysiloxane modified by 2-phenylethyl, 2-phenylpropyl or 6–30C alkyl groups and 90–50 parts by weight of a metal oxide, such as silica, diatomaceous earth, zinc oxide, alumina, or titanium oxide (Yamada 2001), and another material is the thixotropic thermally conductive material which comprises an oily organosilicone carrier, a thermal conductivity-providing filler powder selected from a group consisting of thin-leaf aluminum nitride, dendrite-form zinc oxide, thin-leaf boron nitride, and a mixture of two or more thereof and a silica fiber acting as an exudation inhibitor (Aakalu 1981). Other materials are the thermally conductive silicone grease composition comprising an organopolysiloxane, SiC, and aerosol silica (Yamada 2001), and the thermally conductive silicone oil compound comprising a hydroxyl group-containing organopolysiloxane having a viscosity of 10–100,000 cs (centistokes) wherein the hydroxyl groups comprise 5–50 mol% of the total end groups and a powder of at least one metal compound selected from a group consisting of zinc white, alumina, aluminum nitride, and SiC (Yamada 2001). However, these grease-state silicone composites have a drawback of being insufficient in thermal conductivity. For further improvement, it has been found that the bleeding liability of a base oil can be controlled by combining a particular organopolysiloxane with a thickener, such as zinc oxide, alumina, aluminum nitride, boron nitride, or Sic, an organopolysiloxane having at least one hydroxyl group attached directly to a silicon atom and an alkylalkoxysilane, thereby forming a thermally conductive silicone composition having high reliability which can steadily display thermally conductive properties over a long period of time, brings no oil stain on the surroundings, and does not cause contact point disturbance. The typical thermally conductive silicone composition comprises (1) 5–30 wt% of a liquid silicone; (2) 50–94.98 wt% of at least one thickener selected from the group consisting of a zinc oxide powder, an alumina powder, an aluminum nitride powder, a boron nitride powder and a SiC powder; (3) 0.01–10 wt% of an organopoly-siloxane having at least one per molecule of hydroxyl group attached directly to a silicon atom; and (4) 0.01–10 wt% of an alkoxysilane. The thermally conductive silicone composition can control effectively the bleeding of the base oil and the thermal conductive properties thereof can persist stably for a long time, so that it is suitable for thermal conductive silicone grease. In addition, novel nonbleeding thixotropic thermally conductive dielectric materials which find particular application in the cooling of electronic components, especially in direct contact with integrated circuit chips. The composition of the composite comprises a liquid silicone carrier, a thermal filler powder selected from the group consisting of lamellar aluminum nitride, dendritic zinc oxide, lamellar boron nitride, or mixtures thereof in combination with silica fibers which function as a bleed inhibiting agent.

Polymer ceramic composites have also been applied in the energy storage, electric, electronic, biomedical, and structural fields. Alumina and zirconia are often selected additives to fabricate high performance electrolyte composites to be used in. Other electronic applications are piezoelectric and dielectric materials where CdO, borium strontium titanate, strontium cesium titanate are used as the dispersed phase. In some cases the dielectric constant of the composite equals that of a pure ceramic material, such is the case of CdO plus polymer composites which show a dielectric constant of 2,200 while for pure $BaTiO_3$ the value is around 2,000. Among thermal materials those composed of $Al_2O_3$, $SiO_2$ and $TiO_3$ compounds where low isobaric expansivities are desired in high performance chips, wiring and electronic packaging.

## *Polymer Matrix Nanocomposites*

CNTs have typical diameters in the range of ~1–50 nm and lengths of many microns (even centimeters in special cases). They can consist of one or more concentric graphitic cylinders. In contrast, commercial (PAN and pitch) carbon fibers are typically in the 7–20 μm diameter range, while VGCFs have intermediate diameters ranging from a few hundred nanometers up to around a millimeter.

The graphite sheet may be rolled in different orientations along any 2-D lattice vector (m, n) which then maps onto the circumference of the resulting cylinder; the orientation of the graphite lattice relative to the axis defines the chirality or helicity of the nanotube. As-grown, each nanotube is closed at both ends by a hemispherical cap formed by the replacement of hexagons with pentagons in the graphite sheet which induces curvature. Single-walled carbon nanotubes (SWNTs) are usually obtained in the form of so-called ropes or bundles, containing between 20 and 100 individual tubes packed in a hexagonal array. Rope formation is energetically favorable due to the van der Waals attractions between isolated nanotubes. Multi-walled carbon nanotubes (MWNTs) provide an alternative route to stabilization. They consist of two or more coaxial cylinders, each rolled out of single sheets, separated by approximately the interlayer spacing in graphite. The outer diameter of such MWNTs can vary between 2 and a somewhat arbitrary upper limit of about 50 nm; the inner hollow core is often (though not necessarily) quite large with a diameter commonly about half of that of the whole tube. Carbon nanofibers (CNFs) are mainly differentiated from nanotubes by the orientation of the graphene planes; whereas the graphitic layers are parallel to the axis in nanotubes, nanofibers can show a wide range of orientations of the graphitic layers with respect to the fiber axis. They can be visualized as stacked graphitic discs or (truncated) cones, and are intrinsically less perfect as they have graphitic edge terminations on their surface. Nevertheless, these nanostructures can be in the form of hollow tubes with an outer diameter as small as ~5 nm, although 50–100 nm is more typical. The stacked cone geometry is often called a herringbone fiber due to the appearance of the longitudinal cross section. Slightly larger (100–200 nm) fibers are also often called CNFs, even if the graphitic orientation is approximately parallel to the axis (Shaffer and Sandler 2006).

As reinforcement in a polymer matrix, the formed functional nanocomposites have exhibited the unique physical properties, such as high thermal or electrical conductivity. Electrically or thermally conductive polymer composites, for example, are used in antistatic packaging applications, as well as in electromagnetic interference shielding or thermal managing components in the electronics, automotive, and aerospace sector. The incorporation of conductive filler particles into an insulating polymer host leads to bulk conductivities at least exceeding the antistatic limit of $10^{-6}$ S/m. Compared with common conductive fillers such as metallic or graphitic particles or fibrous, the incorporation of CNTs allows for a low percolation threshold, a high quality surface finish, a robust network, and good mechanical properties–a combination not obtained with any other filler. The use of CNTs/CNFs as a conductive filler in thermoplastics is their biggest current application, and is widespread across the automotive and electronic sectors (Shaffer and Sandler 2006).

Figure 5.6 illustrates the electrical resistivity in a metal particle-filled functional polymer composite with increasing filler loading fraction. In general, the electrical conductivity of a particulate composite reveals a nonlinear increase with the filler concentration, passing through a percolation threshold. At low filler concentrations, the conductive particles are separated from each other and the electrical properties of the composite are dominated by the matrix. With increasing filler concentration local clusters of particles are formed. At the percolation threshold, $\varphi_c$, these clusters form a connected 3-D network through the component, resulting in a jump in the electrical conductivity. Close to the percolation threshold, the electrical conductivity follows a power-law of the form (Shaffer and Sandler 2006):

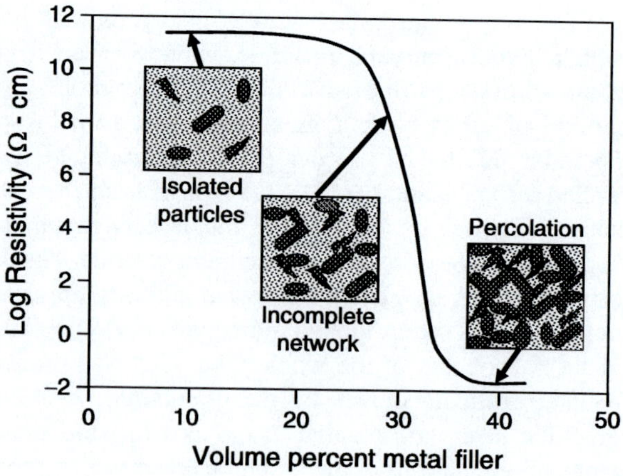


**Fig. 5.6** Variation of electrical resistivity of polymer composite as a function of metal filler volume, with approaching the percolation threshold (Tong 2009)

$$\sigma_0 \propto (\phi_v - \phi_c)^t, \tag{5.22}$$

where $\phi_v$ is the volume fraction of the filler. The exponent $t$ in this equation was found to be surprisingly uniform for systems of the same dimensionality. For 3-D percolating systems $t$ varies between 1.6 and 2. The percolation threshold is reduced on increasing the aspect ratio, but the maximum conductivity is limited by the contact resistance between neighboring particles. A related percolation behavior is observed in the rheological properties, at the point when the filler particles begin to interact. In many cases, the electrical percolation threshold of bulk composites corresponds to the rheological threshold. The electrical properties of nanofibere–thermoplastic composites exhibit characteristic percolation behavior. In the case of untreated CNFs, the critical volume fraction is between 5 and 10 vol%, but depends on the processing technique and resulting degree of CNF dispersion and alignment. At a higher filler content of 15 vol%, even drawing of nanofiber-filled polymer composite fibers does not destroy the conductive network. However, a comparative study of bulk injection-molded nanofiber and PAN-based short carbon fiber reinforced polymer composites showed a lower percolation threshold and higher maximum bulk conductivity for the macroscopic filler (Shaffer and Sandler 2006).

The electrical performance of the composite depends on the intrinsic conductivity as well as the dispersion and alignment of the filler. Oxidation encourages interaction with the polymer, increasing the contact resistance, whereas graphitization both reduces polymer interactions and improves the intrinsic conductivity. As an alternative to graphitization, electro-deposition of copper on nanofiber surfaces has been shown to improve the maximum bulk composite conductivity. Similarly, the electrical percolation threshold of thin MWNT thermoplastic films also depends on the type of nanotube and surface treatment. Threshold values from around 5 wt% for oxidized catalytic MWNTs in polyvinyl alcohol (PVA) to around 0.06 and 0.5 wt% for arc discharge MWNTs in PVA and polymethylmethacrylate, respectively. Interestingly, the optical transparency of such conductive thin film composites for antistatic applications can be significantly improved by using SWNTs, even in bundled form. These very low percolation thresholds are far below the expected values for randomly distributed fibers, and are the result of active aggregation processes (Bal and Samal 2007).

A significant reduction in the critical nanotube volume fraction for electrical percolation can be achieved by exploiting the concept of double percolation through the formation of a cocontinuous morphology in nanotube-filled polymer blends. As in the case of the thermosetting systems, however, the thermal conductivity of nanofiber–thermoplastic composites does not show a percolation transition, even at higher filler volume fractions. A linear increase in thermal conductivity is observed, although the magnitude depends to some extent on the alignment of the filler, in agreement with data for short carbon fiber composites. The overall performance increased for nanofibers was similar to that observed for short carbon fibers in a similar system (Shaffer and Sandler 2006).

For instance, the thermal conductivity of aligned, vapor grown carbon nanoscale fiber reinforced polypropylene composite has been measured in the longitudinal and transverse directions for 9, 17, and 23% fiber reinforcements by volume. The values of thermal conductivity are 2.09, 2.75, 5.38 W/m K for the longitudinal directions and 2.42, 2.47, 2.49 W/m K for the transverse direction respectively, while the thermal conductivity of unfilled polypropylene is 0.24 W/m K (Tavman 2004).

The effective thermal conductivity of liquid with nanoparticle inclusions can be much higher than the normally used industrial heat transfer fluid, such a fluid has terminologized as nanofluid, and considered to be a novel enhanced heat transfer fluid. The possible mechanisms of enhanced thermal conductivity may mainly depend upon the size effect, the clustering of nanoparticles and the surface adsorption, while the Brownian motion of nanoparticles contributes much less than other factors. In particular, it was demonstrated that solid nanoparticle colloids (i.e., colloids in which the grains have dimensions of ≈10–40 nm) are extremely stable and exhibit no significant settling under static conditions, even after weeks or months. Furthermore, the enhancement of thermal-transport properties of such nanofluids was even greater than that of suspensions of coarse-grained materials. For example, the use of $A1_2O_3$ particles ≈13 nm in diameter at 4.3% volume fraction increased the thermal conductivity of water under stationary conditions by 30%.Use of somewhat larger particles (≈40 nm in diameter) only led to an increase of less than ≈10% at the same particle volume fraction; more in accord with theoretical predictions. An even greater enhancement has been reported for Cu nanofluids, where just a 0.3% volume fraction of 10 nm Cu nanoparticles led to an increase of up to 40% in thermal conductivity, a result that is more than an order of magnitude above the increase predicted by macroscopic theory (Tavman 2004).

Metals undergo considerable property changes by size reduction and their composites with polymers are very interesting for functional applications. Some new properties observed in nanosized metals are produced by quantum-size effects, such as electron confinement. These quantum effects arise from the fact that there are a big number of surface atoms compared to a normal bulk metal. These properties are size-dependent and can be tuned by changing the dimension; thus, the same material may show different sets of properties by changing its size. Particularly interesting are the dependence on the size of ferromagnetism and the superparamagnetism characterizing all metals; chromatism observed with silver, gold, and copper metals due to plasmon absorption; the photo and thermoluminescence; and the supercatalytic effect due to the very large superficial area of very fine particles. These materials are highly chemically reactive, highly absorbent, and show very different thermodynamic parameters. For example, they melt at much lower temperatures. Many of these unique chemical and physical characteristics of nanosized metals remain unmodified after embedding in polymers (i.e., optical, magnetic, dielectric, and thermal transport properties), and therefore they can be used to provide special functionalities to polymers (Bal and Samal 2007).

CNTs and nanofibers may not produce practical replacements for existing high-performance materials in the near future. However, there is a continuing market for electrically conducting polymer compounds, and immediate potential to develop the reinforcement of delicate composite structures such as thin films, fibers, and the matrices of conventional fiber composites. Although the full potential of nanotube composites remains to be realized, much progress has been made, and these nanocomposite systems have a bright future once the fundamental questions are resolved.

## Summary

High thermal conductivity polymer matrix composites have been used increasingly for electronic packaging and thermal management. Composites with continuous reinforcements such as fibers, woven or cloths, are usually used as substrates, heat sinks, and enclosures. The polymer matrix in advanced thermally conductive composites includes both thermosetting and thermoplastic types. Different kinds of fillers or reinforcements have been developed to process composite materials with desired thermal, mechanical, and electrical properties. Fillers may be in the form of fibers or in the form of particles uniformly distributed in the polymer matrix material. The properties of the polymer composite materials are strongly dependent on the filler properties as well as on microstructural parameters such as filler diameter, length, distribution, volume fraction, and the alignment and packing arrangement of fillers. It is evident that thermophysical properties of fiber filled composites are anisotropic, except for the very short, randomly distributed fibers, whereas thermophysical properties of particle filled polymers are isotropic.

Reinforcement fillers have an important role to play in maximizing polymer performance and production efficiency. Cost reduction, density control, optical effects, thermal conductivity, magnetic properties, flame retardancy, and improved hardness and tear resistance have been increased the demand for high-performance fillers. Several types of reinforcements, especially nanoparticulate fillers, have been used in polymer matrix composites: VGCF, carbon foam, CNT, and other thermal conductive particles, such as ceramic, carbon, metal, or metal-coated particles, as well as metal or carbon foams. Nanoparticles of carbides, nitrides, and carbonitrides can be used to reinforce polymer matrix nanocomposites with desirable thermal conductivity, mechanical strength, hardness, corrosion, and wear resistance. To achieve these of desirable properties polymer matrix and layout or distribution of nanoparticles need to be optimized.

## References

Aakalu NG, Rittmiler LA (1981) Non-bleading thixotropic thermally conductive material. US Patent 4265775.

Advani SG (2001) Process molding. In Miracle DB and Donaldson SL (eds.) Composites, ASM Handbook, vol. 21: p. 423–433. ASM International, Ohio.

Agari Y, Uno T (1986) Estimation on thermal conductivities of filled polymers. J. Appl. Polym. Sci. **32**: 5705–5712.

Alvino WM (1994) Plastics for electronics. McGraw-Hill, New York.
Bal S, Samal SS (2007) Carbon nanotube reinforced polymer composites-A state of the art. Bull. Mater. Sci. **30** (4): 379–386.
Bootle J, Burzesi F, Fiorini L (2001) Design guidelines. In Miracle DB and Donaldson SL (eds.) Composites, ASM Handbook, vol. 21: p. 388–395. ASM International, Ohio.
Cheng SC, Vachon RI (1969) The prediction of the thermal conductivity of two and three phase solid heterogeneous mixtures. Int. J. Heat Mass Transfer **12**: 249.
Chung DDL (2003) Composite materials: functional materials for modern technologies. Springer, New York.
Davis LC, Artz BE (1995) Thermal conductivity of metal-matrix composites. J. Appl. Phys. **77**: 4954–4960.
Hasselman DPH, Johnson LF (1987) Effective thermal conductivity of composites with interfacial thermal barrier resistance. J. Compos. Mater. **21**: 508–515.
Hatta H, Taya M (1986) Thermal conductivity of coated filler composites. J. Appl. Phys. **59**: 1851–1860.
Hou T-H, Jensen BJ 2007. Double vacuum bag process for resin matrix composite manufacturing. US patent 7186367.
Karayacoubian P (2006) Effective thermal conductivity of composite fluidic thermal interface materials. MS thesis. University of Waterloo, Waterloo, Ontario, Canada.
Lewis T, Nielsen L (1970) Dynamic mechanical properties of particulate-filled polymers. J. Appl. Polym. Sci. **14**: 1449.
Maxwell JC (1954) A treatise on electricity and magnetism. Dover (3rd Ed.), New York, Ch. 9.
Mejia OFO (2007) Micro and nanocomposites composed of a polymer matrix and a metal disperse phase. Ph.D. Dissertation. University of North Texas.
Nordic (2010) The Nordic Electronics packaging guideline – Chapter C: conductive polymers. http://www.extra.ivf.se/ngl/documents/ChapterC/chapterC.pdf. Accessed 28 May 2010.
Nutt SR (2001) Constituent materials. In Damiel R and Micle B (eds.) Composites, ASM handbook, vol. 21. International Handbook Committee, Materials Park.
OSHA (2010) Technical Manual Section III, Chapter 1, Polymer matrix materials: advanced composites. http://www.osha.gov/dts/osta/otm/otm_iii/otm_iii_1.html. Accessed 12 April 2010.
Pilling J (2005) Composite materials design. http://www.mse.mtu.edu/~drjohn/my4150/index.html. Accessed 28 March 2010.
Prasher R (2006) Thermal interface material: histrical perspective, status, and future directions. Proceedings of the IEEE **94**: 1571–1586.
Rightley MJ et al (2007) Advancement in thermal interface materials for future high-performance electronic applications: Part 1. http://prod.sandia.gov/techlib/access-control.cgi/2007/070417.pdf. Accessed on 09 April 2010.
Roberts JC, Luesse MH (1999) Heat sink for increasing through-thickness thermal conductivity of organic matrix composite structures. US Patent 5975201.
Shaffer MSP, Sandler JKW (2006) Carbon nanotube/nanofiber polymer composites. In Advani SG (ed.) Processing and properties of nanocomposites. World Scientific, Hackensack.
Stauffer D, Aharony A (1992) Introduction to percolation theory, Taylor and Francis, Washington, DC.
Tavman IH (2004) Thermal conductivity of particle reinforced polymer composites. In Guceri S et al. (eds.) Nanoengineered nanofibrous materials. Kluwer Academic Publishers, The Netherlands.
Ting J-M (2002) Polymer matrix composite and method for making the same. US Patent 6428890.
Tong XC (2009) Advanced materials and design for electromagnetic shielding. CRC Press, Boca Raton, USA.
Trostyanskaya EB (1995) Polymeric materials. In Shalin RE (ed.) Polymer matrix composites. Chapman & Hall, Norwell.
Yamada K, Isobe K, Takahashi T (2001) Heat-reducing silicon grease composition and semiconductor device using the same. US Patent 6174841.

# Chapter 6
# High Thermal Conductivity Metal Matrix Composites

**Abstract** Metal matrix composites (MMCs) are composed of a metal matrix and a reinforcement, which confers excellent thermally conductive and mechanical performance. High thermal conductivity MMCs have special advantages for particular electronic packaging and thermal management applications because of their combination of excellent thermal conductivity, relatively low density, and tailorable coefficient of thermal expansion (CTE) to match the CTE of semiconductor materials such as silicon, gallium arsenide, or alumina. The optimal design of MMC components is based on appropriate selection of matrix materials, reinforcements, and layer orientations to tailor the properties of a component to meet the needs of a specific design. The specific factors that influence the characteristics of MMCs include reinforcement properties, shape of the dispersed phase inclusions (particles, flakes, fibers, laminates), and orientation arrangement of the dispersed phase inclusions, such as random or preferred; reinforcement volume fraction; matrix properties, including effects of porosity; reinforcement-matrix interface properties; residual stresses arising from the thermal and mechanical history of the composite; and possible degradation of the reinforcement resulting from chemical reactions at high temperatures, and mechanical damage from processing, impact, etc. In the absence of ductility to reduce stress concentrations, joint design becomes a critical design consideration. Numerous methods of joining MMCs have been developed, including metallurgical and polymeric bonding and mechanical fasteners. This chapter will give a brief review about the processing of these composites and their performance and applications in the thermal management of electronic packaging. The contents will include typical processing methods for MMCs, the principal high conductivity for MMCs such as aluminum, copper, and their alloy-based MMCs, and to a lesser extent, other MMCs based on beryllium and silver, etc., as well as low CTE composite solders and advanced multifunctional laminates.

## Introduction

Conventional metallic materials used in manufacture of electronics packaging typically include aluminum, Kovar, molybdenum, and copper–molybdenum–copper. For instance, electronic packages made out of Kovar comprise a Kovar housing and

X.C. Tong, *Advanced Materials for Thermal Management of Electronic Packaging*, Springer Series in Advanced Microelectronics 30,
DOI 10.1007/978-1-4419-7759-5_6, 

Kovar feedthrough pin or wire material and glass seals between the housing and feedthroughs. The CTE of the sealing glass and the Kovar components are matched to provide a hermetic seal. Furthermore, the sealing glass is selected to melt below the critical temperature for the Kovar feedthrough material. A critical transgranular and intergranular oxide are formed and controlled to create the final hermetic seal. However, such conventional electronic packaging materials would not meet the high performance system requirements such as a low CTE, high thermal conductivity, high stiffness, and low density. As a result, composite materials including MMCs have been proposed in an attempt to meet these high-performance system requirements. High thermal conductivity MMCs are especially advantageous for particular electronic packaging and thermal management applications because of their combination of excellent thermal conductivity, relatively low density, and tailorable CTE to match the CTE of semiconductor materials such as silicon, gallium arsenide, or alumina. Aluminum-graphite MMC is an example of these MMCs exhibiting a weight savings of 65% when used in lieu of conventional Kovar as the package housing material. Additionally, aluminum-graphite MMC materials have a thermal conductivity that is six to ten times greater than that of Kovar material and can be produced at relatively low cost casting operations to shapes that are readily machinable (Smith 1999).

MMCs are typically processed with a metal matrix and one or more different reinforcements which confers excellent thermally conductive and mechanical performance, and can be classified according to whether the reinforcement is continuous (monofilament, multifilament, or multilayer laminate structure) or discontinuous (particle, whisker, short fiber, or other). The principal matrix materials for high conductivity MMCs are aluminum, copper, and their alloys. To a lesser extent, other matrices such as beryllium, magnesium, silver, and cobalt are also used for several specialized applications. The optimal design of MMC components is based on appropriate selection of matrix materials, reinforcements, and layer orientations to tailor the properties of a component to meet the needs of a specific design. The specific factors that influence the characteristics of MMCs include reinforcement properties, shape of the dispersed phase inclusions (particles, flakes, fibers, laminates), and orientation arrangement of the dispersed phase inclusions, such as random or preferred; reinforcement volume fraction; matrix properties, including effects of porosity; reinforcement-matrix interface properties; residual stresses arising from the thermal and mechanical history of the composite; and possible degradation of the reinforcement resulting from chemical reactions at high temperatures and mechanical damage from processing, impact, etc. Particulate-reinforced MMCs such as monolithic metals, tend to be isotropic. The presence of brittle reinforcements and perhaps of metal oxides, however, tends to reduce their thermal conductivity, ductility, and fracture toughness. The properties of materials reinforced with whiskers depend strongly on their orientation. Randomly oriented whiskers produce an isotropic material. However, processes such as extrusion can orient whiskers resulting in anisotropic properties. Whiskers also reduce ductility and fracture toughness. MMCs reinforced with aligned fibers have anisotropic properties. They have higher thermal conductivity and are stronger and stiffer in the direction of the fibers than perpendicular to them. The transverse thermal conductivity, strength, and stiffness of unidirectional MMCs (materials having all fibers oriented parallel to one axis), however, are frequently sufficient enough for use. This is one of the major advantages of MMCs over polymer matrix composites (PMCs), which can rarely be used without transverse reinforcement. Another factor that has a significant effect on the behavior of fiber-reinforced metals is the frequently large difference in coefficient of expansion between the two constituents. This can cause large residual stresses in composites when they are subjected to significant temperature changes. In fact, during cooling down from processing temperatures, matrix thermal stresses are often severe enough to cause yielding. Large residual stresses can also be produced by mechanical loading. Fibrous MMCs are essentially brittle materials, as are PMCs. In the absence of ductility to reduce stress concentrations, joint design becomes a critical design consideration. Numerous methods of joining MMCs have been developed, including metallurgical and polymeric bonding and mechanical fasteners.

## Processing of Metal Matrix Composites

Fabrication methods for MMCs can be divided into two major categories, primary and secondary. Primary fabrication methods are used to create the MMC from its constituents. The resulting material may be in a form that is close to the desired final configuration, or it may require considerable additional processing, called secondary fabrication, such as forming, rolling, metallurgical bonding, and machining. The processes used depend on the type of reinforcement and matrix. In general, the primary fabrication methods can be classified as solid state method, liquid state method, in situ fabrication method, and vapor deposition process.

### *Solid State Methods*

Solid state fabrication of MMCs is a process in which MMCs are processed through bonding matrix metal and dispersed phase as a result of mutual diffusion occurring between them in solid states at elevated temperature and under pressure. One of the advantages of the low temperature of solid state fabrication process over liquid state fabrication of MMCs is that some undesirable reactions can be depressed on the interface between the matrix and dispersed reinforcing phases. There are two principal groups of solid state methods for fabricating MMCs: diffusion bonding and powder metallurgy.

Diffusion bonding is common for fabrication of laminate composites with multilayer structures, as shown in Figure 6.1. The matrix can be in the form of foils and the dispersed phase can be in the form of foils or layers of long fibers/wires which are stacked in a particular order, and then pressed at an elevated temperature.

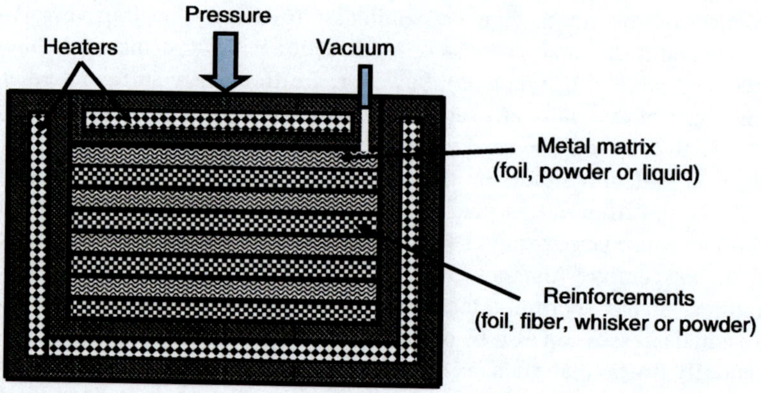


**Fig. 6.1** Illustration of diffusion bonding process for metal matrix composite (MMC) fabrication

The process is dependent on a number of parameters, in particular, time, applied pressure, bonding temperature, and the method of heat application. Diffusion bonding itself can be categorized into a number of processes, dependent on the form of pressurization, the use of interlayers, and the formation of a transient liquid phase.

Bonding in the solid phase is mainly carried out in vacuum or a protective atmosphere, with heat being applied by radiant, induction, direct or indirect resistance heating. Pressure can be applied uniaxially or isostatically. In the former case, a low pressure (3–10 MPa) is used to prevent macrodeformation of the parts (i.e., no more than a few percent). This form of the process therefore requires a good surface finish on the mating surfaces as the contribution to bonding provided by plastic yielding is restricted. Surface finishes of better than 0.4 μm Ra (arthmetic average of the roughness profile) are typically recommended and in addition, the surfaces should be as clean as practical to minimize surface contamination. In hot isostatic pressing, much higher pressures are possibly required (100–200 MPa) and therefore surface finishes are not so critical, finishes of 0.8 μm Ra and greater can be used. A further advantage of this process is that the use of uniform gas pressurization allows complex geometries to be bonded, against the generally simple laminates possible with uniaxial pressurization. For MMC fabrication where dissimilar materials usually need to be joined in the solid phase, it is possible to introduce single or multiple interlayers of other materials to aid the bonding process and to modify postbond stress distribution.

Liquid-phase diffusion bonding/diffusion brazing is applicable only to dissimilar material combinations during MMC fabrication. Solid state diffusional processes lead to a change of composition at the bond interface and the bonding temperature is selected as the temperature at which this phase melts. Alternatively, with the dissimilar metal insert, it melts at a lower temperature than the parent material. Thus, a thin layer of liquid spreads along the interface to form a joint at a lower temperature than the melting point of either of the parent materials. A reduction in bonding temperature leads to solidification of the melt, and this phase can subsequently be diffused into the parent materials by holding at temperature.

Superplastic forming/diffusion bonding is usually used for MMC fabrication, in which the matrix material being one that exhibits superplastic properties at elevated temperatures within defined strain rate conditions. These conditions of temperature and pressure coincide with the conditions required for bonding, and therefore the two processes have been combined into one manufacturing operation either in sequence or together.

Other variants of diffusion bonding are roll bonding, wire/fiber winding, and/or tape casting. Roll bonding is a process of combined rolling (hot or cold) strips of two different metals, such as the matrix and the reinforcement, resulting in the formation of a laminated composite material with a metallurgical bonding between the two layers.

Wire/fiber winding and/or tape casting is a kind of process of combined winding continuous reinforcing fibers and metallic wires, and metallic powders applied to the fiber winding mat with tape casting process, followed by pressing at elevated temperature, as shown in Figures 6.2 and 6.3. For the fiber winding and tape casting process, a monolayer fiber reinforcement for the unconsolidated tape is first formed with drum winding process by disposing a long, substantially continuous reinforcing fiber adjacent an outer surface of a substrate that is coated with a film of

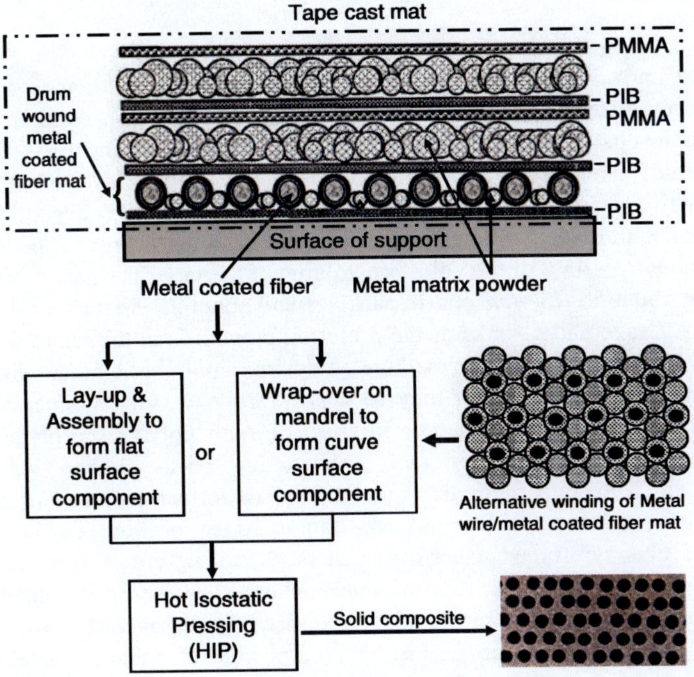


**Fig. 6.2** Illustration of wire/fiber winding and tape casting processes for fabrication of fiber-reinforced MMCs. *PIB* polyisobutylene; *PMMA* polymethylmethacrylate

**Fig. 6.3** Illustration of hot isostatic press for cylinder or curved surface component of MMCs

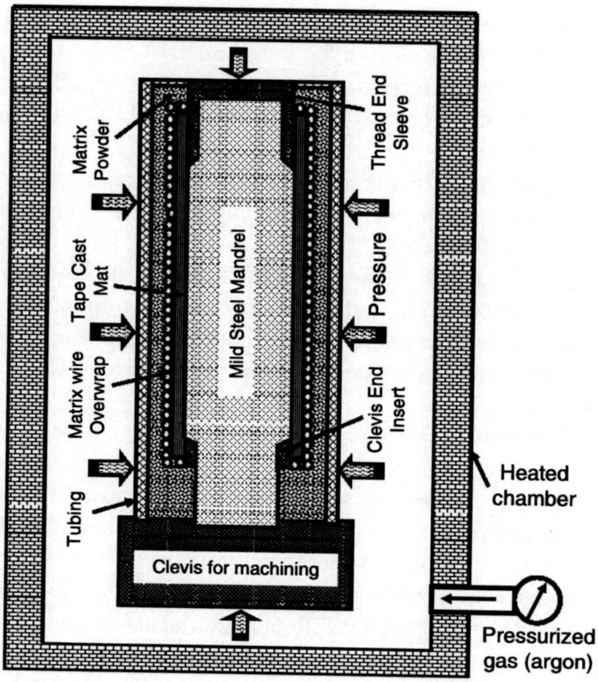


the binder and arranged to cover a cylindrical body. A particularly preferred substrate comprises polyethylene terephthalate. The fiber is wound in a helical pattern on the coated substrate and covered with a film of the binder to form wound fiber reinforcement. This reinforcement is then coated with a layer of slurry of the mixture with powdered matrix material and binder to form the unconsolidated tape casting. A doctor blade can be used to control the coating thickness. The fibers may make up about 25–45% of the void-free volume of the tape. The slurry comprises a mixture of about 50–75 wt% matrix particles and about 25–50 wt% of the organic medium. Average particle size of the preferred powder is in the range of approximately 80–120 μm. The organic medium comprises a polymeric binder dissolved in an organic solvent. The binder may be a polycarbonate, polystyrene, acrylic, or polyisobutylene or a copolymer or mixture of such polymers. Polyisobutylene is particularly preferred. The organic solvent may be an aliphatic or aromatic hydrocarbon. Toluene is particularly preferred. The formed tape casting mat has a width ranging up to several feet and length up to 10 feet or more. Furthermore, the monolayer fiber reinforced unconsolidated tapes can be cut or trimmed to size, separated from the substrate, stacked, encapsulated, degassed and consolidated at high temperature and pressure to make the desired shapes of MMC articles.

Powder metallurgy is a process in which a powder of a matrix metal is mixed with a powder of dispersed phase in the form of particles, short fibers, or whiskers for subsequent compacting and sintering in solid state (sometimes with some presence of liquid). Sintering is a high temperature process used to develop the final properties of the component. It involves heating to temperatures below the melting point of the major constituent in an inert atmosphere to protect against oxidation. Typically, temperatures of approximately 80% of the melting point are used. During the sintering process, adjacent particles bond together by solid state diffusion processes. In contrast to the liquid state fabrication of MMCs, the sintering method allows obtaining materials containing up to 50% of the dispersed phase.

When sintering is combined with a deformation operation, the MMCs are formed into shapes using pressure-based techniques such as uniaxial pressing, hot isostatic pressing, and hot forging.

Periodically, secondary operations are used to improve surface finish or to make sure components fall within required tolerances. Such operations may include drilling, grinding, or repressing. Density and strength can also be increased by a secondary sintering and depressing operation such as rolling, forging, drawing, or extrusion. Deformation of sintered composite materials with the dispersed phase in the form of short fibers results in a preferred orientation of the fibers and anisotropy of the material properties, enhanced thermal conductivity, and strength along the fibers orientation.

Porous metal components are periodically deliberately produced so that they can be infiltrated with secondary materials. Furthermore, they are produced such that they have interconnected pores or capillary pores. These can be infiltrated with coolant or other lower melting point metals. For this to work properly, oxide-free pores are usually required, with processing in an inert atmosphere.

## *Liquid State Methods*

Liquid state fabrication of MMCs involves incorporation of dispersed reinforcing phase into a molten matrix metal, followed by its solidification. Interfacial bonding (wetting) between the dispersed phase and the liquid matrix is crucial to achieve desired performance of the MMC. Proper coating on the dispersed phase particles or fibers not only reduces interfacial energy, but also prevents chemical interaction between the dispersed phase and the matrix, and therefore becomes a common method to improve the wetting and interfacial bonding between the reinforcement and the matrix.

Stir casting is one of simplest and most effective liquid state methods of composite materials fabrication in which dispersed phase particles, or short fibers are mixed with a molten matrix metal by means of mechanical stirring. In the stir casting process, the alloy is melted at a controlled temperature and the desired quantity of reinforcing phase is added to the molten aluminum alloy. The molten alloy is stirred continuously to create a vortex to force the reinforcing phase particles into the melt. Stirring continues to disperse the reinforcing particles as uniformly as possible in a short time. The liquid composite material is then cast by conventional casting methods and may also be processed by conventional metal forming technologies.

Stir casting has some typical features: (1) Content of dispersed phase is limited, usually not more than 30 vol%; (2) distribution of dispersed phase throughout the matrix is not perfectly homogeneous; (3) there are local clouds or clusters of the dispersed particles or short fibers; (4) there may be gravity segregation of the dispersed phase due to a difference in the densities of the dispersed and matrix phase. Distribution of dispersed phase may be improved if the matrix is in semisolid condition, which is called rheocasting. High viscosity of the semisolid matrix material enables better mixing of the dispersed phase.

Infiltration is another common liquid state method of composite materials fabrication, in which a reinforcing preform of dispersed phase (particles, fibers, or woven) is soaked in a molten matrix metal, which fills the space between the dispersed phase inclusions. The motive force of an infiltration process may be either capillary force of the dispersed phase, called spontaneous infiltration; or an external gaseous, mechanical, electromagnetic, centrifugal, or ultrasonic pressure applied to the liquid matrix phase, which is called forced infiltration. For instance, tungsten–copper composites can be made by infiltration process. The tungsten powder is first coated with nickel that is about 0.04% content of the plated powder, mixing the Ni-plated tungsten powder with polymer binder, and then compacting the powder by a molding method, such as metal injection molding, die pressing, or isostatic pressing. Compaction should provide the predetermined porosity level (apparent density) of the tungsten structure. After solvent debinding, the green compact is sintered at 2,200–2,400°F (1,204–1,315°C) in hydrogen atmosphere for 2 h. Placing the sintered part on a copper plate (powder) in the infiltration/sintering furnace, the sintered tungsten skeleton porous structure is infiltrated with copper at 2,100–2,300°F (1,100–1,260°C) in either hydrogen atmosphere or vacuum for 1 h.

Gas pressure infiltration is a forced infiltration method of liquid phase fabrication of MMCs, using a pressurized gas for applying pressure on the molten metal and forcing it to penetrate into a preformed dispersed phase. In contrast to the methods using mechanical force, gas pressure infiltration results in low damage of the fibers. Gas pressure infiltration usually includes several steps: evacuation of pressure vessel (some materials, such as Mg are heated under protective gas because of the evaporation which can damage equipment), heating under vacuum until matrix material is melted, immersion of porous preform into the melt, infiltration under high pressure and withdrawal of the sample from the melt (in the case of some material systems where wettability is not sufficient, sample solidifies together with matrix in crucible). In some cases, bonding between constituents is established as a result of the chemical reaction at the interface (for example formation of carbides), but sometimes this bond is only mechanical interlocking.

Taking the advantages of the attractive fabrication economics and processing flexibility of casting technologies, pressurized liquid metal infiltration of reinforcing particles or chopped fiber preforms has been recognized as one of the competitive routes to produce near-net shape MMC structures. However, the presence of the preform and its physicochemical interactions with the matrix melt make the quality of the fabricated material very sensitive to the infiltration conditions. The success of infiltration largely relies on optimal control of thermal parameters and the associated solidification. Melt solidification during pressurized infiltration of a preform can take place during and postmelt infiltration of a reinforcement preform. Only melt solidification during infiltration, the so-called premature melt solidification thereafter, can change melt flow behavior and associated infiltration quality. During linear gas pressurized infiltration free of solidification, the melt infiltration depth $z$ can be approximately related to the gas pressure $P_{gas}$ applied, the geometrical features of fiber preform $f(V_f)$, and the viscosity of the melt $\mu$ as follows (Long 1996; Long et al. 1997):

$$P_{gas} = P_{cap.\,min} + f(V_f)(\mu \cdot dP/dz)^{0.5} z + P_{back}. \tag{6.1}$$

This equation indicates a linear relationship after gas pressure overcomes the minimum capillary resistance of the preform $P_{cap.min}$ to initiate infiltration. Considering the pressure dominant nature of gas pressurized infiltration, the retainment of linear $P$–$z$ relationship during infiltration accompanied by premature melt solidification can be attributed to the preferential penetration of the melt flow through large interspaces between fibers in the microscopically nonuniform fiber preform, the well maintained melt fluidity by the release of solidification potential heat, and the partition behavior of the alloying elements (Long et al. 1997).

Squeeze casting infiltration is a another forced infiltration method of liquid phase fabrication of MMCs, using a movable mold part (ram) for applying pressure on the molten metal and forcing it to penetrate into a performed dispersed phase, placed into the lower fixed mold part. As shown in Figure 6.4, squeeze casting infiltration involves several steps: a preheated preform of dispersed phase (particles or fibers) is

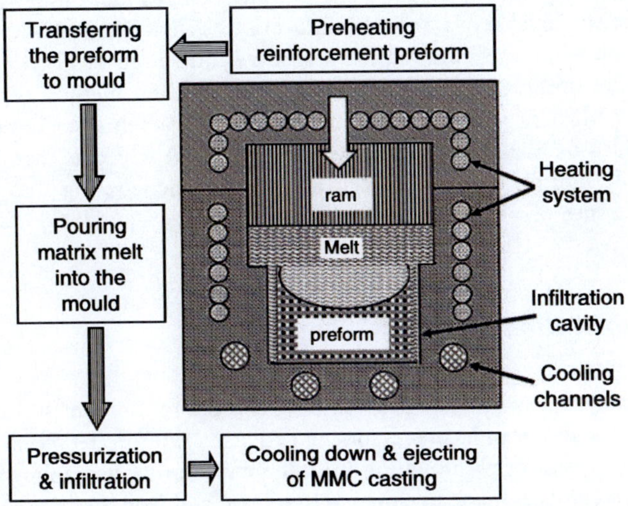


**Fig. 6.4** Illustration of squeeze infiltration casting of MMCs

placed into the lower fixed mold half; a molten metal in a predetermined amount is poured into the lower mold half; the upper movable mold half (ram) moves downward and forces the liquid metal to infiltrate the preform; the infiltrated material solidifies under the pressure; the part is removed from the mold by means of the ejector pin. During solidification free infiltration of preforms under squeeze pressurization, where a constant flux remains throughout the infiltration, the Infiltration Pressure–Ram Displacement–Infiltration Time (*P–D–t*) relationship can be expressed as follows (Long et al. 1997):

$$P_{\text{squ}} = P_{\text{cap.min}} + f(V_{\text{f}})\mu V_{\text{ram}} z + P_{\text{back}}, \tag{6.2}$$

which indicates the infiltration consisting of three stages: initiation of infiltration, linear stable infiltration, and applied force compression. Premature melt solidification is expected to increase the pressure gradient in three ways: increasing the melt viscosity as melt temperature is reduced down below its liquidus; reducing the effective dimension of interspaces and even blocking them due to the presence of solidified matrix phase; and deforming the preform under the significantly increased infiltration pressure. Obviously, all the above mentioned phenomena and their kinetic nature will divert the *P–D–t* relationship during stable infiltration stage away from linear. The existence of a short linear infiltration stage post initiation of infiltration indicates that slight premature melt solidification does not affect melt flow at the beginning of the stable infiltration. Then, the melt solidification begins to play its role, giving rise to a rapid increase of melt flow resistance. To maintain the constant melt flux determined by constant rate of ram displacement, the hydraulic system correspondingly increases its pressure, resulting in a nonlinear rapid pressure increase and rise in the value of pressure to penetrate through the preform. This may result in permanent preform deformation if the increased infiltration pressure exceeds the ultimate compressive strength of the preform before the melt penetrates through the preform (Long et al. 1997).

Pressure die infiltration is a similar forced infiltration method of liquid phase fabrication of MMCs, using a die casting technology, when a preformed dispersed phase (particles or fibers) is placed into a die or mold which is then filled with a molten metal entering the die through a sprue and penetrating into the preform under the pressure of a movable piston or plunger.

### *In Situ Fabrication Methods*

In situ fabrication of MMC is a process in which the reinforcement made in place inside the composite from its precursors, or dispersed reinforcing phase is formed in the matrix as a result of precipitation from the melt during its cooling and solidification. Reinforcements made in situ tend to be fine and well distributed, in addition to having good bonding with the matrix, as the reinforcement-matrix interface tends to be cleaner for in situ than ex situ reinforcements (Chen and Chung 1996). Furthermore, equipment and technologies for in situ fabrication of MMCs are generally less expensive. On the other hand, the making of an in situ reinforcement requires the use of an appropriate reaction, so the choice of in situ reinforcements is much more limited than that of ex situ reinforcements, as their precipitation in particular matrix depends on thermodynamic behavior, and the size of dispersed phase particles is determined by solidification conditions.

Different types of MMCs can be prepared by in situ fabrication method: (1) particulate composite reinforced by in situ synthesized dispersed phase in form of particles, such as aluminum matrix reinforced by titanium boride ($TiB_2$) particles, magnesium matrix reinforced by $Mg_2Si$ particles. (2) Short-fiber composite reinforced by in situ synthesized dispersed phase in form of short fibers or whiskers (single crystals grown in form of short fibers), such as titanium matrix reinforced by titanium boride ($TiB_2$) whiskers, aluminum matrix reinforced by titanium aluminide ($TiAl_3$) whiskers. (3) Long-fiber composite reinforced by in situ synthesized dispersed phase in form of continuous fibers, such as nickel-aluminum (NiAl) matrix reinforced by long continuous fibers of Mo (NiAl-9Mo alloy). Dispersed phases of in situ fabricated MMCs may consist of intermetallic compounds, carbides, borides, oxides, one of eutectic ingredients. Unidirectional solidification of a eutectic alloy (alloy of eutectic composition) may result in formation of eutectic structure, in which one of the components has a form of long continuous filaments, as shown in Figure 6.5, crucible with an eutectic alloy moves downward (or alternatively the induction coil moves upwards). This movement results in remelting followed by resolidification of the alloy under controlled cooling conditions. Value of heat transfer through the crucible bottom together with the crucible speed ($v$) and the power of the heating elements (induction coil) determine particular temperature gradient, which provides unidirectional solidification with flat solidification front. The alloy acquires eutectic structure directed along the solidification direction with eutectic components in form of long mono-crystals (fibers). A distance between the fibers ($d$) is determined by the solidification speed ($v$) according to the formula:

$$d^2 \sim v. \tag{6.3}$$

This indicates that the relative size and spacing of the reinforcement phase can be controlled by controlling the solidification speed, although the volume fraction of the reinforcement will always be constant. In practice, the solidification speed is limited to a range of 1–5 cm/h in order to maintain a stable growth front which requires a high temperature gradient (Chawla 1999).

## *Codeposition*

Codeposition is a process in which matrix metal is deposited together with the dispersed phase by one of the deposition techniques, typically including electrolytic

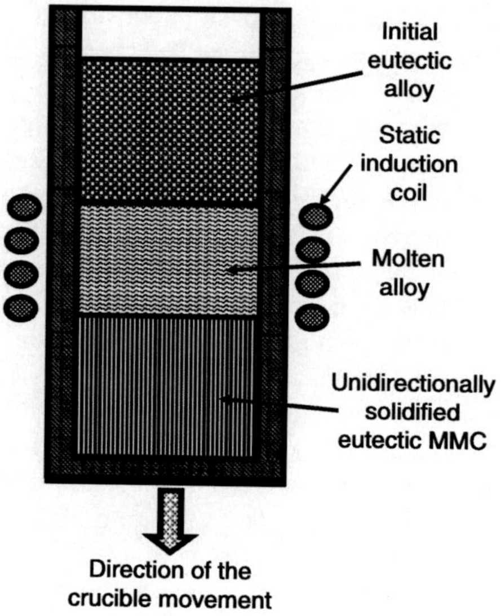


**Fig. 6.5** Illustration of in situ formed MMC by unidirectional solidification

codeposition, spray codeposition and vapor codeposition. Electrolytic codeposition involves electroplating process, in which electrolyte solution of matrix metal ions contains suspended particles of dispersed phase. When the matrix metal is deposited on a substrate, the dispersed phase particles are entrapped by the coating, reinforcing the matrix material. Typical MMCs made with electrolytic codeposition include: (1) Nickel matrix composite materials with various dispersed phases are fabricated by electrolytic codeposition from Nickel Sulfamate and Watts electrolyte, such as Ni–$Al_2O_3$–oxidation resistant nickel matrix composite; Ni–SiC–wear resistant nickel matrix composite; Ni–PTFE, Ni–C, Ni–$MoS_2$–antifriction nickel matrix composites. (2) Antifriction coating of engine bearings consisting of lead–tin–copper alloy and reinforced by alumina ($Al_2O_3$) is fabricated by electrolytic codeposition from electrolyte solution of lead, tin and copper with alumina particles. (3) Aluminum matrix material reinforced by silica ($SiO_2$) is prepared from $AlCl_3$–dimethylsulfone electrolyte containing fine silica particles.

Spray codeposition implements thermal spraying process for atomizing molten matrix metal, droplets of which are delivered to a substrate in a high velocity gas stream together with dispersed phase particles supplied to the stream from a separate container. The method allows fabrication of near-net-shape forming of MMCs. For example, Aluminum matrix material reinforced by silicon carbide (SiC) is produced by spray codeposition followed by rolling. High velocity oxyfuel spraying method is used for fabrication tungsten carbide–cobalt (WC–Co) composite material, which is conventionally manufactured by more expensive technology of sintering fabrication of MMCs.

Vapor codeposition is a group of various methods, utilizing materials in vapor state: physical vapor deposition, chemical vapor deposition (CVD), and direct vapor deposition. In these methods coating of solid material is formed as a result of vapor condensation or chemical reaction on a substrate surface. Vapor codeposition is used for coating fibers, creating multilayer depositions, and fabricating nanostructure.

## Aluminum Matrix Composites

Aluminum matrix composites (AMCs) have received great interest because of the combination of their low weight, good thermal and electrical conductivities, high mechanical strength, and excellent wear properties, potentially becoming a material for many engineering applications, such as sporting goods, electronic packaging, armors, and automotive industries. The reinforcement in AMCs could be in the form of continuous/discontinuous fiber, whisker or particulates, in volume fractions ranging from a few percent to 70%. Properties of AMCs can be tailored to the demands of different industrial applications by suitable combinations of matrix, reinforcement and processing route. When selected reinforcements are well combined with an aluminum matrix through an appropriate process, the resulting material has significant increases in elastic modulus (stiffness), wear resistance, and, in some cases, strength and fatigue resistance. In addition, the CTE of aluminum is reduced by the addition of the reinforcement, while the material retains the high thermal or electrical conductivity and low density inherent in the aluminum alloy. AMCs were first developed to meet very high performance defense and aerospace needs. Continuous fiber reinforced aluminum was used in the Space Shuttle and Hubble Space Telescope. As material cost became a more significant consideration, the emphasis shifted toward particulate-reinforced materials, with the goal of a lower-cost, high-volume product that could be used in automotive and commercial aerospace applications. AMCs reinforced with SiC particles is one of the most suitable and versatile materials among the MMCs for use in aerospace and electronic applications, particular for thermal management material like heat sinks. Other AMCs include boron reinforced aluminum, continuous carbon fiber reinforced aluminum, discontinuous carbon fiber reinforced aluminum, carbon flake reinforced aluminum, diamond particle reinforced aluminum, and silicon reinforced aluminum. The application of these AMCs in electronics packaging market is fast growing, primarily for thermal management applications in which the ability to match the CTE of the electronic materials is a key attribute. These applications will continue to grow as processing methods for forming, machining, and plating continue to develop and further experience is gained.

### *Aluminum–Boron*

Aluminum–boron is the earliest developed aluminum MMC based on CVD boron filament, which combines the outstanding strength, stiffness, and low density of the boron fiber with fabrication and engineering reliability of an aluminum alloy. The overall improvement in modulus to density ratio of the boron fiber is almost six times that of any of the standard engineering materials, including steel, aluminum, molybdenum, and magnesium. This is advantageous in the prevention of micro-buckling of the fibers in the matrix under compression. Other important physical and mechanical properties of aluminum–boron composites include high electrical and thermal conductivity, ductility, and toughness, nonflammability and the ability to be coated, formed, heat treated, and joined. This MMC has been used in tubular structure in the Space Shuttle. Aluminum–boron may be used for thermal management of electronic packaging, for example, it can be a candidate heat sink material. However, although boron fibers have a relatively low CTE, they are poor thermal conductors. As a consequence, increasing the fiber volume fraction to lower the CTE has the undesirable effect of reducing thermal conductivity. Therefore, it is necessary to trade off CTE and conductivity, for instance, the thermal conductivity, a fiber volume fraction of 20% B reinforced aluminum can keep around 180 W/m K. From this point of view, the aluminum–boron probably would be superseded by CVD SiC or SiC-coated boron reinforced aluminum, which has the advantages of much higher thermal conductivity, and greater resistance to attack by liquid aluminum. The increased attack resistance also amplifies composite fabrication process and improves reinforcement/matrix compatibility at elevated temperature.

### *Aluminum–Graphite*

The aluminum–graphite composites are very attractive because the composite can be designed with the CTE approaching zero. The extremely high stiffness of the graphite fibers makes a composite possible that is ideal for applications where precise pointing and tracking are required. These are well suited for strut assemblies, especially in space structures that are subject to a wide temperature range across them. Compared with boron fibers, carbon fibers are more suitable for thermal management because the carbon fibers are more heat resistant, of lower specific gravity, less expensive, and more suitable for mass production. Therefore, aluminum–graphite is particularly applicable for stable instrument platforms, electronics, and thermal control devices such as heat pipes. Stiffness to weight is high because the material is 30% stiffer than aluminum with no thermal expansion.

To obtain dense, high-thermal-conductivity composites, attempts have been vigorously conducted to improve the wettability of graphite with metals. For instance, a aluminum composite material comprising gas-phase-grown carbon fibers having high thermal conductivity, the carbon fibers being coated with a silicon dioxide layer to have improved wettability to the aluminum matrix. However, the use of carbon fibers makes the production cost high. And the silicon dioxide layer having as low thermal conductivity as 10 W/m K limits the thermal conductivity to reach a high level.

Another method has been used to produce a heat sink material comprising firing carbon or its allotrope (graphite, etc.) to form a porous-sintered body, impregnating the porous sintered body with a metal, and cooling the resultant metal-impregnated, porous-sintered body, the metal containing a low-melting-point metal (Te, Bi, Pb, Sn, etc.) for improving wettability in their boundaries, and a metal (Nb, Cr, Zr, Ti, etc.) for improving reactivity with carbon or its allotrope. This method also has a high production cost because a porous-sintered body of carbon or its allotrope is impregnated with a metal. Further, the impregnating metal has reduced thermal conductivity because it contains the low-melting-point metal and the reactivity-improving metal that have relatively low thermal conductivity.

A thermally conductive carbon fiber reinforced aluminum composite is made with the thermal conductivity of at least 300 W/m K, the carbon fibers being plated with nickel. A graphite-particles-dispersed aluminum composite has been produced by compacting graphite particles coated with a high-thermal-conductivity aluminum, the graphite particles having an average particle size of 20–500 μm, the volume ratio of the graphite particles to the metal being 60/40 to 95/5, and the composite having thermal conductivity of 150 W/m K or more in at least one direction. The compacting of the aluminum-coated graphite particles is preferably conducted by at least one of a uniaxial pressing method, a cold-isostatic-pressing method, a rolling method, a hot-pressing method, a pulsed-current pressure sintering method, and a hot-isostatic-pressing method (Fukushima 2009).

## *Aluminum–Diamond*

The Al–diamond composite consists of an aluminum matrix which contains finely dispersed diamond particles, which exhibits enhanced thermal conductivity and the thermal expansion is adapted to commonly used semiconductor materials. The thermal conductivity is 440–530 W/m K at 20°C, the CTE is 7.0–9.0 ppm/K at 20°C, which is well adapted to semiconductors and ceramics. Figure 6.6 shows some application examples of Al–diamond composites for thermal management, including GaAs laser diode or power electronic modules (IGBT) mounted on Al–diamond heat spreader, Ni/Au metalized Al–diamond substrate, Al–diamond heat sink, and Al–diamond insert in heat sink case. Al–diamond composites can be fabricated by many different techniques. Among these fabrication methods, gas pressure infiltration and squeeze casting infiltration are most common.

In addition, coating the diamond particles with a protective layer before contacting the diamond particles with molten aluminum forming the aluminum composite can prevent the reaction to form $Al_4C_3$. A loose bed of industrial diamond powder, with a particle size of 40–50 μm, was coated with SiC using a CVD process of the diamond particle array. In this process, a distinct SiC coating is applied or deposited on the surface of the diamond particles. The thickness of the SiC coating varied between 0.41 and 1.6 μm, depending on process conditions. Total SiC content of the coated diamond particles can be at 3–11% by volume. The preform

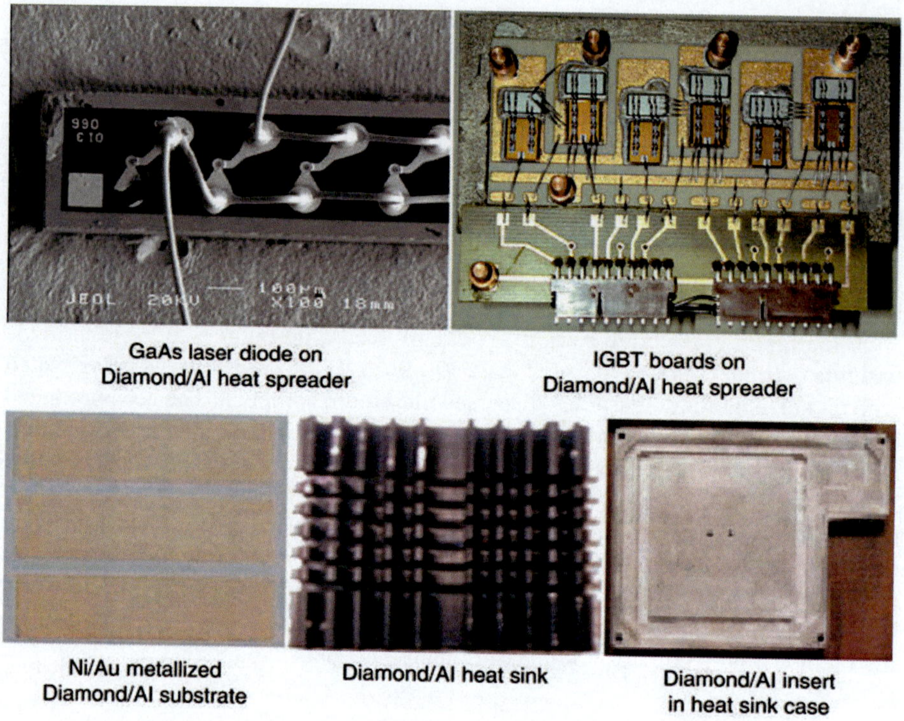


**Fig. 6.6** Application examples of Al–diamond composite for thermal management

particle arrays were observed to have stiffened by the SiC coating. The SiC-coated diamond preforms can be infiltrated by a pressureless metal infiltration process, using an Al alloy.

## *Aluminum–Silicon Carbide*

Al–SiC MMC materials have a unique set of material properties that are well suited for all electronic packaging applications requiring thermal management. The Al–SiC CTE value is compatible with direct integrated circuit (IC) device attachment for the maximum thermal dissipation through the 170–200 W/m K thermal conductivity value material. Additionally, the low material density of Al–SiC makes it ideal for weight sensitive applications such as portable devices.

The SiC is readily wet by molten aluminum and aluminum alloys when it comes in contact with the filler particles. The Al–SiC composites can achieve strengths of 400 MPa at a filler loading of more than 40 vol%, indicating a good bond was formed between the SiC particles and Al. The Al–SiC microstructure is composed of a continuous Al-metal phase with discrete SiC particulate phase. The Al–SiC composite microstructure can be fully dense with no void space creating a hermetic material, which allows fabricated Al–SiC packages to provide environmental protection of functional components. The unique Al–SiC material properties result from a combination of the constituent material properties; Al–SiC properties are tailored by varying the ratio of these constituents. The IC CTE compatible Al–SiC compositions have a SiC particulate content between 50 and 68 vol%. Al–SiC composite materials have thermal conductivity values that are similar to aluminum metal, and CTE values that are similar to alumina. These attributes make Al–SiC packages ideal for direct active device attachment to maximize thermal dissipation and improve product reliability. Al–SiC strength and stiffness compare favorably to traditional packaging materials. The ultimate bending strength of Al–SiC is two to three times greater than Al and its alloys. The Al–SiC Young's modulus–a measure of a material's stiffness–is three times greater than Al and its alloys, and two times greater than Cu and its alloys. The high stiffness to low density ratio is structurally desirable for larger parts with thin cross sections. This attribute permits designs to incorporate feature like fins to maximize cooling surface area. Al–SiC packages and substrates can be plated using similar plating processes as traditional packaging materials that include Ni, Ni–Au, anodization and plasma flame spray coating. Al–SiC packages can also be assembled with sealrings, feedthrus, and substrates using traditional low temperature eutectic brazing and soldering techniques.

In general, however, the thermal conductivity of the Al–SiC MMCs does not meet desired expectations for thermal management of high heat flux electronic packaging. Thermal conductivity of pure aluminum is ~200 W/m K, and the thermal conductivity of pure crystalline SiC particles is ~320 W/m K. Values of thermal conductivity for Al–SiC MMCs are generally less than 200 W/m K, and typically less than 180 W/m K. These Al–SiC MMCs were consolidated by processes such as stir casting, powder metallurgy, or low pressure and pressureless melt infiltration. These methods are relatively slow, and have a considerable residence time when the aluminum is in the molten state, allowing the SiC to react with the molten Al forming aluminum carbide. For that reason, these processes generally require the use of Al–Si alloys which decrease the activity of the Si and reduce the kinetics of the adverse carbide reaction during long contact times with the molten aluminum. These Al–Si alloys generally have a lower thermal conductivity than pure aluminum, thus reducing the thermal conductivity of the Al–SiC. Alternatively, the use of rapid high pressure metal infiltration or squeeze casting to consolidate the particulate reinforced aluminum composites results in a much faster consolidation of the composite. Exposure times of the particles to the molten aluminum are generally seconds as opposed to hours for the non-pressure processes described above. As a result of the rapid consolidation with squeeze casting, pure aluminum can be used, and thermal conductivities of up to about 225 W/m K would be expected for SiC loadings of ~55 vol% in the composite.

Comparably, aluminum-matrix composites are reinforced with SiC whiskers, which are more costly than particles due to the following advantages. (1) Single crystal whiskers have high tensile strength than other discontinuous reinforcements such as particles, polycrystalline flakes and chopped fibers. (2) Standard metal working methods can orient whiskers in the metal-matrix. (3) Extrusion, rolling, forging and superplastic forming can impart directional strength.

Relatively large-diameter monofilament fibers, such as boron and silicon carbide, have been incorporated into metal matrices by hot pressing a layer of parallel fibers between foils to create a monolayer tape. In this operation, the metal flows around the fibers and diffusion bonding occurs. The same procedure can be used to produce diffusion-bonded laminates with layers of fibers oriented in specified directions to meet stiffness and strength requirements for a particular design. In some instances, laminates are produced by hot pressing monolayer tapes in what can be considered a secondary operation.

Monolayer tapes are also produced by spraying metal plasmas on collimated fibers, followed by hot pressing. Structural shapes can be fabricated by creep and superplastic forming of laminates in a die. An alternate process is to place fibers and unbonded foils in a die and hot press the assembly. Composites can be made by infiltrating liquid metal into a fabric or prearranged fibrous configuration called a preform. Frequently, ceramic or organic binder materials are used to hold the fibers in position. The latter is burned off before or during infiltration. Infiltration can be carried out under vacuum, pressure, or both. Pressure infiltration promotes wetting of the fibers by the matrix and reduces porosity.

Cast Al–SiC MMCs now consistently offer net or net–net shape, improved stiffness and strength, and compatibility with conventional manufacturing techniques. They are also consistently lower in cost than those produced by other methods, are available from a wide range of fabricators, and offer dimensional stability in both large and small parts.

Furthermore, studies were undertaken to evaluate the tensile behavior of low-cost discontinuous Al–SiC composites, containing SiC-whisker, -nodule, or -particulate reinforcement. The effects of reinforcement type, matrix alloy reinforcement content, and orientation were determined by analysis of stress-strain curves and by SEM examination. These investigations led to the following conclusions:

(1) Discontinuous Al–SiC composites offer a 50–100% increase over the modulus of unreinforced aluminum and offer a modulus equivalent to that of titanium, but at a third less density. The Al–SiC composites had modulus/density ratios of up to almost twice those of titanium and aluminum structural alloys. The modulus of Al–SiC composites tended to be isotropic and was controlled by the amount of SiC reinforcement.
(2) The yield and tensile strengths of Al–SiC composites demonstrated up to a 60% increase over those of the unreinforced matrix alloys. Yield and ultimate tensile strengths of the composites were controlled by the type and temper of the matrix alloy and by reinforcement content. In general, these properties were independent of the type of reinforcement.
(3) Ductility of Al–SiC composites, as measured by strain to failure, was dependent upon reinforcement content and matrix alloy. Composites with ductile matrix alloys and lower reinforcement contents exhibited a ductile shear fracture with a 5–12% failure strain. As reinforcement content increased, the

fracture progressed through a transition and became brittle, reaching a less than 1–2% failure strain, at higher reinforcement contents. The increase in ductility over that reported previously was probably attributable to cleaner matrix alloy powders, better mixing, and increased mechanical working.

(4) A fine dimple network was observed in the fracture surfaces of composites with higher strains. At lower fracture strains, a coarser dimple network was observed. Composites failing in a brittle manner showed increasing amounts of cleavage fracture.

(5) The SiC-whisker reinforcement was generally oriented in the extrusion direction. Composites with a higher degree of preferred orientation tended to have higher ultimate tensile strength in the direction of whisker orientation. Composites with a more random whisker orientation tended to be isotropic in strength.

Methods that have been used for the manufacture of Al–SiC MMCs can be classified as follows:

(1) Liquid state processes in which the ceramic particulate is added to the molten matrix alloy, agitated to prevent SiC settling due to density differences, and cast to shape. However, this method is limited to SiC levels up to 30 vol%, because of problems with fluidity, and to certain aluminum alloy matrices which do not generate excessive aluminum carbide which is detrimental to mechanical properties and corrosion resistance.

(2) Solid state processes in which the matrix alloy is blended with ceramic particulate and pressed. In a practical sense these processes are limited to SiC reinforcements of approximately 60 vol% placing a limit on the CTE reduction available.

(3) Infiltration processes in which a green SiC compact is infiltrated with the molten matrix alloy. Very high SiC particle loadings can be achieved by infiltration processes (up to 75 vol%). To well bond SiC with Al, infiltration is assisted using pressure infiltration or by conditioning the SiC particle surfaces to assist infiltration by capillary action. While successful in achieving the SiC loading required for CTE matching with semiconductor materials, the MMCs are extremely difficult to machine. In addition, because of the multimodal distribution of SiC particle sizes required to form the green compacts, the MMC is also difficult to metallize, to join, and can be prone to gas leakage at the high vacuums required for certain applications, e.g., helium leakage in satellite applications.

(4) Another route is spray forming, in which an aluminum-based alloy is melted and inert gas atomized and ceramic particulate is injected into the metal spray which codeposits to form a MMC. However, it is not possible to introduce the required high loading of SiC (65–75 vol%) by this method. Such attempts have resulted in excessive porosity, particle clumping, and poor SiC loading reproducibility. Another method for the production of aluminum alloys with a low coefficient of expansion is spray forming of hypereutectic aluminum-silicon alloys in which silicon acts as the reinforcing phase. Instead of introducing an

exogeneous reinforcing ceramic phase or phases in the form of particulate or fibers, the reinforcing phase, silicon, is formed "in situ" by nucleation and growth from the liquid during solidification of the spray formed deposit. The advantage of this method is that the endogeneous reinforcing phase, silicon, is in perfect atomic contact with the aluminum matrix because the solid-state phases which constitute the alloy are derived from the same source, namely, the alloy in the liquid state.

Using net-shape fabrication process, high volume Al–SiC components have been manufactured, including flip-chip lids for microprocessor, heat sink, flip-chip, DSP application, power applications (IGBT base plates, four power substrates), optoelectronic and microwave packaging. Different variants are offered with different thermal expansion coefficients to suit application requirements. Typical ratings (isotropic) are CTE 8 ppm/K, thermal conductivity ~200 W/m K and density 3.01 g/cm$^3$. The high loading of SiC imparts high modulus and forms a strong interface with the matrix, and the resulting high strength-to-weight ratio suits it for many weight-critical applications. It is also used commercially for power (IGBT) base plates (Young et. al 2006).

## *Aluminum–Silicon*

Al–Si alloys can be produced by melting and casting. However, in the compositional range of interest (51–90 wt% Si) the as-cast microstructure is characterized by mainly large, discrete, faceted, high aspect ratio primary silicon crystals which impair mechanical properties and machinability. The primary silicon particle size, which appears as single acicular crystals, is of the order of millimeters which results in a very anisotropic microstructure. This makes it quite unsuitable for the application to electronic packaging. For example, the plates which go to make up electronic packages are typically 1–5 mm in thickness, with a chill cast material it would be possible to have one single silicon crystal passing through the whole plate thickness. This would make the material extremely difficult to machine to fine surface finishes required for metallization, because the silicon crystal would be susceptible to fracture in a single direction along a preferred crystal lographic plane. In addition, because of the large silicon crystal size, the local CTE and thermal conductivity of the material may vary widely depending upon whether aluminum or silicon is in contact with the chip or chip carrier. It is generally considered that these alloys have no engineering application and are only used as master-alloys for liquid metal processing industries or steel deoxidation.

To avoid the issues mentioned above, silicon-based alloy have been made by inert gas atomization to form either a spray-formed deposit or gas atomized powder, which comprises the steps of melting a silicon alloy containing greater than 50 wt% Si and inert gas atomizing such alloy to produce a product in which the silicon forms a substantially continuous phase made up of fine, randomly oriented crystals in the microstructure of the alloy material. The product may be a coherent spray-formed deposit or powder and the alloy preferably includes aluminum. Powder would be used for subsequent compaction and/or sintering using traditional powder metallurgy routes or, preferably, by hot isostatically pressing above the solidus temperature. A coherent spray formed deposit would preferably be hot isostatically pressed at a temperature above solidus to produce a product of substantially 100% density.

If desired, ceramic particles may be introduced into the atomizing spray. The ceramic particles may be injected into the spray or, when making powder, may simply be added to the powder material. The method may comprise the further steps of hot isostatically pressing the material. This may be in the semisolid condition, with or without encapsulation. If the alloy material includes ceramic particulate additions, such as silicon carbide, up to 35 vol% the MMC material may be encapsulated, evacuated, and hot isostatically pressed in the semisolid condition. Suitably, alloys or MMCs can be tailored to have a CTE in the range 4.5–11 ppm/K and thermal conductivity greater than 100 W/m K. The maximum value of thermal conductivity that could be achieved would depend upon the reinforcing particulate, e.g., aluminum-nitride or diamond.

By atomizing the silicon alloy, silicon crystal growth begins during atomized droplet flight and, in the deposit, a large number of nucleative sites are created which grow and impinge into one another to form a network structure in which silicon crystals are randomly oriented and not discrete and highly orientated as in a cast structure and therefore present a substantially continuous phase making the deposit structurally coherent and capable of being machined to fine surface finishes. Therefore, advantages of the alloy materials are that they are machinable with conventional carbide tools, are weldable, brazable, and coatable with materials such as gold.

## Copper Matrix Composites

Copper-based composite materials have been widely used when superior thermal and electrical conductivities are required. For instance, carbon fiber reinforced copper composites offer the unique complex of properties, e.g., high specific stiffness, strength and fracture toughness, low thermal expansion, relatively low density, and excellent thermal and electrical conductivity. The property can be tailored into the required values by changing the volume fraction and the arrangement of the fibers. Copper matrix composites have been used for thermal management in electronic packaging, electric contact with high dimensional stability, substrates for high-power semiconductors with similar thermal expansion as silicon, cooling tubes and components for fusion or space application where high stiffness together with the ability to withstand high thermal loads are demanded (Deve and Skildum 2002).

Copper matrix composites are usually reinforced by (1) continuous fibers of carbon, silicon carbide, tungsten, stainless steel, and diamond coated fibers;

(2) carbon or metal whiskers, carbon foam and nanofibers; and (3) particles such as tungsten, SiC, and diamond. Powder metallurgy (sintering) and infiltration technique can be used for fabrication copper matrix composites, which typically possess low coefficient thermal expansion, high stiffness and elastic modulus, good electrical conductivity, high thermal conductivity and good wear resistance. The copper matrix composites that can be used for thermal management of electronic packaging mainly include Cu–carbon, Cu–nanofiber, Cu–SiC and Cu–diamond.

## *Copper–Graphite*

Copper–graphite composite combines the advantages of high thermal and electrical conductivity from the copper and low thermal expansion coefficient and lubricating properties from the graphite. Addition of graphite to copper also reduces density, increase stiffness, and raises the service temperature. According to the type of reinforcement, copper–graphite composites can be classified into three types: (1) particulate reinforced; (2) whisker/short fiber reinforced; and (3) continuous fiber reinforced composites. Particulate reinforced copper–graphite composites indicate that particulate or particles with aspect ratio less than 5 are reinforced in metal matrix to have a dispersed phase of roughly equiaxed particles. Copper–graphite composites reinforced with graphite particles from 5 to 40% in the volume fraction can be used for structural and wear resistance applications. In general, the manufacturing of these composites is cost effective compared with continuous fiber reinforced composite. Mechanical properties are inferior to whisker/continuous fiber reinforced copper matrix but superior to unreinforced copper alloys, in addition to tailored thermal expansion coefficient. These composites were isotropic in general. In order to improve the mechanical properties, these composites are subjected to post processing operations such as rolling, forging, extrusion and heat treatment, etc. Whisker reinforced copper–graphite composites refer that whiskers are described as either elongated particles or short fiber with aspect ratio grater than 5. Simple-shaped components of short fiber composites can be manufactured by direct squeeze casting method. The application of indirect squeeze casting makes possible to manufacture more complex parts, but it results in more expensive die and tools for casting. Whisker reinforced composites can also be manufactured by infiltration method. Mechanical and structural properties of whisker-reinforced composites are superior to particle reinforced ones. Continuous fibers reinforced copper matrix composites indicate that copper matrix contains long fiber reinforcement that is available as tows of several hundred or thousands of fiber. The diameters of the fiber range from 5 to 20 μm. Continuous fibers are exceptionally strong and they have higher stiffness in the direction of the fiber, which results in anisotropy. These composites were used in thermal management in aero structure for directional heat transfer. Copper matrix having fiber volume fraction up to 40% is produced by squeeze infiltration and or pressure infiltration route (Sebo and Stefanik 2003; Rajkumar and Aravindan 2007).

When continuous fibers reinforced copper composites are used for thermal management, a low linear thermal expansion coefficient and no directional property are usually desired. Carbon fibers have high strength and high elastic modulus, and are characterized by a desirable low thermal expansion coefficient. However, carbon fibers are insufficient in electric conductivity and usually have nonuniform or various thermal conductivities. Combined in certain proportions with carbon fibers, copper matrix–carbon fiber composites exhibit low thermal expansion, high electrical conductivity, and high thermal conductivity and can be applied to various equipments. For example, if such a composite is used as a substrate for a silicon semiconductor element, a semiconductor device having a high capacity will be prepared with ease at a low cost. When the fiber direction is fixed in a copper–carbon fiber composite, however, the thermal expansion coefficient is low in the longitudinal direction of the fiber but is high in the direction perpendicular to said direction. In other words, if carbon fibers have a unidirectional characteristic, the thermal expansion characteristic of the composite is anisotropic. Accordingly, when the composite is applied to a semiconductor device or the like, it is necessary to render fiber directions random to thereby expel the anisotropic characteristic of the thermal expansion coefficient. As one of means for attaining this object, there can be mentioned a method in which carbon fibers are arranged in a net-like form. This method, however, is defective in that production of composites is troublesome and difficult. When carbon fibers are randomly arranged in the copper matrix, the anisotropic characteristic of the thermal expansion coefficient is completely expelled. However, it has been found that if the composite is subjected to a temperature higher than the softening point of the copper matrix, an abnormal volume change takes place and this abnormal volume change results in deformation and breakdown of the composite. This undesirable phenomenon takes place frequently, especially when the volume ratio of carbon fibers is increased to reduce the thermal expansion coefficient of the composite. Furthermore, in connection with the influences of temperatures, deformation and breakdown are caused by exposing the composite to a high temperature for a short time as in case of brazing, and further, even if the exposure temperature is relatively low, deformation and breakdown similarly take place when the composite is exposed for a long time. It has been confirmed that breakdown is due to elastic deformation of carbon fibers. More specifically, in order to increase the amount of carbon fibers in the copper matrix, it is necessary to compress a mass of fibers entangled in random directions, and if such carbon fibers in which elastic deformation has been caused by this compression are sealed into the copper matrix, a large inner stress is left in the resulting composite and so far as the strength of the matrix is sufficiently stronger than the compression stress left in the fibers, deformation or breakdown is not caused in the composite but if there is a weak portion in the matrix in the composite, the stress is concentrated to this weak portion and finally overall breakdown takes place in the composite. Accordingly, it has been found that a risk of such deformation or breakdown is greatly reduced when the fiber amount is small or the ratio of the length to the diameter in the used fibers is low even if the fiber amount is relatively large. Therefore, it is desirable to provide a carbon fiber–copper matrix composite having no directional characteristic in either mechanical properties or linear thermal expansion coefficient. This can be attained by a carbon fiber–copper matrix composite formed by mixing homogeneously (Arakawa et al. 1978): (1) carbon fibers nondirectionally entangled with one another; (2) an additive element capable of forming a carbide and bonding the carbon fibers together; (3) a copper matrix, and molding integrally the mixture by heating the mixture under pressure in a non-oxidizing atmosphere at a temperature lower than the melting point of copper.

The application of copper–graphite fiber composites has been explored for thermal management of electronic packages, many of which are installed on various satellites and defense platforms. One approach is to use CVD coating graphite fiber with a thin layer of molybdenum. There are three reasons for this (Young et al. 2006): (1) it produces a surface which is wettable by infiltrant metal (compared with untreated) without requiring application of pressure; (2) molybdenum is effectively insoluble in copper so the intrinsically high conductivity of oxygen-free copper is not degraded; and (3) molybdenum coating is required in order to develop a carbide interface which has low interfacial thermal resistance. This is important for "coupling" the graphite and the matrix for optimized thermal performance. By adjusting the construction the CTE and thermal conductivity can be set to predefined values within a range. The main benefit is the provision of a material with low expansion with enhanced in-plane conductivity and good through-thickness thermal and electrical conductivity with reduced density compared to copper. Using this technique, fibers having an axial conductivity of 600 W/m K are combined with the copper matrix to deliver a material with thermal expansion controllable within the range 2–10 ppm/K with in-plane thermal conductivity substantially equivalent to copper (400 W/m K) and *Z*-plane conductivity of 200 W/m K. This combination is of particular interest in power electronic applications and especially as substrates or base plates for large-area die attaches (Young et al. 2006).

In addition, graphite foam–copper composites have been developed as a replacement for copper–tungsten (Cu–W). Carbon foams are an attractive alternative to fiber composites as it is easier to control volume loadings, and high conductivity values (600–1,200 W/m K) are available in the ligaments, so should also be substantially lower cost. An important difference is that the material is substantially isotropic. Developmental materials have thermal conductivity of $\sim$342 W/m K and CTE 7.4 and 5.7 ppm/K. Compared to baseline Cu–W composites, improved thermal management is expected with easier machining and lower mass. Applications targeted include LDMOS microwave power transistor substrates, high power laser diodes, various power applications and high-power transmit-receive module packages. However, control of surface roughness can be a problem on machined surfaces (Young et al. 2006).

Another option is to use laminates in which one or more of the layers consist of flexible graphite sheets, and metal-like copper sheets that are bonded between sheets of flexible graphite. The laminate structures can be fabricated by cold-working a flexible graphite sheet on both sides of a metal net and then press-adhering the graphite to the metal net. In addition to the utility in gasket materials, graphite copper laminates can also be used for heat transfer or cooling apparatus. Graphite layered thermal management components offer several advantages in electronic applications and can help eliminate the potential negative impacts of heat generating components in computers, communications equipment, and other electronic devices. The laminates of the graphite/copper can provide a machinable structure comprising layers of epoxy impregnated flexible graphite together with optional layers of metals for thermal management of electronic packaging (Norley et al. 2004).

## *Copper–Carbon Nanofiber*

During the 1970s the carbon fibers grown from vapor phase hydrocarbons were developed showing outstanding mechanical and thermophysical properties, which had thermal conductivity as high as 1,950 W/m K (Heremans and Beetz 1985). Traditionally, there are two methods for the production of these vapor grown carbon fibers (VGCFs) named as "Fix bed" and "Flying seed." The former is based on the growth from catalytic particles and is used to obtain high quality of VGCFs. The main disadvantage is its low production yield. The latter is based on the continuous feed of catalytic seeds in a furnace and subsequently collection of the fibers at the bottom furnace exit. It is being used for mass production of novel carbon nanofibers (VGCNFs), typically some 10 μm in length and diameter in the 30–150 nm range. Compared with carbon nanotubes (CNTs), VGCNFs provide much better availability and performance-cost ratio for industrial production of electronic packaging components. Due to their tailorable CTE, low density and high thermal conductivity, VGCNFs reinforced MMCs constitute an ideal material for heat sinks in high power and high frequency electronic devices working at much higher temperatures, such as those based on new GaN semiconductors. From thermal conductivity point of view, VGCNFs can be considered as an intermediate material between carbon fibers and CNTs. One of the main drawbacks in VGCNFs handling and mixing with a matrix is dispersion. By its nature, with high aspect ratio (L/d) and van der Waals attracting forces, VGCNFs tend to keep entangled in big clusters difficult to disperse in any media. However, some other difficulties, such as absence of wettability, solubility and reactivity between carbon and copper or other metals, or density mismatches with most of metallic matrices, makes it very difficult to apply production routes of MMCs in liquid state (stir-casting, rheocasting or liquid infiltration of porous preforms). Therefore, the most interesting manufacturing routes are those based on powder metallurgy and its many variants, in which the matrix is handled in powder state and composites are consolidated in solid state. However, dispersion of nanoreinforcements inside the matrix, and more in particular VGCNFs, is a problem that has to be addressed and solved before arriving to encouraging results, that means that the lowest porosity values in the composites are consolidated. In order to promote a good contact between matrix and reinforcement, VGCNFs were coated by copper through electroless plating process. Prior to the coating process it is necessary to catalyze the VGCNF surface. Thus, the fiber is immersed in two catalyzing baths composed by hydrochloric acid and tin (sensitizing bath), followed by a bath containing hydrochloric acid and palladium(activation bath). After catalyzing, nanofibers are soaked into the coating solution, in which the red-ox reaction is carried-out. VGCNFs are maintained into the solution until the bath turns into transparent. At this point all copper from the metallic salt is converted into metallic copper on the nanofiber surface. The whole catalyzing process can be aided by the application of ultrasounds and mechanical stirring into the baths. VGCNFs were rinsed and filtered at the end of each step and finally dried under vacuum at 100°C. Another way is to use polyacrylic acid (PAA) to increase VGCNFs dispersion, in the co-deposition with copper by electrodeposition. In this way, PAA was added to the plating bath in 140 μm/L concentration. The cover thickness, and as a result the vol% of the copper matrix in the final composite, were controlled by means of VGCNFs concentration into the plating bath. In this way different samples containing 20 and 40 vol% of reinforcement have been assessed. Finally, copper coated VGCNFs are consolidated into dense samples by hot pressing. The consolidation was performed in the semisolid state (The melting temperature of copper is 1,083°C). Thus, composite powders were inserted in a high strength graphite die (25 mm inner diameter) and hot-pressed under high vacuum with different pressure and temperature parameters: 25 and 35 MPa, and 700, 900 and 1,000°C for 2 h. Consolidation at 900°C and higher pressures is beneficial for porosity and VGCNFs dispersion in the matrix. Minimum oxide amount is formed at the end of the process, which will be beneficial for the final thermophysical properties (CTE and thermal conductivity). Microstructure observed and physical properties measured suggest that the manufacturing processes here addressed are both suitable for the production of copper matrix composites reinforced by VGCFs and other type of nanoreinforcements (CNTs, nanoparticles, etc.) although a further reduction of remaining porosity at high vol % and avoid oxide formation would lead to better thermophysical properties. Further work will also address some other critical characteristics of composites, such as interface modification by coating and/or functionalization of fibers without degrading their physical properties in order to promote a better coupling of matrix and reinforcement, thus leading to improved physical and mechanical characteristics (Barcena et al. 2006).

In addition, molecular level mixing is essential to get homogeneous distribution between the phases. This is quite a challenge as the CNTs are heavily entangled after CVD-production and form a quite stable felt. By the application of wet chemical methods, the felt could be easily penetrated which results then in a homogeneous distribution of the respective phases. Additions of alloying elements are also essential to address the interface.

## *Copper–Silicon Carbide and Copper–Diamond*

Cu–SiC is promising used for heat sink and base plate materials of electronic devices as it can provide high thermal conductivity (>200 W/m K), tailored CTE

(8–12 ppm/K), high stiffness (>200 GPa), and high bending strength (>300 MPa). Due to its high wear resistance, however, the Cu–SiC composite is difficult to machine. This creates some limitations as to the method of manufacture and causes problems with machine time and tool wear. Figure 6.7 shows microstructure, thermal conductivity and CTE of several SiC particle reinforced Cu composites. This material could be useful where weight and/or CTE issues are important. By increasing the SiC fraction, the CTE and density are decreased. Unfortunately, the thermal conductivity is also reduced. This means the benefit of lower weight and tailorable CTE have to be weighed against reduced thermal conductivity.

As an application example, the development of new heat sink materials with high thermal conductivity and sufficient strength under service loading is required for the efficiency of future fusion reactors. In the case of the divertor, the maximum allowable working temperature under neutron irradiation for the currently utilized copper alloy is 350°C. To increase efficiency, it is necessary to develop heat sink materials for operating temperatures up to 550°C. The maximum allowable heat sink temperature determines, among other parameters, the efficiency of the fusion reactor process. The heat sink in the region of the divertor has to be capable of withstanding 10–20 MW/m$^2$ heat flux. Because the efficiency of the steam turbine process depends on the temperature of the cooling agent in the steam generator, the heat sink temperature should be as high as possible. To increase the efficiency of the entire process it is necessary to develop materials with a thermal conductivity of more than 200 W/m K for operating temperatures up to 550°C. Copper meets the

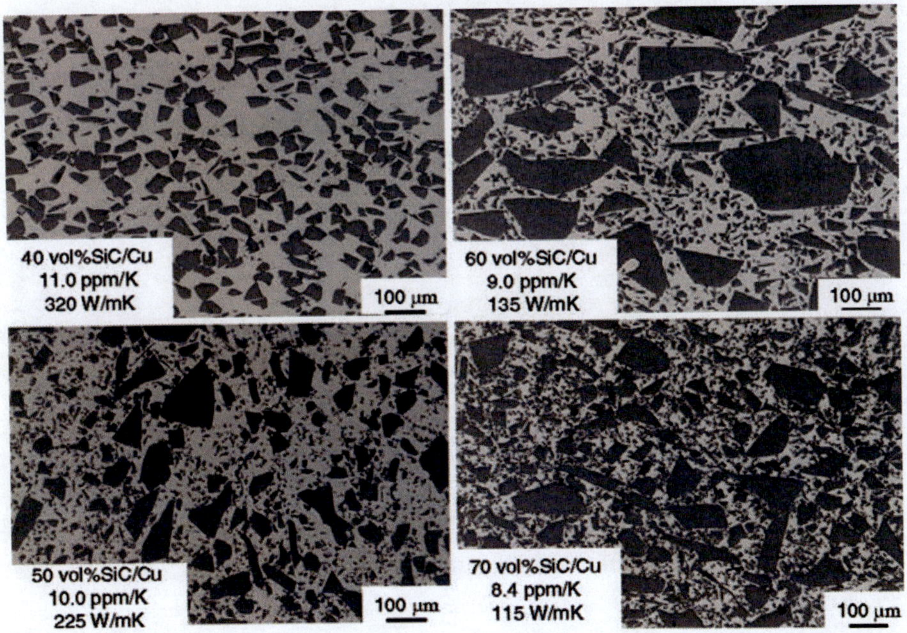


**Fig. 6.7** Illustration of Cu–SiC composites

requirements in thermal conductivity (400 W/m K), but not in mechanical strength at elevated temperatures. High strains occurring at the interface between plasma facing material and heat sink due to the different CTE. This zone should be strengthened by reinforcing copper with SiC long fibers. Even under intense neutron irradiation the copper should remain ductile at elevated temperatures. An additional advantage of the SiC fibers is the comparatively low swelling under neutron irradiation (lower than 1%). A composite made of a copper matrix for high thermal conductivity reinforced with SiC fibers should resist high stresses at the interface between plasma facing material and heat sink even at 550°C under neutron irradiation. Figure 6.8 shows microstructure of the SiC-fiber reinforced copper composite, compared with diamond particles reinforced copper composite. SiC-fiber reinforced copper combines both high thermal conductivity and mechanical strength. SiC fibers with a diameter of 140 μm were coated with an 80 μm

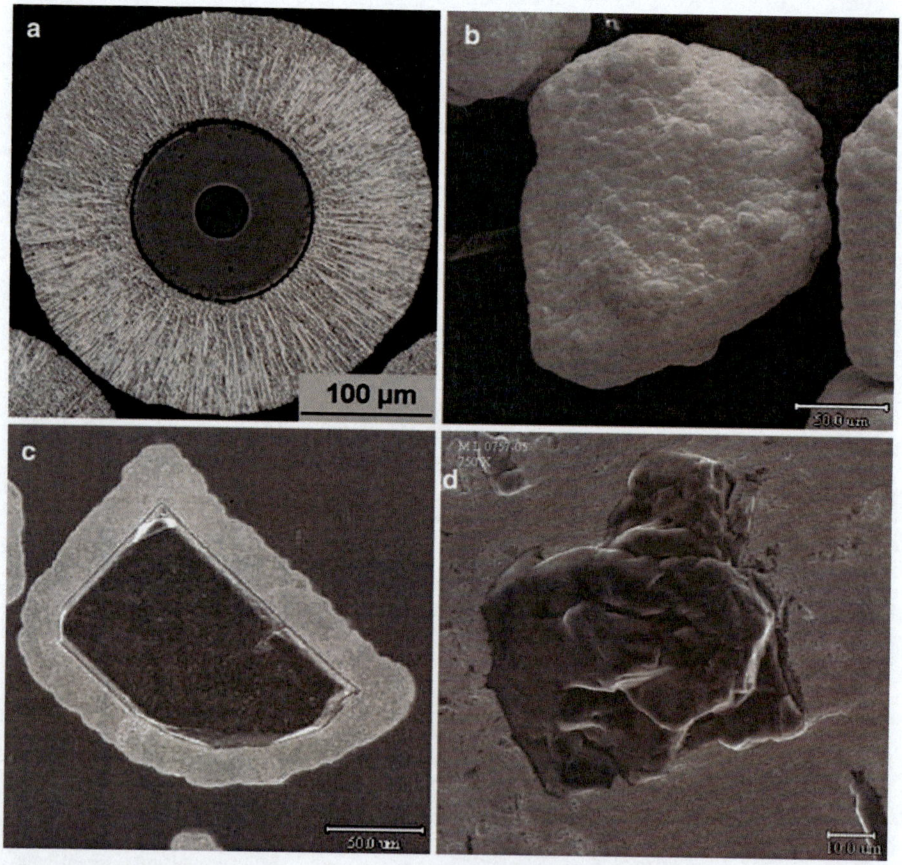


**Fig. 6.8** Illustration of Cu–Si and Cu–diamond composites. **(a)** Copper coated SiC fiber; **(b)** copper coated diamond particle; **(c)** cross-section of **(b)**; **(d)** thermally compressed Cu–diamond composite

copper layer by electroplating. Coated fibers were heat treated at 550°C for 1 h and hot-isostatic pressed in a copper capsule at 650°C and 1,000 bar to form a composite material. From push out tests, interfacial shear strength of 7.2 MPa and interfacial friction stress of 4.3 MPa were calculated. Both values characterize the interface between fibers and matrix as poorly bonded. One possibility to increase the bonding strength is to deposit a 100 nm thin titanium interlayer at the fibers carbon surface. During the heat treatment the titanium reacts with the carbon and forms TiC. With this interlayer the fibers are chemically and mechanically bonded with the matrix. The interfacial shear strength increases up to 54 MPa, approximately ten times higher in comparison to fibers without titanium interlayer. Titanium leads to an improvement of the interface properties in a Cu–SiC composite, deposited on the fibers by magnetron sputtering to increase the bonding strength (Brendel et al. 2004).

Compared with Cu–SiC, diamond composites are quite promising materials for thermal management applications to gain highest thermal conductivity. Interface engineering is essential to gain full potential of thermal conductivity: solved by coatings on diamonds with Ti, Ni, Mo, W; on SiC with Ni; on graphite flakes with Ni, Cu; or by addition of active alloying elements to the matrix.

Copper–diamond composites consisting of a high volume fraction diamond provide a fitting solution to existing thermal management issues with their excellent thermal conductivity (>450 W/m K) and a tailorable CTE matching that of the semiconductor material. These composites are available with a machinable surface for further surface operations such as milling, lapping for realizing surfaces of optical quality. Metallization of high purity elements or combinations, such as Ni, Ni/Au, Au, Ag, Ru, etc. is offered for various applications. Advantages of copper–diamond composites can be summarized as (a) excellent thermal conductivity (470 W/m K or above at 20°C); (b) low CTE (6.7 or tailored value at 20°C); and (c) well adapted to semiconductors and ceramics.

Typical thermal management applications of copper–diamond composites include (1) heat spreaders and heat sinks for microprocessor packages; (2) substrates, submounts and heat spreaders for high power laser diodes and diode arrays; and (3) base plates for automotive and power electronics. A great many copper–diamond composite components with various fabrication processes have been explored. For example, heat sink made from the copper–diamond composite material comprises 40–90% by volume of diamond and 7–59% by volume of copper or a copper-rich solid solution where Cu > 80 at.%. Heat sinks with a polycrystalline diamond layer on the side facing the semiconductor chip have been produced. Diamond layers produced by way of CVD have a thermal conductivity of 1,000–1,500 W/m K. However, cracks can occur in the diamond layer even during cooling from the coating temperature, on account of the lack of plastic deformability of the diamond layer. One diamond composite material with a thermal conductivity of >1,700 W/m K, in which loose, shaped diamond particles are converted into a stable shaped body by way of subsequent vapor deposition of diamond. The diamond composite produced in this way is too expensive for commercial use in mass-produced parts (Herb et al. 1993). A process for producing a diamond–silicon carbide composite material has been developed. The composite material is produced by infiltrating a diamond skeleton with silicon or a silicon alloy. With the high melting point of silicon and the resulting high infiltration temperature, diamond is partially converted into carbon and/or then into silicon carbide. Due to its high brittleness, this material is very difficult and expensive to machine. A kind of sintered diamond material is produced using a transition metal, boron and silicon which are deposited on diamond powder by means of physical coating processes. Then, the coated diamond grains are joined to one another by means of a solid-phase sintering process at high pressure (Kume et al. 1990). A drawback of this process is that the product which is formed has a high porosity. Moreover, the production process is highly complex. Another process in which diamond grains are coated with W, Zr, Re, Cr or titanium, then the coated grains are compacted and the porous body is infiltrated for example with Cu, Ag or Cu–Ag alloy melts (Colella et al. 1998). The high coating costs restrict the fields for which composite materials produced in this way can be used. In addition, the transition metals are all strongly carbide-forming elements. Consequently, the carbide layers formed are relatively thick. This combined with the low thermal conductivity of these carbides (10–65 W/m K) reduces the thermal conductivity boosting effect of the diamond phase considerably.

For further improvement, the composite material has been produced to make a heat sink, comprising 40–90% by volume diamond grains; 7–59% by volume copper or a copper-rich solid solution with Cu more than 80 at.%; and 0.01–20% by volume boron or a boron-rich phase with B more than 50 at.%. The heat sink has an excellent bonding strength between the diamond grains and the copper-rich phase. Boron enrichment in the atomic layer range through to the formation of a boron–carbon compound at the interfaces between the diamond grains and the copper-rich phase. The thickness of the boron–carbon compound is preferably in the range from 1 nm (corresponding to a boron–carbon compound content by volume of approximately 0.001%) to 10 μm, preferably up to 0.05 μm. Even incomplete coverage of the diamond grains is sufficient to achieve enough bonding. It is also possible for atoms of other elements to be included in the boron–carbon compound. Boron–carbon compounds have a sufficiently high thermal conductivity (e.g., $B_4C$, approx. 40 W/m K). Since boron is a relatively weakly carbide-forming agent, the carbide-forming element is not depleted too quickly in the copper melt during an infiltration process, with the result that a highly homogeneous material can be achieved. Furthermore, the carbide layers which form are in relative terms very thin, which is likewise of benefit to the thermal conductivity. Because the solubility of boron in copper is very low (between 0.005 and 0.3 at.%, depending on the cooling rate) in the solid state, the thermal conductivity of the copper matrix deteriorates only slightly. Boron is present in precipitated form in the copper matrix, forming between 0.01 and 20% by volume. Boron-rich compounds with a boron content of >50 at.% may also precipitate, depending on further alloying elements in the copper. Further microstructure constituents, such as for example amorphous carbon, do not have an unacceptably adverse effect on the properties, provided that they do not exceed 10% by volume and the thermal conductivity of these microstructural constituents exceeds 50 W/m K. The crucial factor in this context is always that the copper matrix should as far as possible remain free of foreign atoms, or dissolved fractions should have the minimum possible negative influence on the thermal conductivity, as is the case with Cu–Ag or Cu–Zn. The machining properties are sufficient, on account of the highly ductile copper-containing microstructural constituents. A further advantage for inexpensive production is that with the high thermal conductivity of the copper-containing microstructural constituents, the diamond content can be reduced compared for example to diamond–SiC materials. By varying the diamond, copper and boron contents, it is possible to produce heat sinks with a thermal conductivity and thermal expansion which has been tailored to a very wide range of requirements. Particularly advantageous boron carbide and copper or copper-rich phase contents are 0.005–1% by volume and 15–40% by volume, respectively. A wide range of processes can be used for production. For example, it is possible for diamond powder to be compacted with a copper–boron alloy under temperature and pressure. For example, this can take place in hot presses or hot isostatic presses. In principle, it is also possible for copper and boron to be introduced separately. In this case, the alloy is formed during the hot-pressing operation. The starting point in this case may also be $B_4C$-coated or boron-coated diamond powder. Infiltration has proved particularly advantageous. This involves producing a precursor or intermediate material which, in addition to diamond powder, may also contain a binder. Diamond powder and binder are mixed and followed by the shaping, which can take place by introducing a bed into a mold or with the assistance of pressure, for example by pressing or metal injection molding. The intermediate material is subsequently heated to a temperature at which the binder is at least partially pyrolyzed. However, the pyrolysis of the binder can also take place during the heating in the infiltration process. The infiltration process can take place in unpressurized or pressure-assisted form. The latter option can be implemented by squeeze casting, die casting, or in a sinter-hot isostatic press (HIP) installation. The infiltration material used is a foil advantageously made from a copper-boron alloy with a boron content of 1–4% by weight. In this case, it is also possible for the alloy formation to take place in situ during the infiltration process. When selecting the composition, it is necessary to take into account the fact that the liquidus temperature of the respective alloy should not be higher than 1,050–1,200°C, otherwise excessively high levels of diamond fraction decompose. Foils with a eutectic or near-eutectic composition, in which near-eutectic compositions encompass alloys with a liquidus temperature of $<$1,050°C, are particularly suitable for the infiltration. In addition to the particularly advantageous use of the components for the dissipation of heat from semiconductor components, the composite material can also be used as a heat sink in other application areas, such as for example in aeronautical or aerospace applications or motor manufacture (Weber 2007).

As shown in Figure 6.8, another process for fabrication of the copper-diamond composite is that the diamond is plated with copper, or coated with SiC conversion first then coated with copper, and finally hot-pressed into components. The effective thermal conductivity of Cu–diamond can be predicted based on the following equation (Hasselman and Johnson 1987; Yoshida and Morigami 2004):

$$k = k_m \left( \frac{2\left(\frac{k_d}{k_m} - \frac{k_d}{ah_c} - 1\right)V_d + \frac{k_d}{k_m} + \frac{2k_d}{ah_c} + 2}{\left(1 - \frac{k_d}{k_m} + \frac{k_d}{ah_c}\right)V_d + \frac{k_d}{k_m} + \frac{2k_d}{ah_c} + 2} \right), \tag{6.4}$$

where $k_m$ is the thermal conductivity of the matrix, $k_d$ is the thermal conductivity of the dispersions, $a$ is the radius of the spherical dispersions, and $h_c$ is the boundary conductance. From (6.4), Figures 6.9–6.12 shows the relationship among thermal conductivity, volume fraction of diamond and the diamond particle size. The interfacial thermal barrier is a crucial factor to improve the composite thermal conductivity.

As continuous improvement of the processing for copper–diamond composite materials, their performance would definitely be modified uninterruptedly. Currently, copper–diamond heat spreaders can typically offer a thermal conductivity of 600 W/m K and an average CTE of 3.0 ppm/k in the range of 20–300°C. Therefore, copper–diamond composite as a heat spreader material is intermediate between CVD diamond and other heat spreading materials, such as AlN and Cu. The copper–diamond heat spreader material is conventionally produced by high temperature, high pressure sintering of small diamond particles to form a homogenous intergrown diamond lattice structure. The fine grain structure and uniform distribution of the binder makes this material ideal for heat spreader applications. It can also

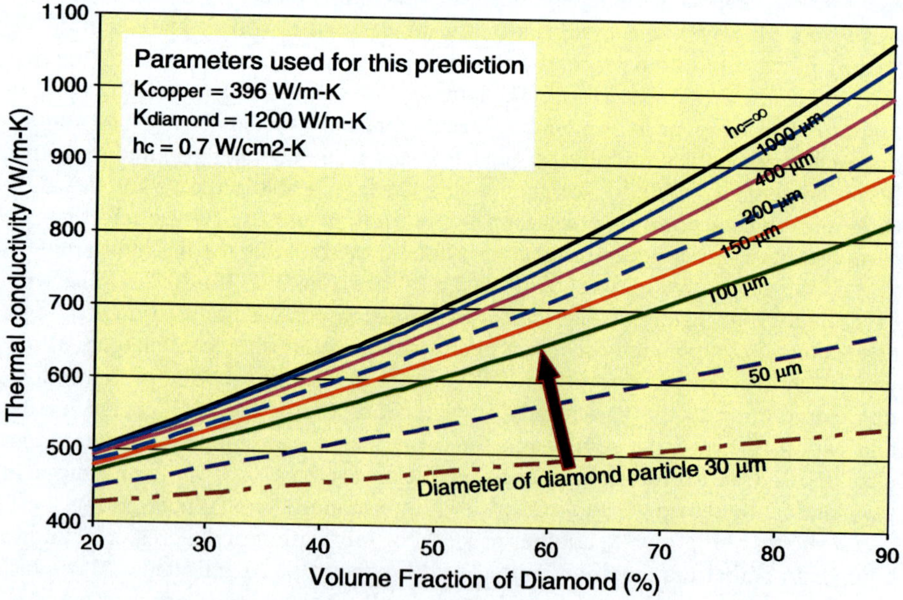


**Fig. 6.9** Illustration of thermal conductivity variation with volume fraction of diamond

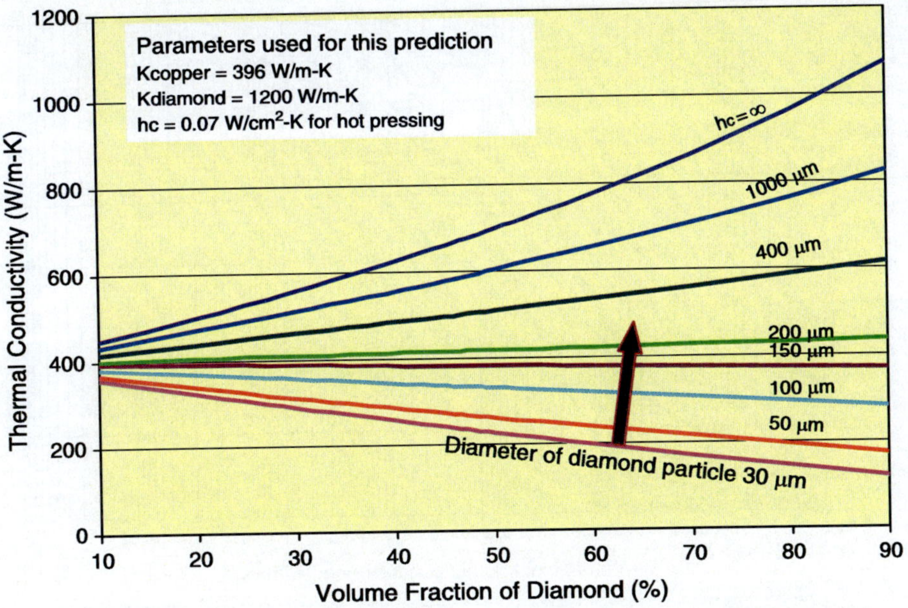


**Fig. 6.10** Illustration of thermal conductivity variation with volume fraction of diamond

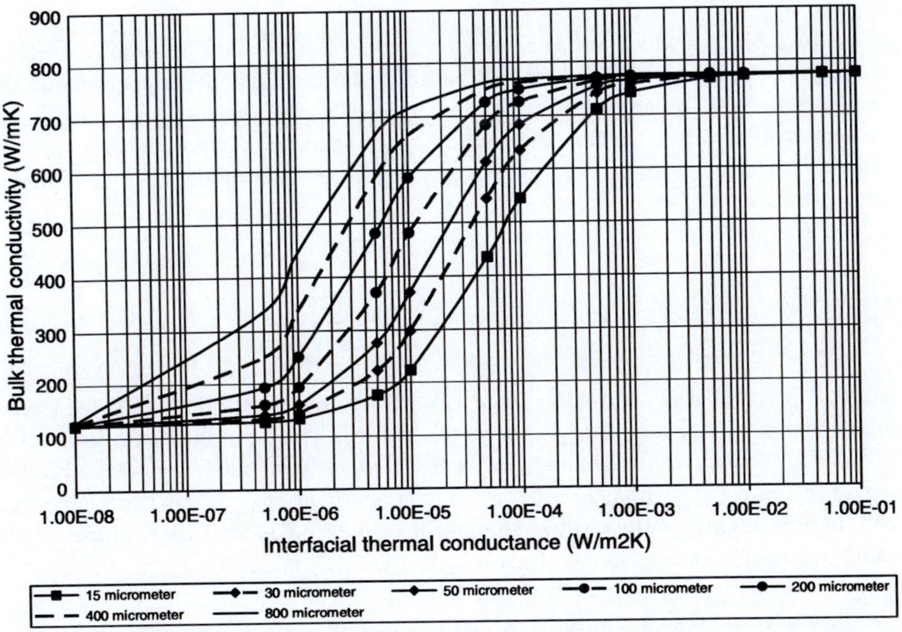


**Fig. 6.11** Illustration of thermal conductivity variation with interfacial thermal conductance when the volume percentage is 60% diamond

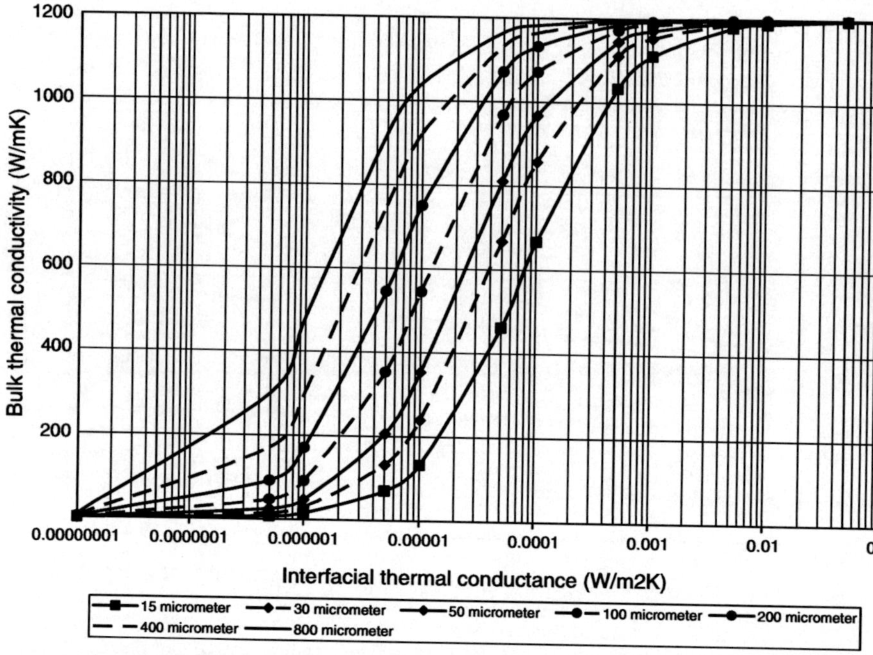


**Fig. 6.12** Illustration of thermal conductivity variation with interfacial thermal conductance when volume percentage is near 100% diamond

be used for devices with small heat sources, such as laser diodes, laser diode arrays, and high power light emitting diodes. In addition, the surface of the copper–diamond heat spreader can be polished to a high degree of surface quality and flatness. If well controlled with high precision laser cutting, excellent edge qualities can be achieved, which is usually required for edge emitting laser diodes and laser diode arrays.

## Other Metal Matrix Composites

Conventionally, $Al_2O_3$ and AlN whose CTE is approximate to that of a semiconductor element have been used as a heat sink material in spite of their comparatively low thermal conductivity because conventional electronic components incorporating semiconductor elements such as semiconductor lasers, microwave elements, etc., have generated only a slight amount of heat. With increasing application of semiconductor elements having high output, heat sink and other thermal management materials having remarkably excellent thermal conductivity are demanded.

The material having the highest thermal conductivity is diamond and c-BN (cubic boron nitride). However, their CTE is very low, wherein that of diamond is 2.3 ppm/°C and that of c-BN is 3.7 ppm/°C, and the Young's modulus of these materials is very high to be 830 through 1,050 GPa. Therefore, a large thermal stress occurs when brazing a heat sink and a semiconductor element together or between the heat sink and semiconductor element when being used as a device, wherein a breakage is likely to occur.

Various types of composite materials such as Al–SiC have been proposed as a heat sink material having a low CTE as well as comparatively high thermal conductivity. However, since the thermal conductivity of Al is as low as approx. 238 W/m K at room temperature, there exists an upper limit in the thermal conductivity of a composite material including Al–SiC. It may be considered that metals having high thermal conductivity such as Cu (395 W/m K at room temperature) and Ag (420 W/m K at room temperature) are used instead of Al. However, since the wettability thereof with SiC is very inferior, the high thermal conductivity that is inherent in Cu and Ag cannot be sufficiently displayed. A Ag–Cu based diamond composite material has been developed as a heat sink material having improved wettability with Cu and Ag. Diamond powder and Ag–Cu–Ti based powder are blended together and molded, and then are heated at a higher temperature than the melting point of the resultant alloy. This allows Ti constituents to diffuse on the surface of diamond grains and to react to form a TiC film on the surface through sintering process. Since the TiC has good wettability with Cu or melted Ag, the phase boundaries of the diamond grains and the metal are adhered close to each other, whereby high thermal conductivity can be obtained. In addition, an infiltration method is used to produce such a diamond–Ag based or diamond–Cu based composite material. With this method, after diamond powder and Ag–Cu–Ti based powder are blended and molded, the molded body is heated at a higher temperature than the melting point of the corresponding alloy to form a TiC layer on the surface of diamond grains. After that, the molded body is further heated to elute and volatilize the Ag constituents and Cu constituents, thereby producing a porous body. Impregnating the porous body with an Ag–Cu alloy produces a composite material having a higher relative density and a higher thermal conductivity than that obtained by the sintering method (Kawai and Nakata 2003). Other MMCs for thermal management of electronic packaging include beryllium composites, Silver–diamond, low CTE composite solders, and multilayer composites.

## *Beryllium Composites*

Beryllium is a steel-gray mineral which is found in igneous rock, particularly in the minerals beryl and bertrandite. Beryllium is relatively scarce and as such it will be of limited use. Beryllium is one of the lowest density metals. It has a remarkable combination of properties found in no other metal. It has six times the specific stiffness of steel. Beryllium has a very high melting point and it maintains useful mechanical properties up to much higher temperatures than most other metals, yet it also has excellent cryogenic properties. Beryllium combines high thermal conductivity with the highest specific heat of any metal for superior thermal diffusivity, making it extremely useful for dissipating thermal transients. Beryllium has excellent damping characteristics. It is a low-atomic-number element, with the highest X-ray transparency of any engineering material. Due to its fascinating properties, this metal has an extensive heritage in space structures, airborne, earth- and space-based optics, and the semconductor, medical and nuclear industries.

On another hand, beryllium oxide (BeO) is unique for oxides as it combines excellent electrical insulating properties with high thermal conductivity. It is also corrosion resistant. The high toxicity of the BeO powders when inhaled, and the high cost of the raw material, has limited its use to applications that exploit its singular properties. BeO is extracted from the naturally occurring minerals beryl and bertrandite, and produced as a powder by the thermal decomposition of Be $(OH)_2$. Powders are commercially available at purity levels of greater than 99%. Components can be made as near net shapes by most of the commonly used fabrication methods, for example, pressing, slip casting, or extruding the powder. Sintering is carried out in the range 1,600–1,800°C. High density components (<5% porosity) can be easily made with commercially pure powders. Near theoretical density (<1% porosity) can be achieved using high purity materials and hot pressing. BeO has an outstanding combination of physical and chemical properties: (1) thermal conductivity is extremely high in comparison with other ceramics, particularly below 300°C. For comparison the thermal conductivity of BeO at room temperature is 300 W/m K, copper is 380 W/m K and alumina is 35 W/m K. (2) Electrical resistivity is high. BeO is classed as an electrical insulator. (3) Mechanical strength is normally lower than alumina, but can reach acceptable levels through control of the fabrication process. (4) BeO has good thermal shock resistance if the component has good strength due to its high thermal conductivity. (5) BeO has lower density than alumina, 3,010 and 3,970 kg/m$^3$, respectively. (6) The thermal expansion of BeO is similar to that of other oxides. BeO is most often used as an electronic substrate, exploiting its high thermal conductivity and good electrical resistivity to give an effective heat sink. The material is found particularly in high power devices or high density electronic circuits for high speed computers.

The composite material beryllium–BeO with 20–60 vol% BeO exhibit low density (2.3 g/cm$^3$ at 40 vol% BeO, compare to 2.9 g/cm$^3$ for Al/SiC at 40% SiC), high thermal conductivity (232 W/m K at 40% BeO, compared to 130 W/m K for Al/SiC at 40 vol% SiC), low CTE (7.5 ppm/K at 40 vol% BeO, compared to 12.1 ppm/K at 40 vol% SiC), and high modulus 317 GPa at 40 vol% BeO, compared to 134 GPa for Al/SiC at 40 vol% SiC) (Chung 2003).

In the composite, the beryllium metal is always present as a continuous phase with the BeO dispersed therein. Most preferred are beryllium alloys having at least about 30% by volume of beryllium. Higher volume fractions of BeO result in lower thermal expansion coefficients and higher thermal conductivities. It should also be appreciated that processing becomes more difficult with volume fractions of greater than about 60%. Preferred volume loadings are in the range from about 20% to about 60% by volume, more preferably in the range of about 40–60% by volume. The beryllium–BeO composite material is fabricated by first providing a beryllium metal powder and BeO powder. Appropriate measures of the powders are placed in a roll blender or V-blender. The ratio of beryllium to BeO is chosen by the material designer according to property requirements. If a higher thermal expansion coefficient or lower thermal conductivity is required, the amount of beryllium metal is increased relative to the BeO. As with conventional processing, the input powders must be dry, inclusion-free and without lumps. The mixture of powders is then blended for a few hours to form a homogeneous composite powder. After the powders are blended, it is preferred that the composite powder be examined to determine if any agglomerations are present. Agglomerated powder is removed by screening or a milling media can be added to the mixture during blending to facilitate deagglomeration. The milling media usually include 2 cm diameter BeO spheres. Another method for deagglomerating the powder is to perform the mixing in a liquid medium. If liquid blending is used, the mixture must be thoroughly dried before processing continues. The composite powder is then formed into a desired shape and densified. Densification is accomplished by conventional hot isostatic pressing techniques, with the resulting billet being further processed into the desired shape with required dimensions. In general, densification is accomplished by first loading a mild steel HIP can with the composite powder. The size and shape of the HIP can is determined by the dimensions of the billet from which the final product is made. The powder may be loaded into the HIP can either manually or with the aid of a mechanical loading device. Conventional processing often includes a vibrating device to facilitate the flow of powder or slip casting thick slurry into a mold. The HIP can is loaded with the composite powder and attached to a vacuum system for evacuation. At this point it is desirable to check the can for leakage. If no leaks are observed, the can is slowly heated under vacuum to drive off residual moisture and gases from the powder. After degassing, the HIP can is sealed and placed into a HIP unit. The composite powder in the can is densified by heating to about 1,000°C at 15 ksi for about 3 h. The composite may be hot isostatic pressed in the temperature range of 900–1,275°C, more generally from about 900°C up to the melting point of the beryllium metal or alloy. The minimum pressure for successful densification at 900°C is about 10 ksi. At higher HIP temperatures, a lower pressure may be used. For example, at about 1,200°C, a HIP pressure of about 5 ksi is sufficient for densification. The maximum HIP pressure is limited generally by the processing equipment. HIP times depend on both temperature and pressure, with HIP time increasing with decreasing temperature and/or pressure. HIP times of between about 2 and 6 h are generally sufficient. hot isostatic pressing preferably is done in an inert atmosphere, such as argon or helium. It should also be noted that the particle size distribution will effect the final density of the hot isostatic pressed article, with narrower distributions yielding denser pieces. However, broader particle size distributions can be accounted for by increasing HIP pressure. The composite may also be densified by hot pressing, although HIP is preferred. The density of the final composition will be generally in the range of about 1.95 g/cm$^3$ to about 2.65 g/cm$^3$. When densification is complete, the sealed can is removed from the dense beryllium–BeO billet by leaching in nitric acid or by other techniques. The beryllium–BeO composite billet can be machined into various shapes. For electronic board applications a sheet configuration is the preferred geometry. To accomplish this geometry the composite billet is rolled at about 1,000°C to a desired thickness. Sheets may also be formed by sawing small sections from the billet and surface grinding to required tolerances. It is also possible to densify by hot isostatic pressing to the sheet morphology. Conventional machining techniques can be used for the composite materials. It is important to note that the composite material is very abrasive and causes tool wear. For example, electrodischarge machining cutting rates are very low when used on the composite material. Once machined to the desired specifications, the composite article can be plated and/or anodized in a fashion similar to that of beryllium. The composites may be stress relieved and flattened with no loss of thermal properties. However, the rolling technique may have a detrimental effect on thermal conductivity and the CTE for the composite material, but to a small degree. The composites may be further processed by rolling to decrease the thickness. Rolling may be performed at temperatures generally between 850 and 1,200°C. The rolling reduction per pass preferably is between 4 and 20%. Rolling may be done under any non-reactive atmosphere, including air. Preferably, rolling is done at about 1,000°C with a reduction per pass of 10% to achieve a total reduction of 90% (i.e., the resulting article has a thickness 10% of the original thickness). Between passes, the article may be annealed at about 760°C. The composites may also be stress relieved, a standard beryllium metallurgical process which removes certain dislocations and makes the composite less brittle (Grensing 1992).

## *Silver–Diamond*

The silver–diamond composite consists of a silver alloy reinforced with a high volume content of diamond particles. The composite material features an excellent thermal conductivity and a relatively low thermal expansion which is well adapted to semiconductors and ceramics. The silver–diamond composite is excellently suited for thermal management applications, such as heat sinks and heat spreaders in high performance micro- and optoelectronics. For precision machining of surfaces and edges, the techniques such as diamond milling or lapping can be used.

The physical and mechanical properties of the silver–diamond composite are dependent on the volume fraction and the grade of the diamond powder. Typical property of silver–diamond composite: density at 20°C, 6.0–6.2 g/cm$^3$; specific heat at 20°C, 0.31 J/gK; thermal conductivity at 20/100°C, 550–650/520–600 W/m K; CTE at 20/200°C, 5.8–6.4/7.0–7.8 ($10^{-6}$/K); specific electrical resistivity 20°C, 8.5–9.0 μΩ cm; Young's modulus at 20°C 250–300 GPa; flexural strength at 20°C, 320–380 MPa.

Silver–diamond composite material can be produced through the following steps: (1) coating the surfaces of diamond powder particles with molten silver nitrate at a temperature of from just above the melting point of silver nitrate to just below the decomposition temperature of silver nitrate; (2) heating the molten silver nitrate coated diamond particles produced in step 1 at a temperature from the decomposition temperature of silver nitrate up to 550°C until the molten silver nitrate decomposes to form a coating of silver metal on the surfaces of the diamond particles; (3) consolidating the silver metal coated particles produced in step 2 into a composite structure having continuous porous between the silver coated diamond particles; (4) infiltrating the continuous pores of the composite structure produced in step 3 with molten silver nitrate at a temperature of from just above the melting point of silver nitrate to just below the decomposition temperature of silver nitrate; (5) heating composite structure at a temperature of from the decomposition temperature of silver nitrate to 550°C until the molten silver nitrate in the pores of the composite structure decomposes to silver metal; and (6) repeating steps 4 and 5 until the pores of the composite structure are essentially filled with silver metal.

This method is realized by coating the diamond particles with molten silver nitrate which is then decomposed to form the silver metal coating on the diamond particles. In one procedure the diamond powder is intimately mixed with silver nitrate powder. The mixture is then heated to a temperature above the melting point of silver nitrate (212°C) but below the decomposition temperature of silver nitrate (444°C) until the silver nitrate melts. The mixture is agitated or stirred to assure that molten silver nitrate coats all the diamond particles. In a second procedure the silver nitrate is dissolved in a suitable solvent such as water or more preferably ethylene glycol. Saturated silver nitrate solutions are preferably used. The silver nitrate solution is mixed into the diamond powder. The solvent is then evaporated causing the silver nitrate to form small crystals on the surfaces of the diamond particles. In removing ethylene glycol the temperature should be kept below 75°C and preferably should be in the range of from 50 to 65°C. The silver nitrate coated diamond particles are then heated to a temperature above the melting point of silver nitrated but below the decomposition point of silver nitrate to form the molten silver nitrate coating on the diamond particles. For both procedures, the next step comprises heating the molten silver nitrate coated diamond particles at a temperature of preferably from the decomposition temperature of silver nitrate to 550°C, more preferably from 450 to 525°C, and still more preferably from 475 to 500°C until the molten silver nitrate decomposes to form a thin silver metal coating. The silver metal coated diamond powder will preferably comprise from about 10 to about 60, more preferably from 15 to 50, and still more preferably from 20 to 40 wt% of silver metal, with the remainder being the diamond powder. The weight of silver requires producing the desired weight percentage of silver for a given weight of diamond material. The amount of silver nitrate needed will be 1.575 times the weight of silver needed (Ferrando 1994).

The silver–diamond composites can also be directly prepared by: (1) coating the diamond particles with silver metal; (2) hot or cold pressing the silver metal coated particles into a preform of the desired shape; and (3) filing the interstices of the preform with silver metal to produce the final composite structure.

## *Low-CTE Composite Solder*

Composite solder is conventionally in the form of a solder preform. A solder preform is solder which has been shaped into a two-dimensional shape such as a sheet or ring. A composite solder is a solder in the form of a composite material. Solder preforms are used to join various parts of an electronic package. For example, they are used to join a ceramic cap and a multilayer ceramic substrate in a multichip module. Because the thermal expansion coefficient of a solder is in general much higher than that of a ceramic, the solder joint suffers from a poor resistance to thermal fatigue. Thus, there is a need for a solder with a low CTE. Composite materials can be tailored to exhibit a chosen thermal expansion coefficient, as the filler species and the filler volume fraction can be judiciously chosen. For a low thermal expansion composite, the filler must have a low thermal expansion coefficient. Moreover, the filler should preferably be a good electrical conductor, because the soldered joint may serve as an electrical connection as well as a mechanical connection. It should also preferably be a good thermal conductor for heat dissipation from the electronic package. In addition, the filler should be wetted well by liquid solder in order to facilitate composite fabrication, and bond well to matrix for maintaining continuity. Further requirements for the reinforcements include: (1) the density of the reinforcement matches with solder matrix to achieve uniform distribution in matrix. And upon remelting and solidification of the solder in the solder-matrix composite, the filler distribution in the composite remains uniform. (2) Minimum solubility to solder to keep stable during reflow. (3) Optimal reinforcement size to stabilize the microstructure of the composite solder. (4) Not prone to coarsening to stabilize microstructure under aging. And (5) not affecting current processing temperature and wettability of the solder matrix to maintain its advantage.

The reinforcements used for composite solders can be classified into several categories: (1) elemental metallic particles; (2) intermetallic particles or intermetallics in situ formed from elemental particles through a reaction with Sn or other elements in the solder matrix during manufacturing or by subsequent aging and reflow processes; and (3) phases that have low solubility in Sn or are nonreactive with Sn. The reinforcements used also varied in size, from micron-size to nano-size, and in shape, such as particles, wires and nanotubes. For instance, various nano-size reinforcements have been chosen to synthesize the composite solders. The nano-size reinforcements are usually preferred over their micron-size counterparts as they could more effectively locate at the Sn–Sn grain boundaries, thus acting as obstacles to restrict dislocation motion (Guo 2007).

For instance, molybdenum particles have been used as filler in solder-matrix composites. However, their distribution in the solder (Sn-40% Pb alloy) became non-uniform after remelting and solidification of the solder in the composite. Thus, molybdenum is not suitable filler in spite of its low CTE. Carbon fibers have nearly zero thermal expansion coefficient. In contrast to ceramic fibers (which also have low thermal expansion coefficients), carbon fibers are a good electrical conductor and a good thermal conductor. Carbon fiber reinforced tin–lead alloys had been previously fabricated by liquid metal infiltration for bearing applications (Old et al. 1974), and by investment casting for fundamental process study (Chung 1992). For both methods, the carbon fibers were electroplated with copper in order to ensure good wetting of the fibers by the alloy. Carbon-fiber reinforced tin (containing 7.5 wt% Sb and 3.5 wt% Cu) have also been fabricated by squeeze casting for fundamental process study wherein the carbon fibers were electroplated with nickel in order to avoid the oxidation of the fibers during casting. Such carbon fiber reinforced materials would have particularly good utility in the formation of solder joints. The thermal expansion coefficient in a direction parallel to the sheet decreases with increasing volume fraction of fibers in the composite. For a volume fraction ranging from 8 to 55%, the thermal expansion coefficient at 25°C in a direction parallel to the sheet ranges from $-1 \times 10^{-6}$ to $11 \times 10^{-6}$/°C. For a volume fraction ranging from 17 to 43%, the thermal expansion coefficient at 25°C in a direction parallel to the sheet ranges from 0.5 to $8.5 \times 10^{-6}$/°C. For a volume fraction of about 29%, the thermal expansion coefficient at 25°C in a direction parallel to the sheet is about $4 \times 10^{-6}$/°C (Chung 1992).

The solder matrix is preferably a metal either elemental or alloy with a solidus temperature below 500°C. It preferably contains one or more of the elements, such as tin, lead, antimony, cadmium, copper, silver, gold, zinc, indium, gallium, bismuth, tellurium, aluminum, mercury, selenium, thallium and arsenic. It is preferably chosen from the group tin–lead, and tin–silver–copper. Tin–lead solders have been used intensively in the electronics industry due to their unique characteristics such as low cost and ease of manufacturing. It is a challenging task to find suitable alternatives for the lead-containing solders. Although a number of lead-free solders have currently been available, there is still no drop-in alternative for eutectic Sn–Pb. Not only must the lead-free alternatives meet health, environment and safety requirements, as well as solder joint reliability and performance expectations, they must also be compatible with the existing soldering processes. When identifying a replacement to the current widely used Sn–Pb solders, it is crucial to ensure that the properties of the replacement solder are comparable to or superior than Sn–Pb solders. For most Pb-free solders, it has been identified that the basic building block element is tin (Sn) and the potential substitutes for lead (Pb) include: silver (Ag), antimony (Sb), zinc (Zn), copper (Cu), bismuth (Bi), indium (In) and gold (Au). There is also a special class of eutectic alloys of Sn and noble metals: Au, Ag and Cu. A eutectic alloy is used as it has a single and low melting point. Hence, partial melting or solidification will not occur and the entire solder joint will melt or solidify at a given temperature. From all these lead-free solders, Sn–Ag–Cu solder alloy wets and forms good quality joints with copper. Its thermo-mechanical property is better than those of the conventional Sn–Pb solder. Furthermore, it is a promising solder in automotive and avionics applications where solder joints are subjected to temperatures up to 150°C (Kariya and Otsuka 1998). Therefore, its usage has been increase in the electronic industry.

Composite solders can be applied either as a preform or a solder paste configuration. In addition to conventional reflow soldering process, faster soldering operation (like laser reflow soldering) and processes which make use of preheating temperature below the melting point of the solder (such as thermal compression bonding or thermosonic bonding) can also be utilized (Chaminde et al. 2006). The properties of the composite solders can mostly be retained when using these bonding approaches.

## Summary

High thermal conductivity MMCs have especial advantages for particular electronic packaging and thermal management applications, because of their combination of excellent thermal conductivity, relatively low density, and tailorable CTE to match the CTE of semiconductor materials such as silicon, gallium arsenide or alumina. The optimal design of MMC components is based on appropriate selection of matrix materials, reinforcements, and layer orientations to tailor the properties of a component to meet the needs of a specific design. MMCs can be fabricated with two major categorized methods, primary and secondary. Primary fabrication methods are used to create the MMC from its constituents. The resulting material may be in a form that is close to the desired final configuration, or it may require considerable additional processing, called secondary fabrication, such as forming, rolling, metallurgical bonding, and machining. The processes used depend on the type of reinforcement and matrix. In general, the primary fabrication methods can be classified as solid state method, liquid state method, in situ fabrication method, and vapor deposition process.

Among the MMCs, AMCs have received great interest since the combination of their low weight, good thermal and electrical conductivities, high mechanical strength and excellent wear properties, potentially becoming a material for many engineering applications, such as sporting goods, electronic packaging, armors and automotive industries. The reinforcement in AMCs could be in the form of continuous/discontinuous fiber, whisker or particulates, in volume fractions ranging from a few percent to 70%. Properties of AMCs can be tailored to the demands of different industrial applications by suitable combinations of matrix, reinforcement, and processing route. When selected reinforcements are well combined with an aluminum matrix through an appropriate process, the resulting material has significant increases in elastic modulus (stiffness), wear resistance, and, in some cases, strength and fatigue resistance. In addition, the CTE of aluminum is reduced by the addition of the reinforcement, while the material retains the high thermal or electrical conductivity and low density inherent in the aluminum alloy. AMCs reinforced with SiC particles is one of the most suitable and versatile materials among the MMCs for use in aerospace and electronic applications, particular for thermal management material like heat sinks. Other AMCs include Boron reinforced Aluminum, Continuous Carbon Fiberre reinforced Aluminum, Discontinuous Carbon Fiber reinforced Aluminum, Carbon Flake reinforced Aluminum, Diamond Particle reinforced Aluminum, and Silicon reinforced Aluminum. The application of these AMCs in the electronics packaging market is fast growing, primarily for thermal management applications in which the ability to match the CTE of the electronic materials is a key attribute. These applications will continue to grow as processing methods for forming, machining, and plating continue to develop and further experience is gained.

Copper matrix composite materials have been widely used when superior thermal and electrical conductivities are required, especially for thermal management in electronic packaging, electric contact with high dimensional stability, substrates for high-power semiconductors with similar thermal expansion as silicon, cooling tubes and components for fusion or space application where high stiffness together with the ability to withstand high thermal loads are demanded. Copper matrix composites are usually reinforced by (1) continuous fibers of carbon, SiC, tungsten, stainless steel, and diamond coated fibers; (2) carbon or metal whiskers, carbon foam, and nanofibers; and (3) particles such as tungsten, SiC, and diamond. Powder metallurgy or sintering and infiltration technique can be used for fabrication copper matrix composites, which typically possess low CTE, high stiffness and elastic modulus, good electrical conductivity, high thermal conductivity, and good wear resistance. The copper matrix composites that can be used for thermal management of electronic packaging mainly include Cu–graphite, Cu–nanofiber, Cu–SiC and Cu–diamond.

Other MMCs for thermal management of electronic packaging include beryllium composites, Silver–diamond, low CTE composite solders, and multilayer composites. In particular, the composite solder is technologically attractive because of its low thermal expansion coefficient. The low expansion results in an increased thermal fatigue life for solder joints between materials with low thermal expansion coefficients, such as ceramics, which are encountered in electronic packages in the form of caps and multilayer substrates in multichip modules. The thermal fatigue of solder joints presently limits the reliability of electronics severely.

## References


Arakawa H, Kuniya K, Namekawa T, Iizuka T (1978) Copper-carbon fiber composites and process for preparation thereof. US Patent 4083719.

Barcena J, Maudes J, Coleto J, Obieta I (2006) Novel copper/carbon nanofibers composites for high thermal conductivity electronic packaging. https://escies.org/GetFile?rsrcid=1691. Accessed 21 April 2010.

Brendel A, Popescu C, Leyens C, Woltersdorf J, Pippel E, Bolt H (2004) SiC – fiber reinforced copper as heat sink material for fusion applications. J. Nucl. Mater. **329–333**: 804–808.

Chaminde C, Fogarassy E, Boisselier D (2006) Diode laser soldering using a lead-free filler material for electronic packaging structures. Appl. Surf. Sci. **252**: 4406–4410.

Chawla KK (1999) Composite Materials: Science & Engineering, 2nd edition, Springer, New York.

Chen Y, Chung DDL (1996) In situ Al-TiB composite obtained by stir casting. J. Mater. Sci. **31**: 311–315.

Chung DDL (1992) Carbon fiber reinforced tin-lead alloy as a low thermal expansion solder perform. US Patent 5089356.

Chung DDL (2003) Composite materials: functional materials for modern technologies. Springer, New York.

Colella NJ et al. (1998) Composite material having high thermal conductivity and process for fabricating same. US Patent 5783316.

Deve HE, Skildum JD (2002) Copper matrix composites. US Patent 6416876.

Ferrando WA (1994) Diamond/silver composites. US Patent H001358.

Fukushima H (2009) High-thermal-conductivity graphite-particles-dispersed-composite and its production method. US Patent 20090035562.

Grensing FC (1992) Method of making beryllium-beryllium oxide composites. US Patent 5124119.

Guo F (2007) Composite lead-free electronic solders. J. Mater. Sci. Mater. Electron. **18**: 129–145.

Hasselman DPH, Johnson LF (1987) Effective thermal conductivity of composites with interfacial thermal barrier resistance. J. Compos. Mater. **21**: 508–515.

Herb JA, Pinneo JM, Gardinier CF (1993) Method for consolidating diamond particles to form high thermal conductivity article. US Patent 5273790.

Heremans J, Beetz CP (1985) Thermal conductivity and thermopower of vapor-grown graphite fibers. Phys. Rev. B **32**(4): 1981–1986.

Kariya Y, Otsuka M (1998) Mechanical fatigue characteristics of Sn-3.5Ag-X (X = Bi, Cu, Zn and In) solder alloys. J. Electron. Mater. **27**(11): 1229–1235.

Kawai C, Nakata H (2003) High thermal conductivity composite material, and method for producing the same. US Patent 6569524.

Kume S et al. (1990) Method for production of a sintered article of diamond. US patent 4902652.

Long S (1996) Cast Fibrous MMCs: Transfer Phenomena and Microstructure Formation, PhD. Thesis, Imperial College, London University.

Long S, Flower HM, Beffort O (1997) Effects of premature melt solidification on pressurized infiltration kinetics and infiltration quality. In Beech J and Jones H (eds) Solidification Processing 1997; Proceedings of the 4th Decennial International Conference on Solidification Processing, July, 1997, Sheffield.

Norley J, Brady JJ, Getz G, Klug J (2004) Laminates prepared from impregnated flexible graphite sheets. US Patent 6777086.

Old CF, Barwood I, Nicholas MG (1974) Practical Metal Composites, Spring Meeting, B47–B50, Institution of Metallurgy, London.

Rajkumar K, Aravindan S (2007) Copper-graphite composites. In International Conference on Advanced Materials and Composites (ICAMC-2007), October 24–26, 2007, CSIR, Trivandrum.

Sebo P, Stefanik P (2003) Copper matrix carbon fiber composites. Int. J. Mater. Prod. Tech. **18** (1/2/3): 141–159.

Smith CR (1999) Graphite aluminum metal matrix composite microelectronic package. US Patent 5998733.

Weber L (2007) Heat sink made from diamond-copper composite material containing boron, and method of producing a heat sink. USPTO Application# 20070042895. http://www.freshpatents.com/Heat-sink-made-from-diamond-copper-composite-material-containing-boron-and-method-of-producing-a-heat-sink-dt20070222ptan20070042895.php. Accessed 6 May 2010.

Yoshida K, Morigami H (2004) Thermal properties of diamond/copper composite material. Microelectron. Reliab. **44**: 303–308.

# Chapter 7
# Thermally Conductive Ceramic Matrix Composites

**Abstract** As a result of increased performance in a wide range of engineered products ranging from computer processors to advanced aerospace vehicles, there is a critical need for improved thermal management systems for transferring heat. The required enhancements include increased thermal conductivity, increased surface area, reduced weight/volume, as well as operability in harsh environments, such as durability under high flow rates, vibrations, stress, elevated temperatures, and oxidative environments. For example, improved thermal management is needed to increase the power density of electronics and more effectively cool electronic enclosures that are envisioned for future aircraft, spacecraft, and surface ships. Typically, heat exchangers must increase in size in order to more effectively dissipate any increased heat loads. This is impossible in many cases, thus new materials and concepts for heat exchanger cores/systems are required. Another high-profile application involves thermal protection systems (TPS) for aerospace vehicles (e.g., the reinforced carbon composite leading edge of the Space Shuttle). Future TPS systems will include a systematic approach where a temperature-resistant, durable exterior composite skin is coupled with a combination of conductive and insulating core materials both of which will need to be capable of withstanding extreme environments. Thermally conductive ceramic composites have been actively developed for meeting these requirements in thermal management of electronic packaging. Fiber, whisker, particulate, or nanotube reinforced composites can be made tougher, stronger, and more effective for thermal management with tailored coefficient of thermal expansion (CTE). The desirable characteristics of ceramic matrix composites (CMCs) include high-temperature stability, high thermal shock resistance, high hardness, high corrosion resistance, light weight, nonmagnetic and nonconductive properties, and versatility in providing unique engineering solutions. The combination of these characteristics makes CMCs attractive alternatives to thermal management of electronic packaging, particularly for high-temperature electronic packaging systems. The aim of this chapter is to discuss and highlight newly developed CMC materials and the associated technologies related to thermal management. Examples of these new materials include diamond or carbon reinforced silicon carbide (SiC) composites, reaction-bonded SiC composites, aluminum-toughened SiC composites, and ceramic-based nanocomposites.

X.C. Tong, *Advanced Materials for Thermal Management of Electronic Packaging*,
Springer Series in Advanced Microelectronics 30,
DOI 10.1007/978-1-4419-7759-5_7, 

## Introduction

CMCs combine reinforcing phases with a ceramic matrix to have been developed to overcome the intrinsic brittleness and lack of reliability of monolithic ceramics and create new and superior properties. In CMCs, in another word, the primary goal of the reinforcement is to provide toughness to an otherwise brittle ceramic matrix. Reinforcements can also be added to the ceramic matrix during processing to enhance or tailor characteristics such as electrical conductivity, thermal conductivity, thermal expansion, and hardness. The desirable characteristics of thermally conductive CMCs include high thermal conductivity, high-temperature stability, high thermal shock resistance, high hardness, high corrosion resistance, light weight, nonmagnetic and nonconductive properties, and versatility in providing unique engineering solutions. The combination of these characteristics makes CMCs attractive alternatives to thermal management of electronic packaging, particularly for high-temperature electronic packaging system. Ceramic matrices can be categorized as either oxides or nonoxides and in some cases may contain residual metal after processing. Some of the more common oxide matrices include alumina, silica, mullite, barium aluminosilicate, lithium aluminosilicate, calcium aluminosilicate, and beryllium oxide. Of these, alumina and mullite have been the most widely used because of their in-service thermal and chemical stability and their compatibility with common reinforcements. Although oxide matrices are often considered more mature and environmentally stable, nonoxide ceramics are rapidly entering the marketplace with superior structural properties, hardness, and, in some environments, corrosion resistance. Some of the more common nonoxide ceramics include SiC, $Si_3N_4$, boron carbide, and AlN. Of these, SiC has been the most widely used, with AlN of increasing interest where high thermal conductivity is required and $Si_3N_4$, where high strength is desired. The reinforcements are available in a variety of forms. Early ceramic composites used discontinuous reinforcements which, when combined with ceramic matrices, could be formed using conventional monolithic ceramic processes (Richerson 1997).

Discontinuous oxide ceramic reinforcements are less prevalent because of their incompatibility with many common ceramic matrices. Advanced composites, which use the fiber in a continuous form to better optimize the structural properties, generally require more costly manufacturing processes. Continuous fiber is available in both monofilament and multifilament tow forms, with multifilament tow fiber being lower in cost on a per-pound basis and, in many cases, easier to process into complex shapes. Some of the more common continuous reinforcements include glass, mullite, alumina, carbon, and SiC. Of these, SiC fibers have been the most widely used because of their high strength, stiffness, and thermal stability. SiC matrix continuous fiber ceramic composites (CFCCs) have been successfully demonstrated in a number of applications where a combination of high thermal conductivity, low thermal expansion, light weight, and good corrosion and wear resistance is desired. SiC matrix CFCCs can be fabricated using a variety of processes, fibers, and interface coatings. Fibers widely used for industrial applications where long life is desired include SiC or mullite. Processes available to fabricate SiC matrix CFCCs and the matrix composition formed include chemical vapor infiltration (CVI) (SiC), polymer infiltration (SiCN, SiC), nitride bonding (Si–SiC–$Si_3N_4$), and melt infiltration (Si–SiC). The interface coating can be either carbon or boron nitride with a protective overcoat of SiC or $Si_3N_4$ (Richerson 1997).

For applications where temperature is lower (<1,100°C) or exposure times limited, mullite fibers have been the most widely used because of their lower cost. Continuous ceramic fibers are of increasing interest because of their ability to provide pseudoductile characteristics to otherwise brittle ceramic materials. Continuous ceramic fibers become especially important in large structures where the probability of processing- or in-service-induced flaws increases with part size and could result in catastrophic failure (Richerson 1997). As an example of applications, the use of CMCs in advanced engines will allow an increase of the temperature at which the engine can be operated and eventually the elimination of the cooling fluids, both resulting in an increase of yield. Further, the use of light CMCs in place of heavy superalloys is expected to yield significant weight saving. Although CMCs are promising thermostructural materials, their applications are still limited by the lack of suitable reinforcements, processing difficulties, sound material data bases, lifetime and cost.

With advent of carbon nanotubes (CNTs) or other naomaterials, the nanoscale reinforcements have been utilized in composites in order to improve their mechanical, thermal, and electrical properties. The combination of the nanotubes with a ceramic matrix could potentially create composites that have high-temperature stability as well as exceptional toughness and creep resistance. Increased thermal and electrical conductivity of the composites even at low volume fractions also might provide the thermal transport needed to reduce material operating temperatures and improve thermal shock resistance in applications like thermal elements and electrical igniters. The high temperature and high reactive environment during conventional fabrication methods of ceramics damage the CNTs and thus, there is a need of alternate methods of processing the composites. There is also a need for control of interface between nanotube and the matrix for better interfacial bonding. However, the research is still in the embryonic stage and there are many challenges to be resolved before it is ready for use in varied industrial applications including thermal management of electronic packaging (Samal and Bal 2008).

## State of the Art in Processing of SiC Matrix Ceramic Matrix Composites

Depending on the discontinuous or continuous reinforcements, SiC matrix ceramic composites can be fabricated by a variety of processes. Discontinuous reinforced ceramic composites are produced using processes originally developed for monolithic ceramics. Processing methods commonly used include slip casting or injection molding followed by sintering to full density in a high-temperature-capable furnace. The shaping and sintering processes can also be combined using unidirectional hot pressing or hot isotatic pressing. Net or near-net shape processing can be achieved with final machining often limited to satisfying high-tolerance dimensions or surface finishes (Richerson 1997).

Ceramic composites containing continuous fiber reinforcements must be processed by methods that accommodate the continuous nature of the reinforcement. Typically, processing involves the formation of a fiber preform that contains an interface coating applied by chemical vapor deposition (CVD) or a particle-filled slurry process followed by impregnation with a second particle-filled slurry mix, preceramic polymer, precursor gases, molten metal, or other raw material that converts to a ceramic matrix when heated. The interface is a very thin layer (<5 μm total thickness), can be multiple layers to achieve the desired result, and is applied to the individual ceramic filaments. The interface serves as protection for the fibers during matrix processing and as a source of debonding during crack propagation in the brittle ceramic matrix. Some typical SiC ceramic composite microstructures are shown in Figure 7.1. Figure 7.1a illustrates the SiC–Si matrix,

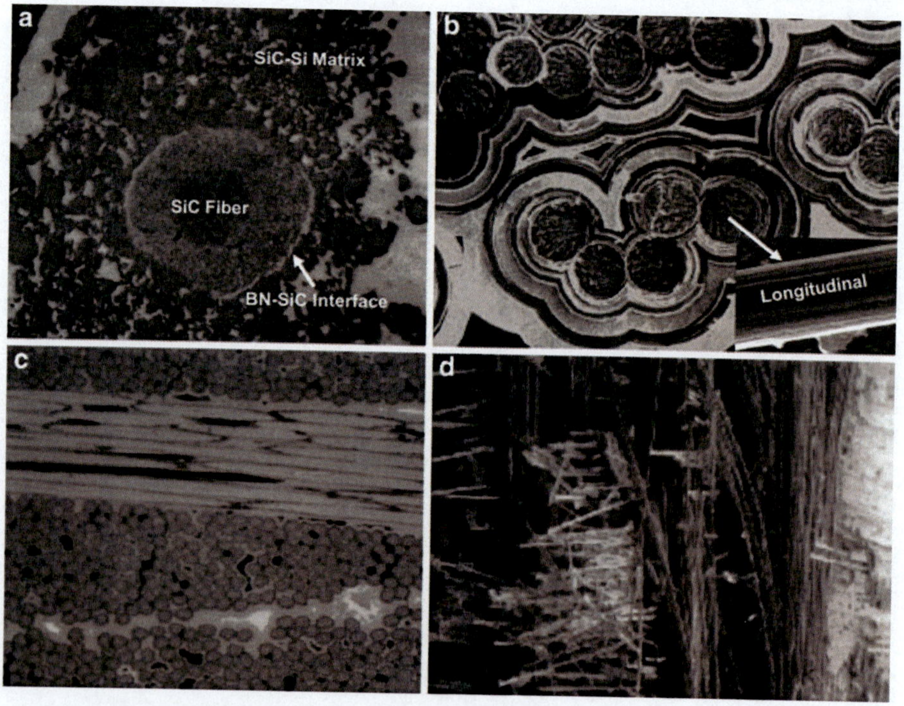


**Fig. 7.1** Typical microstructure of silicon carbide (SiC) matrix composites. (**a**) SiC +Si matrix, SiC fiber and BN–SiC interface phases; (**b**) Multilayered self-healing matrix ceramic matrix composite (CMC), including crack arrester layers and glass-forming layers; (**c**) Carbon fiber reinforced SiC composite; and (**d**) Fracture surface of (**c**)

SiC fiber, and BN–SiC interfaces. Figure 7.1b shows a highly tailored composite structure: multilayered self-healing interphases and matrices that combine crack arrester layers and glass former layers. Figure 7.1c and d show the microstructure and fracture surface of carbon fiber reinforced SiC composite, separately.

With the CVD process, a ceramic interface and the SiC matrix are both formed by deposition from a gas precursor at elevated temperature. Multiple cycles of the matrix-forming process are required to achieve the desired final density, with each cycle taking up to a week in many cases. The number of cycles varies from two to five, depending on the part geometry and size (Richerson 1997).

The infiltration process for CMC fabrication can be clarified gas-phase routes or liquid-phase routes, each of them having advantages and drawbacks. In gas-phase routes, i.e., the so-called CVI processes, the reinforcements (usually as a multidirectional preform) are densified by the matrix deposited from a gaseous precursor, e.g., a hydrocarbon for carbon or a mixture of methyltrichlorosilane and hydrogen for silicon carbide. It is now well established that a fiber coating, i.e., the interphase, has to be deposited on the fiber prior to the infiltration of the matrix in order to control the fiber-matrix bonding and the mechanical behavior of the composite. Pyrocarbon (PyC), boron nitride or (PyC–SiC)n and (BN–PyC)n multilayers, with an overall thickness ranging from about 0.1 μm to about 1 μm, displaying a layered crystal structure (PyC, BN) or a layered microstructure (multilayers), are the most common interphase materials in nonoxide CMCs. The main role of the interphase is to deflect the microcracks which form in the matrix under loading and hence to protect the fiber form notch effect (mechanical fuse function). There are several versions of the CVI process. The most commonly studied and used version is isothermal/isobaric CVI (or I-CVI). It is a relatively slow process because mass transfer in the preform is mainly by diffusion and it yields some residual porosity and density gradient. Conversely, I-CVI is a clean and flexible process (it can be used to densify simultaneously a large number of preforms, eventually of different shapes). For these reasons, it is the preferred process at the plant level. It is well suited to the fabrication of relatively thin parts. In order to increase the densification rate and hence to reduce the processing times, temperature or/and pressure gradients can be applied to the preform. In temperature gradient CVI, or forced CVI, the processing time can be reduced by one order of magnitude with respect to I-CVI. A similar processing time lowering has also been reported for the film-boiling (or calefaction) process, in which the heated fiber preform is directly immersed in a liquid matrix precursor. Finally, pressure pulsed-CVI has been presented at the micrometer (or even nanometre) scale, either the interphase or the matrix. Based on this technique, multilayered self-healing interphases and matrices (combining crack arrester layers and glass former layers) have been designed and produced, through a proper selection of chemical composition of the layers (Fig. 7.1b) (Naslain 2010).

The methods of fabrication through liquid phase routes include (1) infiltration of molten ceramic; (2) slurry infiltration process (SIP); (3) reactive melt infiltration (RMI); and (4) polymer infiltration and pyrolysis (PIP). Infiltration of molten ceramic into a fiber preform is limited by low viscosity of molten ceramics and by high temperature causing chemical interaction between the molten matrix and the dispersed phase (fibers). This process is similar to liquid state fabrication of metal matrix composites (MMCs), and sometimes used for fabrication glass matrix composites. SIP involves the following operations: (1) passing fibers (tow, tape) through a slurry containing particles of the ceramic matrix; (2) winding the fibers infiltrated by the slurry onto a drum and drying; (3) stacking of the slurry impregnated fibers in a desired shape; and (4) consolidation of the matrix by hot pressing in graphite die at high temperature. The RMI process is used primarily for fabrication of SiC matrix composites. The process involves infiltration of carbon (C) containing preform with molten silicon (Si). Infiltration is usually capillary forced. Carbon of the impregnated preform reacts with liquid silicon, forming SiC. Resulting matrix consists of SiC and some residual silicon. When liquid aluminum (Al) is used for infiltration of a preform in oxidizing atmosphere, alumina–aluminum ($Al_2O_3$–Al) matrix is formed. The RMI method is fast and relatively cost effective. Materials fabricated by the RMI method possess low porosity and high thermal conductivity and electrical conductivity (Naslain 2010).

PIP involves the following operations: (1) fiber preform (or powder compact) is soaked with a soft (heated) polymer, forming polymeric precursor. (2) The polymer is cured (cross-linked) at 480°F (250°C). (3) The polymer precursor is then pyrolyzed at 1,100–1,830°F (600–1,000°C). As a result of pyrolysis the polymer converts to ceramic. Pyrolysis causes shrinkage of the matrix material and formation of pores (up to 40 vol.%). (4) The pyrolyzed polymeric cursor may be hot pressed for densification. Hot pressing is limited by possible damage of fibers. (5) Infiltration pyrolysis cycle is repeated several times until the desired density is achieved. Matrices consisting of carbon, silicon carbide, silicon oxycarbide, silicon nitride and silicon oxynitride may be fabricated by PIP method. The following materials are used as polymers in the PIP method include thermosets (thermosetting resins); Pitches or other carbon-containing liquids for fabrication of carbon matrix; polycarbosilane, polysilastyrol, dodecamethylcyclohexasilan for fabrication of SiC matrix. PIP methods are simple low temperature methods, which allow production of intrinsic parts.

The infiltration of fiber preform with preceramic polymer or particle slurry is illustrated in Figure 7.2. The interface material is applied onto the fiber, cloth, or perform by CVI or particle-filled slurry processes before coating with a matrix-forming second slurry or preceramic polymer. Once coated with a polymer or slurry, the shape is formed and heat treated at low temperatures to rigidize the shape. The rigidized shape is then heat treated at elevated temperatures to convert the polymer to a ceramic matrix or bond particles in the ceramic-filled slurry. For preceramic polymers, multiple cycles are required to achieve the desired final density, with each cycle taking 1–3 days. The number of cycles used varies from 4 to 15 depending on the polymer and the desired final density. When using particle-filled slurries to form the matrix, porosity remaining after rigidizing the preform is either filled with molten silicon metal to form a matrix of Si–SiC or Si contained in the slurry reacts during high-temperature heat treatment in nitrogen to form a matrix of Si–SiC–$Si_3N_4$. Variants of the melt infiltration process can also be used in which the molten silicon reacts with carbon present in the matrix to form

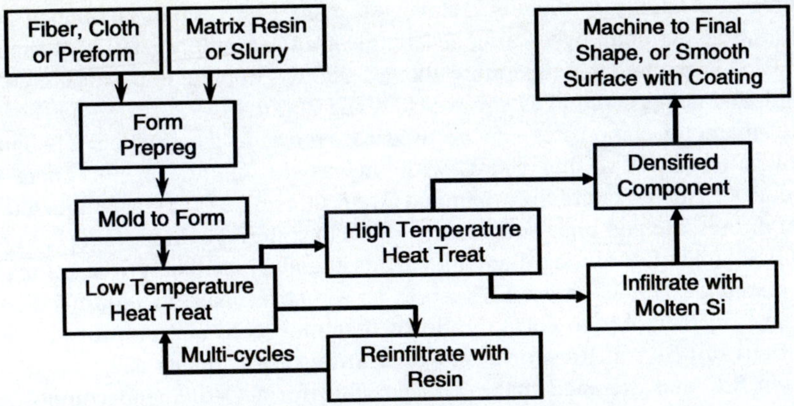


**Fig. 7.2** Flowchart of liquid phase route processing for SiC based CMC

additional SiC. Shaping to near-net dimensions occurs during preforming with final machining limited to high tolerance attachments to avoid surface fiber damage. Additional machining may be required for complex shapes and is generally performed early in the process when the composite is less dense and easier to machine. External coatings can be used to smooth the contact surface for assembly or enhance environmental resistance with some of the more common environmental barrier coatings including mullite, SiC, and $Si_3N_4$ (Richerson 1997).

## SiC–Diamond Composites

The inherent brittleness of single-crystal diamond limits its application in harsh environments of dynamic impacts and high-stress concentrations. Therefore, more flaw-tolerant diamond composites have been developed through the design of their microstructures. In order to manufacture bulk materials containing diamond powder, which starts with single-crystal diamond or diamond powder converted directly from graphite, the composites must be fabricated by using a suitable bonding agent under appropriate conditions. It is difficult to directly sinter diamond because some diamond may transform into graphite because of the inhomogeneous distribution of pressure inside the specimen. To avoid this problem, two classes of diamond composites were developed, polycrystalline diamond (PCD) and SiC–diamond composites. For PCD composites, cobalt usually serves as the bonding agent. Under high temperature and high pressure, cobalt melts and infiltrates the voids among the diamond particles. Then cobalt maintains the hydrostatic condition inside the sample and thus prevents the diamond graphitization during the sintering process. The next step is to leach out cobalt with a solvent, then the PCD composites mainly consist of diamond crystallites growing together around their contacting area, and the remaining strong carbon–carbon (C–C) bonds retain about 90%

modulus of elasticity of diamond. However, even a very small amount of cobalt remaining in the composites after leaching, promotes graphitization of diamond when PCD is exposed to temperature above 1,000 K. Graphite impurities reduce the strength and degrade the PCD compacts (Wang 2006).

SiC–diamond composites can be used to overcome the problem of diamond graphitization by replacing cobalt with SiC as the binding agent. There exist techniques to manufacture SiC–diamond composites, like CVD, vacuum sintering, and high-pressure and high-temperature (HPHT) sintering. Under HPHT, Si melts and fills out cavities among diamond particles, and immediately it starts reacting with diamond. The as-grown SiC serves as a matrix phase to bind all diamond particles together. At the correct ratio of diamond to Si and optimized reaction conditions, all Si will transform into SiC. Strong Si–C bonds and well bonding between SiC and diamond make it is possible for SiC–diamond composites to provide excellent thermal conductivity and mechanical properties. Besides the grain size of the precursors, the overall thermal conductivity and mechanical properties of SiC–diamond composites depend on the concentration of SiC, its structure, density of dislocation, and grain boundaries (Wang 2006).

However, conventional HPHT infiltration sintering is not suitable when the starting materials consist of submicron or nano-size powders due to the so-called self-stop effect. During the sintering process, the newly formed SiC may decrease the open space of tiny pores or block the entrance to a void. This phenomenon is apt to limit or even completely block the infiltration of liquid Si and hence hinder the entire sintering process. To overcome this problem, an alternative method has been developed, which is based on the simultaneous ball milling of diamond and silicon powder mixtures prior to HPHT sintering. After ball milling Si became amorphous and uniformly coated on diamonds. The synthesized composites had a unique nanostructure of diamond embedded into a matrix of nanocrystalline SiC, with large improvement in fracture toughness. These composites showed as much as 50% enhanced fracture toughness while maintaining superhard and superabrasive properties when SiC matrix size decreased from 10 μm to 20 nm. It is in contradiction with the regularly held belief of an inverse correlation between hardness and fracture toughness of materials. Further reduction of SiC grain sizes by the ball milling technique seems very unlikely. Another method of producing nanostructure SiC matrix is to use CNT in SiC–diamond composites manufacturing. This novel composite will consist of diamond imbedded into SiC/CNTs matrix obtained by sintering/infiltration technique under HPHT. Ideally the CNTs will be distributed along the grain boundary of diamond to form a network structure. These materials have great potential to promote thermal management of electronic packaging.

For instance, ceramic heat spreaders (e.g., AlN) and MMC heat spreaders (e.g., SiC/Al) have been used to cope with the increasing amounts of heat generation. However, such materials have a thermal conductivity that is not greater than that of Cu, hence, their ability to dissipate heat from semiconductor chips is limited. A typical semiconductor chip contains closely packed metal conductor (e.g., Al, Cu) and ceramic insulators (e.g., oxide, nitride). The thermal expansion of metal is typically five to ten times that of ceramics. When the chip is heated to above 60°C, the mismatch of thermal expansions between metal and ceramics can create microcracks. The repeated cycling of temperature tends to aggravate the damage of the chip. As a result, the performance of the semiconductor will deteriorate. Furthermore, when temperatures reach more than 90°C, the semiconductor portion of the chip may become a conductor so the function of the chip is lost. In addition, the circuitry may be damaged and the semiconductor is no longer usable (i.e., becomes "burned out"). Thus, in order to maintain the performance of the semiconductor, its temperature must be kept below a threshold level (e.g., 90°C). A conventional method of heat dissipation is to contact the semiconductor with a metal heat sink. A typical heat sink is made of aluminum that contains radiating fins. These fins are attached to a fan. Heat from the chip will flow to the aluminum base and will be transmitted to the radiating fins and carried away by the circulated air via convection. Heat sinks are therefore often designed to have a high heat capacity to act as a reservoir to remove heat from the heat source. The above heat dissipation methods are only effective if the power of the central processing unit (CPU) is less than about 60 W. For CPUs with higher power, more effective means must be sought to keep the hotspot of the chip below the temperature threshold. Alternatively, a heat pipe may be connected between the heat sink and a radiator that is located in a separated location. The heat pipe contains water vapor that is sealed in a vacuum tube. The moisture will be vaporized at the heat sink and condensed at the radiator. The condensed water will flow back to the heat sink by the wick action of a porous medium (e.g., copper powder). Hence, the heat of a semiconductor chip is carried away by evaporating water and removed at the radiator by condensing water. Although heat pipes and heat plates may remove heat very efficiently, the complex vacuum chambers and sophisticated capillary systems prevent designs small enough to dissipate heat directly from a semiconductor component. As a result, these methods are generally limited to transferring heat from a larger heat source, e.g., a heat sink. Thus, removing heat via conduction from an electronic component is a continuing area of research in the industry. One promising alternative that has been explored for use in heat sinks is diamond-containing materials. Diamond can carry heat away much faster than any other material. The thermal conductivity of diamond at room temperature (about 2,000 W/mK) is much higher than either copper (about 400 W/mK) or aluminum (250 W/mK), the two fastest metal heat conductors commonly used. Additionally, the thermal capacity of diamond (1.5 J/cm$^3$) is much lower than copper (17 J/cm$^3$) or aluminum (24 J/cm$^3$). The ability for diamond to carry heat away without storing it makes diamond an ideal heat spreader. In contrast to heat sinks, a heat spreader acts to quickly conduct heat away from the heat source without storing it. In addition, the thermal expansion coefficient of diamond is one of the lowest of all materials. The low thermal expansion of diamond makes joining it with a low thermally expanding silicon semiconductor much easier. Hence, the stress at the joining interface can be minimized. The result is a stable bond between diamond and silicon that does not delaminate under the repeated heating cycles (Sung 2007).

As a result, diamond heat spreaders have been used to dissipate heat from high power laser diodes, such as that used to boost the light energy in optical fibers. However, large area diamonds are very expensive; hence, diamond has not been commercially used to spread the heat generated by CPUs. In order for diamond to be used as a heat spreader, its surface must be polished so it can make intimate contact with the semiconductor chip. Furthermore its surface may be metallized (e.g., by Ti/Pt/Au) to allow attachment to a conventional metal heat sink by brazing. Many current diamond heat spreaders are made of diamond films formed by CVD. The raw CVD diamond films are now sold at over \$10/cm$^2$, and this price may double when it is polished and metallized. This high price would prohibit diamond heat spreaders from being widely used except in those applications (e.g., high-power laser diodes) where only a small area is required or no effective alternative heat spreaders are available. In addition to being expensive, CVD diamond films can only be grown at very slow rates (e.g., a few micrometers per hour); hence, these films seldom exceed a thickness of 1 mm (typically 0.3–0.5 mm). However, if the heating area of the chip is large (e.g., a CPU), it is preferable to have a thicker (e.g., 3 mm) heat spreader. As such, cost-effective devices that are capable of effectively conducting heat away from a heat source, continue to be sought through ongoing research and development efforts (Sung 2007).

Because of their exceptionally high hardness, excellent wear resistance, and thermal stability, SiC–diamond composites have been conventionally used in various industrial applications such as machining, grinding, drilling, and mining. SiC–diamond composites have been prepared by a variety of methods that include CVD, HPHT liquid-phase sintering, and low vacuum liquid-phase infiltration. Most available SiC–diamond composites are composed of microcrystalline diamond held together by microcrystalline SiC. Figure 7.3 shows an examples of SiC–diamond heat spreaders as fabricated or with Si coating for improving surface roughness. The thermal conductivity of the SiC–diamond spreader is approximately 600 W/m K, CET is 1.8 ppm/K. These SiC–diamond heat spreaders are now in production for IBM-server heat spreaders (Zweben 2006).

Despite their extraordinary hardness and wear resistance, SiC–diamond composites have relatively low fracture toughness (<6 MPa·m$^{1/2}$), which limits their potential applications. Fracture toughness of SiC–diamond composites has been improved by incorporating nanocrystalline diamond into the composites. It is believed that the nanocrystalline diamond and SiC hinder dislocation growth and

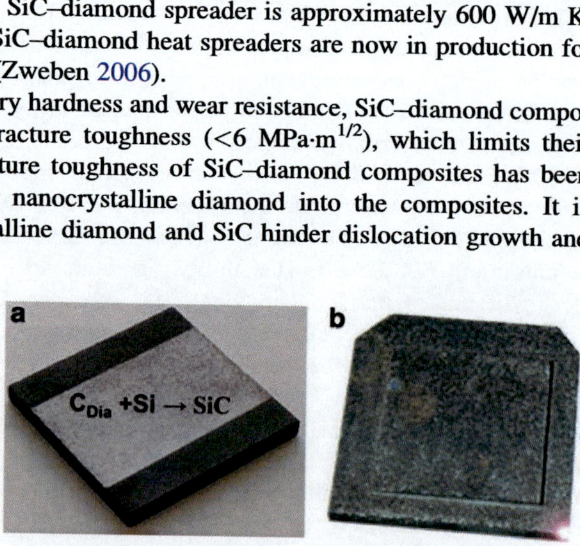


**Fig. 7.3** Diamond particle-reinforced SiC composite heat spreader (**a**) with silicon coating that improves surface roughness; (**b**) without silicon surface coating (Zweben 2006)

microcrack propagation in the composite better than microcrystalline diamond and SiC do, which improves fracture toughness (Ekimov et al. 2000). The composite was prepared by the liquid silicon infiltration of nanocrystalline diamond powder under HPHT conditions HPHT (7.7 GPa, 1,700–2,300 K). The composite displayed high fracture toughness (10 MPa·m$^{1/2}$) but was only partially densified; infiltration depth was only 1–2 mm because the pores closed very quickly during infiltration due to the "self-stop process;" as silicon infiltrates through the pores, it reacts rapidly with diamond to form a silicon carbide phase that seals the pores and prevents further infiltration. Alternative methods that are not limited by the self-stop process may be required to overcome problems relating to the self-stop process in order to provide fully dense composites with high fracture toughness. Such a method would also minimize graphitization of the nanocrystalline diamond, which has also been a problem in the past. Fully dense, SiC–diamond composites having high fracture toughness have been fabricated with a method involving ball-milling microcrystalline diamond powder and microcrystalline silicon powder to form a powder mixture of microcrystalline diamond and amorphous silicon, then sintering the ball-milled powder mixture at a pressure of about 5 GPa to about 8 GPa and a temperature of about 1,400 K to about 2,400 K for a sufficient period of time to form a fully dense diamond-silicon carbide composite of microcrystalline diamond and nanocrystalline silicon carbide having a fracture toughness of at least 10 MPa·m$^{1/2}$ and with minimal graphitization (Qian and Zhao 2005).

## SiC–Carbon Composites

C–C composites are resistant to heat and chemicals and lightweight, and have a high strength. Accordingly, C–C composites are useful as heat-resistant materials used in nonoxidizing atmosphere, and are particularly used for brake discs and pads or the like of airplanes and automobiles because of their superior resistance to heat attack. Unfortunately, carbon materials are generally oxidized at about 500°C, and therefore, are not suitable to use in high-temperature atmospheres except for being used for a very short time. In order to prevent the physical or chemical degradation of the C–C composite resulting from oxidation, or, if it is used as a sliding member, in order to prevent the reduction of mechanical strength resulting from oxidation and abrasion, the reduction of frictional properties at low temperatures, and the reduction of frictional properties caused by rain drops and other water attached to the sliding member, the C–C composite is impregnated with melted Si to react with the Si and thus to transform the carbon of the C–C composite into C/SiC. In general, polyacrylonitrile (PAN)-based carbon fiber is used as the carbon fiber of the C–C composite that is to be impregnated with melted Si to form SiC. One method to produce C–C composite includes a step for preparing a material comprising a bundle of carbon fiber wherein a powdery binder pitch acting as a matrix in the material comprising the carbon fiber bundle and ultimately acting as free carbon being free from the carbon fiber bundle is added to the carbon fibers aligned in a single direction, and then phenol resin powder, for instance, is added thereto. The carbon fibers thus prepared are covered with a flexible coating made of a resin such as a thermoplastic resin to produce a preformed yarn used as a flexible intermediate material. The preformed yarn is formed into sheets. The C–C composite is produced by stacking a desired number of the sheets one on top of another in such a manner that the directions of the carbon fiber orientations are perpendicular to each other, and subsequently performing predetermined process steps. The resulting C–C composite is impregnated with molten silicon, and thus a SiC–C/C composite material having a porosity of 5% is produced. Since the SiC–C/C composite material is a stack of carbon fiber sheets whose fiber orientations are perpendicular to each other, it is considered that the composite material has a high tensile strength in the longitudinal direction of the fibers (in the 0° direction or the 90° direction), but a low tensile strength in directions of 45° with respect to the perpendicular carbon fibers. In addition, in a SiC–C/C composite prepared by impregnating a C–C composite using PAN-based carbon fibers with a melted Si for transformation into SiC, the reinforcing effect of the carbon fibers is degraded, so that the SiC–C/C composite has no more than a tensile strength close to that of carbon materials not reinforced with carbon fibers. Another modified method to produce the carbon fiber-reinforced SiC composite is by impregnating a carbon fiber-reinforced carbon composite with melted Si. The C–C composite includes short carbon fibers prepared from a pitch-based carbon fiber. The pitch-based short carbon fibers are oriented in two-dimensional random directions in the carbon fiber-reinforced SiC composite (Fukagawa and Kubo 2009).

The methods described above are all aimed at enhancing the toughness of SiC ceramics by the addition of C or transformed SiC fibers to ceramics. Especially for the impartation of high toughness, the production of continuous SiC fiber reinforced SiC ceramics is carried out. None of the continuous SiC fibers heretofore developed is capable of retaining its strength intact at elevated temperatures exceeding 1,400°C. In the circumstances, there has developed a need for practicable fibers capable of retaining strength at such elevated temperatures. Carbon fibers are capable of retaining strength at high temperatures (2,000°C) as compared with SiC fibers. When such carbon fibers are combined with SiC matrix, the strength of the carbon fibers is degraded by a reaction which occurs between the matrix and the fibers. When continuous carbon fibers are used, it is necessary to curb the reaction of the carbon fibers with the matrix. To eliminate the degradation, a method for the production of a continuous carbon fiber reinforced SiC composite, which comprises impregnating continuous carbon fibers coated with at least one member selected from the group consisting of SiC, TiC, $TiB_2$, and TiN with a slurry mixture which comprises at least one member selected from the group consisting of SiC, $Si_3N_4$, $SiO_2$, and Si with a thermosetting resin or a high-carbon caking agent thereby obtaining impregnated continuous carbon fibers, then molding the impregnated continuous carbon fibers thereby obtaining a shaped article, curing the shaped article, subsequently heating the shaped article in an inert gas to make a carbon fiber–carbon composite having the interstices between the carbon fibers filled with carbon and the aforementioned silicon or silicon compound, further impregnating the carbon fiber–carbon composite with melted Si, and thereafter heat treating the impregnated composite (Nakano et al. 1998).

On the other hand, carbon fiber-reinforced silicon carbide matrix (SiC–C) composites processed by CVI are candidate materials for aerospace thermal structures. Carbon fibers can retain properties at very high temperatures, but they are generally known to have poor oxidation resistance in adverse, high-temperature environments. Nevertheless, the combination of CVI process and SiC matrix with higher stiffness and oxidation resistance, the interfacial coating, and additional surface-seal coating provides the necessary protection to the carbon fibers, and makes the material viable for high-temperature space applications operating under harsh environments. Furthermore, SiC–C composites, like other CMCs, exhibit graceful noncatastrophic failure because of various inherent energy-dissipating mechanisms. The material exhibits nonlinearity in deformation even at very low stress levels. This is the result of the severe matrix microcracking present in the as-processed composite because of large differences between the coefficients of thermal expansion of the fiber and the matrix.

In addition, a comingled blend of carbon and SiC fibers is used to eliminate process-induced matrix microcracking that is typical in SiC–fiber reinforced SiC–matrix composites. The blending of thermoelastically dissimilar fibers allows for tailoring the fiber preform thermal expansion to be more favorably compatible with the matrix, while improving the high-temperature mechanical performance and lowering cost over C–fiber reinforced SiC–matrix composites. Silicon carbide matrix composite thrust chambers utilizing hybrid fiber preforms were shown to exhibit virtually zero through-thickness permeability over typical C–fiber reinforced SiC–matrix composites resulting from a hybrid preform comprising an optimized blend of carbon and SiC fiber reinforcement.

Often, some of the desirable property characteristics allow composites to offer advantages over conventional structural materials, such as tailoring of composite properties. However, the complexity is in fact responsible for their greater statistical variability and the requirements for more characterization tests. Composite properties are anisotropic as well, having different properties in different directions. This means that characterization of a property such as stiffness, which will vary greatly depending on the orientation of the fiber relative to the direction of the testing, must be repeated for several different directions and loading conditions. The fabrication process for composites also introduces statistical variations in properties and geometry. A composite part is produced in a number of steps, each of which introduces statistical variability. The matrix is usually produced from a combination of raw materials; and the fiber, which has its own set of properties, is often coated or surface treated, introducing yet another source of variability. The processes are usually performed by various vendors and are not under the control of the fabricator of the composite part. Additional irregularities are introduced by the influence of temperature and moisture. Composites are usually more susceptible to environmental conditions. Changes in environmental conditions produce a significant change in properties, leading not only to a source of property variability, but also to a requirement for additional testing to characterize the effects of these variables. In general, CMCs are complex, have brittle constituents, and are potentially flaw-sensitive materials. They inherently have considerable scatter in their properties. It is important to note that because of the flaw-sensitivity of brittle materials, additional characterization is required to characterize the volume effect. Thus, the advantages that composites bring must be weighed against increased material testing costs. Any CMC material characterization effort based solely on a large test matrix is simply impractical because of time and cost considerations.

## Reaction-Bonded SiC Composites

Although SiC can be densified with high temperature and pressure, the process is not a viable commercial process. Reaction bonded SiC is made by infiltrating compacts made of mixtures of SiC and carbon with liquid silicon. The silicon reacts with the carbon forming SiC. The reaction product bonds the SiC particles. Any excess silicon fills the remaining pores in the body and produces a dense SiC–Si composite. The ratio of SiC to carbon and particle size distribution varies widely in practice. Articles are produced with a wide range of compositions and properties. At one extreme, carbon fiber felt or cloth can be infiltrated with liquid silicon, while at the other extreme, an impervious SiC body can be made with a small amount of carbon. Most reaction bonded SiC is made with formulations that contain an organic plasticiser, carbon, and SiC particles. This mixture is ideally suited to near net formation by pressing, injection molding, or extruding. Furthermore, because the reaction process typically gives a dimensional change of <1%, manufacturers have excellent control of component tolerances (Ceram 2001).

Reaction bonded SiC ceramics combine the advantageous properties of high performance traditional ceramics, with the cost effectiveness of net shape processing. Table 7.1 shows properties of some typical reaction bonded SiC compared

**Table 7.1** Properties of some typical reaction bonded SiC compared with CVD SiC composites

| Material | Density (g/cc) | Coefficient of thermal expansion (ppm/K) | Thermal conductivity (W/m K) | Modulus (GPa) | Specific heat (J/kg K) | Polishability |
|---|---|---|---|---|---|---|
| Chemical vapor deposition silicon carbide (SiC) | 2.95 | 2.4 | 175 | 364 | 700 | Slow |
| Reaction-bonded SiC (70%)–Si | 2.95 | 2.9 | 170 | 350 | 680 | Fast |
| Reaction-bonded SiC (80%)–Si | 3.03 | 2.9 | 185 | 380 | 670 | Fast |
| Reaction-bonded SiC–Cf[a] | 2.4 | −0.5 to 0.5 | 100–200 | 100–250 | | Fast |

[a]Tailorable by choosing fiber and interface

with CVD SiC composites (Karandikar 2003). These materials provide high surface hardness, very high specific stiffness, high thermal conductivity, and very low CTE. The processing consists of two steps. First, a carbon containing near net shape porous preform is fabricated; and second, the preform is reactively infiltrated with molten Si to form a primarily SiC body. Reaction bonded SiC ceramic offers extremely high levels of mechanical and thermal stability. It possesses low density (similar to Al alloys) and very high stiffness (~70% greater than steel). These properties lead to components that show little deflection under load, allow small distances to be precisely controlled with fast machine motion, and do not possess unwanted low frequency resonant vibrations. In addition, due to the high stiffness and hardness of the material, components can be ground and lapped to meet stringent flatness requirements. Furthermore, as a result of very low CTE and high thermal conductivity, reaction bonded SiC components show little motion with temperature changes and are resistant to distortion if localized heating occurs. Also, due to an excellent CTE match with Si, reaction bonded SiC ceramics are well suited as substrates for Si handling operations. Furthermore, both Si and SiC possess refractory properties, which yield a composite with good performance in many high-temperature and thermal shock applications. Finally, dense, high purity SiC coatings can be applied when extremely high purity and/or superior resistance to corrosion are required (Evans et al. 2003).

Many processes have been developed to form reaction-bonded SiC. Figure 7.4 shows a schematic of this process. In this process, the preform containing the reinforcement and a carbon precursor or binder is "carbonized" in an inert atmosphere above 600°C to convert the precursor to carbon. Next, the preform is placed in contact with Si metal or alloys of Si in an inert or vacuum atmosphere and heated to above the melting point of the alloy. As a result of the spontaneous wetting and reaction between carbon and molten Si, the preform is infiltrated completely. The carbon in the perform reacts with the Si forming SiC and in the process bonds the reinforcement together. Some residual Si remains (Karandikar et al. 2007).

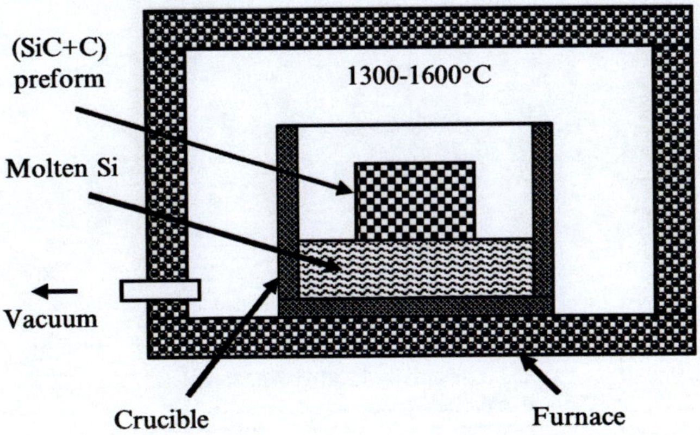


**Fig. 7.4** Schematic of a reaction bonding process for reaction-bonded SiC composite (Karandikar et al. 2007)

Typically, a molten infiltrant containing silicon and one or more sources of boron are contacted to a porous mass that contains at least some boron carbide, and also containing at least some free carbon. The molten infiltrant infiltrates the porous mass without a pressure or vacuum assist to form a composite body of near theoretical density. The silicon component of the infiltrant reacts with the free carbon in the porous mass to form in situ SiC as a matrix phase. Furthermore, the tendency of the molten silicon to react with the boron carbide component can be suppressed or at least greatly, attenuated by the alloying or doping of the silicon with the boron source. The resulting composite body thus comprises boron carbide dispersed or distributed throughout the SiC matrix. Typically, some residual, unreacted infiltrant phase containing silicon and boron is also present and similarly distributed or interspersed throughout the matrix (Si–SiC–$B_4C$). Reaction formed SiC composites featuring a boron carbide reinforcement possess stiffness (e.g., elastic or Young's Modulus) comparable with their counterparts featuring the usual SiC reinforcement, but exhibit a lower specific gravity for the same volumetric filler loading. Accordingly, such $B_4C$ reinforced SiC composites will find utility in applications requiring low mass and high stiffness, such as equipment requiring precise motion control, often at high accelerations. Additionally, because of the extreme hardness and low specific gravity of boron carbide, such composites are attractive armor material candidates (Evans et al. 2003).

Figure 7.5 shows typical microstructure and property of reaction-bonded SiC with laser sintered process (SLS). Three main manufacturing phases: preform SLS,

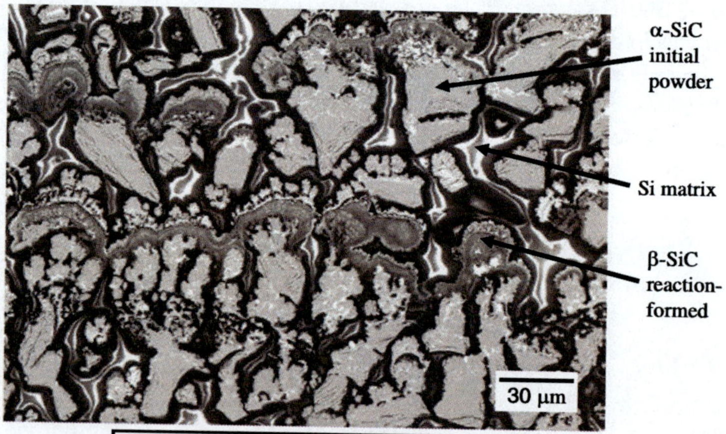


| |
|---|
| 70-80% SiC, 20-30% Si |
| Density 2.9-3.04 kg/m$^3$ |
| Elastic modulus 350-400 GPa |
| Flexural strength 400-450 MPa |
| Thermal conductivity 110-145 W/mK |
| Thermal expansion coefficient 4.3 ppm/K |

**Fig. 7.5** Microstructure and properties of reaction-bonded SiC composite by laser sintering process (Evans et al. 2003)

binder burnout, and reactive infiltration. Preform formation utilizes SiC powder of an appropriate average particle size mixed with a multicomponent binder. The preform or green part is then placed in a vacuum furnace to carbonize the binder. The binder chemistry must support accurate component shapes and acceptable surface roughness, a strong green part and maintenance of the part shape during the first furnace operation. Finally, the physics and chemistry of the infiltration process, based on the microstructure of the initial green preform, determine the viability of the manufacturing process and the characteristics of the final composite material. The governing principles of binder function, carbonization, and infiltration must be established to move beyond functional prototyping and into manufacturing arenas of thermal managing components in electronic packaging industry (Evans et al. 2003).

The properties of SiC components depend on the material grade. In the case of a fully dense SiC–Si composite, the material demonstrates good bend strength at room temperature (typically 400 MPa), which is maintained to the melting point of silicon (1,410°C) where it decreases to around 250 MPa. Young's modulus is typically in the range 350–400 GPa. The properties that lead to selection of the material are resistance to wear, resistance to corrosion; the material tolerates a wide range of acids and alkalis, resistance to oxidation, abrasion resistance, good thermal shock resistance due to low thermal expansion coefficient and high thermal conductivity, strength at high temperature, and good dimensional control of complex shapes. The negligible volume change after reacting with liquid silicon means that components can be formed with complex shapes and to exacting tolerances (Evans et al. 2003). Examples of components made by this route include laser mirror blanks, wafer handling devices, optical benches, and active cooling components such as heat pipe. The components are light weight and stiff with excellent thermal stability.

## Aluminum-Toughened SiC

Reaction bonded SiC ceramics have many outstanding properties, including high-specific stiffness, low CTE, and high thermal conductivity. However, they have low fracture toughness. As an alternative, Al-toughened SiC composite maintains many of the advantageous properties of reaction bonded SiC while providing higher toughness, higher thermal conductivity, and more tailorable CTE. The composite is produced using a reactive infiltration process, which allows near-net-shape components to be fabricated.

Figure 7.6 shows typical microstructures of aluminum-toughened 12%Si/70% SiC (a); and 70%SiC (b) (Karandikar 2003). Al-toughened SiC provides a nominally 37% increase in fracture toughness relative to standard reaction bonded SiC. In addition, the composite can be used in thin-walled component designs that would be difficult to produce with a low toughness ceramic. The addition of the aluminum to the microstructure does lead to a slight decrease in stiffness; however, the

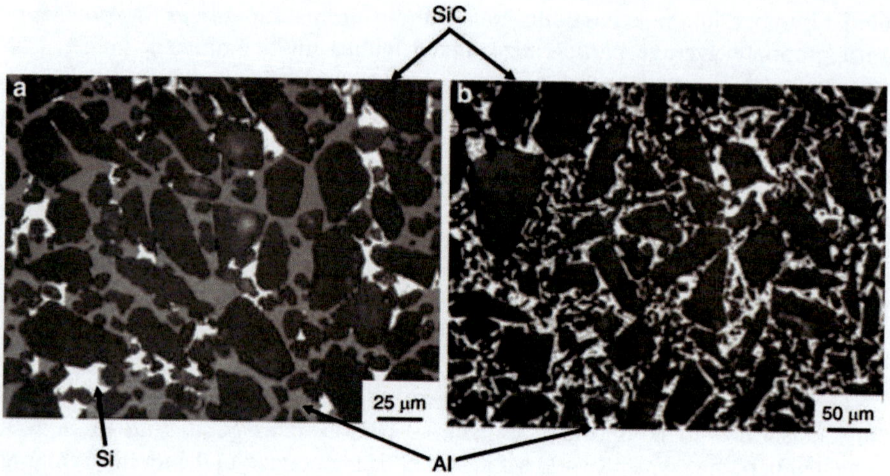


**Fig. 7.6** Microstructure of aluminum-toughened 12%Si/70%SiC (**a**); and 70%SiC (**b**) (Karandikar 2003)

Al-toughened SiC is still well stiffer than comparable Al–SiC MMCs. Thus, the composite is still well suited to applications where very high levels of mechanical stability are required. The presence of the aluminum results in an increase in thermal conductivity relative to the standard SiC product, which is valuable in heat sink applications or in components where localized heating can occur. In addition, the thermal conductivity is in excess of that of most MMCs because no metallic alloying elements are used. The CTE of Al-toughened SiC is low (between those of SiC ceramic and Al–SiC MMCs), and thus provides a high level of thermal stability. In addition, it very closely matches the CTE of AlN, and thus has utility in applications where interfacing with AlN components occurs. Finally, Al-toughened SiC ceramic represents a family of materials. By changing the respective volume fractions of Al, Si, and SiC, the properties can be tailored to meet specific application needs (MMMT 2004).

Al-toughened SiC components can be fabricated with a near-net-shape reactive infiltration process. These manufacturable ceramic-metal composites offer high levels of mechanical and thermal stability, with higher fracture toughness relative to conventional ceramic materials, and will be promising for future thermal management applications.

## Ceramic Nanocomposites

Since the discovery of CNTs, vast areas of research have opened which also include nanoscale reinforcements in composites in order to improve their mechanical, thermal, and electrical properties. Although the focus of the research in nanotube-based composites has mostly been on polymer-based composites, the unique properties of CNTs have also be exploited in CMCs. The combination of these nanotubes with a ceramic matrix could potentially create composites that have high temperature stability as well as exceptional toughness and creep resistance. Increased thermal and electrical conductivity of the composites even at low volume fractions might also provide the thermal transport needed to reduce material operating temperatures and improve thermal shock resistance in applications such as thermal elements and electrical igniters. The high temperature and high reactive environment during conventional fabrication methods of ceramics damage the CNTs and thus, there is a need of alternate methods of processing the composites. There is also a need for control of interface between nanotube and the matrix for better interfacial bonding. However, the research is still in the embryonic stage and there are many challenges to be resolved before it is ready for use in varied industrial applications (Samal and Bal 2008).

The physical properties of material critically depend on the size of grains due to the fact that for different grain sizes a significantly different fraction of atoms forms the surface of the grain. When the grain size is reduced to the nanoscale, the corresponding nanostructure shows remarkably different properties from those of microsize or bulk materials. For instance, the strength of the one-dimensional (1-D) SiC nanostructure is substantially greater than that of larger SiC structures, and it approaches theoretical values. Consequently, as the reinforcing phase, 1-D nanosize SiC is a promising candidate to fabricate superhard nanocrystalline composites. SiC nanorods were first synthesized successfully in 1995, in the process in which CNTs were converted to SiC nanorods by reaction with SiO vapor. Subsequently, a number of methods of synthesis of 1-D SiC nanostructures have been developed. They fall broadly into the following categories: CNTs confined reaction; carbothermal reduction of SiO by C nanocapsules or amorphous activated C; CVD; reaction between $SiCl_4$ and $CCl_4$ and metal Na as coreductant; and annealing CNTs covered with Si. However, all of the above methods either operated at high temperature via vapor–vapor or solid–vapor reactions, or needed some kind of metal, e.g., Fe, Cu, Ni, as a catalyst to cause and speed up the reaction (Guo et al. 2002; Wang 2006).

The most promising solution is the dispersion of nanoparticles, especially CNTs into the ceramic matrix. Apart from structural applications, nanocomposite technology can also be applied for fabricating functional ceramics like ferroelectric, piezoelectric, varistor, and ion-conducting materials. Incorporating a small amount of ceramic and metallic nanoparticles into barium titanate, zinc oxide or cubic zirconia can significantly improve their strength, hardness, and toughness, which are very important in creating highly reliable electric devices operating in severe environmental conditions. In addition, dispersing conducting metallic nanoparticles or nanowires can enhance the electrical properties. For example, the range of electroceramic nanocomposites for information and charge storage has revolutionized the electronic industry. In the nanoscale, quantum effects can be utilized to modify energy states and electronic structures of components. Conducting nanosized particles dispersed in a dielectric matrix (i.e., Ni in PZT) can improve dielectric properties. The solution sol–gel technique, in particular, offers an opportunity to produce not only ultrahomogeneous materials but also heterogeneous or nanocomposite materials. After crystallization and densification, those materials are appropriate for numerous applications, such as electronic or structural materials. The most common methods for processing nanocomposites are mechanical alloying, sol–gel synthesis, and by thermal spray synthesis. Mechanical alloying occurs as a result of repeated breaking up and joining of the component particles which can prepare highly metastable structures such as amorphous alloys and nanocomposite structures with high flexibility. Scaling up of synthesized materials to industrial quantities is easily achieved in this process, but purity and homogeneity of the structures produced remains a challenge. In addition to the erosion and agglomeration, high-energy milling can provoke chemical reactions that are induced by the transfer of mechanical energy, which can influence the milling process and the properties of the product. This is used to prepare magnetic oxide-metal nanocomposites via mechanically induced displacement reactions between a metal oxide and a more reactive metal. Sol–gel is another process in which aerogels are the precursors. Aerogels are made by solgel polymerization of selected silica, alumina or resorcinol-formaldehyde monomers in solution and are extremely light but very porous, mostly nano-sized pores. In the nanocomposites derived from aerogels, the product consists of a substrate like silica gel and one or more additional phases with one of the phases with dimension of the order of nanometers. Thermal spray coating is another commercially relevant, proved technique for processing nanostructured coatings. Thermal spray techniques are effective because agglomerated nanocrystalline powders are melted, accelerated against a substrate and quenched very rapidly in a single step. This rapid melting and solidification promotes retention of a nanocrystalline phase and even amorphous structure. Retention of nanocrystalline structure leads to enhanced wear behavior, greater hardness, and sometimes a good coefficient of friction compared to conventional coatings. As the research in nano-reinforced ceramics is still in its infancy, it is not clear which, if any, of the above toughening and strengthening mechanisms apply to ceramic nanocomposites (Ajayan et al. 2003).

Figure 7.7 shows typical microstructures of ceramic nanocomposites reinforced with dispersed and aligned CNTs. The CNT reinforcements promise to increase the fracture toughness of the composites by absorbing energy through their highly flexible elastic behavior during deformation, which will be especially important for nanotube-reinforced CMCs. However, materials fabrication difficulties have limited research on the composites. The major concern is to obtain a uniform dispersion of nanotubes in the matrix. Damage or destruction of the nanotubes is often observed because of the high temperatures and highly reactive environments associated with many methods of processing ceramic and metal matrices. Lack of optimized fabrication and processing of nanotubes has also restricted research efforts, although this seems to have become less of a problem in recent years as more groups produce their own nanotubes with catalyzed CVD methods. Powder processing methods have been used to fabricate ceramic matrices. The small diameter and large aspect ratio of the nanotubes can make it difficult to obtain a

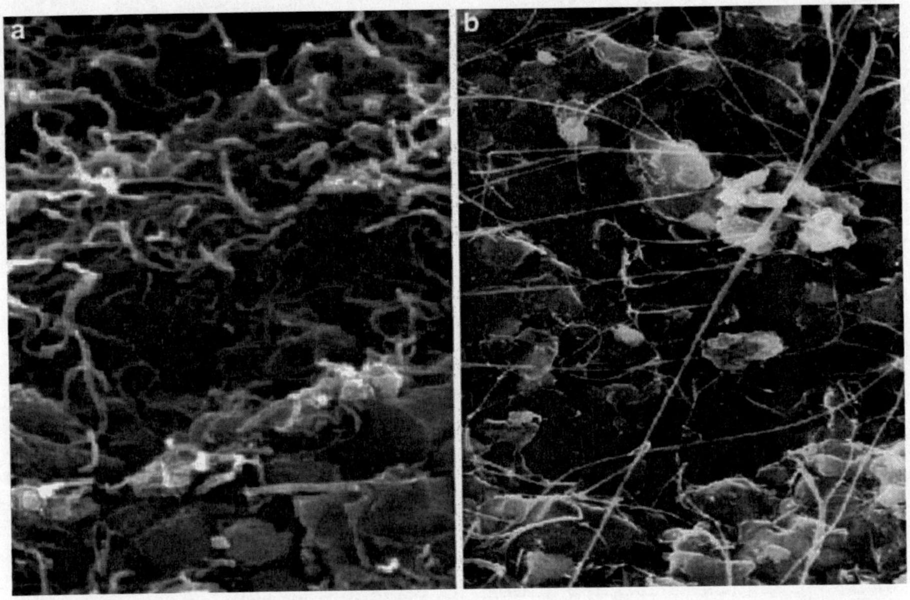


**Fig. 7.7** Typical microstructure of ceramic nanocomposites reinforced with (**a**) dispersed carbon nanotubes (CNTs); and (**b**) aligned CNTs

good mixture of the two phases prior to sintering or hot pressing. Some success has been achieved with conventional milling techniques, primarily with the use of low to moderate nanotube volume fractions. By minimizing the ceramic particle size, large surface CNT materials can be achieved. Nano-SiC ceramic-based CNT composites have been successfully fabricated by mixing SiC nanoparticles with 10 wt% CNTs and hot pressing at 22,730 K. The CNTs played a strengthening and toughening role in this CMC. Both bend strength and fracture toughness of this SiC–10% CNTs composite increase by about 10% over monolithic SiC ceramic. Another interesting approach is to use a mixture of ceramic powder and catalyst for nanotube growth (Fe, Ni, or Co), followed by a CVD process that grows nanotubes inside the particulate perform. This can produce a uniform mixture of the two phases as long as mass transport of the carbon-containing gas (e.g., $C_2H_2$, $CH_4$, etc.) into the powder bed does not produce significant gradients. In some cases, the nanotubes are damaged by high-temperature sintering. However, spark-plasma sintering has been used to produce materials where the nanotube structure is retained. The success of this method apparently results from the use of lower temperatures and shorter firing times. While the sintering of traditional whisker and fiber reinforced composites is typically plagued by differential densification rates, this may be less of a problem with nanotubes because of their small dimensions. Conventional hot pressing and hot isostatic pressing methods have also been used to produce nanotube-reinforced composites. Tape casting and the lamination method have been used to achieve better dispersion of CNTs in an alumina matrix. The tribological studies of the fabricated composites showed that tape casting helped in achieving higher dispersion. The presence of CNTs is also expected to inhibit grain growth during high-temperature processing. This beneficial effect is also known to occur during high-temperature processing of a variety of other composite materials. Several methods have been used to form nanotubes inside preexisting porous matrices. One method of producing nanotubes inside a porous ceramic is to use anodic alumina templates with controlled pore structures. This approach is generally limited to coatings or membranes up to ~100 μm thick. The resulting materials consist of aligned nanotubes whose diameter is determined by the preexisting cylindrical (or nearly cylindrical) pores in the alumina. The dimensions of these materials make them potentially applicable as coatings. Also, the unidirectional microstructure is an advantage for fundamental investigations of mechanical properties. Nanoscale powders with CNTs provide another opportunity for creating dense ceramic-matrix powders with enhanced mechanical properties. The strength and fracture toughness of hot pressed α-alumina is typically much greater than that of conventional grain size polycrystalline alumina. The addition of CNTs to the alumina results in lightweight composites with even greater strength and fracture toughness. The mechanical properties of such composites depend strongly on the processing methods and surface treatment of the CNTs. The most interesting and challenging applications for ceramic-CNT composites are as tough materials. Toughening in CMCs is typically achieved by a weak fiber–matrix interface coupling that permits debonding and sliding of the fibers within the matrix. The closing forces exerted by fibers on matrix cracks that propagate around the fibers constrains the crack growth, and the work required to pull broken fibers out against residual sliding friction at the fiber–matrix interface imparts significant fracture toughness to CMCs. Weak interfaces between a CNT and many ceramic matrices is expected, and pyrolytic carbon is the most common interface material deposited onto fibers to induce such debonding. Thus the mechanisms of interface debonding and sliding, and associated toughening should be operative in CNT-CMCs unless the sliding resistance is too low (Curtin and Scheldon 2004).

CNTs have been successfully used to enhance toughness of reaction bonded SiC ceramics. Fracture toughness of reaction bonded SiC was increased from 4 to 7 MPa $m^{1/2}$ (a 73% increase) using CNT reinforcement. Table 7.2 gives a

**Table 7.2** Comparison of properties between reaction-bonded Carbon nanotube (CNT) reinforced silicon carbide (SiC) and standard reaction-bonded SiC

| Material | Density (g/cc) | Coefficient of thermal expansion (ppm/K) | Thermal conductivity (W/m K) | Modulus (GPa) | Toughness (MPa√m) | Bend strength (MPa) | Electrical resistivity (Ohm cm) |
|---|---|---|---|---|---|---|---|
| Standard reaction-bonded SiC | 3.02 | 2.9 | 188 | 359 | 4 | 288 | 1.097 |
| Reaction-bonded SiC–CNT | 3.06 | 2.72 | 185 | 374 | 6.9 | 285 | 0.245 |

comparison of properties between reaction-bonded SiC–CNT and standard reaction-bonded SiC (Karandikar et al. 2007).

Ceramic-CNT composite systems continue to hold promise but with significant challenges to real success. Given the costs of the materials and processes involved, it is not sufficient to obtain marginal improvements in properties over traditional micron-scale composites or virgin matrices. Yet, with few exceptions, notable enhancements have not been observed. The traditional interplay of careful processing and evaluation, coupled with mechanistic assessment of properties, remains a valid paradigm at the nanoscale and should be assiduously applied to future research in CNT composite systems (Samal and Bal 2008).

## Ceramic Matrix Composite Thermal Protection System

TPS and hot structures are required for a range of hypersonic vehicles ranging from ballistic reentry to hypersonic cruise vehicles, both within Earth's atmosphere and non-Earth atmospheres. There are multiple options for dealing with the severe thermal environments encountered during hypersonic flight. Passive, semipassive, and actively cooled approaches can be utilized. The differences between rockets and air-breathers, for instance, can have a significant impact on the TPS and hot structures. As moving toward air-breathing hypersonic vehicles, the severe thermal structural challenges require a new approach to thermal management, one that includes both TPS and hot structures (Glass 2008).

Thermal structural challenges can be quite severe on air-breathing vehicles. One of the primary thermal structural challenges results from large thermal gradients. For example, in a cryogenic tank containing liquid hydrogen as a fuel, the liquid hydrogen will be −423°F and the outer surface of the TPS may be between 2,000 and 3,000°F. With different materials operating at a wide range of temperatures, attaching the various components (tank, insulation, structure, TPS, etc.) that are growing and shrinking at different magnitudes is challenging. Control surfaces are often hot and are often connected to an actuator inside the vehicle which is much cooler. On some structures, there are thin cross sections (due to the need for low drag) at high mechanical loads. These high mechanical loads are often imposed at elevated temperatures. The stability of the outer mold line is important because it can impact performance. For example, sharp nose leading edges generate shocks which are necessary to maximize airflow into the engines. As a result, leading edges should not ablate, both to generate the desired shocks, and to enable reuse. Steps and gaps can be a problem. Gaps may potentially allow sneak flow, where hot plasma leaks into the structure. Forward facing steps may result in local hot spots, thus locally increasing the surface temperature. Thermal expansion of the propulsion system creates other issues. On air breathers, the propulsion system includes much of the undersurface of the vehicle, is very long, and can grow several inches. It must be attached to the airframe and allow for the differential growth between the propulsion system and the airframe. Other vehicle and structural challenges include affordability. Production costs, life cycle cost, and inspection and maintenance costs are all important. Other issues to consider are damage tolerance, low speed impact such as tool drop, foreign object damage on a runway, hypervelocity impact, weather, and reuse potential. All of these requirements lead to a new approach to thermal protection that move away from the Shuttle Orbiter-type of insulated airframe. The Space Shuttle Orbiter is an aluminum "airplane" with ceramic tiles and blankets, which were developed to enable this type of vehicle to be flown. The required material attributes for hypersonic air-breathing vehicles are high-temperature capability (2,000–4,000°F), high strength at those elevated temperatures, high toughness, light weight, and environmental durability. High specific strength at elevated temperature is the goal, which is obtained with high strength and low density. Metallic options include MMC, super alloys, and titanium. These all have good specific strength, but it drops off by the 2,000°F range. Grouped together as CMC, the SiC–C material, advanced carbon–carbon, and SiC–SiC provide high strength at elevated temperatures which are key for air-breathing vehicles. A CMC is illustrated in Figure 7.8a, where the matrix is reinforced with fibers. The fibers carry the load and the matrix transfers the external load to the fibers. Depending on the processing and fabrication approaches, the fiber–matrix interface could have an interface coating which, if present, is the key to the toughness of the structure. It provides a weak mechanical interface for increased toughness and graceful failure. One of the key components of the CMC is the environmental barrier coating which protects the materials from oxidation at high temperatures. The CMC hot structure control surface approach provides the lowest

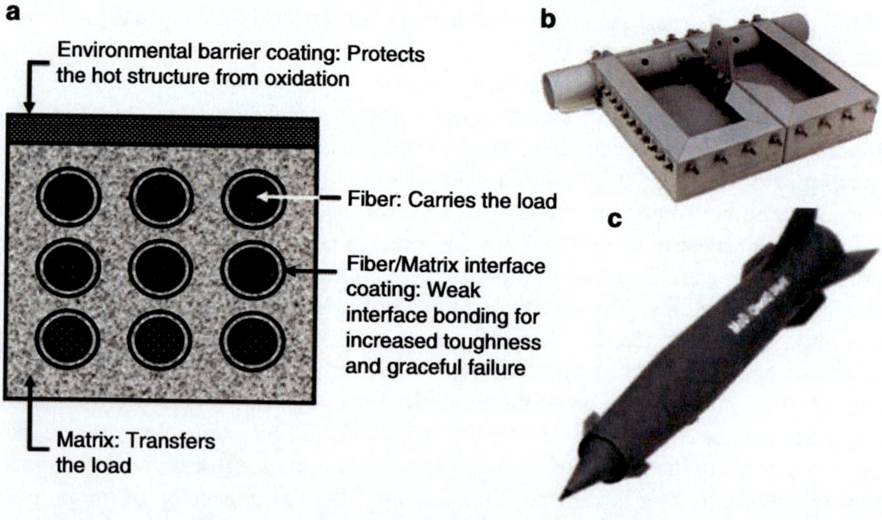


**Fig. 7.8** CMC for thermal protection system and hot structure: (**a**) Schematic of a CMC reinforced with fibers; (**b**) mechanically assembled SiC–$C_f$ body flap including SiC–$C_f$ fastened joints and a thin ply torque tube and box structure along with gusset members for load transfer; and (**c**) SiC–$C_f$ load bearing aeroshell utilized on sustained hypersonic flight missile

weight and thinnest cross section, minimal thermal expansion mismatch problems, and a good thermal margin. This approach is also advantageous because of its sufficient strength and stiffness and there is no external insulation required. Disadvantages to this approach include the high manufacturing and tooling costs for the box structure and scale up. This approach may not be recommended for very large structures. Its repair capability is limited and the manufacturing risk is high in cases of production failure or damage. Access for coating, inspection, and maintenance of internal areas is also a disadvantage. A mechanically assembled CMC control surface can be fabricated and assembled using multiple smaller parts, as shown in Fig. 7.8b. This control surface used SiC–C fastened joints and a thin ply torque tube and box structure along with gusset members for load transfer. Advantages to this approach are that the tooling is relatively simple and damaged components can be replaced without a complete scrap of the entire control surface. A disadvantage to this approach is that tolerance buildup can be problematic in assembly of numerous separate parts. High part count is another disadvantage of this approach. Figure 7.8c shows a SiC–$C_f$ load bearing aeroshell that carries the aerodynamic and vehicle axial loads. It is to design and fly a prototype ramjet capable of sustained hypersonic flight. The vehicle weighs approximately 30 kg, and is 1.5 m long, with a diameter of 7 in. The aeroshell is being utilizing CVI. Fuel flows in an annulus between two SiC–$C_f$ tubes. The fin roots are bonded to the aeroshell. Insulation may either be incorporated in the structure or separate, below the aeroshell. This type of aeroshell has the potential for reduced weight (Glass 2008).

CMCs can be used to enable many of the components on air-breathing hypersonic vehicles, such as leading edges, acreage, hot structures, and in the propulsion system. For most CMC structures, there are two primary materials and structures technical challenges: fabrication and environmental durability. There are several processes for fabricating SiC–C, and SiC–SiC, each having their own challenges. Small coupons can be fabricated with relative consistency. However, a coupon built today may or may not be similar to one that is built 6 months from now. The key point is that a state-of-the-art material is not the same as a state-of-the-art structure. Just because the material can be fabricated, does not at all guarantee that a structure can successfully be fabricated and that it would survive the required mission life. The following fabrication challenges, again process dependent, arise in going from a small material coupon to a large flight structure, which include thickness (density uniformity through out the structure), complex curvature, large scale, low interlaminar properties, delaminations, critical flaw size, nondestructive inspection, tooling, assembly methods and tolerances, reproducibility, fabrication modeling, manufacturability of structures design, and affordable (cost and schedule) fabrication techniques. Meeting the flight requirements is another challenge that impacts the structure maturity on a component. Operation has a significant impact on the ability to use these materials as structures on flight vehicles. Challenges to consider for flight requirements include thermal loads, thermal gradients, mechanical loads, acoustic and vibration loads, pressure (oxidation), combined loads, and number of cycles. During flight, many of the loads are combined, thus generating potentially severe thermal-structural loading. In addition, the primary environment durability challenge is oxidation resistance, which has a major impact on mission life. The number of cycles required under combined loads, inspection and repair, and the ability to predict mission life all present challenges. Hypersonic air-breathing vehicles will require moving beyond an insulated aluminum "airplane," such as the Space Shuttle Orbiter, to a vehicle with multiple TPS and hot structure approaches. The ability to build and fly these vehicles successfully will depend on the ability to use multiple types of CMC structures, but only after solving the environmental durability and fabrication challenges (Glass 2008).

## Summary

Ceramic composites are being actively developed in many research establishments primarily for structural and load-bearing applications. Fiber, whisker, or particulate reinforced composites can be made tougher and stronger than traditional unreinforced ceramics. Among the key factors responsible for the improved stress resistance are differences in Young's moduli between the phases and the nature of the interaction at interfaces between matrix and the reinforcement. As user confidence continues to grow and as energy savings, increased productivity, and reduced maintenance are confirmed, the need has emerged for advanced ceramics with improved toughness. Discontinuous reinforced ceramic composites have partially filled this need, but their application is limited in both size and geometry, and their toughness is less than desired for a risk-adverse industry. Continuous fiber reinforced ceramic composites are viewed as the ultimate solution with many applications rapidly becoming commercially viable. Thermally conductive ceramic composites have been actively developed thermal management of electronic packaging. Fiber, whisker, particulate, or nanotube reinforced composites can be made tougher, stronger, and more effective for thermal management with tailored CTE. The desirable characteristics of CMCs include high-temperature stability, high thermal shock resistance, high hardness, high corrosion resistance, light weight, nonmagnetic and nonconductive properties, and versatility in providing unique engineering solutions. The combination of these characteristics makes CMCs attractive alternatives to thermal management of electronic packaging, particularly for high temperature electronic packaging system.

Of these, SiC fibers have been the most widely used because of their high strength, stiffness and thermal stability. SiC matrix CFCCs have been successfully demonstrated in a number of applications where a combination of high thermal conductivity, low thermal expansion, light weight, and good corrosion and wear resistance is desired. SiC matrix CFCCs can be fabricated using a variety of processes, fibers, and interface coatings. Fibers widely used for industrial applications where long life is desired include SiC or mullite. Processes available to fabricate SiC matrix CFCCs and the matrix composition formed include Sic, polymer infiltration (SiCN, SiC), nitride bonding (Si–SiC–$Si_3N_4$), and melt infiltration (Si–SiC). The interface coating can be either carbon or boron nitride with a protective overcoat of SiC or $Si_3N_4$.

SiC–diamond composites are composed of microcrystalline diamond held together by microcrystalline SiC. The thermal conductivity of the SiC–diamond composite spreader can reach around 600 W/m K, CET is 1.8 ppm/K. These SiC–diamond heat spreaders have been commercialized. Reaction bonded SiC ceramics combine the advantageous properties of high performance traditional ceramics, with the cost effectiveness of net shape processing. These materials provide high surface hardness, very high specific stiffness, high thermal conductivity, and very low CTE. Al-toughened SiC composite maintains many of the advantageous properties of reaction bonded SiC while providing higher toughness, higher thermal conductivity and more tailorable CTE. The composite is produced using reactive infiltration process, which allows near-net-shape components to be fabricated. CNTs have been successfully used to enhance toughness of reaction bonded SiC ceramics. Fracture toughness of reaction bonded SiC was increased from 4 to 7 MPa $m^{1/2}$ (a 73% increase) using CNT reinforcement.

Hypersonic air-breathing vehicles will require moving beyond an insulated aluminum "airplane," such as the Space Shuttle Orbiter, to a vehicle with multiple TPS and hot structure approaches. CMCs provide desirable characteristics to meet these requirements, with high-temperature stability, high thermal shock resistance, high hardness, high corrosion resistance, light weight, nonmagnetic and nonconductive properties, and versatility in providing unique engineering solutions. The ability to build and fly these vehicles successfully will depend on the ability to use multiple types of CMC structures, but only after solving the environmental durability and fabrication challenges.

## References


Ajayan PM, Schadler LS, Braun PV (2003) Nanocomposite science technology. Wiley-VCH GmbH, Weinheim.

Ceram (2001) Silicon Carbide-Reaction bonded silicon carbide (RBSC). http://www.azom.com/details.asp?ArticleID=147. Accessed 03 May 2010.

Curtin WA, Scheldon BW (2004) CNT-reinforced ceramics and metals. Materialstoday 7(11): 44–49.

Ekimov EA et al (2000) High-pressure, high-temperature synthesis of SiC-diamond nanocrystalline ceramics, Applied Physics Letters 77(7): 954–956.

Evans RS, Bourell DL, Beaman JJ, Campbell (2003) Reaction bonded silicon carbide: SFF, Process refinement and applications. http://www.me.utexas.edu/~campbell/pubs/conf/SFF2003paper_SiC.pdf. Accessed 03 May 2010.

Fukagawa T, Kubo S (2009) Carbon-fiber-reinforced SiC composite material and slide member. USPTO Application# 20090029163.

Glass DE (2008) Ceramic matrix composite (CMC) thermal protection systems (TPS) and hot structures for hypersonic vehicles1. 15th AIAA Space Planes and Hypersonic Systems and Technologies Conference. http://ntrs.nasa.gov/archive/nasa/casi.ntrs.nasa.gov/20080017096_2008016802.pdf. Accessed 12 May 2010.

Guo YH, Bando Y, Kurashima K, Sato T (2002) SiC nanorods prepared from SiO and activated carbon. Journal of Material Science **37**: 2023–2029.

Karandikar PG (2003) Rapid processing method for large, low-expansion, ligh-weight mirrors. http://optics.nasa.gov/tech_days/tech_days_2003/docs/42MCubedRapidProcessingMethod.pdf. Accessed 12 May 2010.

Karandikar PG, Evans G, Aghajanian MK (2007) Carbon nanotube (CNT) and carbon fiber reinforced high toughness reaction bonded composites. Ceramic Engineering and Science Proceedings **28**(6): 53–63, Proceedings of the 30th International Conference on Advanced Ceramics and Composites, Daytona Beach, January 2007.

MMMT (2004) Aluminum-toughened SiC ceramic. http://www.mmmt.com/Application Notes/ToughenedSiCNote.pdf. Accessed 06 May 2010.

Nakano K, Hayashi H, Ohnishi N, Nagasaki S (1998) Method for production of continuous carbon fiber reinforced SiC composite. US Patent 4722817.

Naslain R (2010) Ceramic matrix composites. http://www.mpg.de/pdf/europeanWhiteBook/wb_materials_213_216.pdf. Accessed 06 May 2010.

Qian J, Zhao Y (2005) Diamond-silicon carbide composite and method for preparation thereof. US Patent 6939506.

Richerson DW (1997) Ceramic matrix composites. In Mallick PK (ed.) Composites Engineering Handbook. Marcel Dekker, New York.

Samal SS, Bal S (2008) Carbon nanotube reinforced ceramic matrix composites – a review. Journal of Minerals & Materials Characterization & Engineering **7**(4): 355–370.

Sung CM (2007) Silicon-diamond composite heat spreader and associated methods. USPTO Application#: 20070170581.

Wang Y (2006) Silicon carbide nanowires and composites obtained from carbon nanotubes. PhD dissertation. Texas Christian University, Fort Worth.

Zweben C (2006) Thermal materials solve power electronics challeges. http://powerelectronics.com/mag/602PET24.pdf. Accessed 09 May 2010.

# Chapter 8
# Thermal Interface Materials in Electronic Packaging

**Abstract** With the continual increase in cooling demand for electronic packaging, there has been an increased focus within the microelectronics industry on developing high performance thermal solutions. Thermal interface materials (TIMs) play a key role in thermally connecting various components of the thermal solution. As electronic assemblies become more compact and there is an increase in processing bandwidth, escalating thermal energy has become more difficult to manage. The major limitation has been nonmetallic joining using poor TIMs. The interfacial, versus bulk, thermal conductivity of an adhesive is the major loss mechanism and normally accounts for an order magnitude loss in conductivity per equivalent thickness. The next generation TIM requires a sophisticated understanding of material and surface sciences, heat transport at submicron scales, and the manufacturing processes used in packaging of microelectronics and other target applications. Only when this relationship between bond line manufacturing processes, structure, and contact resistance is well understood on a fundamental level, will it be possible to enhance interfacial thermal conductance and advance the development of miniaturized microsystems. TIMs are widely needed to improve thermal contacts for facilitating heat transfer in electronic packaging, such as that associated with the flow of heat from a microprocessor to a heat spreader or a heat sink in a computer. A TIM is commonly in the form of paste, solder, or a resilient sheet that serves to fill a gap between the two adjoining surfaces. The resiliency helps the conformability. The performance of a TIM is enhanced by conformability of the interface material to the topography of the mating surfaces because the air residing in the valleys in the surface topography is thermally insulating and should be displaced by the interface material. This chapter will discuss interfacial conduct mechanism and review current status and future trends of commercial and advanced TIMs, including metallic, organic, graphite, hybrid, and nanotechnology-based TIMs.

X.C. Tong, *Advanced Materials for Thermal Management of Electronic Packaging*, Springer Series in Advanced Microelectronics 30,
DOI 10.1007/978-1-4419-7759-5_8, © Springer Science+Business Media, LLC 2011

## Thermal Joint Conductance and Selection of Thermal Interface Materials

Thermal management is meant to ensure a long-term lifetime and functioning of electronic components such as power semiconductors in electronic circuits. It all begins with a profound knowledge of the occurring power dissipation and of the whole thermal path. Any engineering surface of the electronic components is rough on a microscopic level due to the presence of microscopic asperities. When two such rough surfaces come into contact during assembly of electronic packaging, the actual contact occurs only at a few discrete spots, usually at the high points of the two mating surfaces. Typically, the ratio of real contact area to apparent contact area is approximately 1–2%. Heat flowing from one body into the other is constricted to flow through the actual contact spots because the thermal conductivity of the solid contact spots is much higher than that of the surrounding gap which is filled with air in most electronic applications. TIMs are often inserted between the surfaces of a contact pair to reduce the thermal contact resistance. Although they typically have lower thermal conductivity than the substrate, they are highly compliant and hence, under the application of relatively small contact pressure, deform to conform to the geometry of the adjacent rough surfaces. A part of the low thermal conductivity gas present is therefore replaced by a higher conductivity material. This leads to a decrease in the constriction of the heat flow lines, and an increase in the contact conductance (Singhal et al. 2004). The heat transfer of these complex joints is governed by contact pressure; number, size, and shape of contact spots and voids; types of fluid in voids; pressure of fluid in voids; hardness and flatness of contact surfaces; modules of elasticity of contact surfaces; surface cleanliness; and property of the TIM used in the joint.

### *Thermal Joint Conductance*

The contact interface configuration can be divided into conforming rough surfaces, and nonconforming wavy, convex, or concave interfaces. At this point the thermal joint conductance will be derived based on the assumption of conforming surfaces in which the voids are so small, thermal energy is mainly transferred by conduction, not by convection. In fact, the assumption of the conforming rough surfaces is close to most cases of the electronic packaging contact interfaces. Because radiation heat transfer at most interfaces is negligible or nonexistent, it will not be included in this analysis. The thermal joint conductance, $h_j$, of the interface formed by two conforming, rough surfaces is the sum of contact conductance and gap conductance, which is given by (Yovanovich et al. 2004)

$$h_j = h_c + h_g = 1.25 k_s \frac{m}{\sigma} \left( \frac{P}{H} \right)^{0.95} + \frac{k_g}{Y + \sigma M}, \tag{8.1}$$

where $h_c$ is contact conductance, $h_g$ is the gap conductance. $k_s$ is the harmonic mean thermal conductivity of the interface, and can be obtained by $k_s = 2\,k_1k_2/(k_1 + k_2)$, $k_1$ and $k_2$ are thermal conductivity of two joint solids, respectively. $\sigma$ is the combined RMS (Root-Mean-Square) roughness and can be obtained by $\sigma = \sqrt{\sigma_1^2 + \sigma_2^2}$, $\sigma_1$ and $\sigma_2$ are the RMS roughness of two joint surfaces, respectively. $m$ is the effective mean absolute asperity slope of the interface and can be obtained by $m = \sqrt{m_1^2 + m_2^2}$, $m_1$ and $m_2$ are the absolute asperity slope of two joint surfaces, respectively.

The surface asperity slope is frequently not given. For the joint interface between aluminum heat sink and ceramic packages, the mean absolute asperity slope can be approximated by the correlation equation (Yovanovich et al. 2004):

$$m = 0.125\ (\sigma \times 10^6)^{0.402}, \tag{8.2}$$

which was developed for the surface roughness range:

$$0.216\,\mu\text{m} \leqslant \sigma < 9.6\,\mu\text{m}, \tag{8.3}$$

$P$ represents the contact pressure, and $H_c$ is the surface microhardness of the softer of the two contacting solids. The microhardness is in general complex because it depends on several geometric and physical parameters, such as the Vickers microhardness correlation coefficients. For the relative contact pressure range:

$$10^{-5} < P/H_c < 2 \times 10^{-2}. \tag{8.4}$$

The gas parameter $M$ accounts for rarefaction effects at high temperatures and low gas pressures. This gas-surface parameter depends on the thermal accommodation coefficients, the ratio of specific heats, the Prandtl number, and the molecular mean free path of the gas. This complex gas surface parameter depends on gas pressure and temperature according to the relationship. Its value is 0.081 for air at 1 atm and 100°C, 0 for liquids (Yovanovich et al. 2004).

Although this conforming rough surface model was developed for bare surfaces, it can also be applied to interfaces with thermal grease, assuming that the grease behaves as a liquid and fills all gaps between the contacting asperities and substituting $M = 0$ and the thermal conductivity of the grease into the gap conductance relationship in equation (8.1). However, when solid interstitial materials are used, such as thermal compounds, elastomers, or adhesive tapes, the joint conductance problem becomes much more complicated because the use of a solid interstitial material introduces an additional interface to the problem. Using thermal resistance concepts, the overall joint conductance for this problem is determined by the series combination:

$$1/h_j = 1/(h_{j,1}) + t/k + 1/(h_{j,2}), \tag{8.5}$$

where $h_{j,1}$ and $h_{j,2}$ refer to the joint conductance between each of the contacting surfaces and the interfacial material, and $t$ and $k$ are the average thickness and thermal conductivity of the interstitial layer. Completing this analysis requires characterization of the relevant surface parameters, such as the slope, roughness, and microhardness for the various interstitial materials. In addition, for elastomeric materials, the layer thickness $t$ is not constant but instead depends on the contact pressure. Additional research is needed before a model can be developed to address this complex phenomenon.

For example, the aforementioned models will be used to calculate the joint resistances for the interface formed by an aluminum 6063-T5 heat sink and alumina ($Al_2O_3$) package. The thermal conductivities of the heat sink and ceramic package are $k_1 = 201$ W/m K and $k_2 = 20.9$ W/m K respectively. The harmonic mean thermal conductivity of the interface is $k_s = 37.85$ W/m K. Because the microhardness of the aluminum alloy is 1,094 MPa, which is much less than that of the alumina, it will be used to compute the contact parameters. Based on a surface roughness for fly cut aluminum of $\sigma_1 = 0.4$ μm and a surface roughness for ground alumina of $\sigma_2 = 1.3$ μm, the effective surface roughness of the interface is calculated as $\sigma = 1.36$ μm. The surface slopes are $m_1 = 0.139$, $m_2 = 0.0865$, respectively. The effective surface slope of the interface is therefore $m = 0.164$. Therefore, the join resistance of a typical aluminium-ceramic interface can be calculated for contact pressures between 0.007 and 0.35 MPa, which includes the practical microelectronic pressure range of 0.07 and 0.17 MPa. The greatest joint resistances are found when air is present in the interstitial gap. In the contact pressure range of 0.007–0.35 MPa, the air joint resistance goes from 2.665 to 1.903 $cm^2$ °C/W. When silicon grease is placed in the gap, the joint resistance is much smaller than the bare interface. The calculated values of the joint resistance lie in the range of 0.335–0.213 $cm^2$ °C/W which is an order of magnitude smaller than the joint resistances of a bare joint. If greases with thermally conductive ceramics are used, the joint resistance can be reduced to values below 0.065 $cm^2$ °C/W. The correlation equations are based on conforming rough surfaces with interstitial substances which perfectly wet all portions of the surfaces which form the gap. Any nonflatness will result in interfaces with larger gaps which will have larger joint resistances. If the interstitial substance does not perfectly *wet* the contacting surfaces, this will also produce a more thermally resistive interface. The correlation equations therefore correspond to the best thermal joints which have the smallest joint resistances. The use of other interstitial materials, such as thermal compounds, elastomers, or adhesive tapes has been shown to increase the complexity of the joint conductance problem significantly. In order to successfully solve this problem, simulation methods provided the way to model the complex interface phenomena. It is common practice for high-power applications to assemble power devices as bare dies in the chip on PCB or to use flip chip assemblies with a directly mounted copper or aluminum heat sink. The thermal resistance of the TIM is the bottleneck of the heat transfer from the active device junction to the cooler. Better TIMs are crucial to even out the localized hot spots (Yovanovich et al. 2004).

## *Criteria for Selection of Thermal Interface Materials*

The TIM acts to connect the different thermal management components in electronic packaging. Formation of a small thermal barrier is an important property of a TIM. The thermal barrier can be described in terms of effective thermal resistance, or reverse effective thermal conductivity through the TIM which is preferably as high as possible. The effective thermal conductivity of the TIM is primarily due to the interfacial heat transfer coefficient as well as the intrinsic bulk thermal conductivity of the TIM. A variety of other properties are also important for a TIM depending on the particular application, for example, an ability to accommodate or avoid thermal expansion stresses when joining two materials, an ability to form a mechanically sound joint that is stable during thermal cycling, a lack of sensitivity to moisture and temperature changes, manufacturing feasibility, and low cost.

Figure 8.1 shows thermal barrier of thermal resistance across an interface with a TIM. After inserting a TIM between two contact solid surfaces, the effective thermal resistance, $R_{\text{TIM}}$, at the interface consists of the bulk resistance, $R_{\text{bulk}}$, of the TIM arising from its finite thermal conductivity, and the contact resistance, $R_{\text{c}}$ between the TIM and the adjoining solids. Therefore, $R_{\text{TIM}}$ can be expressed as (Prasher et al. 2003):

$$R_{\text{TIM}} = \frac{BLT}{k_{\text{TIM}}} + R_{\text{c1}} + R_{\text{c2}}, \tag{8.6}$$

where BLT is the bond-line thickness of the TIM, $k_{\text{TIM}}$ is the thermal conductivity of the TIM, and $R_{\text{c1}}$ and $R_{\text{c2}}$ are the contact resistances of the TIM with the two adjoining surfaces. The sum of contact resistance of the TIM with the two adjoining substrates can be expressed as (Prasher et al. 2003):

$$R_{\text{c}} = R_{\text{c1}} + R_{\text{c2}} = \left(\frac{\sigma_1 + \sigma_2}{2k_{\text{TIM}}}\right)\left(\frac{A_{\text{nominal}}}{A_{\text{real}}}\right), \tag{8.7}$$

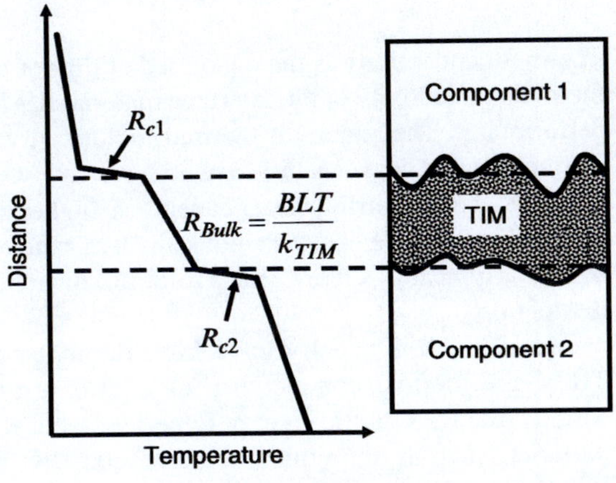


**Fig. 8.1** Thermal resistance across an interface with a thermal interface material (TIM)

where $\sigma_1$ and $\sigma_2$ are the surface roughness of the substrates, $A_{nominal}$ is the nominal area of heat transfer, and $A_{real}$ is the real area of heat transfer. Therefore, minimization of contact resistance can be done through increasing pressure applied on the interface, decreasing surface roughness, increasing thermal conductivity of the TIM, and increasing capillary force by changing the surface chemistry.

One of the goals of thermal design is to reduce the thermal resistance $R_{TIM}$, which can be accomplished by reducing the BLT, increasing the thermal conductivity and reducing the contact resistances $R_c$ ($R_{c1}$ and $R_{c2}$). As the reduction of BLT may be limited in some cases, the two most desirable properties of a TIM are high thermal conductivity and high compliance in order to increase the thermal conductance and reduce the contact resistances. Only few homogeneous materials possess both these properties, such as indium and other solder alloys, therefore, TIMs are typically composite materials with metallic or ceramic fillers in a polymeric matrix. Typically used fillers such as alumina ($Al_2O_3$) or boron nitride (BN) are characterized by relatively high thermal conductivity and low compliance. Most matrix materials, e.g., silicone, have low thermal conductivity but high compliance. In view of practical applications, optimal volume fractions and geometric distributions of filler and matrix materials are sought at which the joint conductance assumes a maximum value. The optimal filler volume fraction is expected to depend on a series of factors, including the relative thermal and mechanical properties of the matrix and filler, the filler shape, its distribution, and orientation. Furthermore, the size of the filler particles relative to the thickness of the TIM layer will also affect the optimal filler volume fraction, as will the boundary resistance between filler and matrix (Singhal et al. 2004). As a result, the optimum property of the TIMs can be achieved through optimizing the volume fraction and the geometric distribution of filler particles for which the joint conductance of a rough surface-TIM-rough surface system takes the maximum value.

In practical applications, the criteria for selection of the TIMs can be summarized as shown below.

### Thermal Conductivity of the Material

Thermal conductivity is the quantified ability of any material to transfer heat. The thermal conductivity of the interface material has a significant impact on its thermal performance. The higher the thermal conductivity, the more efficient the material is at transferring heat. Materials that have a lower thermal conductivity are less efficient at transferring heat, causing a higher temperature differential to exist across the interface. To overcome this less efficient heat transfer, a better cooling solution (typically, a more costly solution) must be used to achieve the desired heat dissipation.

Metallic TIMs usually have high thermal conductivity. For polymer based TIMs, the thermal conductivity of a TIM is typically enhanced by loading a soft, sometimes liquid-like polymeric material matrix with conducting solid particles, such as aluminum, alumina, and BN. If the design requirement is that the TIM should be thermally conducting but electrically insulating then the ceramic-based filler particles are more typically chosen. The bulk thermal conductivity of particle laden polymer-based TIMs can be functionally expressed as (Prasher et al. 2003):

$$k_{\mathrm{TIM}} = f(k_{\mathrm{f}}, k_{\mathrm{m}}, \varphi, R_{\mathrm{f}}), \tag{8.8}$$

where $k_{\mathrm{f}}$ is the filler thermal conductivity, $\varphi$, filler volume fraction, and $R_{\mathrm{f}}$ is the contact resistance between the fillers and the polymeric matrix. When $k_{\mathrm{f}}/k_{\mathrm{m}} \gg 1$, (8.8) can be expressed as (Prasher et al. 2003):

$$\frac{k_{\mathrm{TIM}}}{k_{\mathrm{m}}} = \frac{1}{(1-\varphi)^{3(1-\alpha)/(1+2\alpha)}}, \tag{8.9}$$

where $\alpha$ is the Biot number and is given as:

$$\alpha = \frac{R_{\mathrm{f}} k_{\mathrm{m}}}{d}, \tag{8.10}$$

where $R_{\mathrm{f}}$ is the interface resistance between the filler and the particle and $d$ is the diameter of the particle. $R_{\mathrm{f}}$ could arise because of imperfect mixing of the particle with the polymer matrix, due to phonon acoustic mismatch, or due to a combination of the two phenomenon. The contribution due to phonon acoustic mismatch at room temperatures or higher is important only when the thermal conductivity of both the materials is very high. In the case of silicon oil-based TIMs, for instance, it is safe to assume the $R_{\mathrm{f}}$ is arising from imperfect wetting or mixing of the filler particles with the silicon oil, as the thermal conductivity of silicon oil is very low. Therefore, the phonon acoustic mismatch component is very low at room temperatures. The reason for higher $R_{\mathrm{f}}$ for 60% volume fraction could be due to difficulty in wetting the surface of the particle with the silicone oil as the volume fraction is so high.

### Electrical Conductivity of the Material

Depending on the application requirement, the TIMs can be provided with electrically conductive or electrically insulated products. Some metallic TIMs such as solders, low melting alloys (LMAs), or metal-based compounds are electrically conductive, whereas ceramic-based compounds are typically not. Polymer-based composites can be both thermally and electrically conductive. Electrically conductive TIMs can be used for application that do not cause circuit shorting or intend to enhance electromagnetic interference (EMI) shielding performance. Electrically insulated TIMs are used to circuit short-sensitive cases. For example, metal-based thermal compounds are not hazardous to the processor die itself, but other elements on the processor or motherboard can be at risk if they become contaminated by the compound. For this reason, much of electronic packaging is not recommended in the use of electrically conductive TIM.

## Spreading Characteristics of the Material

The spreading characteristics of the TIM determine its ability, under the pressure of the mounted heat sink, to spread and fill in or eliminate the air gaps between the processor and the heat sink. Because air is a very poor thermal conductor, the more completely the interface material fills the gaps, the greater the heat transference.

During spreading of the TIM, reaching a minimum BLT is often another parallel goal of thermal design. BLT is a function of various parameters such as application pressure (i.e., pressure applied in bringing the two contact surfaces together) and particle volume fraction. An empirical expression for the BLT of particle laden polymeric TIMs is (Prasher et al. 2003):

$$\mathrm{BLT} = 1.31 \times 10^{-4}\left(\frac{\tau_y}{P}\right)^{0.166}, \tag{8.11}$$

where $\tau_y$ is the yield stress of the TIM and $P$ is the applied pressure. This correlation was validated in the pressure range of 25–200 psi application pressures. Because the yield stress of the TIM increases with increasing filler loading, BLT is higher for higher volume fraction. Therefore, there are two competing effects with regard to filler loading for the thermal resistance of the TIM: $k_{\mathrm{TIM}}$ increases and BLT also increases with increasing filler volume fraction at the same pressure, which leads to an optimal filler loading for the minimization of $R_{\mathrm{TIM}}$ (Prasher et al. 2003).

## Long-Term Stability and Reliability of the Material

The long-term stability and reliability of the TIM is its ability to provide a sufficient thermal conductance even after an extended time or extensive use of the electronic equipment (for example, servers or personal computers that work 24 h a day, 7 days a week). Low-quality compounds may harden or leak out over time (the pump-out effect), leading to overheating or premature failure of the processor. High-quality compounds provide a stable and reliable TIM throughout the lifetime of the processor. Thermal greases with higher viscosities are typically more resistant to pump-out effects on lidless processors. Another typical requirement of TIMs is that of reworkabilityBecause the heat sink is attached to the device by the OEM in a number of applications, reworkability is a requirement to avoid yield loss due to heat sink attach. Reworkability implies that the heat sink should be easily removed and that the TIM should be easily cleanable so that the heat sink may be reattached if needed. This requirement has led to certain classes of materials seeing continued popularity; e.g., filled greases, filled phase change materials (PCMs), and certain gels.

## Ease of Application

A spreadable TIM-like grease requires the installer to carefully use the appropriate amount of material. Too much or too little material can cause problems. The PCM, for instance, has a fixed size and is therefore easier to apply in a consistent manner.

During materials design, apart from thermal requirements, the specific needs for the electrical insulation of electronic parts and other parameters, e.g., minimum distances and computer telephony integration have considerable impacts on the product design-in of TIMs. TIMs are therefore sometimes classified into electrically insulating or electrically noninsulating materials. Furthermore, parameters such as chemical consistence, proper surface properties for compensating surface deviations or mechanical tolerances, softness and flexibility, tensile strength, processability, easy handling and delivery shapes, and last but not least the economic efficiency among others must be considered when doing the product design-in. Environmental compatability, reworkability, suitability to adhesive coating, chemical, temperature and aging resistance, as well as life time and low outgasing of silicone are other important influences.

Table 8.1 shows some typical TIMs which have been developed and utilized in electronic packaging (Mahahan 2004). TIMs can basically be classified as metallic TIMs, conductive elastomer TIMs, thermal grease, phase-change materials, and emerging TIMs such as polymer solder hybrid (PSH), composite, or smart TIMs (Tong 2009).

TIMs are usually used on the following application areas of an electronic package: (1) to bring a bare die package into contact with heat sink hardware; (2) to bring the die into good thermal contact with an integrated heat spreader, and then to bring the integrated heat spreader into contact with system thermal solutions, such as chassis, heat sink, or OEM applied hardware. Conventionally, the TIM between the die or the die package and the heat spreader is called a TIM1, and the TIM between the heat spreader and the heat sink hardware is called a TIM2.

In addition, many applications call for TIM that can easily be placed on a chip, on a lid, or perhaps just against a heat source and a cooling solution contact plate. Metal foil with/without galvanic compatible surface plating, and compressible solder perform have been developed for such an application. The metal foil and compressible TIMs also fin the burn-in head to optimize their performance. The pressure range is around 35 to 100 psi. For instance, an indium foil preform offers uniform thermal resistance at lower applied stresses in compressed interfaces. The malleability of indium minimizes surface resistance and increases heat flow. This kind of TIMs delivers superior performance over time, they cannot experience

**Table 8.1** Properties of typical thermal interface materials

| Material type | Typical composition | Advantages | Drawbacks | Bond line thickness (mm) | Thermal conductivity (W/m K) |
|---|---|---|---|---|---|
| Grease | AlN, Ag, ZnO, Silicon oil | High bulk thermal conductivity, conforms to surface irregularities, no cure needed, reusable | Pump out and phase separation, migration | 0.05 | 3–5 |
| Gel | Al, Ag, Silicon oil, Olefin | Good bulk thermal conductivity, Conforms to surface irregularities before cure, no pump out or migration, reusable | Cure needed, lower thermal conductivity than grease | 0.025–0.04 | 3–4 |
| Phase Change Materials | Polyolefins, epoxies, polysters, acrylics, BN, Alumina, Al, Carbon nanotubes | Conforms to surface irregularities, no cure needed, no delamination, easy handling, reusable | Lower thermal conductivity than grease, no uniform bold line thickness | 0.05–0.05 | 0.5–5 |
| Phase change metallic alloy | Pure In, In/Ag Sn/Ag/Cu In/Sn/Bi | High thermal conductivity, reusable | Complete melt possible, voiding | 0.05–0.125 | 30–50 |
| Solder | Pure In, In/Ag Sn/Ag/Cu In/Sn/Bi | High thermal conductivity, easy handling, no pump out | Reflow needed, stress crack, delamination, Voiding possible Not reusable | 0.05–0.125 | 30–86 |

pump out problems even under power cycling, and also resist bake out due to its solid state.

## Metallic Thermal Interface Materials

The thermal conductivity of conventional polymeric TIMs generally does not exceed about 5 W/m K and is typically less than about 1 W/m K. However, metallic TIMs that form thermal interfaces with effective thermal conductivities of about 50 W/m K or greater will be the alternative to meet increasing thermal management requirement of high performance electronics. Most metallic TIMs may also exhibit a favorable solder or wetting behavior upon reflow which facilitates a low thermal interfacial resistance. During reflow, the solder and substrate are heated, the solder melts and wets by surface tension and/or local surface alloying. The interfaces consist of intermetallics or interdiffused metals with thermal properties that are frequently less desirable than those of the bulk TIM metal but much better than polymeric TIMs. In addition, the reflow or active bond process can also help to form reliable thermal interfaces. Metallic TIMs, however, can fail in certain applications as a result of the relatively large differences between the coefficients of thermal expansion (CTE) of the TIM and the semiconductor and/or heat sink components and the lack of compliance (Sreeram et al. 2003). To take advantages of the metallic TIMs and avoid their disadvantages, a large number of metallic TIMs have been developed and offered different characteristics for selection. Such metallic TIMs typically include reflow solders and active bond process, nonreflow solders and LMAs, composite solders, and hybrid metallic solder materials as well as gold–gold interconnection.

## *Reflow Solders and Active Bond Process*

Reflow soldering is a process in which a solder TIM preform or TIM solder paste is used to attach one or several electrical components to each other, after which the assembly is subjected to controlled heat, which melts the solder, permanently connecting the interfacial joint. Heating may be accomplished by passing the assembly through a reflow oven or under an infrared lamp or by soldering individual joints with a hot air pencil. A thermal reflow process profile typically undergoes five transitions: (1) Preheat–brings the assembly from 25°C to preheat zone and evaporates solvents from solder perform or solder paste. A slow ramp-up rate will prevent any damage due to thermal shock. The time and temperature to evaporate the solvents will depend upon the solder perform or solder paste that is used. (2) Flux activation (preheat)–dried solder perform or solder paste is heated to a temperature in which the flux will react with the oxide and contaminants on the surfaces to be joined. The time and temperature should be long enough to allow the flux to fully clean these surfaces but not too long that the flux may be exhausted before soldering takes place. (3) Thermal equalization–achieves temperature equalization approximately 20–40°C below the peak reflow temperature. Time and temperature will depend on the mass and materials. (4) Reflow–in this stage, the assembly is briefly brought to the temperature sufficient to produce reflow of solder. (5) Cool down–this is the final stage in the solder process. Gradual cooling should be used. The end result should be as fast as possible without causing thermal shock to the components. Cool down in this manner will produce a finer grain structure in the solder joint, which will yield a more fatigue-resistant solder joint.

Although each offers different thermal and mechanical properties, reflow solders are very attractive for use as a TIM because they have high thermal conductivities ranging from 30 to 86 W/m K. The key to using solder TIMs is to reduce voiding during reflow when assembling or mounting the components together. Unlike polymer- or grease-based materials used for thermal interfacing, solder voids will not propagate later during use due to pump out or migration, which assures a higher end of life performance. Thermal solders provide the best physical connection of all TIMs. Using solder eliminates the issue of bleed-out from the thermal greases while providing very good adhesion. This mechanical attachment is a particular advantage when external clamping is not feasible. To take advantage of these properties, interfaces and assembly are designed to ensure proper wetting of the contact surfaces, which minimizes voiding. Furthermore, thermal expansion of components in a thermal assembly may stress solder TIMs. For this reason, brittle alloys should not be used as TIMs. Bismuth-containing alloys are especially poor choices because of their low thermal conductivity and brittle nature. Au–Sn alloys handle stress well, although they may not be soft enough for some applications where the heated materials expand at markedly different rates. It is important to choose a solder that will be able to compensate for CTE mismatch. Design of the solder thermal interface requires understanding the interconnect materials to minimize the effects of intermetallic formation and performance degradation. Additionally, the solder must provide for the compliance of the thermal expansion mismatch between the semiconductor and the heat sink. Regarding interface compliance, indium may be the best solder to use as a thermal interface. Pure indium metal provides a unique combination of high conductivity (86 W/m °C) and compliance (60 psi shear flow stress and no work hardening). Indium is applied in the mechanical sealing of metallic and nonmetallic surfaces with excellent integrity. Because indium will wet to nonmetallic substrates, it is especially well suited for use against many materials that are commonly used in semiconductor packages. This mixture of properties makes indium uniquely suited to thermal interface solutions with or without reflowing. Indium can also be cold welded in applications where normal soldering temperatures are detrimental to the electronic packaging.

In fact, as a Pb-free solder-based TIM, indium has been successfully and commercially used to meet the increasing demand for thermal cooling capability of electronic packaging. Figure 8.2 gives an example of indium as a TIM1 to connect the silicon die and the heat spreader together (Renavikar et al. 2008). To assemble the indium TIM1 appropriately, the following issues need to be paid attention: (1) effectively reduce the thermodynamically stable native indium oxide on the indium TIM1 perform; (2) to control solder joint voiding post joint formation; (3) to control interfacial reactions with surface finishes on the heat spreader lid and the back side metallization on the silicon die; and (4) to deal with reliability issues faced in small and large die products, such as thermal fatigue cracking of the indium during thermal cycling. The assembly process, including the reflow of the indium TIM1 to form uniform intermetallic compounds post assembly, is a key challenge. The indium oxide on the surface of the indium needs to be effectively reduced in order to form uniform and defect-free intermetallic layers at both the die/indium and the heat spreader lid plating (Ni/Au) and the indium (In). Indium oxide is an extremely tenacious and thermodynamically stable oxide. The presence of voiding in the joint can potentially lead to an increase in local thermal resistance and consequently lead to the degradation of the thermal performance of the joint. Additionally, excessive spallation of the

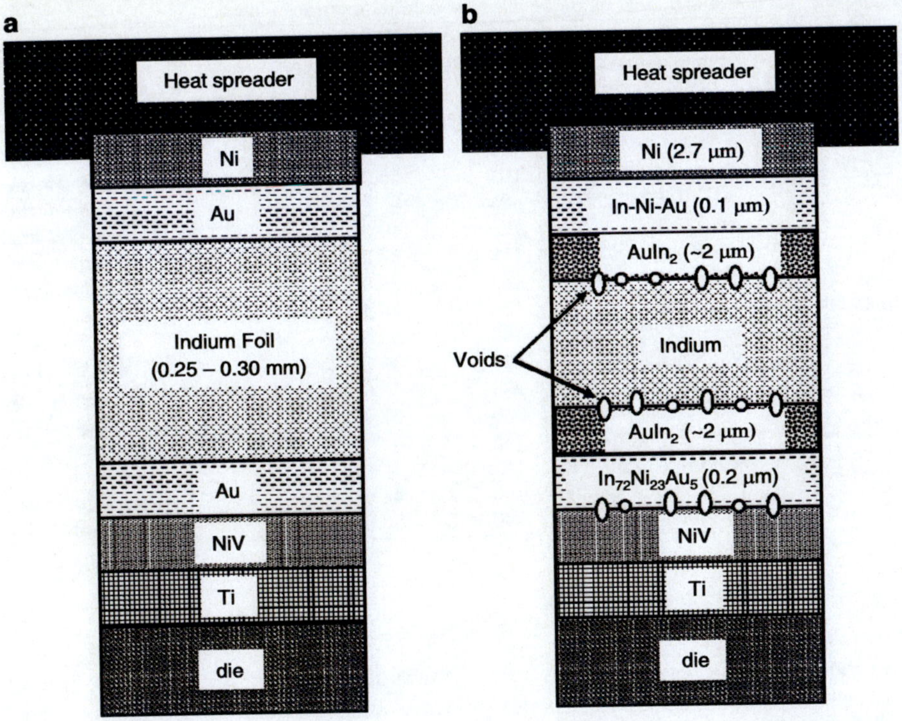


**Fig. 8.2** Use of indium as TIM1 and interactions with surface finish on heat spreader and back side metallization on the die (**a**) pre- and (**b**) post-reflow assembly

binary Au–In intermetallic compound as well as the formation of excessive Kirkendall voiding due to relatively different diffusion coefficients of In–Au and Ni can result in an increase in the thermal resistance of the joint. The reliability performance of the joint can be modulated by the relative thickness and morphology of the binary and ternary intermetallic compounds as influenced by the fluxing ability of the flux used and the reflow profile used, as well as mechanical design attributes of the heat spreader dimensions/die size, package stiffness, and preform dimensions (Renavikar et al. 2008).

Other reflow solders that might be used for TIMs are shown in Table 8.2. Generally, next-generation solder TIMs should meet the following requirements: High reliability over a wide application and product design range; high process yields with existing processing equipment; compatibility with the fabrication needs for various solder forms; low toxicity; and low cost. One thing which will probably never change is the need for good solder wetting and spreading. This requires that the solder exhibit a low viscosity and low surface tension, and a sufficiently high dissolution rate of base metal or metallized surface finishes into the molten solder. Rapid formation of intermetallic compounds is also desirable, at least in the initial stage because it tends to lower the free energy of the system,

**Table 8.2** Reflow solders that might be used for thermal interface materials

| Type of solder alloys | Composition (brand name) | Melting point (°C) | Thermal conductivity (W/m K) | CTE (ppm/°C) | TIM format |
|---|---|---|---|---|---|
| In and its alloy | Indalloy 4 | 157 | 86 | 29 | Paste/preform |
| | 100In | 157 | 84 | 32.1 | Paste/preform |
| | 21.5In16Sn62.5Ga | 10.7 | 35.0 | – | Paste/preform |
| | 51In16.5Sn32.5Bi | 60 | – | – | Paste/preform |
| | 66In34Bi | 72 | 40.0 | – | Paste/preform |
| | 52In48Sn | 118 | 34.0 | – | Paste/preform |
| | 97In3Ag | 143 | 73.0 | – | Paste/preform |
| Sn and its alloy | 42Sn58Bi | 138 | 18.5 | 14 | Paste |
| | 63Sn37Pb | 183 | 50.0 | – | Paste/preform |
| | 96.5Sn3.5Ag | 221 | 33.0 | – | Paste/preform |
| | 100Sn | 223 | 73.0 | 22 | Paste/preform |
| | 80Au20Sn | 280 | 57.0 | – | Paste/preform |
| Active bond solders | SB115 (In/Sn+Ti) | 115–120 | – | – | Preform |
| | SB140 (Bi/Sn+Ti) | 135–150 | – | – | Preform |
| | SB220 (Sn/Ag+Ti) | 190–232 | 48.0 | 19 | Preform |
| | SB400 (Zn/Al+Ti) | 390–415 | 80.0 | 32 | Preform |

and thereby promotes the further spread of solder. High-tin alloys, for instance, are adequate for viscosity, dissolution rate, and intermetallic formation rate requirements, but unfavorable due to the relatively high surface tension of tin. The latter may be improved with doping of some surface-active, low-surface energy elements. Good oxidation resistance of the solder will also be needed because formation of the oxide during soldering immediately impedes the spreading process. The oxidation resistance can often be enhanced by doping with an element such as P, Ge, or Ga.

As shown in Table 8.2, a group of novel active bond solders have been developed and enable wetting directly to metal and ceramic surfaces without the need for flux. This is achieved through the modification of conventional solders by the addition of titanium (0–5%) and/or rare earth elements. These active elements migrate to the interface and react with the opposing material surface to remove oxides and nitrides and transport them into the bulk of the solder as an inert material. This process can be conducted in an air environment but requires the application of a low-level mechanical shearing action to break a thin oxide (Sn–Ti) layer on the molten solder and initiate the reaction with the component interfaces. The level of shear is claimed to be small and can be delivered by brushing or scraping the surface, sliding the joining surfaces relative to one another, or by the application of high frequency vibration to the parts to be joined. Once the oxide layer has been disrupted, the bulk solder reacts very rapidly with the substrate surfaces to form either a metallurgical (metals) or atomic attraction/van der Waals bond. The bonding procedure is typically as follows (Young et al. 2006):

(1) Heat substrate surface to 250°C (solder material dependent), typically the process is conducted at 10–20°C above the melting point of the solder. The molten active solder would retain the ability to self align components
(2) Attach the solder preform and establish layer of material on the surface
(3) Use ultrasonic tool or mechanical action to apply enough solder on one surface to fill gap between the die and the heat spreader lid
(4) Bring lid and die together to minimize trapped air and load with ~2 lb/in$^2$
(5) Cool to below 190°C. The optimum bond line thickness for these active solders is 50–100 μm compared with 200–250 μm for conventional die attach solders

Thicker layers result in degraded performance. The mechanical performance of active solder material is similar to comparable conventional solders, although the addition of titanium is to improve the high temperature creep performance of Sn–Ag–Cu alloys. Mechanical strength will be dependent on the substrate materials and the joint design. In general, metallurgical bonds achieve 20–55 MPa (3,000–8,000 psi), and those employing van der Waals forces 6–12 MPa (1,000–2,000 psi). Silicon bonded to substrates has been shown to survive thermal cycling tests (1,000 cycles for -55°C to +125°C) and multiple solder reflow cycles (~260°C). The ability to join graphite and aluminum foams to metal-based plates opens up the potential of replacing some fin-plate heat exchangers. Thermal property (heat transfer coefficient in closed loop water) measurements showed graphite-foam samples to possess superior heat transfer coefficients compared with commercial aluminum fin-plate heat exchangers, having heat transfer coefficients between 10,000 and 20,000 W/m$^2$ K compared with 500–1,200 W/m$^2$ K for commercial aluminum fin-plate designs, thus demonstrating active solder joined graphite-foam cores to be a factor of over 15 times better. Comparative investigations on active-solder joined aluminum-foam cores resulted in coefficient values of 1,500–2,000 W/m$^2$ K. The potential benefits of active soldering are no flux, eliminates cleaning and residual corrosive material, no preplating, reduction in thermal impedance layers, reduction in process steps/cost, ability to solder a wide range of materials, good thermal/electrical joint conductivity, reworkable joint, high surface tension (low flow), and low penetration into foam materials and pipes/channels (Young et al. 2006).

Active soldering is a relatively new process and consequently is still developing a market presence. The active bond solder is potentially an attractive product for electronic thermal management systems. Two key features are: (1) ability to directly join combinations of high thermal conductivity materials with a reduced number of interface (thermal impedance) layers; and (2) clean (no-flux) processing. The latter is becoming increasingly important as package density increases, systems become more complex, and environments more severe. The process is relatively immature and consequently new applications will need to be fully tested to establish the performance of the joint and its impact on the application and product reliability. The need to precoat materials prior to soldering may limit some high-volume applications unless alternative metallization techniques, such as sputtering can be developed.

## *Nonreflow Solders and LMAs*

Nonreflow solders indicate phase change metallic alloys, also called LMAs, typically including alloys of indium, bismuth, gallium, tin, or silver. They are in liquid phase at operating temperatures; typical-phase change occurs between 60 and 80°C. Because LMAs are all-metal they contain no organic materials and hence require no curing during application. The superior performance of LMAs as TIMs is a result of the thermally conductive nature of metals and the high degree of wetting they enjoy. Good wetting means low contact resistance and allows relatively thin bond lines. These alloys are well suited as a thermal interface between materials of dissimilar CTEs, however, reliability problems in production implementations and resulting catastrophic failure have retarded their adoption. Mechanical problems include shake-out due to mechanical shock and drip-out due to low viscosity, especially with vertically oriented interfaces. Thermal problems include dry-out caused by thermal cycling leading to the creation of voids in the interface and catastrophic failure; the growth of thermally resistive intermetallic compounds through interfacial contact, and oxidation/corrosion in nonhermetically sealed applications, both leading to a steady decrease in performance and eventual thermal failure (Macris et al. 2004).

Many alloy combinations can be classified as LMAs for TIM applications. Alloys which possess a liquidus point below the operating temperature of the electronic component are needed to allow the LMA to adequately flow into all surface asperities of the interface. In general, alloys comprised of the fewest number of constituents possess the highest thermal conductivities. Common LMA candidates include bismuth-based or gallium-based alloys as depicted in Table 8.3. Liquid metals, such as alloys of bismuth, gallium, and indium potentially offer both low interfacial resistance and high conductivity. Several alloys of gallium with very low melting points have also been identified as potential liquid metal interface materials. Thermal performance of such an interface would be more than one order of magnitude greater than many adhesives typically in use. Several very low melting point gallium-based alloys are liquid at room temperature. These gallium-based alloys are finding increased use in various applications as a replacement for toxic mercury, which has a high vapor pressure at room temperature. These alloys have reduced toxicity and lower vapor pressure than mercury.

Alloy systems that are liquid at room temperature have a high degree of thermal conductivity far superior to ordinary nonmetallic liquids. This results in the use of these materials for specific heat conducting and/or dissipation applications. Other advantages of these liquid alloy systems are their inherent high densities and electrical conductivities. These alloys will wet most metallic surfaces once oxides have been sufficiently removed from the substrate surface. However, gallium is very reactive with some metals, even at room temperature. At high temperatures, gallium dissolves most metals, although a number of them, including Na, K, Au, Mg, Pb, Ni, and interestingly Hg are only slightly soluble at moderate temperatures. Gallium and the gallium alloys, like gallium–indium, have the ability to wet too

**Table 8.3** Typical low melting alloys

| Low melting alloys | Liquidus (°C) | Solidus (°C) |
|---|---|---|
| 61Ga/25ln/13Sn/1Zn | 7.6 | 6.5 |
| 62.5Ga/21.5ln/16.0Sn | 10.7 | 10.7 |
| 75.5ga/24.5ln | 15.7 | 15.7 |
| 95Ga/5ln | 25.0 | 15.7 |
| 100Ga | 29.78 | 29.8 |
| 44.7Bi/19.1ln/6.3Sn/22.6Pb/5.3Cd | 47.2 | 47.2 |
| 49Bi\21ln\12Sn\18Pb | 58 | 58 |
| 32.5Bi\51ln\16.5Sn | 60 | 60 |
| 49Bi\18ln\5Sn\18Pb | 69 | 58 |
| 33.7Bi/66.3ln | 72 | 72 |
| 57Bi/26ln\17Sn | 79 | 79 |
| 54.02Bi\29.68ln\16.3Sn | 81 | 81 |

many nonmetallic surfaces such as glass and quartz. Gently rubbing the gallium alloy into the surface may help induce wetting.

Typical applications for these materials include thermostats, switches, barometers, heat transfer systems, and thermal cooling and heating designs. Uniquely, they can be used to conduct heat and/or electricity between nonmetallic and metallic surfaces. Alloys are packaged in polyethylene bottles and shipped in accordance with applicable US federal regulations. Unopened bottles have a guaranteed 1-year shelf life. As the alloy is removed from the bottle, the volume can be replaced with dry aragon, which will minimize the possibility of oxidation at the surface of the alloy. If the alloy has been stored below its melting point and has solidified, it should be remelted and thoroughly shaken or mixed before use. Care should be taken in reheating the alloy in the original packaging provided. Temperatures should not exceed 65.6°C.

Although LMAs offer both low interfacial resistance and high conductivity, they have historically suffered from various reliability issues including: corrosion/oxidation, intermetallic formation, drip-out, dewetting, and migration. Unless mitigated, these mechanisms will continue to degrade the interface, resulting in a thermally related catastrophic failure of the actual electronic component.

Corrosion, considered to be the main failure mechanism in LMA materials, is driven primarily by moisture, oxygen, and heat. The liquid phase of these alloys at operating temperatures facilitates rapid diffusion, accelerating the corrosion process. Corrosion of an LMA is primarily an electrochemical process in which surface atoms of the alloy react with a substance in contact with the exposed surface. Electricity passes from a negative region to a positive region (both on the LMA) through an electrolyte or corroding medium. In the typical electronics environment, the corroding medium is liquid or vapor phase moisture. Moisture films on an LMA may contain dissolved substances which affect corrosion such as oxygen, various oxides/dioxides, sulfates, chlorides, and metal ions. Corrosion may also be accelerated by galvanic effects which operate by a difference in potential between dissimilar metals. As most LMA combinations lie near the anodic end of the galvanic series, the LMA will preferentially corrode when in contact electrically with a metal of higher cathodic potential and an electrolyte. Many current flip chip packages do not contain a lid seal, but instead use a partial bead of adhesive to allow venting of the package cavity. The die underfill provides all the moisture protection needed. To provide adequate LMA moisture protection, the package cavity must be of a closed configuration. The longest package lifetimes will be obtained with epoxies, or alternatively sealants being developed for near hermetic microelectromechanical system packaging. By providing a near hermetic seal between the package lid and package laminate, corrosion due to moisture films can be significantly reduced (Macris et al. 2004).

Mechanical stability should be considered due to the fact that liquid metals flow quite well. This characteristic alone poses some special challenges to overcome if they are to be used successfully as a TIM. A wire mesh has been incorporated in the thermal interface to control excess LMA, and making a paste of the LMA by incorporating various materials in particulate form to increase viscosity and reduce migration. The solid structures or phases proposed for incorporation within the thermal interface address the basic problem of getting an LMA to stay put when in service. These structures increase the surface contact area with the LMA in the thermal interface. As long as the total solid–liquid interface energy is less than the interface energy of the liquid–gas and solid–gas interfaces it replaces, the LMA will minimize its surface energy by wetting the surfaces within the interface. The LMA may still wet the surface adjacent to the thermal interface if it is the same wettable surface used under the die, particularly when acted upon by an additional force. Additional forces will arise from shock, vibration, and CTE mismatches between the LMA and other components. Once the LMA has wet the surface adjacent to, but outside the thermal interface, it is doubtful surface tension alone will retain it within the interface when acted upon by external forces. Others have proposed addressing this difficulty with gaskets to contain the LMA and fillers or noneutectic (slushy) compositions to increase its viscosity. Another simple way is to deploy the liquid metal in closed cavities where no opportunities for shorts or adverse reactions with other metals exist (Macris et al. 2004).

Designing an LMA thermal interface system has similarities to designing a solder joint system. As with a solder joint system, no deleterious intermetallics must form at the liquid solid interfaces of the LMA system. For the LMA system, deleterious intermetallics will be any having poor thermal conductivity that grow thick enough to decrease performance during the life of the component. The possibility of intermetallic formation for a given metals system can be determined from the binary equilibrium phase diagrams for the various metals present, but finding thermal conductivities for most intermetallics in the literature will be unlikely, and finding parameters necessary to calculate growth rates for them is even less likely. Consequently, for a system where intermetallics are likely to form as indicated by the phase diagram, systems will need to be aged at elevated temperature, thermal performance will need to be tested periodically, and if a decrease in thermal performance is observed, cross sections will need to be prepared to identify the offending intermetallic so it may be eliminated. Systems containing gallium pose additional challenges. Gallium metal is quite corrosive to other metals (including aluminum and copper) because of the rapidity with which it diffuses into the crystal lattices of metals. The few metals that tend to resist attack are molybdenum, niobium, tantalum, and tungsten. To prevent amalgamation with another metal, a nonreactive coating such as Teflon, siloxane, hydrogenated carbon, or thin ceramic layers may be applied to the interfaces. Additionally, refractory metals possessing good thermal performance and chemical inertness may be deposited. Anodized/oxide or conversion coatings can also provide the necessary chemical stability for various LMA compositions (Macris et al. 2004).

These challenges should be mitigated by applying a multidisciplinary approach. Therefore, LMA alloys can be applied as a TIM offering superior thermal performance due to their high thermal conductivities and low contact resistance, resulting from excellent surface wetting. Reworkability, ease of handling, and a lack of cure make this attractive in a high-volume setting.

## *Composite Solders and Hybrid Metallic Thermal Interface Materials*

Next-generation electronic packages need highly conductive TIMs dominated by solder or metallic TIM1. Some solder TIMs, such as AuSn and indium, have been developed and commercialized successfully. The eutectic gold–tin alloy is 80Au/20Sn with eutectic of 280°C (556°F) and is reflowed in a pure nitrogen or forming gas atmosphere with fluxless soldering, normally onto gold-based metallizations. The peak reflow temperature is usually 320–340°C in a bell-shaped profile, with a dwell time at peak temperature of 2–3 min, and a total cycle time of 20–30 min, depending on the thermal load in the furnace/chamber. The solder provides excellent wetting characteristics (zero wetting angle with Au metallization), great joint strength (~275 MPa), superior resistance to corrosion, and high thermal conductivity (57 W/m K), making it the material of choice in many high-reliability applications, such as microwave, medical, and aerospace systems. However, the high cost and high reflow temperature have limited its commercial applications as a TIM.

As one of most promising TIMs, indium has been successfully commercialized. However, its limitation is also apparent: high cost; limited supply (80–100 tons/year); poor corrosion resistance; best used to temperature sensitive components that do not require high joint strength; and will not be exposed to harsh or high-stress environments.

With enhanced thermal performance, hybrid or composite metallic TIM1 is promising to replace indium or other TIMs for high-power microprocessors. The design concept of advanced hybrid metallic TIM1 materials would focus on: (1) low cost and ease of use; (2) compatible with current and future heat spreading systems; (3) improved thermal performance and reliability superior to current TIM1 materials, such as thermal grease, PSH, phase change metal alloy, and indium solder. Advanced metallic TIM1 can be formed via combination of solder pastes with thermal and hybrid reinforcements (such as powder, fiber, cloth, and screen or foam of metals, shape memory alloys, ceramics, and carbon materials, etc.). It will be suitable for high volume production, the metal or other highly conductive reinforcement will provide high thermal conductivity, and the reliable metallurgical bonding interface will result in low-contact resistance.

Materials selection of hybrid TIM1 depends on application requirements: performance, reliability, manufacturing process, cost, and flexibility. Major factors include:

(1) Melting point–satisfy assembly constraints.
(2) Thermal conductivity–the higher the better.
(3) Corrosion resistance–critical for nonhermetic packages.
(4) Strain-hardening exponent–influence thermal fatigue life due to CTE difference.
(5) Solubility of surface finish elements–relative high solubility benefits good bonding and controllable formation of intermetallics.
(6) Oxidation effect–oxidation both to solder and surface finish is negative for bonding and thermal fatigue life.
(7) Selection of surface finishes including barrier and bonding layers: the selection of barrier layer needs to consider dissolution time, wetting time, durability against corrosion, selective etchability, and the bonding layer is ensuring good adhesion, oxidation resistance, and high solubility in solder.
(8) TIM1 thickness (BLT) determination: (a) based on thermal and mechanical requirements; (b) thin BLT is good for heat transfer, while the thick BLT reduces stresses and increases compliance; (c) optimum BLT based on the balance of the thermal and mechanical requirements.
(9) Fatigue resistance can be enhanced by (a) reduce TIM1 area; (b) increase TIM1 thickness; (c) match Si die CTE with that of heat spreader.

The possible fabrication and attachment assembly processes of hybrid TIMs are shown in Figure 8.3. Figures 8.4 and 8.5 give examples of Option 1. The TIM format can be paste fabricated with a mixture of solder, flux, and filler. The solder powder is chosen with a melting point of 70–200°C. The flux can be (1) no-clean (NC) flux, which consists of rosin, solvent, and a small amount of activator; (2) water soluble (WS) flux that consists of organic acids, thixotrope, and solvent. The fillers include highly conductive metals, ceramics, or graphite and other materials; low CTE materials or shape memory alloys; and active interface bonding materials.

Figures 8.6 and 8.7 provide the constitutional sketch and the example of option 2 hybrid materials. The TIM format can be paste; paste plus screen/cloth preform; pad or tape; or combospreader (spreader attached with TIM1). The constitution includes (1) solder powder with melting point of 70–200°C; (2) flux can be (a) NC flux consists of rosin, solvent, and a small amount of activator, (b) WS flux consists of organic acids, thixotrope, and solvent; (3) selectable fillers: highly conductive metal, ceramics or other materials; low CTE

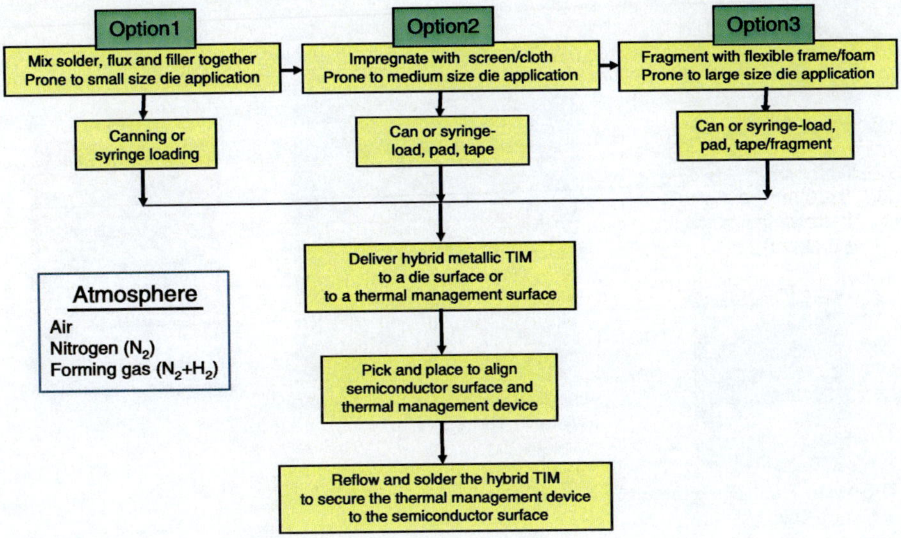


**Fig. 8.3** Sketch of possible fabrication and attachment assembly processes for hybrid thermal interface materials

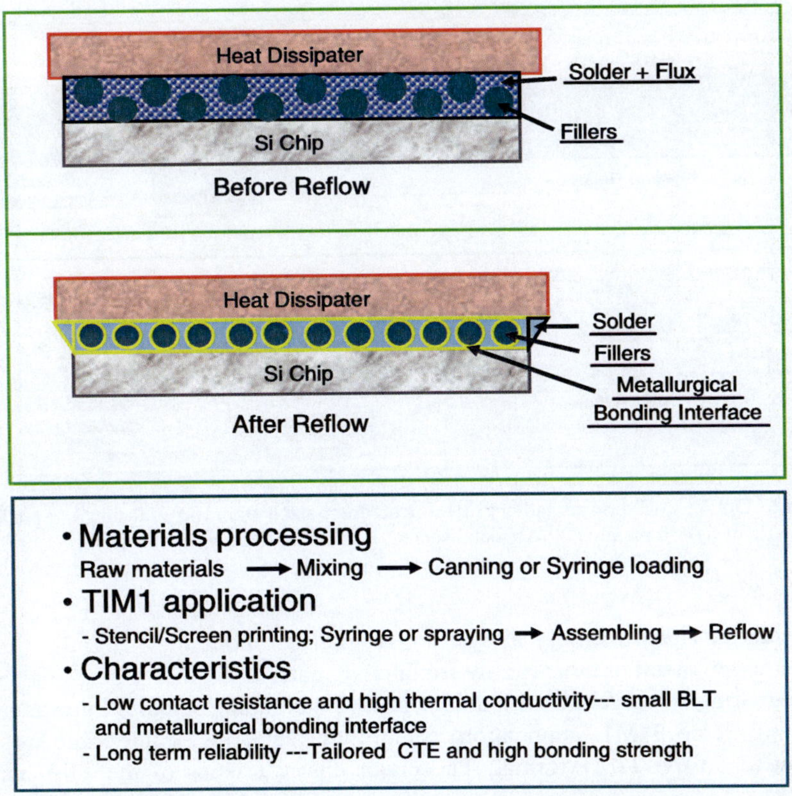


**Fig. 8.4** Option 1–Hybrid metallic TIM1 paste for small die (<100 mm$^2$) application

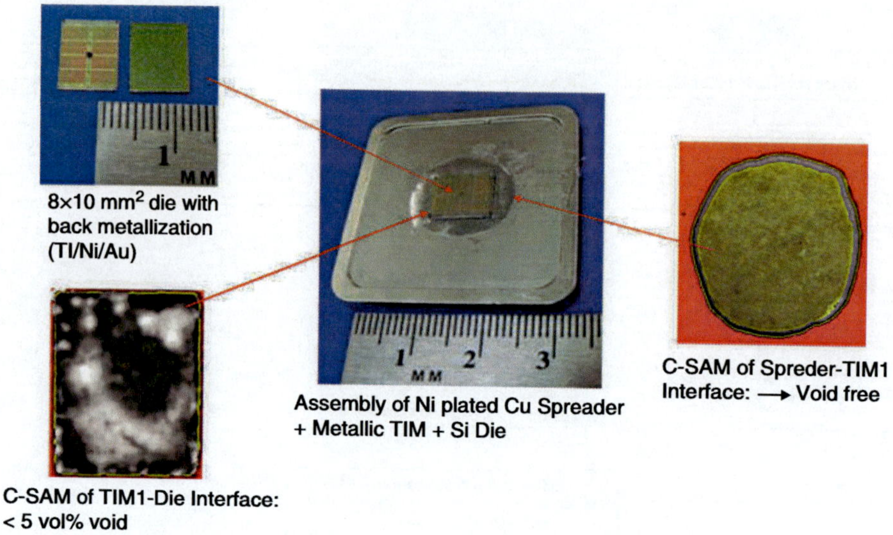


**Fig. 8.5** Option 1–Example of metallic TIM1 for small die (<100 mm$^2$) application

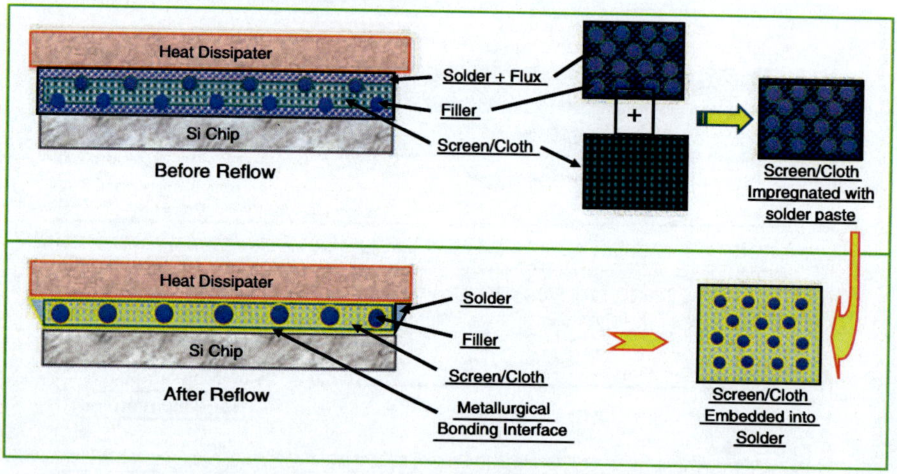


**Fig. 8.6** Option 2–Hybrid metallic TIM1 for medium-sized die (100–200 mm$^2$) application–schematic of hybrid metallic TIM1(metal screen + solder with or without filler)

materials or shape memory alloys; active interface bonding materials; and (4) screen/cloth–metal or other highly conductive materials. TIM1 processing can be impregnating screen/cloth with solder paste by brushing, dipping, printing, and spraying, etc. TIM1 application is through selecting appropriate process depended on the TIM1 format. The major characteristics of the TIM include

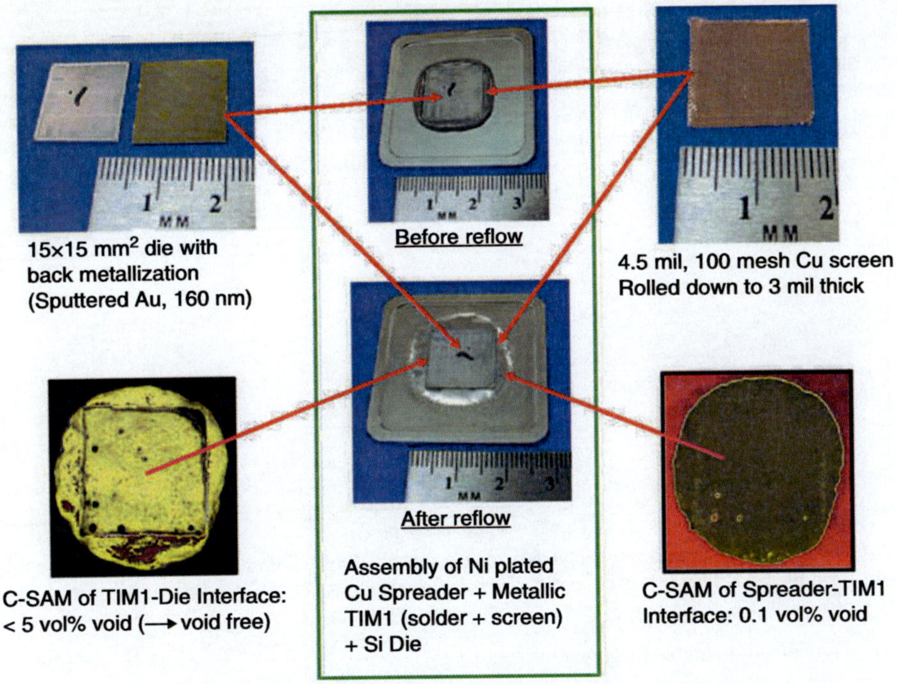


**Fig. 8.7** Option 2–Example of hybrid metallic TIM1 for medium-sized die (100–200 mm$^2$) application–schematic of hybrid metallic TIM1 (metal screen + solder with or without filler)

(1) low contact resistance and high thermal conductivity with highly conductive screen/cloth and metallurgical bonding interface; (2) long-term reliability with tailored CTE, medium BLT, and high interface bonding strength.

The illustration of Option 3 hybrid thermal materials is shown in Figures 8.8 and 8.9. The TIM format can be paste plus the conductive frame; paste plus conductive screen/cloth preform and plus frame; pad or tape; combospreader (spreader attached with TIM1). Materials can be chosen from solder powders with melting point of 70–200°C. Flux can be chosen from NC flux consists of rosin, solvent, and a small amount of activator; or WS flux consists of organic acids, thixotrope, and solvent. Selectable fillers can be highly conductive metal, ceramics, or other materials such as low CTE materials, shape memory alloys, and active interface bonding materials. Selectable screen/cloth can be metal or other highly conductive materials. Selectable flexible frame/foam can made of polymer, carbon, ceramic, metal, composite, or other flexible frame/foam. TIM1 processing includes framing and impregnate screen/cloth with solder paste by brushing, dipping, printing, and spraying, etc. TIM1 can be applied through selected appropriate process depended on the TIM1 format. Major characteristics of the TIM include low contact resistance and high thermal conductivity; adjustable BLT, highly conductive filler,

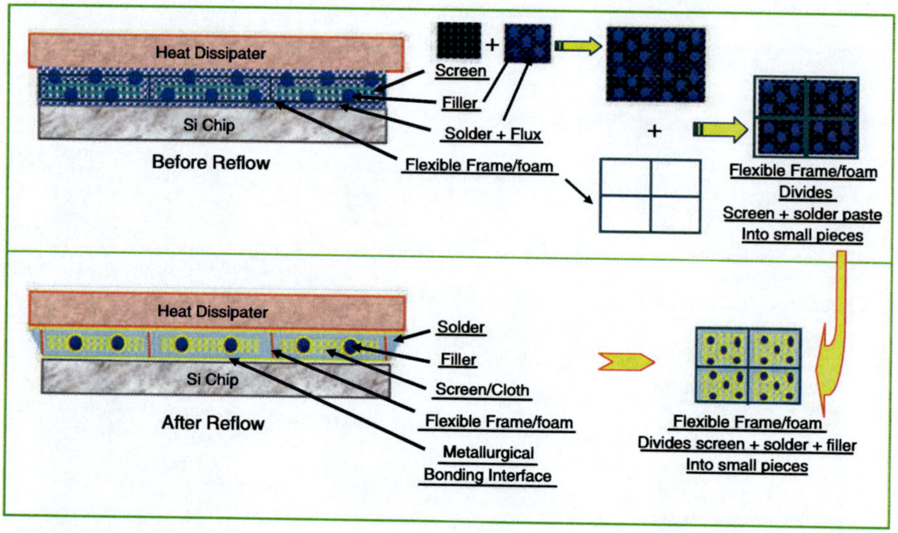


**Fig. 8.8** Option 3–Hybrid metallic TIM1 for large-sized die (>200 mm$^2$) application–schematic of hybrid metallic TIM1 (frame + solder + filler + screen)

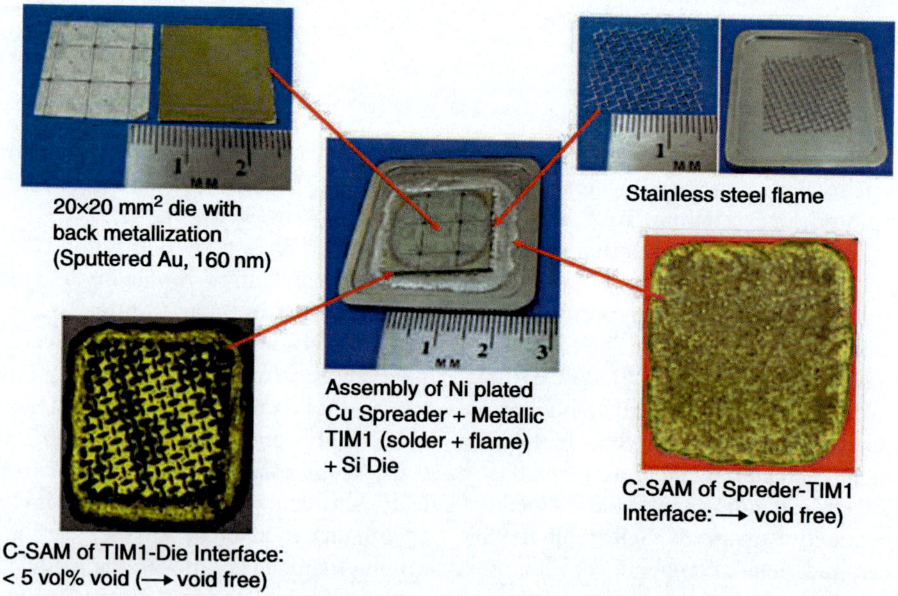


**Fig. 8.9** Option 3–Example of hybrid metallic TIM1 for large-sized die (>200 mm$^2$) application–schematic of hybrid metallic TIM1 (frame + solder + filler + screen)

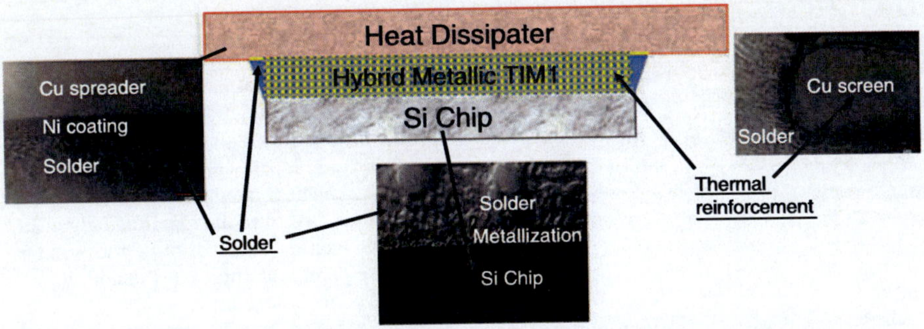


**Fig. 8.10** Structure of hybrid thermal interface material and attachment assembly

screen/cloth, as well as metallurgical bonding interface; long-term reliability; framing TIM1 into smaller pieces to reduce interface shear strength, tailored CTE, adjustable BLT, and high interface bonding strength.

Figure 8.10 shows another option using rolling-cladding of solder and metal wire screen to form the hybrid metallic TIM and its constitutional structure attachment to the components of electronic packaging. The goal is to enhance bulk thermal conductivity and thermal fatigue life through construction of a solder system reinforced with highly conductive and high-reliability network. Flexible solder formulation provides low cost and good manufacturability. Near void-free metallurgical bonding interface maintains low contact resistance and strong adhesion. The basic design criteria include (1) thermal resistance competes with indium: TI < 0.025 °C cm$^2$/W at 4–5 mil; (2) capability to fabricate into multiple formats, such as dispensable paste, preform or tape and combo heat spreader attached with metallic TIM1; (3) reliability to meet or exceed industry standards: thermal cycling, highly accelerated stress test (HAST), high-temperature aging; and (4) Flexibility to optimize formulations for different package applications with the conformability to different die sizes and board materials or components.

The candidate solders used for the hybrid TIMs are shown in the Table 8.4. The Sn–Bi system is most competitive in cost and cost volatility, and particularly interesting with doping of Zn, Cu, and Sb elements. The Sn–Bi phase diagram is shown in Figure 8.11; the Sn–Bi system has the good melting point match for TIM1 application. Sn–Bi doped with Zn, Cu, Sb, etc. would refine grain size, induce uniform dispersion of fine precipitates, and slow coarsening of eutectic Sn–Bi. A combination of high mechanical strength and good ductility is likely to yield improved fatigue resistance properties; also a benefit to manufacturability when utilizing thermomechanical processes, such as rolling, pressing, etc. Figure 8.12 shows microstructure of casting and rolled Sn–Bi solders. The solders showed ductility that was quite good. One Cu wire screen was placed between two rolled solder sheets, and then rolled or pressed together to bond and form into a kind of hybrid metalic TIM, as shown in Figure 8.13. The C-mode scanning acoustic microscope showed that no apparent porosity presented in the laminate before and after reflow, which exhibited a good bonding between the solder and the copper wire

**Table 8.4** Candidate solders used for the hybrid thermal interface materials

| Type of solder alloys | Melting point (°C) | Remark |
|---|---|---|
| Indium | 156.6 | Baseline and current commercial TIM1 |
| Sn30In5Zn0.5Cu | 114–156 | Indium reduces melting point and provides conductivity. Zn, Cu refine microstructure. |
| Sn3.5Ag8In3.5Bi1Zn | 203 | Eutectic Sn–Ag doped with Zn would exhibit good mechanical strength and creep resistance, due to refined microstructure. In reduces melting point. Bi further lowers the melting point and improves wettability |
| Sn58Bi | 139 | Low cost, simple eutectic |
| Sn45Bi1Zn0.5Cu | 142–158 | Sn–Bi doped with Zn, Cu has refined equiaxed grain structure. It can be quite ductile and may exhibit superplastic behavior. |
| Sn50Bi1Zn0.5Cu | 138 | Sn–Bi doped with Zn, Cu has refined equiaxed grain structure. It can be quite ductile and may exhibit superplastic behavior. |
| Sn35Bi1Zn0.5Cu | 139–178 | Sn–Bi doped with Zn, Cu has refined equiaxed grain structure. It can be quite ductile and may exhibit superplastic behavior. |
| Sn56Bi2Al | 144 | Al may provide strength and conductivity. |
| Sn45Bi1Zn1Mg | 142–173 | Mg may improve wettability, for flux-free application. |
| Sn45Bi1Zn1Sb | 142–170 | Zn, Sb effectively refine microstructure. |

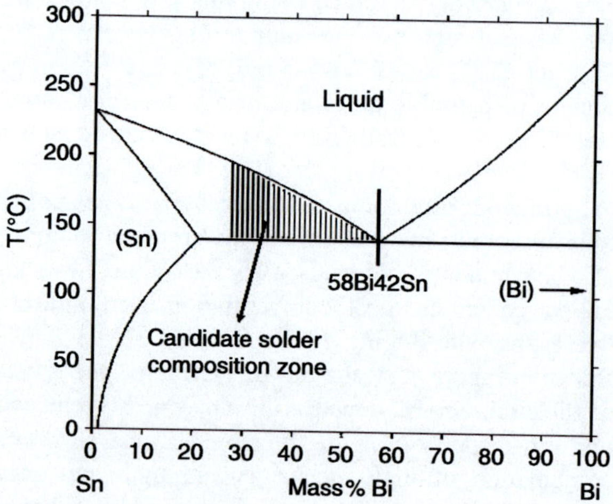


**Fig. 8.11** Phase diagram of Sn–Bi

screen, and the forming process is feasible. The thermal resistance of the hybrid metallic TIM was tested using method as shown in Figure 8.14. The joint resistance is defined by

$$R_j = R_{c,1} + R_{bulk} + R_{c,2}. \tag{8.12}$$

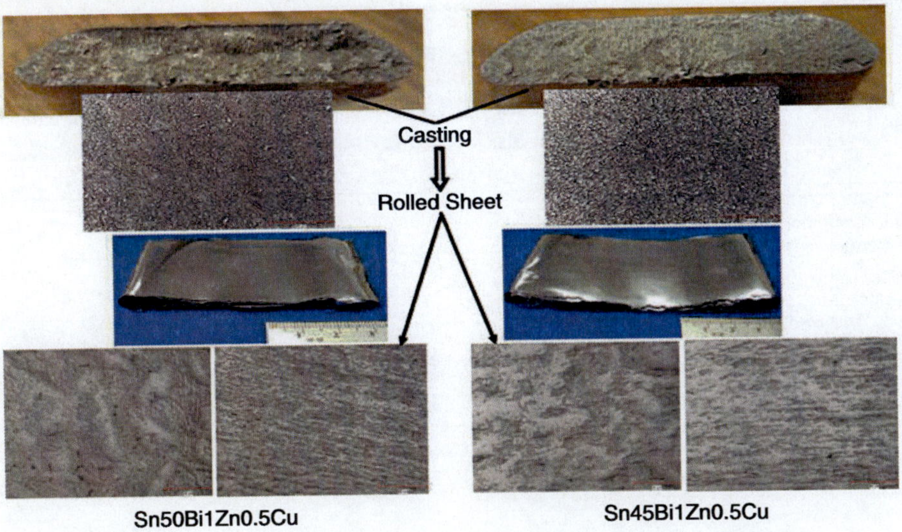


**Fig. 8.12** Microstructure of casting and rolled Sn–Bi solders refined by Zn and Cu

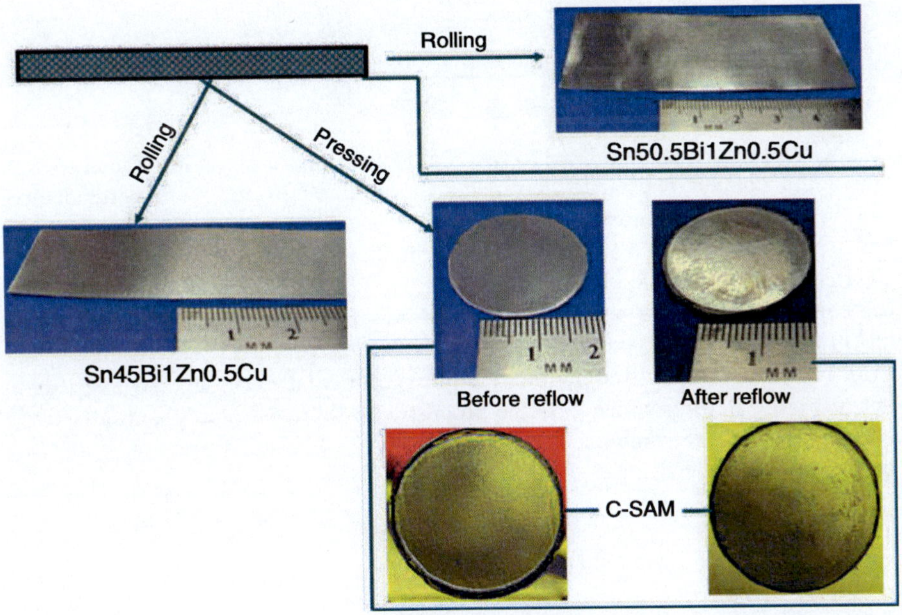


**Fig. 8.13** Rolled/pressed ductile TIM1 preform/tape exhibited good manufacturability

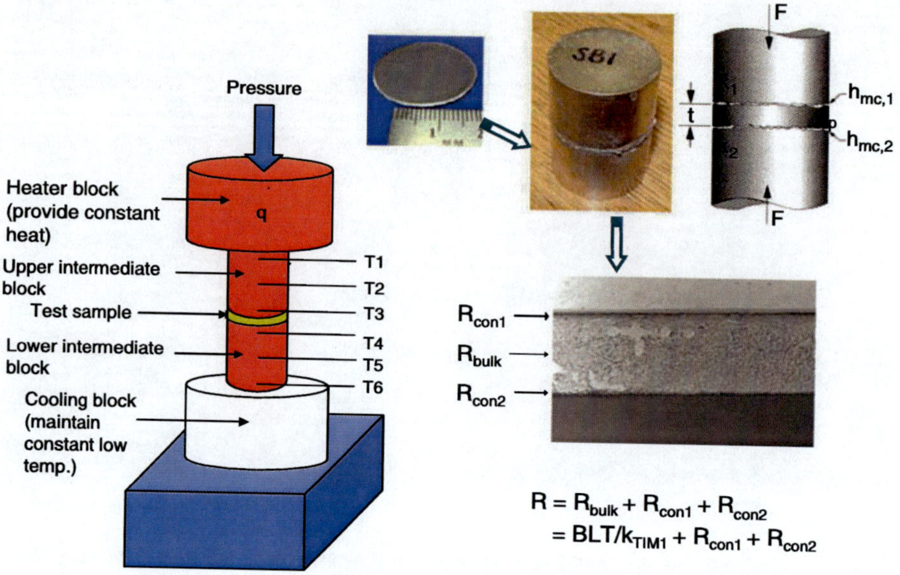


**Fig. 8.14** Test method for thermal resistance that is the sum of the contact resistance and the bulk resistance of the TIM1

And the joint conductance is determined by

$$h_j = \frac{1}{1/h_{c,1} + 1/h_{bulk} + 1/h_{c,2}}. \tag{8.13}$$

The joint resistance is related to two regions: (1) microscopic interfacial region, where the contact resistance dominates the solder/substrate joint conductance. Good metallurgical bonding would eliminate or minimize the contact resistance, and make the contact resistance is much smaller than bulk resistance. (2) Bulk TIM1 region, where the bulk resistance of the TIM1 dominated by thermal reinforcement, such as Cu screen. The tested thermal resistance of typical hybrid metallic TIM is 0.015–0.025 °C cm$^2$/W, which is very competitive among the TIMs. Figures 8.15–8.19 show modeling, experimental example, and assessment result. Figure 8.20 provides possible processes to fabricate, apply, and attach the hybrid TIMs to electronic packaging assembly.

## *Gold–Gold Interconnection*

Gold–gold interconnection (GGI) with gold stud have been developed as a niche segment of the flip chip bonding that emerges as one of the high growth areas of semiconduct assembly. GGI is a lead-free process where the Au bumps and

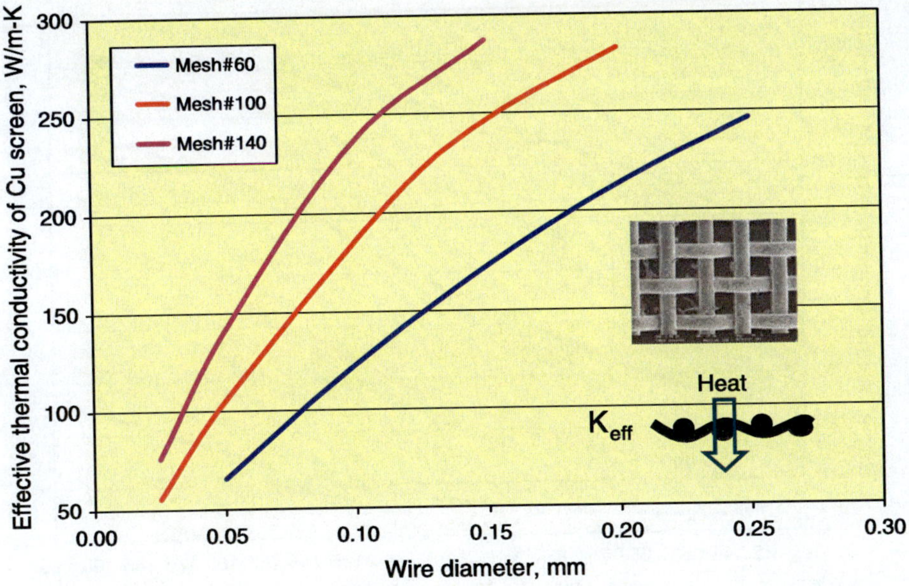


**Fig. 8.15** Cu screen selection: Mesh#-Wire size-Thermal Conductivity-$K_{eff}$ of Cu screen depends on wire and mesh number, assuming the wire is thermally isostatic

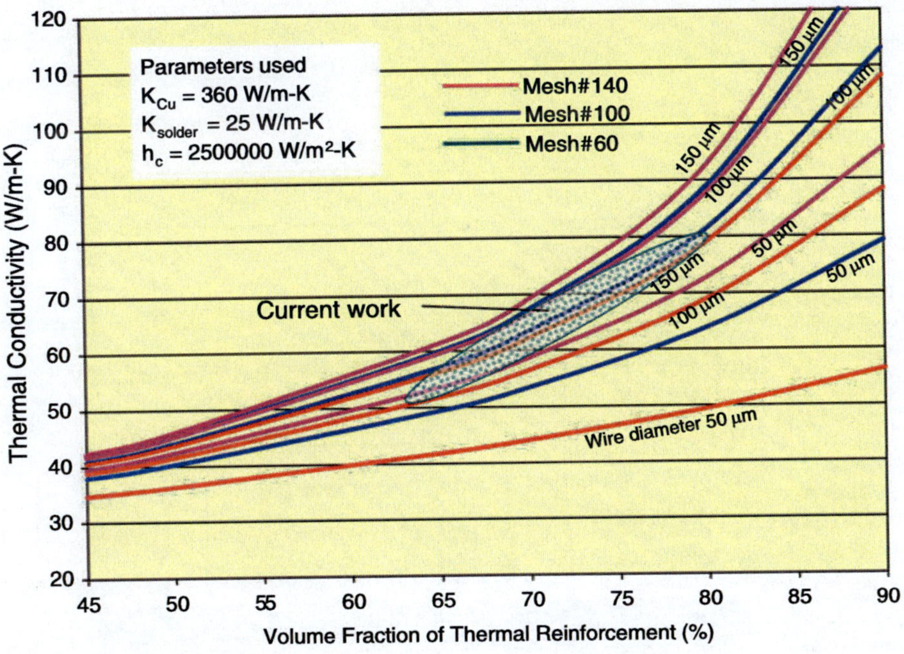


**Fig. 8.16** Predicted thermal conductivity of TIM1 with increasing vol% of reinforcement

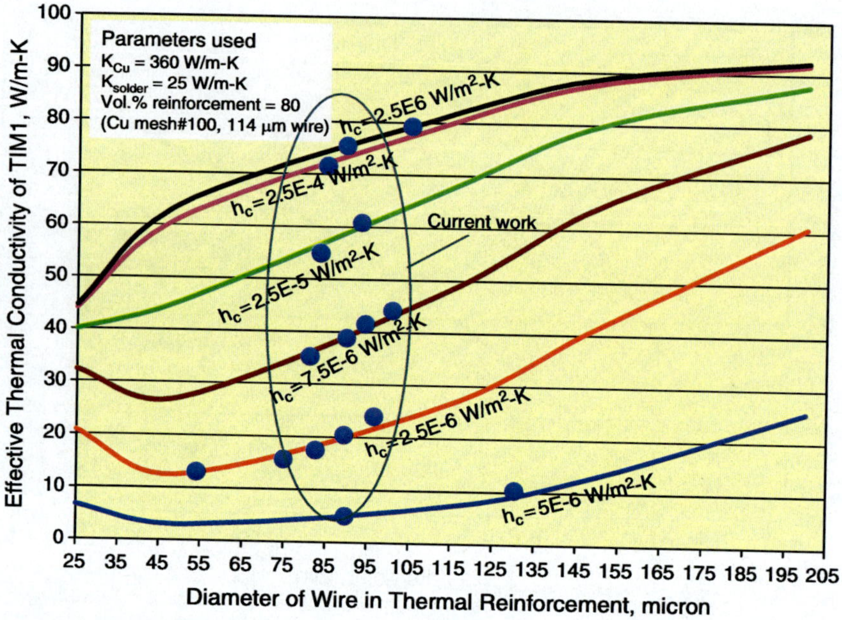


**Fig. 8.17** How interface thermal barrier affect thermal conductivity of TIMI–good interface bonding is the key to enhance the thermal conductivity

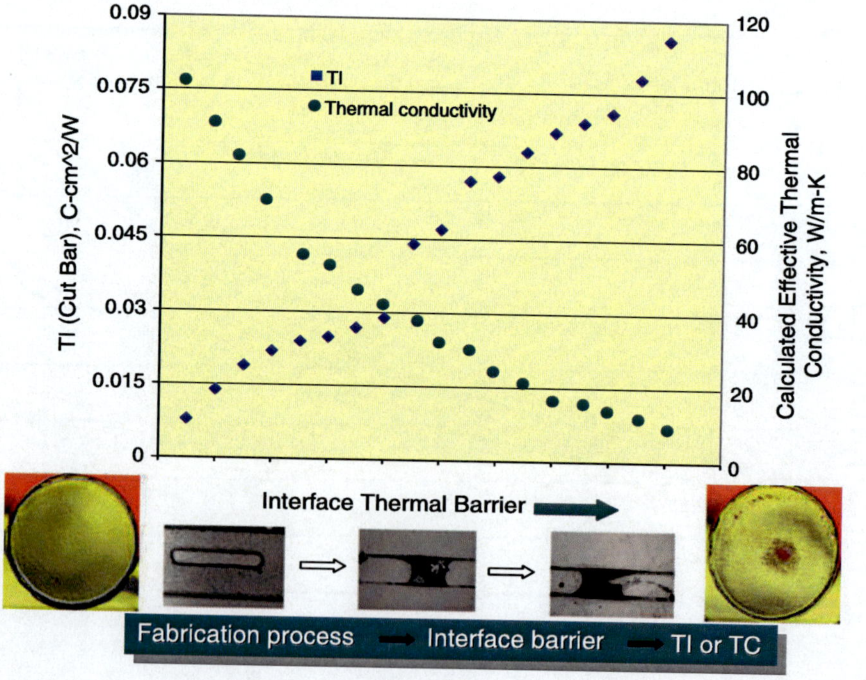


**Fig. 8.18** Effect of interface barrier on thermal impedance (TI) and effective thermal conductivity of TIM1

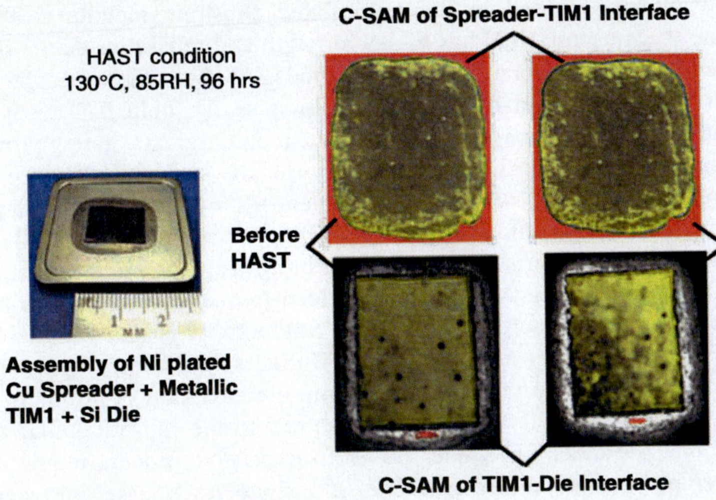


**Fig. 8.19** Highly accelerated stress test (HAST)–no visual change from C-mode scanning acoustic microscope before and after HAST

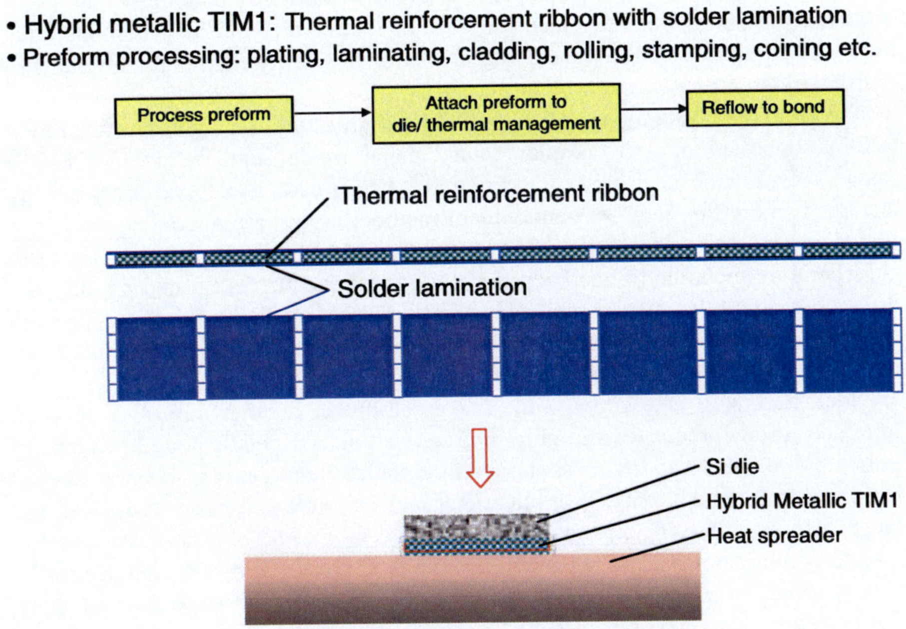


**Fig. 8.20** Application flow map–preform attachment

Au bond pads are joined together by heat and ultrasonic power under a pressure head. Gold bumping uses a variation of traditional wire bond technology to generate gold bumps on a wafer. After bumping, a wafer is diced and flipped, then thermosonically welded to the gold-plated substrate metallurgically, a monometallic thermosonic weld has higher strength and reliability than a solder joint produced by conventional flip chip methods. At the same time, the GGI joining also functions as an excellent TIM because the gold has a superior thermal conductivity of up to 318 W/m K, and the metallurgical bonding provides extremely small interface resistance. Furthermore, the use of a GGI flip chip assembly process will help to eliminate equipment parts and processing steps of the traditional flip chip assembly process and hence will shorten the overall cycle time. The materials characteristics of the GGI joining provide facet benefits. Gold resistivity is 80–85% lower than leaded and lead-free solder alloys, providing better current-carrying capacity. The thermal conductivity is much higher than solders, aiding head transfer and dissipation. Unlike solder bumped packaging, gold stud and GGI does not require under bump metallization or an interposer. When dice are small and the CTE is well matched to the substrate, GGI often does not require an under fill. Under fill is an expensive process required for solder flip chip, because solder is prone of fatigue fracture during thermal cycling. Gold stud bumping can produce small, inexpensive bumps at 50-μm fine pitch that are inherently taller than plated bumps. Plating processes have difficulty achieving this capability without additional expensive masking operations. Therefore, GGI joining can provide a highly thermally conductive interface infrastructure, it is not only a lower cost of ownership, but also demonstrates higher strength and conductance, and a greater flexibility than conventional flip chip methods.

The Gold–gold joining has been applied for high-brightness light-emitting diode (LED) construction, and complementary metal oxide semiconductor (CMOS) image sensors and other areas. Traditionally, LEDs used wire bond processes by changing to the flip chip GGI attachment method, high thermal conductivity, and low electrical resistance of the GGI are superior to solder bump flip chip. GGI flip chip bonding technology has been developed to bond the integrated circuit (IC) drive chip on the IC suspension used in hard disk drives. With the IC suspension design, it becomes possible to assemble the IC drive chip close next to the magneto-resistive head slider on the suspension (Luk et al. 2002).

Gold stud with conductive adhesive assembly is another emerging thermal interface infrastructure for flip chip interconnection. Originally found principally in low bump count and low-cost applications such as smart cards, adhesive assembly has now moved upscale to applications such as implanted medical devices and large imaging (infrared and X-ray/gamma-ray) detector arrays. While indisputably lead free, many other advantages of adhesive over solder bump flip chip assembly clearly foretell a wider role for adhesive flip chip assembly in tomorrow's demanding products. It is not only lead that vanishes with adhesive assembly. Because adhesives require no fluxing, the postcleaning required to remove flux residues also vanishes. The problems that flux residues cause in underfill delamination and in long-term reliability similarly vanish with the fluxes. Adhesive assemblies emerge from the curing oven ready for underfill, if required. The more compliant gold bumps and adhesive can tolerate higher thermal expansion differences between chip and substrate without underfill. Large arrayed silicon detectors with thousands of connections often do not require any underfill, since they are assembled to thermally matched silicon readout chips. Adhesives are suitable for extremely high density interconnects, presently dominated by indium bump bonding, also known as hybridization. Hybridization requires indium bumps on both mating surfaces, cold-welded by applying high pressure. Adhesive assembly arrays with bump pitches as low as 50 μm are much less costly than indium hybridization. Common adhesive cure temperatures of 150°C and below are far lower than today's eutectic solder, and far lower than candidate lead-free solders. For temperature-sensitive materials such as compound semiconductors or polymer-based devices such as pyroelectric detectors, adhesive cure temperatures may be less than 80°C. Special low curing temperature adhesive formulations allow room temperature curing when needed. Gold stud bumps commonly are combined with adhesive attachment. Although direct gold-to-gold interconnection, with no adhesive, is a reliable flip chip technique, the required high thermocompression forces limit it to lower bump counts. In contrast, gold stud bumps placed into conductive adhesive bumps require very low normal forces, making this method suitable for large chips with thousands of connections. Stud bumps can be placed directly on an IC's standard aluminum bond pads with no special preparation, although a light plasma cleaning may be used to remove any contamination on the pads. The costly and complicated multistep process of under bump metallization required for solder bumping is eliminated. Whole wafers, partial wafers, or even individual diced chips can be easily gold stud bumped, eliminating the dependency on processing full wafers which is typical of most solder bumping processes (Clayton 2005).

Adhesive attachment may relax the process window for coplanarity. The adhesive itself compensates for small variations in stud bump height. The adhesive also can be used to identify bumps that may be excessively low by visually inspecting for adequate epoxy transfer onto all bumps before curing the adhesive. If high pin count die makes sequential stud bumping of an entire wafer unacceptably slow, electroplated gold bumps offer an alternative. The main advantages of electroplated bumps are the bumping time per wafer, and a higher overall bump height if thick photoresist is used. The isotropic conductive adhesives (ICA) commonly combined with gold stud bumps are epoxies loaded with up to 80% silver flakes, ranging from 5 to 30 μm in diameter. The flakes tend to lie atop one another like a pile of wet leaves. Cured ICA has the same electrical conductivity in all directions (isotropic). More expensive gold, palladium, or platinum flakes are used in those applications, such as implanted medical devices, which are not compatible with silver. Carbon nanotubes (CNTs) currently are being tested as a potential alternative to silver flakes, although at present, they are rather expensive. In addition, nanotubes have a strong tendency to agglomerate, preventing a uniform distribution of the conductors when mixed into a resin matrix. In summary, the emerging trend toward a gold-bump/adhesive system brings with it a demonstrated, well-tested solution for low-temperature, flux free, lead-free, flip chip assembly. However, beyond those obvious advantages, the low-temperature assembly and manufacturing flexibility

Fig. 8.21 Gold nanolawn for electronic interconnection (Bock 2009)

of gold bump adhesive assembly link it closely to the latest trends in semiconductor and sensor device development. Looking toward a future world of plastic ICs, polymer memory chips, and a variety of sensors fashioned from compound semiconductors, gold bump conductive adhesive may become the only practical, economical choice for low temperature flip chip assembly (Clayton 2005).

Gold nanolawn is a completely new interconnection technique, as shown in Figure 8.21, which promises substantially higher integration densities using techniques such as flip chip. The nanolawn is produced by means of nanoscaled lithography, through-mask electroplating on bond pads, and cold welding. Daisy chains and Kelvin structures were used to characterize the interconnections electrically. The results showed that the resistance of a single $25 \times 25\ \mu m$ connection is about 1 m$\Omega$. Daisy chains with 360 elements confirmed high reliability and excellent electrical properties (Bock 2009).

## Organic Thermal Interface Materials

If the metallic TIMs are mainly suitable to TIM1 applications, organic TIMs would mostly be used for TIM2 application, although some metallic or organic TIMs have been developed for both TIM1 and TIM2 applications (Mahajan et al. 2002). With advances in packaging technology, the proper selection of a TIM has become more complicated as each application will have different design requirements. The final selection will encompass material properties and process considerations beyond the thermal performance of the material. Compliance, thickness, processing, and response to reliability conditioning are key considerations when engineering the final package. To conform the diversified requirements of advanced processor and packaging technology, different formats of organic TIM have been developed and commercialized such as various thermally conductive elastomers, thermal greases, PCM, and PSH materials. Despite the maturity of the organic TIM market, there continues to be a need for higher performance and value-added materials. The performance requirement for a TIM encompasses more than thermal performance but a total solution: low thermal resistance across a bond line, high thermal conductivity, reworkability, sufficient thickness to accommodate nonplanarity, compliance for thermal mismatch, preapplication to heat spreader (TIM1) or a heat sink (TIM2) for speed of assembly, and low cost. For instance, both grease and PCMs possess low thermal resistance due to their liquid nature at operating temperature and thin bond line, however, each possess drawbacks. The reliability and messy nature of grease is a concern, whereas the reworkability and low thermal conductivity for PCMs are drawbacks. Pad and film formats offer many advantages from a processing and packaging perspective, but traditionally, their performance is not as high as grease or PCM. There exists a strong need for high-performance organic TIMs as the strength of those TIMs makes them useful for both TIM2 and TIM1 applications. The emerging high performance organic TIMs include PSH and advanced thermally conductive elastomers. This section will give a review on the status and future trend of various organic TIMs.

## *Thermally Conductive Elastomer Materials*

Thermally conductive elastomers are elastomeric materials which contain a thermally conductive fillers. Depending on loaded fillers, the thermally conductive elastomers can mainly be classified into insulated TIM and noninsulated TIM. Insulated TIM is usually loaded with thermally conductive ceramic fillers. They are primarily used in electronics applications in instances where good thermal conduction and electrical insulation are needed in the same material. For example, when the material is used as TIM2, thermally conductive elastomer material is used as a thermally conductive, electrically insulating interface between the electronic component and the metal heat sinks. Because of their high electrical conductivity, metal heat sinks cannot be in direct contact with electronic components. The thermal conductivity of thermally conductive elastomers is generally much less than that of the metal heat sink. Thus, the thermal conductivity of the thermally conductive elastomer limits the overall ability of heat dissipation.

Noninsulated TIMs are loaded with high thermal conductivity metallic or graphite fillers and aim to replace some solders for electronic assembly. These metal-loaded elastomers pay a penalty in viscosity and adhesive quality to incorporate the higher conductivity materials. As discussed previously, efforts to create high conductivity metallic TIMs have been progressing with some success. These metallic TIMs require fundamentally different application processes and can sometimes strain thermal budgets for the attached devices. As a result, there is still a vigorous effort to improve the filled polymer systems (Mahajan et al. 2002).

Typical thermally conductive elastomer contains ceramic filler such as BN or alumina in an elastomer matrix. The elastomers used are usually urethane or silicone based. While these materials are adequate in many instances, there is a constant demand for thermally conductive elastomers with improved thermal conductivity and electrical insulating properties. For electrical insulating TIM, the thermally conductive electrical insulating filler may be any conventional filler such as boron nitride, alumina, magnesia, or aluminum nitride. Hexagonal BN is the preferred filler. The proportion of filler in the composition is preferably about 45–70 vol%. For BN fillers, the filler preferably forms about 65–80 wt% of the composition. Any conventional particle size may be used. Preferably, the filler particle size is about 5–200 μm, and the filler will contain a mixture of coarse (100–200 μm) and fine (10–50 μm) particle sizes. High bulk density fillers generally provide better thermal conductivity. For alumina fillers, any conventional α-alumina can be used, however, tabular alumina is preferred. The elastomer may be any thermoplastic elastomer, preferably a block copolymer. Preferred block copolymers are styrenic copolymers and olefin copolymers. Particularly preferred are styrene-ethylene-butylenestyrene block copolymers. Examples of these block copolymer elastomers are Kraton® G-1652-M and Kraton® G-657-MX sold by the Shell Chemical Co. These Kraton® elastomers have styrene/rubber ratios of 29/71 and 13/87, respectively. If desired, a cross-linkable elastomer such as Shell Chemical's Kraton® 1901-X or RP-6501 may be used as the elastomer. Cross-linkable elastomers have functional groups such as carboxyl groups or ethoxy silanol groups which are reacted to cross link the elastomer molecules. Any suitable cross-linking agent and a cross-linking catalyst may be combined with the cross-linkable elastomer to bring about the cross linking. Cymel® 303 (hexamethoxymethylmelamine) sold by American Cyanamid Co., is a preferred cross-linking agent for Kraton® 1901-X; Cycat® 600 (aromatic sulfonic acid) sold by American Cyanamid Co., is a preferred cross-linking catalyst for Kraton® 1901-X; Dibutyl tin dilaurate is a preferred cross-linking catalyst for Kraton® RP-6501. The relative proportions of cross-linkable elastomer, cross-linking agent, and cross-linking catalyst may vary depending on the compounds used and the degree of cross linking desired. For hexagonal BN fillers, the filled elastomers preferably have a thermal conductivity of at least about $12 \times 10^{-3}$ cal/s cm K, more preferably at least about $15 \times 10^{-3}$ cal/s cm K. The filled elastomers may be made in any conventional form such as blocks, sheets, or films using any conventional method. A preferred method for making the filled elastomers employs a solvent-containing precursor composition. The elastomer is dissolved in an appropriate solvent to form a solution. The amount of solvent can be any amount which effectively dissolves all the elastomer. Typically, the solvent forms about 50–85 wt% of the elastomer-solvent solution. The thermally conductive filler is then added to form the precursor composition. The composition is dried to remove the solvent. The composition is shaped before and/or after the drying step. Before drying, the composition may be cast in a mold or as a film or sheet. After drying, the composition is pressed at elevated temperatures. The drying temperature may vary depending on the solvent and elastomer involved as well as the drying time. The drying is performed for about an hour at about 70–120°C. The pressing is performed at 500–10,000 psi, more preferably about 1,500 psi. The pressing temperature is about 150–175°C. If a cross-linkable elastomer is used, a cross-linking agent and catalyst would be added to the precursor mixture. The solvent may be any suitable solvent. Toluene is a preferred solvent. The amount of solvent used is the minimum needed to form a workable precursor composition (Block et al. 1993).

The typical format of thermally conductive elastomers are silicone elastomer pads filled with thermally conductive ceramic particles, often reinforced with woven glass fiber or dielectric film for added strength. These elastomers are available in thickness from about 0.1–5 mm and hardness from 5 to 85 Shore A. Unlike compounds and greases, elastomer pads provide electrical insulation and can be used between surfaces that are at different electrical potential. They are typically used under discrete power devices where electrical isolation is required.

Elastomers do not flow freely like greases or compounds, but will deform if sufficient compressive load is applied to conform to surface irregularities. At low pressures, the elastomer cannot fill the voids between the surfaces and the thermal interface resistance is high. For example, when the thermal conductivity of highly conductive elastomer reaches 5 W/m K ($R_{c\text{-}s}$ is 1.8 °C/W), twice that of grease, inspection of the disassembled interface after the test showed that there was less than 30% contact between the material and the two surfaces. As pressure is increased, more of the microscopic voids are filled by the elastomer and the thermal resistance decreases. For most high durometer materials, mounting pressures around 300–500 psi eliminate the interstitial voids and reduce interface resistance to a minimum. Mounting pressure must be permanently maintained by using fasteners or springs to hold the two surfaces together (deSorgo 1996).

As one of elastermic noninsulating TIMs, thermally conductive adhesive tapes are double-sided pressure sensitive adhesive films filled with sufficient ceramic powder to balance their thermal and adhesive properties. The adhesive tape is usually supported either with an aluminum foil or a polyimide film for strength and ease of handling. Polyimide support also provides electrical insulation. Adhesive tapes perform much like the elastomeric films in that they also require some initial mating pressure to conform to irregularities in the mating surfaces. They are also unable to fill large gaps between nonflat surfaces. However, once the joint is formed, the adhesive tapes require no mechanical support to maintain the mechanical or thermal integrity of the interface. Adhesive tapes provide convenience in attaching a heat sink to a semiconductor package because, unlike liquid adhesives, no cure time is required. The film is applied to one of the surfaces, usually to the heat sink, and it is then forced into contact with the semiconductor package to complete the thermal joint. The application pressure is typically 10–50 psi for a few seconds' duration. The bond thus formed can be considered permanent and the heat sink is reliably attached to the semiconductor.

However, this convenience comes at a price. $R_{c\text{-}s}$ for tapes is only slightly better than a dry joint. This is because the thermal tapes do not fill gaps as well as liquids, and thermal joints made with tapes will normally include considerable interstitial air gaps. For the most part, the quality of the two joining surfaces will determine the amount of contact that can be achieved and the thermal performance that can be expected. The high shear strength of these thermal tapes means that reliable joints between heat sinks and semiconductors can be achieved, even with poor surfaces and no mechanical fasteners (deSorgo 1996).

Comparably, conductive elastomeric thermal interface pads can be compressed between the heat generator and the heat receiver. Pads are either electrically conductive or nonconductive, based on application. Pads are generally thicker than tapes and adhesives because the added thickness helps to fill larger gaps.

## *Thermal Grease and Compound*

Thermal conductive greases, usually with a form of paste, are typically made up of silicone or hydrocarbon oils with a heavily loaded suspension of thermally conductive ceramic or metal fillers. The emulsion fillers increase thermal capacity and provide body to minimize flow out of the interface. The actual bulk thermal conductivity of the resulting material is low; however, because the paste-like consistency of the thermal grease allows the final bond line thickness to be very thin, the resulting thermal resistance across the interface can be quite low. As a two-part suspension, thermal greases can separate to form voids and dry out over time, especially in service with high temperatures or extensive thermal cycling. Therefore, the paste-like nature can lead to inconsistent and messy application issuers (deSorgo 1996).

Sufficient grease is applied to one of the mating surfaces such that when pressed against the other surface, the grease flows into all voids to eliminate the interstitial air. Excess grease flows out past the edges and the thinnest possible thermal joint is formed as the two surfaces come into contact at their high points. Joint integrity must be maintained with spring clips or mounting hardware.

Thermal greases are notoriously user unfriendly, but provide very low thermal resistance between reasonably flat surfaces. Grease does not provide electrical insulation between the two surfaces, and excess grease that flows from the joint should be cleaned up to prevent contamination problems. Greased joints can also dry out with time, resulting in increased thermal resistance. The effect of thermal grease and other TIMs on heat transfer through an interface can be seen in Table 8.5. These data were generated using a P54C Pentium Thermal Test Die powered to 6 W in an AMP Socket 5. A Wakefield 799–80 AB pin fin heat sink was attached with a spring clip. Without an interface material, $R_{c\text{-}s}$ was 2.9°C/W. The addition of thermal grease to the joint reduced the resistance to 0.9°C/W. Inspection of the joint after testing showed that the grease covered over 90% of the area and the thickness was less than 0.07 mm (deSorgo 1996).

**Table 8.5** Case-sink and junction ambient thermal resistance for various interface materials (deSorgo 1996)

| Thermal interface joint | Thickness in/mm | Thermal conductivity (W/m K) | Thermal resistance $R_{c\text{-}s}$ (°C/W) | Thermal resistance $R_{j\text{-}a}$ (°C/W) |
|---|---|---|---|---|
| Dry joint | N/A | N/A | 2.9 | 9.9 |
| Thermal grease | 0.003/0.076 | 0.7 | 0.9 | 8.1 |
| Thermal compound | 0.005/0.127 | 1.2 | 0.8 | 7.9 |
| Conductive elastomer | 0.01/0.254 | 5.0 | 1.8 | 8.9 |
| Adhesive film | 0.009/0.229 | 0.7 | 2.7 | 9.6 |

Thermally conductive compounds are an improvement on thermal grease as these compounds are converted to a cured rubber film after application at the thermal interface. Initially, these compounds flow as freely as grease to eliminate the air voids and reduce the thermal resistance of the interface. After the interface is formed, the compounds cure with heat to a rubbery state and also develop secondary properties such as adhesion. Formulations with adhesive properties do not require mechanical fasteners to maintain the integrity of the joint. Because the binder cures to a rubber, these compounds do not have the migration or the dry joint problems associated with thermal greases. Compounds can be used to fill large gaps where greases would bleed from the joint on account of their migratory nature. Clean up is also simple as excess material is easily removed after it has been cured to a rubber. The thermal performance of a typical compound is shown in Table 8.5. Because the compound behaves like a grease, the joint is nearly void free and the improvement in thermal resistance over a dry joint is similar to that of grease, 0.8 vs. 2.9°C/W (deSorgo 1996).

In addition, thermal conductive gels are similar to thermal grease with a new difference. The gel material is cured to form cross-linked polymer chains which provides lateral stability to minimize the problem associated with liquid TIM materials. Gels are reusable, however, they generally have a slightly lower thermal conductivity than thermal grease.

## *Phase Change Materials*

PCM is a substance with a high heat of fusion and capable of absorbing and releasing heat when the material changes from solid to liquid and vice versa. In fact, latent heat storage of PCMs can be achieved through solid–solid, solid–liquid, solid–gas and liquid–gas phase change. However, the only phase change used for PCMs is the solid–liquid change. Liquid–gas phase changes are not practical for use as thermal storage due to the large volumes or high pressures required to store the materials when in their gas phase. Liquid–gas transitions do have a higher heat of transformation than solid–liquid transitions. Solid–solid phase changes are typically very slow and have a rather low heat of transformation (Wikipedia 2008).

When PCMs reach the melting temperature at which they change phase, they absorb large amounts of heat at an almost constant temperature. The PCM continues to absorb heat without a significant raise in temperature until all the material is transformed to the liquid phase. When the ambient temperature around a liquid material falls, the PCM solidifies, releasing its stored latent heat. A large number of PCMs are available in any required temperature range from -5 up to 190°C. Typical PCMs include organic [paraffin ($C_nH_{2n+2}$) and fatty acids ($CH_3(CH_2)_{2n}COOH$)], inorganic [salt hydrates ($M_nH_2O$) and eutectic materials (organic–organic, organic–inorganic, inorganic–inorganic compounds). The advantages of organic PCMs include availability in a large temperature range, freezing without much super cooling, ability to melt congruently, self nucleating properties, compatibility with conventional material of construction, no segregation, chemical stability, high heat of fusion, safe and nonreactive, and recyclable. Their major disadvantages are low thermal conductivity in their solid state, high heat transfer rates are required during the freezing cycle, volumetric latent heat storage capacity is low, and they are flammable. Due to cost considerations, only technical grade paraffins may be used which are essentially a paraffin mixture and are completely refined of oil (Wikipedia 2008).

The characteristics of inorganic salt hydrates ($M_nH_2O$): high volumetric latent heat storage capacity, low cost and easy availability, sharp melting point, high thermal conductivity, high heat of fusion, low volume change, nonflammable, change of volume is very high, super cooling, and nucleating agents are needed and they often become inoperative after repeated cycling.

Eutectics have a sharp melting point similar to pure substance. Volumetric storage density is slightly above organic compounds. Only limited data are available on thermophysical properties as the use of these materials are very new to thermal storage application.

Thermal-composites have been developed by combinations of PCMs and other (usually solid) structures. A simple example is a copper mesh immersed in a paraffin wax. The copper mesh within paraffin wax can be considered a composite material, dubbed a thermal-composite. Such hybrid materials are created to achieve specific overall or bulk properties. Thermal conductivity is a common property which is targeted for maximization by creating thermal composites. In this case the basic idea is to increase thermal conductivity by adding a highly conducting solid (such as the copper-mesh) into the relatively low conducting PCM thus increasing overall or bulk (thermal) conductivity. If the PCM is required to flow, the solid must be porous, such as a mesh. Solid composites such as fiberglass or kevlar-pre-preg for the aerospace industry usually refer to a fiber (the kevlar or the glass) and a matrix (the glue which solidifies to hold fibers and provide compressive strength). A thermal composite is not so clearly defined, but could similarly refer to a matrix (solid) and the PCM which is of course usually liquid and/or solid depending on conditions (Wikipedia 2008).

In order to use for TIMs in electronic packaging, the PCM should possess:

1. Thermodynamic properties
   (a) Melting temperature in the desired operating temperature range
   (b) High latent heat of fusion per unit volume
   (c) High specific heat, relatively high density and high thermal conductivity
   (d) Small volume changes on phase transformation and small vapor pressure at operating temperatures to reduce the containment problem
   (e) Congruent melting
2. Kinetic properties
   (a) High nucleation rate to avoid super cooling of the liquid phase
   (b) High rate of crystal growth, so that the system can meet demands of heat recovery from the storage system
3. Chemical properties
   (a) Chemical stability
   (b) Complete reversible freeze/melt cycle
   (c) No degradation after a large number of freeze/melt cycle
   (d) Noncorrosive, nontoxic, nonflammable, and nonexplosive materials
4. Economic properties
   (a) Low cost
   (b) Large-scale availabilities

Based on this baseline, some special form of phase change TIMs have been developed for electronic packaging applications. The solid–liquid phase change is used for TIMs because a liquid is associated with high conformability, which is needed for minimizing interfacial air pockets, thereby reducing interface thermal resistance. The PCM should melt at a temperature slightly above room temperature (e.g., 45°C) so that the material is a liquid while it functions as a TIM. The liquid is attractive in that it is conformable, but it is disadvantageous in the possibility of seepage and consequent contamination of the surrounding electronic components. By having the interface material be a solid during operation of the electronics at temperatures below the melting temperature, the seepage problem is alleviated. Furthermore, the absorption of the latent heat of fusion during melting provides an additional mechanism of heat dissipation. Both the organic (such as paraffin wax) and inorganic ones (most commonly metal salt hydrates, e.g., disodium hydrogen phosphate dodecahydrate) can be used for TIMs. The organic materials are attractive in their low reactivity, stability in the phase change characteristics under thermal cycling and low supercooling, but they tend to be poor in thermal conductivity. The inorganic materials are attractive in their relatively high thermal conductivity, but they suffer from relatively high reactivity, high supercooling, and poor stability in the phase change characteristics under thermal cycling. For application as TIMs, the organic PCMs are preferred, such as paraffin wax or silicon-based TIMs. The thermal conductivity of a PCM can be increased by using a filler (particles, fibers, or bars) that is thermally conductive. The filler does not melt, but its presence can affect the phase change characteristics, including the melting temperature and the heat of fusion. Due to the low thermal conductivity of the organic PCMs, the use of thermally conductive filler is important. An alternate method of increasing the thermal conductivity involves impregnating porous graphite with paraffin wax, but this method suffers from the inability of the resulting composite to conform to the topography of the surface from which it is to absorb heat. Furthermore, the volume fraction of the component that undergoes phase change is limited. The choice of the organic matrix of a PCM for use as a TIM depends on the melting temperature, the conformability in the molten state, the extent of undercooling during solidification, the latent heat of fusion, and the ability to withstand elevated temperatures. The conformability strongly affects the effectiveness as a TIM, but it is an attribute that is difficult to measure. For instance, carbon black can be a secondary filler used in the TIM because of its exceptional conformability, which is a consequence of its being in the form of porous agglomerates of nanoparticles. Due to its conformability, carbon black is even more effective as a filler in thermal pastes than highly conductive fillers when the mating surfaces are sufficiently smooth (such as 0.05 μm).

A polycaprolactone diol (molecular weight = 2,000 amu) filled with hexagonal BN particles (4 vol%) is an effective phase-change TIM. It exhibits a melting onset temperature of 37°C and a solidification onset temperature of 34°C. The thermal contact conductance across copper mating surfaces (15 μm roughness) is $8 \times 10^4$ $W/m^2$ °C at 70°C and a pressure of 0.69 MPa. The combination of high conductance and high resistance to elevated temperatures makes this material superior as a phase-change TIM to BN-filled paraffin wax, polyvinyl ether, polytetramethylene ether glycol, tetradecanol, lauric acid, polyester diol, polyester triol, or polycaprolactone diol (molecular weight = 1,000 amu). However, the heat of fusion is lower for the polyols than many other organic materials. The conductance attained by BN-filled diols is higher than that attained by commercial phase-change TIMs. BN is more effective as a filler than carbon black for providing high conductance, but carbon black reduces the heat of fusion by a smaller amount than BN does (Liu and Chung 2006).

While novel PCMs are continuously emerging, commercial PCMs provide a combination of grease-like thermal performance with pad-like handling and installation convenience. They have compositions that transform from a solid at room temperature to a mixture of solid and liquid phases at various operating temperatures. The liquid from the mixture makes intimate contact with the contact surfaces while the solid retains the integrity of the gap. At this operating temperature, the PCM is able to act like thermal grease, allowing it to form a thin bond line. Sometimes the installation process requires some compressive force to bring the two surfaces together and cause the material to flow until the two surfaces come into contact at a minimum three points, or the joint becomes so thin that the viscosity of the material prevents further flow. Excess liquid should then flow to the perimeter of the interface and solidify so it remains away from the other components. Solid at room temperature, PCMs are melted (i.e., undergo a phase change) as the temperature rises to the 40–70°C range. This makes the material (0.13 mm thick in its dry film form) as easy to handle as a pad, while assuring that it will, when subjected to heat during the assembly process, flow into voids between mating surfaces as effectively as thermal grease. Ordinarily, applying power to the electronic component introduces the needed heat for the phase change to occur, establishing a stable thermal joint. These materials consist of organic binders (i.e., a polymer and a low-melt-point crystalline component, such as a wax), thermally conductive ceramic fillers, and, if necessary, a supporting substrate, such as aluminum foil or woven glass mesh (deSorgo 1996).

Typical phase-change TIMs have the following characteristics:

(1) Compressible phase-change interface materials that compensates for large area and extreme height tolerances and gaps.
(2) In situ curing thermal interface adhesive materials that provides up to 300 psi as applied and cure to over 600–1,200 psi over time at operating temperatures of the device.
(3) Available as electrically conductive or electrically insulating thermal adhesives, thermal greases, thermal gels, and thermal pads.
(4) Available as dry, single-sided, or double-sided pressure sensitive tacky thermal interface pads and thermal adhesives film tapes.
(5) Available in preforms of thermal PCM pads and thermal adhesive films and tapes.
(6) Phase-change thermal interface pad materials that flow at 55, 90, 130, and 180°C respectively.
(7) PCM of all thickness with outstanding thermal conductivitity and low thermal interface resistance to help cool and manage power devices as adhesives, greases, gels, and gap filling compressible thermal pads.

In summary, PCMs provide a combination of grease-like thermal performance with pad-like handling and installation convenience. They have compositions that transform from a solid at room temperature to a mixture of solid and liquid phases at various operating temperatures. The liquid from the mixture makes intimate contact with the contact surfaces while the solid retains the integrity of the gap. At this operating temperature, the PCM is able to act like thermal grease, allowing it to form a thin bond line. Sometimes the installation process requires some compressive force to bring the two surfaces together and cause the material to flow, until the two surfaces come into contact at a minimum three points, or the joint becomes so thin that the viscosity of the material prevents further flow. Excess liquid should then flow to the perimeter of the interface and solidify so it remains away from other components.

## *Polymer Solder Hybrid Materials*

As one special form of the PCMs, PSHs provide superior long-term reliability performance, and exhibit the lowest thermal impedance of the phase-change family. For optimum performance, the material must be exposed to temperatures above the melting point of contained solder(s) during operation or by a burn-in cycle to achieve lowest thermal impedance and highest thermal performance. Upon reaching the required burn in temperature, the material will fully change phase and attain minimum bond-line thickness less than 0.001 in. or 0.0254 mm and maximum surface wetting. They are typically used in microprocessors, graphics processors, chipsets, memory modules, power modules, and power semiconductors. Material may flow when oriented vertically, especially at higher temperatures. This does not affect thermal performance, but should be considered if appearance is important (Chomerics 2009).

Conventionally, as shown in Figure 8.22, PSHs comprise a siloxane-based polymer and a solder such as indium–silver, indium–tin, tin–bismuth, gallium–tin–indium, gallium–indium–tin–zinc, indium–bismuth, gold–tin or other solder material. Additionally, thermally conductive fillers such as aluminum, silver, copper, or graphite may also be added to the PSH TIM. The combination of conductive filler and solder allows for the formation of thermally conductive pathways or chains throughout the PSH TIM. In this manner, heat generated within the die may be readily transferable across such chains to the attached thermal management device for heat dissipation.

In order to form the indicated chains and properly secure the thermal management device to the die, the PSH TIM is heated by a reflow process to at least the melting point of the solder. 183°C is considered the industry standard dividing line between "low melt" and "high melt" solders. Where conventional solders, such as those mentioned above, are used, the PSH TIM may be a high melt solder requiring reflow at temperatures in excess of about 260°C, for the solder to melt and allow the filler to diffuse there into. For example, a temperature of 260°C may be required where a gold–tin solder is used in the PSH TIM. However, even where other

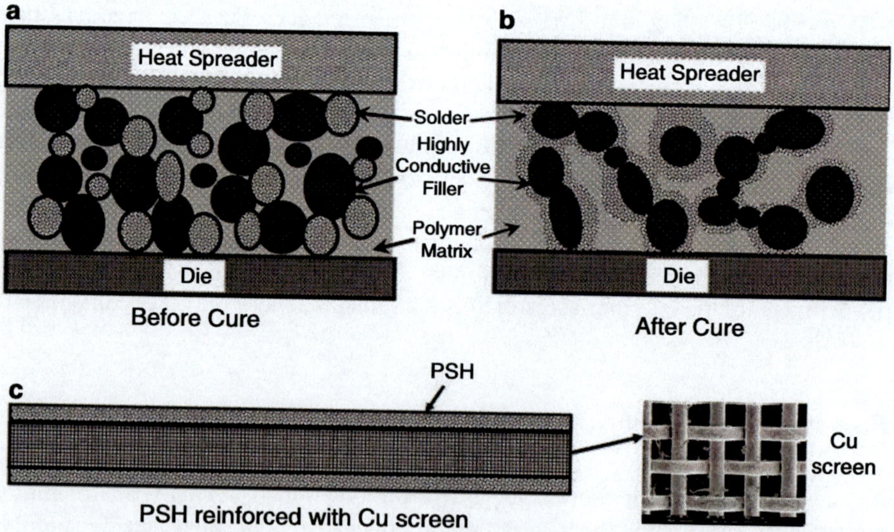


**Fig. 8.22** Illustration of polymer solder hybrid (PSH) thermal interface material **(a)** Before cure; **(b)** After cure, a continuous highly conductive network is formed in the polymer matrix for heat conduction, benefiting of solder conductivity without the compliance issues; **(c)** The thermal conductivity of PSH can be enhanced with Cu screen

conventional solders are used, it is not uncommon for a temperature in excess of about 183°C to be employed.

For TIM1 application, the design target for the PSH performance requirement usually includes: (1) Thermal resistance of the PSH between values of indium and grease: TI < 0.10 C $cm^2/W$, preferably <0.05 °C $cm^2/W$ at BLT 2 mil. (2) Dispensable paste format-syringe dispensable. (3) Pot life > 8 h (full shift). (4) Reliability to meet or exceed industry standards–thermal cycling, HAST, high temperature aging. (5) Solder partially melts at less than 100°C for self healing of microcracks at device operating temperature, and for increasing conductivity at hotspots. A PSH TIM may be used to secure a thermal management device to a temperature-sensitive assembly. For example, larger mobile or bare die assemblies incorporated into an already assembled computing box, such as a laptop, may require use of a PSH TIM. Such assemblies may include a die requiring attachment of a heat management device such as a heat pipe. These assemblies include more than just a die or individual semiconductor package. Rather, these assemblies are closer to being a completed computer, such as the indicated laptop, with added temperature sensitive features.

For temperature-sensitive assemblies, low melt solders are available having melting points lower than about 183°C. However, a PSH to serve as a TIM that employs a low melt solder tends to be highly susceptible to degradation when subjected to standard industry testing. Industry tests may include a HAST which includes PSH exposure to more than 100°C (e.g., about 130°C) with relative humidity of more than 50% (e.g., about 85%) for an excess of about 75 h (e.g., about 100 h), or a standard bake test which includes PSH exposure to even higher temperature (e.g., 150°C for about 100 h) without degradation. Unfortunately, a PSH employing a low melt solder tends to degrade in between about 24 and 48 h when subjected to such tests (Koning and Hua 2004).

While a low melt solder, such as indium–tin–bismuth, would normally degrade or lack the robustness necessary to remain intact when subjected to a HAST, the addition of some fillers prevents such an occurrence. These fillers include finely distributed copper, gold, or silver particles or copper-, gold-, or silver-coated particles. The fillers provide a sufficient degree of robustness to the PSH TIM following reflow, such that the PSH TIM remains intact following exposure to conditions such as HAST conditions (Koning and Hua 2004).

The polymer matrix may be a material that can be applied as a paste such as a dispensable syringe or by screen printing. The polymer matrix may also act as an adhesive to bond the two mating parts together. The nonfusible particles, such as most metals, benefit from a high thermal conductivity, however, a thermal flow path through the TIM is limited by the point-to-point contact of the particles as shown in Figure 8.22. Nonfusible particles refer to particles that will not melt and flow during packaging assembly process, reliability testing, and product operation and so remain as point contacts with each other. This provides thermal conductivity through the TIM that is limited to point-to-point percolation, resulting in a thermal bottleneck through the nonfusible particles (Jayaraman et al. 2007).

The phenomenon of percolation describes the effects of interconnections present in a random system, here the number of filler particles that are randomly in point contact with each other to allow thermal conduction. Normally, to improve conduction limited by percolation, the amount of filler could be increased until a threshold amount is reached and heat conduction, due to the filler, transitions to a sufficiently high value. The volume fraction of filler required to reach this transition level may be too high and can overpower the properties desired from the polymer binder such as low-contact resistance. Another problem is that for some metal particles in contact with some polymer binders, the bare particle filler can poison the polymer cure such as by hindering or blocking the curing agent (Koning et al. 2006).

To address these concerns, a PSH TIM has been developed that includes fusible particles as well as filler particles in a silicone polymer matrix material. The fusible particles melt during the assembly process and can therefore wet the filler particles or self coalesce. Thereby the average particle size grows creating long continuous heat transfer pathways that alleviate the thermal bottleneck of percolation. The fusible particles, e.g., solder material, may be materials such as solder-like materials that melt below approximately 300°C. Solder materials include pure metals, such as indium and tin, alloys of several metals such as In, Cu, Ag, Sn, Bi, Pb, Cd, Zn, Ga, In, Te, Hg, Tl, Sb, Se, Po, or mixtures of any two or more thereof, or another metal or alloy having a melting point lower than that of the metal powder in component. The filler particles may be nonfusible materials with melting points well above 300°C, such as aluminum at 660°C, silver at 961°C, copper at 1,084°C, gold at 1,064°C, etc. Figure 8.22c shows that copper (Cu) mesh can be a effective high thermal conductivity reinforcement for PSHs with very thin BLT PSHs on either side of the copper mesh, providing excellent thermal conductivity and extremely low thermal resistance (0.05°C $cm^2/W$ or below). In addition, Cu mesh can be a carrier for creating extremely thin BLT PSHs.

Silicone exhibits certain characteristics (e.g., low glass transition temperature and low moisture absorbency) that make it suitable as a binder matrix for PSH TIMs. Its low glass transition temperature is approximately 25°C and low moisture absorbency is approximately 1% or less by weight. Other materials may exhibit such characteristics and therefore may likewise be suitable as a binder matrix for PSH TIMs. Such materials may provide better adhesion and lower contact resistance than silicones (Jayaraman et al. 2007). Other polymers usable in the curable matrix include siloxanes, olefins, and epoxies. Polymeric resins usable in the PSH include any thermosetting resin (either monomeric or polymeric) that is cross linkable by the curing agent, a metal catalyst, or a hydroxyl group-bearing agent. Resins which meet this requirement include epoxies, phenolics, novalacs (both phenolic and creosotic), polyurethanes, polyimides, bismaleimides, maleimides, cyanate esters, polyvinyl alcohols, polyesters, polyureas, acrylics, polyamides, polyacrylates, polysiloxanes, cyanoacrylates, and the like. Other resin systems are modifiable to be cross linkable by the curing agent, a metal catalyst, or a hydroxyl group-bearing agent. Examples of such resins include acrylics, rubbers (butyl, nitrile, etc), polyamides, polyacrylates, polyethers, polysulfones, polyethylenes, polypropylenes, polysiloxanes, polyvinyl acetates/polyvinyl esters, polyolefins, cyanoacrylates, polystyrenes, and the like (Matayabas and Dani 2003).

Typical conventional PSH TIMs have the following characteristics:

(1) Storage–Store frozen at –40°C. Shelf life at –40°C is 1 year.
(2) Thaw–Allow to warm to room temperature prior to use.
(3) Pot life–8 h at room temperature (below 30°C).
(4) Dispense–Dispense enough material on the die to give a 50–75 μm thick bond line. 0.4 g of PSH per $cm^2$ of die area.
(5) Cure cycle–150°C for 30–45 min in a preheated oven at a pressure of 30 psi to hold the parts together during the cure.

## Graphite-Based Thermal Interface Material

Flexible graphite sheet materials have become an attractive TIM because of their excellent resilience and flexibility to conform well to the mating surfaces, relatively high thermal conductivity (>3 W/m K) in thickness direction, low thermal expansion, thermal stability, and chemical inertness (Luo et al. 2002).

In general, the flexible graphite is prepared from flakes of natural graphite that are intercalated in an acid solution. After the flakes are intercalated they are washed and dried and then exfoliated by exposure to a high temperature for a short period of time. This causes the flakes to expand or exfoliate in a direction perpendicular to the crystalline planes of the graphite. The exfoliated graphite flakes are vermiform in appearance and are therefore commonly referred to as worms. The worms may be compressed into sheets or foils with a density approaching theoretical density although a density of about 1.1 $g/cm^3$ is considered typical for most applications. In addition, for TIM application, the graphite is usually optimized by reducing ash content and choosing proper graphite flake source, flake size, treatment chemistry, expansion thermal conditions, and sheet density. The sheets of flexible graphite can be cut into any desired configuration to suit a particular TIM application (Smalc et al. 2003).

Flexible graphite sheets are porous and have a density that various between 0.60 and 1.40 $g/cm^3$, considerably lower than the theoretical density of graphite of 2.25 $g/cm^3$ because the pores are open and interconnected, it is possible to impregnate flexible graphite sheets with various materials to improve their performance for TIM application. In particular, polymeric materials, including mineral oils and synthetic oils, can be added to lower the thermal resistance of the TIMs. Flexible graphite TIMs are ideal for TIM2 application. The typical characteristics of the flexible graphite TIMs include light weight; high thermal conductivity 300–400 W/m K; high heat resistance (in nonoxidizing environment above 3,000°C); and excent EMI shielding efficiency. Materials with adhesive coatings are designed for use in applications requiring low contact resistance, high thermal conductivity at low clamping loads, and ease of application. One example is the TIMs that are manufactured from natural graphite and a polymer adhesive. A pressure sensitive adhesive (PSA) can be applied to one surface of the TIM to adhere the TIM pad to heat sinks, spreaders, etc. This kind of TIM is an excellent replacement for thermal grease PCMs with advantages, such as not separate, dry out or pump out, and excellent contact can be maintained for the life of the assembly (Luo et al. 2002).

## Advanced Thermal Interface Materials

In order to manage the escalated need for heat dissipation, advanced TIMs, which promote heat transfer within a package or from a package to a heat sinking device, have a strong need for improvement of thermal dissipation as a result the thermal resistance of interfaces, as a fraction of total junction to ambient resistance, has grown dramatically. Conventional TIMs call for filling a polymeric carrier (grease, PCM, silicone, etc.) with highly conductive particles to produce a material with low to moderate thermal conductivity. These materials work well when gaps between components can be minimized (e.g., bond line thickness of 0.001 in. or less). When gaps are larger, conduction of heat within the interface material (i.e., from particle to particle) dominates the interface thermal resistance and poor performance results. A new class of fiber reinforced interface materials have been developed which overcome this limitation. These materials feature an aligned array of conductive fibers. This structure allows one fiber to span the gap between the two surfaces, resulting in very high thermal performance, even for large gaps. Because the fibers have a high aspect ratio, the material has a high degree of mechanical compliance, making it ideal for application such as interfaces within multichip modules, mobile products, or gap filling applications where variations in interface thickness can be absorbed with little or no thermal penalty. The material has been shown to be very robust with respect to thermal and mechanical cycling, and will not pump out or dry out from the interface during actual use. Meanwhile, nanotechnology-based TIMs also caused great attention and would play a great role in next generation high performance electronic packaging.

### *Gelvet and Fiber-Reinforced Thermal Interface Materials*

Conventional polymer-based TIMs wet out surfaces quite well and provide low thermal contact resistance. To achieve a significant degree of thermal conductivity, however, the material must be highly loaded with thermally conductive particulate or flake. Loading in excess of 60–70 vol% conductive fillers such as alumina, silica, BN, graphite, silver, or other metals is common. Forcing particle to particle contact creates a heat transfer path and increases the thermal conductivity of the polymeric system to approximately 1 W/m K for typical systems and up to 4–8 W/m K for very specialized materials. These conventional interface materials are limited by relatively low thermal conductivity, as the resistance to heat transport generated by particle to particle contact limits the efficiency of the conductive filler used. An alternate to filling a carrier with flake or particulate material has been developed to use an aligned structure of thermally conductive fibers, commercially called Gelvet™. As shown in Figure 8.23, these fibers create a continuous thermal path from one surface to another, and hence give rise to a composite interface material with an inherently high conductivity. As this high conductivity can be achieved with a comparatively low volume fraction of conductive fibers, the TIM can be made very compliant, especially if fibers of suitable aspect ratio (fiber length/fiber diameter) are chosen. The fibers are held in place by some thin layer of adhesive and encapsulated in a gel or elastomeric material to prevent broken fibers from escaping the interface. The encapsulant material is chosen to wet the fibers and result in a tent-like structure at the free fiber tips which gives some degree of wetting out at the fiber tips, while not inhibiting the inherent compliance of the velvet-like fiber structure. Because the thermal conductivity of this structure does not depend on particle to particle contact, some degree of drying out of the encapsulant does not affect thermal performance. Also, this material is easily reworkable and will remain in place and not be subject to pumping out of material

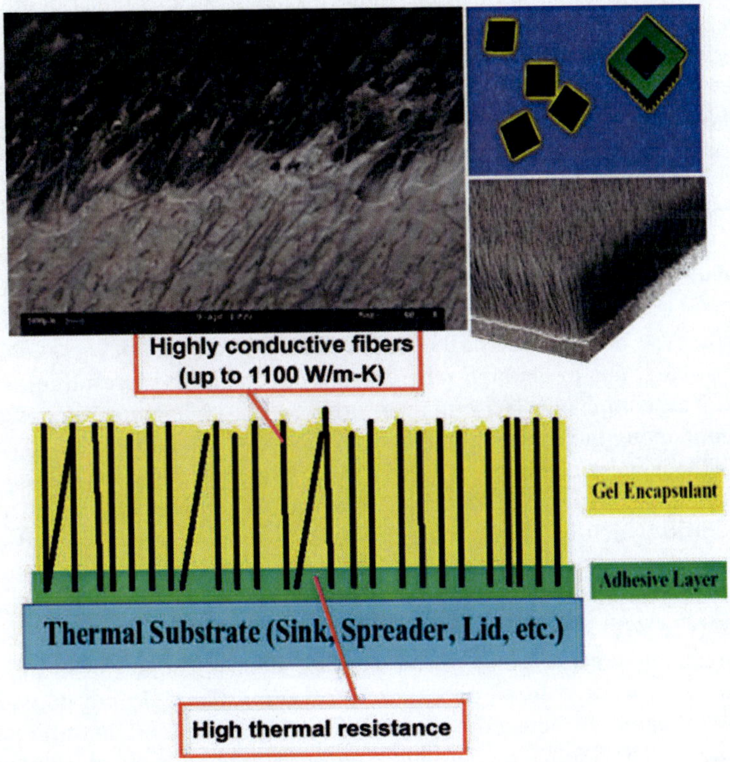


**Fig. 8.23** Structure of Gelvet thermal interface material

because all the conductive elements (fibers) are anchored in place. The materials set of choice for this type of interface material will be a high thermal conductivity fiber which has a small diameter, is readily available, and relatively inexpensive. In general, carbon or graphite fibers meet all of these characteristics. For example, 10 μm diameter carbon (graphite) fibers, with axial thermal conductivities of 800 and 1,100 W/m K have been used for the aligned fibers (Dean and Pinter 1999).

The key feature of an aligned fibrous interface material is its inherent high thermal conductivity. When the fibers are aligned in the direction of heat transport (i.e., spanning the interface gap), the theoretical thermal conductivity of the resulting material can be given by (Dean and Pinter 1999)

$$K_{\mathrm{TIM}} = k_{\mathrm{fiber}}A_{\mathrm{fiber}} + k_{\mathrm{polymer}}(1 - A_{\mathrm{fiber}}), \tag{8.14}$$

where $k$ represents thermal conductivity and $A$ the area fraction occupied by the aligned fibers. For a typical structure with 10% of the area occupied by carbon fibers having a conductivity of 1,100 W/m K, this gives a TIM thermal conductivity of approximately 110 W/m K. Actual measurements give a thermal conductivity of 50–90 W/m K for this material (Dean and Pinter 1999).

As the pressure increases, the thermal resistance can decrease dramatically. Benchmarking has indicated that the thinner versions of this material are able to perform on par with typical greases and PCMs for moderate bond line thicknesses and interface pressures. As the mating surfaces deviate from planarity, the effect an increased interface thickness does not significantly affect the thermal resistance of these aligned materials, while the thermal resistance of most other interface materials increases dramatically. This effect is due to the differences in bulk thermal conductivity between the aligned interface materials and conventional materials (Dean and Gettings 1998).

The aligned fibrous structure presents a very compliant interface compared with highly loaded structures. When the fiber aspect ratio is high, it takes very little force to bow, or elastically buckle, the fiber. Compliance of a material may be tailored by varying the type of fiber used or the length of fiber (material thickness), with little thermal penalty, due to the high inherent thermal conductivity resulting from this structure. This compliance is useful in absorbing stack-up tolerances for assemblies or accommodating die of differing height in multichip modules.

This TIM is produced in sheet form and may be preapplied to any component surface or shape. As it resembles a sheet or gasket and maintains this form, the component to which it is attached may be disassembled and reworked without necessitating replacement of the interface material. Numerous assembly cycles for a heat sink application and thousands of cycles for a burn-in application have been observed with little degradation to the interface material. This material has been demonstrated not to pump out in thermal, mechanical, or power cycling, as the fibers are anchored in an adhesive. The carbon fibers do not degrade with temperature increases, hence the properties of the material are not affected by a bake test for an assembly. Through appropriate choice of encapsulant and adhesive, low outgassing, residue, and high temperature applications may also be serviced.

In summary, the features of the Gelvet TIMs include:

- Highly oriented fibers
- Free tips contact mating surface
- Highly conductive graphite fibers (up to 1,100 W/m K) utilized to create a thermal pad with conductivity 6–20 times higher than conventional interface pads (composite $k = 30$–$100$ W/m K)
- Material may be attached directly to any substrate
- Aligned fiber structure results in a mechanically compliant material
- Thickness of pads may be tailored from 0.020 to 0.100 in.

There are some limitations to this type of material. The thickness cannot be reduced much below 0.015 in, as it becomes difficult to properly orient fibers that are shorter than this. As the carbon fibers used are very strong in tension and compression, but brittle in shear, this material does not exhibit good abrasion resistance and is not suitable for applications that require shear resistance. The temperature limits of the preferred adhesives and encapsulating material restrict their use to temperatures under 110°C, although substitutions for these materials may be made to extend the temperature range if a thermal penalty can be absorbed. Also, the selection of carbon fibers produces a material that is electrically conductive, which may be unattractive for some applications.

To modify and improve the performance of the gelvet TIM, the reduction of the thermal resistance at fiber-gel junction can be attempted with a chemical trimming process as shown in Figure 8.24 and aim to meet:

- A preferably orientalized fiber structure can be utilized to create a thermal interface pad able to used for TIM1 or TIM2
- Fiber tips can be exposed out of the surface of the adhesive layer and the gel encapsulant
- The length of the fiber tips can be effectively controlled and matched with the roughness of the mating surface to ensure a good contact
- Exhibits the higher thermal conductivity than that of current Gelvet
- Exhibits relatively low thermal resistance
- Modified TIMs could be generated when using high performance raw materials or modified processes

This requires the proper selection of fiber, adhesive and encapsulant materials:

- Fibers–high thermal conductivity with maximum density
  - Graphite
  - Diamond
  - CNT (if available) or other nano-fibers
  - Metal wires
  - Others

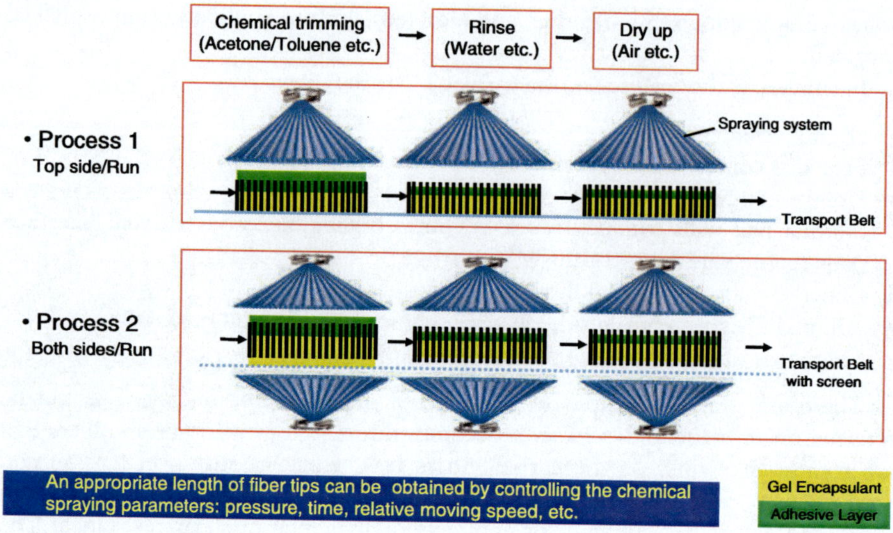


**Fig. 8.24** Schematic surface trimming process to get an appropriate length of fiber tips for minimizing contact thermal resistance at fiber–gel junction

- Adhesive layer–sticky and highly conductive
  - Gel with high thermal conductivity fillers
  - Liquid crystal polymer or its composites
  - Graphite paints or other adhesives with high thermal conductivity
- Encapsulant–balance between compliance and stiffness
  - Gel with stiff filler
  - Liquid crystal polymer or its composites
  - Graphite paints or other adhesives with high thermal conductivity

In fact, Gelvet TIM has been used for thermal interface material between heat sink to die interface, heat sink to package interface, and heat sink to multichip module interface.

In addition to Gelvet™, AF Inter Connector material manufactured by Shin Etsu, for instance, also uses carbon fibers in a silicone matrix with the carbon fibers disposed normal to the interface surface. The AF material acts as an electrical conductor between two devices, while the Gelvet material is designed to conduct heat between heat sinks and heat generating devices. Both of these solutions are designed for electrical nonspace applications and do not use a low- or non-outgassing polymer matrix material. It has also been found that the Gelvet material may also exhibit fiber damage during cutting of the graphitized carbon fiber and may have a less consistent distribution of fiber length (Suzuki et al. 2009).

Accordingly, an improved TIM that uses carbon-based fiber in a low- or non-outgassing polymer matrix have been used for use in contamination-sensitive cases, such as spacecraft applications. The TIM comprises highly thermally conductive (carbon-based) fibers contained, embedded, or immersed, in a low- or substantially non-outgassing polymer matrix material or gel encapsulant. The low- or non-outgassing polymer matrix material comprises high thermal conductivity, small diameter fibers, such as carbon fibers for example, that are disposed substantially normal to interface surfaces, and protrude above at least one of the surfaces of the low- or non-outgassing polymer matrix material. The thermally conductive fibers adjust to and contact surfaces to which they are attached and conform to macroscopic non-planarity of the surfaces. The thermally conductive fibers act as continuous conductors of heat from surfaces to which they are attached. For some applications, the thermally conductive fibers may be cut in a manner that reduces damage. This is achieved by cutting green state ungraphitized fibers to length, instead of cutting graphitized thermally conductive fibers. Heat treatment of the carbon based fibers may be done after the green state thermally conductive fibers are cut to length. The thermally conductive fibers, such as carbon fibers, for example, in the green state (ungraphitized) are cut to a desired length and are then graphitized. The graphitized thermally conductive fibers are then embedded in one or more layers of low- or non-outgassing polymer matrix material having a desired thickness. The graphitized thermally conductive fibers protrude from an interface surface of the low- or non-outgassing polymer matrix material by a predetermined amount. The graphitized thermally conductive fibers and the polymer matrix material securing them are then cured for a prescribed time period and temperature depending on the selected polymer matrix material and the desired end use properties. The polymer matrix material used in the TIM is chosen to pass stringent spacecraft environmental requirements. For example, the polymer matrix material is substantially non-outgassing which means that it contains volatile condensable material of less than 0.1% and has a total mass loss of less than 1.0% per an ASTM E595 procedure. Methods for introducing the graphitized thermally conductive fibers into the polymer matrix material include flocking, mixing, infiltration, electrostatic, preimpregnation, wicking, or vacuum transfer, for example. By using green state thermally conductive fibers, cost is reduced because the green state fibers are significantly less expensive to purchase and are easier to cut to length with less damage. The graphitized thermally conductive fibers may be heat treated to enhance the properties that provide high thermal conductivity. The TIM improves heat transfer between components and/or devices that generate heat and an external environment, such as is provided by a radiator panel, or other heat radiating device, for example, employed on a spacecraft. Also, the polymer matrix material used in the TIM exhibits low outgassing, thus enabling the TIM to be used in a space environment (Bonneville et al. 2001).

## *Nanotechnology Based Thermal Interface Materials*

A new class of nanotechnology based TIMs have been developed, such as CNT-based TIMs, Graphene TIMs, and nano-TIM with adhesion functions that has low thermal resistance, high thermal conductivity and high mechanical strength using electrospinning process. With the electrospinning process, polymer nanofibers with nanoscale diameter can be formed. Nanoparticles such as nano-silver particles, nano- CNT and nanosilicon carbide particles can be embedded into the nanofibers to enhance the thermal conductivity and to reduce the thermal resistivity. This new class of the nanofiber-based composite TIMs can offer high thermal conductivity, low thermal resistivity, similar operation temperature range, and similar degradation behavior, two to five times higher ultimate tensile strength, in comparison with commercially available TIMs. Incorporating adhesion functions into the developed nano-TIMs has been carried out using lamination or direct mixing methods. To achieve this, hot melt adhesive is planed to be directly laminated into the nano-TIM materials to achieve adhesion function. Another method is to directly dissolve thermosets into solutions to achieve the nano-TIMs with adhesion function in the same operation.

### Carbon Nanotube Adhesive

CNTs are increasingly finding applications in thermal management materials and are also being considered as potential interface and attachment techniques. Figure 8.25 shows typical CNT thermal interface options. (1) CNT array growths on

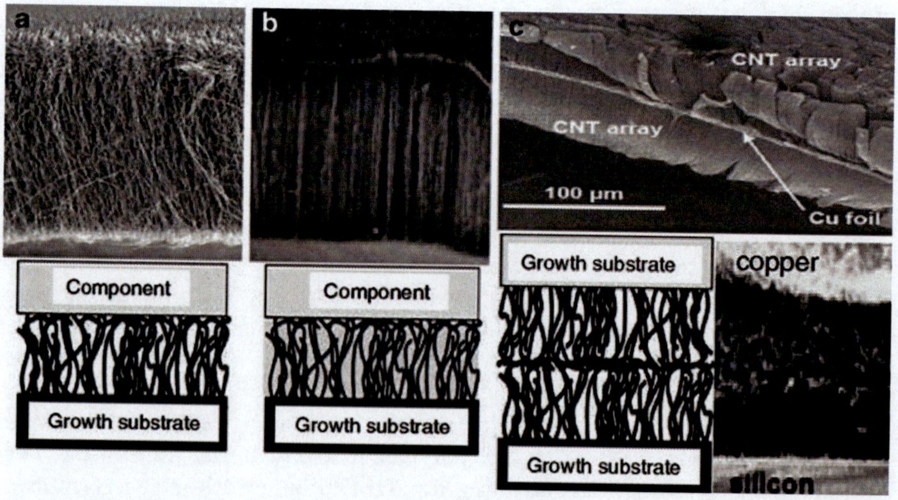


**Fig. 8.25** Carbon nanotube (CNT) thermal interface options. (**a**) CNT array growths on one substrate; (**b**) CNT array growths one substrate with filler; (**c**) CNT arrays grow on both substrates and butt against each other with or without copper foil between

one substrate; (2) CNT array growths one substrate with filler; (3) CNT arrays grow on both substrates and butt against each other with or without copper foil between. When grown on a surface in an array, CNTs can provide an extremely large surface-to-volume ratio and bind to each other and to surfaces through van der Waals (vdW) interactions. When acting collectively, vdW forces can provide significant adhesive strength (~12 N/cm$^2$) regardless of the hydrophobicity of the surfaces. In conjunction with their electrical and thermal conductivity, this adhesive characteristic makes CNTs potentially attractive as a device or substrate attachment material. Typical nanotube dimensions: MWNT diameter 20–30 nm; multiwalled carbon nanotube (MWNT) height 5–500 μm; Spatial density 1,010–1,011 tubes/cm$^2$; Contact area 4–20 mm$^2$. With a preload of $< 2$ kg in the normal direction, the maximum measured adhesive strength for 5–10 μm height arrays to a glass surface was 11.7 N/cm$^2$ in the normal direction (apparent contact area of 4 mm$^2$) and 7.8 N/cm$^2$ in shear (apparent contact area of 8 mm$^2$). The adhesive energy between MWNT and a glass surface was found to be 20–80 mJ/m$^2$. The adhesive strength over repeated adhesion and separation cycles degraded, due to the relatively poor adhesion of the MWNTs to their silicon growth substrate. This was improved significantly by adding molybdenum to the catalyst underlayer (Zhang et al. 2008).

Thermal interface measurements for the silicon/MWNT-glass system identified that the dominant interface is between the MWNT layer and the glass substrate ($10^{-5}$–$10^{-6}$ K m$^2$/W) which is at least one order of magnitude lower than that of the chemical vapor deposition (CVD) grown MWNT-silicon interface. Basically, adhesion performance is dependent on the tube length, for example, 5–10 μm tubes gave best adhesion. A relatively large preload (2 kg) is necessary to affect a joint. Adhesive strengths were related to apparent contact areas (4–20 mm$^2$), e.g., smaller contact areas had higher strengths, possibly due to coplanarity or surface embed vertically aligned MWNT in soft polymer films for improved conformity to surfaces. Adhesive energy and strength can be increased through the use of conformable substrates, single-walled CNTs, solder coated fibers, such as indium, or adhesive on the target surface. The vertically aligned nanotubes (15–250 nm diameter) were grown using microwave-enhanced CVD on silicon and copper substrates using a titanium adhesion layer (~10 nm), aluminum buffer layer (~10–20 nm) and nickel or iron catalyst layer (~16–10 nm). For MWCNT-coated silicon to copper interface (CNT array height 7–13 μm, CNT diameter 15–25 nm, CNT density 20–30 $\times$ $10^7$ CNTs/mm$^2$, substrate area 10 $\times$ 10 mm), thermal interface resistance of 20–37 mm$^2$ K/W at a pressure of 0.445 MPa. A combination of a CNT array and a PCM (load 0.35 MPa) produced a minimum resistance of 5.2 mm$^2$ K/W. With 3-ω measurements of vertically aligned CNTs on silicon substrate (CNT array height 13 μm, CNT diameter 10–80 nm), effective thermal conductivity reaches 74–83 W/m K in the temperature range 295–323 K, one order of magnitude higher than the test thermal greases or phase-change materials. In general, the major thermal impedance is at the tube tip substrate interface. The main issues with the current technique are ensuring that the silicon is kept below 350°C during CNT processing (this is not an issue with silicon carbide devices) and the need to apply a clamping pressure (1–4 atm) to effect a joint (Young et al. 2006).

By using vertically aligned CNT array as dry interface material to bridge the interfaces, this technology could perform as a thermally conductive tape that provides not only good thermal conducting ability but also strong mechanical bonding between the mating surfaces. It has been shown that a small fraction of CNTs as filling material could induce a 125% enhancement in thermal transport. This interface material can be directly applied to any interface that needs high thermal conductance. A preload of a few kilograms would be sufficient to create a good contact and adhesion. The bonding of the two mating surfaces is based on the van der Waals interaction between the CNTs and the surfaces which is inversely proportional to the feature size of the structure. Because of the high density and small diameter of the CNTs, the contact of the CNT array with the mating surface could generate a fairly good adhesion force in the order of few Newton per centimeter square ($N/cm^2$). Benefits of using CNTs as a TIM include: (1) not messy; (2) saves time through ease of installation; (3) eliminates problem of applying the exact amount of grease with each application; (4) CNTs have extraordinary mechanical properties besides the ultra high thermal conductivity. The fine structure of CNTs ensures the filling-in of the cavities at the interface, and ensures direct contact at mating surfaces. (5) Different from other approaches the vertically aligned CNT arrays form parallel paths across the mating surfaces with each path containing one CNT and two junctions at surfaces. This allows a better minimization of the number of junctions and therefore maximization of the thermal conductivity. (6) The high density of CNT array ($>10^{10}$ $cm^{-2}$) enables a high effective thermal conductance at interface. Different from traditional thermal greases, this interface material could be reusable, detachable, and mechanically robust. Furthermore, it can provide much better thermal performance than the general thermal conductive tapes that are commercially available. A more important advantage of this technology over thermal greases or thermal tapes is that it can work under some extreme environments, such as vacuum, very low and very high temperatures (Zhang et al. 2008).

Potential applications include TIMs of microprocessors and power electronics especially for thermal devices in space, because traditional TIMs could not effectively perform in the space environment.

### Graphene-Based Thermal Interface Materials

Graphene-based TIMs have been developed with vertically grown graphene or individual graphene sheets that align perpendicular to the plane of microchips and heat spreaders or heat sinks, providing a superior thermal conduction path from chip to sink. Graphene has extraordinary thermal conductivity ~3,000–5,000 W/m K, and the graphene based TIMs have the advantages of high packing density, rich shapes/geometry, easily functionalized, and possibilities to bond to surface (Chen et al. 2008).

Figure 8.26 shows the concept of TIM with vertically grown graphene sheets between (and bonded to) substrates. Microwave plasma enhanced CVD can be used

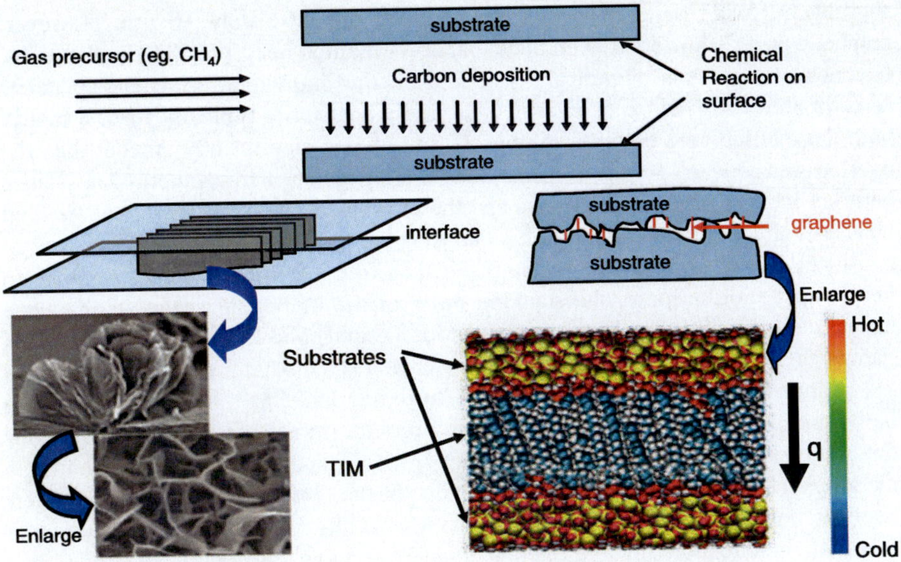


**Fig. 8.26** Concept of chemical vapor deposition grown vertical graphene between two substrates as TIM (Chen et al. 2008)

to grow vertically aligned graphene sheets: no catalyst needed, works on almost any substrate, and graphene bonded to substrate surface can have very high filling/packing density. Theoretically predicted thermal conductivity of the vertically grown graphene would be around 1,500 W/m K (Chen et al. 2008).

In addition, graphite oxide is a layered material consisting of hydrophilic oxygenated graphene sheets (graphene oxide sheets) bearing oxygen function groups on their basal planes and edges. Graphite oxide can undergo complete exfoliation in water, yielding colloidal suspensions of almost entirely individual graphene oxide sheets with a mean lateral dimension of approximately 1 μm. Such sheets can be chemically functionalized, dispersed in polymer matrices, and deoxygenated to yield novel composites. Furthermore, graphene paper and graphene oxide paper are synonymous with each other with the only difference being that the oxide in graphene oxide paper is reduced using hydrazene before the individual sheets are stacked together as graphene paper. These graphene oxide sheets can be assembled into a paper-like material under a directional flow, similar to the way buckypaper is formed from CNTs. The graphene oxide paper was made by filtration of colloidal dispersions of graphene oxide sheets through an Anodisc inorganic membrane filter that is fabricated from a unique form of aluminum oxide with a highly controlled, uniform capillary pore structure. The graphene oxide sheets were then air dried and peeled from the filter. The thickness of each graphene oxide paper sample was controlled by adjusting the volume of the colloidal suspension. The thickness of the free-standing graphene oxide paper ranged between about 1 μm and about 30 μm. Well-packed layers through almost the entire cross section of a graphene oxide paper having a thickness of approximately 10 μm. However, graphene paper is different from buckypaper for thermal heat spreading applications. Graphene paper has significant overlap of the individual graphene platelets as compared to buckypaper in which the individual tubes typically form a matrix with the overlap area between individual tubes being significantly smaller than the overlap area between graphene sheets. This overlap results in lower thermal resistance between individual graphene sheets and therefore better heat transport. The thermal resistance at the overlap region of nanotubes is the biggest contributor to the thermal resistance in buckypaper. Graphene paper therefore has a significant advantage to buckypaper. Furthermore, the stacking order of the individual sheets of graphene paper, would give the graphene paper superior thermal conductivity, is not possible in buckypaper because CNTs, being one-dimensional (1-D) structures cannot be stacked like graphene, which are 2-D structures. Third, interlamellar water present between individual sheets of graphene paper, which provides graphene paper with superior thermal conductivity, is not present in buckypaper. The combination of these makes the graphene paper a superior thermal conductor than buckypaper. Given that interlamellar water is present between individual sheets of graphene paper, the individual sheets are held together by hydrogen or covalent bonding. By contrast, the individual sheets in commercially available graphite paper are held together by van der Walls bonding. This distinction between graphene oxide paper and graphite paper indicates that graphene paper may exhibit superior mechanical properties, and in particular superior thermal conductivity properties, as compared with commercially available graphite paper. Specifically, the hydrogen or covalent bonding of the individual sheets in graphene paper would decrease phonon scattering, and therefore increase the thermal conductivity of the graphene paper. The measured thermal diffusivity of graphene paper having a thickness of about 14 μm (0.014 mm) using a laser flash technique apparatus is about 30 $cm^2/s$ (2,929.72 $mm^2/s$). The specific heat was measured to be about 0.8 J/gK using a differential scanning calorimetric technique. The density was measured to be about 1.83 $g/cm^3$ using He Pycnometry. Subsequently, the thermal conductivity of the graphene paper was calculated to be about 44 W/cm K. Based on the measurements of the thermal conductivity, it was envisioned that individual sheets of the graphene paper could be aligned to be perpendicular to the plane of a heat source and heat spreader or heat sinks, thus providing a superior conduction path from the heat source to the heat sink (Tilak et al. 2010).

## Thermal Interface Materials Selection and Application

The essential purpose of TIMs is to maintain effective transfer of heat from hot chips to dissipating devices such as heat sinks or spreaders. As heat flows, it encounters thermal resistances that impede overall heat transfer. TIMs reduce the most problematic of these, the contact resistance between the mating parts (heat source–heat sink). Air gaps significantly limit heat flow from the hot component into the sink or spreader. An effective TIM replaces the gaps created by the nonsmooth mating surfaces with a material whose thermal conductivity is much greater than that of air. Basically, it replaces poor conduction from point contacts and air to enhanced conduction through solids. The effectiveness of a thermal-interface material can enhanced by high thermal conductivity filler or reinforcement and low thickness of the interface material and low thermal-contact resistance between the interface material and each mating surface.

## *Commercial TIMs Selection and Application*

As the mating surfaces are not perfectly smooth, the interface material must be able to flow or deform in order to conform to the topography of the mating surfaces. If the interface material is a fluid, grease, or paste, it should have a high fluidity or workability so as to conform and have a small thickness after mating. On the other hand, the thermal conductivity of the grease or paste increases with increasing filler content, and this is accompanied by decrease in the workability. Without filler, as in the case of oil, the thermal conductivity is poor. A thermal-interface material in the form of a resilient thermal-conductor sheet (e.g., a felt consisting of conducting fibers clung together without a binder, a resilient polymer-matrix composite containing a thermally conducting filler, or flexible graphite) usually cannot be as thin or conformable as one in the form of a fluid, grease, or paste, so its effectiveness requires a very high thermal conductivity within it. Therefore, it is important to balance all factors when choosing a proper TIM.

Most TIMs are polymer-based composites loaded with heat-conducting filler particles. Common fillers include aluminum oxide (alumina), BN, aluminum nitride, and magnesium oxide. Metal fillers, such as silver, can be used where electrical isolation is not needed. Some level of pressure is usually needed between the mating surfaces to compress filler particles and make the material flow into the surface irregularities to reduce contact resistance. Once in place, a TIM's effective thermal resistance comprises the bulk resistance of the material and the contact resistance between the TIM and its mating surfaces. For instance, thermal-interface materials used in commercial computing systems are typically made up of a polymer matrix in combination with highly thermally conductive fillers (metal or ceramic), such as thermal epoxies, PCM, thermal greases, and gels. At small interface thicknesses, it is seen that the choice of the interface material is a function of both the contact resistance and the bulk conductivity. The particular material chosen depends on the relative magnitude of these two entities at the target design thickness. Performance considerations as well as cost and manufacturability concerns, inevitably result in tradeoffs that are made during interface material selection. Matrix selection is typically driven by its compatibility with filler, its mechanical properties, its ability to wet the mating surfaces, and its viscosity. The maximum filler loading that can be achieved is dictated by the thermodynamic wetability of the filler by the matrix and by the polymer viscosity. The polymer matrix also allows tailoring of the desired mechanical properties of the interface material under use conditions. Epoxy resins are used when high modulus and adhesion are targeted, whereas silicones are used for low-modulus and stress-absorbing applications. Lower surface energy materials are used to act as a matrix since they improve the wetability at the mating surfaces. Common examples of the use of silicones are the polymer materials. The key ingredient in the interface material is the filler, which is responsible for heat conduction. The fillers are dispersed in a polymeric matrix, which typically has poor thermal conductivity, for handling and processability. The important filler properties are bulk thermal properties, morphology (size and shape), and distribution. To reduce the contact resistance, filler surface treatments (coating) are also critical for ensuring optimum filler and matrix thermodynamic wetting. Ceramic powders such as alumina and magnesium oxide are commonly used due to their lower cost and dielectric properties. Further thermal enhancement is achieved through more conductive particles such as aluminum nitride or BN. These fillers provide a five- to tenfold improvement in bulk thermal conductivity, but due to more elaborate manufacturing techniques, cost 10 to 100 times that of their ceramic counterparts. For higher performance, metal particles such as silver and aluminum are used. Silver is chosen mainly for its very high bulk thermal conductivity. Aluminum provides a balance between the bulk thermal conductivity and density (high volume loading can be achieved because of low specific gravity). Issues such as physical design tolerances, positive pressure at the interface, warpage, and tilt and flatness of the mated surfaces have a direct influence on the thickness of the interface as well as the degree of wetting. Of particular interest is the tradeoff between the flatness (macroscopic) of the surface and the cost of machining. Warpage issues can be alleviated by the choice of assembly materials to lower the processing temperature as well as the CTE mismatch between materials. Interface pressure becomes a key factor with collapsible materials; this is controlled through design. The interaction of filler particles with microstructural asperities at the mated surfaces determines the degree of compaction and wetting at the interface. Materials in a semisolid or liquid state need to be dispensable. A typical tradeoff is that a higher degree of filler loading (to reduce thermal resistance) translates into increased viscosity of the material, which in turn affects the manufacturing throughput (Viswanath et al. 2000).

While thermal interfaces and TIMs are often considered well into the design process, several factors can be summarized as below when choosing a TIM (NeoEase 2010):

(1) *Thermal impedance* is the single most important specification measured in degrees K $m^2$/W. Thermal impedance is an application-specific measure of the ratio of the temperature difference between two mating surfaces to the steady state heat flow through them. Thermal impedance usually decreases with added mounting pressure and contact area, but increases with the thickness of the TIM.
(2) *Thermal conductivity*, in W/m K, measures a material's ability to conduct heat regardless of its thickness. A bulk measurement, thermal conductivity values can be used for comparing TIMs, but it does not describe a TIM's ability to minimize contact resistance in an application.

(3) *The gap space* between the heat source and the heat spreader. As a rule, the thinner the TIM the better, but because mating surfaces are never perfectly flat, a minimum material thickness may be needed to accommodate nonflatness issues.
(4) *Surface flatness* of mating surfaces is important for determining the type of material. If both surfaces are flat, grease or thin films would be ideal choices, but that is seldom the case. Plastic ICs are typically concave in the center and if the heat sink is extremely flat, the contact area would be limited to the periphery leaving an air pocket in the center.
(5) *Electrical isolation*, measured in kV, is sometimes needed. Silicone-based TIMs provide this property, along with thicker materials such as gap fillers. Thinner PCMs and, greases may not be reliable electrical insulators. Solder and graphite are electrically conductive.
(6) *Compressibility* is important when working with irregular surface and when covering a number of components. If heat and excess pressure are applied to a silicone-based TIM, silicone can escape and migrate along the printed circuit board. Without sufficient pressure there may be excess thermal resistance across the interface.
(7) *Temperature range* in the interface determines which materials can be used. Silicone TIMS, e.g., gap fillers are rated to higher temperatures than silicone free interface materials.
(8) *UL flame class rating*. A UL flammability rating requirement is needed for many TIM applications. Most of these materials are available with V-0 ratings, which will meet most needs.
(9) *Silicone or silicone free*. Silicone is an excellent thermal material with a high temperature range but some applications, e.g., in space, cannot use it due to outgassing.
(10) *Ease of application*. The method of attachment is a cost and performance decision. Most small heat sinks are attached with a double sided thermal adhesive tape. Larger heat sinks require mounting hardware. Adhesives can be added to both or one side of the thermal material. However, with a layer of adhesive, thermal impedance will be increased.
(11) *Utility*. How easy are the materials to work with in a manufacturing environment? How easy are they to rework when heat sinks must be removed? Some gap fillers can be reused, but PCMs and grease must be replaced.
(12) *The long-term stability* of the material depends on such factors as the usage temperature, time, application, and material properties.

The characteristics of typical TIMs would be described and compared as below (NeoEase 2010).

PCMs undergo a transition from a solid to a semisolid phase with the application of heat from the operating processor and a light clamping pressure. The semisolid PCM readily conforms to both surfaces. This ability to completely fill the interfacial air gaps and surface voids, usually under light clamping pressure, allows performance comparable to thermal grease. While less "runny" than grease, PCMs contain wax and once the melt-on temperature is reached, they may flow out of tight areas. Recently introduced phase-change type materials are not wax-based and will not drip. At room temperature these materials are firm and easy to handle. This allows more control when applying the solid pads to a heat sink surface. After installation, some phase-change pads create a strong adhesive bond between the processor and the heat sink. Exercise caution when removing the heat sink from the processor. A slight twisting or rotating movement should help to remove the heat sink. Using strong force to remove the heat sink can damage the processor.

Thermal greases typically are silicones loaded with thermally conductive fillers. They do not need curing and they can flow and conform to interfaces. They also offer reworkable thermal interface layers. It is important to ensure that the proper amount of paste or grease is dispensed prior to installing the heat sink. Too little grease may leave gaps between the heat sink and processor; too much might also cause air gaps and leak material outside the interface. On extended operation and over time, some greases can degrade, pump out, or dry out, which affects thermal transfer performance. Despite these drawbacks, greases are the interface materials of choice in high performance processor applications. Thermal conductivity of high performance thermal greases is in the order of 10 W/m K, which is superior to other TIMs.

One of the largest segments of the thermal interface market, gap fillers are supplied in different thicknesses and can cover large segments of a board. Effective materials can fill gaps up to one-quarter inch with a soft, highly thermally conductive interface. Gap fillers can blanket over multiple components of varying height to conduct heat into a common heat spreader. These pads are often used when low compression forces are required, therefore, high compressibility is an important feature. Gap fillers can be custom molded, and new form-in-place gap filler compounds are an option for high volume automation.

Thermal films provide electrical isolation along with thermal transfer. Their film carriers give superior resistance to tear and cut-through from burrs on heat sinks. This category includes silicone, silicone-free (e.g., ceramic-filled polyurethane), and graphite materials with a wide range of thermal performance and price points.

Thermal pads usually are fabricated by molding nonreinforced silicone with conductive fillers. Reinforcements for thermal pads can include woven glass, metal foils, and polymer films. Thermal pads are typically precut in sizes to accommodate different size components. From a performance standpoint, they are inferior to PCMs and thermal grease, but offer a practical, low cost TIM solution in many applications with less cooling requirements.

Graphite films are electrically conductive, have a low cost, and have been used for a long period of time. Graphite films are effective in very high temperatures (up to 500°C). The oriented fibers can be laminated in a horizontal plane resulting in very different thermal conductivity measurements. For example, the thermal conductivity can be 7.0 W/m K on the $x$-axis and 150.0 W/m K on the $y$-$z$ axis.

Most small heat sinks are attached to components with a double-sided PSA thermally conductive tape. Factors for tapes include peel strength, lap- and die-shear strength, holding power, and thermal resistance. Thermally conductive adhesive tapes are considered to be convenient for heat sink attachment with mid-range thermal performance. While they replace mounting hardware, thermal tapes often have problems with the lack of flatness on component surfaces. Plastic ICs usually are concave in the center and heat sink surfaces vary as well. This can result in air gaps in the interface. One thermal adhesive tape consists of a finely woven nickel coated copper mesh that conforms closely to irregular mounting surfaces varying up to 50% of its thickness.

Thermal adhesives are one- or two-component systems containing conductive fillers. They are typically applied via dispensing or stencil printing. These adhesives are cured to allow for cross linking of the polymer which provides the adhesive property. The major advantage of thermal adhesives is that they provide structural support, therefore eliminating the need for mechanical clamping.

Thermal gels are low modulus, paste-like materials that are lightly cross linked. They perform like grease with respect to their ability to conform to surfaces, while displaying reduced material pump out.

Metallic TIMs can be made in many different forms and are no longer limited to solder applications. In some applications metallic TIMs are totally reworkable and recyclable. The need for better performing TIMs in such devices as power amplifiers and insulated-gate bipolar transistor (IGBT) modules have prompted exploration of other types of metallic TIMs such as liquid metals, phase change metals, and soft metal alloys. The soft or compressible metal TIM is the most easily adopted metallic TIM because it does not need to be reflowed or contained in a gasket like a solder or liquid metal. Metallic TIMs are very thermally conductive, reliable, and in the case of compressible metals, easily adopted. A new hybrid material consists of a thermally conductive silicone film on one side bonded to a copper film. The advantage of this material is that it can be used to manufacture flex circuits as well as provide electromagnetic interference and RFI (Radio Frequency Interference) noise protection.

Thermal interfaces are often considered late in the design stages of cooling systems. This is not the best practice as TIMs can be the limiting factors in the expense of thermal management designs. With more and more excess heat to be dealt with, there is a steady demand for higher performing TIMs. Used effectively, TIMs can help reduce the size of heat sinks and the need for larger fans. The extended benefit is that an effective TIM is a faster, easier applied, and less costly solution than changing heat sinks or redesigning a chassis.

## *Future Directions*

The focus of future research and application should be on understanding the reliability and performance degradation of TIMs. Current commercial TIMs are capable of providing a thermal resistance between 0.03 and 0.1°C $cm^2/W$; however, due to degradation at large exposures to high temperatures, the thermal performance can degrade severely depending on the temperature of the processor and time of exposure. There is no mechanistic understanding of these degradations. Fundamental physics-based modeling is needed to relate the degradation of the polymer properties to thermal properties of the polymer composites. Use of nanoparticles and nanotubes is almost inevitable; however, any research in this area should take the performance of the current commercially available TIMs as benchmark. Research should also focus on minimizing the total thermal resistance rather than just increasing the thermal conductivity. A physics-based model for the contact resistance between the particle-laden TIMs and the substrate is still incomprehensive. Contact resistance will become important for thin highly conducting TIMs. Modeling of the CNT concepts will be also needed in the future due its promise as interface material (Prasher 2006).

## Summary

TIMs play a key role in thermally connecting various components of the thermal management solutions. As electronic assemblies become more compact and increase in processing bandwidth, escalating thermal energy has become more difficult to manage. The major limitation has been nonmetallic joining using poor TIMs. The interfacial, versus bulk, thermal conductivity of an adhesive is the major loss mechanism and normally accounts for an order magnitude loss in conductivity per equivalent thickness. The next generation TIM requires a sophisticated understanding of material and surface sciences, heat transport at submicron scales, and the manufacturing processes used in packaging of microelectronics and other target applications. Only when this relationship between bond line manufacturing processes, structure, and contact resistance is well understood on a fundamental level will it be possible to enhance interfacial thermal conductance and advance the development of miniaturized microsystems. TIMs are widely needed to improve thermal contacts for facilitating heat transfer in electronic packaging, such as that associated with the flow of heat from a microprocessor to a heat spreader or a heat sink in a computer. A TIM is commonly in the form of paste, solder, or a resilient sheet that serves to fill a gap between the two adjoining surfaces. The performance of a TIM is enhanced by conformability of the interface material to the topography of the mating surfaces because the air residing in the valleys in the surface topography is thermally insulating and should be displaced by the interface material.

Any engineering surface of the electronic components is rough on a microscopic level, due to the presence of microscopic asperities. When two such rough surfaces come in contact during assembly of electronic packaging, the actual contact occurs only at a few discrete spots, usually at the high points of the two mating surfaces. Typically, the ratio of real contact area to apparent contact area is approximately 1–2%. Heat flowing from one body into the other is constricted to flow through the actual contact spots because the thermal conductivity of the solid contact spots is much higher than that of the surrounding gap which is filled with air in most electronic applications. TIMs are often inserted between the surfaces of a contact pair to reduce the thermal contact resistance. Although they typically have lower thermal conductivity than the substrate, they are highly compliant and hence under the application of relatively small contact pressure, deform to conform to the geometry of the adjacent rough surfaces. A part of the low thermal conductivity gas present is thus replaced by a higher conductivity material. This leads to a decrease in the constriction of the heat flow lines, and therefore, an increase in the contact conductance. The heat transfer of these complex joints is governed by contact pressure; number, size, and shape of contact spots and voids; types of fluid in voids; pressure of fluid in voids; hardness and flatness of contact surfaces; modules of elasticity of contact surfaces; surface cleanliness; and property of the TIM used in the joint.

The thermal conductivity of conventional polymeric TIMs generally does not exceed about 5 W/m K and is typically less than about 1 W/m K. However, metallic TIMs that form thermal interfaces with effective thermal conductivities of about 50 W/m K or greater will be the alternative to meet increasing thermal management requirement of high performance electronics. Most metallic TIMs may also exhibit a favorable solder or wetting behavior upon reflow which facilitates a low thermal interfacial resistance. During reflow, the solder and substrate are heated, the solder melts and wets by surface tension and/or local surface alloying. The interfaces consist of intermetallics or interdiffused metals with thermal properties that are frequently less desirable than those of the bulk TIM metal but much better than polymeric TIMs. In addition, the reflow or active bond process can also help to form reliable thermal interfaces. Metallic TIMs, however, can fail in certain applications due to the relatively large differences between the coefficients of thermal expansion of the TIM and the semiconductor and/or heat sink components and the lack of compliance. To take advantage of the metallic TIMs and avoid their disadvantages, a great many metallic TIMs have been developed and offered different characteristics for selection. Such metallic TIMs typically include reflow solders and active bond process, nonreflow solders and LMAs, composite solders and hybrid metallic solder materials, as well as gold–gold interconnection.

If the metallic TIMs are mainly suitable to TIM1 applications, organic TIMs would mostly be used for TIM2 application, although some metallic or organic TIMs have been developed for both TIM1 and TIM2 applications. With advances in packaging technology, the proper selection of a TIM has become more complicated as each application will have different design requirements. The final selection will encompass material properties and process considerations beyond the thermal performance of the material. Compliance, thickness, processing, and response to reliability conditioning are key considerations when engineering the final package. To conform the diversified requirements of advanced processor and packaging technology, different formats of organic TIM have been developed and commercialized such as various thermally conductive elastomers, thermal greases, PCMs, and PSH materials. Despite the maturity of the organic TIM market, there continues to be a need for higher performance and value-added materials. The performance requirement for a TIM encompasses more than thermal performance but a total solution: low thermal resistance across a bond-line, high thermal conductivity, reworkability, sufficient thickness to accommodate non-planarity, compliance for thermal mismatch, preapplication to heat spreader (TIM1) or a heat sink (TIM2) for speed of assembly, and low cost. For instance, both grease and PCMs possess low thermal resistance due to their liquid nature at operating temperature and thin bond-line, however, each possess drawbacks. The reliability and messy nature of grease is a concern, whereas the reworkability and low thermal conductivity for PCMs are drawbacks. Pad and film formats offer many advantages from a processing and packaging perspective, but traditionally, their performance is not as high as grease or PCM. There exists a strong need for high-performance organic TIMs as the strength of those TIMs makes them useful for both TIM2 and TIM1 applications. The emerging high performance organic TIMs include PSH and advanced thermally conductive elastomers. Meanwhile, flexible graphite sheet materials have become an attractive thermal-interface material because of their excellent resilience and flexibility to conform well the mating surfaces, relatively high thermal conductivity (>3 W/m K) in thickness direction, low thermal expansion, thermal stability, and chemical inertness.

In addition, a new class of fiber reinforced interface materials have been developed, which feature an aligned array of conductive fibers. This structure allows one fiber to span the gap between the two surfaces, resulting in very high thermal performance, even for large gaps. Because the fibers have a high aspect ratio, the material has a high degree of mechanical compliance, making it ideal for applications such as interfaces within multichip modules, mobile products, or gap-filling applications where variations in interface thickness can be absorbed with little or no thermal penalty. The material has been shown to be very robust with respect to thermal and mechanical cycling, and will not pump out or dry out from the interface during actual use. Furthermore, nanotechnology-based TIMs have also caused great attention and would play a great role in next generation high performance electronic packaging.

## References

Block J, Rice RW, Morgan CR (1993) Thermally conductive elastomer. US Patent 5194480.

Bock IK (2009) Gold nano-lawn: a new interconnection technique. http://www.izm.fraunhofer.de:80/EN/fue_ergebnisse/system_integration/KontaktierverfahrenmittelsGoldNanorasen.jsp. Accessed 26May 2010.

Bonneville SW, Cooney JE, Peck SO (2001) Thermal interface materials using thermally conductive fibers and polymer matrix materials. US Patent 6311769.

Chen YP, Ruan X, Fisher TS (2008) Graphene-based thermal interface materials. www.physics.purdue.edu/quantum/.../grapheneCTRC_oct2008a.ppt. Accessed on 3 June 2010.

Chomerics (2009) Thermal management products & custom solutions catalog. http://www.darcoid.com/images/uploads/pdfs/THERMALMP.pdf. Accessed on 30 May 2010.

Clayton JE (2005) Adhesive interconnect flip chip assembly. http://www.polymerassemblytech.com/corp/papers/AdvancedPackaging.pdf. Accessed on 8 June 2010.

Dean NF, Gettings AL (1998) Experimental testing of thermal interface materials with non-planar surfaces, Proceedings of the Institute of Electrical and Electronics Engineering SemiTherm Conference, San Diego, CA, 1998, IEEE.

Dean N, Pinter M (1999) Novel thermal interface material with aligned conductive fibers, Proceedings of Annual meeting, International Microelectronics and Packaging Society (IMAPS), Chicago, IL, 1999.

deSorgo M (1996) Thermal interface materials. http://www.electronics-cooling.com/1996/09/thermal-interface-materials-2/ Accessed 30 May 2010.

Jayaraman S, Koning PA, Dani A (2007) Polymer matrices for polymer solder hybrid materials. US Patent 7252877.

Koning PA, Hua F (2004) Polymer solder hybrid. US Patent 6813153.

Koning PA, Hua F, Deppish CL (2006) Polymer with solder pre-coated fillers for thermal interface materials. US patent 7036573.

Liu Z, Chung DDL (2006) Boron Nitride Particle filled paraffin wax as a phase-change thermal interface materials. Journal of Electronic Packaging, 128, 319–323.

Luk CF, Chan YC, Hung KC (2002) Development of gold to gold interconnection flip chip bonding for chip on suspension assemblies. Microelectronics Reliability, 42, 381–389.

Luo X, Chugh R, Biller BC, Hoi YM, Chung DDL (2002) Electronic applications of flexible graphite. Journal of Electronic Materials, 31 5, 535–544.

Macris CG, et al. (2004) Performance, reliability, and approaches using a low melt alloy as a thermal interface material. http://www.enerdynesolutions.com/downloads/imaps_2004_man.pdf. Accessed on 12 May 2010.

Mahajan R et al. (2002) Emerging directions for packaging technologies. Intel Tecnology Journal 6, 62–75.

Matayabas JC, Dani AA (2003) Polymer solder hybrid interface material with improved solder filler particle size and microelectronic package application. US Patent 7030483.

NeoEase (2010) Choosing thermal interface materials-knowing your application needs is key to picking the best TIM. http://www.mhw-thermal.com/blog/?p=3. Accessed on 3 June 2010.

Prasher R (2006) Thermal interface materials: historical perspective, status, and future directions. Proceedings of the IEEE, 94, 1571–1586.

Prasher R, Shipley J, Prstic S, Koning P, Wang J-L (2003) Thermal resistance of particle laden polymeric thermal interface materials. Journal of Heat Transfer, 125, December 2003.

Renavikar MP et al. (2008) Materials technology for environmentally green microelectronic packaging. Intel technology Journal 12, 1–16.

Singhal V, Siegmund T, Garimella SV (2004) Optimization of thermal interface materials for electronics cooling applications. IEEE Transactions on Components and Packaging Technologies, 27 2, June 2004.

Smalc M, Norley J, Reynolds III RA, Pachuta R, Krassowski DW (2003) Advanced thermal interface materials using natural graphite. IPACK03 – International electronic packaging technical conference and exhibition, Maui, HI, July 6–11, 2003.

Sreeram AN, Lewis B, Hozer L, Liberatore MJ, Minogue G (2003) Thermal interface material and heat sink configuration. US Patent 6653741.

Suzuki Y, Miura Y, Horikoshi K, Yada M (2009) Carbon fiber-reinforced resin composite materials United States Patent 7585558.

Tilak V et al. (2010) Thermal management system with graphen-based thermal interface material. US Patent 0128439.

Tong XC (2009) Advanced Materials and Design for Electromagnetic Shielding. CRC Press, Roca, Raton USA.

Viswanath R, Wakharkar V, Watwe A, Lebonheur V (2000). Thermal performance challenges from silicon to systems. Intel Technologies Journal Q3, 2000. http:/www.intel.com/technology/itj/q32000/pdf/thermal_perf.pdf. Accessed on 03 June 2010.

Wikipedia (2008) Phase change material. http://en.wikipedia.org/wiki/Phase_Change_Material. Accessed on 29 May 2010.

Young R et al. (2006) Developments and trends in thermal management technologies – a mission to the USA. www.lboro.ac.uk/research/iemrc/documents/. . ./CB2007.pdf. Accessed 26 February 2010.

Yovanovich MM, Culham JR, Teertstra P (2004) Calculating interface resistance. http://www.mhtl.uwaterloo.ca/pdf_papers/mhtl97–4.pdf. Accessed on 03 may 2010.

Zhang K, Chai Y, Yuen MMF, Xiao DGW, Chan PCH (2008) Carbon nanotube thermal interface material for high-brightness light-emitting-diode cooling. Nanotechnology, 19, 215706 (8pp) doi:10.1088/0957–4484/19/21/215706.

# Chapter 9
# Materials and Design for Advanced Heat Spreader and Air Cooling Heat Sinks

**Abstract** Air cooling is the simplest and principal method of thermal management most widely used for a variety of electronic systems ranging from portable electronics to large business systems. The advantages of air cooling are its ready availability and ease of application. Air cooling usually dissipates heat through heat spreaders and heat sinks. Low-power electronic systems are conveniently cooled by natural convection and radiation. When natural convection is not adequate, the forced convection is adopted by a fan or blower to blow the air through the enclosure that houses the electronic components. Various heat spreading methods have been widely used in electronic packaging. Heat spreaders are usually used in die level packaging to spread heat from the microprocessor chip into the package. Spreaders are commonly used within electronic enclosures to move heat from discrete components to the walls of the enclosure. Materials and process technologies to make heat spreaders have been involved in metallic heat spreaders (<400 W/m K); composite spreaders (<1,200 W/m K); and advanced heat spreaders including high conductivity plates (500–5,000 W/m K), vapor chambers (>5,000 W/m K), and oscillating flow heat spreaders (up to 120,000 W/m K). Heat sinks can usually be divided into active and passive. Active heat sinks utilize power and are usually a fan type, or liquid cooling or some other Peltier cooling device. Passive heat sinks are fully reliable, usually made of a metal such as aluminum-finned radiator that dissipates heat through convection. Next generation heat sinks and leading-edge materials in air-cooled heat removal have been developed and are increasing, such as high thermal conductivity polymers, carbon-based composites, sintered metallic powders, phase-change compounds, synthetic diamonds, and strand-oriented graphite materials. Composite heat sinks are emerging and getting increased attention since nanotechnology can be applied. This chapter will review materials and design for advanced heat spreaders and air cooling sinks, including the concept of air cooling, spreading and constriction resistance analysis, heat spreader classification and materials selection, and overview of design, materials, and processing of air cooling heat sinks.

X.C. Tong, *Advanced Materials for Thermal Management of Electronic Packaging*, Springer Series in Advanced Microelectronics 30,
DOI 10.1007/978-1-4419-7759-5_9, 

## Overview of Air Cooling

Air cooling is a method of dissipating heat that works by making the object to be cooled have a relatively larger surface area or have an increased flow of air over its surface, or both. Typical examples using air cooling are heat spreaders, and heat sinks with or without blowing air. Compared with other cooling approaches, air cooling is the simplest and principal method of thermal management used most widely for a variety of electronic systems ranging from portable electronics to large business systems. The advantages of air cooling are its ready availability and ease of application. Before 1964, all IBM computers were cooled solely by forced air. In many cases, air moving devices are installed at the bottom or top of a column of boards to provide sufficient cooling. For high heat flux, a push–pull air flow arrangement with air moving devices at both the bottom and top of the column of boards was used to provide high pressure drop capability. Low-power electronic systems are conveniently cooled by natural convection and radiation. When natural convection is not adequate, the forced convection is adopted by a fan or blower to blow the air through the en closure that houses the electronic components (Anandan and Ramalingam 2008).

### *Passive Air Cooling*

Passive air cooling is the simplest form of air cooling in which heat is transferred to the air by conduction, natural convection, and radiation through heat spreaders and heat sinks. Circuit boards that dissipate up to about 5 W of power can be cooled effectively by natural convection. Passive cooling is well suited to certain products, especially small electronic devices with low heat output, such as MP3 players and mobile phones. It is familiar in consumer electronics such as TV, VCD, etc. by providing a sufficient number of vents on the case to enable the cooled air to enter and the heated air to leave the case freely. An alternative cooling technique that enhances the natural convection heat transfer can be done from discrete heat sources. The combination of an appropriately placed cross flow opening and a strategically positioned transversely vibrating plate is studied by varying the parameters and geometric configurations. The combined effects cause significant improvement in the thermal conditions over pure natural convection. In some applications, such as rotary machines, guided missiles, and space-based manufacturing process involves natural convection in rotating condition. All portable electronic cooling ultimately depends on the passive air cooling, which can be accomplished directly at a hot component, or by conducting the heat to a distant radiating surface, such as a heat spreader or a heat sink, where it is transferred to the air. Due to their inherent simplicity, reliability and low long-term costs, natural convection heat sinks have proved to be instrumental in cooling single or multiple chip circuit boards.

However, passive air cooling cannot keep up with the thermal demands of high-performance electronics. Even if a notebook computer were made out of solid diamond, it could only passively dissipate 10–15 W before overheating. By contrast, a typical notebook computer requires 25–60 W or more to be dissipated.

## *Active Air Cooling*

When natural convection cooling is not adequate, active air cooling is provided by external means such as a fan, a pump, a jet of air, or microscale electronic or ion-driven flows. In electronic systems cooling, the fan is a popular means of circulating air over hot surfaces. For active air cooling with force convection, the hot surfaces are characterized by their extended surfaces such as fins in heat sinks. The use of microjet of air to cool hotspots is more attractive. The fan selection is the important aspect in forced convection. The primary considerations in the selection of the fan include (1) the static pressure head of the system, which is total resistance, an electronic system offer to air as it passes through; and (2) the volume flow rate of air required for cooling. Piezoelectric fans are preferred as alternatives for conventional fans to cool low-power electronics owing to their low-power consumption, minimal noise emission, and small dimensions (Anandan and Ramalingam 2008).

In many instances, thermal enhancement techniques such as heat sinks with fans or blowers are required to cool high density microelectronic packages. It increases the effective surface area for heat transfer and lower thermal resistance between source and sink. Heat sinks can be operated under free or forced convective modes depending on the cooling load requirement. The diverse mix of geometric configuration, thermophysical properties and flow conditions present in microelectronic applications must be factored into thermal modeling tools used for design or reliability assessment. The geometries encountered in heat sink assemblies are difficult to model using analytical techniques because of the complex fluid flow around and between the various components of the heat sink. Therefore, computing simulation is usually used for the heat sink and fan blowing system design.

Modern electronics not only need the kind of thermal efficiency that forced air cooling provides, but they also need to function in a smaller and thinner space than ever before. Smaller fans need to work harder and run faster to move as much air as large ones. Higher speeds mean more noise and faster wear out, which makes them far from ideal for projectors, ultrathin laptops, and other devices that need quiet thermal management in a compact space. Clearly, the industry needs a new way to cool electronics, i.e., one that is compact, quiet, efficient, and cost effective. No such technology is yet commercially available, but Figure 9.1 shows promising candidate examples of novel active air cooling: (1) Microscale ion-driven air flow resulted from electrical charge imbalances between the nanotubes (electrodes) to cause tiny lightning bolts that ionize the surrounding air and propel cooling winds over electronics. (2) Quantum electron tunneling as a cool chip is placed such that

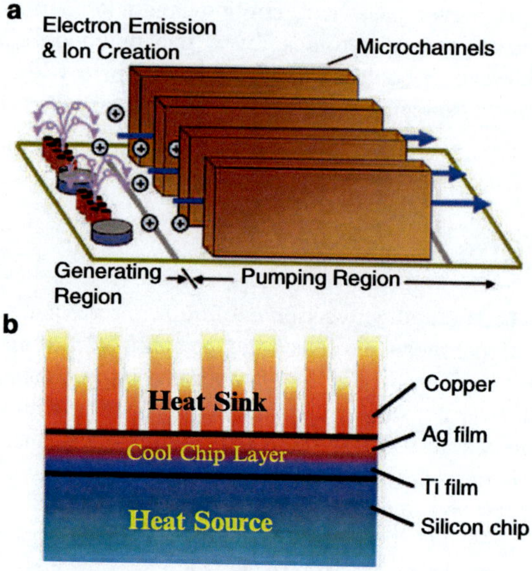


**Fig. 9.1** Examples of novel active air cooling: **(a)** microscale ion driven air flow resulted from electrical charge imbalances between the nanotubes (electrodes) to cause tiny lightning bolts that ionize the surrounding air and propel cooling winds over electronics. **(b)** quantum electron tunneling as a cool chip is placed such that electrons tunnel across the gap (formed by low adhesion between Ag and Ti films) and do not return back and make the heat is effectively dissipated

electrons tunnel across the gap (formed by low adhesion between Ag and Ti films) and do not return back and make the heat is effectively dissipated.

As shown in Figure 9.1a, in silent air cooling technology, airflow is created by ionizing air molecules and setting them in motion with an electric field. High-intensity electric field ionizes air molecules surrounding the corona electrode tip. As ions move from the corona electrode to the collector electrode, they collide with neutral air species, transferring charge and momentum and generating air flow. The ionized molecules transfer their momentum to neutral air molecules, resulting in a steady airflow that cools components with minimal noise. The noise generated by this system can be as little as 1 dB above the background level of a semianechoic chamber, making it essentially inaudible to the human ear in a normal environment. With optimizing airflow and reducing thickness, the system has shown the ability to remove equal amounts of heat as a stock cooling system, but with only half the power consumption.

One of the most intriguing applications for ion driven air flow cooling is light-emitting diode (LED) lighting, the energy-efficient successor to incandescent bulbs. Incandescent light bulbs, operating at 800 cd (candela), dissipate most of their heat through radiation. In contrast, LEDs only reach temperatures between 80 and 120°C and do not give off a significant amount of infrared energy. However, as LEDs heat up, they experience undesirable shifts in color and drops in luminescence. Because conventional light sockets were not designed to conduct appreciable heat away from the bulb, heat in retrofit LEDs must be dissipated via convection. Passive air cooling, with large, heavy, and expensive heat sinks is the most common solution for LED cooling and has the advantage of not creating any noise or vibration, which is unacceptable in many lighting applications. However, forced convection can remove far more heat in a smaller and lighter form factor than can passive convection, making silent act a promising choice for LED lighting applications (Yang et al. 2003).

As a developing cooling technology, as shown in Fig. 9.1b, thermotunnel cooling has been developed for fast moving electrons carry heat across a gap, but cannot return due to a voltage difference. The problem with using thermal electrons to carry heat is the fact that, due to the high work function of metals, which are the only practical emitters, the lowest cooling temperature is around 600°C–clearly not useful except in the most unusual applications. Thermotunnel cooling avoids this problem by making the gap narrow enough that electrons can tunnel across the gap, carrying the heat with them. The problem with this approach has been getting two surfaces near enough that they can tunnel over a large area, yet not touch at any point, which would short out the device preventing it from doing any useful cooling.

In fact, various component materials, including thermal managing materials and processing methods, have been explored to allow them to be scaled down and/or combined to accommodate the demands for smaller electronic components, as electronic devices become smaller and operate at higher speeds, energy emitted in the form of heat increases dramatically with heat flux often exceeding 100 W/cm$^2$. Thermal management components such as heat sinks, heat spreaders, and various novel passive and active air cooling devices have been utilized to decrease potential negative impacts of heat-generating components in a wide range of electronic devices by aiding in the transfer of heat to the ambient environment.

Figure 9.2 shows a conventional heat spreading and heat sink system. Heat spreaders have typically been utilized in flip-chip technology to provide a lower thermal resistance pathway between the chip and ultimate heat sink. Various materials such as copper and aluminum alloys have been utilized for flip-chip heat spreader applications. In particular instances, materials such as carbon–carbon (C–C) composites or diamond can be advantageously utilized for heat spreader applications due to their exceptional thermal conductivity. Diamond and C–C composite heat spreaders can have greatly enhanced thermal transfer rates relative

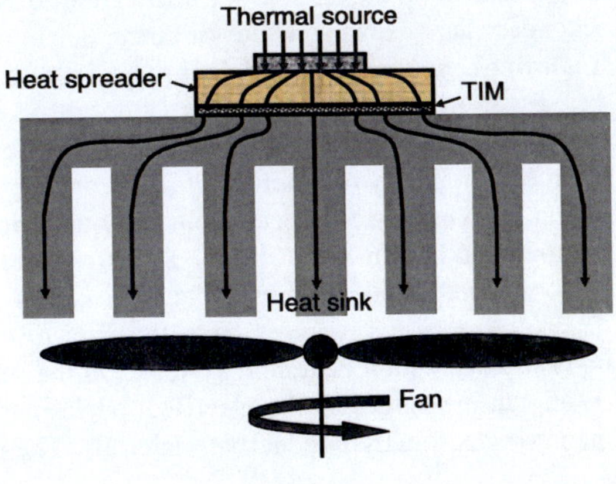


**Fig. 9.2** Typical heat spreading and heat sink system

to alternative materials having lower thermal conductivity. Diamond heat spreaders can also allow a better thermal expansion match between the chip and packaging components. However, due to the expense of diamond materials and the relative difficulty in fabricating conventional heat spreader configurations, heat spreaders for flip-chip and other microelectronic applications fabricated from these materials can be cost prohibitive. Thus, the continuous demand in the thermal management materials, design, and processing of electronic packaging has been focused on: (1) designing and producing thermal interconnects and thermal interface materials, layered materials, head spreaders and heat sink components and products that meet specification requirements while minimizing the size of the device and number of layers; (2) developing more efficient heat spreader and heat sink materials, products, and/or components with respect to the compatibility requirements of the material, component, or finished product; (3) producing materials and layers that are more compatible with other layers, surfaces, and support materials at the interface of those materials; (4) effectively reducing cost of manufacture of heat spreaders, heat sinks, and other thermal transfer components.

## Spreading and Constriction Resistance

A heat spreader is a primary heat exchanger (most often in the form of a simply highly thermal conductive plate) which moves heat between a heat source and a secondary heat exchanger such as a heat sink. The secondary heat exchanger is always larger in the cross sectional area, surface area, and volume. The heat flow is the same in both heat exchangers, but the heat flux density is less in the secondary, so it can be made of a less expensive material such as aluminum, and is a better match to an air heat exchanger because the low heat transfer coefficient for air convection is adequate for a low heat flux. A heat spreader is generally used if and only if the heat source tends to have a high heat flux density (high heat flow per unit area), and for whatever reason, heat cannot be conducted away effectively by the secondary heat exchanger. For instance, this may be because it is air cooled, giving it a lower heat transfer coefficient than if it was liquid cooled. A high enough heat exchanger transfer coefficient is often sufficient to avoid the need for a heat spreader. The use of a heat spreader in electronic packaging is an important part of an economically optimal design for transferring heat from high to low heat flux media, such as a microprocessor in air cooling integrated circuits, and air cooling a photovoltaic cell in a concentrated photovoltaic system. To achieve good thermal performances in electronic packaging, a heat spreader is usually used to help dissipate the heat generated from the chip to the ambience through the package body usually with a heat sink system. On the other hand, the use of heat spreader could cause changes to the electrical performances of the system as it is usually made of electrically conductive materials. The ever-increasing processing speeds in high-speed digital systems and frequency in radio frequency and microwave circuits makes the presence of a heat spreader much more complex in terms of its impact on system's electrical behaviors. For example, mitigation of electromagnetic interference (EMI) by grounding the heat spreader or by other solutions may need to be considered during design of the heat spreader system.

Thermal spreading is used to describe the case whenever heat leaves a source of finite dimensions and enters a larger cross-sectional area. Thermal spreading resistance has been evaluated to design and compare the performance of various thermal management components, such as heat spreaders and heat sinks. As shown in Figure 9.2, considering a square source with zero thickness of size $A_s$ centrally located on a square heat spreader plate of size $A$, thickness $d$ and thermal conductivity $k$ dissipates heat flow $q$ (W). The top and sides of the plate are adiabatic (insulated), the bottom "sees" a uniform heat transfer coefficient $h$ (W/m$^2$ K). The remarkable thing is that even for this simple configuration, no explicit solution is known for the description of heat spreading. Observing the exact implicit solution of the governing differential equations reveals the source of the complexity of heat spreading: it is not possible to separate the convection and conduction parts. In other words, changing the heat transfer coefficient also changes the value of the spreading resistance. Consequently, it is not possible to write the problem in terms of one conduction resistance describing the heat spreading inside the solid and one convection resistance describing the boundary condition because the two are dependent. There is one exception: the analysis becomes much more straightforward when the temperature gradients over the area that is in contact with the environment can be neglected. In other words, a uniform temperature may be assumed. Such is often the case with relatively small heat sinks and spreaders. A final complexity stems from the fact that decreasing the thickness of the thermal spreader plate does not automatically result in a decrease in temperature, caused by the fact that a smaller thickness also implies a decrease in spreading capability. Hence, for a certain combination of thickness and thermal conductivity, given the boundary conditions and the dimensions, a minimum in the total thermal resistance may be found (Lasance 2008).

In practice, there is one source with multiple layers that is most often to be dealt with; when a single source on a submount is itself connected to a second spreader. Closer observation reveals two heat spreading effects: one from the submount to the spreader, and one from the source to the submount. In first order, this problem can be handled as a single-layer issue, provided that the boundary condition of the first layer is replaced by the spreading plus the boundary condition of the second layer. Multiple sources add another layer of complexity because the coupling between the sources is not only dependent on the dimensions and physical properties but also on the boundary conditions and, worst of all, on the dissipation of the sources themselves. Using superposition techniques is the recommended approach (Lasance 2008).

There have been three approximate solutions for the single source with single layer system.

(a) One-dimensional [1-D]-Series-Resistance Approach. As shown through (9.1)–(9.2), the temperature rise $\Delta T$ of the source due to dissipation of heat flow $q$ via a total thermal resistance $R_{\text{total}}$ from junction to ambient, which is split up into $R_{\text{spreding}}$ and $R_{\text{convection}}$ (Lasance 2008):

$$\Delta T = qR_{\text{total}}, \tag{9.1}$$

$$R_{\text{total}} = R_{\text{spreader}} + R_{\text{convection}} = g\frac{d}{kA} + \frac{1}{hA}, \tag{9.2}$$

with $g = f(A_s, A)$ some geometrical factor representing the spreading, e.g., the 45° rule. The 1-D approximations should not be used at all in practice.

(b) Assuming that a surface source can be replaced by a volume source extending over the thickness and that a square plate can be replaced by a flat cylinder, by approximating the Bessel-function solution for the cylinder by algebraic equations, one-sided heat transfer for a single source on a plate with uniform heat transfer coefficients at both sides, mainly for flat heat sinks can be obtained (Lasance 2008):

$$R_{thj-a} = \frac{1}{hA_2} + \frac{\ln(A_2/A_1)}{4\pi kd} - \frac{\gamma}{2\pi kd}, \tag{9.3}$$

where $\gamma$ is Euler's constant, 0.577. With some care, the first term of the right-hand side could be interpreted as a convective term, the second as a spreading or conductive term, and the third as a correction term. Due to the negative sign, this equation does not represent a simple series resistance network. To guarantee an accuracy of better than 90%, the following inequalities should be obeyed: $b/a > 2$, $ma < 0.5$, $mb < 3$, $md < 0.15$. $m$ is the so-called fin factor, and $a$ and $b$ equivalent radii for respectively source and base area, defined as: $m = \sqrt{h/kd}$, $a = \sqrt{A_1/\pi}, b = \sqrt{A_2/\pi}$. This equation can be used, provided $h/k{\cdot}d < 0.025$; e.g., $h = 2{,}500$ W/m$^2$K, $k = 100$ W/m K, $d = 1$ mm: $h/k{\cdot}d = 0.025$. While this equation is acceptable for a certain range, its use is not recommended because too many conditions have to be met to guarantee accurate values.

(c) The maximum spreading resistances is defined as (Lasance 2008):

$$R_{\text{max}} = \frac{T_{\text{max}} - T_{\text{average}}}{q}. \tag{9.4}$$

The temperature rise $\Delta T_{\text{max}}$ from junction to ambient is calculated using (Lasance 2008):

$$\Delta T_{\text{max}} = q(R_{\text{max}} + R_{\text{convection}}). \tag{9.5}$$

At first sight this expression looks like an ordinary series resistance network. However, it should be understood that $R_{max}$ is a function of the boundary conditions. It showed that the errors stay within 5% or less for the majority of cases of practical interest. Equations (9.4) and (9.5) can still be used over a very large range but errors increase when $h/k > 10$, $k_{subm} > 100$ W/m K and $d < 1$ mm.

More practically, consider an example problem as shown in Figure 9.3. A device is mounted on a heat spreader which is attached to a heat sink with a known average external sink-to-ambient thermal resistance. Because the majority of the heat dissipated by the device will flow through the heat spreader and into the heat sink, the heat loss through the other exposed surfaces may be assumed negligible, and the thermal system can be approximated by a simple serial resistant network. The chip and heat spreader can be considered as a combination of 1-D material and spreading resistances and, in the case of the heat sink, the total resistance consists of the sink-to-ambient resistance and the spreading resistance in the heat spreader. Many semiconductor devices are square or rectangular. However, the spreading resistances associated with the different parts of the system can be determined by considering the axisymmetric problem, provided that the square root of the area is used as the characteristic length, and the same area ratio is used. Heat enters the top surface of the heat spreader over a concentric circular surface of radius $a$ and leaves

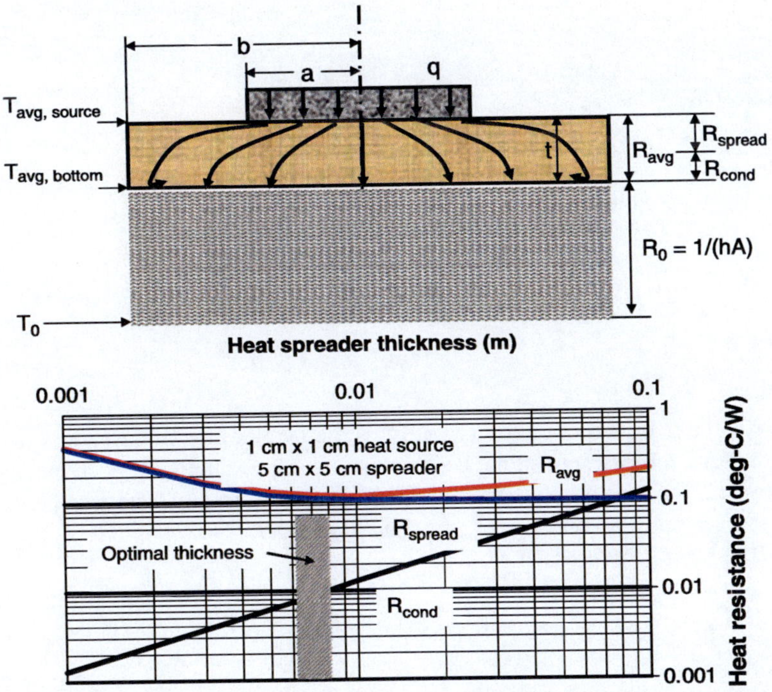


**Fig. 9.3** Sketch of thermal spreading modeling and optimization of the heat spreader thickness

the heat spreader of radius $b$ through the bottom surface over which a uniform heat transfer coefficient, or an external resistance is prescribed. The remaining top and side surfaces are assumed to be adiabatic. For noncircular devices, the equivalent contact and heat spreading radii are obtained as follows (Lee et al. 1995):

$$a = \sqrt{A_s/\pi}, \tag{9.6}$$

$$b = \sqrt{A_p/\pi}, \tag{9.7}$$

where $A_s$ is the contact area of the heat source, and $A_p$ is the area of the heat spreader. The governing differential equation for this problem is Laplace's equation in a 2-D cylindrical coordinate system (Lee et al. 1995):

$$\frac{1}{r}\frac{\partial}{\partial r}\left(r\frac{\partial T}{\partial r}\right) + \frac{\partial^2 T}{\partial z^2} = 0. \tag{9.8}$$

The boundary conditions are

$$\text{at } r = 0,\quad \frac{\partial T}{\partial r} = 0, \tag{9.9}$$

$$\text{at } r = b,\quad \frac{\partial T}{\partial r} = 0, \tag{9.10}$$

$$\text{at } z = 0,\quad \frac{\partial T}{\partial z} = hT, \tag{9.11}$$

$$\text{at } z = t,\quad k\frac{\partial T}{\partial z} = q \quad \text{for} \quad 0 < r \le a, \tag{9.12}$$

$$k\frac{\partial T}{\partial z} = 0 \quad \text{for} \quad a < r \le b, \tag{9.13}$$

where $k$ is the thermal conductivity, and $T$ is the local temperature excess over the ambient fluid temperature. The heat source is represented by $q$, denoting the uniform heat flux over the contact area, and $h$ is the constant heat transfer coefficient at the bottom surface of the heat spreader. One popular solution of (9.8)–(9.13) can be expressed as (Lee et al. 1995):

$$R_{\text{spread}} = R_{\text{avg}} - R_{\text{cond}} = (T_{\text{ave, source}} - T_{\text{ave, bottom}})/q = \Delta T/q, \tag{9.14}$$

$$R_{\text{cond}} = \frac{t}{k(\pi b^2)}, \tag{9.15}$$

$$R_{\text{avg}} = \frac{\Psi_{\text{avg}}}{ka\sqrt{\pi}}, \tag{9.16}$$

$$\Psi_{\text{avg}} = \frac{\varepsilon\tau}{\sqrt{\pi}} + \frac{1}{2}(1-\varepsilon)^{3/2}\Phi_{\text{c}}, \tag{9.17}$$

$$\Phi_{\text{c}} = \frac{\tanh(\tau\lambda_{\text{c}}) + \frac{\lambda_{\text{c}}}{B_i}}{1 + \frac{\lambda_{\text{c}}}{B_i}\tanh(\tau\lambda_{\text{c}})}, \tag{9.18}$$

$$\lambda_{\text{c}} = \pi + \frac{1}{\varepsilon\sqrt{\pi}}, \tag{9.19}$$

where

$$B_i = \frac{1}{\pi\kappa bR_0},$$

$$\varepsilon = a/b$$
$$\tau = t/b.$$

Using (9.14)–(9.19), the optimum thickness of chosen heat spreaders can be estimated. As shown in Figure 9.3, for a 1 cm by 1 cm heat source, and 5 cm by 5 cm heat spreader, its optimum thickness is 6–8 mm. Typical experienced rule-of-thumb suggests that the heat spreader thickness should be approximately 1/4 to 1/2 the lateral chip dimensions. If it is too thick, the heat spreader performance is not significantly improved because of the added thermal resistance produced by the additional thickness. If it is too thin, there is insufficient material to spread the heat laterally away from the source. Typical lateral dimensions for heat spreaders are approximately five times the heat source device dimension for centrally mounted chips, and approximately 2.5 times the device dimension for edge mounted chips. Figure 9.4 compares the spreading resistance of typical desktop heat spreaders made with aluminum, copper, silicon carbide (SiC)/diamond and CVD diamond, which shows that different materials can be used to get the same spreading resistance with different spreading thickness. In fact, heat spreading is a complex phenomenon that can be addressed by analytical formulas only for geometrically simple cases for which no explicit solution exists. For situations where double-sided heat transfer plays a role, or multiple sources or multiple layers, the problem becomes intractable from an approximate analytical point of view and computer codes must be used. Implicit solutions are known for multilayer cases with multiple sources and uniform boundary conditions, even when time is a parameter. User friendly software has been developed that is based on these solutions, with the additional advantage that no mesh generation is required. The big advantage of such software is that even people with little background in heat transfer can get insight in

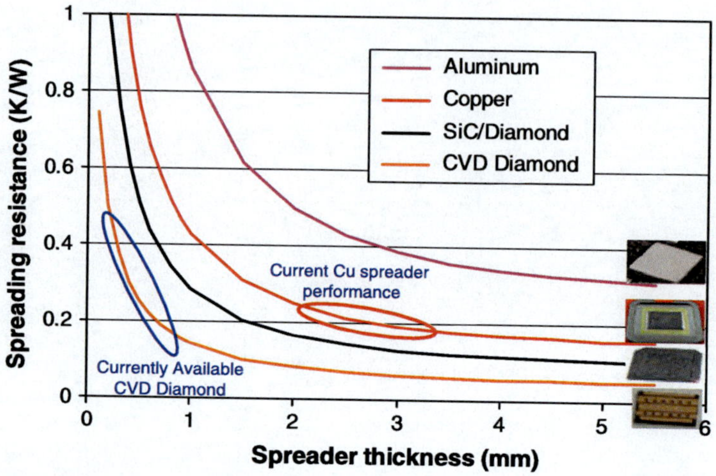


**Fig. 9.4** Spreading model – typical desktop estimate, spreader area 38 × 38 mm; hot spot area 7 × 7 mm. Bottom $R_0 = 0.2$ K/W. Different materials can be used to get the same spreading resistance with different spreading thickness

the physics underlying heat spreading by simply changing a few parameters. For the more practical cases for which layers consist of more than one material or for which the boundary conditions cannot be considered uniform, more advanced conduction-only codes should be used (Lasance 2008).

## Type of Heat Spreaders and Their Materials Selection

A typical electronic packaging consists of various levels of cooling requirements, including heat spreading within a semiconductor device package, heat transfer, or spreading from device package to chassis, heat transport from chassis to system heat exchanger such as heat sink, and heat dissipation through system heat exchanger. Heat is generally first dissipated by spreading heat out, and then by sinking heat away. Thermal conductivity of a heat spreader material indicates the ease of thermal transport in the heat spreader whereas heat capacity of a heat sink identifies how much heat is required to raise its temperature 1 K. In other words, a heat spreader resembles a water hose, it can transmit heat like dispensing water. On the other hand, a heat sink is similar to a reservoir; it can store heat as holding water. A heat spreader does not retain heat, so it requires the combination of high thermal diffusivity and low heat capacity. In contrast, a heat sink may store heat so it needs high thermal diffusivity and high heat capacity. During design of a heat spreader, effective thermal conductivity, tailored thermal expansion coefficient, area in the plane parallel to the lead frame, thickness, and physical separation from the chip are major parameters and determine the effectiveness of a heat spreader, while high heat flux and miniaturization are two dominant design concerns in current electronic packages.

Various heat spreading methods have been widely used in electronic packaging. Heat spreaders are usually used in die level packaging to spread heat from the microprocessor chip into the package. For instance, multichip modules have been built to incorporate a copper heat spreader. On a component level, active spreaders such as heat pipes are used to move heat from chip packages to heat sinks. Heat sinks themselves can incorporate spreaders such as a copper base or heat pipes embedded in an aluminum base. Passive heat spreaders have been incorporated into plasma display panels. Spreaders are commonly used within electronic enclosures to move heat from discrete components to the walls of the enclosure. Materials and process technologies to make heat spreaders have been involved in metallic heat spreaders made of solid copper, aluminum, copper–tungsten or cupper–moly (<400 W/m K); composite spreaders including ceramic, metallic, and carbon composites (<1,200 W/m K); and advanced heat spreader technologies including high conductivity plates (500–5,000 W/m K), vapor chambers (>5,000 W/m K), and oscillating flow heat spreaders (up to 120,000 W/m K). Table 9.1 shows typical materials for fabrication of heat spreaders. Metallic materials have a high concentration of electrons and their thermal conductivity is dominated by electrons, while nonmetallic materials where electrons are not available for thermal conduction, thermal transport typically occurs through phonons, such as diamond or other carbon or ceramic materials. When some applications, such as laser diodes (LDs), and many semiconductor integrated circuits, require both electrical isolation and thermal conduction, dielectric heat spreaders that are made with nonmetallic materials are ideal.

## *Dielectric Heat Spreader Materials*

A dielectric heat spreader formed from a thermally conductive, electrically insulating substrate is used to mount high power semiconductor devices to a printed circuit board, such as RF (Radio Frequency) and microwave devices in which the dimensions of conductors near the device need to be carefully controlled, as these conductors affect performance of the system. The dielectric heat spreader may contain additional components and features such as passive components or transmission lines. The dielectric heat spreader provides thermal transfer between the semiconductor device and a printed circuit board (PCB). Thermal performance of the PCB may be improved by providing thermal vias which provide additional heat transfer from the dielectric heat spreader. The dielectric heat spreaders are made from a thermally conductive, electrically insulating material such as beryllium oxide (BeO), sapphire ($Al_2O_3$), aluminum nitride (AlN), high resistivity silicon, chemical vapor deposited diamond, SiC, and boron nitride. The properties of some typical heat spreading materials are shown in Table 9.1.

**Table 9.1** Typical materials for heat spreaders

| Materials type | Composition | Thermal conductivity (W/m K) | Coefficient of thermal expansion (ppm/K) | Density ($g/cm^3$) |
|---|---|---|---|---|
| Dielectric heat spreading materials (electrical inslulators) | Alumina ($Al_2O_3$) | 30 | 7.3 | 3.9 |
| | AlN | 170–200 | 4.3 | 3.3 |
| | BeO | 250 | 8–8.5 | 2.9 |
| | SiC | 250 | 2.4 | 3.2 |
| | Si | 150 | 2.6 | 2.3 |
| | Chemical vapor deposition diamond | 1,200–1,700 (*Z*); 1,000–1,300 (*X* & *Y*) | 0.8–1.5 | 3.5 |
| | SiC/diamond | 600 | 5–10 | 3.3 |
| Metallic materials (electrical conductors) | Al | 237 | 24 | 2.7 |
| | Cu | 401 | 17 | 8.9 |
| | Metal–diamond composites | ~600 | 3.0–7.5 | 3.3–5.8 |
| | Copper tungsten | | | |
| | 90 wt% W + 10 wt % Cu | 200 | 6.4 | 17.3 |
| | 85 wt% W + 15 wt % Cu | 216 | 7.2 | 16.6 |
| | 80 wt% W + 20 wt % Cu | 219 | 16.2 | 7.6 |
| | 75 wt% W + 25 wt % Cu | 228 | 8.1 | 15.7 |
| | Molybdenum copper | | | |
| | 85 wt% Mo + 15 wt % Cu | 155 | 6.9 | 10 |
| | 80 wt% Mo + 20 wt % Cu | 165 | 7.5 | 9.9 |
| | 75 wt% Mo + 25 wt % Cu | 175 | 8.0 | 9.8 |
| | Al/SiC | | | |
| | 37 vol% Al + 63 vol % SiC | 180 | 8.8 | 3 |
| | 45 vol% Al + 55 vol % SiC | 180 | 10.6 | 2.96 |
| | 63 vol% Al + 37 vol % SiC | 170 | 11.5 | 2.9 |
| | Kovar | | | |
| | 29 wt% Ni + 17 wt % Co + 53 wt% Fe | 17.3 | 5.1 | 8 |
| Carbon materials (electrical conductors) | Natural graphite | 140–500 (*a*, *b*) 3–10 (*c*) | 50 (L); −0.5 (T) | 1.1–1.7 |
| | Annealed pyrolytic graphite (APG) | 1,700 (*x*, *y*) 8 (*Z*) | 20 (L); −1 (T) | 2.24–2.25 |

The supreme thermal conductivity of diamond makes it the natural choice for a range of semiconductor devices operating at high current densities such as LDs, laser diode arrays (LDAs), RF transistors, or high-power electronics. For example, a diamond heat spreader placed between the device active region and a metal or ceramic carrier provides optimum heat transfer by conduction, rapidly removing heat from the active region and spreading it into the much larger volume of metal or ceramic beneath. The majority of the diamond films have been grown on Si wafers, primarily due to availability, low cost, and material characteristics. Crystal growth at the nucleation (Si) surface (800–1,100°C substrate temperature) is exposed to a plasma of hydrogen and a carbon containing reagent such as methane. The ionized hydrogen first attaches to the substrate then combines with the methane radicals leaving diamond, diamond-like-carbon (DLC), and graphite crystals behind. Crystal growth at the nucleation surface occurs first by the accumulation of DLC crystals, then by growth in the planar direction, and finally by epitaxial growth of the film normal to the substrate plane (Z axis). This results in much smaller grain sizes and smaller spaces between the grains on the nucleation side of the diamond film and larger grains and fewer, yet larger, spaces between the crystals on the growth side. Such a coarsening effect will result in a columnar structure of crystals. The columnar structure will continue with increasing thickness of the diamond film, thus, the thicker the film the coarser the diamond grains on the top. The surface finish on the growth side is typically estimated at 5% of the film thickness and the nucleation side is that of the silicon wafer. Polishing the growth side is slow and costly, two times by some estimates. The thermal path is less impeded by grain boundaries along the growth direction (Z axis) and is most impeded in the *X* and *Y* directions. The down side of chemical vapor depositon (CVD) diamond heat spreaders is the low coefficient of thermal expansion (CTE). The semiconductor device must be maintained in a compressive state at all times. The temperature excursions of the device will impact the long-term reliability due to ambient temperature variations, high rate modulation, or CW operation.

The typical commercialized CVD diamond heat spreaders include all sides coated, four sides coated, and patterned CVD diamond heat spreaders. The common metallization used for all and four sides coated heat spreaders are Ti/Pt/Au coatings and eutectic AuSn coatings with sub-layers of TiN and Ti functioning as barrier layers and adhesion and wetting promoters respectively. The patterned CVD diamond heat spreaders can be produced through patterned Ti/Pt/Au and/or Au/Tin metallization for LD, LDA, or multichip module applications. The photolithography process has been used to achieve pattern sizes down to 10 mm in size with high precision. Metallization pads to the edge can be added to the process. Diamond can be bonded onto metal (Cu or Kovar) or ceramic (AlN) submounts using thermal compression bonding, vacuum soldering or die bonding with AuSn or/and AuGe to form CVD diamond heat spreader assemblies. RF power transistors such as those used in GSM (Global System for Mobile Communications) cellular base stations are an ideal application for CVD Diamond substrates. The enhanced thermal performance of CVD diamond with respect to the traditional BeO substrates ensures a better linearity and RF performance and increases the lifetime of the device. In addition, the size of the base station can be reduced significantly. Ultraflat diamond film substrates are particularly useful as heat spreading substrates for high-power LDAs. A CVD diamond heat spreader mounted between the laser bar and a Cu or Si microcooler can reduce the overall thermal resistance of the package by 30%. This opens the door to power levels of up to 100 W from a single bar, with lifetimes well in excess of 10,000 h.

CVD diamond is the ultimate heat spreader material. A tuneable CTE is currently not available; however, experiments are underway to ascertain adhesion characteristics of various metals, to the polished surface, and dielectric stresses associated with thick Cu and Au plating. Very high volume applications in a multiplicity of industries will reduce the cost. Performance applications will dominate the immediate future.

BeO still has an unjustifiable cloud, worldwide, over its application domain. Machining, grinding, and dicing in a dry dusty environment is the basis of the toxicity reports. Reputable vendors process BeO in a wet, high exhaust, filtered, environment that precludes dust inhalation. The thermal characteristics of BeO heat spreaders, on a comparative basis, are quite attractive. They have good CTE match for GaAs devices.

Aluminum Nitride (AlN) is ranked third in the hierarchy of heat spreader dielectric materials. The CTE match to Si is considered near optimum with the cost less than half of BeO. However, the CTE match to GaAs is unsuitable. With its thermal conductivity, high electrical resistivity, and good thermal expansion coefficient, AlN demonstrates high performance as a heat dissipating substrate or as a heat spreader for low power LDs, light-emitting diodes and Si RF microwave devices. With a slight modification most of the CVD diamond processes can be used for AlN heat spreader fabrication, such as all side coated AlN heat spreaders using a thin film metallization process of Ti/Pt/Au, patterned AlN heat spreaders with Ti/Pt/Au pattern and/or additional AuSn solder layer. AlN heat spreader assemblies using thermal compression bonding or soldering processes with AuSn and AuGe for laser submounts.

As a result, dielectric materials provide a great deal of flexibility in a wide variety of, heat spreader, high power applications. The dielectric heat spreaders are particularly suitable for using in the RF and microwave arts includes devices such as power amplifiers, Gallium Arsenide (GaAs) devices, mixers, or high speed digital logic devices, which is mounted to the dielectric heat spreader using 80Au/20Sn solder, or other adhesives (Hutchison et al. 2002).

### Metallic and Composite Heat Spreader Materials

Traditionally, CuW and CuMo are the leading materials in CTE matched heat spreaders. The W and Mo typically constitute 60–90% of the composite, which imparts the lowered CTE but reduces the thermal conductivity to less than half of Cu alone. When the source of heat is a point source, such as laser diodes with

0.25 × 0.25 mm area, the reduced thermal conductivity of composite heat spreaders are unsuitable. Therefore, metal matrix composites (MMCs) are finding widespread use in electronic packaging where strength, weight, thermal conductivity, and thermal expansion are important. Because of their composite nature, the properties of MMCs can be tailored to suit the particular application based on the contribution of individual materials. Applied to electronics cooling applications, the primary advantage of MMCs is the ability to tailor the CTE to match bonded and mating materials, such as copper, braze material, ceramic substrates, and silicon chip. This reduces interfacial shear stresses that result from relative expansion or contraction, and improves fatigue and failure rate characteristics. For instance, Al–SiC metal matrix composite heat spreader lids for system-in-package (SiP) applications provide high thermal and structural reliability along with improved packaging functionality and flexibility. Al–SiC offers device-compatible CTEs to reduce stresses due to differences in device and assembled materials thermal expansion rates and high thermal conductivity to improve SiP reliability. The density of Al–SiC is one-third that of Cu and its stiffness is one-third greater. The low weight reduces the weight per solder ball to improve yield and vibration tolerance, and the weight/stiffness combination makes larger lids possible.

A variety of advanced materials have been proposed for use as heat spreaders, such as the use of multilayer diamond heat spreader substrates bonded to silicon chips; composite heat spreader with carbon fibers dispersed in a carbon or metal matrix; 2-D, composite heat spreaders made from C–C and carbon–silicon carbide (C–SiC) materials. The thermal performance of these materials was found to be superior to that of metal alloy spreader materials with similar CTE such as copper-tungsten. Advanced materials have also been used as heat spreaders in electronic assemblies and housings. The conformal metallic layers are applied to the inside of a molded plastic electronics enclosure. The coating provides both heat dissipation and EMI shielding. A graphite loaded polymer (GLP) has been used as a material for the housing of cell phone handset. The thermal performance of GLP handsets were compared with those of handsets made with conventional acrylonitrile–butadiene–styrene and incorporating aluminum or copper heat spreaders. C–C heat spreader has been applied in a laptop computer. This heat spreader was used to move heat from laptop motherboard to the back of the liquid crystal display (LCD). Carbon fibers in various orientations were used to form heat transfer conduits to move heat from an absorption plate and spread the heat uniformly across the back of the LCD. Two-phase heat spreaders are incorporated microfabricated structures to enhance boiling (Smalc et al. 2010).

## *Graphite Heat Spreader*

Graphite material is highly anisotropic, with a thermal conductivity ranging from 140 to 500 W/m K in and along the "*a*" and "*b*" axes (parallel to the layer planes) and from 3 to 10 W/m K along the "*c*" axis (perpendicular to the planes). By comparison, conventional isotropic materials used for spreading heat, such as 1100 series aluminum and 11000 series copper, have a thermal conductivity of 200 and 387 W/m K in all three directions. Natural graphite has a density ranging from 1.1 to 1.7 g/cm$^3$ compared with 2.7 and 8.89 g/cm$^3$ for aluminum and copper respectively. Also, because of their excellent flexibility, natural graphite materials are able to conform well to surfaces under low contact pressures. This combination of properties makes natural graphite a potential substitute for aluminum and copper materials as heat spreaders. In particular, the highly anistropic thermal conductivity of natural graphite implies that a natural graphite sheet can function as both a heat spreader and an insulator and can be used to eliminate localized hot spots in electronic components (Smalc et al. 2005).

Graphite spreaders have been used to cool a hard drive in a typical laptop computer. Figure 9.5 shows some examples of graphite spreaders (Young et al. 2006): (1) graphite spreaders and inserts embedded in flip-chip lid; (2) graphite spreader used for notebook computer; (3) graphite spreader retrofitted in LCD TV. The in-plane thermal conductivity, through-thickness thermal conductivity, specific thermal contact resistance, and resistance to bending are the four most significant material properties that could affect the performance of natural graphite heat spreaders in the laptop application. The in-plane thermal conductivity directly affects the ability of natural graphite to spread heat away from a point source, while the through-thickness thermal conductivity affects heat transfer through the material. The specific thermal contact resistance influences how well heat is

**Fig. 9.5** Examples of graphite spreaders (Young et al. 2006): **(a)** Graphite spreaders and inserts embedded in flip-chip lid; **(b)** graphite spreader used for notebook computer; **(c)** and **(d)** graphite spreader retrofitted in liquid crystal display TV

transferred between the spreader and the internal components of the laptop. Finally, the resistance to bending shows how easily the spreader can conform to differences in the height of the heat sources and sinks. Installed in a laptop, as shown in Figure 9.5b, the natural graphite spreader was able to conform easily to the height differences of the components it contacted. The spreader reduced the overall temperature gradient within the laptop and on the outer case. Heat was transferred from the hard drive, which generated an estimated 16 W of heat, to other areas of the laptop and reduced the temperature rise of the hard drive above ambient by 21%, from 27.8°C above ambient to 21.9°C. The corresponding temperature rise of the left palm rest, directly above the hard drive, was reduced by 35% from 16.2 to 10.5°C, while the rise in temperature of the right palm rest increased from 2.9 to 7.3°C. In addition, natural graphite is an electrical conductor with a typical in-plane electrical resistivity of 6.8 mW m. Because the natural graphite spreader contacts components of the laptop that may be electrically charged, an electrically insulating sheet of polyethylene terephthalate (PET) film was affixed to the spreader to prevent shorting. PET materials are commonly used as low-cost electrical insulators in electronic applications. The thickness of a PET insulating film can vary from as little as 2.5 mm to 0.36 mm. The 2.5 mm has a dielectric breakdown voltage of 490 V and is used in capacitors. As the thickness increases, both the dielectric breakdown voltage and the thermal resistance of the material also increase. PET films are available in roll form and can economically be applied to natural graphite materials in a continuous process. Where low thermal resistance is critical, such as in thermal interface materials, 2.5 mm PET films have been attached to natural graphite to provide electrical insulation. The PET had a relatively minor effect on contact resistance as the thermal power passing through the spreader is relatively low and there is no single highly concentrated heat source. Heat transfer in and out of the spreader occurs over a relatively wide area. For practical purposes, the increases in temperature due to the PET layer are negligible compared with the overall changes in temperature achieved by the spreaders. However, these thin PET films tend to be costly and are not practical for large spreaders where thicker, more durable films are required. However, the net effect of the graphite spreader on the changes in touch temperatures was to reduce the temperature gradient between the left and right palm of the laptop rests from 11.1°C to only 2.9°C. This reduction in case temperature gradient results in more comfortable operating conditions for the user (Smalc et al. 2005).

In addition, graphene or few-layer graphene (FLG) has been explored to use for lateral heat spreaders in silicon-on-insulator-based chips. It is shown that the incorporation of graphene or FLG under the insulating layer can lead to substantial reduction in the temperature of the hotspots. The efficiency of the hotspot removal with graphene depends on the specifics of the device structure and geometry. Numerical experiments suggest that FLG heat spreaders may be technologically more feasible than single layer graphene. The development of graphene can lead to a new type of high heat flux thermal management of nanoelectronic circuits (Balandin et al. 2008).

## *Advanced Heat Spreaders*

Progress in the field of power electronics is guided by miniaturized package design, selection of new materials with high thermal conductivity, manufacturing feasibility, and production cost. Frequently cited applications for new substrate technologies with enhanced thermal performance are, e.g., power semiconductor lasers where optical characteristics are affected by temperature gradients inside of the component, and insulated-gate-bipolar transistors for power control assemblies where power loss densities of more than 100 W/cm$^2$ must be removed, frequently not exceeding a junction temperature of 125°C. In either case the highest possible heat transfer coefficient between component and cooling agent is required. For this purpose, packaging concepts based on a new substrate technology are needed. Figure 9.6 shows some examples of advanced heat spreaders: (1) Si–microwisker heat spreader; (2) vapor chamber with planar heat pipes; and (3) Oscillating flow heat spreader–metallic plate with oscillating channels. The highest heat transfer rates are obtained using liquid coolants. For best thermal performance even substrates with microchannels or cooling pipes are used, such as vapor chamber and oscillating flow heat spreaders. Drawbacks of such rather complex structures are costly manufacturing techniques and serious limitations of the flow rate because the flow resistance rises dramatically with decreasing cross section. The heat transfer rate is limited by the onset of boiling. A promising attempt to increase the possible heat transfer rate per

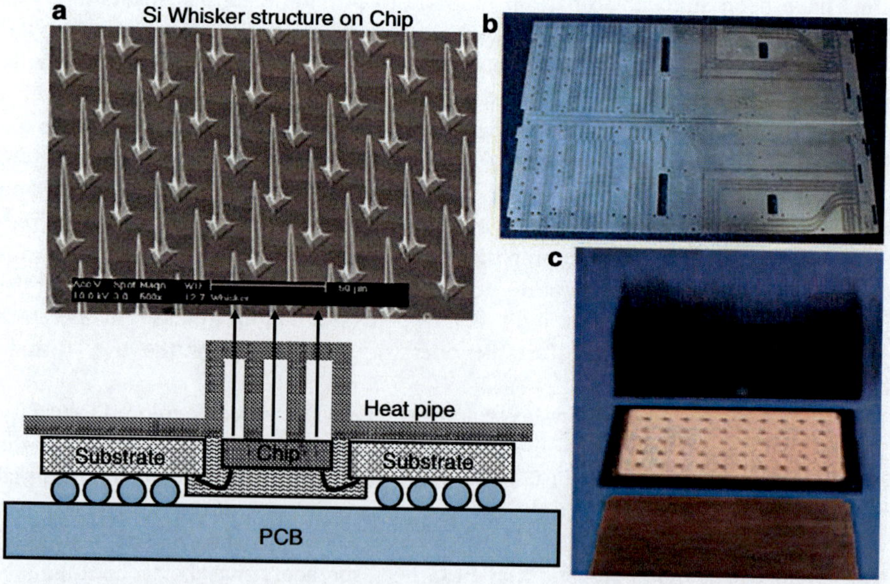


**Fig. 9.6** Examples of advanced heat spreaders: **(a)** Si–microwisker heat spreader; **(b)** vapor chamber with planar heat pipes; and **(c)** oscillating flow heat spreader–metallic plate with oscillating channels

unit of area is significantly the use of silicon substrates structured with needle-like microwhiskers, as shown in Figure 9.6a (Hanreich et al. 2000).

As shown in Figure 9.6b, vapor chambers are flat, thin heat pipes that are typically used as heat spreaders for cooling of high power, high heat flux electronics. A concentrated high heat flux source is attached to one surface of the vapor chamber. The heat input vaporizes the working fluid. The vapor spreads to the entire inner volume and condenses over a much larger, cooler surface of the vapor chamber. The condensed liquid is transported back to the heat input area in the wick structure lining the vapor chamber inner wall. In some cases, the vapor chamber is referred to as a "heat flux transformer" because of its ability to convert higher heat fluxes into lower heat fluxes. Vapor chambers may be used to cool a single microprocessor or multiple processors in a single plane. The vapor chamber can accept heat from each source and transfer the heat to an integral air cooled heat sink or water cooled edge rails. The vapor chamber can be bonded to the base of the heat sink. It accepts concentrated heat inputs and transfers the heat uniformly over the entire heat sink base. An isothermal heat sink base eliminates the temperature gradients due to conduction and improves the fin efficiency of the heat sink. Some of the benefits of using vapor chambers include (1) effective thermal conductivity can be orders of magnitude greater than copper spreaders; (2) direction independent (isotropic) heat spreading; and (3) lighter than equivalent size copper spreaders. The effective thermal conductivity of a vapor chamber is dependent on the ratio of the heat removal area to the heat input area. This ratio is sometimes referred to as the area aspect ratio. The effective thermal conductivity of the vapor chamber increases with the area aspect ratio. Typical effective thermal conductivities range from 5,000 to 200,000 W/m K.

Oscillating flow heat spreaders remove heat from a concentrated heat source and spread the heat over a larger area by mechanically oscillating fluid in a channel, as shown in Figure 9.6c. The oscillating fluid has no net motion but simply oscillates back and forth. Heat diffusion (effective thermal conductivity) is enhanced by the axial fluid mixing caused by rapidly changing velocity profiles in the fluid. Heat transfer coefficients are also enhanced by the disruption of boundary layers near the walls of heat exchanging surfaces. Oscillating flow heat spreaders offer several benefits. First, they do not need excess fluid reservoirs, and therefore conserve valuable space in electronics packages. Second, they tend to have less thermal gradients across heat sources compared with traditional pumped liquid loops. Pumped liquid accumulates heat as it travels past heat sources causing large temperature gradients across heat sources. Because heat diffuses along oscillating flow, smaller temperature gradients can be maintained across heat sources. Finally, the largest benefit is their ability to absorb heat fluxes in excess of 1,000 W/cm$^2$ because of the elimination of the critical heat flux limit. Oscillating flow channels can be integrated into the heat sink base to spread heat from the heat source located at the center of the heat sink. The temperature gradient along the heat sink base was reduced by 70% as compared with the baseline heat sink. The maximum junction temperature was decreased by 30%. Table 9.2 summarizes the comparisons between the oscillating flow heat spreader technology and traditional liquid cooling and air cooling (the details will be discussed later).

**Table 9.2** Comparison of oscillating flow heat spreading and traditional aircooling and liquid cooling system

| Parameter | Oscillating flow heat spreader | Air cooling heat sink assembly | Liquid cooling cold plate |
|---|---|---|---|
| Heat flux limit | >1,300 W/cm$^2$ | <10 W/cm$^2$ | <100 W/cm$^2$ |
| Thermal conductivity | ~120,000 W/m K | <1,200 W/m K | ~5,000 W/m K |
| Heat transfer coefficient | >60,000 W/m$^2$ K | <30 W/m$^2$ K | <10,000 W/m$^2$ K |
| Volume | Very compact | Very large | Very compact |
| Power consumption | Moderate | Very large | Moderate |
| Integrate-ability | Easy to integrate | Easy to integrate | Difficult due to flow imbalance |

## Air Cooling Heat Sink

Heat sinks are devices that enhance heat dissipation from a hot surface, usually the case of a heat generating component or a heat spreader, to a cooler ambient, usually air. In most situations, heat transfer across the interface between the solid surface and the coolant air is the least efficient within the system, and the solid–air interface represents the greatest barrier for heat dissipation. A heat sink lowers this barrier mainly by increasing the surface area that is in direct contact with the coolant. This allows more heat to be dissipated and/or lowers the device operating temperature. The primary purpose of a heat sink is to maintain the device temperature below the maximum allowable temperature specified by the device manufacturers.

### *Type of Air Flow Heat Sinks*

Heat sinks can usually be divided into two types: active and passive. Active heat sinks utilize power and are usually a fan type or liquid cooling or some other peltier cooling device. Passive heat sinks are fully reliable because they have no mechanical components. Passive heat sinks are usually made of a metal such as aluminum-finned radiator that dissipates heat through convection. For passive heat sinks to work to their full capacity, it is recommended that there is a steady air flow moving across the fins. Active heat sinks are most common in electronic devices. Through a large number of fins, it extracts as much heat as possible from the air flowing through the fan and heat sink. The complexity of the air flow inside the electronics cooling system results in a nonuniform flow throughout the active heat sink. This approximation of the air flow through the fins is further increased by the use of fan curves that can oversimplify the effect of the fan, neglecting the impact of the system geometry. An explicit modeling of the fan within the active heat sink, including the entire blade details, as an integral part of the global system model, has been developed dramatically. A more realistic description of the system design leads to more accurate evaluation of the heat sink performance, which will result in a better system design.

The most common types of air cooled heat sinks can be classified as shown below in terms of manufacturing methods and their final form shapes (Lee 1995):

(1) *Stampings*. Stamped heat sinks are usually light-gauge stampings of aluminum and copper with desired shapes. Because they are stamped from a sheet of metal, these heat sinks have low fin density. The mounting surface is usually not flat, requiring high mounting torque or a conformable interface material. They are used in traditional air cooling of electronic components and offer a low-cost solution to low-density thermal problems. They are suitable for high volume production because advanced tooling with high-speed stamping would lower costs. Additional labor-saving options, such as taps, clips, and interface materials, can be factory applied to help to reduce the board assembly costs.
(2) *Extrusion*. Extruded heat sinks are either bidirectional or omnidirectional. Bidirectional sinks let air flow either way along the extrusion direction, which allows the formation of elaborate 2-D shapes capable of dissipating large heat loads. They may be cut, machined, and have options added. A crosscutting will produce omnidirectional, rectangular pin fin heat sinks, and incorporating serrated fins improves the performance by approximately 10–20%, but with a slower extrusion rate. One advantage of the omnidirectional type is that the pin-fin arrangement allows for more precise positioning of the heat sink using clips. Extrusion limits, such as the fin height-to-gap fin thickness usually dictate the flexibility in design options. Typical fin height-to-gap aspect ratio of up to 6 and a minimum fin thickness of 1.3 mm are attainable with a standard extrusion. A 10 to 1 aspect ratio and a fin thickness of 0.8 in can be achieved with special die design features. However, as the aspect ratio increases, the extrusion tolerance is compromised.
(3) *Bonded/Fabricated Fins*. Most air cooled heat sinks are convection limited, and the overall thermal performance of an air cooled heat sink can often be improved significantly if more surface area can be exposed to the air stream. These high performance heat sinks utilize thermally conductive aluminum-filled epoxy to bond planar fins onto a grooved extrusion base plate. This process allows for a much greater fin height-to-gap aspect ratio of 20–40, greatly increasing the cooling capacity without increasing volume requirements. In a bonded-fin heat sink, for instance, an extruded aluminum base holds plate fins in place, usually with an epoxy adhesive. Typical fin aspect ratios are around 10 or more; in many applications, the performance of a bonded-fin heat sink compares with that of a good extruded heat sink. The advantage of the bonded fin's construction method is that its heat-sink base can hold much better dimensional tolerances, and, if the heat sink is very wide, the bonded fin's cost is comparable with that of an extrusion. In addition, fins taller

than 3 in are available, and their height can vary along the length of the heat sink. This height variation can even extend to leaving gaps in the fins where fins are not needed. In this case, bonded-fin technology can offer potential savings over an extrusion because it is unnecessary to remove any fin material.

(4) *Castings*. Sand, lost core, and die casting processes are available with or without vacuum assistance, in aluminum or copper/bronze. This technology is used in high density pin fin heat sinks which provide maximum performance when using impingement cooling.

(5) *Folded Fins*. Corrugated sheet metal in either aluminum or copper increases surface area and, hence, the volumetric performance. The heat sink is then attached to either a base plate or directly to the heating surface via epoxy-bonding or brazing. These heat sinks pack a lot of heat-transfer surface area into a small space, but at the price of air pressure drop. It is not suitable for high profile heat sinks on account of the availability and fin efficiency. Hence, it allows high performance heat sinks to be fabricated for applications. It is needed to duct the air flow into the heat sink to get good performance from the sink. As with the bonded fins, the fan cannot move as much volume because of the pressure drop. A typical fan curve shows that as pressure increases, air flow decreases. Another consideration is that the convoluted fins do not perform well in natural convection, which is a potential problem if the system fan fails.

Figure 9.7 shows the typical rage of cost functions for different types of heat sinks in terms of the required thermal resistance. The comparison of cost, suitable application, thermal resistance, advantages, and disadvantages for different types of heat sinks are shown in Table 9.3.

## *Heat Sink Design Constraints and Design Parameters*

Commercial air cooling heat sinks are typically fabricated of extruded aluminum lamella plates that are compressed together leaving air flow ducts between the plates. The copper-layered heat sink is manufactured by compressing extruded copper plates on the base of the heat sink. A coat of protective finish is applied on the copper and aluminum junction. Assuming (1) the effect of the edge surfaces of the heat sink is ignored; (2) the heat source is assumed to be generating uniform heat flux; (3) the heat source is centrally mounted on the base plate, and the heat sink is cooled uniformly over the exposed finned surface; (4) all heating power is conducted to the fins so that external heat losses are legible; (5) the contribution of radiation is neglected because it is rather small for forced convection and the polished surfaces of the aluminum reduce radiation, the heat sink thermal resistance $R_T$ that includes fin resistance $R_f$, flow resistance $R_{flow}$ and spreading resistance $R_{spd}$ can be expressed as (Behm and Huttunen 2001):

$$R_T = R_f + R_{flow} + R_{spd}. \tag{9.20}$$

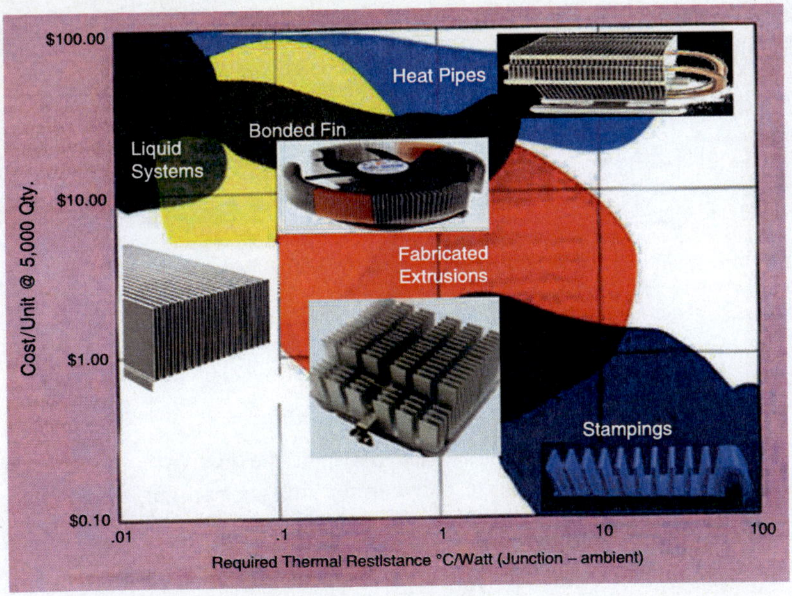


**Fig. 9.7** Types of heat sinks: cost vis. required thermal resistance

For a single fin,

$$R_{\mathrm{fin}} = \frac{1}{\eta_{\mathrm{f}} h A_{\mathrm{fin}}}, \tag{9.21}$$

where $h$ is average heat transfer coefficient (W/m$^2$K); $A_{\mathrm{fin}}$ is surface area of a single fin (m$^2$); $\eta_{\mathrm{f}}$ is fin efficiency. When the heat convection coefficient is uniform over the fin surface area, the fin efficiency is given as

$$\eta_{\mathrm{f}} = \frac{\tanh(mL)}{mL}, \tag{9.22}$$

where $L$ is length of the fin ($m$). When the width of a fin is much larger than its thickness,

$$m = \sqrt{\frac{2h}{kt}}, \tag{9.23}$$

where $k$ is effective thermal conductivity of the fin (W/m K), $t_{\mathrm{fin}}$ is fin thickness (m).

The thermal resistance of the base between the fins is

$$R_{\mathrm{base}} = \frac{1}{hA_{\mathrm{base}}}, \tag{9.24}$$

**Table 9.3** Comparison of different types of heat sinks

| Type | Cost index | Most suitable application | Thermal resistance | Pros | Cons |
|---|---|---|---|---|---|
| Stamped sink | 0.5 | Low power | 1–100°C/W | Inexpensive | Low performance |
| Extruded sink | 1 | Most applications | 0.1–12°C/W | Versatile | Limited size |
| Bonded fin | 2 | Large applications | 0.01–5°C/W | Close tolerances | Expensive |
| Conviluted fin | 3 | Ducted air | 0.01–0.15°C/W | High heat-flux density | Expensive, needs ducting |
| Fan sink | 5 | In a pinch | 0.01–10°C/W | Band-aid solution | Reliability, cost, warm-air circulation |

where $A_{\mathrm{base}}$ is surface area of the base between the fins (m$^2$). As the thermal resistances of the fins and the base between the fins are parallel, the total thermal resistance for the fin system with $N$ fins would be

$$R_{\mathrm{f}} = \frac{1}{\frac{N}{R_{\mathrm{fin}}} + \frac{1}{R_{\mathrm{base}}}}. \tag{9.25}$$

Heat flow is transferred to the coolant fluid around the fins and is calculated as

$$Q = q_{\mathrm{m}} C_p (T_{\mathrm{f,out}} - T_{\mathrm{f,in}}), \tag{9.26}$$

where $q_{\mathrm{m}}$ is mass flow of coolant (kg/m$^3$); $C_p$ is specific heat capacity (J/kg K). Maximum temperature difference in the fluid divided by total heat dissipation derives to the flow thermal resistance

$$R_{\mathrm{flow}} = \frac{1}{q_{\mathrm{m}} C_p}. \tag{9.27}$$

Thermal conductivity is calculated for the copper-layered heat sink base plate. The base plate of the heat sink can be treated as a homogeneous material with two effective thermal conductivities, the in-plane conductivity $k_{\mathrm{pl}}$ and the through-plane conductivity $k_z$. The equations for calculating conductivities of the copper-layered base plate are given by

$$k_{\mathrm{pl}} = \frac{\sum_{i=1}^{N} k_i t_i}{t}, \tag{9.28}$$

$$k_{\mathrm{pl}} = \frac{t}{\sum_{i=1}^{N} k_i t_i}. \tag{9.29}$$

Effective conductivity $k$ can be calculated by taking an average value of the results of these equations. In cases where the footprint of a heat sink is much larger than the size of the heat source, the contribution of the spreading resistance to the overall device temperature rise is significant. Therefore heat spreading in the base of the heat sink is calculated analytically by using verified correlations for the spreading resistance. The spreading resistance can be determined from (9.30) to (9.35). This set of correlations results in an accuracy of approximately 10% compared with measurements. Because fluid is flowing through heat sink, pressure loss occurs. The pressure loss is a sum of the kinetic and viscous losses.

$$\Delta p_{\text{drop}} = \frac{1}{2}\left(\sum \zeta + f\frac{L}{D_{\text{h}}}\right)\rho v^2, \tag{9.30}$$

where $f$ is friction factor, $\Sigma\zeta$ is sum of kinetic loss coefficient. The friction factor can be obtained from the chart drawn after Moody (Bejan 1995). Kinetic loss coefficient at entrance is expressed as

$$\zeta_{\text{in}} = \frac{0.681}{1 + 1.116\left(\frac{A_{\min}}{A_{\max}}\right)^{0.21}} - 0.182, \tag{9.31}$$

and at exhaust

$$\zeta_0 = 1 - 2\left(\frac{A_{\min}}{A_{\max}}\right) + \left(\frac{A_{\min}}{A_{\max}}\right)^2, \tag{9.32}$$

$$\frac{A_{\min}}{A_{\max}} = \frac{1}{1 + \frac{t_{\text{bp}}}{s}}, \tag{9.33}$$

where $s$ is width of flow channels between plates (m).

The common fans used for heat sink and their characteristics are shown in Table 9.4, including (1) propeller fans: a basic fan, consisting of a motor and a propeller, which tip vortices can be a source of turbulent and noise; (2) tube axial fans are most commonly used in electronic cooling, which are similar to propeller fans, but with a venture around the blades, reducing the tip vortices; (3) vane axial fans improve flow rate while reducing the pressure loss by means of vanes or stators incorporated into the fan housing to straighten the air flow; and (4) centrifugal blowers are often used in applications with confined configurations such as laptop computers, typically draw air in axially and blows out radially, and provide less air flow but with higher air pressure than fans. Combining fans in series or parallel can sometimes achieve the desired airflow without greatly increasing the system package size or fan diameter. Parallel operation is defined as having two or more fans blowing together side by side. The performance of two fans in parallel will increase

**Table 9.4** Type of fans and their characteristics

| Fan type | Wheel type | Static pressure (in water gauge) | Wheel diameter (in) | Air flow (cubic feet per minute, CPM) | Break horse power (bhp) |
|---|---|---|---|---|---|
| Propeller axial | Direct drive | 0–1 | 10.0–50.0 | 50–50,000 | 1/6–10 |
| Tube axial | Direct drive | 0–1 | 20.0–50.0 | 2,500–50,000 | 1/3–15 |
| Vane axial | Direct drive | 0–5 | 20.0–60.0 | 1,200–150,000 | 1/3–150 |
| Centrifugal blower | Backward inclined | 0–12 | 10.0–75.0 | 500–125,000 | 1/3–200 |
| | Backward inclined airfoil | 0–14 | 20.0–120.0 | 1,500–450,000 | 1/3–1,500 |

the volume flow rate (double at maximum delivery). The best results for parallel fans are achieved in systems with low resistance. In series operation, the fans are stacked one upon the other, resulting in an increase of static pressure (doubling at shut-off, but less elsewhere). The best results for series fans are achieved in systems with high resistance. In both series and parallel operation, particularly with multiple fans (5, 6, 7, etc.), certain areas of the combined performance curve will be unstable and should be avoided. This instability is unpredictable and is a function of the fan and motor construction and the operating point. It is also important to consider fan placement in the enclosure. Pressurizing the enclosure is the preferred method because incoming air can be readily filtered. In addition, a pressurized enclosure will prevent dust from entering through cracks or crevices. The fan is also handling cooler, denser air, and it will therefore have a slightly higher pressure capability (this may be a very slight advantage for low heat dissipating systems). An important feature of a pressurized system is that the fan life and reliability are increased due to the fan ambient temperature being lower. The disadvantage of pressurization is that heat generated by the fan is dissipated into the enclosure.

The fan laws are generated and used to determine the output of a given fan under other conditions of speed or density, or to convert the known performance of an air mover of one size to that of another geometrically similar unit of a different size. Geometrically and dynamically similar fans can be characterized by the following four equations (Turner 1996):

$$\text{Volumetric flow rate}: G = K_{\mathrm{q}} N D^3, \tag{9.34}$$

$$\text{Mass flow rate}: \dot{m} = K_{\mathrm{m}} \rho N D^3, \tag{9.35}$$

$$\text{Pressure drop}: P = K_{\mathrm{p}} \rho N^2 D^2, \tag{9.36}$$

$$\text{Fan power}: \mathrm{HP} = K_{\mathrm{HP}} \rho N^3 D^5, \tag{9.37}$$

$$\text{Fan noise}: L = a \ln(G) + b, \tag{9.38}$$

$$\text{Thermal resistance of fan}: R_{\mathrm{HS}} = cG^m + d, \tag{9.39}$$

where *K, a, b, c, d, m* are constant for geometrically and dynamically similar operation; *G* is volumetric flow rate; $\dot{m}$ is mass flow rate; *N* is fan speed in RPM; *D* is fan diameter; HP is power output; ρ is air density.

In selecting an appropriate heat sink that meets the required thermal criteria, various parameters need to be examined, including air temperature rise, pressure drop, volume flow rate, heat sink size/weight (fin height, length, thickness, density, profile/shape, cross-cut pattern, flow management, base plate thickness and materials selection), orientation, appearance and cost, which affect not only the heat sink performance itself, but also the overall performance of the system. The choice of a particular type of heat sink depends largely to the thermal budget allowed for the heat sink and external conditions surrounding the heat sink. It should be emphasized that there can never be a single value of thermal resistance assigned to a given heat sink because the thermal resistance varies with external cooling conditions (Lee 1995).

When selecting a heat sink, it is necessary to classify the air flow as natural, low flow mixed, or high flow forced convection. Natural convection occurs when there is no externally induced flow and heat transfer relies solely on the free buoyant flow of air surrounding the heat sink. Forced convection occurs when the flow of air is induced by mechanical means, usually a fan or blower. There is no clear distinction on the flow velocity that separates the mixed and forced flow regimes. It is generally accepted in applications that the effect of buoyant force on the overall heat transfer diminishes to a negligible level (under 5%) when the induced air flow velocity is in excess of 1–2 m/s (200–400 lfm). The next step is to determine the required volume of a heat sink. Table 9.5 shows approximate ranges of volumetric thermal resistance of a typical heat sink under different flow conditions (Lee 1995).

The volume of a heat sink for a given low condition can be obtained by dividing the volumetric thermal resistance by the required thermal resistance. Table 9.5 is to be used only as a guide for estimation purposes in the beginning of the selection process. The actual resistance values may vary outside the above range depending on many additional parameters, such as actual dimensions of the heat sink, type of heat sink, flow configuration, orientation, surface finish, altitude, etc. The smaller values shown above correspond to a heat sink volume of approximately 100–200 $cm^3$ (5–10 $in^3$) and the larger ones to roughly 1,000 $cm^3$ (60 $in^3$). The above tabulated ranges assume that the design has been optimized for a given flow condition. Although there are many parameters to be considered in optimizing a heat sink, one of the most critical parameters is the fin density. In a planar fin heat sink, optimum fin spacing is strongly related to two parameters: flow velocity and fin length in the direction of the flow. The average performance of a typical heat sink is

**Table 9.5** Approximate ranges of volumetric thermal resistance of a typical heat sink

| Flow condition m/s (lfm) | Volume resistance $cm^3$ °C/W ($in^3$ °C/W) |
|---|---|
| Natural convection | 500–800 (30–50) |
| 1.0 (200) | 150–250 (10–15) |
| 2.5 (500) | 80–150 (5–10) |
| 5.0 (1,000) | 50–80 (3–5) |

linearly proportional to the width of a heat sink in the direction perpendicular to the flow, and approximately proportional to the square root of the fin length in the direction parallel to the flow. For example, an increase in the width of a heat sink by a factor of 2 would increase the heat dissipation capability by a factor of 2, whereas and increase the heat dissipation capability by a factor of 1.4. Therefore, if the choice is available, it is beneficial to increase the width of a heat sink rather than the length of the heat sink. Also, the effect of radiation heat transfer is very important in natural convection, as it can be responsible for up to 25% of the total heat dissipation. Unless the component is facing a hotter surface nearby, it is imperative to have the heat sink surfaces painted or anodized to enhance radiation (Lee 1995).

The performance of different heat sink types varies dramatically with the air flow through the heat sink. To quantify the effectiveness of different types of heat sinks, the volumetric heat transfer efficiency can be defined as (Lee 1995):

$$\eta = \frac{Q}{\dot{m}c\Delta T_{\text{sa}}}, \tag{9.40}$$

where, $\dot{m}$ is the mass flow rate through the heat sink, $c$ is the heat capacity of the fluid, and $\Delta T_{\text{sa}}$ is the average temperature difference between the heat sink and the ambient air. The heat transfer efficiencies have been measured for a wide range of heat sink configurations, and their ranges are listed in Table 9.6 (Lee 1995). The improved thermal performance is generally associated with additional costs in either material or manufacturing, or both.

Generally, air cooling heat sink is traditionally used to cool down electronic packaging systems. Originally, thermal design was only considered when laboratory tests failed or a component turned the PC (Personal Computer) board brown. As a heat transfer agent, air is a poor media with the following limitations: (1) low density and low specific heat, resulting in low heat carrying capacity; (2) low thermal conductivity. However, due to its availability, low cost, ease of maintenance, and "no problem if there's a spill" characteristics, air cooling is still the preferred choice. Potential enhancements in air cooled heat sinks can be achieved in either of two ways: (1) Increasing the heat transfer coefficient, $h$; (2) Adding more exposed surface. The heat transfer coefficient is a measurement of how effectively heat is removed from a surface. It is expressed in terms of Watts dissipated per square meter per °C of temperature of rise above the cooling medium. Newton's law of cooling states that heat dissipated is equal to the heat transfer coefficient ($h$) times area times the temperature rise between the heat sink and the cooling air or (Soule 2001):

**Table 9.6** Heat transfer efficiencies of different heat sinks

| Heat sink type | Heat transfer efficiency (%) |
|---|---|
| Stamping and flat plates | 10–18 |
| Finned extrusions | 15–22 |
| Impingement flow fan heat sinks | 25–32 |
| Fully ducted extrusions | 45–58 |
| Ducted pin fin, bonded and folded fins | 78–90 |

$$q = hA(T_{\text{heat sink}} - T_{\text{air}}). \tag{9.41}$$

Only "*A*," the number of square centimeters of surface area of the fins, and "*h*," the effectiveness of the heat removal from each square centimeter of surface can be controlled during design. Both of these parameters are bounded by the limits of cost and marketability. The heat transfer coefficient "*h*" can be increased by a number of methods (Soule 2001):

(1) Increasing airflow velocity is most common. Increased airflow past the heat sink fins generally increases "*h*." It also increases the mass flow rate, reducing the overall temperature of the exit air. Increasing air velocities past 8–10 m/s adds significantly to the backpressure and the acoustic noise.
(2) Cross cutting of flat fins into multiple short sections. This improves the heat transfer coefficient at the fin surface while slightly reducing the surface area exposed to the air stream. The drawback of this method is the resultant additional pressure drop. Typically, crosscutting is best employed where airflow is from an unpredictable direction.
(3) Augmentation of fins is similar to cross cutting but adds a separate "twist" in the leading and trailing edge of a fin. This curvature of the metal into the air flow reduces the "*h*" by scrubbing the dead air molecules away from the surface. Augmentation works best in air velocities of 3 m/s or higher.
(4) Impingement or jet cooling of heat sinks is achieved using a high-speed air flow directed at the fin tip toward the base. This cooling layout can increase heat removal but must be designed with an eye toward the system as a whole and tested in situ to be effective.

Increased heat removal from any of these heat transfer enhancement schemes comes at the price of increased pressure drop. The heat sink, fan, and system as a whole must be examined in conjugate analysis to determine if any of these techniques will produce lower overall thermal resistance than just increased surface area through increased numbers of thin, flat fins. A real limitation in this area is the ability to control contamination in the air stream. Closely spaced fins can cause clogging of the fins and a reduction in thermal performance. The technology of fans is also under scrutiny in the electronics cooling industry. If a fan with higher performance (significantly increased air flow at equivalent pressure drop) becomes available, many designs will be built around that fan. The difficulty in achieving this higher performance comes back to fan blade technology and motor design, as well as acoustic noise limitations. Future developments are expected in this market. In convective heat transfer, the increase in heat removal is based on available cooling surface area. In many cases, the use of an increased number of tall, thin, flat fins will give better performance and less pressure drop than any modification of the fin shape. Based on air flow resistance, the flat fins maintain the lowest pressure drop and hence the greatest mass flow of air. Augmentation of extended surfaces used to dissipate heat increases the overall effectiveness of a heat sink and increases the heat removed per unit volume. This amount of increase depends on the number of augmentations, air flow velocity, and overall length of the cooling channels. A great deal of research is being put into low-cost manufacturing methods of producing extended surfaces that incorporate more fins in smaller heat sink package sizes (Soule 2001).

## *Heat Sink Materials Selection and Fabrication Process*

Heat sinks have been most commonly made by (Keller 1998): (1) extruding, used most often in higher powered and amplifier type cooling; (2) stamped, used in low power discrete component cooling; and (3) machined, used in medium powered discrete component cooling, especially high value items like processors; (4) cast. Cast heat sinks do not have the limitations imposed by extrusions from a 2-D profile nor the backpressure inherent in the saw cut, nonaerodynamic front end of an extrusion. In forced air systems, the squared shape entrance of a saw cut creates a large pressure build up region creating for much of the back pressure in the heat sink.

### Aluminum Extrusions

Aluminum has many characteristics that make it an excellent heat sink material. It is easily shaped in fine detail with inexpensive tooling, but the height and number of fins, or extrusion ratio, limit surface area. The extrusion or aspect ratio of the fins is the limit that the extrusion tool can produce without failure. The aspect ratio of fin height to spacing has increased over the past decade from 6:1 up to as much as 15:1 depending on size and extrusion press operator. The objective of increasing those ratios is to create more fins in a smaller volume. Serrations on extruded fins have for years been believed to increase surface area and hence heat transfer. In most cases, however, they have no added thermal effect. Serrations can increase the heat removal rate but, as typically seen, are not tall enough to provide any increase in cooling over flat fins. Three-dimensional extrusions formed by impact extrusion offer an alternative to the standard 2-D parts. Using the same high-conductivity alloy, impact extrusions are commonly seen as pin fin or elliptical fin parts. Extrusion or draw ratios exceed the capability of linear extrusions, offering an increase in the amount of surface area per unit volume.

Most aluminum extrusion heat sinks used in cooling of electronics use flat fins or extended heat transfer surfaces to dissipate waste heat to the ambient air. Flat fins are simple to manufacture but can only remove a given amount of heat per square inch of surface. This performance decreases dramatically as the length of the fins increases. Staggered and augmented fins have been shown to increase performance by 15–25% or more depending on air speed. The nature of air flow on extended cooling surfaces is to form a layer of stagnant, insulating air over any surface exposed to the air stream. This insulating layer results in a heat sink temperature rise and is caused by the viscous nature of air, or any other fluid, used to transfer heat in a convective mode. The build-up of this surface film or boundary layer decreases the amount of heat energy that can be dissipated from each square centimeter of cooling area by acting as an insulation layer between the heated surface and the cooling air. The thickness, and its ability to transfer heat, of this insulating layer is dependent on many parameters, such as the velocity of the air over the extended surface, the turbulent intensity of the incoming air, the length of the cooling surface in the direction of air travel, and the roughness of the heat transfer surface.

Many methods of increasing the heat transfer characteristics of a flat surface have been used to promote turbulence to help decrease temperature rise. Any increase in heat transfer at the air to metal (usually aluminum) interface will increase overall ability of the heat sink to remove overall heat from the semiconductor and thus reduce temperature rise at a given heat load. This desirable characteristic will help reduce the junction temperatures of the semiconductor devices being cooled or offer a potential reduction in the volumetric requirements of the heat sink. This will allow smaller system package sizes as well as decreased weight and, if correctly designed, reduce the cost of the entire assembly.

The insulating boundary layer near the leading edge of a flat surface is generally thin and the airflow within the layer is relatively fast. This provides an increase in the heat energy the film can remove. As the flow continues, uninterrupted, down the length of the fin the boundary layer thickness or air grows and consequently reduces the amount of heat that can be removed from a given surface area. The thickness of this layer continuously grows as the length of the fin increases and only comes back to the original state when the fin breaks and allows mixing of the cooler duct flow air with the heated molecules from the boundary layer. The boundary layer thickness, and hence its inability to remove heat from a solid surface, increases continuously in the direction of air flow. Depending on air speed, thickness of the boundary layer can increase to the point of merging layers between closely spaced fins, further reducing the heat transfer efficiency.

Extruded fins with minute serrations offer little or no increase in the additional heat transfer when compared with a flat fin of equal height. The thickness of the boundary layer is, in all but the fastest air speeds, thicker than the height of the amplitude of the serrations. This closes off the valleys between the serration's peaks to any additional heat removal, resulting in a stagnant recirculation zone over the valleys and effectively limits any added surface area increase. The result is that the effective surface area of the fin is essentially equal to the straight line height of the fin without the added length of the serrations. The heat transfer coefficient of the fin can be expressed as (Kraus and Bar-Cohen 1983):

$$h = (J/t)^{0.66} \times (0.323) \times (D \times V \times L/D_v)^{0.5} \times D \times C_p \times J, \qquad (9.42)$$

where $h$ is heat transfer coefficient; $J$ is velocity of the air past the fin; $D$ is air density; $L$ is length of the cooled surface in the direction of air flow; $D_v$ is dynamic viscosity; $t$ is thickness of the boundary layer; and $C_p$ is specific heat of the cooling air. This expression applies to both laminar air flow as well as turbulent air flow over flat plate surfaces typical of those found on an aluminum heat sink extrusion. The expression has a strong dependence on the air velocity, physical properties of the cooling air, as well as the length of the surface in the direction of air travel. This equation shows that multiple shorter cooled surfaces will remove more heat than one single surface of the total equivalent combined length. This is due to the heat transfer being directly proportional to the thickness of the boundary layer and to the thinning or reattachment of this layer at every leading edge in the flow pattern. In another word, if the boundary layer of stagnant air along the heat dissipation surface is relatively thin, the heat transfer is proportionately greater than if the boundary layer is thick. The longer the uninterrupted cooled surface the thicker the boundary layer will be. Heat transfer can be improved by reattachment of the boundary layer at every leading edge. Whenever a cooling fin offers a leading edge or new projection into an air stream or when the heat transfer surface is interrupted the film layer becomes thinner and heat transfer is increased. In both natural and forced convection heat removal increases with more surface area exposed to the cooling air stream, as expressed by (Sathyamurthy et al. 1996)

$$Q = hA\Delta T, \tag{9.43}$$

where the heat input is equal to the heat transfer coefficient ($h$) times the surface area exposed to the cooling air flow ($A$) times the differential temperature ($DT$). However, after a certain channel length the addition of more length adds no cooling effect. In effect, in the direction of air flow the efficiency of heat removal decreases as the length of the heat sink increases. This is due primarily to the stagnant layer of air molecules which, through the inherent viscosity of the air stick to the cooling surfaces and act as insulation resisting the flow of heat from the surface to the cooling air stream. This is depicted mathematically as the "$h$" value. When the heated surface is interrupted in the direction of air flow the insulation layer is also interrupted and causes the thickness to decrease. These interruptions cause the boundary layer to become thinner or disappear altogether, allowing increased heat transfer rates to the cooler air passing the surface. The properly designed heat sink limits the length between flow interruptions sufficient to maximize heat removal but still minimize the effect of additional pressure drop. The cost of this additional heat removing capacity is at the price of increased resistance to air flow. To optimize the number of leading edges in a design pressure drop must also be considered. These must not present the cooling air with too much resistance to the air flow, but be enough to be effective. A large number of leading edge points could decrease velocity and the subsequent mass flow of the cooling air to the point that the augmentation techniques would reduce the total thermal resistance of a heat sink to less than the original single piece fin design. This would choke air flow and slow heat removal (Sathyamurthy et al. 1996).

In classic fluid flow analysis the total pressure drop through a heat sink can be broken up into four segments: (1) the entrance effect which analyzes the air as it goes from free steam velocity to channel flow; (2) the flow acceleration, which is based on Bernoulli principles; (3) the core friction effect or the drag that the extended fin surface has on the passing air; and (4) the exit effect as the air again reduces velocity. This section actually decreases the total pressure drop. The friction factor of air flowing in constricted channels between fins will account for a large portion of the total overall system pressure drop. Any effect of inlet and outlet pressure drops and flow acceleration will be small (typically less than 10%) relative to the overall. To help improve the overall thermal performance of finned heat sink, some manufacturing methods have been developed that will modify a flat finned part and create several smaller, more efficient finned sections. This modification can be done using one piece of fin to reduce manufacturing costs and increase in fin heat transfer (Soule 1996).

### Bonded Fin Heat Sinks

Bonded fin heat sinks use separate base and fins assembled to form a heat sink. For over a decade these parts have offered an alternative to the limitations of the extrusion process. Increases in cooling surface area of 200–300% allow significant decreases in heat sink volume. Many different assembly styles are available on the market, such as (1) epoxy bonding of fins into a heat-spreading base; (2) brazed assemblies; (3) cold formed or swaged; (4) welded-ultrasonic or resistance; (5) stacked fins (fin and base extruded individually and formed together); and (6) diffusion bonding.

### Folded Fin

Folded fins were originally developed for the military and aerospace market to provide large amounts of surface area with very little weight. This method of making cooling surface area using corrugated metal sheet has been adopted by the commercial cooling market and is now used in every application from personal computers to radio transmitters. Bonded, brazed, or soldered to heat spreading surfaces, these folded lengths of metal come in many styles and sizes. Manufacturing capability offers fin heights up to 4.0 in with densities of 8–10 fins per inch of width. Flat fins are the most common style.

### Castings

Castings are commonly used as low-cost heat sinks integrated into electronic packaging. Due to cost, castings are limited to high-volume production programs that use tens of thousands of parts. Castings typically provide near net-shaped parts that require little or no machining prior to assembly. Traditionally, the thermal conductivity of aluminum casting alloys has been significantly lower than extruded alloys. That, coupled with the potential for air pores and inclusions in some casting styles, can yield poor thermal designs. Newly developed aluminum casting materials offer conductivities as high as 170 W/m K. These materials offer conductivity close to extruded aluminum but still are limited by fin height to spacing ratios noted here. Also, die life using this new cast material is reduced to one-third of normal. Most cast fin sections are limited in height and density to an aspect ratio of ~4:1. This limits heat dissipation levels or requires additional surfaces be added to the casting; i.e., bonded fin or folded fins.

Although high thermal conductivity is an important factor to consider when choosing heat sink materials, however, this is not always necessary. For instance, in the moderate power ranges, it is the surface area and the heat sink geometry where most of the heat sink efficiency is gained. The perceived problem of resistance of the heat sink material being a major factor of natural or forced air cooling at this power level is only about 5% of the total resistance. Most is in the exchanges from fin surfaces to passing air. Much cost savings in the heat sink can be attained by using less expensive materials or manufacturing methods that only affect a few percent of the entire heat sink efficiency (Keller 1998).

Aluminum has been most commonly used for manufacturing air cooled heat sinks mainly due to the following reasons: (1) it has an exceptionally high thermal conductivity; (2) it is easy to form and fairly easy to machine, although, not nearly as easy as free machining aluminum which has more silicon and other alloys to increase strength and machinability but with a substantial penalty to thermal conductivity; (3) aluminum is also quite light for a highly thermal conductive material. The drawback is that secondary processing is required on all aluminum heat sinks used in electrical packaging. This is a plating required for UL approval. Raw Aluminum would oxidize creating an insulating film that can be easily rubbed down onto the conductive surface. This leaves a danger of rubbing wires causing potential shorts. For casting, aluminum has some other inherent disadvantages. These mostly stem from the poor die castability of pure aluminum. A couple percent of silicon is usually added to the aluminum to allow the casting part to slip out of the mold easily. Unfortunately, this causes the internal structure of aluminum to be nonhomogeneous which reduces its thermal conductivity to about one-third of pure aluminum. In addition, porosity forms inside the casting resulting in poor material contact, greatly increasing the thermal resistance inside the casting. Hot isostatic pressing (HIP) can be used to alleviate this porosity problem but this still has only half the conductivity of raw aluminum; the cost of this method is two to four times higher than standard die casting. However, it may be economically advantageous in systems with strict weight limitations–specifically avionics and portable electronic equipment (Keller 1998).

Brass and bronze are two materials not as frequently used for thermal dissipation. Pure copper has an extremely high thermal conductivity but, even to a more extreme than aluminum, the impurities needed to make casting or machining possible drop its conductivity dramatically. The most common copper heat sink used is a pure copper slug used as a heat spreader on chip carriers. Many medium- and high-powered chip carriers utilize this effective method. On top of this slug often times a heat sink will be mounted to add more surface area for additional cooling. The heat sink-slug-die contains an intermediate copper slug. This slug is normally plated to prevent oxidation. For large heat sinks in corrosive environments where abrasion can occur or strength is required, bronzes (thermal conductivity

60–80 W/m K) are sometimes used, most prevalently in marine environments on commercial gear.

Zinc, with its good thermal conductivity (112 W/m K) (but not great compared with pure aluminum [204 W/m K]) can often be overlooked as a heat sink material. In weight sensitive equipment, this oversight may be acceptable but in most other electronic cooling applications, especially medium to high volume, zinc alloys may have some added benefits over aluminum. Several zinc alloys have unusual properties which make them rather attractive for cast heat sinks. Even with alloys added to zinc to help in strength, durability and castability, zinc can be maintained porosity-free in the casting process. This results in thermal conductivity loss of a much less degree than aluminum or copper. In fact, its cast thermal conductivity exceeds that of most cast aluminum and copper alloys. Low cost zinc, sometime referred to as pot metal, is known as a low-cost material for casting. In the electronic manufacturing industry, it is used extensively in high tolerance motor castings, electronic mounts, and brackets and even switch housings. Places where aluminum castings are also used are a premium but at a penalty of two to three times the cost of a zinc alloy die cast. Low pouring temperature die casting always entails much more tooling costs up front than an extrusion or machined heat sinks. Aluminum die costs can be quite high because the mold must be machined out of steel. Lower melting temperature zinc alloys (the high thermal conductivity ones that we are interested in) have the mold made out of easily machined aluminum, saving about 50% on the mold cost. Greater savings can be made by utilizing carbon molds; although these have an effective life of only about 60,000 parts (Keller 1998).

As can be seen when selecting a zinc alloy, ZA-8 has a thermal expansion match close to many ceramic and copper backed chip carrier packages. Even though its thermal conductivity is less than that of ZA-27, or not shown ZA-12, the problems associated with thermal expansion mismatch can cause stress problems down the road from this expansion difference. Multiple cavity molds should be looked at when very large volumes of heat sinks are required or when the process times require such. When casting aluminum, there is a certain amount of cooling time required to let the core solidify further, which results in a better thermally conductive heat sink. This cooling period takes up a significant amount of time on very expensive die casting machines. Because of this, it is economical to produce multicavity molds even for runs of as low as 20,000 parts. The cooling of zinc alloy in the die is relatively quick partly due to its lower melting temperature–smaller temperature differential change. This results in a much faster die casting time, benefiting in reduced costs per part. Single cavity molds are most economical for most zinc alloy castings until casting volumes pass 60,000 or so. Zinc and its alloys are quite heavy, and as mentioned previously, do not make good heat sinks for portable equipment. For portable equipment, look into cast 380 series aluminum and where very high power densities are required, HIP can be used with 357 or similar series aluminum (Keller 1998).

Cast heat sinks give discrete chip cooling an economical and efficient heat sink. The shapes allow for manufacturing cues, reduced back pressure in the system, and reduce the cost of heat sinks to about 30% of standard extruded or machined ones.

Die casting materials for heat sinks come in many different variations that need to be optimized for their particular use. Aluminum is the most common material because of lower cost, ease of manufacturing, and the existing infrastructure. However, a higher conductivity material is preferred. Copper is getting more popular to be used for both heat sinks and chip integrated circuit (IC) lead frames. A material that was more conductive than aluminum, but not as costly as copper would be viable, especially if easily shaped, such as laminate materials Cu/W and Cu/Mo, graphite materials, and composite materials. Zinc with a moderate amount of aluminum doped in provides an extremely good heat sink material with compatible thermal expansion characteristics with most chip packages at a low per piece cost. Poured aluminum castings create complicated, lightweight shapes less expensively than machined versions and are better tailored to portable equipment (Keller 1998).

## Material Thermal Conductivity

Next generation heat sinks and leading-edge materials in air cooled heat removal have been developed and are increasing, such as sintered heat sinks, metal injection molding, high (5X) performance fans, aluminum extrusions at 25:1 ratio, and self-fanning piezoelectric fin heat sinks. Thermal conduction comes into play in two areas, heat spreading in the base, and fin efficiency. Pure aluminum has a conductivity of 230 W/m K. Its ability to conduct heat in thin sections is not sufficient to cool many high heat flux components. To overcome this conduction limitation, many alternative materials and techniques are being explored, including pure copper as a base material, heat pipes, vapor chambers, and thermosiphons (further details would be given in Chap. 10), increased conductivity of heat spreaders, and copper- or carbon-based materials, laminate, and composite materials. Each of these materials or systems will increase heat conduction compared with an aluminum extrusion. However, they will all be more costly and possibly heavier than aluminum. At 390 W/m K, copper offers a 70% increase in conduction over aluminum, and has been widely used for heat spreading materials. For heat sink material, the down side of copper is that it weighs three times more than aluminum, costs the same on a per pound basis, and is more difficult to machine. Due to limited high temperature formability, a copper extrusion will not yield the same detail as aluminum. Also, machining copper takes more time and wears cutters at a much higher rate. However, when an application is limited in conduction, copper is a commonly used alternative.

Figure 9.8 shows heat pipe based heat sinks: (1) a heat sink is installed in an open space of the PC and the heat is transferred from the central processing unit to the heat sink using heat pipes; (2) the heat from the processor is transferred to the LCD panel of a notebook through microheat pipe based heat sink. A heat pipe is a closed-loop, phase-change heat transfer system that, by itself, does not dissipate heat. It does move heat from one point to another with a very low temperature rise. An average heat pipe has an equivalent thermal conductivity of up to 1,000 W/m K. Although it is not strictly correct to quote thermal conductivity of a heat pipe, it is

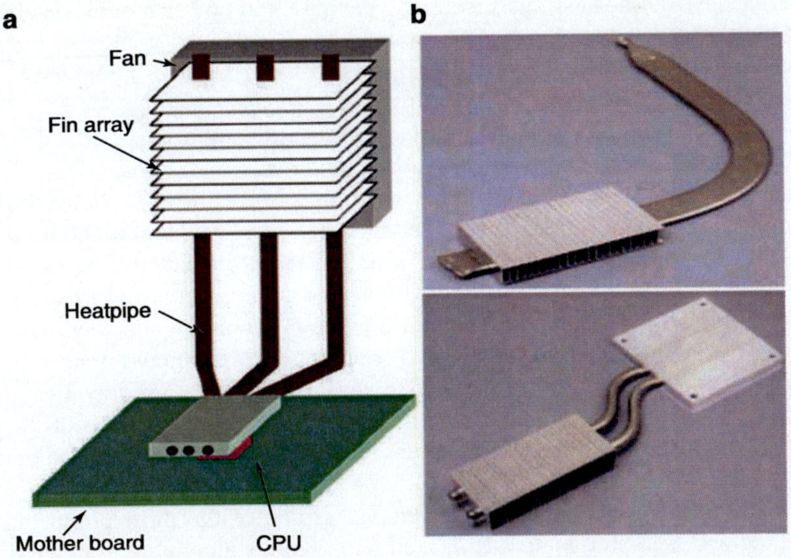


**Fig. 9.8** Heat pipe based heat sinks. **(a)** A heat sink is installed in an open space of the PC and the heat is transferred from the central processing unit to the heat sink using heat pipes; **(b)** the heat from the processor is transferred to the liquid crystal display panel of a notebook through microheat pipe based heat sink

used here as a comparison to solid materials. The heat transport mechanism of a heat pipe is latent heat transport via vapor flow and not thermal conduction. When a value of thermal conductivity is given for a heat pipe, it is usually the equivalent thermal conductivity that a solid conductor of the same cross-sectional area and length would require to conduct the same amount of heat from end-to-end with the same temperature rise. The addition of heat pipes to an aluminum heat spreader can improve the conductivity and reduce spreading resistance with a minimum of weight increase. Heat pipes come in various sizes and styles, typically round but also in square or rectangular cross sections. Most heat pipes used in electronic cooling are copper with water or alcohol as the working fluid. Heat pipes with sintered wicks are the current state of the art and provide the highest heat flux density control. They are also best at returning liquid against gravity by capillary flow. Thermosiphons use buoyancy effects caused by a temperature difference in the working fluid to move heat inside a closed system. These systems can be either single phase (liquid only) or two phase (evaporative and condensing). Loop heat pipes are similar to conventional heat pipes in that they use a phase change liquid to move heat. They are different in that the system forms a circulating loop with hot vapor and cooled liquid in separate lines to and from the condenser. Vapor chambers, or flat plate heat pipes also use a change of phase liquid to move heat. The vapor chamber provides a uniform temperature across the heat dissipation surface with low mass. This technology allows direct attachment of heat dissipation fins and can be used for air cooling of applications including microprocessors,

RF frequency transmitters, thermoelectric devices, and power modules. Advances in vapor chamber technology will revolve around improved wick structures and working fluids. However, water still appears to have the best characteristics for moving heat.

High thermal conductivity polymers, carbon-based composites, sintered metallic powders, phase-change compounds, synthetic diamonds, and strand-oriented graphite materials are all part of leading edge materials for heat sinks. The most desirable properties include controlled levels of conductivity, high machinability, low weight, low CTE, low toxicity, and above all a per pound cost that will be lower than aluminum. Composite heat sinks are emerging and getting increasing attention since nanotechnology can be applied. Composites with nanomaterials such as carbon nanotubes (CNTs) are under way. Metal and nanomaterials may possibly be used together as composites. Future heat sink materials will likely use metallic composites and nanomaterials such as CNTs that have extremely high (~3,000 W/m K) thermal conductivity, especially because considerable funding is available for nanotechnology research.

For instance, Al–SiC has been developed as one of the most promising heat spreading and heat sink materials. Mixed with various aluminum alloys it can be tailored to specific physical properties. Controlled thermal expansion, high conductivity, and significant strength make Al–SiC attractive. Due to cost, this material is typically used in base sections and as substrates for direct die attachment in power module bases. This material has been used in some applications to form the entire heat sink, although because of cost issues, an aluminum extrusion is typically used.

On the other hand, liquid cooling, typically a water and glycol mixture, overcomes most of the heat removal and transport limitations of air cooled systems. Smaller heat sink volumes, elimination of audible noise, the ability to move rejected heat away easily from the user area, and significantly increased system reliability are all liquid cooling characteristics. The problem is that only a few markets currently accept liquid-cooled electronics. The primary resistance is customer acceptance. The concept of a liquid-cooled personal computer has yet to gain general market acceptance. However, as heat fluxes exceed 100 W/in.$^2$, the need for phase change heat transfer or liquid cooling are the only alternatives (Soule 2001).

## Nanostructure Enhanced Heat Sink and Complex Spreader Sink

Nanostructures and nanomaterials have been used to enhance the performance of heat sink and hybrid cooling systems. The efforts are mainly focusing on reducing contact resistance between solid–solid interfaces, and convective resistance from heat exchanger surface to the atmosphere; increasing capillary limit and boiling limit of interfacial cooling structures; and enhancing thermal performance of working fluids.

Contact resistance between solid–solid interfaces is a major concern for assembling the heat sink. A convectional approach is to employ a clamping device that exerts pressure on two attached components of which the interface is filled with thermal grease with high thermal conductivity for alleviating the large thermal contact resistance caused by the air layer. CNT has been used to improve the contact thermal resistance between solid–solid interfaces. The interface material can be produced by using a soft matrix material mixed with CNTs and/or nanofibers. In particular, the CNTs or nanofibers are preferred to be aligned and perpendicular with the contact surface of the electronic component so that the thermal conductivity of the thermal interface material is effectively elevated but highly directional. Comparably, the thermal interface material formed by injection molding may cause the CNTs to be disposed in the matrix material randomly and multidirectionally, which can not sufficiently utilize the heat conduction capability of the CNTs (Chen 2008). A specific example of the thermal interface material is composed of silver colloid and CNTs. The nanotubes are aligned parallel in the interface material and normally are attached to two solid surfaces. Meanwhile, nanometer-scale silver is filled into the empty spaces inside nanotubes to enhance the effective thermal conductivity (Leu et al. 2006). The composite material with the CNT-filler can also be designed by allowing one end of the CNT exposed in the atmosphere, and then pressed against a solid surface that transports heat through these nanotubes. A heat dissipated component packed with an integrated circuit device and a heat sink, which is filled with CNTs that can help to transport heat from the circuit device to the heat sink. Nanofibers can be initially implanted on one surface and adhered to a second surface through van der Waals forces when the second surface is moved to attach the nanofibers. This phenomenon is similar to the way that a gecko with foot hairs (setae) can hang from a ceiling. The empty space between two contacted solid surfaces can be significantly reduced by the presence of adhesive nanofibers (Chen 2008).

Convective resistance from the heat exchanger surface to the atmosphere can be reduced through arrayed nanorods that are immobilized on a substrate surface. The heat exchange area of substrate is increased significantly with the nanorods attached on the substrate. For a radiator of which the surface is covered with aligned CNTs normal to the radiator surface, if the substrate with an area of 1 cm$^2$ is uniformly covered by nanotubes with a length of 1 mm, a diameter of 40 nm, and spacing between centers of neighboring nanotubes of 100 nm, the heat exchanger area is increased 12 times (Chen 2008). Further research is needed to explore how air flow is driven into the narrow gaps between nanotubes.

Capillary limit and boiling limit of wick structures can be increased for heat pipes that combines with a heat sink system. As will be discussed in Chapter 10, the maximum heat that a convectional heat pipe can dissipate is dominated by sonic limit, capillary limit, and boiling limit. If the heat transport capacity of heat pipe is constrained by the capillary limit, the heat pipe does not have wick structures small enough on the wall surface to create enough capillary force to drive condensed liquid from the condenser section back to the evaporator section for the case that applied heat is over the capillary limit. The evaporator section is finally dried out.

If the heat transport capacity of heat pipe is constrained by the boiling limit, it is directly related with the critical heat flux of deionized water on the heat pipe wall. When the applied heat is over the critical heat flux, a vapor film is formed on the wall of evaporator section. The capillary limit is linearly proportional to the reciprocal of effective capillary radius of wick pore at the liquid–vapor interface. The reduction in wick pore can also result in an increase in critical heat flux. In commercially available heat pipes used in electronic devices, the capillary radius is at an order of around one hundred micrometers. Therefore, nanostructure could be a preferred choice to replace microwick structures on the wall surface of heat pipes for raising both the capillary and boiling limits (Chen 2008). For instance, a nanostructured composite wick comprises a channel, a plurality of nanostructures, wherein the nanostructures have a differentially spaced apart gradient along the length of the channel so as to promote capillary fluid flow. The formation of the wick can be one or a combination of the following processes (Chen 2008): depositing a solid film on a substrate and then electrochemically etching the solid film forming arrays of nanopores; and/or electrochemically growing nanowires or nanotubes in the nanopores and annealing the nanowires to form polycrystalline wires. By using the nanobristle array organized by nanowires as the evaporator, for example, the nanobristle heat pipe showed a significant 25–30% reduction in thermal resistance over the sintered metal powder heat pipe (Habib 2008).

In addition, thermal performance of working fluid can be enhanced by adding suspended nanoparticles into a base fluid to increase the thermal conductivity and convective heat transfer coefficient of base fluid. This is so-called nanofluid. However, the sediment time of suspended nanoparticle is a key factor to affect the reliability of the nanofluid used in a liquid cooling system (Chen 2008). One example of thermally conductive nanofluid is composed of a heat transfer medium and an additive with or without stabilized surfactant. The size of additive material ranges from 1 to 100 mm. The additive materials can be carbon nanoparticle with or without stabilized surfactant, diamond nanoparticle, or other nanoparticles. The heat transfer medium can be soy-based oil or other carrying fluids. A ferrofluid-base pump has been developed to drive working fluid with ferro-nanoparticles for transporting heat from a heat generator to a heat sink (Cheng 2009).

Miniaturizations of electronic devices along with improved performances have led to the development of integrated spreader sinks involving phase change and/or forced liquid convection. Spreader sink combines the functions of both heat spreader and heat sink to improve the performance of a cooling system. For example, uses of a combined liquid cooling system can inherit high convective coefficients such as sprays and the enhanced surfaces with pins, ribs and grooves which trip shear layers to augment turbulent heat convection as well as the adoption of nanofluid as the primary working fluid (Cheng 2009). In particular, the use of nanofluid can elevate the overall convective capacity and performance for the complex spreader sink system. The increase of the Prandtl number for a nanofluid generally increases the Nusselt numbers from the counterpart conventional coolant flows at a constant pumping power; while the increase of thermal conductivity for nanofluid elevates the convective heat transfer coefficient at a fixed Nusselt number

(Chen 2008). An integrated forced convective liquid cooling system comprises a heat spreader in contact with the electronic component, a pump and a heat sink exchanging section that is thermally connected to the cooling section to radiate heat convected from the heat spreader to the atmosphere (Kobayashi 2008). A low cost boiling heat sink has been developed to utilize the enhanced microporous coating to significantly augment the nucleate boiling heat transfer and the critical heat flux and integrate vapor chambers and radiators. The boiling cooler comprises a vessel that is partially filled with a liquid coolant, consolidated with the vapor chambers or condensers for the vapor to spread heat around the extended surfaces around which the secondary coolant flows or air flows convect latent heat away to convert vapors into liquids. A layer of microporous coating for boiling heat transfer enhancement is applied on the thermally conductive wall, which is at least partially submerged in the liquid within the vessel. As the pumping force to circulate vapor and condensate within the heat sink is driven by free convection, cooling performances for such a heat sink are orientation dependent. Another integrated heat sick system involves phase change processes, in which a thermoelectric module (TEC) or a thermosyphon is adopted to transfer heat in the lateral direction from the central portion of the heat plate to its periphery and then to the fin assembly for heat exchanges (Batscha et al. 2008). The TEC establishes a differential potential of thermal energy between the cold and hot sides to electrically pump heat from the cold to hot sides by using the electrical energy. The TEC is assigned to transfer heat from the central portion of an electronic chip toward its peripheral portion that smoothes out the nonuniform heat fluxes over the electronic chip. In addition, such integrated heat sink can be comprised by a thermosyphon cooling system within which a refrigerant is disposed in the lower portion of thermosyphon for liquid-to-vapor transformations. Extended cooling fins are disposed in the lower portion of the thermosyphon for transforming heat from the electronic component to the refrigerant that vaporizes the refrigerant. Upper portions of the thermosyphon are equipped with fins in contact with the externally supplied cooling airflows as the condenser to recover vapors into liquids (Chen 2008).

In fact, various complex spreader sink systems have been developed with different combinations. In addition to the wide range of heat transfer augmentation methods developed for the single-phase convection, the efficient wick structures for two-phase components as well as the methods for reducing the thermal contact resistances have utilized the nanotechnology to improve the thermal physical properties and/or the surface topology for capillary actions. The emerging design proposals with various combinations of the interdependent cooling components such as "vapor chamber + heat sink", "TEC + heat pipe + heat sink", and "thermosyphon + heat pipe + heat sink" have led the future developments of electronic cooling concepts toward the system development (Chen 2008). In this respect, the interdependency between the cooling components in an electronic cooling system involving two-phase devices and the performance improvements for the worst components in an electronic cooling system have drawn attentions for future developments. These techniques developed for electronic cooling systems also keep finding their alternative applications to the optical and electronic communication systems where the high cooling power density is essential for their normal operations.

As an example, "heat pipe or vapor chamber + heat sink" is a combined system with a heat pipe or vapor chamber directly integrated to a heat sink (Chen 2008). This system can be constructed by forming a boundary structure by sealing together two formed plates: contact plate and cover plate. Contact plate and cover plate are sealed together at their peripheral lips and by conventional means, such as soldering or brazing, to form heat pipe. The interior of heat pipe is, however, constructed unconventionally. While contact plate is essentially flat with the exception of peripheral lip, the cover plate includes multiple depressions. Depressions are formed and dimensioned so that when the contact plate and the cover plate are joined, the flat portions of depressions are in contact with the inner surface of the contact plate. Depressions thereby assure that the spacing between the contact plate and the cover plate will be maintained even through pressure differentials between the inside volume of the heat pipe and the surrounding environment might otherwise cause the plates to deflect toward each other. Vapor chambers make use of the same principle as the heat pipe, which can be used to improve the base conductivity of the heat sink. If the area of the cooling surfaces is much larger than the evaporating surface, the spreading of heat can be achieved effectively because the phase-change mechanism occurs near the isothermal condition. Consequently, the heat flux can be reduced inversely proportionally to the area ratio of heating surface to the cooling surface. In addition, boiling enhancement features can be adapted into the vapor chamber through a boiling enhanced multiwick structure. With this structure, the condensate is collected from the condensation sites using a wicking structure with a spatially varying wicking power. Various boiling enhancement structures can be adapted at the heating zone to simultaneously provide wicking power and boiling enhancement. In this manner, the boiling enhancement structure is not totally submerged inside a pool of liquid, and thus can operate in antigravity orientations. Furthermore, this boiling enhancement structure may act as a 3-D bridging wick, which may also provide a structure supporting function. The boiling enhancement structure is a protruded wick having a wicking power greater than that at the condensation site. The protruded wick can be in the form of fins so that the liquid can be wicked between the fins toward the tips of the fins. Besides fins, the protruded wick can also be an array of pins. Interlinking structures between fins or pins can also be used to increase the boiling surface area. Foam/porous structures can also be used in the protruded wick to provide the larger boiling surface area with the added heat diffusion in the porous media. With all of these structures, a heat conduction path can be provided from the heating source toward a larger boiling surface, and to saturate this boiling surface (without total immersion) with condensate that is continually supplied by the complex wicking system. Similarly, a vapor chamber can serve as a heat spreader for the microelectronics die. A TEC serves to cool the vapor chamber and maintain proper functioning of the vapor chamber, thus keeping the microelectronic die cooled. A controller can be used to receive inputs from temperature sensors, and these temperature measurements can be utilized to control the current through the TEC and the voltage/current to a motor that drives a fan and provides additional cooling. The vapor chamber contacts the microelectronic die tightly and has a larger footprint than the die, serving as a heat spreader (Chen 2008).

Furthermore, a "vapor chamber + heat pipe + heat sink" system can be a thermal management device utilizing varied methods of heat transfer augmentation to cool a heat generating component. At least one heat pipe is used to transfer heat from the condensation portion of a vapor chamber to a colder location on a finned heat sink which is also attached to the condensation portion of the vapor chamber (Chen 2008). "Thermosyphon + heat pipe + heat sink" is designed to use a "heat pipe + thermosyphon" overcomes the low cooling capacity of the "conventional fan + heat sink" unit. A heat pipe mounted on a predetermined location of a heat emitting unit in electronic device transfers heat from the device. A vaporizing unit for vaporizing liquid inside the thermosyphon with the heat transferred via the heat pipe contacts a predetermined location of the heat pipe. A condensing unit to discharge the heat carried by the vapor to the environment is connected to the vaporizing unit through a vaporization line and a condensation line. A small capacity fan provides cooling air to the condensing unit. This thermpsyphon does not need a separate power, but uses the heat generated by the unit. By way of employing the high-density fin stack around the condensing unit, the size of such system can be reduced with the same heat discharge capability; or the capacity burden on the fan can be reduced (Chen 2008). In addition, "TEC + heat pipe or vapor chamber + heat sink" system has been designed using a plurality of heat pipes directly inserted between the plates of the TEC unit. The heat pipes can be freely allocated, wherein the space around the heat exchanger can be used more efficiently with a high-density fin stack around the heat pipes. Thus, the size of the system can be reduced with the same heat discharge capability (Chen 2008).

Without question, the increased power density for electronic chipsets and assemblies would continuously promote the developments of cooling units for high power electronic components toward the two-phase systems which employ various heat transfer augmentation designs. The hybrid technology for cooling devices such as the combination of phase-change process with nanofluid and the utilization of CNTs for wick structure or for thermal interface materials has emerged as one of the future trends of developments for thermal managements of electronic systems. Although the air cooled heat sink is a subject of long-term development, this device, which convects heat to the surrounding ambience appears to be essential for all different types of electronic cooling systems. Design concepts for optimizing the cooling effectiveness of the subassembly in an electronic cooling system would shift from the increased cooling area by using fins or pins toward the more effective conductive and convective measures due to the everlasting needs for miniaturization of electronic devices. From this point of view, nanotechnology can be used to reduce the thermal contact resistance between different subassemblies with temperature differences, to increase the capillary limit of interfacial device by improving the microstructures of wicks and/or inner surfaces in two phase components, and to upgrade the thermal physical properties of working fluid. Cooling performances of the thermally driven, self-circulating phase change devices depend greatly on their orientations in respect to gravity, which can be combined with the forced liquid cooling system in order to upgrade their cooling performances (Chen 2008). The detail design and optimization of the liquid cooling and thermoelectric cooling systems would be discussed in the following chapters.

## Summary

Air cooling is the simplest and principal method of thermal management used most widely for a variety of electronic systems ranging form portable electronics to large business systems. The advantages of air cooling are its ready availability and ease of application. Air cooling usually dissipates heat through heat spreaders and heat sinks, and works by making the object to be cooled to have a relatively larger surface area or have increased air flow over its surface, or both. For high heat flux, a push–pull air flow arrangement with air moving devices at both the bottom and top of the PCBs is used to provide high pressure drop capability. Low-power electronic systems are conveniently cooled by natural convection and radiation. When natural convection is not adequate, the forced convection is adopted by a fan or blower to blow the air through the enclosure that houses the electronic components.

A heat spreader is a primary heat exchanger most often in the form of a highly thermal conductive plate, which moves heat between a heat source and a secondary heat exchanger such as a heat sink. The secondary heat exchanger is always larger in the cross-sectional area, surface area, and volume. The heat flow is the same in both heat exchangers, but the heat flux density is less in the secondary, so it can be made of a less expensive material such as aluminum, and is a better match to an air heat exchanger because the low heat transfer coefficient for air convection is adequate for a low heat flux. A heat spreader is generally used if and only if the heat source tends to have a high heat flux density, and for whatever reason, heat cannot be conducted away effectively by the secondary heat exchanger. A high enough heat exchanger transfer coefficient is often sufficient to avoid the need for a heat spreader. The use of a heat spreader in electronic packaging is an important part of an economically optimal design for transferring heat from high to low heat flux media, such as a microprocessor in air cooling ICs, and air cooling a photovoltaic cell in a concentrated photovoltaic system. To achieve good thermal performances in electronic packaging, a heat spreader is usually used to help dissipate the heat generated from the chip to the ambience through the package body usually with a heat sink system. On the other hand, the use of a heat spreader could cause changes to the electrical performances of the system as it is usually made of electrically conductive materials. The ever-increasing processing speeds in high-speed digital systems and frequency in RF and microwave circuits makes the presence of a heat spreader much more complex in terms of its impact on a system's electrical behaviors. For example, mitigation of EMI by grounding the heat spreader or by other solutions may need to be considered during design of the heat spreader system. Various heat spreading methods have been widely used in electronic packaging. Heat spreaders are usually used in die level packaging to spread heat from the microprocessor chip into the package. For instance, multichip modules have been built to incorporate a copper heat spreader. On a component or board level, active spreaders such as heat pipes are used to move heat from chip packages to heat sinks. Heat sinks themselves can incorporate spreaders such as a copper base or heat pipes embedded in an aluminum base. Passive heat spreaders have been incorporated into plasma display panels. Spreaders are commonly used within electronic enclosures to move heat from discrete components to the walls of the enclosure. Materials and process technologies to make heat spreaders have been involved in metallic heat spreaders made of solid copper, aluminum, copper-tungsten or cupper-moly with thermal conductivity of less than 400 W/m K; composite spreaders including ceramic, metallic and carbon composites (<1,200 W/m K); and advanced heat spreader technologies including high conductivity plates (500–5,000 W/m K), vapor chambers (>5,000 W/m K), and oscillating flow heat spreaders (up to 120,000 W/m K). When some applications, such as LDs, and many semiconductor ICs, require both electrical isolation and thermal conduction, dielectric heat spreaders that are made with nonmetallic materials are ideal.

Heat sinks can usually be divided into two types: active and passive. Active heat sinks utilize power and are usually a fan type, or liquid cooling or some other peltier cooling device. Passive heat sinks are fully reliable, as they have no mechanical components. Passive heat sinks are usually made of a metal like aluminum-finned radiator that dissipates heat through convection. For passive heat sinks to work to their full capacity, it is recommended that there is a steady air flow moving across the fins. Active heat sinks are most common in electronic devices. Through a large number of fins, it extracts as much heat as possible from the air flowing through the fan and heat sink. The complexity of the air flow inside the electronics cooling system results in a nonuniform flow throughout the active heat sink. This approximation of the air flow through the fins is further increased by the use of fan curves that can oversimplify the effect of the fan, neglecting the impact of the system geometry. An explicit modeling of the fan within the active heat sink, including the entire blade details, as an integral part of the global system model, has been developed dramatically. A more realistic description of the system design leads to more accurate evaluation of the heat sink performance, which will result in a better system design.

Next generation heat sinks and leading-edge materials in air cooled heat removal have been developed and are increasing such as sintered heat sinks, metal injection molding, high performance fans, aluminum extrusions at 25:1 ratio, and self-fanning piezoelectric fin heat sinks. High thermal conductivity polymers, carbon-based composites, sintered metallic powders, phase-change compounds, synthetic diamonds, and strand-oriented graphite materials are all part of leading edge materials for heat sinks. The most desirable properties include controlled levels of conductivity, high machinability, low weight, low CTE, low toxicity and above all the cost will be expected lower than aluminum. Composite heat sinks are emerging and getting increasing attention since nanotechnology can be applied.

## References

Anandan SS, Ramalingam V (2008) Thermal management of electronics: A review of literature. Thermal Science **12** (No. 2): 5–26.

Balandin AA, Kotchetkov D, Ghosh S (2008) Lateral graphene heat spreaders for electronic and optoelectronic devices and circuits, UC Case No.: 2008-781-2; U.S.A. Patent Appl. No. 1192-01-PA-H.

Batscha D, Argas S, Leshem A (2008) Direct heating organic rankine cycle. US Patent 0289313.

Behm J, Huttunen J (2001) Heat spreading conduction in compressed heatsinks. 10th International Flotherm User Conference, Amsterdam, May 14–15, 2001.

Bejan A (1995) Convection Heat Transfer, Second Edition. John Wiley & Sons, Inc., New York, p. 347.

Chen PH et al. (2008) High power electronic component: review. Recent patent on Engineering **2**: 174–188.

Cheng L (2009) Nanofluid heat transfer technologies. Recent patent on Engineering **3**: 1–7.

Habib YM et al. (2008) Heat pipe with nanostructured wicking material. WO Patent 08016725.

Hanreich G, Hauser H, Riedling K, Rieder W (2000) Silicon-Microwhisker heat-spreader for enhanced heat transfer in power electronics. Research report. Vienna University of technology, Vienna, Austria.

Hutchison BR, Schwiebert M, Thompson RJ (2002) Integral dielectric heatspreader. US Patent 6414847.

Keller K (1998) Cast heat sink design advantages. http://www.cs.unc.edu/~pxfl/papers/heatsink.pdf. Accessed 07 June 2010.

Kobayashi S (2008) Cooling device of electronic device. US Patent 7337830.

Kraus A, Bar-Cohen A (1983) Thermal Analysis and Control of Electronic Equipment. McGraw-Hill, New York.

Lasance CJM (2008) Heat spreading: Not a trivial problem. Electronics Cooling. http://www.electronics-cooling.com/2008/05/heat-spreading-not-a-trivial-problem/. Accessed 01 June 2010.

Lee S (1995) How to select a heat sink. http://www.aavidthermalloy.com/technical/papers/pdfs/select.pdf. Accessed 03 June 2010.

Lee S, Song S, Au V, Moran KP (1995) Constriction/spreading resistance model for electronics packaging. ASME/JSME Thermal Engineering Conference **4**: 199–206.

Leu C, Yu TC, Chen GL (2006) Thermal interface with silver-filled carbon nanotubes. US Patent 7148512.

Sathyamurthy P, Rundstadler PW, Lee S (1996) Numerical and experimental evaluation of planar and staggered heat sinks. Fifth Annual ITHERM Conference, May 29, 1996.

Smalc M, Shives G, Chen G, et al (2005) Thermal performance of natural graphite heat spreaders. Proceedings of IPACK, ASME InterPACK'05, San Francisco, CA, USA, July 17–22.

Smalc MD et al (2010) Heat spreader for display panel. US patent 7666270.

Soule C (2001) Future trend in heat sink design. http://electronics-cooling.com/articles/2001/2001_feb_a1.php. Accessed 06 June 2010.

Soule CA (1996) Augmentation improves thermal performance of air cooled heat sinks, 5 pages, from Nov. 1996 issue of PCIM Magazine, reprinted from http://www.aavid.com/atp/articles/agmtThrmIHS/. Accessed 09 June 2010.

Turner M (1996) All you need to know about fans. Electronics Cooling. http://www.electronics-cooling.com/1996/05/all-you-need-to-know-about-fans/. Accessed 06 June 2010.

Yang F et al (2003) Corona driven air propulsion for cooling of electronics. XIIIth International Symposium on High Voltage Engineering, Netherlands, 2003, Smit (ed.) Millpress, Rotterdam, ISBN 90-77017-79-8.

Young R et al (2006) Developments and trends in thermal management technologies – A mission to the USA. http://www.lboro.ac.uk/research/iemrc/documents/.../CB2007.pdf. Accessed on 26 February 2010.

# Chapter 10
# Liquid Cooling Devices and Their Materials Selection

**Abstract** Microprocessor power density has been steadily increasing over the past decade due to increases in microprocessor power dissipation and reduction in feature size of the processing central processing unit (CPU) core, where most of the power on a die is generated. This trend is expected to continue into the future, leading to next generation electronics with a power dense core covering a fraction of the total die surface area bounded by regions of reduced power density cache, with localized power densities exceeding 100 W/cm$^2$. Conventional cooling technologies in the electronic industry have limitations on removing these nonuniform, high heat fluxes from the surface of microprocessors. The combination of high heat fluxes with the nonuniformity of heat dissipation requires technologies able to remove large amounts of heat in a spatially and temporally variable manner. These heat removal technologies, which can meet the challenging cooling requirements of next generation electronics, include direct liquid cooling, indirect liquid cooling, heat pipe, liquid–vapor phase change, nonrefrigeration phase change techniques, pool boiling, jet impingement cooling, evaporative spray cooling, and embedded droplet impingement for integrated cooling of electronics (EDIFICE). This chapter will give a brief review on these methods with single or/and two-phase cooling.

## Introduction

Thermal management of nearly all high performance electronic packaging products devolved to the aggressive use of air cooling in the late 1990s under the influence of market forces, exploiting technology that was a natural outgrowth of the air-cooled multichip modules of the 1980s. Thus, by the end of the decade, a renaissance in thermal packaging produced heat sinks for high-end commercial workstations and servers that were routinely dissipating 60–70 W, with chip heat fluxes of some 26 W/cm$^2$. Although air cooling is widely used in cooling methods for electronic packages, liquid cooling can accommodate significantly higher heat fluxes due to higher specific heat and thermal conductivity. With power dissipation approaching 200 W, cost-effective thermal management of large silicon chips requires exploration

X.C. Tong, *Advanced Materials for Thermal Management of Electronic Packaging*, Springer Series in Advanced Microelectronics 30,
DOI 10.1007/978-1-4419-7759-5_10, 

of nonconventional thermal packaging techniques, including direct cooling with dielectric liquids. The high dielectric strength and low dielectric constant of these liquids, as well as their chemical inertness, make it possible to immerse most electronic components directly in these fluorochemical liquids (perfluorinated chamicals and Novec fluids). Pool boiling on the chip surface transfers the dissipated heat directly into the liquid and thus, overcomes the barrier posed by the interface resistance otherwise encountered at the solid interface between the chip and the cooling hardware. With saturated pool boiling, component heat fluxes in the range of 10–20 W/cm$^2$ could be removed at surface superheats of typically less than 20 K. The critical heat flux (CHF) places an upper limit on this highly efficient heat transfer process. Heat fluxes in excess of CHF result in the formation of an insulating film of vapor on the heat transfer surface and superheats typically higher by an order-of-magnitude than encountered in nucleate pool boiling. The development of liquid cooling has been focused on (Arik and Bar-Cohen 2003): (1) The fundamentals of the boiling heat transfer and CHF based on the analysis of hydrodynamic phenomena leading to the "choking off," or interruption, of the liquid supply to the heater; (2) Selection of dielectric liquids and their effects on the CHF (pressure, subcooling, surface treatments, binary mixtures, thermal effusivity etc.); and (3) Design and materials selection of passive and active cooling devices.

In fact, liquid cooling of electronic packaging has been utilized for many years, in some segments more than others. The acceptance for applying liquid cooling varies from business segment to business segment. For example, the automotive industry has always used liquid cooling for cooling of combustion engines, so the use of liquid cooling for power electronics in an automotive application is not frightening. In other segments, the idea of having liquid flowing through power electronic assemblies is most disturbing. Maybe the biggest impediment for applying liquid cooling has been the relative high cost and the limitations in performance that confines the usability to more exotic applications where no alternative exists – these limitations being large temperature gradients across the cooling areas and high pressure drops in the coolers that make large/expensive pump systems necessary. Standard liquid cooled power modules suffer from two major drawbacks: (1) High cost. Standard liquid cooled power modules feature copper baseplates with quite complicated structures, e.g., pin fins, which are necessary in order to achieve sufficient cooling. A typical baseplate for a large module gets five times more expensive with pin fins compared with the flat baseplate. (2) Inhomogeneous cooling. Standard liquid cooling is the inherent temperature gradient which arises through the module due to the warming up of the coolant. In order to reduce cost and to improve the cooling performance, a series of cooling systems has been developed, such as indirect single-phase liquid cooling, high-flux cooling with phase-change heat transfer, heat pipe cooling, direct immersion cooling (passive and active), liquid jet/spray cooling, microjet cooling, refrigeration cooling, boiling and condensation heat transfer, and liquid encapsulated heat spreader. A distinction of direct liquid cooling and indirect cooling is that there is intimate contact between the coolant and the electronics; while for the latter, the electronics are physically separated from the liquid. In the 1980s, indirect cooling systems were devised to exceed the heat dissipation capabilities of air cooling, while avoiding difficulties with direct liquid cooling. As applied to multichip modules, indirect liquid cooling involves attachment of a liquid-cooled cold plate to the module; therefore, interfacial and solid–solid contact resistances are inherent to such schemes. If these resistances are to be eliminated, resort must be made to indirect liquid cooling schemes. Direct liquid cooling can be achieved by immersing the electronic components in an inert, dielectric liquid or by spraying the liquid onto the electronics, or circulating the liquid through microchannels machined in the chip.

Historically, liquid cooling methods for computers have been regarded as frivolous despite their efficacy. However, it is readily apparent that the future of computing lies in miniaturization. Examples of miniaturized computers are ubiquitous in everyday life, whether the scale is from a former supercomputer to a desktop computer, a desktop computer to a laptop computer, or a laptop computer to an iPhone. These computers will only follow the trend and become even smaller, and the limited capacity of air cooling will soon undoubtedly reach its limit. Liquid cooling methods for microelectronics can be categorized as either indirect or direct. Indirect liquid cooling prevents any contact between the microelectronic components and the coolant. In these cases it is necessary to provide a good thermal conduction path from the microelectronic heat sources to the liquid. Water is the preferred coolant for indirect liquid cooling because of its superior thermophysical properties, whereas fluorocarbon liquids are more suitable for direct cooling on account of its chemical characteristics. Direct and indirect liquid cooling methods can each be further categorized into single-phase and two-phase liquid cooling. As an example, jet impingement cooling may be utilized as a two-phase direct method, in contrast to the two-phase indirect technology of heat pipes and thermosyphons. Two-phase cooling methods are more desirable because of high heat transfer coefficient; however the systems tend to be more complex. In the mid 1980s, IBM employed indirect liquid cooling technology using water for mainframes and supercomputers. This technology then became the norm for high performance computers which easily accommodated the large cold plates and heat conduction devices. In large mainframe computers, the cold plates (containing water) can be sufficiently separated from the electronics and thermally connected with heat conduction devices (Yarin et al. 2009). Two-phase cooling would be considered for notebook computers although most microprocessors have been sufficiently cooled with metallic heat sinks and fans.

## Direct Liquid Cooling

The inexorable rise in chip power dissipation and emergence of on-chip hotspots with heat fluxes approaching 1 kW/cm$^2$ has turned renewed attention to direct cooling especially with dielectric liquids. Use of dielectric liquids in intimate contact with the heat dissipating surfaces eliminates the thermal interface materials and deleterious effects of solid–solid interface resistances, and harnesses the highly efficient phase-change processes to the critical thermal management of advanced integrated circuit (IC) chips. Direct liquid cooling techniques allow for direct contact between a dielectric liquid and the surface of the chip, holding great promise for solving hotspot-driven thermal management issues. Furthermore, use of emerging and phase-change processes, including pool boiling, gas-assisted evaporative cooling (GAEC), jet impingement, and spray cooling, exploit the latent heat of these liquids to reduce the required mass flow rates and can provide the added advantage of inherently high heat transfer coefficients (Bar-Cohen et al. 2006). This section would define, quantify, and then compare these direct liquid-cooling techniques.

## *Immersion Cooling*

Direct liquid immersion cooling brings the coolant in direct contact with the back of the chip, there are no physical walls separating the microelectronic chips and the surface of the substrate from the liquid coolant. This form of cooling offers the opportunity to remove heat directly from the chip(s) with no intervening thermal conduction resistance, other than that between the device heat sources and the chip surfaces in contact with the liquid. Interest in direct liquid immersion as a method for cooling IC chips may be traced back as early as the 1960s. Figure 10.1 illustrates

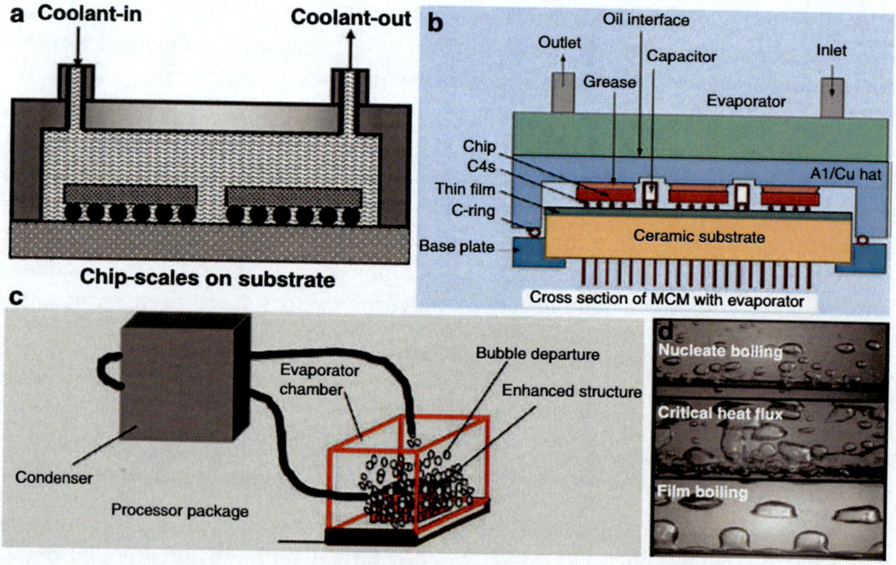


**Fig. 10.1** Illusions of direct liquid emersion cooling techniques: (**a**) natural convection; (**b**) assisted with evaporation; (**c**) two-phase loop thermosyphon; and (**d**) enhanced with pool boiling

some typical direct liquid immersion cooling techniques that have been developed: (1) natural convection; (2) assisted with evaporation; (3) two-phase loop thermosyphon; and (4) enhanced with pool boiling. An electronic package, such as a power supply, can be filled with a dielectric fluid that transfers heat from electronic components to an external heat sink through conduction/convection. A larger volume of fluid may be required depending on the size of the flow channel(s); but nozzles, plumbing, a pump, and filter required by spray or jet impingement methods are eliminated. A variation of direct immersion is a flow through/channel flow configuration that directs coolant over the entire surface of an electronic circuit card and components. Nozzles are not needed, plumbing is minimized or eliminated, a pump circulates the coolant, and a filter traps contamination. The liquid may be moving passively due to natural convection or be driven by a pump. The liquid may also undergo partial-phase change, evaporation, or two-phase change in which case much higher heat fluxes may be attained. Bubbles are generated by the phase change as heat is transferred from components to the coolant. In the saturated flow, large bubbles are generated. Large bubbles may inhibit liquid contact with electronic parts; therefore reducing heat transfer. Typically, excessive boiling should be avoided in order to reduce the creation of large bubbles that would lead to local hotspots. Instead, the preferred condition is that of subcooled boiling where the bubbles are small enough to recondense into the main flow. The subcooled flow illustrates the preferred situation where very small bubbles (microbubbles) are produced by the phase change that naturally condenses back into the bulk coolant flow. Subcooling is achieved exactly as the term implies. The dielectric fluid is cooled some amount below the phase change point based on the specific system requirements. An important fact is that a variety of dielectric fluids exist that can be mixed to achieve specific cooling performance.

Direct liquid immersion cooling offers a high heat transfer coefficient which reduces the temperature rise of the chip surface above the liquid coolant temperature. However, the relative magnitude of a heat transfer coefficient is affected by both the coolant and the mode of convective heat transfer, i.e., natural convection, forced convection, or boiling (Hwang and Moran 1981). Water is the most effective coolant and the subcooled boiling mode offers the highest heat transfer coefficient. Direct liquid immersion cooling also offers greater uniformity of chip temperatures than is provided by air cooling. Historically, the coolant considered was a dielectric (e.g., fluorocarbon coolants) primarily because it would also be in direct contact with the chip interconnects and substrate top metallurgy surface. For this exercise, it is assumed that water can be brought in direct contact with the back of the chip. Obviously, a seal or barrier preventing water from contacting interconnects or the substrate top metallurgy surface is needed. The selection of a liquid for direct immersion cooling cannot be made on the basis of heat transfer characteristics alone. Chemical compatibility of the coolant with the chips and other packaging materials exposed to the liquid must be a primary consideration. There may be several coolants which can provide adequate cooling, but only a few will be chemically compatible. Water is an example of a liquid which has very desirable heat transfer characteristics, but which is generally unsuitable for direct immersion cooling on account of its chemical characteristics. Fluorocarbon liquids (e.g. FC-72, FC-86, FC-77, etc.) are generally considered to be the most suitable liquids for direct immersion cooling, in spite of their poorer thermophysical properties. As shown in Table 10.1, the thermal conductivity, specific heat, and heat of vaporization of fluorocarbon coolants are lower than water (Danielson et al. 1987). These coolants are clear, colorless per-fluorinated liquids with a relatively high density and low viscosity. They also exhibit a high dielectric strength and a high volume electrical resistivity. The boiling points for the commercially available fluorinert liquids range from 30 to 253°C. Other candidate fluids are hydrofluoroethers (HFEs). In general, the direct cooling of microelectronic components imposes stringent chemical, electrical, and thermal requirements on the liquids to be used in this thermal control mode. Direct liquid cooling of microelectronic components requires compatibility between the liquid coolant and a system-specific combination of the chip, chip package, substrate, and printed circuit board (PCB) materials, e.g. silicon, silicon dioxide, silicon nitride, alumina, o-rings, plastic encapsulants, solder, gold, and epoxy glass. A liquid coolant must possess the dielectric strength needed to provide electrical isolation between adjacent power/ground conductors and signal lines operating at a potential of approximately 1–5 V and spaced as close as 0.05 mm. It is also desirable that the liquid's dielectric constant be close to unity to avoid introducing significant propagation delays. Furthermore, such coolants must be nontoxic and chemically inert (Bar-Cohen et al. 2006).

The convective heat transfer processes upon which liquid immersion cooling depends may be classified as natural convection, forced convection, or boiling modes. As in the case of air cooling, natural convection is a heat transfer process in which mixing fluid motion is induced by coolant density differences caused by the heat transferred to the coolant. This mode of heat transfer offers the lowest heat flux or cooling capability for a given wall superheat. Nonetheless, the heat transfer rates attainable with liquid natural convection can easily match or exceed those attainable with forced convection of air. Natural convection would typically be

**Table 10.1** Thermophysical properties of dielectric coolants and water at atmospheric pressure

| Property | Coolant FC-87 | FC-72 | FC-77 | FC-40 | HFE-7100 | HFE-7200 | $H_2O$ |
|---|---|---|---|---|---|---|---|
| Boiling point at 1 atm (°C) | 30 | 56 | 97 | 156 | 61 | 76 | 100 |
| Density (kg/m$^3$) | 1,633 | 1,623 | 1,780 | 1,870 | 1,500 | 1,430 | 957.8 |
| Specific heat (J/kg K) | 1,088 | 1,088 | 1,172 | – | 1,180 | 1,210 | 4,217 |
| Thermal conductivity (W/m K) | 0.0551 | 0.0545 | 0.057 | – | – | – | 0.613 |
| Dynamic viscosity (N s/m$^2$) | 0.00042 | 0.00045 | 0.00045 | – | – | – | 0.000855 |
| Heat of vaporization (kJ/kg) | 87.9 | 84.97 | 83.7 | 711.6 | 125.6 | 122.6 | 2,257 |
| Surface tension (N/m) | 0.0089 | 0.0084 | 0.008 | 0.016 | 0.014 | 0.014 | 0.0589 |
| Thermal coefficient of expansion (1/K) | 0.0016 | 0.0016 | 0.0014 | – | – | – | 0.0002 |
| Dielectric constant | 1.71 | 1.72 | 1.75 | – | – | – | 78.0 |

employed within a closed container to transfer heat from chips or modules to liquid, and then from the liquid to the walls of the container, as shown in Figure 10.1a. Heat could then be transferred from the walls to outside air by natural or forced convection. For forced convection, higher heat transfer rates may be attained by utilizing a pump to provide forced circulation of the liquid coolant over the chip or module surfaces, as shown in Figure 10.1b–d. As with air cooling, its allowable heat flux for a given surface-to-liquid temperature difference can be increased by increasing the velocity of the liquid over the heated surface. Depending upon the surface geometry and the nature of the flow (i.e., laminar or turbulent), the heat transfer coefficient will be proportional to the velocity to a power between 0.5 and 0.8. The price to be paid for increasing cooling performance in this way will be a higher pressure drop. This can mean a larger pump and higher system operating pressures. Although forced convection requires the use of a pump and the associated piping, it offers the opportunity to remove heat from high-power modules in a confined space; and then transport the heat via the liquid coolant to a remote heat exchanger to reject the heat to air or water. In addition, as shown in Figure 10.1a, b, when liquid in a constant-pressure vessel is uniformly heated and reaches its evaporation temperature, vapor and liquid are present as coexisting phases in equilibrium. The addition of heat increases the vapor fraction, and hence the volume, without altering the temperature of the two-phase mixture. When the liquid has vaporized, further heating results in a temperature rise of the vapor along a locus appropriate to the vapor constant-pressure specific heat. The termination of the all-liquid state and the beginning of the all-vapor state are generally referred to as the liquid and vapor saturation points, respectively. Because of the higher internal energy of a vapor molecule relative to a liquid molecule of the same species and at the same temperature, evaporation of a fluid, under conditions far removed from the critical point, generally results in the absorption of a very substantial amount of heat. Rapid evaporation thus facilitates the transfer of a very high heat flux from the surface wetted by the fluid. The migration or diffusion of vapor, generated at the free liquid–vapor interface by this evaporation process, is governed by the difference in vapor pressure or vapor concentration between the interface and the ambient. Along the interface, it is generally assumed that the vapor and the liquid are in thermodynamic equilibrium; therefore, the vapor pressure at the interface corresponds to the saturation value at the local liquid temperature. Based on the evaporation principle, GAEC has been developed. With high velocity gas–liquid flow in the narrow channels between populated substrates, GAEC offers distinct advantages in the thermal management of compact three-dimensional packaging of microelectronics and has been successfully incorporated into high-performance computer modules (Bar-Cohen et al. 2006).

Boiling is a complex convective heat transfer process depending upon liquid-to-vapor phase change by the formation of vapor bubbles at the heated surface. In contrast to evaporation, boiling generally refers to a process that occurs along solid surfaces submerged in the liquid and is characterized by the dominant influence of vapor bubbles. Boiling typically commences when the surface temperature exceeds the liquid saturation temperature (the boiling point) by 3–10°C, and vapor bubbles are then found to grow and issue from minute cavities in the surface. It is commonly characterized as either pool boiling (occurring in a stagnant liquid) or flow boiling. The pool boiling heat transfer rate usually follows a relationship of the form (Simons 1996):

$$Q = C_{sf}A(T_{wall} - T_{sat})^n, \tag{10.1}$$

where $C_{sf}$ is a constant depending on each fluid–surface combination, $A$ is the heat transfer surface area, $T_{wall}$ is the temperature of the heated surface, and $T_{sat}$ is the saturation temperature (i.e., boiling point) of the liquid. The value of the exponent $n$ is typically about 3. If chip power is gradually increased in small steps, as shown in Figure 10.1d, cooling occurs first by natural convection (point A). Eventually, a power level is reached at which sufficient superheat is available to initiate the growth of vapor bubbles on the surface and boiling starts (point B). As power is increased, more nucleation sites become active and the frequency of bubble departure increases (point C). The region between B and C is termed the nucleate boiling regime. In the nucleate boiling regime, the bubbles increase in size with increasing heat flux at the heater surface and begin to coalesce to form larger bubbles. Vigorous agitation of the hot boundary along the heated surface, and gross fluid circulation caused by the motion of the vapor bubbles, provide the ability to accommodate substantial increases in heat flux with minimal increases in surface temperature. As power is increased to point C, the CHF condition is reached. The CHF places an upper limit on nucleate boiling heat transfer. At CHF, typically occurring at 15–20 W/cm$^2$ for saturated FC (Fluorocarbon) liquids at atmospheric pressure, the vapor generation rate is so high that the heated surface becomes partially or completely blanketed by vapor and the boiling mechanism changes to film boiling (Bar-Cohen et al. 2006). In film boiling, the vapor blanket on the heater may periodically thin or even break, thus allowing the liquid to touch the heater surface and temporarily resume nucleation on a small fraction of the heated surface. In this regime, heat transfer from the surface to the liquid is dependent on thermal conduction through the vapor and it is very poor. In most electronic cooling applications, transition to film boiling will result in failure due to high temperatures. To take advantage of boiling to cool electronic devices, it is desirable to operate in the nucleate boiling regime (B–C). A problem often associated with pool boiling of fluorocarbon liquids is that of temperature overshoot. This behavior is characterized by a delay in the inception of nucleate boiling (i.e., beyond point B), such that the heated surface continues to be cooled by natural convection; with increased surface temperatures unless a sufficient superheat is reached for boiling to occur. This behavior is a result of the good wetting characteristics of the fluorocarbon liquids and the smooth nature of silicon chips. There is usually little or no temperature overshoot associated with flow boiling cooling applications. The typical CHFs encountered in saturated (i.e., liquid saturation temperature) pool boiling of fluorocarbon liquids ranges from about 10–15 W/cm$^2$, depending upon the nature of the surface (i.e., material, finish, geometry). The allowable CHF may be extended by subcooling the liquid below its saturation temperature. For example, experiments conducted at IBM demonstrated that it is possible to increase the CHF to as much as 25 W/cm$^2$ by dropping the liquid temperature to −25°C. Higher CHFs may be achieved using flow boiling. For example, heat fluxes from 25 to over 30 W/cm$^2$ have been reported for liquid velocities of 0.5–2.5 m/s over the heated surface. Heat fluxes in excess of 100 W/cm$^2$ have been obtained with a FC-72 liquid jet impinging upon a 6.5 × 6.5 mm chip at a flow rate of 2.2 cm$^3$/s (Chrysler et al. 1995).

Therefore, while boiling heat transfer encompasses a variety of thermal transport phenomena, it is the highly efficient nucleate boiling regime that is of primary interest for thermal control of high heat flux electronic and microelectronic components. This regime lies between boiling incipience, associated with the initial generation of a steady stream of vapor bubbles from distinct nucleation sites, and the peak nucleate boiling heat flux, or CHF, associated with vapor blanketing of the heated surface. In established nucleate boiling, large variations in heat flux can be accommodated with small variations in surface temperature. To maintain operating chip temperatures in the preferred range of 65–125°C, the saturation temperature of the immersion cooling liquids, at atmospheric pressure, must be moderate. Several perflourinated (FCs) and HFE fluids possess such properties, as shown in Table 10.1. These properties combine to yield nucleate boiling superheats similar to those of conventional coolants (5–20 K) but result in anomalous boiling incipience phenomena and in relatively modest peak heat fluxes. The low surface tension of these liquids contributes to near-zero contact angles on most known surfaces and interferes with the trapping and preservation of vapor and/or gas in surface cavities essential to heterogeneous nucleation. As a result, incipience superheats of 20–30 K and even as high as 72 K, have been encountered in ebullient heat transfer to these liquids. Alternately, their high gas solubility, approaching 50% by volume at standard temperature and pressure, can lower the saturation temperature of the gassy liquid by as much as 20 K. In view of the temperature constraints imposed on individual chips and package arrays, the widespread implementation of ebullient immersion cooling for electronic systems appears to require a detailed understanding of the controlling thermal transport mechanisms and predictive relations for boiling incipience, established nucleate boiling, and the peak nucleate boiling heat flux for immersed components (Bar-Cohen et al. 2006).

In spite of prolonged interest in direct immersion liquid cooling as a means to cool high heat flux microelectronics, there have been only a limited number of applications. As with indirect liquid cooling, these applications have been almost exclusively in the large mainframe and supercomputer arena. This is not surprising, because this has been the microelectronics technology sector with the highest packaging densities and concentration of heat. The liquid encapsulated module (LEM) developed at IBM in the 1970s provides an example of a package utilizing pool boiling. A substrate with integrated circuit chips was mounted within a sealed module-cooling assembly containing a fluorocarbon coolant (FC-72). Boiling at the exposed chip surfaces provided high heat transfer coefficients (1,700–5,700 W/m$^2$ K) to meet chip cooling requirements. Internal fins provided a means to condense the vapors and remove heat from the liquid. Either an air-cooled or water-cooled cold plate could be used to cool the module. Using this approach, it was possible to cool 4 W chips (4.6 × 4.6 mm) and module powers up to 300 W. Direct liquid immersion cooling has been used within IBM for over 20 years as a means to cool high-powered chips on multichip substrates during electrical testing prior to final module assembly. An example of a large scale forced convection fluorocarbon cooling system is provided by the CRAY-2 supercomputer (Danielson et al. 1986). Stacks of electronic module assemblies were cooled by a forced flow of FC-77 in parallel across each module assembly. Each module assembly consisted of eight PCBs on which were mounted arrays of single chip carriers. A total flow rate of 70 gpm was used to cool 14 stacks containing 24 module assemblies each. The power dissipated by a module assembly was reported to be 600–700 W. Coolant was supplied to the electronics frame by two separate frames containing the required pumps and water-cooled heat exchangers to reject the total system heat load to customer supplied chilled water. In addition, several other considerations should be kept in mind when considering direct liquid immersion for cooling electronics. Because fluorocarbon liquids are expensive, they should only be considered for use in closed systems. Whether the application is in a self-contained module like the LEM or a forced flow scheme, care must be taken to ensure that the seal materials chosen are compatible with the liquid. Information or guidance in this regard may sometimes be obtained from the manufacturer of the coolant. If boiling is to take place, then the design must incorporate a means to condense the resulting vapors. A finned surface may be designed for this purpose as in the LEM example, or a remote finned condenser surface cooled by air or water might be used. In flow systems, care must be taken in selecting a pump. The relatively high vapor pressure of the low boiling point fluorocarbons generally requires that a higher suction head be provided to prevent cavitation in the pump. Whether using a self-contained boiling module or a circulating flow system, care should be taken to make sure all internal surfaces in contact with the coolant are clean. This will ensure that manufacturing process residues or unclean surfaces do not introduce a contaminant into the liquid which could be carried to the heated chip surfaces and interfere with the boiling process. In forced circulating liquid systems, it may be desirable to add a particulate and a chemical filter to ensure the long-term purity of the coolant. By selecting the appropriate liquid coolant and the mode of heat transfer, and by giving appropriate attention to these other considerations; direct liquid immersion cooling can be used successfully to provide an effective solution for cooling high-heat flux chips and packages (Simons 1996).

In summary, direct immersion can be utilized with natural convection and forced convection. Natural convection requires no fluid moving hardware, directing contact with parts, which can be used in either sensible or phase change methods that can provide close temperature control. However, it has the lowest heat removal capability of the fluorocarbon mechanisms. To increase thermal transfer coefficient, forced convection direct immersion can be used. With phase change and channel flow phase change, high rate of heat removal results in approximate constant part temperatures, and subcooled inlet liquid can cause complete condensation of the vapor microbubbles within the bulk fluid flow, thus requiring conditioning only a single-phase liquid. This leads to a more efficient mode of heat exchanger action and system operation. It is a more simple method than spray and jet methods, yet affording high-heat flux values. The major disadvantages of the method are that it requires pumping hardware and associated maintenance/cost, fluid reconditioning. CHF of the liquid coolant must also be matched to the system requirements to avoid exceeding nucleate boiling regime and leading to overheating and catastrophic failure. This is not necessarily a disadvantage, just a design requirement.

## *Liquid Jet Impingement*

Jet impingement is an attractive cooling mechanism due to its capability of achieving high-heat transfer rates. Impingement cooling may involve a single jet directed at a single component or an array of electronic components, multiple jets directed at a single component, arrays of jets directed at an array of chips on a common substrate, or an array of jets directed at chip packages on a PCB. The jets may be formed by circular slot-shaped orifices or nozzles of various cross sections. A fluid jet issuing into a region containing the same fluid is characterized as a submerged jet while a fluid jet issuing into a different, less dense, fluid is characterized as a free-surface jet. As a final distinction, jet impingement cooling of electronic components may involve forced convection alone or localized flow boiling, with or without net vapor generation (Bar-Cohen et al. 2006). A typical configuration for single-phase submerged jet impingement cooling is illustrated in Figure 10.2.

### Concept of Free-Surface and Submerged Jet Impingement

When a jet impinges on a surface, very thin hydrodynamic and thermal boundary layers form in the impingement region due to jet deceleration and increase in pressure. Consequently, extremely high heat transfer coefficients are obtained within the stagnation zone. Because the peak heat transfer only occurs within the stagnation zone, a single impinging jet can provide effective heat transfer when highly localized heating or cooling is required. As shown in Figure 10.2, the flow in a jet impinging perpendicularly on a plate surface is commonly divided into three separate regions: the free region, the impingement region, and the radial flow region. The flow in the free jet zone is mainly in the axial direction and is not overly affected by the presence of the impingement surface. Within this free jet zone, there are two subregions, the potential core with velocity equal to the jet exit velocity and a free surface jet flow with a lower velocity shear layer, which is slowed by the drag and entrainment of the surrounding fluid. Downstream from the nozzle, the shear layer progressively expands into the potential core, eventually reaching the jet centerline. In the stagnation flow region, the flow impinges on the surface and then turns, flowing parallel to the surface. The parallel flow portion is called the wall-jet region. Depending on the temperature of the liquid in the jet

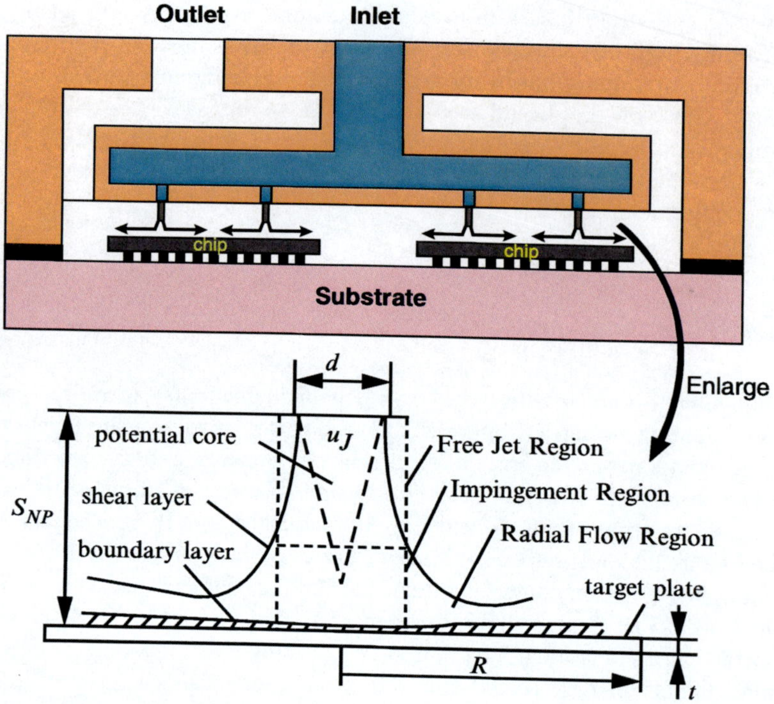


**Fig. 10.2** Typical configuration of submerged jet impingement cooling

(saturated or subcooled) and the temperature difference between the target wall temperature and the saturation temperature of the jet liquid, impingement heat transfer may provide either single-phase or two-phase cooling (Bar-Cohen et al. 2006).

Higher heat transfer coefficients result from submerged jet conditions than from free-surface jet conditions for Reynolds number $Re \geq 4{,}000$. An impinging jet can be identified confined or semiconfined if the radial spread is confined in a narrow channel, usually between the impingement surface and the orifice plate. The presence of a confining top wall in jet impingement causes lower heat transfer coefficients, believed to be caused by the recirculation of fluid heated by the target plate. Confinement promotes a more uniform heat transfer distribution for the area enclosed by a nondimensional radial distance from the stagnation point ($R/d$) of 5. The key parameters determining the heat transfer characteristics of a single impinging jet are the Reynolds number, Prandtl number, jet diameter, and jet-to-target spacing. Nozzle geometry can also have a significant influence on heat transfer. For a constant jet diameter, heat transfer increases with increasing Reynolds number. For a constant Reynolds number, decreasing the jet diameter yields higher stagnation and average heat transfer coefficients. This can be attributed to the higher jet velocities created by the smaller nozzles. Additionally, the jet-to-target spacing ($S_{NP}$) has a much greater influence on heat transfer for submerged jets than for free-surface jets. Little change in stagnation and average heat transfer for $S_{NP}/d < 4$, then a decrease in heat transfer as $S_{NP}/d$ increases beyond this point. The relative consistency of heat transfer for $S_{NP}/d < 4$ can be explained by the jet impingement taking place within the potential core with its nearly uniform velocity, while the decrease in heat transfer at higher $S_{NP}/d$ values is attributed to complete degradation of the potential core prior to impingement. For jet impingement, the coolant is sprayed at a higher rate than in the spray cooling case, and does not necessarily need to evaporate, although it is convenient if evaporation still occurs because in that case, no draining mechanism would be necessarily required. As with spray cooling, the fluid pours in direct contact with electronic connections without an effective sealant, water is still not an option; in addition, with operating temperatures being below 100°C, water might require a draining mechanism to ensure proper flow. For the case of F-72, heat flux of up to 120 W/cm$^2$ can be supported on an operating temperature of 60°C higher than that of the liquid. If a sealant were available, a theoretical heat flux of up to 460 W/cm$^2$ would be feasible using water as the operating fluid (Stull and Herz 2007).

Because impinging jets can provide very high local heat transfer rates, this cooling method has been used in a wide range of industrial applications such as quenching of metals and glasses, cooling of gas turbine blades, cooling in grinding processes, and cooling of photovoltaic cells. Jet impingement has also become a viable candidate for high-powered electronic and photonic thermal management solutions.

### Convective Jet Impingement

In general, the surface–convection resistance is the smallest in the stagnation-flow region and increases in the wall-jet region. Average heat transfer coefficient over the impingement surface depends on parameters such as the jet Reynolds number *Re*, nozzle-to-plate distance $S_{NP}$, nozzle geometry *d*, the impinging wall geometry *R*; and the inlet coolant Prandtl number *Pr*. One of the most widely used correlations for the average Nusselt number in a single-jet impingement is expressed as (Bar-Cohen et al. 2006):

$$Nu/Pr^{0.42} = 2\left(\frac{d}{r}\right)\frac{1-1.1d/r}{1+0.1(S_{NP}/d-6)d/r}\sqrt{Re_J\left(1+\frac{Re_J^{0.55}}{200}\right)}, \tag{10.2}$$

where $Re_J$ and *Nu* are the Reynolds number and the Nusselt number based on the nozzle diameter $Re_J = u_J d/\nu, Nu = hd/k_f$ and where *h* is the average heat transfer coefficient based on the average temperature difference between the target and the coolant. The general form of the equation for the local *Nu* number at the stagnation zone both for submerged and free surfaced jets is given by (Bar-Cohen et al. 2006):

$$Nu_0 = 1.25Pr^{1/3}Re^{1/2}. \tag{10.3}$$

The exponent of the *Re* number clearly indicates the laminar characteristic of impingement flow in the stagnation zone. The heat transfer rate for multiple-jet impingement can be estimated from the single-jet impingement case by allocating a unit cell on the heated surface to each one of the jets. If the interaction between adjacent jets within the representative area and the influence of the spent fluid flow is neglected, the heat transfer data inferred from a single jet can approximately represent the actual situation. Therefore, the heat transfer coefficient for an array of nozzles can be expressed by (Bar-Cohen et al. 2006)

$$\left(\frac{Nu}{Pr^{0.42}}\right)_{\mathrm{AN}} = Re_J^{2/3}\left[1+\left(\frac{S_{\mathrm{NP}}/d}{0.6/\sqrt{\alpha_J}}\right)^6\right]^{-0.05}\frac{\sqrt{\alpha_J}(1-2.2\sqrt{\alpha_J})}{1+0.2(S_{\mathrm{NP}}/d-6)\sqrt{\alpha_J}}. \quad (10.4)$$

The above correlation is valid in the range of $2{,}000 < Re_J < 100{,}000$, $0.004 < \alpha_J < 0.04$; and $2 < S_{\mathrm{NP}}/d < 12$ (Martin 1977).

In the absence of boiling, a free jet forms a radial wall jet that emanates from the impingement zone while remaining mostly in contact with the heated wall. Unlike the situation with single-phase jet impingement cooling, during boiling along the heated surface, the vigorous, at times explosive, generation of vapor bubbles within the wall jet splashes away a significant portion of the wall jet liquid flow. Further increases in heat flux result in the formation of dry patches in the outer circumference of the wall jet, as much of the wall jet liquid is splashed away in these outer regions. Eventually, this dry out propagates inward toward the impingement zone, causing separation of the wall jet from much of the heated wall and resulting in dry out/CHF. Due to the existence of different boiling regimes, which depend on surface temperature and geometry as well as coolant flow conditions and subcooling, boiling from high temperature surfaces experiencing liquid jet impingement can be quite complex. There are two types of behavior in jet impingement boiling (Bar-Cohen et al. 2006): (1) nucleate boiling, in which bubbles are formed by nucleation at the solid surface washed by the impinging jet. In saturated boiling, these bubbles grow, detach, and join the main two-phase flow. In highly subcooled boiling, they collapse rapidly while heating the main liquid flow towards the saturation temperature. (2) Convective boiling, or thin-film evaporation, in which heat is transferred by conduction and convection to the liquid/vapor interface, sometimes assisted by bubble dynamics. Two-phase jet impingement on a flat hot plate can be further divided into two modes, the free film flow and the stagnation jet flow. In steady-state jet impingement boiling, the dryout or the CHF generally occurs at the downstream location furthest from the stagnation point, and are typically correlated in terms of the heat source dimension 2R (Bar-Cohen et al. 2006).

In applying correlation (10.4) to the submerged jet cooling of electronic components, the heat transfer coefficient produced by impinging liquid jet(s) can be obtained (Bar-Cohen et al. 2006):

$$h \propto [k^{0.58}\rho^{0.67}\mu^{-0.25}]\,[(n/A)^{0.35}d^{0.67}]\,[h_{-0.3}v^{0.67}], \quad (10.5)$$

where $n$ would be determined by experimental data. In this equation, the first bracketed term represents a fluid figure-of-merit for submerged-jet heat transfer;

the second term constitutes a thermal figure-of-merit for the jet plate, and the third, the operating conditions of an impingement cooling system. Clearly, to maximize the jet heat transfer rate, it is desirable to choose a liquid with high thermal conductivity and density but relatively low viscosity. The ideal jet manifold would contain many large-diameter nozzles per component. For 16-jet arrays for instance, the heat transfer coefficient increases from 16.5 kW/m$^2$ K for a jet diameter of 0.27 mm to 17.8 kW/m$^2$ K at 0.32 mm. Due to the strong dependence of the heat transfer rate on the jet Reynolds number, maximization of the heat transfer coefficient also requires increasing the liquid velocity at the nozzle and decreasing the distance of separation between the nozzle and the component. Alternately, if a liquid has been selected and if the jet Reynolds number is to remain constant, a higher heat transfer coefficient can only be obtained by increasing $n/A$ or decreasing $S_{NP}$. In addition, intense small-scale turbulent jets have been developed with periodic entrainment and expulsion of fluid by microfluidic devices immersed in the liquid. The jets can be made to impinge upon electronic components, thereby providing forced convection impingement cooling. The small size of these devices, accompanied by the high exit velocity of the fluid, provides an opportunity to significantly reduce the volume of the hardware used for the thermal management of electronics (Bar-Cohen et al. 2006).

Consequently, jet impingement offers excellent cooling performance in low to moderately high heat applications. Production of components is quite easy because precision manufacturing is not required. A simple jet impingement system could be constructed that delivers and directs dielectric liquid through holes (nozzles) in a plate. Rectangular or circular shapes can be used and tailored for size, flow rate, and velocity to match the heat dissipating components. Precise positioning is not required. Contamination, corrosion, and wear of jet impingement nozzles are of very little concern, even though the dielectric fluid should be filtered and conditioned in both jet impingement and spray configurations. Major advantages of jet impingement include very high heat fluxes being achieved if desired and critically designed/manufactured nozzles not being required. Plates with machined openings can be used; flow can be localized if only a single jet is used; multiple jet placement is not as critical with respect to closeness to/from parts; jet hardware is more repeatable and durable due to less precision required in jet openings; will not splash and separate away from parts. There is no cone effect requiring specific locations with respect to parts. Thermal transfer can also take place in sensible regime better than with spray. This could benefit overall cooling and offer some redundancy. The disadvantages are that it requires complex fluid handling and reconditioning (condensation, heat exchange with ultimate sink, etc.) hardware. This adds weight and associated penalties to platform systems.

## *Spray Cooling*

Spray cooling involves spraying the fluid directly onto the chip and allowing it to evaporate, cooling the chip down by the same basic mechanism as human sweat. Figure 10.3 illustrates typical spray processes (1) photograph of spray; (2) light

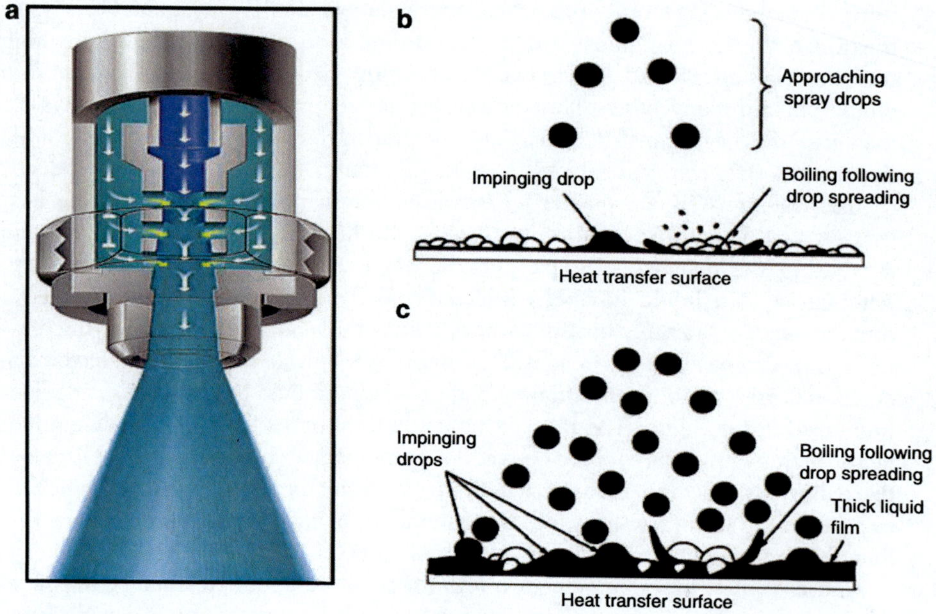


**Fig. 10.3** Illustration of spray processes (**a**) photograph of spray; (**b**) light spray (lower Weber number); and (**c**) dense spray (high Weber number)

spray (lower Weber number); and (3) dense spray (high Weber number). The breakup of a droplet upon impingement on the surface is described by the droplet Weber number, $We_d$, which is defined as the ratio of droplet inertia forces to surface tension forces, using the droplet diameter $d$ as the length scale and $u_d$ the characteristic velocity of the droplet normal to the surface (Bar-Cohen et al. 2006).

$$We_d = \frac{\rho_f u_d^2 d}{\sigma}. \tag{10.6}$$

As an impinging droplet contacts a hot solid surface, heat is transferred from the solid to the liquid phase by conduction, convection, and radiation, increasing the temperature of the liquid or alternatively vaporizing liquid from the base of the droplet. The droplet Weber number has a strong influence on the spreading characteristics and integrity of the droplet and several distinct dynamical regimes of droplet impact associated with specific ranges of the Weber number have been recognized. Thus, both the Weber number and the surface superheat can affect the behavior of the impinging droplets and are the spray cooling heat transfer rates.

As shown in Figure 10.3b, c, two spray cooling regimes can be recognized (Bar-Cohen et al. 2006): a light spray (a small volumetric flux) and a dense spray (a high volumetric flux). In a light spray, the frequency of drop impingement upon the heated surface is low, leaving much of the surface covered with fairly stagnant liquid within which vapor bubbles can easily nucleate and aid the evaporation process. Evaporation efficiency in light sprays is, therefore, very high. Because the fluid will directly contact the chip without a sealant, only dielectric fluids can be used. A commonly used fluid is the fluorocarbon FC-72, which is capable of supporting heat fluxes up to 50–60 W/cm$^2$ while limiting chip junction temperatures to 85°C. Jet impingement of a dielectric liquid on the chip surface may also be considered. When chip heat fluxes approach 120 W/cm$^2$ and a chip to liquid temperature difference of 60°C while cooling a 6 × 6 mm silicon chip with a single jet of FC-72 liquid at a flow rate of 2.2 × 10$^{-4}$ m$^3$/s. Ultimately, given that an adequate chip to liquid seal can be provided, direct jet impingement of water on the back of the chip may be considered. For these various options, the chip heat removal capability for a 60°C chip to liquid inlet temperature difference is estimated to be 60, 120, 300, and 460 W/cm$^2$ for spray cooling, dielectric jet impingement cooling, forced convection dielectric cooling with a heat sink attached to the chip, and water jet impingement, respectively.

In general, spray cooling heat transfer displays three distinct domains of behavior at low, middle, and high surface temperatures, corresponding to the nucleate, transition, and film boiling regions. Spray heat transfer with dielectric liquids appears to be much more effective than saturated pool boiling, achieving peak heat fluxes that can be several times higher than saturated pool boiling CHF, although spray cooling does require the investment of significant pumping power. The temperature overshoot encountered with boiling incipience seems to be entirely eliminated by the use of liquid sprays, and spray cooling can provide a relatively uniform surface temperature. However, the cooling rates achieved in spray cooling are dependent on the liquid droplet properties and behavior. The maximum heat transfer per drop impinging on hot surfaces ($q_{\text{max}}$) is a function of the fluid properties and the normal component of the impact velocity $u_{\text{d}}$ for superheats of about 165°C for water, acetone, alcohol, and some freons. As much as 50% of the droplet mass would evaporate during the short time interval associated with impact and bouncing. A correlation of experimental data was given as (Bar-Cohen et al. 2006):

$$\frac{q_{\text{max}}}{\rho_{\text{d}} d^3} = 0.00183 \left( \frac{\rho_{\text{d}}^2 u_{\text{d}}^2 d}{\rho_{\text{v}} \sigma g} \right)^{0.341} \left[ h_{\text{fg}} + C_{p,v} \left( \frac{T_{\text{w}} - T_{\text{sat}}}{2} \right) \right]. \tag{10.7}$$

Sprays can be classified into either pressure sprays or atomized sprays, depending upon the method used to accomplish the liquid breakup. Despite their superior cooling performance, atomized sprays are difficult to incorporate in a closed loop electronic cooling system because of the complexity of separating air from dielectric liquid coolants. The droplet sprays can have the form of a mist and impinge on the surface with a random pattern. After hitting the surface, the liquid droplets spread and often merge to form a thin liquid film. If the wall superheat is above the Leidenfrost point, a thin vapor layer is present underneath the droplets or the liquid film (Bar-Cohen et al. 2006).

Basically, spray cooling offers good performance in applications with low to moderate heat removal requirements. Figure 10.4 illustrates an inkjet-assisted spray cooling device. Dielectric liquid vaporizes on impact, cooling the chip, and the vapor is then pumped back into a reservoir that feeds the spray device. The individual spray cooling components, such as nozzles, plumbing, pumps, filters, and heat exchangers are commercially producible. However, it should be noted that special machining capabilities are required to manufacture spray nozzles to the precision and tolerance required. It is revealed that variation in nozzle performance due to contamination, corrosion, and long-term wear. Placement of spray nozzles is fairly critical to assure adequate cooling. The major advantages of spray cooling is that it can provide high heat flux, give good coverage due to atomizing action, may use less fluid than forced convection, bulk flow mechanism, and may use slightly less fluid than jet impingement. However, it requires special spray nozzles, complex fluid handling and reconditioning (condensation, heat exchanger with ultimate sink, etc.) equipment. This adds weight and associated penalties to platform systems. Its quality control is critical; the nozzles are very sensitive to manufacturing tolerances and quality. Nozzle action can change in time due to erosion, corrosion build up, and contaminants. Spray distance to components is critical for development of spray cone. Spray velocity and momentum can be critical. If too great, it can lead to part erosion, and splashing away without proper wetting of the part surfaces, leading to poor cooling. Proper distance and some degree of confinement have been found necessary to avoid separation from surfaces, leading to poor wettability. In addition, it must be used only as phase change, has poor sensible cooling action, and need high pump pressure.

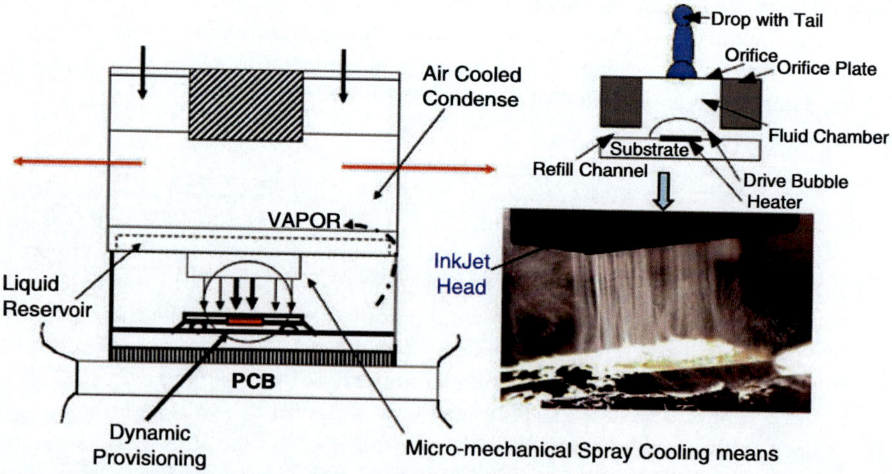


**Fig. 10.4** Illustration of inkjet-assisted spray cooling. Dielectric liquid vaporizes on impact, cooling the chip, and the vapor is then pumped back into a reservoir that feeds the spray device

## Indirect Liquid Cooling

Indirect liquid cooling schemes are characterized by the presence of distinct physical barriers between the chip and the liquid. Several possible implementations are illustrated in Figure 10.5. The most conservative implementation is to attach a separable cold plate to a lidded module (Figure 10.5a). When greater thermal performance is desired, the interface between the cold plate and the module can be eliminated by integrating the cold plate with the module (Figure 10.5b). Taking the concept to the limit, chip-scale high performance cold plates can be attached directly to the chips (Figure 10.5c). For three indirect cooling schemes, the liquid volumetric flow is in proportion to the heat dissipated at a ratio of $6.31 \times 10^{-8}$ $m^3/s/W$. Under these conditions, the module cold plate thermal performance is estimated to be as low as 12.8°C $mm^2/W$, when cold plate has an active area of 42.5 × 42.5 mm and channel widths approaching the microchannel scale of less than 0.2 mm. Additionally, a lid-to-cold plate unit resistance of 25°C $mm^2/W$ (a PAO (Polyalphaolefin) oil interface with an average bond line of 0.0045 mm) is assumed for the separable cold plate. Assuming a metal interface (i.e., solder) between the chip and the cold plate with microchannels less than 0.1 mm in width, the chip heat removal capability for a chip junction to liquid inlet temperature difference of 60°C is 227, 320, and 397 $W/cm^2$ for the separable module cold plate, integral module cold plate, and individual chip-scale cold plate configurations, respectively (Ellsworth and Simons 2005).

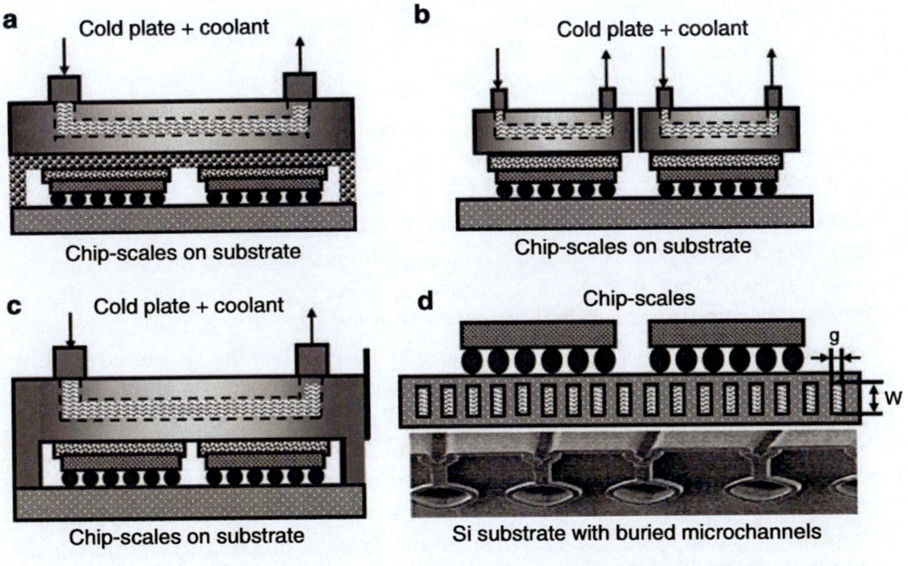


**Fig. 10.5** Indirect liquid cooling schemes: (**a**) separable module cold plate; (**b**) integral module cold plate; (**c**) individual chip-scale cold plates; and (**d**) microchannels through Si substrate

Microchannels have most commonly been used for indirect liquid cooling of electronic packaging. They can be machined into a heat spreader or heat sink and then attached to a chip or array of chips, or directly etched onto the surface of a chip die itself or a substrate in the shape of rectangular grooves, as shown in Figure 10.5d. Water is a normal fluid due to the fact that it is cheap and has superior thermal properties. When laminar water flows through a rectangular channel, the heat transfer coefficient can be expressed as (Kandlikar and Grande 2004)

$$h = N_u k / D_{\mathrm{h}}, \tag{10.8}$$

where $D_{\mathrm{h}} = 4gw/2(g + w)$, $Nu = 8.235\,[1–2.0421\,(g/w) + 3.0853\,(g/w)^2 – 2.4765\,(g/w)^3 + 1.0578\,(g/w)^4 – 0.1861\,(g/w)^5]$.

Therefore, microchannels are based on a very simple heat transfer concept: the heat transfer coefficient for laminar flow (very slow flow through a narrow channel) is inversely proportional to the hydraulic diameter. This means that the smaller the channel is, the higher the ability to draw heat from the source. Microchannels typically have sizes in the 5–100 μm range leading to a heat transfer coefficient of that may reach 80,000 W/m$^2$ K. There are commonly two main problems when designing a system of microchannels: pressure drop and flow uniformity across the channels.

The pressure drop due to forcing a fluid through a small channel may produce design limitations, which include pumping power, and mechanical stress limitation of the chip or substrate materials. The pressure drop can be predicted by the following equation (Kandlikar and Grande 2004):

$$\Delta p = \frac{2fL\rho V^2}{D_{\mathrm{h}}}, \tag{10.9}$$

where $V$ is the mean flow velocity; $L$ is the flow length; $\rho$ is the fluid density; and $f$ is the friction factor, depends on the aspect ratio $g/w$. The smaller hydraulic diameter results in a higher heat transfer coefficient on the one hand but higher pressure-drop on the other hand. This would require higher pumping power. In order to improve the performance, the channels should be deep so that the hydraulic diameter is small but the channel surface area is large. However, making the channels too small may result in unreasonable pressure drop. The other issues related to microchannels may include possible liquid leaking, potential fouling leading to clogging, and critical requirements for mini-pumps: (1) ability to move liquid through the channel at a required rate; (2) ability to produce large pressure heads to overcome the large pressure drop associated with the small channels; and (3) traditional rotary pumps cannot be used to their large size and power consumption. One solution for keeping a high heat transfer coefficient and reasonable pressure drop that is using stacking–instead of having a single layer of microchannels on top of the heat source–two, three, or more stacks can be built. In most cases, two or three stacks are a good compromise between heat transfer behavior and pressure drop. Flow nonuniformity across the microchannels would result in nonuniform cooling, which may have implications on both the performance and reliability.

In order to improve the heat transfer performance of single-phase convection in microchannels, the following efforts have been made. (1) Increase surface area by adding protrusions to the channels; or by adding and arranging fins in a manner that is similar to a compact heat exchanger, such as staggered fins, posts, and T-shaped fins. (2) Produce other enhanced geometries. When microchannel diameters become small, the classical theory for friction and heat transfer coefficients may have some extent of deviation. A good analytical model is needed to describe the heat transfer process and evaluate the effect of design parameters such as surface roughness. In addition, the method to accurate measurements of system parameters needs also to be developed.

Other forms of indirect liquid techniques are shown in Figure 10.6. The examples of indirect liquid cooling schemes include: (1) Liquid cooling through nanograss–water forms a nearly perfect ball, suspended on the tips of tiny blades of nanograss (Schweber 2005). This method is to control the behavior of tiny liquid droplets by applying electrical charges to specially engineered silicon surfaces that resemble grass blades. The specially engineered silicon surface is covered with a nonstick, water-repellent material. The individual blades of the nanograss are so small that liquid droplets sit on top and can be easily maneuvered by applying voltage. A droplet could be sent to a hotspot where it would sink in and absorb the

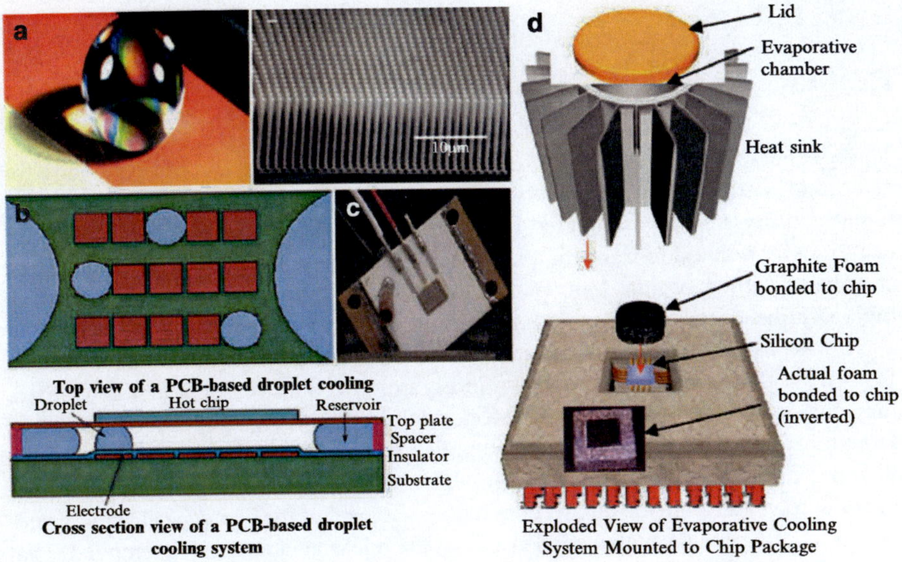


**Fig. 10.6** Examples of indirect liquid cooling schemes: (**a**) liquid cooling through nanograss – waters a nearly perfect ball, suspended on the tips of tiny blades of nanograss (*right*); (**b**) droplet-based microfluidics; (**c**) heat spreader filled with water as working fluid; and (**d**) evaporative cooling system using graphite foam

heat, and then go on its way. With a predicted shelf life of 25 years, the nanograss liquid cooling system can dissipate powers greater than 100 W/cm$^2$. (2) Droplet-based microfluidics (Pamula and Chakrabarty 2003); this method is based on pumping nanoliters of water droplets in a single-phase forced convection. The actuation mechanism is based upon surface tension gradients induced in a droplet by voltage-induced electrowetting. Hot areas also attract droplets and are returned to their reservoirs by electrowetting resulting in a self-contained and self-regulated system. Liquid flow increases with increasing temperature ensuring that local hot areas have increased cooling applied locally without the need for external sensors. (3) Heat spreader filled with water as working fluid; the system uses a new heat spreader which can be integrated at the interconnect substrate level. It is operated on a principle similar to heat pipes and an active fluid pumping system based on a piezo-driven micropump. Initial heat spreader fabricated and successfully tested at the level of 10 W power. The target is to develop an integrated nonmoving-part pump, and to transfer heat fluxes of around 200 W/cm$^2$. (4) Evaporative cooling system using graphite foam (Klett 2002). The technology uses a flip-chip design where the silicon chip is inverted and graphite foam is used to replace the material currently used to dissipate heat. The cell walls are made of oriented graphitic planes and exhibit thermal conductivities greater than 150 W/m K. The thermal response time of the foam is similar to that of diamond yet exhibits a surface area thousands of time greater than diamond. Using this technology, a power density of 100 W/cm$^2$ was attained–more than 350% greater that previously attainable.

## Heat Pipe Cooling

The concept of heat pipe was put forward by R.S. Gaugler as a cooling strategy for electronic equipment and presents a promising alternative compared with traditional cooling schemes (Gaugler 1944). Heat pipe provides a passive two-phase heat transfer method to transmit a large quantity of heat from a confined space to a condenser with minimum temperature drop. This method offers the possibility of high local heat removal rates with the ability to dissipate heat uniformly.

A classical heat pipe consists of a metallic vessel in which its inner walls are lined with a wick structure, as shown in Figure 10.7. The vessel is first vacuumed, then charged with a small amount of working fluid (typically water or methanol), and hermetically sealed. When the heat is applied to one end of the metallic vessel, the working fluid vaporizes (phase change). The resulting pressure gradient forces the vapor to travel through the hollow core to the other end of the heat pipe at sonic speed, where it condenses through a condenser and the released latent heat is removed. That is, the vapor condenses back to liquid and releases heat at the same time. The liquid is then pumped back to the original hot end via wick by capillary action, completing the continuous evaporation/condensation cycle. The types of wick structures, such as extruded grooves, screen mesh, and powdered metal, are crucial in providing the

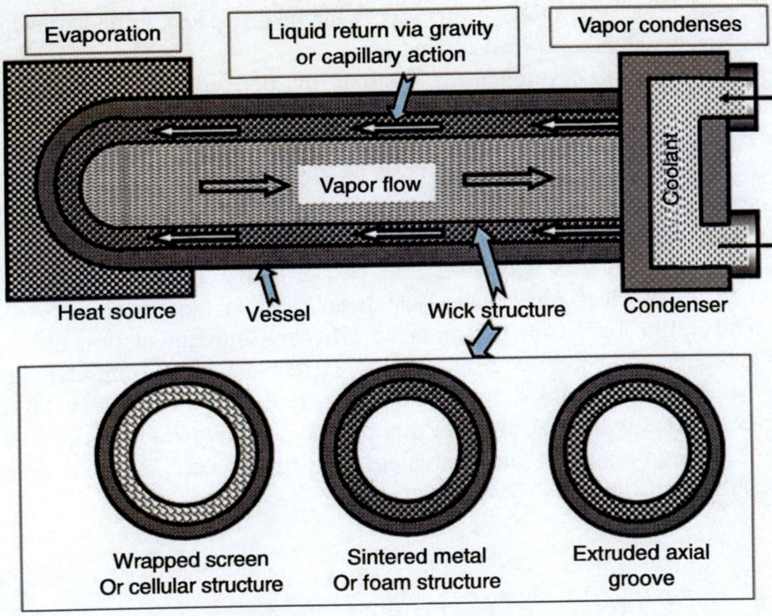


**Fig. 10.7** Schematic structure of heat pipe

capillary pumping action to circulate the fluid. The means by how the condensate is being transferred back from the condenser to the evaporator is the major distinction between different heat pipe designs. Different forces can be utilized, e.g., capillary forces (surface tension), gravitation, acceleration (centrifugal) forces, thermally induced pressure differences, or some combinations of them. Copper tubing is the most common material used to make heat pipes due to the material's compatibility with common working fluids, high thermal conductivity, and manufacturability. When assembling with electronic components, flattened heat tubes allows heat pipes to interface with flat surfaces and minimize thermal contact resistance. Structure, shape, and length are also factors that affect the heat pipe's performance. The general advantages of heat pipes include: (1) Very high effective thermal conductance–comparative thermal conductivity is up to 90 times greater than copper for the same size component. Less temperature difference is needed to transport heat than traditional conductive materials; therefore, low thermal resistance can be achieved. (2) Power flattening. A constant condenser heat flux can be maintained while the evaporator experiences variable heat fluxes. (3) Efficient transport of concentrated heat. (4) An isothermal surface of low thermal impedance can be formed. The condenser surface of a heat pipe will tend to operate at uniform temperature. If a local heat load is applied, more vapors will condense at this point, tending to maintain the temperature at the original level. In other words, the evaporator and condenser temperature can remain nearly constant while heat flux into the evaporator may vary. (5) The cylindrical geometry heat pipe suitable for many applications but other geometries can be adopted to meet special requirements. The condenser and evaporator can have different areas to fit variable area spaces. High heat flux inputs can be dissipated with low heat flux outputs only using natural or forced convection.

Heat pipes have been developed for a variety of applications, covering almost the entire spectrum of temperatures encountered in heat transfer processes. Heat pipes are used in a wide range of products like air conditioners, refrigerators, heat exchangers, transistors, and capacitors. Their application in the field of cryogenics is very significant, especially in the development of space technology. Heat pipes have been commercially available since the mid 1960s. The electronic industry has just embraced heat pipe as a reliable and cost-effective solution for high-end cooling application. Heat pipes have also been used in laptops to reduce the working temperature for better performance. The first implementation of a heat pipe in a notebook computer occurred in 1994. Earlier notebook computers relied on simple metallic heat sinks, but by the late 1990s, approximately 60% of notebook computers used heat pipes in thermal management solutions (Ali et al. 2000). It is now common to find heat pipes in different configurations used in thermal solutions for portable computers.

## *Fundamental Principles of Heat Pipe Cooling*

As shown in Figure 10.7, a typical heat pipe consists of three main sections which include an evaporator section, an adiabatic section, and a condenser section. Heat added at the evaporator section vaporizes the working fluid, which is in equilibrium with its own vapor. This creates a pressure difference between the evaporator section and the condenser section, which drives the vapor through the adiabatic section. At the condenser section, heat is removed by condensation and is ultimately dissipated through an external heat sink. The capillary effect of the wick structure will force the flow of the liquid from the condenser to the evaporator section.

In principle, heat pipes use two-phase flow, latent heat of vaporization, and capillary action to circulate a working fluid between heated and cooled regions via a wick. The capacity of a heat pipe, increase in heat transfer must be consistent with capillary driven circulation of the working fluid. Difference in pressure between the vapor phase, $p_v$, and the liquid phase $p_l$, must be balanced by the surface tension in the capillary structure (Shafii et al. 2001).

$$p_v - p_l = \Delta p_v - \Delta p_l = \frac{2\gamma \cos\theta}{r_c}, \tag{10.10}$$

$$\Delta p_l = p_l(O) - p_l(L) = -\rho_l g L \sin\varphi - \frac{b\eta_l Q_e L}{2\pi(r_w^2 - r_v^2)\rho_l \varepsilon r_c^2 l_e}, \tag{10.11}$$

$$r_c = \frac{b\eta_l Q_e L}{4\pi(r_w^2 - r_v^2)\rho_l \varepsilon l_e \gamma \cos\theta}. \tag{10.12}$$

Maximum heat transfer

$$Q_e = \frac{\pi r_w^2 l_e \gamma \cos\theta}{3L}\sqrt{\frac{\varepsilon\rho_v\rho_l}{3b\eta_v\eta_l}} \quad \text{(Reynolds number} \ll 1), \tag{10.13}$$

$$Q_e = \frac{4\pi r_w^2 l_e}{3}\sqrt{\frac{2\rho_v\rho_l\varepsilon\gamma^2\cos^2\theta}{(\pi^2-4)bL\eta_v\eta_l}} \quad \text{(Reynolds number} \gg 1), \tag{10.14}$$

where $r_w$ is inner radius of wick; $r_v$ is inner radius of the metal vessel tube; $l_e$ is length of evaporation regime; $L$ is the length of the metal tube; $r_c$ is radius of capillary pore; $\theta$ is contact angle; $b$ is constant, if pores are not interconnected, b ~ 8; if the pores are interconnected, b ~ 10–20.

Heat pipes operate on a closed two-phase cycle and utilize the latent heat of vaporization to transfer heat with a very small temperature gradient. Heat pipes consist of three main parts, which are the vessel, wick structure, and working fluid. The vessel or a container is normally constructed from glass, ceramic, or metal, while wick structure is constructed from woven fiberglass, sintered metal powders, screen, wire meshes, or grooves. Finally, typical working fluid used varies from nitrogen or helium (for low temperature heat pipes) to lithium, potassium, or sodium (for high temperature). In order to fabricate a working heat pipe, all three parts are given important consideration to the material type, thermophysical properties, and compatibility. Heat pipes are capable of creating its own capillary pressures at the evaporator end. This would cause a continuous flow of liquid to the wick and replenish the liquid at the evaporator zone. Heat flow through the evaporator section and condenser section are assumed to be adiabatic. Due to this reason, the vapor experiences a negligible temperature drop. Heat pipes generally exhibit thermal characteristics that are even better than a solid conductor of the same. As for wick structure, the working fluid travels from the condenser section to the evaporator section. The working fluid should be evenly distributed over the evaporator section. In order to provide a proper flow path with low flow resistance, an open porous structure with high permeability is desirable. This is to ensure that the working fluid returns from the condenser to the evaporator.

### *Heat Pipe Design and Type of Heat Pipes*

The main types of heat pipes are shown in Table 10.2. Apart from standard heat pipes, the thermosyphon is also an attractive device, which has successfully implemented two-phase liquid cooling by indirect contact with electronics. A two-phase thermosyphon basically consists of an evaporator and a condenser, which are connected through a passage, or a loop. Heat is transferred from the source through an interface to the evaporator, where the fluid vaporizes, taking the latent heat of vaporization. The vapor then moves to the condenser through gravity force, where it

**Table 10.2** Type of heat pipes

| Type of heat pipes | Method of condensate return |
|---|---|
| Standard heat pipe | Capillary force |
| Loop heat pipe | Capillary force |
| Thermosyphon | Gravity |
| Inverse thermosyphon | Bubble pump |
| Rotating or revolving heat pipe | Centripetal or centrifugal force |
| Electrohydrodynamic heat pipe | Electrokinetic force |
| Electro-osmotic heat pipe | Electro-osmotic force |
| Magnetohydrodynamic heat pipe | Magnetic force |
| Magnetic fluid heat pipe | Magnetic force |
| Osmotic heat pipe | Osmotic forces |

condenses. The released heat is dissipated to the ambient from the condenser and the condensed liquid is returned to the evaporator, thus completing a closed loop. The density difference between the liquid and vapor creates a pressure head, which drives the flow through the loop, and as such no outside driving force is needed. Thermal resistance at the interface between the heat source and the evaporator is a key design parameter. This must be minimized in order to get the benefit of low thermal resistance of the evaporator (Garner and Patel 2001).

Standard heat pipes can be designed into different shapes or structures depending on application requirements, such as (1) Flat plate–much like traditional cylindrical heat pipes but are rectangular. Used to cool and flatten temperatures of semiconductor or transistor packages assembled in arrays on the top of the heat pipe. (2) Microheat pipes–small heat pipes that are noncircular and use angled corners as liquid arteries. Characterized by the equation: $r_c/r_h \geq 1$ where $r_c$ is the capillary radius, and $r_h$ is the hydraulic radius of the flow channel. Employed in cooling semiconductors (improve thermal control), laser diodes, photovoltaic cells, medical devices. (3) Variable conductance–allows variable heat fluxes into the evaporator while evaporator temperature remains constant by pushing a noncondensable gas into the condenser when heat fluxes are low and moving the gas out of the condenser when heat fluxes are high, thereby, increasing condenser surface area. They come in various forms such as excess-liquid or gas-loaded form. (4) Capillary pumped loop heat pipe–for systems where the heat fluxes are very high or where the heat from the heat source needs to be moved far away. In the loop heat pipe, the vapor travels around in a loop where it condenses and returns to the evaporator.

There are many factors to consider when a heat pipe is designed. Compatibility of materials, operating temperature range, length and diameter of the heat pipe, power limitation, heat transport limitation of the heat pipe, thermal resistance, effect of bending and flattening of the heat pipe, and operating orientation are given high importance. However, the design issues are reduced to certain major considerations by limiting the selection to copper/water heat pipes for cooling electronics. The main consideration is the amount of power the heat pipe is capable of carrying. Another aspect is the temperature range that the particular working fluid can operate. This working fluid needs a compatible vessel material to prevent corrosion or any chemical reaction. Table 10.3 illustrates the typical characteristics of heat pipes.

**Table 10.3** Typical characteristics of heat pipe

| Applied temperature range (°C) | Working fluid | Vessel material | Axial heat flux ($kW/cm^2$) | Surface heat flux ($W/cm^2$) |
|---|---|---|---|---|
| −200 to −80 | Liquid nitrogen | Stainless steel | 0.067 at −163°C | 1.01 at −163°C |
| −70 to 60 | Liquid ammonia | Nickel, aluminum, stainless steel | 0.295 | 2.95 |
| −45 to 120 | Methanol | Copper, nickel, stainless steel | 0.45 at 100°C | 75.5 at 100°C |
| 5 to 230 | Water | Copper, nickel | 0.67 at 200°C | 146 at 200°C |
| 400 to 800 | Potassium | Nickel, stainless steel | 5.6 at 750°C | 181 at 750°C |
| 500 to 900 | Sodium | Nickel, stainless steel | 9.3 at 850°C | 224 at 760°C |
| 900 to 1,500 | Lithium | Niobium + 1% zirconium | 2.0 at 1,250°C | 207 at 1,250°C |
| 1,500 to 2,000 | Silver | Tantalum + 5% tungsten | 4.1 | 413 |

In standard heat pipes, wick provides structure and force that transports the condensate liquid back to the evaporator. It also ensures that working fluid is evenly distributed over the evaporator surface. The driving force that transports the condensed working liquid through the wick to the evaporator is provided by capillary pressure. Working fluids that are employed in heat pipes have concave facing menisci (wetting liquids) as opposed to convex facing menisci (nonwetting liquids). Contact angle is defined as the angle between the solid and vapor regions. Wetting fluids have angles between 0 and 90°. Nonwetting fluids have angles between 90 and 180°. The shape of a fluid's meniscus is dependent on the fluid's surface tension and the solid–fluid adhesion force. If the adhesion force is greater than the surface tension, the liquid near the solid will be forced up and the surface tension of the liquid will keep the surface intact causing the entire liquid to move up. When the liquid in the evaporator vaporizes, the radius of curvature of the menisci in the wick will decrease. As the vapor condenses in the condenser, the radius of curvature of the menisci in the wick will increase. The difference in the radius of curvature results in capillary pressure. Capillary pressure is also due to body forces and phase-change interactions. The capillary pressure created by two menisci of different radii of curvature is given by Young–Laplace Equation (Peterson 1994):

$$P_{cap} = \sigma\left(\frac{1}{R_{\mathrm{I}}} + \frac{1}{R_{\mathrm{II}}}\right) \tag{10.15}$$

where $R_{\mathrm{I}}$ and $R_{\mathrm{II}}$ are radii of curvature and $\sigma$ is the surface tension. To maximize capillary pressure, the minimum radius is needed. For a circular capillary the minimum radius is:

$$(R_{\mathrm{I}}, R_{\mathrm{II}})_{\min} = \frac{r}{\cos\theta}. \tag{10.16}$$

Substituting (10.16) values into the formula for capillary pressure:

$$P_{\text{cap}} = \frac{2\sigma\cos\theta}{r}. \tag{10.17}$$

For maximum capillary pressure the $\theta$ must be zero.

$$P_{\text{cap, max}} = \frac{2\sigma}{r}. \tag{10.18}$$

Wetting fluids have a cos $\theta$ value that will be positive. This results in a positive capillary pressure that creates a pushing force on the liquid in the wick near the condenser; this forces the liquid to move to the evaporator. Nonwetting fluids will have cos $\theta$ values that are negative, resulting in a negative capillary pressure that creates a suction force on the liquid in the wick. The liquid is prevented from moving to the evaporator. For this reason, the working liquid in heat pipes must be a wetting liquid.

Two typical types of wicks, homogeneous and composite, have been used, as shown in Figure 10.8. Homogeneous wicks are made from one type of material or machining technique, and tend to have either high capillary pressure and low permeability or the other way around. They are also simple to design, manufacture, and install. Composite wicks are made of a combination of several types or porosities of materials and/or configurations. Capillary pumping and axial fluid transport are handled independently and tend to have a higher capillary limit than homogeneous wicks but cost more. During wick design, three properties should be evaluated: (1) High pumping pressure–a small capillary pore radius (channels through which the liquid travels in the wick) results in a large pumping (capillary) pressure. (2) Permeability–large pore radius results in low liquid pressure drops and low flow resistance. Design choice should be made that balances large capillary pressure with low liquid pressure drop. Composite wicks tend to find a compromise between the two. (3) Thermal conductivity–a large value will result in a small temperature difference for high heat fluxes. (4) The wick materials compatible with working fluid.

Heat pipes work on a cycle of vaporization and condensation of the working fluid which results in the heat pipe's high thermal conductivity. When choosing a working fluid for a heat pipe, the fluid must be able to operate within the operating temperature range of the heat pipe. For instance, if the operating temperatures are too high, the fluid may not be able to condense. However, if the operating temperatures are too low, the fluid will not be able to evaporate. Watch the saturation temperature for your desired fluid at the desired heat pipe internal pressure. In addition, the working fluid must be compatible with the wick and container material. Some typical working fluids are shown in Table 10.4.

In general, as the operating temperature range of the working fluid increases, the heat transport capability increases. Selection of working fluid should also incorporate the fluid's interactions with the heat pipe container and wick. A parameter

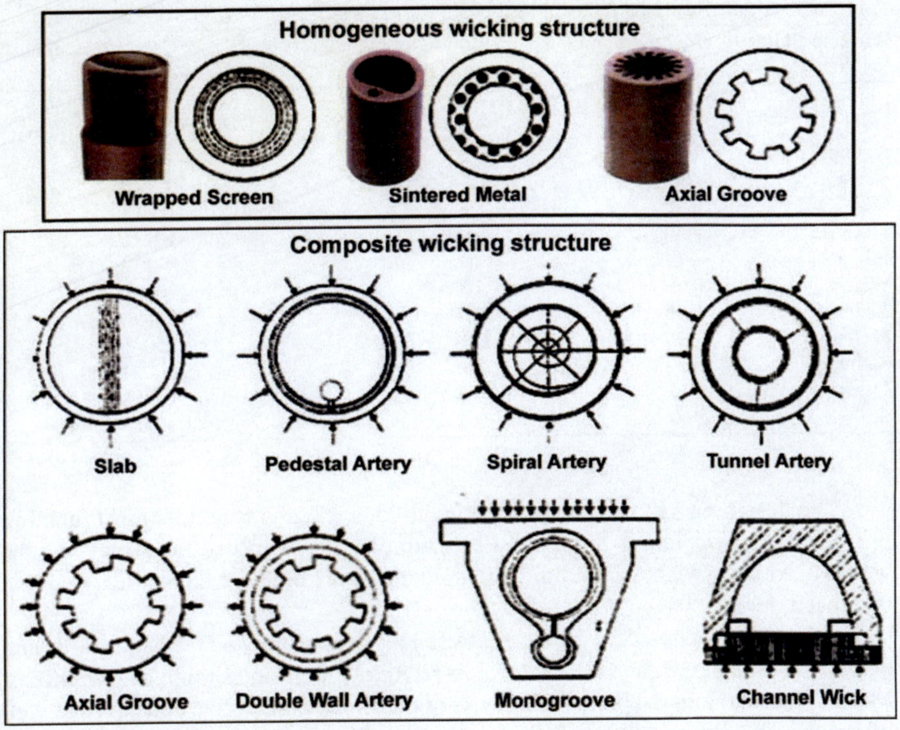


**Fig. 10.8** Typical wicking configurations and structures of heat pipe

called the liquid transport factor to gauge the effectiveness of a working fluid has been developed (Chi 1976):

$$N_l = \frac{\rho_l \sigma \lambda}{\mu_l}, \tag{10.19}$$

where $\lambda$ is the latent heat of vaporization and $\sigma$ is the surface tension. Subscript l refers to the liquid. For electronics cooling applications, occurring in low to moderate temperatures, water is the liquid with the highest liquid transport factor. Another common fluid is ammonia.

Parameters that should be considered for container design are operating temperature range of the heat pipe; internal operating pressure and container structural integrity; evaporator and condenser size and shape; possibility of external corrosion; prevent leaks; and compatibility with wick and working fluid. Typical materials include aluminum, stainless steel, copper, composite materials, and high temperature heat pipes may use refractory materials or linings to prevent corrosion. Because the heat pipe is like a pressure vessel, it must satisfy ASME (American Society of Mechanical Engineers) pressure vessel codes. Typically, the maximum allowable stress at any given temperature can only be one-fourth of the material's maximum tensile strength.

**Table 10.4** Typical working fluids of heat pipe

| Working fluid (medium) | Melting point (°C) | Boiling point at atmosphere pressure (°C) | Useful temperature range (°C) |
|---|---|---|---|
| Helium | −271 | −261 | −271 to −269 |
| Nitrogen | −210 | −196 | −203 to −160 |
| Ammonia | −78 | −33 | −60 to 100 |
| Acetone | −95 | 57 | 0–120 |
| Methanol | −98 | 64 | 10–130 |
| Flutec PP2 | −50 | 76 | 10–160 |
| Ethanol | −112 | 78 | 0–130 |
| Water | 0 | 100 | 30–200 |
| Toluene | −95 | 110 | 50–200 |
| Mercury | −39 | 361 | 250–650 |
| Sodium | 98 | 892 | 600–1,200 |
| Lithium | 179 | 1,340 | 1,000–1,800 |
| Silver | 960 | 2,212 | 1,800–2,300 |

When designing a heat pipe, the working fluid, wick, and container must function properly when operating together. For example, the working fluid may not be wettable with the wick; or the fluid and container may undergo a chemical reaction with each other.

In addition, the contact resistance between the evaporator and the heat source and between the condenser and the heat sink is relatively large and should be minimized. Methods used to join the parts include use of thermally conductive adhesives as well as brazed or soldered techniques. Cooling of electronics has one primary goal: maintain a component's temperatures at or below the manufacturer's maximum allowable temperature. As the temperature of an electronic part increases, the rate of failure increases. Heat pipes are excellent candidates for electronics cooling because of their high thermal conductivity and high heat transfer characteristics, and they provide constant evaporator temperatures with variable heat fluxes and variable evaporator and condenser sizes. Therefore, they are good alternatives to large heat sinks, especially in laptops where space is limited. They are a good alternative to air cooling because of their better heat transport capabilities. Air cooling may still be used to combine with the heat pipe to remove heat from its condenser.

## *Limitations on Heat Transport Capacity*

Heat pipe performance and operation are strongly dependent on shape, working fluid, and wick structure. Certain heat pipes can be designed to carry a few watts or several kilowatts, depending on the application. The effective thermal conductivity of the heat pipe will be significantly reduced if the heat pipe is driven beyond its capacity. Therefore, it is important to assure that the heat pipe is designed to transport the required heat load safely. However, during steady state operation, the maximum heat transport capability of a heat pipe is governed by several limitations which must be clearly known when designing a heat pipe. There are five primary heat pipe transport limitations:

1. *Viscous* Viscous force will prevent vapor flow in the heat pipe. This causes the heat pipe to operate below the recommended operating temperature. The potential solution is to increase the heat pipe operating temperature or operate with an alternative working fluid.
2. *Sonic* Vapor will reach sonic velocity when exiting the heat pipe evaporator resulting at a constant heat pipe transport power and large temperature gradient. The main reason is the power and the temperature combination. In other words, the heat pipe is operating at low temperature with too much of power. This is a normal problem during a start-up. The potential solution for this limitation is to create large temperature gradient so that heat pipe system will carry adequate power as it warms up.
3. *Entrainment/Flooding* This is where high velocity vapor flow prevents condensate vapor from returning to the evaporator. The main reason is due to low operating temperature or high power input that the heat pipe is operating. To overcome this, the vapor space diameter or the operating temperature needs to be increased.
4. *Capillary* This is the combination of gravitational, liquid, and vapor flow and pressure drops exceeding the capillary pumping head of the heat pipe wick structure. The main cause is the heat pipe input power exceeds the design heat transport capacity of the heat pipe. The problem can be resolved by modifying the heat pipe wick structure design or reducing the power input.
5. *Boiling* Described as a film boiling in a heat pipe evaporator that typically initiates at 5–10 W/cm$^2$ for screen wick and 20–30 W/cm$^2$ for power metal wicks. This is caused by high radial heat flux. It will lead toward film boiling resulting in heat pipe dryout and large thermal resistances. The potential solution is to use a wick with a higher heat capacity or spread out the heat load efficiently.

Each limit has its own particular range in which it is important. However, in practical operation, the capillary and boiling limits are the most important. For a heat pipe to function properly, the capillary pressure must be greater or equal to the sum of the pressure drops due to inertial, viscous, and hydrostatic forces, as well as pressure gradients. That is (Peterson 1994).

$$(\Delta P_c)_{max} \geq \int_{L_{eff}} \frac{\partial P_v}{\partial x} \mathrm{d}\, x + \int_{L_{eff}} \frac{\partial P_l}{\partial x} \mathrm{d}\, x + \Delta P_{ph,e} + \Delta P_{ph,c} + \Delta P_{\perp} + \Delta P_{//}, \tag{10.20}$$

where $(\Delta P_c)_{max}$ is maximum capillary pressure difference generated within capillary wicking structure between wet and dry points; $\int_{L_{eff}} \partial P_v/\partial x \,\mathrm{d}\, x$ is the sum of inertial and viscous pressure drops occurring in vapor phase; $\int_{L_{eff}} \partial P_l/\partial x \,\mathrm{d}\, x$ is the sum of inertial and viscous pressure drops occurring in liquid phase; $\Delta P_{ph,e}$ is pressure gradient across phase transition in evaporator; $\Delta P_{ph,c}$ is pressure gradient across phase transition in condenser; $\Delta P_{\perp}$ is normal hydrostatic pressure $\Delta P_{//}$ is axial hydrostatic pressure drop.

If it is not, then the working fluid is not supplied rapidly enough to the evaporator to compensate for the liquid loss through vaporization. If this occurs, there is dryout in the evaporator. The boiling limit is due to excessive radial heat flux; all the other limits are due to axial heat flux. The maximum heat flux beyond which bubble growth will occur resulting in dryout is given by (Peterson 1994).

$$q_{\mathrm{b,e}} = \left(\frac{2\pi L_{\mathrm{e}} K_{\mathrm{eff}} T_{\mathrm{v}}}{\lambda \rho_{\mathrm{v}} \ln(r_{\mathrm{i}}/r_{\mathrm{v}})}\right)\left(\frac{2\sigma}{r_{\mathrm{n}}} - \Delta P_{\mathrm{c,m}}\right), \quad (10.21)$$

where $\Delta P_{\mathrm{c,m}}$ is maximum difference in capillary pressure, $\sigma$ is surface tension which is a function of fluid and temperature, $r_{\mathrm{n}}$ is nucleation site radius (assumed to be between 0.254 and 25.4 μm), $T_{\mathrm{v}}$ is vapor temperature, $L_{\mathrm{e}}$ is length of evaporator, $r_{\mathrm{v}}$ is vapor space (radius of open heat pipe allowing vapor to pass through), $r_{\mathrm{i}}$ is inner radius of heat pipe, $\rho_{\mathrm{v}}$ is density of vapor, $\lambda$ is latent heat of vaporization, and $K_{\mathrm{eff}}$ is effective conductivity, which is given by (Chi 1976):

$$\text{For wick and liquid in series } K_{\mathrm{eff}} = \frac{k_{\mathrm{l}} k_{\mathrm{w}}}{\varepsilon k_{\mathrm{w}} + k_{\mathrm{l}}(1-\varepsilon)}, \quad (10.22)$$

$$\text{For wick and liquid in parallel } K_{\mathrm{eff}} = \varepsilon k_{\mathrm{l}} + k_{\mathrm{w}}(1-\varepsilon), \quad (10.23)$$

$$\text{For wrapped screen } K_{\mathrm{eff}} = \frac{k_{\mathrm{l}}[(k_{\mathrm{l}} + k_{\mathrm{w}}) - (1-\varepsilon)(k_{\mathrm{l}} - k_{\mathrm{w}})]}{(k_{\mathrm{l}} + k_{\mathrm{w}}) - (1-\varepsilon)(k_{\mathrm{l}} - k_{\mathrm{w}})}, \quad (10.24)$$

$$\text{For packed spheres } K_{\mathrm{eff}} = \frac{k_{\mathrm{l}}[(2k_{\mathrm{l}} + k_{\mathrm{w}}) - 2(1-\varepsilon)(k_{\mathrm{l}} - k_{\mathrm{w}})]}{(2k_{\mathrm{l}} + k_{\mathrm{w}}) - (1-\varepsilon)(k_{\mathrm{l}} - k_{\mathrm{w}})}, \quad (10.25)$$

$K_{\mathrm{l}}$ is thermal conductivity of liquid, $k_{\mathrm{w}}$ is thermal conductivity of wick, and ε is wick porosity. The radial and axial resistances can be determined from traditional resistance equations for cylindrical shapes and flat plates depending on the shape of the heat pipe.

## *Applications of Heat Pipe for Electronic Cooling*

There are many advantages in the application of heat pipe compared with other thermal management devices. First, the heat transfer capacity of a heat pipe may be several orders of magnitude greater than even the best solid conductors because it operates on a closed two-phase cycle. These characteristics may result in a relatively small thermal resistance and allow physical separation of the evaporator and condenser without high penalty in overall temperature drop. Furthermore, the increase in the heat flux in the evaporator may increase the rate at which the working fluid is vaporized, without significant increase in the operating temperature. Thus, the heat pipe can operate as a nearly isothermal device, adjusting and maintaining a relatively constant source temperature. Another advantage of the heat pipe is that both the evaporator and condenser operate independently. It needs only common liquid and vapor stream. Because of this, the area where heat is added can differ in size and shape from the area over which heat is rejected, provided the rate at which the liquid is vaporized does not exceed the rate at which it is condensed. In this case, heat fluxes generated over relativity small areas can dissipate over larger areas with reduced heat fluxes. This characteristic is useful in the thermal control of electronic components. Because it allows the high fluxes generated at the component level to be reduced and allows convection to be used to dissipate the heat. Finally, its thermal response time is less than other heat transfer devices such as solid conductors. The main reason is the closed two-phase cycle. In addition, the heat pipe requires a very small temperature difference because of the usage of the latent heat of vaporization. This will cause the effective thermal conductivity to be few times greater than the best solid conductors.

Figure 10.9 shows some application examples of heat pipes, such as laptop heat pipe, desktop retrofitted with heat pipe, heat pipe integrated with heat sink, and flat

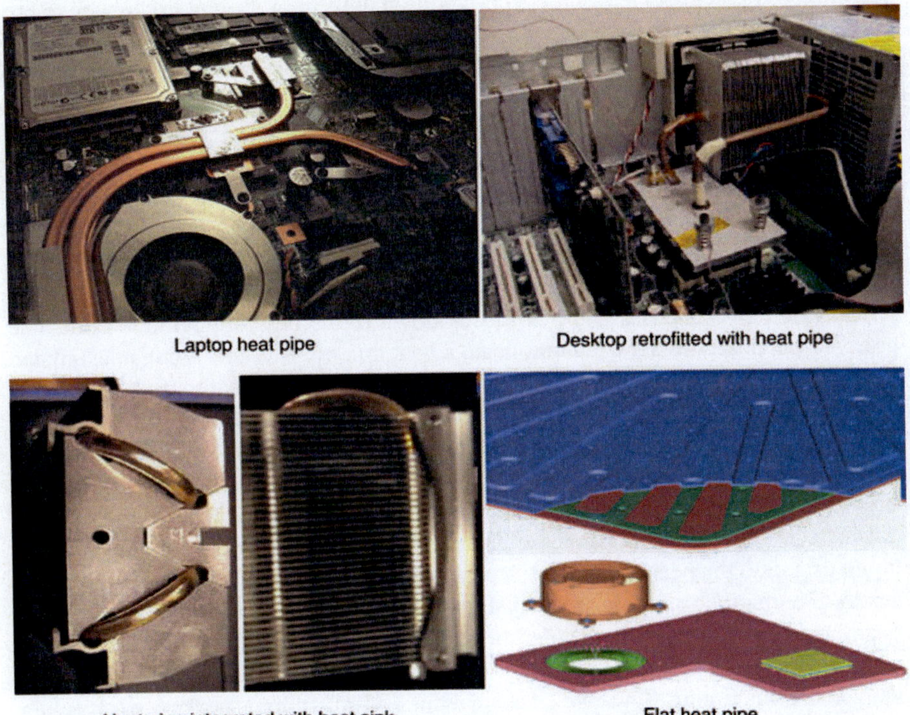


**Fig. 10.9** Application examples of heat pipes

heat pipes. For instance, portable electronic devices such as cellular phones increasingly offer more electronic features. The operating conditions of cellular phones along with high-powered components present enormous challenges to the cooling system. To overcome this, heat spreaders are used to dissipate heat in cellular phones. One large heat spreader is perceived to be more efficient compared with several smaller ones. Heat pipes can further enhance the heat dissipation by transporting the heat away from the source. Higher efficiency is achieved when a heat spreader is incorporated with a heat pipe having good contact between them. A keypad can be used as a heat spreader, which dissipates heat to the environment. This approach is used to reduce hotspots on cellular phones.

Basically, a heat spreader is a thin planar heat pipe whose thickness can be as low as 500 μm. It can be used to replace the outer shell of portable electronic device. Heat pipes typically transfer heat away from the CPU and dissipate the heat through larger areas such as components located in the base of the laptop (keyboard and chassis). However, there is a limit to the amount of power dissipation achievable through a sole heat pipe in the base. Another common mobile thermal solution is a remote heat exchanger system: a combination of heat pipe, heat sink, and fan. In the mid 1990s Intel used the heat pipe with spreader alone, but later replaced it with the remote heat exchanger system implemented in Intel Mobile Pentium III cooling technology with an average TDP (Thermal Design Power) of 20 W. The Toshiba Portege 3480CT notebook computer uses a different configuration of the remote heat exchanger where a magnesium heat sink draws heat away from the chip surface. Water vaporization occurs in the hot end of the copper tube (attached to heat sink), and heat is spread through the aluminum plate as the vapor cools. Chip powers in excess of a few tens of Watts are beyond the capacity of conventional heat pipes even with remote heat exchangers. The limiting factors in the heat pipes are the wick thickness (capillary forces) and the cross-sectional area. In 1994, Aavid Engineering demonstrated its two-phase passive cooling system, known as Oasis technology, for notebook computers. Fluid contained in a sealed unit is heated by the chip and vaporizes. Natural convection effects force the vapor to travel to a thin chamber behind the display screen where it condenses and flows back down to the high temperature chip. The passive system reportedly accommodated a heat flux of 29 W/cm$^2$ with an estimated junction temperature of 110°C. In an effort to provide an alternative to heat pipes, a closed-loop two-phase microchannel cooling system has been developed using electroosmotic pumping for the working fluid (1 mM buffered deionized water). The cooling scheme for VLSI (Very-Large-Scale Integration) chip application consists of a microchannel heat exchanger above the chip, a heat rejector placed downstream of the heat exchanger and an electroosmotic pump which returns cooled liquid back into the heat exchanger. A pressure transducer provides the pressure drop across the microchannel heat exchanger while the heat rejector consists of an aluminum heat sink with embedded flow channels and a fan. The pump was characterized by a linear relationship between flow rate and back pressure, where it delivered a maximum flow rate of 7 mL/min and maximum pressure of 160 kPa. The two-phase system accommodated up to 38 W/cm$^2$ of input power giving a junction-fluid thermal resistance value of about 0.1 K/W. The chip temperature was maintained below 120°C, with a pressure drop of 30 kPa at 38 W and an estimated saturation temperature of 110°C. The Hitachi Flora 270 W silent model is a liquid-cooled notebook computer, which also utilizes the backside of the display screen and a micropump. A liquid cooling jacket removes heat from the chip and the liquid is then pumped to the backside of the display screen where heat is dissipated from a radiator panel. The radiator panel is connected with flexible tubes to a reserve tank that supplies cooled liquid back to the chip. The radiator panel dissipated 30 W of heat, while an additional 10 and 20 W were dissipated through the keyboard and laptop chassis respectively. The fanless system eliminated the acoustic noise associated with fans found in conventional forced air-cooled systems. In 2002 Hitachi Ltd commercialized the first long-life, water-cooling system for a high performance Mobile Intel Pentium Processor-M (Ali 2004).

Therefore, one of the highest volume applications for heat pipes is cooling the microprocessors in notebook computers. Due to the limited space and power available in notebook computers, heat pipes are ideally suited for cooling the high power chips. Fan-assisted heat sinks require electrical power and reduce battery life. Standard metallic heat sinks capable of dissipating the heat load are too large to be incorporated into the notebook package. Heat pipes, on the other hand, offer a high efficiency, passive, compact heat transfer solution. Three or four millimeter diameter heat pipes can effectively remove the high flux heat from the processor. The heat pipe spreads the heat load over a relatively large area heat sink, where the heat flux is so low that it can be effectively dissipated through the notebook case to ambient air. The heat sink can be the existing components of the notebook, from electromagnetic interference shielding under the key pad to metal structural components. Typical thermal resistances for these applications at 6 to 8 Watt heat loads are 4–6°C/W. High-power mainframe, minimainframe, server, and workstation chips may also employ heat pipe heat sinks. High end chips dissipating up to 100 W are outside the capabilities of conventional heat sinks. Heat pipes are used to transfer heat from the chip to a fin stack large enough to convect the heat to the supplied air stream. The heat pipe isothermalizes the fins eliminating the large conductive losses associated with standard sinks. The heat pipe heat sinks can dissipate loads in the 75–100 W range with resistances from 0.2 to 0.4°C/W, depending on the available air flow (Garner 1996).

In addition, other high-power electronics including silicon controlled rectifiers, insulated gate bipolar transistors (IGBTs), and thyristors often utilize heat pipe heat sinks, which are capable of cooling several devices with total heat loads up to 5 kW. These heat sinks are also available in an electrically isolate diversion where the fin stack can be at ground potential with the evaporator operating at the device potentials of up to 10 kV. Typical thermal resistances for the high-power heat sinks range from 0.05 to 0.1°C/W. Here, the resistance is also predominately controlled by the available fin volume and air flow (Garner 1996).

Although heat pipes themselves do not actually dissipate significant amounts of heat, they do effectively transfer heat without a large increase in temperature. This unique transfer capability allows them to transport or spread heat to a point remote from the heat generator. A variety of basic heat sink technologies benefit from integrating heat pipes which improve conduction paths, reduce overall weight, and raise thermal performance without increasing volume. The availability of a wide range of heat pipe sizes and power handling capabilities make them suitable for integration in heat sinks for 50 W processors to multi kilowatt IGBTs. Due to the large variety and diverse requirements, most heat sinks using integrated heat pipes are developed specifically for the application. Heat pipe assemblies have been used to solve thermal problems in desktop, notebook, computer servers, telecommunication, motor drives, UPS (Uninterruptible Power Supply), and transportation applications.

## Refrigeration Cooling

High heat flux cooling approaches have been explored to meet the demand of increasing dissipation rates while satisfying the required reliability and cost considerations; these methods include heat pipes, liquid immersion, jet impingement and sprays, thermoelectrics, and refrigeration. Among these approaches, refrigeration is one of the only methods which can work in high-temperature ambient, and even result in negative values of thermal resistance. As a result, the advantages of refrigeration cooling include maintenance of low junction temperatures while dissipating high heat fluxes, potential increases in microprocessor performance at lower operating temperatures, and increased chip reliability. However, these advantages must be balanced against the increased complexity and cost of the cooling system, possible increases in cooling system volume, and uncertainties in the system reliability due, for instance, to moving parts in the compressor. Therefore, minimization and cost reduction may need to be taken into account to make it feasible to use this approach in electronics cooling. Refrigerated cooling of electronic equipment has been classified into four major categories: refrigerated cooling of air or liquid, refrigerated heat sinks, liquid nitrogen baths, and thermoelectric coolers. The key difference between the first two of these is that a refrigerated heat sink results in a lower temperature at the surface of the chip compared with a refrigerated system using a cooling fluid, because it avoids use of a secondary fluid loop to carry heat from the chip to the refrigeration loop. In addition, the evaporator of the refrigerated heat sink is mounted directly to the chip leading to greater compactness. A liquid nitrogen bath may be used for cryogenic applications; its efficiency decreases significantly in the temperature range typical of electronics cooling. The liquid nitrogen bath is a batch-cooling mode operation and liquid nitrogen needs to be provided from an external cryogenic refrigeration system (Kirkconnel 2008). The need for insulation of the bath and other implementation difficulties renders this approach impractical for electronics applications (Trutassanawin et al. 2006).

Vapor compression refrigeration has long been used to cool telecommunications equipment and some high performance computers. On the whole, however, its usage has been confined to high-value, relatively large, and stationary applications. The advantages of vapor compression cooling (VCC) systems typically include providing heat sinks at below ambient temperatures, removing large amounts of heat using relatively low power, and protecting electronic components from overheating. Additionally, at low temperature, reliability improves and microprocessors run at faster speeds. A number of refrigeration cooling systems for electronics are available commercially. For example, a vapor-compression refrigeration system has been used to cool the processor (a multichip module) in the IBM S/390 G4 CMOS server system, as shown in Figure 10.10, which was the first IBM system to employ refrigeration cooling: all other components of the server were air cooled (Schmidt et al. 2000). Two modular refrigeration units were connected to the evaporator, with one of these serving as a backup. The size of the cooling unit was 267 × 267 × 711 mm, with a weight of 27 kg. The average processor temperature for the G4 server was maintained at 40°C, which was approximately 35°C lower than the temperature that could be achieved with a conventional air cooling design. The apparent advantages of subambient cooling when using refrigeration include faster switching of complementary metal oxide semiconductor (CMOS) transistors, increased carrier mobility, improved subthreshold operating characteristics and interconnect conductivity, and reduced junction leakage. However, one of the critical problems with this approach is moisture condensation on exposed surfaces at temperatures below the dew point of the surrounding air. Hence, any exposed cold surfaces must be insulated and sealed. In addition, the bulk and weight of the compressor, as well as a lack of robust interactive capacity control, are the reasons limiting widespread acceptance of refrigeration cooling for electronics. As a result, miniature-scale refrigeration systems (MSRSs) for

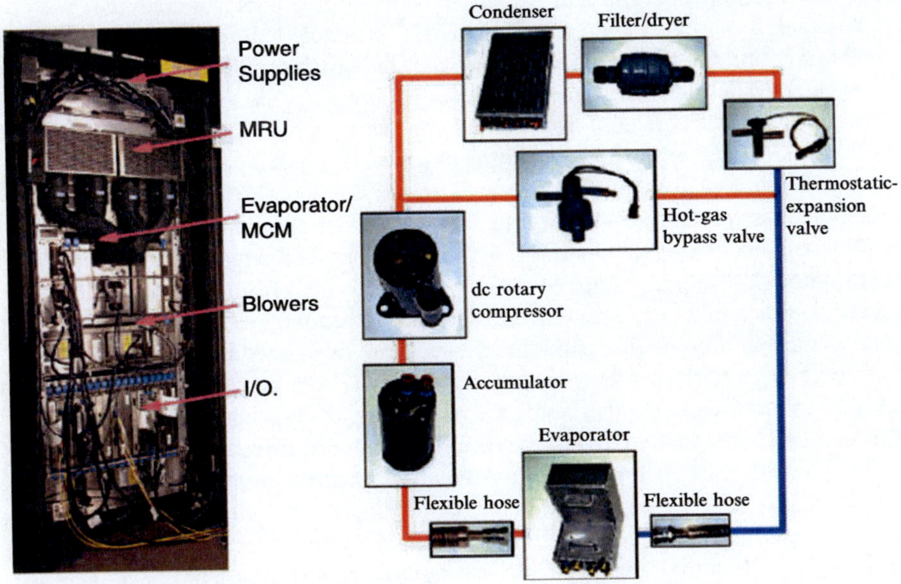


**Fig. 10.10** Schematic of vapor compression refrigeration components–an evaporator-cooled IC module is enclosed within an insulated housing which is surrounded by an atmosphere of dehumidified air

electronics cooling have caused attention. The MSRS for electronics cooling usually consists of four main components: an integrated microchannel cold plate evaporator–heat spreader, a compressor, a microchannel condenser, and an expansion device. An optimum heat spreader thermal conductivity of 2,000 W/m K for a heat spreader thickness of 3–4 mm is suggested. A cold plate evaporator microchannel length of 30 mm is recommended to achieve low overall system thermal resistance (Trutassanawin et al. 2006).

Although historically, VCC systems have had difficulty fitting within the space limitations of a small electronics chassis. The challenge has been to create a small but powerful enough system for electronics systems. A major obstacle to building small VCC systems was the absence of commercially available and affordable miniature refrigeration compressors in the fractional-kilowatt range that would fit within a small space. As a result, such a compressor (measuring only 5.6 cm in diameter × 7.6 cm high and weighing only 630 g) has been developed and become available (Deming et al. 2008). The compressor is a rotary design with a rolling piston driven by a sensorless brushless DC motor running at 24 VDC. Utilizing the miniature compressor, a compact VCC electronics cooling unit (ECU) system can slide into a standard 48 cm wide and 2U high electronics rack. The replacement 2U rack-mounted ECU was to deliver 500 W (J/s) of cooling and 800 W (J/s) of heating to satisfy the requirements in all operating conditions. It can handle high heat loads with a high coefficient of performance, and is a robust miniaturized refrigeration system, having a unique size to fit a 2U rack with a cooling capacity of 500 W (Deming et al. 2008). That is, the MSRS can be selected from commercially available components to fit into the space available in a personal computer.

Another miniature refrigeration model is assembled directly into a portable cooler or cooling box for cooling the inside of the box by utilizing the Peltier effect of an electronic type thermomodule using thermoelectric cooling elements. A conventional electronic cooling type box consists of a heat insulating housing and a lid. In a side wall of the heat insulating housing, an electronic type thermomodule having a plain plate shape and formed of thermo-electric cooling elements is assembled such that a heat absorbing side of the thermomodule is oriented inside the housing and the heat radiating side thereof is oriented outside the housing. The thermomodule is a plain shape module, which is formed from plurality of thermo-electric cooling elements assembled in series. Each cooling element is formed such that p-type and n-type semiconductor elements are connected between two ceramic base boards. A heat absorbing member with fins is heat-conductively fixed to the heat absorbing side of the thermoelectric module, and a heat radiating member with fins is heat-conductively fixed to the heat radiating side thereof. A wind channel is formed outside the housing to surround the fins of the heat radiating member, and an outside fan is attached to the wind channel to supply outer air to the fins. A direct current electric source, such as a battery, is used to supply electricity to the thermo-module. Based on the same principle, the refrigeration structure has been modified and consists of the following: (1) The refrigerator includes at the inner side of a heat absorbing member with fins heat-conductively connected to a heat absorbing side of a thermoelectric module, and an inner fan for forcibly supplying air in the refrigerator to the fins of the heat absorbing member arranged in rows. (2) The inner fan is assembled inside to the fins of the heat absorbing member. In a preferred embodiment, the thermomodule is vertically arranged and is installed in a housing wall of the refrigerator, and the fins of the heat absorbing member connected to the heat absorbing side of the thermomodule are arranged vertically. Also, a drain receiver is provided at a lower side of the heat absorbing member for receiving dew droppings from the fins arranged in the rows and draining outside the housing. Furthermore, a plurality of thermoelectric modules heat-conductively laminated and assembled in the heat insulating housing may be provided. Furthermore, air inside the housing is forcibly supplied by the inner fan to the rows of the fins of the heat absorbing member projecting into the housing so that the heat exchange rate between the heat absorbing member and the air inside the housing is increased several times when comparing the natural convection system. Thus, it is possible to sufficiently exhibit the cooling ability of the electronic type thermomodule. Because the inner fan is installed inside the fins of the heat absorbing member arranged in the rows, the limited space inside the housing is effectively utilized. Also, a plurality of thermoelectric modules is laminated and arranged together as the thermomodule, so that the difference of temperature between the heat absorbing side and the heat radiating side is effectively increased. Thus, the cooling ability can be increased. Also, the fins of the heat absorbing member are vertically arranged, and the drain receiver is provided at the lower side of the heat absorbing member for receiving the dew flowing along and dropping from the fins and discharging the dew outside the housing. Thus, it is possible to easily discharge the dew formed on the surfaces of the heat absorbing fins to the outside of the housing through the drain receiver (Higashi and Sawano 1997).

Therefore, microrefrigerators and coolers have a unique characteristic for electronics cooling that differentiates them from all other thermal management technologies: the ability to generate cooling temperatures below ambient or ultimate sink temperature. Thermoelectric or Peltier coolers are the most common refrigeration devices used for chip and board level electronics cooling, and have been scaled to the microdomain. Unfortunately, thermoelectrics have low efficiencies resulting in high power requirements and added waste heat. Meso- and/or microscale devices operating on the Stirling cycle are an attractive potential alternative based on the high efficiencies realized for Stirling machines. A further attraction of microscale Stirling devices is the ability of the Stirling cycle to be used for generating power where a temperature difference is maintained; or, in the "reverse" refrigeration cycle with power input. However, attempts to miniaturize such a device for application in electronics cooling have been scale-limited by the use of traditional components (e.g., pistons, linkages, and pressure vessels) and traditional fabrication methods. As a result, Stirling coolers have been impractical for most electronic packaging applications. Stirling machines face two significant dilemmas at meso- and/or microscale: piston frictional losses and axial thermal conduction losses. Frictional losses scale unfavorably relative to performance in the microdomain. As surface forces, frictional losses diminish only as the square of unit length; whereas performance diminishes roughly as volume or the cube of unit length. In other words, a Stirling device that has been reduced in overall size by a factor of 1,000 will have a vastly reduced performance by a factor of 1,000,000,000 (as measured by output power or refrigeration capacity), while frictional losses will only be reduced by a factor of 1,000,000.

The regenerator of a Stirling device maintains the temperature differential between the hot temperature source and cold temperature sink, while simultaneously storing and releasing heat from/to the working gas on every cycle. Axial thermal conduction through the regenerator introduces a direct loss from the device performance. At traditional scales, this loss mechanism is manageable. However, as the conduction length is reduced to the order of millimeters or less, minimizing axial conduction losses becomes paramount. Figure 10.11 shows schematic of

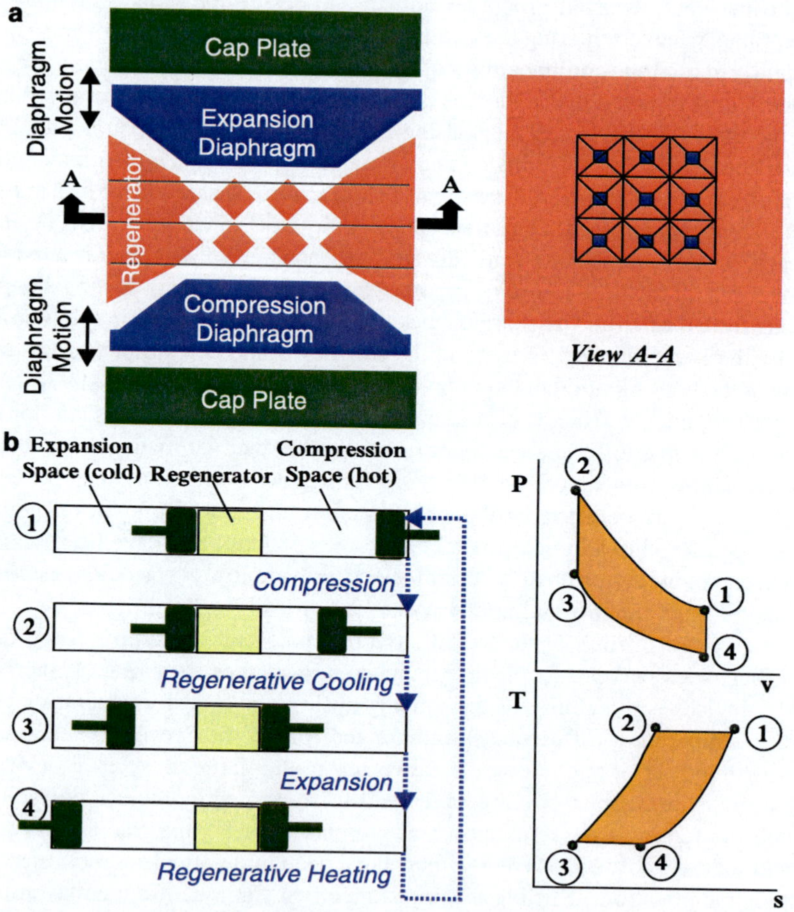


**Fig. 10.11** Schematic of Stirling cooler that uses diaphragm instead of piston to produce a Stirling refrigeration cycle, and is fabricated with semiconductor processing techniques. (**a**) Crossection of one cell of the microelectromechanical system Stirling cooler, which illustrates the key components of the device; (**b**) the Stirling refrigeration cycle creates a cold (expansion) region and a hot (compression) region

Stirling cooler that uses diaphragm instead of piston to produce a Stirling refrigeration cycle, and is fabricated with semiconductor processing techniques (Moran et al. 2004): (1) The cross section of one cell of the microelectromechanical system (MEMS) Stirling cooler, which illustrates the key components of the device. (2) The Stirling refrigeration cycle creates a cold (expansion) region and a hot (compression) region. Frictional losses are eliminated by replacing conventional pistons and associated linkages with electrostatically driven solid silicon diaphragms. The device is fabricated with semiconductor processing techniques to produce a device with planar geometry. The result is a flat cold surface for extracting heat and an opposing flat hot surface for thermal dissipation. A thin film temperature sensor deposited on one of the surfaces provides control feedback. This sensor, along with the ability to switch hot and cold ends by altering the cycle with control software, permits the device to be used for precise thermal control as well as for active cooling. The expansion and compression diaphragms are the only moving parts, and are deflected toward and away from the regenerator region in phase-shifted sinusoidal fashion to produce the Stirling refrigeration cycle. Expansion of the working gas in the space directly adjacent to the expansion diaphragm in each cycle creates a cold end for extracting heat, while compression at the other end creates a hot region for dissipating heat. Heat is transferred to and from the working gas as it is forced through the regenerator region by the moving diaphragms. The slanted geometries of the diaphragm and regenerator surfaces are characteristic of the wet etching process used to create the structure, and advantageously increase the potential swept volume in the expansion and compression regions. The regenerator can alternatively be made up of constant cross-section passages that are created after the regenerator layers are bonded. Issues associated with the breakdown of continuum behavior in the working gas have been examined, and the initial design was intentionally sized to be above the continuum scale limits. Future components may be selectively scaled to take advantage of noncontinuum behavior, e.g., higher voltage breakdown in the electrostatic-driven diaphragms (Moran et al. 2004).

Consequently, a vapor compression refrigeration system is a promising electronics cooling method because it can achieve negative values of the overall system thermal resistance. In other words, the refrigeration system is an active cooling system that can be used to dissipate high heat flux from the chip even if the ambient air temperature is higher than the chip surface temperature. This feature is unique of refrigeration applied to electronics cooling, that is, high heat dissipation electronic devices are guaranteed to work continuously at low stable temperatures even in warm surrounding environments. However, some of the main challenges of microrefrigeration systems include the design and development of efficient, compact, and reliable small components. The diaphragm compressor design is quite new in the field of microrefrigeration and it seems to offer a suitable solution in every application where volume, weight, and energy-saving considerations are of concern. Another important characteristic for the design of microrefrigeration systems is the design of the cold plate evaporator and a better understanding of the heat transfer mechanisms. With the growth of research and development, microrefrigeration systems will be able not only to be competitive with other advanced electronics cooling techniques, but they will be one of the most promising techniques to provide the cooling of future generations of microprocessors (Cremaschi et al. 2007).

## High-Flux Cooling with Phase-Change Heat Transfer

High heat flux removal is a major consideration in the design of a number of systems, such as high-performance computer chips, laser diodes, and future nuclear fusion and fission reactors.

Phase-change processes, such as pool and flow boiling, are generally very effective modes of heat transfer. However, the demands of modern thermal systems have required the development of methods to enhance boiling systems. While heat fluxes above $10^8$ W/m$^2$ have been accommodated in carefully controlled situations, the required fluid and the convective conditions usually dictate maximum heat fluxes several orders of magnitude lower. Therefore, the passive techniques, enhanced surfaces for pool boiling and enhanced surfaces and inserts for forced convection boiling/vaporization, as well as active techniques and compound techniques, such as EDIFICE, have been developed for exploring effective methods to dissipate increasing high heat fluxes.

The highest steady heat flux removal, $3.7 \times 10^8$ W/m$^2$, has been obtained (Ornatskii and Vinyarskii 1965) by high velocity and highly subcooled flow of water in a small diameter tube with one-sided heating. A small percentage of alcohol was added to enhance heat transfer. This heat flux removal would, of course, satisfy virtually any system requirement; however, it is difficult to achieve this. Furthermore, there are many applications where water is neither an acceptable coolant nor an appropriate working fluid. Such is the case with direct immersion cooling of microelectronic devices. An array of microelectronic chips, representing a memory or central processing component of a computer, is simply submerged in the liquid. The devices are readily exposed to the coolant, and boiling occurs at higher power levels. Such phase-change cooling is desirable from the standpoint of the electronics because the devices are then subject to rather small changes in temperature, even for large changes in power. The fluids used for immersion cooling must be dielectric and inert, as shown in Table 10.1. Unfortunately, all of these fluids have poor heat transfer characteristics compared with water because of low thermal conductivity and low latent heat of vaporization. There is also the problem of boiling-curve hysteresis due to the extreme wettability of these fluids. There is a particular need to elevate the peak nucleate or burnout heat flux for these fluids. Some enhancement techniques are indicated that accomplish this and, additionally, reduce the wall superheat. Depending on the coolant and the convective conditions, maximum heat fluxes may vary widely, and in most cases these fluxes are several orders of magnitude below the ultimate limit that has been recorded to date (Bergles 2003).

Enhanced heat flux removal techniques can be classified either as (1) passive, which require no direct application of external power, such as treated surfaces, rough surfaces, extended surfaces, displaced enhancement devices, swirl flow devices, coil tubes, and surface-tension devices; or as (2) active, which require external power, such as mechanical aids, surface vibration, fluid vibration, electrostatic fields, suction or injection, and additives for fluids. The effectiveness of both types of techniques is strongly dependent on the mode of heat transfer, which may range from single-phase free convection to dispersed-flow film boiling. Two or more of the above techniques may be utilized simultaneously to produce an enhancement that is usually larger than the individual techniques applied separately. This is called compound enhancement (Bergles 2003).

## *Enhancement of Pool Boiling*

Some passive and active enhancement techniques have been shown to be effective for pool boiling. Most techniques apply to nucleate boiling; however, some techniques are applicable to transition and film boiling. It should be noted that phase-change heat transfer coefficients are relatively high. The main thermal resistance in a two-fluid heat exchanger often lies on the nonphase-change side. (Fouling of either side can, of course, represent the dominant thermal resistance.) On the other hand, the overall thermal resistance may then be reduced to the point where significant improvement in the overall performance can be achieved by enhancing the two-phase flow. Two-phase enhancement is also important in double-phase change (boiling/condensing) heat exchangers, such as thermosyphon reboilers.

Surface material and finish have a strong effect on nucleate and transition pool boiling. However, reliable control of nucleation on plain surfaces is not easily accomplished. For many years, this was accomplished simply by area increase in the form of low helical fins. The subsequent tendency was to improve the nucleate boiling characteristics by fundamental changes in the boiling process. Many of these advanced surfaces are being used in commercial shell-and-tube boilers. Furthermore, they are being considered for use in other areas, such as immersion cooling of microelectronic chips. This passive technology has found much more favor than active technology. Several manufacturing processes have been employed: machining, forming, layering, and coating. The geometry formed can be low-fin tubing, tunnel and pore arrangement produced by rolling, upsetting, and brushing, and the porous metallic matrix produced by sintering or brazing small particles to the base surface. The advantage is seen to be not only high nucleate boiling heat transfer coefficients, but also the fact that boiling can take place at very low temperature differences. Thus, these surfaces can be characterized as low temperature difference surfaces as well as high heat flux surfaces. In all cases, a complex liquid–vapor exchange is involved. The liquid flows in at random locations around the helical aperture or through selected pores to the interior of the structure, where thin film evaporation occurs over a large surface area. The vapor is then ejected through other paths by bubbling. The latent heat transport is complemented by agitated free convection from the exposed surfaces. The behavior of structured surfaces is not yet understood to the point where correlations are available to guide custom production of the surfaces for a particular fluid and pressure level. Some manufacturers have accumulated sufficient experience to provide optimized surfaces for important applications, such as flooded evaporators for refrigerant dry-expansion chillers and thermosyphon reboilers for hydrocarbon distillation columns. In spite of the difficulty, it can be said that considerable progress has been made in the analytical/numerical descriptions of some of the enhanced pool boiling surfaces (Webb 1994a).

Structured surfaces are not exempt from temperature overshoots and resultant boiling curve hysteresis. However, the superheats required for incipient boiling with the highly wetting liquids, are generally lower than for plain surfaces due to the greater probability of finding active sites over the larger surface area. The practical problem is to provide, at least on a transient basis, a trigger to initiate boiling. In some cases, the stimulus can be provided by injected vapor, either naturally as with a dry-expansion chiller or intentionally through sparging–the artificial introduction of vapor. The latter method has been used to improve the boiling performance of simulated microelectronic chips (Bergles and Kim 1988).

Structured boiling surfaces developed for process and refrigeration evaporation have been used as heat sinks for microelectronic chips. A small section of the surface is attached to the chip with a thin layer of industrial thermal grease or epoxy. The heat sinks can be bonded to the foil heater so that the heat transfer is well defined. It is seen that the order of performance improvement is: microfin, microhole, High Flux (Linde), and Thermoexcel-E. The boiling curve shift relative to the plain heater is substantial, especially at lower heat fluxes. The enhanced heat sinks permit higher heat fluxes than those expected for the plain surface; however, the wall superheat advantage is less at high heat fluxes. This seems to be observed with many enhanced boiling surfaces. The greatest improvements in boiling performance have been reported with testing of the massive heat sink stud (Bergles 1981); however, very high heat fluxes are attained. Unfortunately, the highest heat fluxes occur at superheats or surface temperatures that are above 85°C, which is usually considered to be the upper level of semiconductor junction temperature for acceptable reliability. The goal, however, is to develop enhanced boiling heat sinks that permit high peak heat fluxes at modest superheats Webb (1994b).

The performance of enhanced tubes in flooded, horizontal bundles has received considerable attention because this is the major application for these tubes. The issue at hand can be characterized as the relationship of separate effects or single-tube performance to the actual bundle performance. As in most other applications of enhancement technology, the understanding of enhanced tube behavior depends on the mechanism of the normal situation. Plain-tube bundles often exhibit substantial increases in average heat transfer coefficients above plain single tubes. This bundle factor is attributed to the convective effect in the upper portion of the bundle when the vapor quality is high. In the case of enhanced tubes, nucleate boiling overshadows the convective effect, with the result that the bundle performance tends to be reasonably close to that of the single tube. The result is a lesser increase in performance than would be expected from simply comparing the single-tube data. The bundle-average heat transfer coefficient (at constant heat flux) can be up to three times greater than for a smooth-tube bundle Yilmaz et al. 1980. Flooded natural circulation, recirculation reboilers (thermosyphons) can be of two types: horizontal (kettle reboilers) and vertical. Enhanced tubes have been applied to both types; however, in the former, outside diameter enhancement is used, whereas in the latter, inside diameter (ID) enhancement is used. Enhancement affects the vapor generation rate, which, in turn, controls the circulation. Enhanced bundles are not immune from the problem of boiling curve hysteresis with highly wetting fluids. While this problem does not occur in all systems, because of normal vapor injection or favorable transients, the consequences of the inability to initiate boiling can be severe. For horizontal bundles where falling-film or spray evaporation is routinely used (absorption refrigeration, chemical processes), there is the possibility of reducing the working-fluid inventory. This is important for the expensive, ozone-friendly refrigerants in large commercial and industrial chillers (Bergles 1989).

Boiling of multicomponent mixtures is the norm, rather than the exception, in the chemical process industry. Also, the air conditioning industry is vigorously investigating mixtures of refrigerants for use in air conditioners and heat pumps. Enhanced tubes maintain their advantage over plain tubes when boiling mixtures. However, the single-tube characteristics of enhanced tubes, as a function of concentration, do not agree with those of plain tubes. In pool boiling, the effect of a mixture is to cause the wall superheat to increase above that for a linear interpolation based on mixture composition between the single-component data points. For smooth tubes, the deviation in wall superheat can be very large. On the other hand, for a sintered porous surface with a mixture of a volatile fluid in a relatively nonvolatile fluid, the deviation is much smaller than that of a smooth tube. The boiling processes for mixtures are different for the smooth and enhanced surfaces. In addition to the T-shaped fins, the interior channels have small notches around the circumference (Uhlig and Thome 1985).

The final issue that is very important with surfaces having small interior channels is fouling. It would be reasonable to expect solids to precipitate, or less volatile components to hide out, in the pores, and render the normal boiling process ineffective. The conventional wisdom is that the performance of enhancement devices is reduced under fouling conditions. Furthermore, it is believed that these devices promote fouling. In fact, the truth is often different from the perception, and some enhanced heat transfer devices may also be antifouling devices. Enhanced heat transfer and reduced fouling are observed with a sintered surface and calcium sulfite scale. Under potential fouling conditions, it is important to control the liquid chemistry and use a pore size that keeps up the liquid circulation. Compressor lubricating oil is a likely contaminant in flooded refrigeration evaporators, thus, this could be considered a fouling situation. Alternatively, it could be considered a mixture problem. Oil also considerably reduced the performance of an enhanced bundle with spray evaporation.

Active enhancement techniques for pool boiling include heated-surface rotation, surface wiping, surface vibration, fluid vibration, electrostatic fields, and suction at the heated surface. Because of its greater potential for practical application, electro-hydrodynamic-enhanced boiling has attracted considerable attention from academic and industrial researchers. Pool boiling enhancements of heat transfer coefficients up to

900% have been reported for refrigerants. Although active techniques are effective in reducing the wall superheat and/or increasing the critical heat flux (CHF), the practical applications may be restricted, largely because of the necessity to develop reliable, low-cost transducers or power supplies and associated enhanced heat exchangers.

Compound enhancement involves two or more of the techniques applied simultaneously to produce an enhancement that is larger than the individual techniques applied separately. For pool boiling, this has been demonstrated with fins and electric fields, electric fields in a bundle of treated tubes, radically grooved rotating disk, extended surfaces that are treated, and rough surfaces with additives (Zaghdoudi and Lallemand 2002).

## *Enhancement of Forced-Convection*

A variety of surfaces and devices for passive enhancement of forced-convection boiling and evaporation are depicted in Figure 10.12. Under certain circumstances, subcooled CHF to water can be improved with a volatile additive. For instance, the addition of about 2% 1-pentanol to water elevated the CHF in plain tubes at high subcooling by as much as 15%. However, the opposite trend was observed at low subcooling. None of the active techniques are especially effective in subcooled boiling. The reason is that subcooled boiling is a very efficient heat transfer process, and a very large disturbance is required to alter it. Finally, one of the few compound techniques that have been used with subcooled flow boiling is water–alcohol plus the twisted tape (Bergles 1997, 1998).

The devices and surfaces depicted in Figure 10.12 have also been tested for passive enhancement of in-tube, forced-convection vaporization. The structured surfaces in Figure 10.12 are generally not used for in-tube vaporization due to the difficulty of manufacture. One notable exception is the porous metallic matrix surface in a vertical thermosyphon reboiler because the coating can be applied to the inside of larger diameter tubes (Thome 1996). The overall heat transfer coefficient was increased at least 80% with the application of these tubes. The behavior of vertical tubes in bundles can be studied with single tubes. The ID enhancement can be very effective because reboilers usually are driven by condensing steam, and the in-tube fluid is typically a hydrocarbon with relatively low heat transfer coefficient. To get the circulation going, it is often necessary to sparge the flow. This artificial introduction of vapor is equivalent to boiling at very low superheat, which is possible with enhanced tubes. Helical repeated ribs and helically coiled wire inserts have been used to increase vaporization coefficients and dryout heat flux in once-through boilers. Numerous tubes with internal fins, either integral or attached, are available for refrigerant evaporators. Original configurations were tightly packed, copper, offset strip fin inserts soldered to a copper tube or aluminum, star-shaped inserts secured by drawing the tube over the insert. Average heat transfer coefficients (based on the surface area of smooth tube of the same diameter) for typical evaporator conditions are increased by

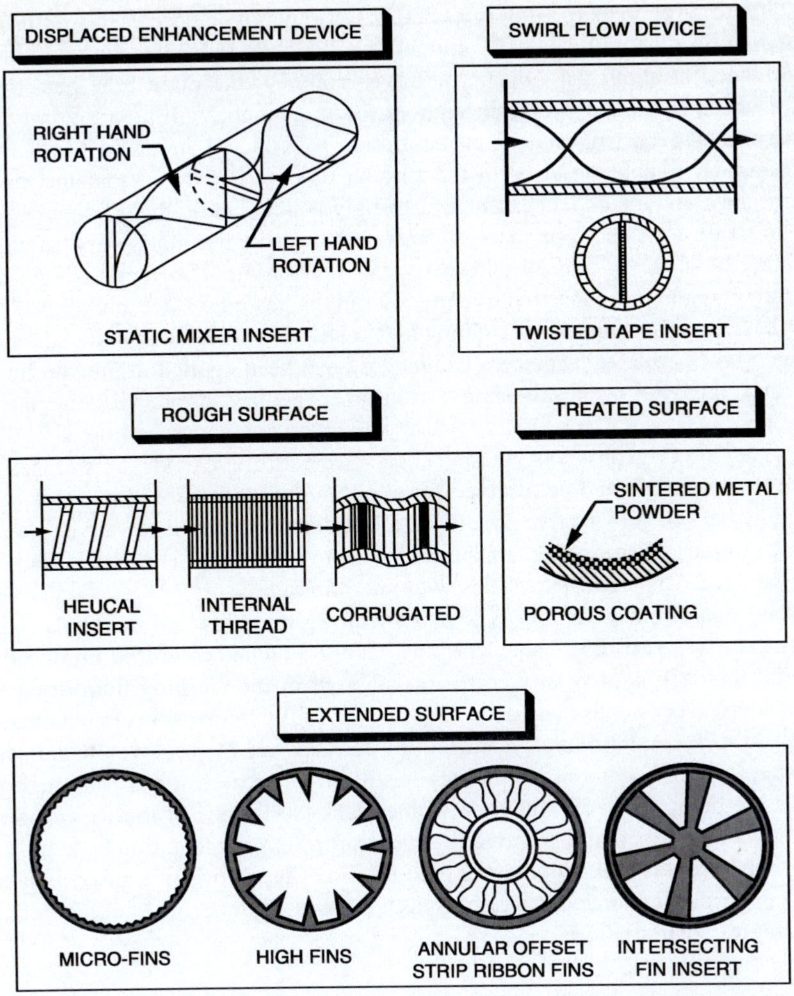


**Fig. 10.12** Examples of passive enhancement techniques for forced-convection boiling and evaporation (Bergles 2003)

as much as 200%. For reasons of high cost and pressure drop, the composite tubes and high fin tubes in Figure 10.12, and even the medium fins have been largely replaced with microfins. They are usually produced inside circular tubes (commonly copper) by swaging. They are widely used for convective vaporization of refrigerants, and improve heat transfer coefficients by about 50–100%. The one variant in commercial microfin tubes is the fin profile. Regarding geometry, the higher fins perform better, flat valleys are preferred, and flat peaks are better. Single-grooved tubes are generally used, but cross-grooved tubes perform better. Cross grooves can be produced by enhancing the pattern on a strip, rolling up the strip, and welding it. The usual contaminant in refrigeration systems is compressor lubricating oil, which may be present in concentrations of up to 10% (Bergles 2003). Oil generally degrades the performance of the low-fin tube while a small amount of oil slightly enhances the microfin tube performance at about 2% concentration. In contrast, the plain tube heat transfer coefficient is enhanced by oil throughout the test range, with a particularly sharp enhancement occurring at 3% concentration. While the inner-fin tubes always enhance heat transfer relative to the smooth tube, it is evident that the absolute performance and the enhancement are strongly dependent on the oil concentration. The effect of a twisted-tape insert in forced convective boiling is very effective in elevating the CHF in boiling with net vapor generation. The swirl-induced radial pressure gradient promotes vapor removal from the heated surface, and directs liquid to the heated surface, thereby permitting higher heat fluxes before dryout of the surface occurs. Some active enhancement techniques have been applied to flow boiling, but the results have not been promising. For instance, neither surface vibration nor fluid vibration (pulsations to ultrasound) has any effect on developed boiling or CHF. The same result has been observed with injection and electrostatic fields. This suggests that active techniques should be reserved for pool boiling, or possibly low velocity flow boiling, where the convective conditions still influence the boiling curve. There are some examples of compound enhancement in flow boiling: (1) rough surfaces and additives; and (2) electrostatic fields with microfin tubes.

When enhancement is applied, manufacturing methods and materials requirements may be overriding considerations. Can the enhancement be produced in a material that will survive any corrosion inherent in the working fluid or environment? Much work needs to be done to define the fouling/corrosion characteristics of enhanced surfaces. Particularly, antifouling surfaces need to be developed. It should be noted that enhancement technology is still largely experimental, although great strides are being made in analytical/numerical descriptions of the various techniques. Accordingly, it is imperative that the craft of experimentation be kept viable. Experimentation is still a vital art needed for direct resolution of transport phenomenon in complex enhanced geometries, as well as benchmarking of computer codes for computing simulation.

## *Embedded Droplet Impingement for Integrated Cooling of Electronics*

Microprocessor power density has been steadily increasing over the past decade due to increases in microprocessor power dissipation and reduction in feature size of the CPU core, where most of the power on a die is generated. This trend is expected to continue into the future, leading to next generation electronics with a power dense core covering a fraction of the total die surface area bounded by regions of reduced power density cache, with localized power densities exceeding 100 W/cm$^2$. Conventional cooling technologies in the electronic industry have limitations on removing these nonuniform, high heat fluxes from the surface of microprocessors. The combination of high heat fluxes with the nonuniformity of heat dissipation requires technologies able to remove large amounts of heat in a spatially and temporally variable manner. These heat removal technologies can meet the challenging cooling requirements of next generation electronics, including direct liquid cooling, liquid–vapor phase change, nonrefrigeration phase change techniques, pool boiling, jet impingement cooling, and evaporative spray cooling. Among these technologies, a droplet impingement-cooling device, EDIFICE, has been developed for removing heat fluxes over 100 W/cm$^2$ for both portable and desktop electronics (Amon 2003). The goal is to integrate the chip cooling solutions with the chip level packaging using MEMS technology, which offers the possibility of miniaturization and inexpensive batch fabrication.

The EDIFICE utilizes latent heat of vaporization of dielectric coolants to obtain high chip heat transfer rates. Cooling is through impingement of micron-sized droplets (50–100 μm) generated through multiple nozzles manufactured with deep reactive ion etching. The objective is to provide adaptive spatial and temporal cooling. Precise design and control of heat transfer rates would be achieved by using on-chip monitoring of temperature, thermal gradients, and dielectric coolant film thickness. EDIFICE employs MEMS-enabled technologies for manufacturing (1) micronozzles, swirl nozzles, and microinjectors for jet break-up, (2) microstructured silicon surfaces for the enhancement of thin-film evaporation, and (3) electrostatic microvalves for on-demand control of dielectric coolant flow rates. A schematic of the EDIFICE design is shown in Figure 10.13 with the overall system (Amon 2003). The droplet impingement mechanism avoids temperature overshoot at boiling incipience and almost eliminates swings in chip temperature and thermal cycling. The system level arrangement transports the removed heat to the periphery of the system, from where it is either dissipated by a condenser/heat sink or stored in a detachable heat storage unit. In this way, the volume near the chip is made available for electronics. In the case of the detachable energy storage unit, it contains an organic-phase change material (e.g., eicosene) with a melting temperature of about 50°C along with thermal conductivity enhancers such as aluminum foams or fins. For the case of the heat sink/condenser arrangement, the vapor generated at the component level is transported through the system to the combined heat sink and condenser thereby minimizing thermal resistance offered by multimaterial interfaces. The EDIFICE combines efficient phase-change heat transfer utilizing latent heat of vaporization of dielectric coolants and on-chip control to provide localized, adaptive, on-demand cooling. To satisfy temporal and spatial heat removal requirements, it contains built-in software to provide on-demand cooling achieved through the control of droplet sizes, impingement frequencies and impingement locations based on the on-chip sensing of temperature, thermal gradients and dielectric film thickness (Amon 2003).

## Summary

Microprocessor power density has been steadily increasing over the past decade due to increases in microprocessor power dissipation and reduction in feature size of the CPU core, where most of the power on a die is generated. This trend is expected to

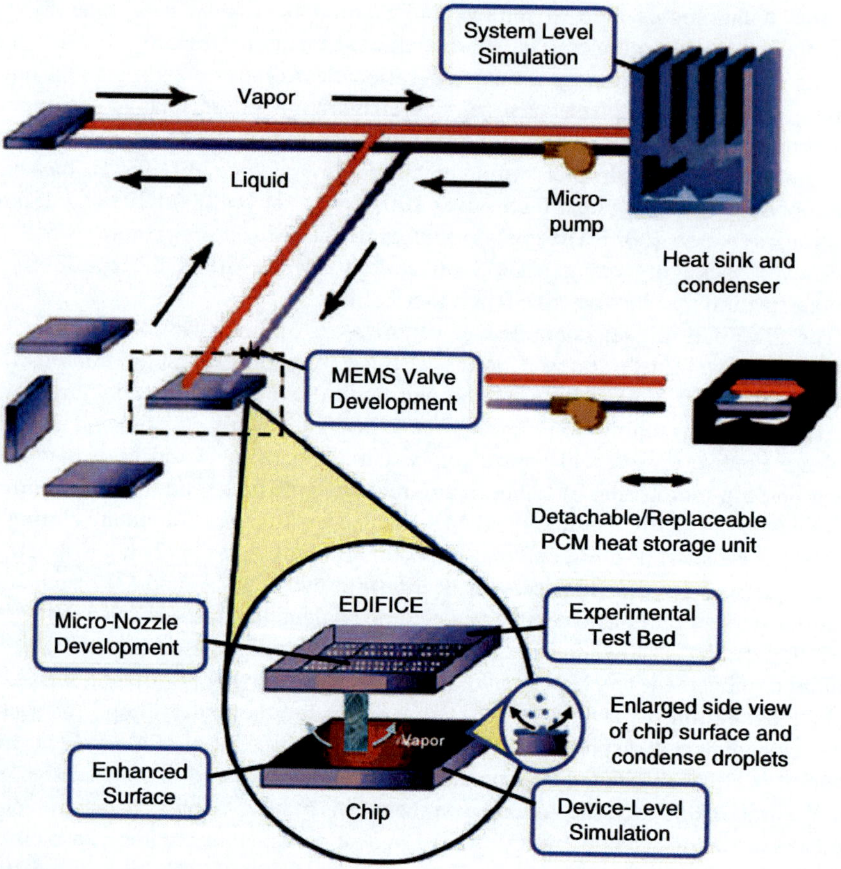


**Fig. 10.13** Schematic of EDIFICE and overall system (Amon 2003). Dielectric liquid coolant sprayed on a chip evaporates as the chip-level heat flux rises. The vapor is then collected and pumped back into the system. Combines efficient phase-change heat transfer utilizing latent heat of vaporization of dielectric coolants and on-chip control to provide localized, adaptive, on-demand cooling

continue into the future, leading to next generation electronics with a power dense core covering a fraction of the total die surface area bounded by regions of reduced power density cache, with localized power densities exceeding 100 W/cm$^2$. Conventional cooling technologies in the electronic industry have limitations on removing these nonuniform, high heat fluxes from the surface of microprocessors. The combination of high heat fluxes with the nonuniformity of heat dissipation requires technologies able to remove large amounts of heat in a spatially and temporally variable manner. These heat removal technologies, which can meet the challenging cooling requirements of next generation electronics, include direct liquid cooling, indirect liquid cooling, heat pipe, liquid-vapor phase change, nonrefrigeration phase change techniques, pool boiling, jet impingement cooling, evaporative spray cooling, and EDIFICE.

The inexorable rise in chip power dissipation and emergence of on-chip hotspots with heat fluxes approaching 1 kW/cm$^2$ has turned renewed attention to direct cooling especially with dielectric liquids. Use of dielectric liquids in intimate contact with the heat dissipating surfaces eliminates the thermal interface materials and deleterious effects of solid–solid interface resistances, and harnesses the highly efficient phase-change processes to the critical thermal management of advanced IC chips. Direct liquid cooling techniques allow for direct contact between a dielectric liquid and the surface of the chip, holding great promise for solving hot spot-driven thermal management issues. Furthermore, use of emerging and phase-change processes, including pool boiling, GAEC, jet impingement, and spray cooling, exploit the latent heat of these liquids to reduce the required mass flow rates and can provide the added advantage of inherently high heat transfer coefficients. Different direct liquid-cooling techniques have been defined, quantified, and compared with each other.

Comparably, indirect liquid cooling schemes are characterized by the presence of distinct physical barriers between the chip and the liquid. The most conservative implementation is to attach a separable cold plate to a lidded module. When greater thermal performance is desired, the interface between the cold plate and the module can be eliminated by integrating the cold plate with the module. Taking the concept to the limit, chip-scale high performance cold plates can be attached directly to the chips. Assuming a metal interface or solder between the chip and the cold plate with microchannels less than 0.1 mm in width, the chip heat removal capability for a chip junction to liquid inlet temperature difference of 60°C is 227, 320, and 397 W/cm$^2$ for the separable module cold plate, integral module cold plate, and individual chip-scale cold plate configurations, respectively. Other promising indirect liquid techniques may include: (1) liquid cooling through nanograss–water forms a nearly perfect ball, suspended on the tips of tiny blades of nanograss; (2) droplet-based microfluidics, which are based on pumping nanoliters of water droplets in a single-phase forced convection; (3) heat spreader filled with water as working fluid; and (4) evaporative cooling system using graphite foam.

Heat pipe cooling is a passive two-phase heat transfer method to transmit a large quantity of heat from a confined space to a condenser with minimum temperature drop. This method offers the possibility of high local heat removal rates with the ability to dissipate heat uniformly. A classical heat pipe consists of a metallic vessel in which its inner walls are lined with a wick structure. The vessel is first vacuumed, then charged with a small amount of working fluid, and then hermetically sealed. When the heat is applied to one end of the metallic vessel, the working fluid vaporizes. The resulting pressure gradient forces the vapor to travel through the hollow core to the other end of the heat pipe at sonic speed, where it condenses through a condenser and the released latent heat is removed. The liquid is then pumped back to the original hot end via wick by capillary action, completing the continuous evaporation/condensation cycle. There are many advantages in the application of heat pipe compared with other thermal management devices. First, the heat transfer capacity of a heat pipe may be several orders of magnitude greater than even the best solid conductors because it operates on a closed two-phase cycle.

These characteristics may result in a relatively small thermal resistance and allows physical separation of the evaporator and condenser without high penalty in overall temperature drop. Furthermore, the increase in the heat flux in the evaporator may increase the rate at which the working fluid is vaporized, without significant increase in the operating temperature. Thus, the heat pipe can operate as a nearly isothermal device, adjusting and maintaining a relatively constant source temperature. Another advantage of the heat pipe is that both the evaporator and condenser operate independently. It needs only common liquid and vapor stream. Because of this, the area where heat is added can differ in size and shape from the area over which heat is rejected, provided the rate at which the liquid is vaporized does not exceed the rate at which it is condensed. In this case, heat fluxes generated over relativity small areas can dissipate over larger areas with reduced heat fluxes. This characteristic is useful in the thermal control of electronic components. Because it allows the high fluxes generated at the component level to be reduced and allows convection to be used to dissipate the heat. Finally, its thermal response time is less than other heat transfer device such as solid conductors. The main reason is the closed two-phase cycle. In addition, a heat pipe requires a very small temperature difference because of the usage of the latent heat of vaporization. This will cause the effective thermal conductivity to be few times greater than the best solid conductors.

Among high heat flux cooling approaches, refrigeration is one of the only methods which can work in high-temperature ambient, and even result in negative values of thermal resistance. As a result, the advantages of refrigeration cooling include maintenance of low junction temperatures while dissipating high heat fluxes, potential increases in microprocessor performance at lower operating temperatures, and increased chip reliability. However, these advantages must be balanced against the increased complexity and cost of the cooling system, possible increases in cooling system volume, and uncertainties in the system reliability due, for instance, to moving parts in the compressor. Therefore, minimization and cost reduction may need to be taken into account to make it feasible to use this approach in electronics cooling. Refrigerated cooling of electronic equipment has been classified into four major categories: refrigerated cooling of air or liquid, refrigerated heat sinks, liquid nitrogen baths, and thermoelectric coolers. The key difference between the first two of these is that a refrigerated heat sink results in a lower temperature at the surface of the chip compared with a refrigerated system using a cooling fluid because it avoids use of a secondary fluid loop to carry heat from the chip to the refrigeration loop. In addition, the evaporator of the refrigerated heat sink is mounted directly to the chip leading to greater compactness. A liquid nitrogen bath may be used for cryogenic applications; its efficiency decreases significantly in the temperature range typical of electronics cooling. The liquid nitrogen bath is a batch-cooling mode operation and liquid nitrogen needs to be provided from an external cryogenic refrigeration system. The need for insulation of the bath and other implementation difficulties renders this approach impractical for electronics applications.

High heat flux removal is a major consideration in the design of a number of systems such as high-performance computer chips, laser diodes, and future nuclear fusion and fission reactors.

Phase-change processes such as pool and flow boiling, are generally very effective modes of heat transfer. However, the demands of modern thermal systems have required the development of methods to enhance boiling systems. While heat fluxes above $10^8$ W/m$^2$ have been accommodated in carefully controlled situations, the required fluid and the convective conditions usually dictate maximum heat fluxes several orders of magnitude lower. Therefore, the passive techniques, enhanced surfaces for pool boiling and enhanced surfaces and inserts for forced convection boiling/vaporization, as well as active techniques and compound techniques, such as EDIFICE, have been developed for exploring effective methods to dissipate increasing high heat fluxes. For example, the EDIFICE combines efficient phase-change heat transfer and on-chip control to provide localized, adaptive, on-demand cooling.

## References

Ali AA (2004) Design and analysis of a compact two phase cooling system for a laptop computer. MS thesis, Georgia Institute of Technology.

Ali A, DeHoff R, Grubb K (2000) Advanced heat pipe thermal solutions for higher power notebook computers. Proceedings of 5th Annual Pan Pacific Microelectronics Symposium, Jan. 2000, pp. 143–148.

Amon CH (2003) MEMS-based thermal management of high heat flux devices edifice: embedded droplet impingement for integrated cooling of electronics. http://www.web.mit.edu/hmtl/www/papers/AMON.pdf. Accessed on 21 June 2010.

Arik M, Bar-Cohen A (2003) Effusivity-based correlation of surface property effects in pool boiling CHF of dielectric liquids. International Journal of Heat and Mass Transfer **46**: 3755–3764.

Bar-Cohen A, Arick M, Ohadi M (2006) Direct liquid cooling of high flux micro and nano electronic components. Proceedings of the IEEE **94**: 1549–1570.

Bergles AE (1981) Two-phase flow and heat transfer, 1756–1981. Heat Transfer Engineering **2 (3–4)**: 101–114.

Bergles AE (1989) The challenge of enhanced heat transfer with phase change. Heat and Technology **7 (3–4)**: 1–12.

Bergles AE (1997) Heat transfer enhancement – the encouragement and accommodation of high heat fluxes. Journal of Heat Transfer **119**: 8–19.

Bergles AE (1998) Techniques to enhance heat transfer, in Handbook of Heat Transfer, Rohsenow WM, Hartnett JP, Cho YI, eds., McGraw-Hill, New York, pp. 11.1–11.76.

Bergles AE (2003) High-flux processes through enhanced heat transfer. http://www.web.mit.edu/hmtl/www/papers/BERGLES.pdf. Accessed on 21 June 2010.

Bergles AE, Kim C-J (1988) A method to reduce temperature overshoots in immersion cooling of microelectronic devices. Proc. InterSociety Conference on Thermal Phenomena in the Fabrication and Operation of Electronic Components, IEEE, New York, pp. 100–105.

Chi SW (1976) Heat pipe theory and practice. Hemisphere Publishing Corp., New York.

Chrysler GM, Chu RC, Simons RE (1995) Jet impingement boiling of a dielectric coolant in narrow gaps, IEEE Transactions on Components, Hybrids and Manufacturing Technology Part A **18 (3)**: 527–533.

Cremaschi L, Groll EA, Garimella SV (2007) Potential performances and challenges of future refrigeration cooling techniques as electronics cooling systems. Extended Abstract Draft THERMES II 2007 Thermal Challenges in Next Generation Electronic System II, January 7–10, 2007, Santa Fe, New Mexico, USA.

Danielson RD, Krajewski N, Brost J (1986) Cooling a superfast computer. Electronic Packaging and Production, pp. 44–45, July 1986.

Danielson RD, Tousignant L, Bar-Cohen A (1987) Saturated pool boiling characteristics of commercially available perfluorinated liquids. Proc. of ASME/JSME Thermal Engineering Joint Conference, 1987.

Deming G, Wysk R, Lee KP (2008) 2U Rack mountable vapor compression cooling system for high power electronics. Electronics Cooling. http://www.electronics-cooling.com/html/2008_may_techbrief.php. Accessed on 20 June 2010.

Ellsworth Jr. MJ, Simons RE (2005) High powered chip cooling – air and beyond. Electronics Cooling. http://www.electronics-cooling.com/2005/08/high-powered-chip-cooling-air-and-beyond/. Accessed on 12 June 2010.

Garner SD (1996) Heat pipes for electronics cooling applications. http://www.electronics-cooling.com/1996/09/heat-pipes-for-electronics-cooling-applications/. Accessed 12 June 2010.

Garner SD, Patel CD (2001) Loop thermosyphons and their applications to high density electronics cooling. Proceedings of the International Electronic Packaging Technical Conference and Exhibition, 2001.

Gaugler RS (1944) Heat transfer device. US Patent 2350348.

Higashi I, Sawano R (1997) Electronic cooling type refrigerator. European Patent EP0645593.

Hwang UP, Moran KP (1981) Boiling heat transfer of silicon integrated circuits chip mounted on a substrate. ASME Heat Transfer Division **20**: 53–59.

Kandlikar SG, Grande WJ (2004) Evaluation of single phase flow in microchannels for high heat flux chip cooling – thermohydrolic performance enhancement and fabrication technology. Heat Transfer Engineering **25 (8)**: 5–16.

Kirkconnell CS (2008) A cryocooler in your laptop computer? Maybe... Electronics Cooling. http://www.electronics-cooling.com/html/2008_nov_a1.php Accessed on 18 June 2010.

Klett J (2002) Graphite foam heat sinks – signature and heat management. http://www.ornl.gov/sci/oetd/documents/graphite_foam_2.pdf. Accessed on 16 June 2010.

Martin H (1977) Heat and mass transfer between impinging gas jets and solid surfaces. Advances in Heat Transfer **8**: 1–60.

Moran ME, Wesolek DM, Berhane BT, Rebello KJ (2004) Microsystem cooler development. NASA/TM – 2004–213307 AIAA–2004–5611. http://www.gltrs.grc.nasa.gov/reports/2004/TM-2004–213307.pdf. Accessed on 18 June 2010.

Ornatskii AP, Vinyarskii LS (1965) Heat transfer crisis in a forced flow of underheated water in small-bore tubes. High Temperatures **3**: 444–451.

Pamula VK, Chakrabarty K (2003) Cooling of integrated circuits using droplet-based microfluidics. Proceedings of the 13th ACM Great Lakes Symposium on VLSI, Washington, DC, USA, pp. 84–87.

Peterson GP (1994) An introduction to heat pipes. John Wiley & Sons, Inc., Canada.

Schmidt RR, Ellsworth MJ, Jr., Chu RC, Agonafer D (2000), Refrigeration cooled computers: application and review. Proceedings of the ASME International Engineering Congress and Exposition, 2000, vol. 28, pp. 277–283.

Schweber B (2005) Nanograss provides effective thermal-transfer surface. http://www.highbeam.com/doc/1G1–133836697.html. Accessed on 16 June 2010.

Shafii MB, Faghri A, Zhang Y (2001) Thermal modeling of unlooped and looped pulsating heat pipes. Journal of Heat Transfer Copyright **123**: 1159–1172.

Simons RE (1996) Direct liquid immersion cooling for high power density microelectronics. Electronic Cooling. http://www.electronics-cooling.com/1996/05/direct-liquid-immersion-cooling-for-high-power-density-microelectronics/. Accessed on 16 June 2010.

Stull C, Herz R (2007) Alternative cooling methods for electronics: utilization of liquid convection and evaporation. http://www.courses.ucsd.edu/rherz/mae221a/reports/Stull_221A_F07.pdf. Accessed on 16 June 2010.

Thome JR (1990) Enhanced boiling heat transfer. Hemisphere, New York.

Trutassanawin S et al. (2006) Experimental investigation of a miniature-scale refrigeration system for electronics cooling. IEEE Transactions on Components and Packaging Technologies **29 (3)**: 678–687.

Uhlig E, Thome JR (1985) Boiling of acetone–water mixtures on smooth and enhanced surfaces. Advances in Enhanced Heat Transfer – 1985. ASME Heat Transfer Division **43**: 49–56.

Webb RL (1994a) Advances in modeling enhanced heat transfer surfaces. Heat Transfer 1994. Proc. of the 10th International Heat Transfer Conference, vol. 1, IChemE, Rugby, U.K., pp. 445–459.

Webb RL (1994b) Principles of enhanced heat transfer. Wiley, New York.

Yarin LP, Mosyak A, Hetsroni G (2009) Fluid flow, heat transfer and boiling in micro-channels. Springer, New York.

Yilmaz S, Hwalck JJ, Westwater JW (1980) Pool boiling heat transfer performance for commercial enhanced tube surfaces, ASME Paper No. 80-HT-41.

Zaghdoudi MC, Lallemand M (2002) Electric field effects on pool boiling, Journal of Enhanced Heat Transfer **9**: 187–208.

# Chapter 11
# Thermoelectric Cooling Through Thermoelectric Materials

**Abstract** Thermoelectric (TE) cooling has been used for thermal management of high-power-dissipating electrical components, with silent, compact, reliable, and durable characteristics and being modulated to maintain a fixed temperature. However, TE coolers currently in use have a coefficient of performance (COP) of only about 0.5. This is quite a low value compared with COPs of other cooling approaches such as air conditioners and refrigerators at levels of 3.0–5.0. With increasing demands for high performance thermoelectric coolers, advanced emerging TE materials provide probability for improving their efficiency. These emerging materials include new families of advanced bulk TE materials based on crystal structures that contain weakly bound atoms or molecules with large vibrational amplitudes at partially filled structural sites acting as effective phonon scatterers, such as skutterudites, clathrates, and oxides; low dimensional materials systems, such as quantum well superlattices, quantum wires, quantum dots, thin film or band engineering structures; as well as nanocomposites, which demonstrates much higher *ZT* values than that of their bulk counterparts. The nanocomposites can be fabricated inexpensively, quickly, and in a form that is compatible with existing TE device configurations. Further research in this field will allow TE cooling to play a significant role in any future thermal management solution. This chapter will review the principle, design, and application of the TE cooling, as well as the effects of the emerging novel TE materials on its efficiency. The main contents include TE effects, design methodology and multistage architecture of TE cooling devices, and advanced TE materials and future development trends.

## Introduction

Thermoelectric cooling uses the Peltier effect to form a solid state heat pump, and works in conjunction with a heat sink to dissipate heat from an electronic packing system. A modern TE cooler is a semiconductor-based electronic component that functions as a small heat pump. By applying a low voltage DC power source to a TE module, heat will be moved through the module from one side to the other. One module face, therefore, will be cooled while the opposite face simultaneously is

X.C. Tong, *Advanced Materials for Thermal Management of Electronic Packaging*,
Springer Series in Advanced Microelectronics 30,
DOI 10.1007/978-1-4419-7759-5_11, 

heated. This phenomenon may be reversed, whereby a change in the polarity (plus and minus) of the applied DC voltage will cause heat to be moved in the opposite direction. Therefore, a thermoelectric module may be used for both heating and cooling thereby making it highly suitable for precise temperature control applications (Ferrotec 2010).

In electronic packaging, as shown in Figure 11.1, if a typical single-stage TE module was placed on a heat sink that was maintained at room temperature and the module was then connected to a suitable battery or other DC power source, the cold side of the module would cool down to Δ, for instance −40°C. At this point, the module would be pumping almost no heat and would have reached its maximum rated temperature difference ($\Delta T$). If heat was gradually added to the cold side of the module, the cold side temperature would increase progressively until it eventually equaled the heat sink temperature. At this point the TE cooler would have attained its maximum rated heat pumping capacity ($Q_{\max}$). Actually, TE coolers work principally in a similar fashion to mechanical refrigerators. In a mechanical refrigeration unit, a compressor raises the pressure of a liquid and circulates the refrigerant through the system. In the evaporator or freezer area the refrigerant boils and, in the process of changing to a vapor, the refrigerant absorbs heat causing the freezer to become cold. The heat absorbed in the freezer area is moved to the

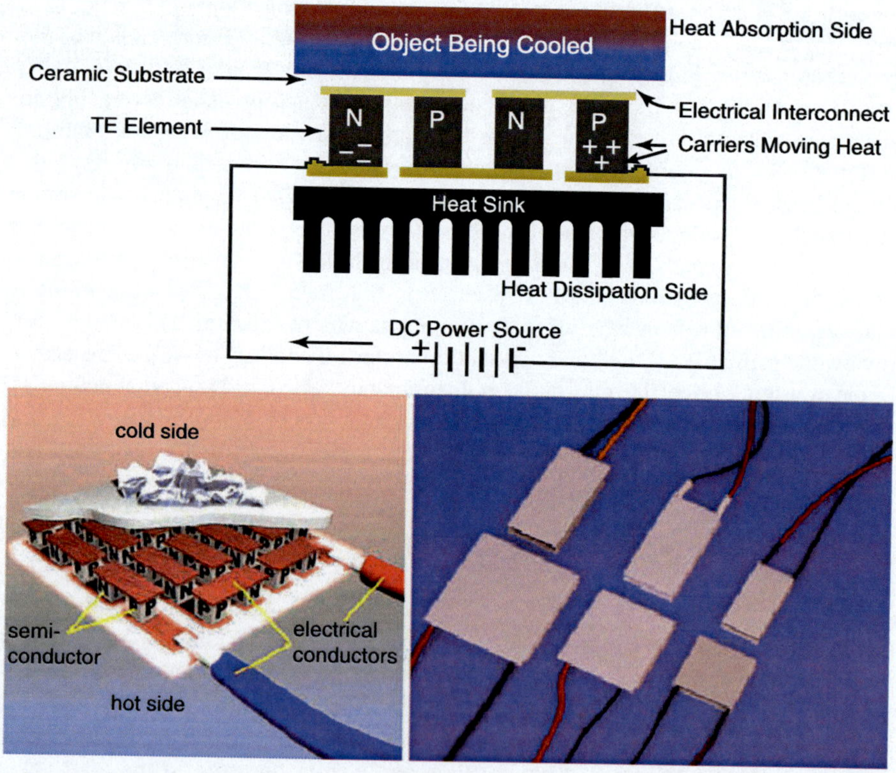


**Fig. 11.1** Schematic of thermoelectric cooling device

condenser where it is transferred to the environment from the condensing refrigerant. In a TE cooling system, a doped semiconductor material essentially takes the place of the liquid refrigerant, the condenser is replaced by a finned heat sink or a liquid cooling system, and the compressor is replaced by a DC power source. The application of DC power to the TE module causes electrons to move through the semiconductor material. At the cold end (or freezer side) of the semiconductor material, heat is absorbed by the electron movement, moved through the material, and expelled at the hot end. Because the hot end of the material is physically attached to a heat sink, the heat is passed from the material to the heat sink and then, in turn, transferred to the environment (Ferrotec 2010).

The physical principles upon which modern TE coolers are based actually date back to the early 1800s, although commercial TE modules were not available until almost 1960. The first important discovery relating to thermoelectricity occurred in 1821 when Thomas Seebeck, a German scientist, found that an electric current would flow continuously in a closed circuit made up of two dissimilar metals provided that the junctions of the metals were maintained at two different temperatures. Seebeck did not actually comprehend the scientific basis for his discovery, however, and falsely assumed that flowing heat produced the same effect as flowing electric current. In 1834, Jean Peltier, a French watchmaker and part time physicist, while investigating the Seebeck effect, found that there was an opposite phenomenon whereby thermal energy could be absorbed at one dissimilar metal junction and discharged at the other junction when an electric current flowed within the closed circuit. Twenty years later, William Thomson issued a comprehensive explanation of the Seebeck and Peltier effects and described their interrelationship. At the time, however, these phenomena were still considered to be mere laboratory curiosities and were without practical application. In the 1930s, Russian scientists began studying some of the earlier TE work in an effort to construct power generators for use at remote locations throughout the country. This Russian interest in thermoelectricity eventually caught the attention of the rest of the world and inspired the development of practical TE modules. Currently, TE coolers make use of modern semiconductor technology whereby doped semiconductor material takes the place of dissimilar metals used in early TE experiments (Ferrotec 2010).

Its applications have been limited because of the low values of efficiency that are inherent in current thermoelectric coolers (TECs). A TEC requires relatively large amounts of electrical power in order to produce a cooling effect. In fact, more conventional vapor compression refrigeration systems have as much as a 3 to 1 advantage in efficiency over TEs. In addition, when using the TEC in a system for cooling, the total heat that generated and needed to be removed from the system might be even more than without the TEC. However, TECs can be useful in certain applications where the advantages outweigh the disadvantages. Actually, TECs have no moving parts and do not require the use of chlorofluorocarbons; therefore, they are safe for the environment, inherently reliable, and virtually maintenance free. They can be operated in any orientation and are ideal for cooling devices that might be sensitive to mechanical vibration. Their compact size also makes them ideal for applications that are size or weight limited where even the smallest compressor would have excess capacity. Their ability to heat and cool by a simple reversal of current flow is useful for applications where both heating and cooling is necessary or where precise temperature control is critical. Despite their small size, TEC devices can all create temperature differences of around 70°C. They can also be stacked on top of each other to create even larger temperature differences (up to 130°C). Because of their small size, TECs can be used for localized cooling, where only a small part of the system needs to be cooled. TECs are also useful as active temperature control devices. A simple feedback loop can be used to implement a TEC in a system where a specific temperature is required, because the temperature of one side of the TEC is directly proportional to the input power. The cold side of the device can be either cooled or heated to the proper temperature, depending on the direction of the electrical current. As long as the temperature variation is within the range of the TEC's capability, and a proper heat sink/heat source is used on the other side, the temperature can be controlled quite accurately. Some of the specific features of TE modules include (Ferrotec 2010):

1. *No Moving Parts*: A TE module works electrically without any moving parts so they are virtually maintenance free.
2. *Small Size and Weight*: The overall TE cooling system is much smaller and lighter than a comparable mechanical system. In addition, a variety of standard and special sizes and configurations are available to meet strict application requirements.
3. *Ability to Cool Below Ambient*: Unlike a conventional heat sink where the temperature necessarily must rise above ambient, a TE cooler attached to that same heat sink has the ability to reduce the temperature below the ambient value.
4. *Ability to Heat and Cool with the Same Module*: TE coolers will either heat or cool depending upon the polarity of the applied DC power. This feature eliminates the necessity of providing separate heating and cooling functions within a given system.
5. *Precise Temperature Control*: With an appropriate closed-loop temperature control circuit, TE coolers can control temperatures to better than ±0.1°C.
6. *High Reliability*: TE modules exhibit very high reliability due to their solid state construction. Although reliability is somewhat application dependent, the life of typical TE coolers is greater than 200,000 h.
7. *Electrically "Quiet" Operation*: Unlike a mechanical refrigeration system, TE modules generate virtually no electrical noise and can be used in conjunction with sensitive electronic sensors. They are also acoustically silent.
8. *Operation in Any Orientation*: TEs can be used in any orientation and in zero gravity environments. Thus they are popular in many aerospace applications.
9. *Convenient Power Supply*: TE modules operate directly from a DC power source. Modules having a wide range of input voltages and currents are available. Pulse width modulation may be used in many applications.
10. *Spot Cooling*: With a TE cooler it is possible to cool one specific component or area only, thereby often making it unnecessary to cool an entire package or enclosure.

11. *Ability to Generate Electrical Power*: When used in-reverse by applying a temperature differential across the faces of a TE cooler, it is possible to generate a small amount of DC power.
12. *Environmentally Friendly*: Conventional refrigeration systems cannot be fabricated without using chlorofluorocarbons or other chemicals that may be harmful to the environment. TE devices do not use or generate gases of any kind.

More exotic TEC devices are being developed that could result in better performance such as superlattice structures, quantum wires and quantum wells, thin films using SiGe/Si, and thermionic cooling.

## Thermoelectric Effects

The TE effect is the direct conversion of temperature differences to electric voltage and vice versa at the junctions of two different conductors. Traditionally, the TE effect encompasses three separately identified effects, the Seebeck effect, the Peltier effect, and the Thomson effect. These three effects, together with several other phenomena, form the basis of functional TE modules.

### *Seebeck Effect*

The Seebeck effect is the conversion of temperature differences directly into electricity. In 1822, Seebeck discovered that a compass needle would be deflected when a closed loop was formed of two metals joined in two places with a temperature difference between the junctions. This is because the metals respond differently to the temperature difference, which creates a current loop, which produces a magnetic field. However, at this time Seebeck did not recognize there was an electric current involved, so he called the phenomenon the thermomagnetic effect, thinking that the two metals became magnetically polarized by the temperature gradient. The Danish physicist Hans Christian Ørsted played a vital role in explaining and conceiving the term thermoelectricity. The Seebeck effect is then used to describe that a voltage is created in the presence of a temperature difference between two different metals or semiconductors. This causes a continuous current in the conductors if they form a complete loop. The voltage developed can be derived from (Ferrotec 2010):

$$V = \int_{T_1}^{T_2} (S_{\mathrm{B}} - S_{\mathrm{A}})\mathrm{d}T. \tag{11.1}$$

$S_{\mathrm{A}}$ and $S_{\mathrm{B}}$ are the Seebeck coefficients (also called TE power or thermopower) of the materials A and B as a function of temperature, and $T_1$ and $T_2$ are the temperatures of the two junctions. The Seebeck coefficients are nonlinear as a function of temperature, and depend on absolute temperature, material, and molecular structure of the conductors. The Seebeck effect is the principle at work behind thermal diodes and TE generators which are used for creating power from heat differentials. Seebeck coefficient of a material, represented by $S = E_x/(dT/dx)$, where $E_x$ is electric field, measures the magnitude of an induced TE voltage in response to a temperature difference across that material. It has units of (V/K), though in practice it is more common to use microvolts per kelvin. Values in the hundreds of μV/K, negative or positive, are typical of good TE materials. The Seebeck effect of a material depends on the material's temperature and crystal structure. Metals typically have small thermopowers because most have half-filled bands. Electrons (negative charges) and holes (positive charges) both contribute to the induced TE voltage thus canceling each other's contribution to that voltage and making it small. In contrast, semiconductors can be doped with excess electrons or holes, and thus can have large positive or negative values of the Seebeck coefficient depending on the charge of the excess carriers.

## *Peltier Effect*

Peltier effect is the reverse of the Seebeck effect, which was discovered by Jean Charles Athanase Peltier in 1834. That is, an applied voltage could create a temperature difference between the two dissimilar metals. Although Peltier is generally credited with the discovery of TE cooling, he did not fully understand the physics of the phenomenon. The full explanation was given four years later by Emil Lenz, who showed that a drop of water on a bismuth-antimony junction would freeze when electrical current was applied one way, and melt again when the current was reversed. As knowledge of TEs increased, the most important discoveries were related to material properties. In 1911, Altenkirch derived the TE efficiency, also called the TE figure of merit Z, which is given as (Mann 2006)

$$Z = \frac{S^2 \sigma}{\kappa}, \tag{11.2}$$

where $S$ is the Seebeck coefficient, $\sigma$ is the electrical conductivity, and $\kappa$ is the thermal conductivity. This expression indicates that ideal TE materials would have a high electrical conductivity to minimize Joule heating and a low thermal conductivity to prevent the backflow of heat from the hot side to the cool side. The TE efficiency can be nondimensionalized by multiplying by the average temperature $T$ $[(T_1 + T_2)/2]$, which yields the most common form of TE efficiency, $ZT$, also known as the dimensionless figure of merit.

Early work in TEs resulted in very small values of $Z$ because the materials being used were mostly metals, which did not possess ideal TE properties. Most traditional materials have a correlation between electrical and thermal conductivity.

That is, a material that conducts electricity well, such as a metal, will also conduct heat well, and a material that insulates heat, such as glass or ceramic, also insulates electricity. Beginning in the late 1930s and continuing into the 1970s, there was a surge of discoveries that showed semiconductors exhibited the best TE properties. Until that time, metals such as bismuth and antimony alloys were the state of the art with a *ZT* value of around 0.1 at room temperature. The implementation of semiconductors such as bismuth telluride ($Bi_2Te_3$) helped to increase that number by tenfold because semiconductors are moderate conductors of heat and electricity. However, a *ZT* value of roughly 3 is needed in order for TE cooling systems to compete with vapor compression refrigerators. A new class of TE materials has been developed and categorized as low-dimensional TEs and they have the potential to greatly increase *ZT* value. These are superlattice thin-films, also called quantum wells, which are structures built up by depositing multiple layers of alternating semiconductor materials with differing electronic band gaps to a thickness in the range of 10–100 Å, on a substrate material. This type of structure provides decoupling of the thermal and electrical conductivities of the material, and further tailoring of the layer materials can yield a higher *ZT* value up to 3 or 4 (Bass et al. 2004).

A typical TE cooling component is an n- or p-type semiconductor (bismuth telluride) sandwiched between two conductors, usually copper. The semiconductor is used because they can be optimized for pumping heat and because the type of charge carriers within them can be chosen. When an electronic field is applied to two ends of the n-type (doped with electrons) semiconductor, the electrons move towards the positive end to release heat, while the negative end absorbs heat. Therefore, the heat is transported in the direction of current flow. When a p-type semiconductor (doped with holes) is used instead, the holes move in a direction opposite the current flow. The heat is also transported in a direction opposite the current flow and in the direction of the holes. Essentially, the charge carriers dictate the direction of heat flow. Electrons can travel freely in the copper conductors but not so freely in the semiconductor. As the electrons leave the copper and enter the hot-side of the p-type, they must fill a hole in order to move through the p-type. When the electrons fill a hole, they drop down to a lower energy level and release heat in the process. Then, as the electrons move from the p-type into the copper conductor on the cold side, the electrons are bumped back to a higher energy level and absorb heat in the process. Next, the electrons move freely through the copper until they reach the cold side of the n-type semiconductor. When the electrons move into the n-type, they must bump up an energy level in order to move through the semiconductor. Heat is absorbed when this occurs. Finally, when the electrons leave the hot-side of the n-type, they can move fieely in the copper. They drop down to a lower energy level and release heat in the process. To increase heat transport, several p type or n type TE components can be hooked up in parallel. However, the device requires low voltage and therefore, a large current which is too great to be commercially practical. The TE components can be put in series but the heat transport abilities are diminished because the interconnectings between the semiconductor creates thermal shorting.

The most efficient configuration is where a p and n TE component is put electrically in series but thermally in parallel. This kind of device is called a couple for electronic cooling. When assembling it into an electronic package, one side is attached to a heat source and the other a heat sink that convects the heat away. The side facing the heat source is considered the cold side and the side facing the heat sink the hot side. There must be an electrical insulator between the heat generating device and the conductor to prevent an electrical short circuit between the module and the heat source. The electrical insulator must also have a high thermal conductivity so that the temperature gradient between the source and the conductor is small. Ceramics such as alumina are generally used for this purpose. The most common devices use 254 alternating p and n type TE devices. The devices can operate at 12–16 V at 4–5 A. These values are much more practical for real-life operations.

## *Thomson Effect*

The Thomson effect is a combination of the Seebeck and Peltier effects which was the phenomenon discovered by William Thomson in 1854, later called Lord Kelvin. It was found that there occurs a reversible transverse heat flow into or out of a conductor of a particular metal, the direction depending upon whether a longitudinal electric current flows from colder to warmer metal or from warmer to colder. Any temperature gradient previously existing in the conductor is thus modified if a current is turned on. Any current-carrying conductor (except for a superconductor) with a temperature difference between two points, will either absorb or emit heat, depending on the material. The Thomson effect does not occur in a current-carrying conductor which is initially at uniform temperature. If a current density $J$ is passed through a homogeneous conductor, heat production per unit volume is (Roberts 1977):

$$q = \rho J^2 - \mu J \frac{\mathrm{d}T}{\mathrm{d}x}, \qquad (11.3)$$

where $\rho$ is the resistivity of the material, $\mathrm{d}T/\mathrm{d}x$ is the temperature gradient along the wire, $\mu$ is the Thomson coefficient. The first term $\rho J^2$ is simply the Joule heating, which is not reversible. The second term is the Thomson heat, which changes sign when $J$ changes direction.

In metals such as zinc and copper, which have a hotter end at a higher potential and a cooler end at a lower potential, when current moves from the hotter end to the colder end, it is moving from a high to a low potential, so there is an evolution of heat. This is called the positive Thomson effect. In metals such as cobalt, nickel, and iron, which have a cooler end at a higher potential and a hotter end at a lower potential, when current moves from the hotter end to the colder end, it is moving from a low to a high potential, there is absorption of heat. This is called the negative

Thomson effect. It is commonly asserted that lead has a zero Thomson effect. While it is true that the TE coefficients of lead are small, they are in general nonzero. The Thomson coefficient of lead has been measured over a wide temperature range and has been integrated to calculate the absolute Seebeck coefficient of lead as a function of temperature. Unlike lead, the TE coefficients of all known superconductors are zero.

Thomson's theoretical treatment of thermoelectricity is remarkable in the fact that it is probably the first attempt to develop a reasonable theory of irreversible thermodynamics (nonequilibrium thermodynamics). This phenomenon is of interest in respect to the principles involved but plays a negligible role in the operation of practical TE modules. For this reason, it is ignored (Ferrotec 2010).

## *Application of Thermoelectric Effects to Thermoelectric Cooling*

The TE semiconductor material used most often in current TE coolers is an alloy of bismuth telluride that has been suitably doped to provide individual blocks or elements having distinct "n" and "p" characteristics. TE materials most often are fabricated by either directional crystallization from a melt or pressed powder metallurgy. Each manufacturing method has its own particular advantage, but directionally grown materials are most common. In addition to bismuth telluride ($Bi_2Te_3$), there are other TE materials including lead telluride (PbTe), silicon germanium (SiGe), and bismuth-antimony (Bi-Sb) alloys that may be used in specific situations. From these materials, bismuth telluride ($Bi_2Te_3$) has highest figure of merit $ZT$ (from 0.8 to 1) within a temperature range of around 50–100°C, and is best suited for most current cooling applications. This is attributed to its specific crystal structure. $Bi_2Te_3$ is highly anisotropic in nature, which results in its electrical resistivity being approximately four times greater parallel to the axis of crystal growth (C-axis) than in the perpendicular orientation. In addition, thermal conductivity is about two times greater parallel to the C-axis than in the perpendicular direction. Because the anisotropic behavior of resistivity is greater than that of thermal conductivity, the maximum performance or figure-of-merit occurs in the parallel orientation. Because of this anisotropy, TE elements must be assembled into a cooling module so that the crystal growth axis is parallel to the length or height of each element and, therefore, perpendicular to the ceramic substrates. In addition, $Bi_2Te_3$ crystals are made up of hexagonal layers of similar atoms. While layers of bismuth and tellurium are held together by strong covalent bonds, weak van der Waals bonds link the adjoining [$Te^1$] layers. As a result, crystalline bismuth telluride cleaves readily along these [$Te^1$][$Te^1$] layers, with the behavior being very similar to that of Mica sheets. Fortunately, the cleavage planes generally run parallel to the C-axis and the material is quite strong when assembled into a thermoelectric cooling module (Ferrotec 2010).

Bismuth telluride material, when produced by directional crystallization from a melt, typically is fabricated in ingot or boule form and then sliced into wafers of various thicknesses. After the wafer's surfaces have been properly prepared, the wafer is then diced into blocks that may be assembled into TE cooling modules. The blocks of bismuth telluride material, which usually are called elements or dice, also may be manufactured by a pressed powder metallurgy process. A practical TE cooler consists of two or more elements of semiconductor material that are connected electrically in series and thermally in parallel, as shown in Figure 11.1. These TE elements and their electrical interconnects typically are mounted between two ceramic substrates. The substrates serve to hold the overall structure together mechanically and to insulate the individual elements electrically from one another and from external mounting surfaces. After integrating the various component parts into a module, TE modules ranging in size from approximately 2.5 to 50 mm (0.1–2.0 in) square and 2.5 to 5 mm (0.1–0.2 in) in height may be constructed (Ferrotec 2010).

Both n-type and p-type bismuth telluride TE materials are used in a TE cooler. This arrangement causes heat to move through the cooler in one direction only while the electrical current moves back and forth alternately between the top and bottom substrates through each n and p element. N-type material is doped so that it will have an excess of electrons (more electrons than needed to complete a perfect molecular lattice structure) and p-type material is doped so that it will have a deficiency of electrons (fewer electrons than are necessary to complete a perfect lattice structure). The extra electrons in the n material and the "holes" resulting from the deficiency of electrons in the p material are the carriers which move the heat energy through the TE material (Ferrotec 2010).

Figure 11.1 shows a kind of typical TE module. The module consists of pairs of p-type and n-type semiconductor thermoelements forming thermocouples which are connected electrically in series and thermally in parallel. In cooling mode, an electrical current is supplied to the module. Heat is pumped from one side to the other (Peltier effect), the result is that one side of the module becomes cold. In generating mode conversely, a temperature gradient is maintained across the module. The heat flux passing through the module is converted into electrical power (Seebeck effect). TE cooling uses the Peltier effect to create heat flux between the junctions of two different types of materials. A Peltier cooler, heater, or TE heat pump is a solid-state active heat pump which transfers heat from one side of the device to the other side against the temperature gradient (from cold to hot), with consumption of electrical energy. Such an instrument is also called a Peltier device, Peltier diode, cooling diode, Peltier heat pump, solid state refrigerator, or TEC. Because heating can be achieved more easily and economically by many other methods, Peltier devices are mostly used for the cases that a single device is to be used for both heating and cooling. However, with the advent of more and more effective TE materials, the applications of the TEC are increasing. Peltier devices have been commonly used in camping and portable coolers and for cooling electronic components and small instruments. Some electronic equipment intended for military use in the field is thermoelectrically cooled. The cooling effect of Peltier heat pumps can also be used to extract water from the air in dehumidifiers. Peltier elements are a common component in thermal cyclers, used for the synthesis of DNA by polymerase chain reaction, which is a common molecular biological technique which requires the rapid heating and cooling of the reaction mixture for denaturation, primer annealing, and enzymatic synthesis cycles. The effect is also used in satellites and spacecraft to counter the effect of direct sunlight on one side of a craft by dissipating the heat over the cold shaded side, whereupon the heat is dissipated by thermal radiation into space. Photon detectors such as charge-coupled devices in astronomical telescopes or very high-end digital cameras are often cooled down with Peltier elements. This reduces dark counts due to thermal noise. A dark count is the event that a pixel gives a signal although it has not received a photon but rather mistook a thermal fluctuation for one. On digital photos taken at low light these occur as speckles or pixel noise (Kim and Ferreira 2008). In electronic packaging, TECs would be used to cool computer components to keep temperatures within design limits without the noise of a fan, or to maintain stable functioning when overclocking. A Peltier cooler with a heat sink or waterblock can cool a chip to well below ambient temperature.

## Design and Architecture of Thermoelectric Cooling Devices

A basic architecture of a TEC with heat being moved as a result of an applied electrical current (I) is shown in Figure 11.1. Most TEC modules are fabricated with an equal number of n-type and p-type elements where one n and p element pair form a TE couple. Heat flux (heat actively pumped through the TE module) is proportional to the magnitude of the applied DC electric current. By varying the input current from zero to maximum, it is possible to adjust and control the heat flow and temperature.

### *Design Methodology*

Peltier TEC performance is a function of ambient temperature, hot and cold side heat exchanger (heat sink) performance, thermal load, Peltier module (thermopile) geometry, and Peltier electrical parameters. Specifically, the performance of a TEC device depends on the following factors: (1) the temperature of the cold ($T_c$) and hot ($T_h$) sides; (2) thermal and electrical conductivities of the device's materials; (3) contact resistance between the TE device and heat source/heat sink; and (4) thermal resistance of the heat sink. The TE performance is usually characterized by COP. The current yielding the maximum COP is given by (Goldsmid 1986):

$$I_\phi = \frac{(S_p - S_n)(T_h - T_c)}{R[(1 + ZT_m)^{1/2} - 1]}. \tag{11.3}$$

The maximum COP is

$$\phi_{\max} = \frac{Q_c}{W_{in}} = \frac{T_1[(1+ZT_m)^{1/2} - T_2/T_1]}{(T_2 - T_1)[(1+ZT_m)^{1/2} + 1]}, \tag{11.4}$$

where $T_m = (T_H + T_C)/2$. The maximum efficiency of a TE device for electricity generation is given by $\eta_{max}$, defined as

$$\eta_{\max} = \frac{T_H - T_C}{T_H} \frac{\sqrt{1+Z\overline{T}} - 1}{\sqrt{1+Z\overline{T}} + \frac{T_C}{T_H}}, \tag{11.5}$$

where $T_H$ is the temperature at the hot junction and $T_C$ is the temperature at the surface being cooled. $Z\overline{T}$ is the modified dimensionless figure of merit which now takes into consideration the TE capacity of both TE materials being used in the power generating device, and is defined as (Goldsmid 1986)

$$Z\overline{T} = \frac{(S_p - S_n)^2 \overline{T}}{[(\rho_n k_n)^{1/2} + (\rho_p k_p)^{1/2}]^2}, \tag{11.6}$$

where $\rho$ is the electrical resistivity, $\overline{T}$ is the average temperature between the hot and cold surfaces, and the subscripts n and p denote properties related to the n- and p-type semiconducting TE materials, respectively. It is worthwhile to note that the efficiency of a TE device is limited by the Carnot efficiency (hence the $T_H$ and $T_C$ terms in $\Phi_{max}$), because TE devices are still inherently heat engines.

The COP of current commercial TE refrigerators ranges from 0.3 to 0.7, only about one-sixth the value of traditional vapor-compression refrigerators (approximately 3). A simplified way of determining the voltage and the heat load are given by (Buit 1980):

$$Q_c = (\alpha_p - \alpha_n)IT_c - K(T_H - T_c) - 1/2I^2R, \tag{11.7}$$

$$V = 2N\left[\alpha(T_h - T_c)\frac{IRL}{A}\right], \tag{11.8}$$

where $V$ is the voltage and $Q_c$ is the heat load, $N$ is the number of couples, and $L$ is the element height.

There are various ways to design a TEC; and different methodology has been used to design one that meets a given specification. Typically, the first step in the design of a TEC system involves making an analysis of the system's overall thermal characteristics. This analysis may be quite simple for some applications or highly complex in other cases, which must never be neglected if a satisfactory and efficient design is to be realized. The active heat load is the actual heat generated by the component to be cooled. For most applications, the active load will be equal to the electrical power input to the article being cooled (Watts = Volts × Amps) but in other situations the load may be more difficult to determine. The passive heat load (sometimes called heat leak or parasitic heat load) is that heat energy which is lost or gained by the article being cooled as a result of conduction, convection, and/or radiation. Passive heat losses may occur through any heat-conductive path including air, insulation, and electrical wiring. In applications where there is no active heat generation, the passive heat leak will represent the entire heat load on the TEC. Determination of the total heat leak within a cooling system is a relatively complicated issue but a reasonable estimate of these losses often can be made by means of some basic heat transfer calculations, however, several fundamental heat transfer equations can be used in evaluating some of the thermal aspects of a design or system. The following equation can be used for estimating heat losses due to convection and conduction of an enclosure (Marlow 1998).

$$Q_{\text{passive}} = (A\Delta T)/(x/k + 1/h), \tag{11.9}$$

where: $Q_{\text{passive}}$ = heat load (W); $A$ = total external surface area of enclosure (m$^2$); $x$ = thickness of insulation (m); $k$ = thermal conductivity of insulation (W/m °C); $h$ = convective heat transfer coefficient (W/m$^2$ °C); $\Delta T$ = Temperature change (°C). Some designs require a set amount of time to reach the desired temperature. The following equation may be used to estimate the time required (Marlow 1998):

$$t = [(\rho)(V)(C_p)(T_1 - T_2)]/Q, \tag{11.10}$$

where: $t$ = time (seconds); $\rho$ = density (g/cm$^3$); $V$ = volume (cm$^3$); $C_p$ = specific heat (J/g °C); $T_1 - T_2$ = temperature change (°C); $Q = (Q_{\text{to}} + Q_{\text{tt}})/2$ (J/s; Note: 1 J/s = 1 W); $Q_{\text{to}}$ is the initial heat pumping capacity when the temperature difference across the cooler is zero. $Q_{\text{tt}}$ is the heat pumping capacity when the desired temperature difference is reached and heat pumping capacity is decreased. $Q_{\text{to}}$ and $Q_{\text{tt}}$ are used to obtain an average value.

Before the cooler or heat sink can be selected, the cooling requirements must be defined. This includes determining the amount of heat to be pumped. Minimizing the heat load allows the cooler to achieve colder temperatures or reduces the power required to reach the defined cooling level. The following describes the techniques used to estimate active and passive heat loads and applies only to steady state heat loads. The heat load may consist of two types; active or passive, or a combination of the two. An active load is the heat dissipated by the device being cooled. It generally equals the input power to the device. Passive heat loads are parasitic in nature and may consist of radiation, convection, or conduction. The general equation for active heat load dissipation is (Marlow 1998):

$$Q_{\text{active}} = V^2/R = VI = I^2R, \tag{11.11}$$

where: $Q_{active}$ = active heat load (W); $V$ = voltage applied to the device being cooled (V); $R$ = device resistance (Ω); $I$ = current through the device (A). When two objects at different temperatures come within proximity of each other, heat is exchanged. This occurs through electromagnetic radiation emitted from one object and absorbed by the other. The hot object will experience a net heat loss and the cold object a net heat gain as a result of the temperature difference. This is called thermal radiation. Radiation heat loads are usually considered insignificant when the system is operated in a gaseous environment since the other passive heat loads are typically much greater in magnitude. Radiation loading is usually significant in systems with small active loads and large temperature differences, especially when operating in a vacuum environment. The fundamental equation for radiation loading is (Marlow 1998):

$$Q_{rad} = FesA({T_{amb}}^4 - {T_c}^4), \tag{11.12}$$

where: $Q_{rad}$ = radiation heat load (W); $F$ = shape factor (worst case value = 1); $e$ = emissivity (worst case value = 1); $s$ = Stefan–Boltzmann constant ($5.667 \times 10^{-8}$ W/m$^2$K$^4$); $A$ = area of cooled surface (m$^2$); $T_{amb}$ = Ambient temperature (K); $T_c$ = TEC cold ceramic temperature (K).

When the temperature of a fluid (in this case, a gas) flowing over an object differs from that of the object, heat transfer occurs. The amount of heat transfer varies depending on the fluid flow rate. Convective heat loads on TECs are generally a result of natural (or free) convection. This is the case when gas flow is not artificially induced as with a fan or pump, but rather occurs naturally from the varying density in the gas caused by the temperature difference between the object being cooled and the gas. The convective loading on a system is a function of the exposed area and the difference in temperature between this area and the surrounding gas. Convective loading is usually most significant in systems operating in a gaseous environment with small active loads or large temperature differences. The fundamental equation which describes convective loading is (Marlow 1998):

$$Q_{conv} = hA(T_{air} - T_c), \tag{11.13}$$

where: $Q_{conv}$ = convective heat load (W); $h$ = convective heat transfer coefficient (W/m$^2$ °C) (typical value for a flat, horizontal plate in air at 1 atm = 21.7 W/m$^2$ °C; $A$ = exposed surface area (m$^2$); $T_{air}$ = temperature of surrounding air (°C); $T_c$ = temperature of cold surface (°C).

Conductive heat transfer occurs when energy exchange takes place by direct impact of molecules moving from a high temperature region to a low temperature region. Conductive heat loading on a system may occur through lead wires, mounting screws, etc., which form a thermal path from the device being cooled to the heat sink or ambient environment. The fundamental equation which describes conductive loading is Marlow 1998):

$$Q_{\text{cond}} = kA\Delta T/L, \qquad (11.14)$$

where: $Q_{\text{cond}}$ = conductive heat load (W); $k$ = thermal conductivity of the material (W/m °C); $A$ = cross-sectional area of the material (m$^2$); $L$ = length of the heat path (m); $\Delta T$ = temperature difference across the heat path (°C) (usually ambient or heat sink temperature minus cold side temperature).

Heat loading may occur through one or more of four modes: active, radiation, convection, or conduction. By utilizing these equations heat loads can estimated. The resulting information can then be used to select a suitable TEC for specific application.

TECs are mounted using one of three methods: adhesive bonding, compression using thermal grease, or solder. In general, for a TEC with a ceramic base of 19 mm or less, solder or adhesive bond can be used without fear of failure due to thermal stresses. If the TEC base is larger than 19 mm, the compression method is recommended because thermal grease is not rigid and does not transfer thermal stresses. A thin layer of copper metallization on the hot and/or cold ceramic allows soldering as a means of attachment. Adhesives and greases are prone to outgassing, therefore, they are not as appropriate for use in a vacuum package (Marlow 1998).

Surface preparation is important when using any of the assembly methods. No matter which method is used, the mounting surface should be flat to less than 0.08 mm over the TEC mounting area. In addition, the surface should be clean and free from oil, nicks, and burrs. When multiple TECs are placed in parallel thermally between common plates, the TEC thicknesses should vary no more than 0.05 mm. An adhesive bonding mounting method should be used when it is necessary to permanently attach the TEC to the heat sink, when mounting with solder is not an option, when the TECs need to be lapped to the same height after mounting, or when moderate thermal conductivity is required. A compression mounting method is needed when a permanent bond is not desired, when multiple TECs are used, or when the TEC is larger than 19 mm. The soldering mounting method is necessary when minimal outgassing is needed, when the TEC is smaller than 19 mm, when need a high-strength junction, or when high thermal conductivity is required.

In general, techniques used to install TE modules in a cooling system are extremely important. Failure to observe certain basic principles may result in unsatisfactory performance or reliability. Some of the factors to be considered in system design and module installation include the following (Marlow 1998): (1) TE modules have high mechanical strength in the compression mode but shear strength is relatively low. As a result, a TEC should not be designed into a system where it serves as a significant supporting member of the mechanical structure. (2) All interfaces between system components must be flat, parallel, and clean to minimize thermal resistance. High conductivity thermal interface material is often used to ensure good contact between surfaces. (3) The “hot” and “cold” sides of standard TE modules may be identified by the position of the wire leads. Wires are attached to the hot side of the module, which is the module face that is in contact with the heat sink. For modules having insulated wire leads, when the red and black leads are connected to the respective positive and negative terminals of a DC power supply, heat will be pumped from the module's cold side, through the module, and into the heat sink. Note that for TE modules having bare wire leads, the positive connection is on the right side and the negative connection is on the left when the leads are facing toward the viewer and the substrate with the leads attached presented on the bottom. (4) When cooling below ambient, the object being cooled should be insulated as much as possible in order to minimize heat loss to the ambient air. To reduce convective losses, fans should not be positioned so that air is blowing directly at the cooled object. Conductive losses may also be minimized by limiting direct contact between the cooled object and external structural members. (5) When cooling below the dew point, moisture or frost will tend to form on exposed cooled surfaces. To prevent moisture from entering a TE module and severely reducing its thermal performance, an effective moisture seal should be installed. This seal should be formed between the heat sink and cooled object in the area surrounding the TE module(s). Flexible foam insulating tape or sheet material and/or silicone rubber RTV (Room-Temperature Vulcanization) are relatively easy to install and make an effective moisture seal. Several methods for mounting TE modules are available and the specific product application often dictates the method to be used.

TECs operate directly from DC power suitable power sources can range from batteries to simple unregulated "brute force" DC power supplies to extremely sophisticated closed-loop temperature control systems. A TEC module is a low-impedance semiconductor device that presents a resistive load to its power source. Due to the nature of the bismuth telluride material, modules exhibit a positive resistance temperature coefficient of approximately 0.5% per °C based on average module temperature. For many noncritical applications, a lightly filtered conventional battery charger may provide adequate power for a TEC provided that the AC ripple is not excessive. Simple temperature control may be obtained through the use of a standard thermostat or by means of a variable-output DC power supply used to adjust the input power level to the TE device. In applications where the thermal load is reasonably constant, a manually adjustable DC power supply often will provide temperature control on the order of $\pm 1$°C over a period of several hours or more. Where precise temperature control is required, a closed-loop (feedback) system generally is used whereby the input current level or duty cycle of the TE device is automatically controlled. With such a system, temperature control to $\pm 0.1$°C may be readily achieved and much tighter control is not unusual (Chein and Huang 2004).

## *Multistage Architecture*

Multistage cooling and low-level signal detection are two applications which may require lower values of power supply ripple. In the case of multistage TE devices, achieving a large temperature differential is the typical goal, and a ripple component of less than 2% may be necessary to maximize module performance. In situations where very low level signals must be detected and/or measured, even though the TE module itself is electrically quiet, the presence of an AC ripple signal within the module and wire leads may be unsatisfactory. The acceptable level of power supply ripple for such applications will have to be determined on a case-by-case basis (Ferrotec 2010).

TEC modules are considered to be highly reliable components due to their solid-state construction. For most applications they will provide long, trouble-free service. There have been many instances where TE modules have been used continuously for 20 or more years and the life of a module often exceeds the life of the associated equipment. The specific reliability of TE devices tends to be difficult to define, however, because failure rates are highly dependent upon the particular application. For applications involving relatively steady-state cooling where DC power is being applied to the module on more-or-less continuous and uniform basis, TE module reliability is extremely high. Mean time between failures (MTBFs) in excess of 200,000 h are not uncommon in such cases and this MTBF value generally is considered to be an industry standard. On the other hand, applications involving thermal cycling show significantly worse MTBFs, especially when TE modules are cycled up to a high temperature. It is important that modules are installed in accordance with general requirements in order to minimize the possibility of premature module failure due to faulty assembly techniques. Temperature control methods also have an impact on TE module reliability. Linear or proportional control should always be chosen over ON/OFF techniques when prolonging the life of the module is required (Ferrotec 2010).

Various methods have been used to improve the performance of TECs which are its major drawback. Thin film coolers or multistage (bulk) coolers are two typical examples. Thin films are material layers of about 1 μm thickness. Alternating layers of $Sb_2Te_3$ and $Bi_2Te_3$ are used to produce thin film TECs. As shown in Figure 11.2, the highest power components are mounted on a diamond substrate which would be the top or cold side substrate of a thin film TEC. Power densities were reported to be above 100 W/cm$^2$. Thin film coolers considerably reduce the size of TE devices. Because the cooling density of a Peltier cooler is inversely proportional to its length, scaling to smaller size is desirable.

When the desired temperature differential between the cold and hot side cannot be obtained with a single-stage module, or when the cold side temperature must be lower than a one-stage cooler will allow, a multistage module may need to be applied. Multistage modules are essentially single-stage modules stacked up in a vertical pyramid-shaped array, as shown in Figure 11.3 (Simons 2001; Bar-Cohen et al. 2005). As the number of stages increases, the minimum cold side temperature will decrease. Also, increasing the number of stages increases the COP for a given cold side temperature (Nolas et al. 2001a, b). The COP of $N$ number of multistage module is given by (Goldsmid 1986):

$$\phi = \left[(1 + \frac{1}{\phi'})^N - 1\right]^{-1}, \qquad (11.15)$$

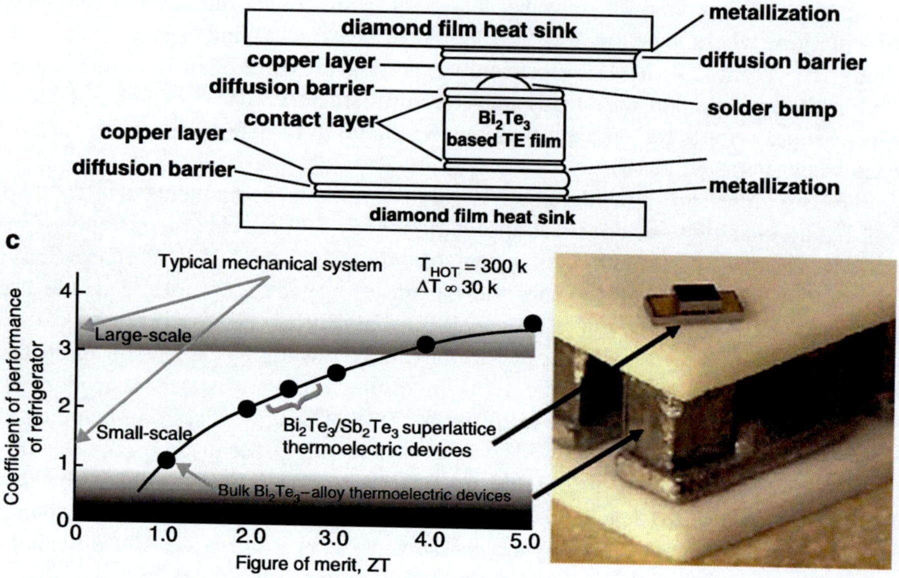


**Fig. 11.2** Thin film cooler

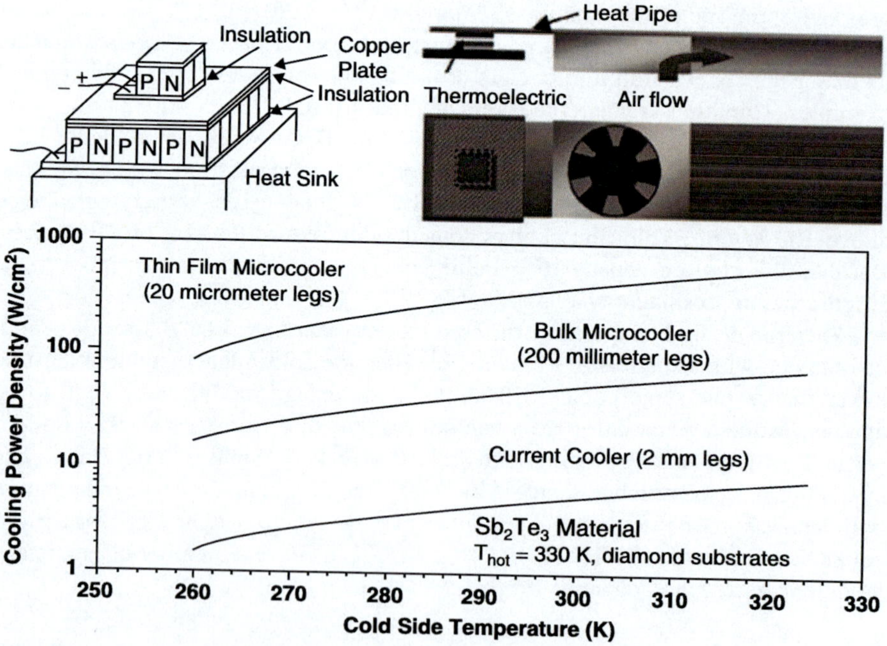


**Fig. 11.3** Multistage thermoelectric cooler and cooling density of different size coolers

where $\phi'$ is the COP of one stage of the module and $N$ is the number of stages. Figure 11.3 also compares the three types of coolers: bulk (multistage), thin film, and current (Simons and Chu 2000). More exotic TE devices are being researched that could result in better performance such as, superlattice structures, quantum wires and quantum wells, thin films using SiGe/Si, and thermionic cooling. Some alternative methods of TEC have been considered, such as (1) the heat produced by a computer chip can be used to provide the electricity to run a fan that cools the chip. The fan uses a TE device operating on the Seebeck effect to convert the heat to electricity. (2) When a laptop is running on batteries, the electricity used to power the fan comes from the battery. Therefore, to conserve battery life, a TE power generator is a good alternative (Bar-Cohen et al. 2005).

## Thermoelectric Materials and Future Development Trends

TECs in use today have a COP of only about 0.5. This is quite a low value compared with COPs of larger scale machines, such as air conditioners and refrigerators at levels of 3.0–5.0. For electronic component cooling, advanced emerging TE materials provide probability for improved efficiency or TECs.

As the fundamental theory of thermoelectricity became well established, the important role of heavily doped semiconductors as good TE materials had been accepted, and the TE material bismuth telluride was discovered and developed for commercialization, the effectiveness of a TE material in terms of the dimensionless TE figure of merit (*ZT*) was established, and the TE industry was launched. From 1960 to 1990, only incremental gains were made in increasing *ZT*, with $Bi_2Te_3$ remaining the best commercial material today, with $ZT \sim 1$. During this three decade period, the thermoelectrics field received little attention from the worldwide scientific research community. Nevertheless, the TE industry grew slowly but steadily by finding niche applications for space missions, laboratory equipment, and medical applications where cost and energy efficiency were not as important as energy availability, reliability, and predictability (Dresselhaus et al. 2007).

In the early 1990s, the US Department of Defense became interested in the potential of TEs for new types of applications, encouraging the research community to reexamine research opportunities for advancing TE materials to the point they could be used more competitively for cooling and power conversion applications. This attempt was successful in stimulating the research community to once again become active in this field and to find new research directions that would have an impact on future developments and would lead to TE materials with better performance. As a result of this stimulation, two different research approaches were taken for developing the next generation of new TE materials: one using new families of advanced bulk TE materials based on crystal structures that contain weakly bound atoms or molecules with large vibrational amplitudes at partially filled structural sites acting as effective phonon scatterers, such as skutterudites, clathrates, and oxides; and the other using low dimensional materials systems, such as quantum well superlattices, quantum wires, quantum dots, thin film or band engineering structures. During the 1990s these two approaches developed independently and mostly along different directions. And then the two approaches seem to be coming together again as the most successful new bulk TE materials are host materials containing nanoscale inclusions that are prepared by chemical approaches (Dresselhaus et al. 2005). Low-dimensional materials systems are now being assembled as nanocomposites containing a coupled assembly of nanoclusters showing short-range low dimensionality embedded in a host material, thereby producing a bulk material with nanostructures and many interfaces that scatter phonons more effectively than electrons (Dresselhaus et al. 2007).

## *Fermi Energy in Thermoelectric Materials*

The Fermi energy, of a substance is a result of the Pauli exclusion principle when the temperature of a material is lowered to absolute zero. By this principle, only one electron can inhabit a given energy state at a given time. At the temperature of absolute zero, all electrons in the solid attempt to get into the lowest available energy level. As a result, they form a sea of energy states known as the Fermi sea. The highest energy level of this sea is called the Fermi energy or Fermi level. At absolute zero no electrons have enough energy to occupy any energy levels above the Fermi level. In metals the Fermi level sits between the valence and conduction bands. The size of the band gap between the Fermi level and the conduction band determines if the metal is a conductor, insulator or semiconductor. Once the temperature of the material is raised above absolute zero the Fermi energy can be used to determine the probability of an electron having a particular energy level. The probability that an energy state will be filled is given by Femi distribution function (Nolas et al. 2001a, b):

$$f_0(E) = \frac{1}{e^{(E-E_f)/(k_B T)} + 1}, \tag{11.6}$$

where $E_f$ is the Fermi energy, $T$ is the absolute temperature, and $k_B$ is Boltzmann's constant. $k_B T$ is also proportional to the thermal energy of the particles. At a temperature of absolute zero, Fermi energy has the value $E_{f_0} = (h^2/2m)(3n/8\pi)^{2/3}$, where $n$ is the total number of free electrons per unit volume, $m$ is the mass of a free electron, $h$ is Planck's constant. As the temperature rises the probability of electrons with energy greater than the Fermi energy increases. These electrons take part in electric conduction which is an important application of the Fermi energy. There is a finite probability for electrons to exist in the band gap between Fermi level and the conduction band, but there is no energy state in the band gap because the density of state is zero. The Fermi energy is important to both electrical and thermal conductivities. Increased temperature leads to increased numbers of electrons jumping into the conduction band from the Fermi level. The electrons in the conduction band bump into each other causing electrical current, and heat from the collisions. With more electrons available, more current, and heat, can flow. Hence, the Fermi energy is intrinsic to both of these properties. For metals, the Fermi energy actually lies within the conduction band; hence, it takes relatively little energy to bump these electrons into the band gap, which is why metals are typically good conductors of both electricity and heat. For semiconductors, the Fermi energy lies between the valence band and conduction band, and more energy is required to excite the electrons sufficiently, and thus both electrical and thermal conductivity are lower than for metals. For insulators, the Fermi energy is considerably below the conduction band, which means that it takes an incredible amount of energy to raise an electron to the conduction band, and thus few electrons do; therefore, heat flow and electrical current are reduced significantly, in some cases to near zero. However, that the correlation between thermal and electrical conductivity is only valid for certain substances, specifically metals and metalloids. Diamond, for instance, has an incredibly high thermal conductivity but very low electrical conductivity. These differences are due to differences in structure, which makes the Fermi energy less important than other properties.

The Fermi energy represents certain minimum "zero-point" energy (the absolute minimum that can be reached) and a "zero-point" pressure. This pressure is due to the electrons moving around, even at 0 K. At high temperatures and low densities, classical pressure dominates. But at low temperatures and incredibly high densities, the Fermi energy and pressure become radically more important. The concept of Fermi energy has been widely used in explanation and development of TE materials.

## *Optimization Criteria of the Thermoelectric Materials*

Because the TE effect was discovered in the 18th century, it had taken many years to find useful applications and materials for this discovery. In the 1950s, Abraham Ioffe found that doped semiconductors showed a much larger TE effect than other materials. This led to research of binary semiconductors as TE materials in the 1950s, with $Bi_2Te_3$ having the greatest TE effect at room temperature. The research led a good understanding of the properties displayed by a good TE material. It was not until the 1990s that interest increased again due to new materials discoveries with the potential for excellent TE properties. Discovery of practical materials with $ZT > 2$ for $\Delta T = 400 \pm 50°C$ can yield fuel savings of 10% in vehicles. Additional fuel savings are possible using TEs to enable new climate control concepts, e.g. localized on demand cooling. The potential uses of TE materials such as Peltier refrigerators, or for power generation are limitless if the dimensionless figure of merit, *ZT* can reach a value of 3 or larger. There are currently a few applications in practice that require high reliability, most notably NASA's use of TE materials to cool deep space probes to temperatures as low as 160 K (Kimmel 1999).

The primary criterion for TE materials is the figure of merit given by $Z = \sigma S^2/k$, which depends on the Seebeck coefficient, *S*, thermal conductivity, $\lambda$, and electrical

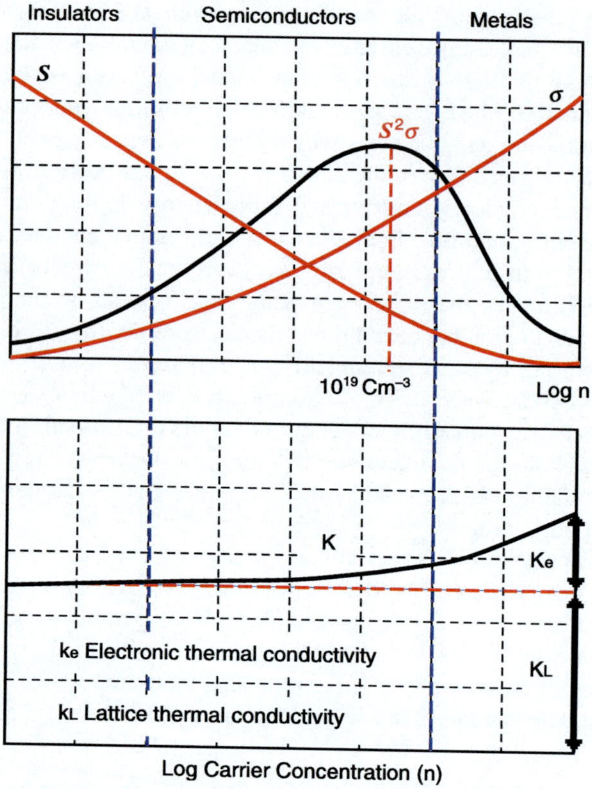


**Fig. 11.4** Sketch of the dependence of $S^2\sigma$ and $k$ on the concentration of free carriers. The relatively moderate values of electrical and thermal conductivity make semiconductors the best known thermoelectric materials (Rowe 1994).

conductivity, σ. In this equation, thermal conductivity and electrical conductivity are typically intertwined. Figure 11.4 shows a rough approximation of the dependence of TE properties on carrier concentration. It can be seen in the figure that semiconductors display the largest value of $\sigma S^2$, and they also have moderate values of $k$, which making the semiconductors the best known TE materials. The efficiency of modern TE materials still cannot compete with more conventional methods because the electrical conductivity is too low and the thermal conductivity is too high. A $ZT$ value of roughly 3 is needed in order for TEC systems to compete with vapor compression refrigerators (Venkatasubramanian et al. 2001).

In order to optimize the figure of merit, phonons which are responsible for thermal conductivity must experience the material as they would in a glass (experiencing a high degree of phonon scattering-lowering the thermal conductivity) while electrons must experience it as a crystal (experiencing very little scattering-maintaining the electrical conductivity). It is through the adjustment of each these properties independently of the other that the figure of merit can be improved. The selection and design rule for TE materials is likely to have high Seebeck coefficient and high *ZT*. Goal is to use these rules in the selection of materials for further investigation by first principles. First principles calculations are used to obtain electronic structure and vibrational properties. Boltzmann transport theory can be applied to first principles band structures to obtain electrical transport quantities, especially Seebeck coefficient, *S*(*T*). Apply linear response and direct methods for lattice vibrations thermal transport functions and thermal conductivity (Madsen and Singh 2006):

$$\text{Conductivity: } \sigma_x(T) = e^2 \int \mathrm{d}\varepsilon N(\varepsilon) {v_x}^2(\varepsilon)\tau(\varepsilon, T)(-f'(\varepsilon)), \tag{11.17}$$

$$\text{Thermopower: } S(T) = (e/T\sigma(T)) \int \mathrm{d}\varepsilon N(\varepsilon) {v_x}^2(\varepsilon)\tau(\varepsilon, T)\varepsilon(-f'(\varepsilon)), \tag{11.18}$$

$$\text{At low } T: S(T) = (\pi^2 k^2 T/3e\sigma)(\mathrm{d}\sigma/\mathrm{d}\varepsilon)|E_{\mathrm{F}}. \tag{11.19}$$

There, high *S* implies high log derivative of σ, i.e. strong energy dependence of electrical transport: band structure and scattering. More practical materials can be found by (1) Metallic conduction can be realized through mixed valence transition element such as Co, and very strong metal–oxygen hybridization. (2) High Seebeck coefficient at high carrier density can be reached with very narrow bands: A consequence of bonding topology (near right angle M–O–M bonds in edge sharing octahedra), and electron count is near crystal field gap. The dimensionless figure of merit *ZT* can be expressed by (Nolas et al. 2001a, b):

$$ZT = \left(\frac{e\alpha}{k_{\mathrm{B}}}\right)^2 \frac{1}{\dfrac{\sigma_0}{\sigma\beta} + \left(\dfrac{e}{k_{\mathrm{B}}}\right)^2 L} = \frac{e^2\alpha^2}{\dfrac{\sigma_0 k_{\mathrm{B}}^2}{\sigma\beta} + e^2 L} = \frac{\alpha^2}{\dfrac{\sigma_0}{\sigma\beta}\left(\dfrac{k_{\mathrm{B}}}{e}\right)^2 + L}, \tag{11.20}$$

where the dimensionless materials parameter $\beta = \sigma_0 T/\lambda_{\mathrm{L}}(k_{\mathrm{B}}/e)^2$.

Although large Seebeck coefficient values are important to good TE materials, other factors are also important. Because charge carriers must move through the material to transport heat, the material should conduct electricity well; otherwise, the deleterious effect of resistive heating will be enhanced. In addition, the material should act as a thermal insulator; the purpose of the device (when operated as a heat pump) is to produce a hot and cold region, therefore, a good thermal conductor will rapidly dissipate the temperature difference established. The best TE materials involve a trade-off among the three factors, combining a high Seebeck coefficient and electrical conductivity with low thermal conductivity. All three parameters are affected by the carrier concentration, *n*, of a solid. Carrier concentrations range from about $10^{14}$ to $10^{21}$ carriers/$\mathrm{cm}^3$ in a semiconductor, and are about $10^{22}/\mathrm{cm}^3$ in a metal. Electrical conductivity, σ, increases with *n*. The thermal conductivity, *k*, has two components, a lattice thermal conductivity $k_{\mathrm{L}}$ and an electronic thermal conductivity $k_{\mathrm{e}}$, such that $k = k_{\mathrm{L}} + k_{\mathrm{e}}$. The lattice component does not vary significantly with *n*; the electronic component increases with *n*. The Seebeck coefficient, $S$, generally decreases with increasing carrier concentration. The greatest $Z$ value is obtained with a carrier concentration between $10^{18}$ and $10^{21}/cm^3$. This implies that the best TE materials will be semiconductors with a relatively high carrier concentration. The choice of carrier type is also important. The direction of both the Seebeck and Peltier effects is reversed depending on whether the carriers are electrons or holes. If both carrier types are present in a material, their effects will work against each other. Semiconductors always contain both carrier types, but often the semiconductor is intentionally laced with impurities ("doped") so that one carrier type is greatly predominant. In this case, the semiconductor is said to be extrinsic. Intrinsic semiconductors, on the other hand, have roughly equal numbers of each type of carrier, causing their performance as TE materials to suffer. Extrinsic semiconductors, then, are the better choice for TE devices. Strategies to further improve the figure of merit of semiconductors generally involve decreasing the lattice thermal conductivity through a number of techniques that affect the microstructure of the material. These include solid-solution alloying of different semiconductors and dispersing inert particles in the semiconductor. Both treatments disrupt the regular ordering of the crystalline grains and decrease the ability of the material to carry heat through lattice vibrations. In another word, the mean free path of phonons (quantized lattice vibrations) in the material decreases. Narrow bandgap semiconductors are generally used for cooling and for power generation applications. Most Peltier coolers are made with alloys of bismuth telluride ($Bi_2Te_3$), antimony telluride ($Sb_2Te_3$), and/or bismuth selenide ($Bi_2Se_3$), the best materials to date for near-room-temperature operation. At higher temperatures, lead telluride (PbTe) is used. For power generation systems, which typically operate at still higher temperatures, silicon–germanium (Si–Ge) alloys are often used.

More specifically, the best TE materials will have a small electronic contribution to the thermal conductivity, $k_e$, thus maximizing the electronic conductivity while minimizing the thermal conductivity. The Weidmann–Franz law states that (Kimmel 1999):

$$k_e/\sigma T = L_0 = 2.44 \times 10^8 V^2/K^2. \tag{11.21}$$

Therefore, the electronic contribution of the thermal conductivity is a constant (Lorenz number $= L_0$) at a given temperature. The total thermal conductivity, is the sum of the electronic and lattice contributions $k = k_e + k_L$. Finding materials that minimize the lattice contribution to the thermal conductivity will maximize the dimensionless figure of merit (Kimmel 1999). Metals are poor TE materials because they have a low Seebeck coefficient, and large electronic contribution to thermal conductivity, so $S$ and $k$ will cancel each other out. Insulators have a high Seebeck coefficient, and a small electronic contribution to thermal conductivity, however their charge density and therefore electrical conductivity are low leading to a low TE effect. The best TE materials are between metals and insulators; i.e., semiconductors with an electronic density of $10^{19}/cm^3$. The next step is finding semiconductors with a low lattice contribution to the thermal conductivity, thus maximizing the dimensionless figure of merit. The materials with the lowest $k_L$

have a large number of atoms in the unit call, a large average atomic mass, and a large average coordination number per atom thus lowering the mean free path. Another method of reducing $k_L$ is finding a unit cell with cages, or voids, with loosely bound atoms that can "rattle" and thus scatter phonons. Based on these principles, four groups of materials have been explored: normal broadband semiconductors, semiconductors with rattling atoms or molecules, "correlated" metals or superconductors, and superlattices, also called multiple quantum well materials. Semiconductors with "rattling" atoms and multiple quantum wells have the greatest potential for improving the *ZT* value, and improved values of *ZT* have already been shown experimentally (Kimmel 1999).

There has been a substantial effort in finding p-type or n-type binary semiconductors with high dimensionless figures of merit. The current state of the art materials include $Bi_2Te_3$, $ZT = 0.9$ at room temperature; PbTe, $ZT = 1.2$ at 700°K; and SiGe, $ZT = 0.6$ at 1,000°K. Large values of electronic conductivity are necessary for good TE materials; thus large values of electron or hole mobility are needed. It can be shown that maximizing *ZT* is equivalent to maximizing $U/k_L$, where $U$ is the weighted mobility for electrons and holes $U = Nvm^{3/2}$, $N$ is the number of equivalent parabolic bands, ν is the electron or hole mobility, and $m$ is the effective mass. Covalently bonded semiconductors will have the highest effective mobility. To maximize *ZT*, superconductors are needed with $U > 0.02\ m^2/V\ s$, and $k_L < 3$ W/m K. These quantities are a good starting point to determine the TE efficiency of semiconductor materials. Ternary materials are one of research directions to maximize *ZT* (Kimmel 1999). The main approach to reducing the lattice contribution of the thermal conductivity has been to lower the mean free path by using heavy atoms within large unit cells. However, the lowest thermal conductivity is seen in amorphous structures. The idea is to find a material that group of the periodic table. To form the loosely bound void, the *R* atom radius has to be smaller than the void radius of the unit cell. By adding La – 1.79 Å, Nd – 1.70 Å, and Sm – 1.69 Å to the $IrSb_3$ binary (not filled) skutterudite unit cell with a void radius of 2.04 Å, an order of magnitude decrease in the $k_L$ was exhibited. The best dimensionless figure of merit at room temperature is only $ZT = 0.25$, but these structures show great promise at higher temperatures (Kimmel 1999).

Correlated metals and semiconductors, or Kondo insulators are described by rare earth intermetallic compounds usually containing Ce. Metals are not good TE materials due to a low Seebeck coefficient; however, this is not true for rare earth intermetallic compound where the 4f level is close to the fermi energy. These compounds have a high density of states near the Fermi energy, thus scattering conduction electrons. A simple power law no longer describes the scattering time, but exhibits resonant scattering. The Kondo effect, which relates to the scattering of conduction electrons, shows compounds with large Seebeck coefficients at low temperatures. Compounds such as $CePd_3$ exhibits the dimensionless figure of merit, $ZT = 1$ (Kimmel 1999).

Superlattices or multiple quantum well (MQW) structures are materials that will only conduct electrons in two dimensions, thus increasing the density of states with decreasing well widths. The increase in density of states will increase a materials figure of merit, assuming no change in the thermal conductivity. Lowering the well thickness and using MQW structures shows a large increase in Z. For instance, 20% improvement in the *ZT* of MQW structures has achieved versus bulk material. At quantum wells <10 Å, phonons will be scattered by the interfaces between quantum wells, thus lowering the thermal conductivity, and further enhancing the *ZT* value (Kimmel 1999).

Consequently, nanostructured materials such as superlattices, quantum dots, nanowires, and nanocomposites have shown promising TE properties. By exploiting nanoscale effects, these materials are able to obtain enhancements in TE properties which cannot be achieved in traditional bulk materials, resulting in large increases in the TE figure of merit. Although a high *ZT* has been reported, many of these materials are not practical for large-scale commercial use because they are fabricated by atomic layer deposition processes such as molecular beam epitaxy, making them slow and expensive to fabricate and restricting the amount of material that can be produced. Another type of nanostructured material, known as a bulk nanostructured material, is a material which is fabricated using a bulk process rather than a nanofabrication process, and has the important advantage of being able to be produced in large quantities and in a form that is compatible with commercially available devices. Thus far, of all the nanostructured materials only bulk nanostructured materials have been produced in enough quantity to be used in this manner; superlattice or quantum dot structures are currently only able to be produced as thin films. This practicality is a key reason why there is significant interest in bulk nanostructured materials and optimism that they can be used in a commercial setting. Indeed, efforts to commercialize the first generation of these materials are ongoing (Minnich et al. 2009).

On the other hand, for bulk TE materials, oxides and chalcogenides promise potentially low cost. For low-dimensional materials, highly textured or single crystal material may be required, and therefore increase costs. Anisotropies are often accompanied by poor mechanical properties. TE nanomaterials may combine the advantages of both bulk and low D materials. In addition, entire doping range with theory can be explored to reduce the need for extensive synthesis characterization to find optimal composition.

## *Bulk Thermoelectric Materials*

There has been substantial renewed research interest in the investigation of new and/or significantly more efficient TE materials for applications in solid-state refrigeration or power generation. Over the past 30 years, bulk TE materials based on the $Bi_2Te_3$ and Si–Ge systems have been extensively studied and optimized to perform a variety of solid-state TE refrigeration and power generation applications. There appears little room for improvement of the figure-of-merit from these materials. Therefore, entirely new classes of compounds need to be investigated if substantial material improvements are to be made. Some of the new materials that have been investigated include skutterudites, oxides, quantum well materials, superlattice structures, and low-dimensional and disordered systems. The enhanced interest in new TE materials has been driven by the need for higher performance and extended temperature regimes for TE devices in many civilian and military applications. As one of bulk TE materials, the skutterudite material system has drawn the most attention for TE applications (Nolas et al. 1999).

Due to the natural superlattice formed by the layered structure in homologous compounds (such as those of the form $(SrTiO3)_n(SrO)_m$ – the Ruddleson-Popper phase), oxides are also being considered for high-temperature TE devices. These materials exhibit low thermal conductivity perpendicular to these layers while maintaining electrical conductivity within the layers. The figure of merit in oxides is yet relatively low (~0.34 at 1,000 K), but the enhanced thermal stability, as compared with conventional high-*ZT* bismuth compounds, makes the oxides superior in high-temperature applications (Senthilkumar and Vijayaraghavan 2009).

## Established Materials: Bismuth Chalcogenides

These materials involve $Bi_2Te_3$ and $Bi_2Se_3$ and comprise some of the best performing TEs at room temperature with a temperature-independent figure of merit, *ZT*, between 0.8 and 1.0. Nanostructuring of these materials to produce a layered superlattice structure of alternating $Bi_2Te_3$ and $Bi_2Se_3$ layers produces a device within which there is good electrical conductivity but perpendicular to which thermal conductivity is poor. The enhanced *ZT* can reach approximately 2.4 at room temperature for p-type. Bismuth telluride, antimony telluride, and bismuth selenide have a nine-layer structure. They are composed of close-packed anions (Te or Se) with cations (Bi or Sb) occupying two-thirds of the octahedral holes. Let *A*, *B*, and *C* represent different relative orientations of the anion close-packed layers; and *a*, *b*, and *c* represent different relative orientations of the cations in the octahedral holes that lie midway between the close-packed layers. The structure can then be written as *AcBAcBaCBaCbACbA*. One of the most important features of $Bi_2Te_3$ is that high rate thermal diffusion of Cu, Ag, and Au can take place in its crystal lattice at relatively low temperature. These elements can act as donor impurities to move into or out of the lattice of $Bi_2Te_3$ in a TE device. For instance, during the manufacturing of TE modules, it is possible for Cu that dissolved in molten solder to pass through the nickel plated contacts into the thermoelements in several minutes. On the other hand, the copper can move out the lattice at ordinary temperature by emerging the thermoelements in a suitable aqueous sink. Because of weak bonding, and relatively large spacing between Te layers, the copper ions can also easily move from one site to another (Nolas et al. 1999).

With the same crystal structure as $Bi_2Te_3$, antimony telluride ($Sb_2Te_3$) and bismuth selenide ($Bi_2Se_3$) are particularly good TE materials, and furthermore, the addition of one or both to $Bi_2Te_3$ will improve its *ZT* by reducing the thermal conductivity. There is complete solid solubility among the three compounds $Bi_2Te_3$, $Sb_2Te_3$, and $Bi_2Se_3$. Within the $Bi_2Te_3$–$Sb_2Te_3$–$Bi_2Se_3$ pseudoternary system, the optimal compositions for TEC are normally $Bi_2Te_{2.7}Se_{0.3}$ and $Bi_{0.5}Sb_{1.5}Te_3$ for the n-type and p-type, respectively (Cui 2009).

Conventionally, many methods have been utilized for fabrication of TE materials, such as (1) a crystal ingot fabrication method in which starting materials are alloyed by melting, converted into an ingot and then sliced; (2) a powder sintering method in which a starting material powder or starting materials which have been alloyed by melting and then powdered, are molded and sintered, and sliced, if necessary; (3) a polycrystallization-zone melting method; (4) an amorphous fabrication method; and (5) a thin or thick film fabrication method. These methods have been used in some limited fields due to intrinsic drawbacks, for example, a long-term treatment of alloying by melting at high temperatures is needed, which resulted a low productivity and high energy consumption. In methods including a slicing step, super-small sized elements are difficult to produce, and in the poly-crystallization-zone melting method, undesirable electrical or mechanical anisotropy result from crystallization. Additionally, the conventional molding methods have limitations to get various desirable molded shapes.

An improved method have been developed to provide a TE material obtained by sintering a molding of a mixture prepared by copulverizing a material containing at least bismuth and a material containing at least tellurium without alloying by melting. The process for producing a TE material comprises copulverizing and intermixing a material containing at least bismuth and a material containing at least tellurium, directly, without alloying by melting, and then molding and sintering. Raw materials (starting materials) for the above TE material are those containing at least bismuth and those containing at least tellurium. In addition to bismuth and tellurium, antimony, selenium, an alloy of tellurium and antimony, can be used. These raw materials are preferably in powder form. The particle diameters of the materials are 100–150 mesh pass. In the case of powders having a large particle diameter, it is preferred to control the particle diameter to be within the above range by such techniques such as grinding. In connection with the type of the raw material and the mixing ratio, various embodiments are considered. Examples are Bi:Te = 2:3 (molar ratio), Bi:Sb:Te = 2:8:15 (molar ratio), or (BiSb):(TeSe) = 2:3 (molar ratio). Particularly, by compounding bismuth (Bi) or bismuth antimony (BiSb) and tellurium (Te) or tellurium selenium (TeSe) in a ratio of about 2:3, a TE material, exhibiting excellent performance at temperatures of 600 K or less can be obtained. As the raw material, metal elements which is not alloyed by melting, or an alloy or a compound which is obtained in the course of scouring can be used as long as it contains the above component. The above raw material preferably contains a suitable amount of dopants. Dopants used conventionally can be added to the raw material by the usual methods. For example, in production of an n-type (negative type) TE material, $SbI_3$, CuTe, $Cu_2$ S, CuI, CuBr, AgBr, etc. can be used. In production of a p-type (positive type) TE material, Te, Cd, Sb, Pb, As, Bi, etc. can be used. Particularly in the case that bismuth and tellurium are added in a ratio of about 2:3, from viewpoints of solubility and stability, $SbI_3$ is preferably used in the n-type, and Te is preferably used in the p-type. The amount of the dopant added is determined appropriately depending on the type and mixing ratio of the raw material, and the type of a substance as the dopant. It is usually 0.01–10 mol% and preferably 0.05–5 mol%. To the raw material or starting material powder which has been copulverized and intermixed, a substance having low thermal conductivity (thermal conductivity reducing agent) can be added. The figure of merit can be improved by adding such thermal conductivity reducing agents. Examples of such thermal conductivity reducing agents are $SiO_2$, $TiO_2$, $ZrO_2$, $B_2O_3$, BN, and $Si_3N_4$. The added amount of the thermal conductivity reducing agent is determined depending on the purpose of use, particularly upon temperatures to which an apparatus is required to cool. It is usually 0.1–50% by weight and preferably 0.5–30% by weight. By adding a thermal conductivity reducing agent, even if the figure of merit of a TE material is the same, the thermal conductivity drops and thus cooling to a lower temperature can be achieved (Ohta et al. 1992).

In the process, the raw materials thus compounded are sufficiently mixed by copulverization and intermixing. In this case, it is desirable that pulverization and intermixing are carried out at the same time to further reduce the particle diameter of the raw materials. The pulverization and intermixing can be carried out simultaneously by various techniques such as a ball mill, an impact fine pulverization machine, a jet pulverization apparatus, and a tower-type friction machine. In the case that a ball mill is used as the pulverization and mixing method, it is preferred that a planetary-type ball mill or similar is used instead of the conventional dropping-type ball mill. Copulverization and intermixing can be carried out either in dry or wet conditions. For example, in the case of a wet condition, alcohols such as ethanol or butanol and various solvents can be used as mixing aids. In connection with the mixing power or mixing time in the above copulverization and intermixing, they are desirably controlled so that the average particle diameter of the starting material powder after copulverization and intermixing is 0.05–10 μm. The starting material powder after copulverization and intermixing is compression molded into a desired form by means of applying pressure such as press molding without the processing of alloying by melting employed conventionally. This compression molding can be carried out by adding a binder component (such as polyvinyl alcohol) if necessary. The pressure at the time of compression molding is usually 0.2–20 ton/cm$^2$ although it varies with the type and particle diameter of the starting material powder. As molding methods, in addition to the above compression molding, any desired molding methods such as hot press, extrusion molding, injection molding, coating, and a screen printing method can be employed. It is necessary to apply sintering after molding. A sintered product obtained exhibits high performance as a TE material. This sintering is applied to a molding obtained by the above mentioned molding, under reduced pressure or atmospheric pressure, or under pressure, specifically in the range of $10^{-3}$ to $10^6$ Torr in an inert gas such as argon or nitrogen. The sintering temperature is chosen appropriately depending on the type and mixing ratio of the starting material, and it is usually 300–600°C. The heating-up rate, particularly at a temperature exceeding 200°C, and especially 400°C is preferably controlled to not more than 10 K/h. If the temperature is raised at a higher rate, the TE material obtained sometimes has reduced performance. If the heating-up rate is too slow, a long time period is needed to reach a predetermined temperature. For example, it is suitable to control the speed to about 3–10 K/h. The time of heating-up varies with the atmosphere under pressure and the composition, and is not necessarily limited to the above range. When the predetermined sintering temperature is reached at the above heating-up rate, the green molding is maintained at the temperature for the time needed to thereby sinter it, whereupon the desired TE material is obtained. This sintering time is usually 0.5–30 h. The figure of merit can be further increased by carrying out the above sintering treatment in a sealed condition, i.e., the green molding is not exposed to the atmosphere during sintering. The sealed condition can be produced by sealing the green molding in an ampule. At the time of sintering, the surrounding gas may comprise only one generated from the molding. The surrounding gas comprises another one, preferably inert gas, and more preferably reducing gas or a mixed gas of reducing gas and inert gas. As the inert gas, Ar (argon), is preferably used, and $H_2$, and CO are preferably used as the reducing gas. In sintering under a pressure of more than atmospheric pressure, it suffices that sintering with gas pressure in a nonoxidizing atmosphere is carried out. When sintering is carried out under a pressure of not more than atmospheric pressure, it suffices that a green molding is placed in a sealed container such as an ampule and after setting it at a predetermined pressure corresponding to a pressure increase generated by raising the sintering temperature, sintering is carried out at a raised temperature. In the case of reduced pressure (i.e., under a pressure of not more than atmospheric pressure), it is preferred for the sintering to be carried out in the presence of reducing gas. The reason why reducing gas is used as a preferred atmosphere is that the copulverized and intermixed starting material powder is relatively small in particle diameter and, as a result, is readily subject to oxidation and, therefore, the reducing gas is used to reduce the material. It is preferred that sintering is carried out under a pressure of not more than atmospheric pressure, thus a TE material obtained by sintering is decreased in density and is reduced in thermal conductivity. The TE material thus sintered is preferably annealed at a temperature of 50–200°C below the sintering temperature for a period of 0.5–30 h, if necessary. This annealing can be carried out in a reducing atmosphere and preferably in a sealed condition. In carrying out the reduction treatment in sintering or annealing, no special additional steps are necessary, and it suffices that the sintering or annealing is carried out in a reducing atmosphere. The figure of merit can be increased by carrying out the reduction treatment in at least any copulverizing and intermixing, sintering and annealing in the course of production of the above mentioned TE material. The reduction treatment usually proceeds by heating in a reducing gas such as $H_2$ or CO atmosphere. In this reduction treatment, a mixed gas of the reducing gas and inert gas such as Ar can be used. The reduction treatment is not critical for operation and can be carried out in various manners. Pressure conditions in the reduction treatment are also not critical. When the reduction treatment is carried out at the time of sintering or annealing, it is preferable that it be carried out in a sealed condition (Ohta et al. 1992).

The TE material can be easily obtained by copulverizing and intermixing metal elements as starting materials (particularly by the use of a planetary-type ball mill), molding, and then sintering. Particularly, because a powder element is used as a starting material, starting materials can be easily prepared. Furthermore, the TE material can be easily produced without a complicated procedure or a special apparatus, and thus the production cost of the TE material can be reduced. Additionally, the figure of merit can be improved by choosing specified sintering conditions. The thermal conductivity κ of the TE element produced by the process is not more than 1.4 W/m K and particularly 0.7–1.3 W/m K. Thus the TE element has the feature that it has lower thermal conductivity than that of the conventional TE element. The reason for this is that grain growth is less likely to occur at the time of sintering and a fine grain structure is obtained. Thus, a TE material is obtained having an increased figure of merit. Undesirable anisotropic properties and production loss do not arise. Moreover, because a material in any desired form can be directly produced, miniaturization is realized and one-body molding of a module is possible. Thus, the TE material can be used in a wide variety of fields such as TE power generation and TE cooling, a temperature sensor, space development, marine development, and electric power generation in the remote areas (Ohta et al. 1992).

## Skutterudite Thermoelectrics

Skutterudite materials have structures in the form of (Co, Ni, Fe)(P, Sb, As)$_3$ and are cubic with space group Im$^3$. Unfilled, these materials contain voids into which low-coordination ions (usually rare earth elements) can be inserted in order to alter thermal conductivity by producing sources for lattice phonon scattering and decrease thermal conductivity due to the lattice without reducing electrical conductivity. Such qualities make these materials behave with phonon-glass electron-single-crystal behavior.

The skutterudite material system possesses the basic conditions for high *ZT* values: a large unit cell, heavy constituent atom masses, low electronegativity differences between the constituent atoms, and large carrier mobilities. In addition, skutterudites form covalent structures with low coordination numbers for the constituent atoms and can thereby incorporate atoms in the relatively large voids formed. Therefore, compounds can be formed with atoms filling the voids of the skutterudite structure. Placing atoms in the interstitial voids of this crystal system would substantially reduce thermal conductivity by introducing phonon-scattering centers. Thus these atomic void-fillers would rattle around in their oversized cages, thereby providing an approach to drastically reduce thermal conductivity, in the highly conductive binary compounds and thereby maximize *ZT*. Another advantage of this cubic material system is that single crystals are not necessary to investigate the electrical and thermal transport properties of the skutterudites, which makes their promise for TE applications more feasible if the appropriate materials parameters are achieved. In addition, these materials are hard and have a relatively low coefficient of thermal expansion (Nolas et al. 2001a, b). These factors make the skutterudites attractive as TE materials.

The basic family of binary semiconducting compounds forming the skutterudite structure consists generally of $CoP_3$, $CoAs_3$, $CoSb_3$, $RhP_3$, $RhAs_3$, $RhSb_3$, $IrP_3$, $IrAs_3$ and $IrSb_3$. The skutterudite structure, indicated by $MX_3$ where M represents a metal atom and X a pnictide atom, belongs to the body-centeredcubic space group Im$^3$.

The crystallographic unit cell consists of eight $MX^3$ units, with the eight M atoms occupying the *c* sites and the 24 X atoms situated on the *g* sites. The structure can be uniquely specified from the lattice constant and the two positional parameters *y* and *z* specifying the *g* site. The skutterudite structure can be considered a distortion of the more symmetric cubic perovskite $ReO_3$ structure (Nolas et al. 1999).

The filled skutterudites crystallize in the cubic space group $Im^3$ with two formula units ($RM_4X_{12}$) per unit cell. There are three unique atomic positions in the normalized unit cell. The rare earth position is (0, 0, 0), the transition metal position is (0.25, 0.25, 0.25), and the pnicogen (P, As, Sb) position (0, *y*, *z*) is variable with $y \approx 0.35$, $z \approx 0.16$. The positions of the remaining 31 atoms in the unit cell are determined by the symmetry operations associated with the $Im^3$ space group (Sales 2003). By alloying with Fe on the Co site or Sn on the pnictide site, compounds with lanthanide-filling fractions between zero and one have been obtained. However, it is also possible to fractionally fill the binary skutterudites without any metal atom substitution. For example, with this approach, $La_{0.2}Co_4P_{12}$ and Ce0.25Co4P12, as well as $Ce_{0.1}Co_4Sb_{12}$ and $La_{0.23}Co_4Sb_{12}$ have been formed. The relationship between the metal atom charge state, the lanthanide valence, and structural and electronic stability of these compounds is subtle and important. A novel approach to skutterudite compound synthesis has resulted in the preparation of many compounds that could not have been successfully formed employing traditional synthesis techniques. The formation of metastable skutterudite compounds in thin-film form was achieved through controlled crystallization of amorphous reaction intermediates formed by low-temperature interdiffusion of modulated elemental reactants. In this approach the elements are deposited sequentially in layers thin enough that the layers interdiffuse before nucleation occurs. The formation of the desired metastable compound, for example $FeSb_3$, is achieved at low-temperatures from the amorphous intermediates that have the same composition. In this way the formation of more thermodynamically stable compounds, such as $FeSb_2$, can be avoided. By employing this technique, in addition to the binary skutterudite $FeSb_3$, filled skutterudites with the general formula $TFe_4Sb_{12}$ were formed, with T being all the lanthanide group elements as well as hafnium. This technique was also used in the synthesis of filled skutterudite with void fillers even smaller than the large mass lanthanides. These included Sn, Al, Ga, In, and Zn (Nolas et al. 1999).

Therefore, a large number of filled skutterudites have been developed and investigated. For instance, $CeFe_4Sb_{12}$ and $YbFe_4Sb_{12}$ are moderately heavy fermion metals. $CeRu_4Sb_{12}$ exhibits nonfermi liquid behavior and may be near a ferromagnetic quantum critical point. $PrOs_4Sb_{12}$ is the first example of a Pr compound that exhibits both superconductivity and heavy fermion behavior. $PrRu_4P_{12}$ and $SmRu_4P_{12}$ each undergo a metal-to-insulator transition and antiferromagnetic order. $PrFe_4P_{12}$ is an extremely unusual material in which quadrupolar order and heavy fermion ground states are extremely close in energy. At low temperatures a magnetic field ($\approx 4$ T) can drive $PrFe_4P_{12}$ between the two ground states. The density of states in $PrFe_4P_{12}$ is remarkably sharp in energy as evidenced by a two order of magnitude change in the Hall coefficient below 5 K and a huge value for $S$ at 5 K of $-130$ μV/K. Most of the La-filled skutterudites and two of the

Pr-filled skutterudites are superconductors with a maximum $T_c$ of 10.3 K for $LaRu_4As_{12}$. Several of the Ce-filled skutterudites are narrow gap semiconductors (also called Kondo insulators) where the gap is created by a strong hybridization between the Ce 4f level and the transition metal and pnicogen states near the Fermi energy. The Nd, Eu, Gd and Tb filled skutterudites order magnetically at temperatures ranging from 2 K for $NdFe_4P_{12}$ to 100 K for $EuFe_4P_{12}$. The coupling between the lanthanide magnetic moments and the conduction electrons is large in these compounds. This large coupling results in unusual peaks in electrical transport data near the onset of magnetic ordering, and in some cases multiple magnetic transitions. Finally, the TE figures of merit for the La, Ce and Yb filled skutterudites are among the highest values reported for any material at elevated temperatures (600–1,000 K). By varying the extrinsic carrier concentration in the filled skutterudites, the yielded maximum value for $ZT$ of 1.4 at 1,000 K, and a maximum $ZT$ value of 0.3 at 300 K. The $ZT$ values of the filled skutterudites are too small for room temperature applications. The relatively large band gap of these compounds ($\approx$0.6 eV) makes it unlikely that further research will result in a skutterudite-based TE material with properties better than the $Bi_2Te_3$-based materials currently in use near room temperature. Only at temperatures in the 600–900 K temperature range are the TE properties of the filled skutterudite antimonides of interest for use in power generation applications (Sales 2003).

The synthesis process of the skutterudites is rather complicated. The skutterudites do not melt congruently and involve pnicogens (P, As, Sb) that generally have high vapor pressures at the formation temperatures of the compounds. The high melting temperatures of Fe, Ru and Os coupled with the reactivity of the lanthanide metals with convenient crucible materials (e.g., $SiO_2$) makes the synthesis of many of these compounds difficult. Small single crystals of most of the lanthanide phosphides can be grown in a molten tin flux. For example, $LaFe_4P_{12}$ crystals were grown using La filings, Fe powder, red phosphorus, and Sn in the atomic ratio 1:4:20:50. The mixture was sealed in an evacuated silica tube, annealed for one week at 1,050 K, and slow cooled (2 K/h) to about 773 K, followed by rapid cooling to room temperature. A 1:1 mixture of HCl and water was then used to dissolve the Sn flux. Single crystals with typical dimensions from 0.1 to 2 mm can be grown by this method. Polycrystalline skutterudite phosphides have also been synthesized directly from the elements using a high pressure (4 GPa) and high temperature (800–1,200°C) wedge-type cubic-anvil apparatus (Shirotani et al. 1996, 1997; Sekine et al. 1997; Uchiumi et al. 1999). This technique avoids contamination from residual Sn flux. Lanthanide arsenides with the filled skutterudite structure were prepared: Lanthanide arsenides (RAs), were first prepared by reacting lanthanide filings with As in a silica tube at 900 K for 2 days. The RAs material was then ground together with the transition metal and excess As (R:T:As = 1:4:20) and sealed again in evacuated silica tubes, rapidly heated to 1,150 K and kept at temperature for 3 h. The samples were then annealed at 1,000 K for 4–7 days. The excess As was removed through sublimation. In spite of this elaborate synthesis procedure, the overall products were only 70–90% single phase. However, small single crystals of the skutterudite phase suitable for X-ray structure refinement could often be isolated from the reaction product. A dense single-phase sample of $CeFe_4As_{12}$ was prepared by a similar procedure followed by the densification of the powder with a hot-press (Watcharapasorn et al. 2002). Single-phase arsenides with a variety of compositions have also been prepared via direct high-pressure high-temperature synthesis from the elements (Shirotani et al. 1997). Synthesis of the lanthanide antimonides with the filled skutterudite structure have been synthesized using a procedure similar to that described above for the synthesis of the arsenides. The phase purity of the antimonides prepared in this manner was only about 80%. A better synthesis procedure was used (Sales et al. 1996, 1997): a thin layer of carbon was deposited on the inside of a round-bottomed silica tube by the pyrolysis of acetone. Stochiometric amounts of high purity lanthanide metal pieces (99.99% electropolished bar), Fe rod (99.9985%), and Sb shot (99.999%) were loaded into the precarbonized tube. The tube was sealed under vacuum at a pressure of $10^{-3}$ Pa and transferred into a programmable furnace. The silica ampoule was heated to 600°C at 2°C/min, left at 600°C for 3 h, and then slowly (0.5°C/min) heated to 1,050°C and left for about 20–40 h. It is important to slowly heat the tube because of the highly exothermic reaction between the lanthanide elements (particularly Ce) and antimony. The silica ampoule containing the homogeneous molten liquid was removed from the furnace and quenched into a water bath. The same ampoule (containing the prereacted elements) was then placed in a furnace and annealed at 700°C for 30 h to form the correct crystallographic phase. The completely reacted solid was removed from the silica tube and cleaned with a wire brush to remove small amounts of carbon from the surface. To form a completely dense polycrystalline solid, the reacted material was ball milled into a fine powder in an argon atmosphere, loaded into a graphite die, and hot-pressed (5,000 psi) in a helium atmosphere at 700°C for 40 min. This procedure results in single-phase and dense polycrystalline samples. Single crystals of the antimonides can also be grown using excess antimony as a flux (Chakoumakos et al. 1999). High purity elements in the ratio R:T:Sb = 1:4:20 are loaded into an evacuated carbon coated quartz tube. The tubes are heated to 900°C for 24 h and then cooled slowly (1–3°C/h) to 600°C, followed by a quench to room temperature. The excess Sb flux can be removed by etching in acid ($HCl:HNO_3$ = 1:1). Small quantities of new compounds with the skutterudite structure can also be synthesized using a clever nonequilibrium thin film method pioneered by D.C. Johnson and colleagues at the University of Oregon (Hornbostel et al. 1997). The new compounds are formed by the low-temperature interdiffusion of multilayer elemental reactants (Sales 2003).

### Oxides

Oxide TE materials have been rapidly developed since high TE performance of layered cobalt oxide $NaCo_2O_4$ reported in 1997. These p-type oxides in single crystal form have been proved to show $ZT \approx 1$. However, performance of n-type oxides had remained at a relatively low level $ZT \approx 0.3$ for Al-doped ZnO in 1996. The efforts have been made to overcome disadvantages of oxides such as low carrier mobility and high lattice thermal conductivity: electron correlation, quantum wells, and bulk nanocomposites, and promising results on n-type ZnO-based bulk nanocomposites showing $ZT \approx 0.65$ (Ohtaki 2002).

Metal oxides have attracted much attention as TE power generation materials at high temperatures on the basis of their potential advantages over heavy metallic alloys in chemical and thermal robustness. To clarify the intrinsic TE properties of oxides, high-quality epitaxial films of several TE oxides, including $Na_xCoO_2$, $Li_xCoO_2$, $Sr_xCoO_2$, $Ca_xCoO_2$, $Ca_3Co_4O_9$, $SrTiO_3$:Nb, $TiO_2$:Nb, and SrO($SrTiO_3$): Nb have been investigated. Two representative oxides, p-type $Ca_3Co_4O_9$ and n-type $SrTiO_3$:Nb, exhibit the best $ZT$ among these oxide TE materials.

Generally, a vapor-phase epitaxy (VPE) method, such as pulsed-laser deposition (PLD), is appropriate for high-quality epitaxial film growth of oxide materials. However, fabrication of a $Na_xCoO_2$ epitaxial film is very difficult by a conventional VPE method (growth temperature, $T_g \sim 700°C$) because revaporization of Na due to its extremely a large vapor pressure of $\sim 10^4$ Pa at 700°C occurs during film growth. A method of reactive solid-phase epitaxy, is a powerful means to fabricate single-crystalline films of layered oxides, and subsequent topotactic ion exchange. First, a high-quality epitaxial film of CoO was deposited on the (0001) face of an $\alpha$-$Al_2O_3$ substrate at 700°C by PLD using a $Co_3O_4$ sintered disk as a target. Then the PLD-deposited CoO film was heated together with a $NaHCO_3$ powder at 700°C in air to give a $c$-axis-oriented $Na_xCoO_2$ ($x \sim 0.8$) epitaxial film. The $Na_xCoO_2$ film can be converted into $Li_xCoO_2$, $Sr_xCoO_2$, $Ca_xCoO_2$, and $Ca_3Co_4O_9$ epitaxial films by the appropriate ion-exchange treatment.

Although very high $ZT$ values ~1 have been reported for p-type $Ca_3Co_4O_9$ and n-type $SrTiO_3$:Nb, reliable $ZT$ values of both materials were clarified to be ~0.05 at 300 K and ~0.3 at 1,000 K. Thus, nanostructural control such as an artificial superlattice or 2DEG may be necessary to obtain a high $ZT$ TE oxide. For example, utilizing a 2DEG in $SrTiO_3$ provide a new route to realizing practical TE materials without employing toxic heavy elements. A two-dimensional electron gas (2DEG) in $SrTiO_3$ demonstrates a Seebeck coefficient $S$ that is enhanced by a factor of ~5 compared with the bulk and an optimized $ZT$ that reaches 2.4, twice that of conventional TE materials (Tervo et al. 2009).

Other oxide materials such as $Na_2CoO_4$, $CaMnO_3$, (ZnO)($In_2O_3$), ZnO and $CuAlO_2$ appear promising because they are chemically stable at high temperatures and they have high oxidation resistance. They are also nontoxic. $Na_2CoO_4$ has especially been reported to show unusually high Seebeck coefficient at room temperature, i.e., 100 μV/K. General problems with oxide TEs, however, were reported to be weak mechanical strength, high contact resistance at interfaces of oxides and electrodes (Tervo et al. 2009).

## Flexible Graphite

Flexible graphite is quite unusual in its combination of compliance and strong TE behavior. Therefore, it can be used as a compliant TE material to reduce thermal stress between a TE cell and the wall of a heat exchanger of a TE energy conversion system, improving the thermal coupling as well as the system's durability. Flexible graphite is much more conductive thermally in the in-plane direction than in the out-of plane direction. The high in-plane thermal conductivity, together with the resilience in the out-of plane direction, helps the attaining of a good thermal contact between flexible graphite and a hot/cold surface. On the other hand, the low out-of-plane thermal conductivity is favorable for the Seebeck effect in the out-of-plane direction. The Seebeck effect in the through-thickness direction of flexible graphite can be used for the generation of electrical energy, as a Seebeck voltage is generated between the two opposite in-plane surfaces of a flexible-graphite sheet when the sheet is placed on a hot object or a cold object. The flexibility of the sheet facilitates the placement on a surface that is not flat. A related application is the sensing of the temperature of the hot or cold object (Luo et al. 2002). Further research would be need for feasible applications.

## *Low-Dimensional Thermoelectric Materials*

Low-dimensional materials, such as quantum wells, superlattices, quantum wires, and quantum dots offer new ways to manipulate the electron and phonon properties of a given material. In the regime where quantum effects are dominant, the energy spectra of electrons and phonons can be controlled through altering the size of the structures, leading to new ways to increase *ZT*. In this regime, the low-dimensional structures can be considered to be new materials, despite the fact that they are made of the same atomic structures as their parent materials. Each set of size parameters provides a novel low-dimensional material that can be examined, to a certain extent, both theoretically and experimentally, in terms of its TE properties. Thus, searching for high *ZT* systems in low-dimensional structures can be regarded as the equivalent of synthesizing many different bulk materials and measuring their TE properties. Because the constituent parent materials of low-dimensional structures are typically simple materials with well-known properties, the low-dimensional structures are amenable to a certain degree of analysis, prediction and optimization. When quantum size effects are not dominant, it is still possible to utilize classical size effects to alter the transport processes, as for example the exploitation of interfaces and boundaries to scatter phonons more effectively than electrons. Investigations on low-dimensional structures have exploited both quantum and classical size effects for electrons and phonons (Yang and Chen 2005).

The field of low-dimensional thermoelectricity started with the introduction of two strategies (Dresselhaus et al. 2007): the use of quantum-confinement phenomena to enhance $S$ and to control $S$ and $k$ somewhat independently, and the use of numerous interfaces to scatter phonons more effectively than electrons and to scatter preferentially those phonons that contribute most strongly to the thermal conductivity. Three additional concepts, including carrier-pocket engineering, energy filtering, and the semimetal–semiconductor transition have further advanced the potential for using low-dimensional materials to enhance TE performance. This was first demonstrated through a two-dimensional (2-D) superlattice consisting of PbTe quantum wells and $Pb_{1-x}Eu_xTe$ barriers. Experiments on cross-plane transport in $Bi_2Te_3/Sb_2Te_3$ superlattices demonstrated that the scattering of phonons by the interfaces reduced the thermal conductivity more than the electrical conductivity, thereby establishing proof-of-principle of this second concept, which has since been shown in practice to yield a great enhancement to *ZT*. Following the experimental demonstration of enhanced TE performance in 2-D superlattices, research moved forward in two different directions. In one direction, advances in superlattice design and growth were pursued, while in the second research direction ordered structures of lower dimensionality, such as 1-D quantum wires and 0-D quantum dots were investigated.

Superlattices have been found to be structures that improve *ZT*. The alternating layers of materials can be manufactured from good TE materials. Superlattice consists of several alternating nanosized layers, each less than 5 nm thick. These layers block the travel of atomic vibrations that produce heat flow but still let the electrons to flow as current. Adding interfacial phonon scattering sites, the thermal conductivity can be reduced. Fabrication methods include lithography and electroplating, which are widely used in semiconductor industry (Tervo et al. 2009). Figure 11.5 shows the example of fabricated micro TEC device with $Bi_2Te_3$ and $Sb_2Te_3$ deposited on Cr/Au/Ti/Pt hot (bottom) or cold connectors (Da Silva et al. 2003).

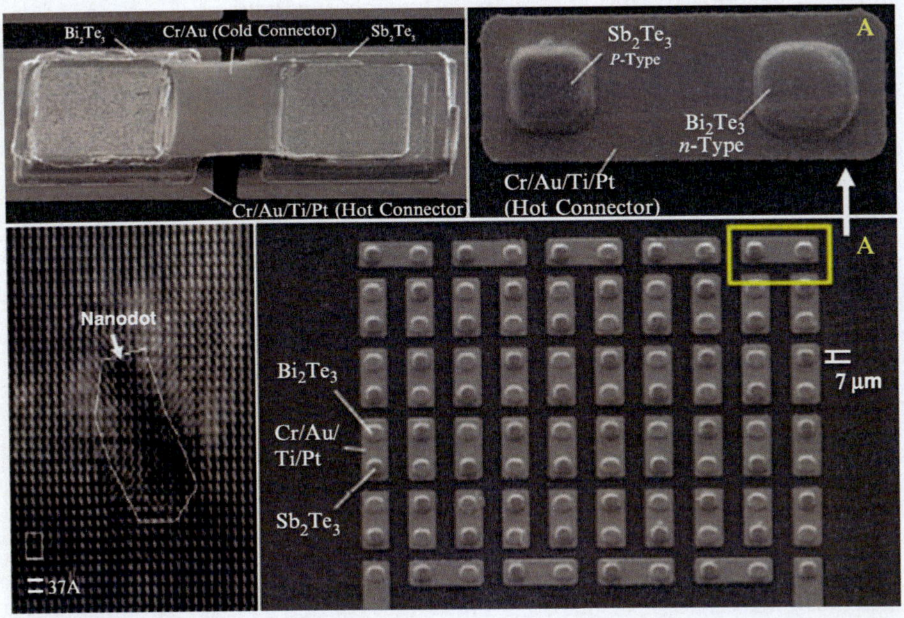


**Fig. 11.5** Example of fabricated micro thermoelectric cooler device with $Bi_2Te_3$ and $Sb_2Te_3$ deposited on Cr/Au/Ti/Pt hot (*bottom*) or cold connectors

In superlattices, the periodicity has three major effects on the phonon spectra: (1) phonon branches are folded due to the new periodicity in the growth direction, (2) mini band-gaps form, and (3) the acoustic phonons in the layer with a frequency higher than that in the other layer become flat, or confined due to the mismatch in the spectrum. The major impact of these changes is the phonon group velocity reduction in the cross-plane direction, which has been proposed as an explanation for the lowered thermal conductivity. For example, superlattices of such sandwich structures over thousands of periods were grown to produce a quantum-dot superlattice (QDSL) of composition $PbTe/PbSe_{0.98}$ $Te_{0.02}$ on top of a $BaF_2$ substrate followed by a relatively thin PbTe buffer layer. Using Bi as n-type dopant for this QDSL, values of $ZT \sim 1.6$ and 3.5 were achieved at 300 K and 570 K, respectively. Encouraging results were also reported for a p-type QDSL-based on using Na as the dopant. The very large $ZT$ values obtained by using this approach show that it is possible to both increase the power factor $S^2\sigma$ and to decrease the thermal conductivity at the same time in QDSLs, and therefore enhancing $ZT$. The superlattice structures appear to be more than twice as efficient as previous bulk TE materials. However, serious difficulties will be faced due to phenomena taking place in nanoscale. The biggest problem is reported to arise with electrical conductivity as a result of contact resistance and difficulties in predicting thermal conductivity. The problem requires mastering of electron-hole-phonon transport in nanoscale (Bulusu and Walker 2008). Thin-film TEC devices based on improved model system may someday be utilized for high-performance characteristics. In addition, the behavior of bulk thin films of SiGe and superlattice thin films of Si/SiGe was compared as microrefrigerators (Ezzahri et al. 2008). Although superlattice has enhanced electrical properties (larger TE power factor), the maximum cooling of thin film refrigerators based on SiGe alloys are comparable with that of superlattices. This is considered to be due to larger thermal conductivity of superlattices as compared to bulk SiGe thin films. Bulk thin films are also easier to grow by using methods such as chemical vapor deposition. Si-based microrefrigerators as well as other devices are attractive for their potential monolithic integration with Si microelectronics (Tervo et al. 2009).

Following the second research direction of going to lower dimensions, the study of quantum wires for TE applications was pursued (Dresselhaus et al. 2007). One material with very high potential for TE applications has for many years been Bi and Bi-related materials, because of the high S of the Bi. But unfortunately, Bi is a semimetal with both electron and hole carriers, and electrons and holes therefore contribute with opposite signs to its total S. To take advantage of the excellent electronic properties of the electron carriers in Bi materials, they would have to be prepared as an n-type semiconductor. The development of Bi-based quantum-well superlattices was, however, impeded by the difficulty in finding a suitable barrier material for Bi quantum wells in preparing 2-D quantum-well superlattices. Therefore, the development of Bi and $Bi_{1-x}Sb_x$ alloys as low-dimensional materials took the form of preparing ordered arrays of 1-D quantum wires inside the pores of anodic Al templates: Al is a well-behaved barrier material. The mechanism by which Bi can be converted into a semiconductor is the size-dependent semimetal–semiconductor transition. When the size of a semimetal nanowire decreases so that there are relatively few quantum states for the direction normal to the axis of the nanowire, then the energy bands split into discrete sub-bands. In this quantum regime, as the wire diameter decreases, the lowest conduction sub-band edge moves up in energy while the highest valence subband edge moves down in energy until these energy levels cross as the material makes a transition from a semimetal (with overlapping energy states for the lowest conduction band and the highest valence band) to a semiconductor with a band gap between the valence and conduction bands. In the semiconducting phase, the material can be doped to have one strongly dominant carrier type. As alloying Bi with Sb changes the electronic structure of the bulk alloy, the semimetal–semiconductor transition for a Bi–Sb nanowire, is dependent on both wire diameter and Sb concentration. By changing both the wire diameter and Sb composition, two variables can be provided for controlling and optimizing nanomaterials for enhanced TE performance (Dresselhaus et al. 2007).

Nanowires can be used for blocking free movement of phonons. As shown in Figure 11.6a, the electrochemically synthesized large-area, wafer-scale arrays of rough Si nanowires are 20–300 nm in diameter (Hochbaum et al. 2008). These nanowires have Seebeck coefficient and electrical resistivity values that are the same as those of doped bulk Si, but those with diameters of about 50 nm exhibit 100-fold reduction in thermal conductivity, yielding $ZT = 0.6$ at room temperature. For such nanowires, the lattice contribution to thermal conductivity approaches the limit for amorphous Si. Bulk Si is a poor TE material. However, by greatly reducing thermal conductivity without affecting much the Seebeck coefficient and electrical resistivity, Si nanowire arrays show promise as high-performance, scalable TE material. The similar result is as shown in Figure 11.6b, efficient TE performance was achieved from the single-component system of silicon nanowires for cross-sectional areas of 10 nm by 320 nm (Boukai et al. 2008). By varying the nanowire

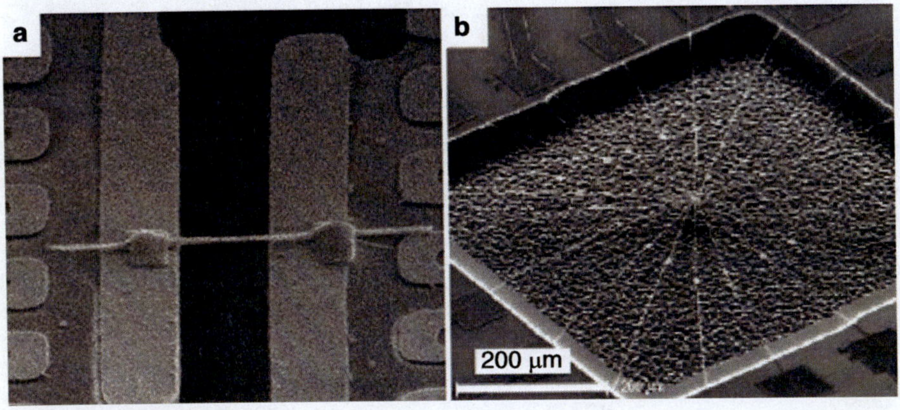


**Fig. 11.6** Examples of Si nanowire thermoelectric materials (Tervo et al. 2009): **(a)** Pt-nanowire; **(b)** overview of the suspension arrangement of Si nanowires

size and impurity doping levels, *ZT* values representing an approximately 100-fold improvement over bulk Si were achieved over a broad temperature range, including $ZT < 1$ at 200 K. Moreover, arrays of quantum dots were synthesized with each only a few nanometers in diameter (Harman et al. 2002). The materials were PbSeTe-based quantum dot superlattice structures grown by molecular beam epitaxy. Improved cooling values relative to the conventional bulk $(Bi,Sb)_2(Se,Te)_3$ TE materials were obtained using an n-type film in a one-leg TE device, which cooled the cold junction 43.7 K below the hot junction temperature of 299.7 K. The typical device consists of a substrate-free, bulk-like (typically 0.1 mm in thickness, 10 mm in width, and 5 mm in length) slab of nanostructured PbSeTe/PbTe as the n-type leg and a metal wire as the p-type leg (Tervo et al. 2009).

### *Thermoelectric Nanocomposites*

Figure 11.7 shows the historic progress of high efficiency TE materials. The maximum *ZT* has stayed stagnant at around 1 for all temperature ranges over 50 years since the important advancements using alloying approach. In the 1990s, two parallel approaches were proposed for the enhancement of the *ZT* of TE materials. The first of these approaches is based on new categories of advanced bulk materials,

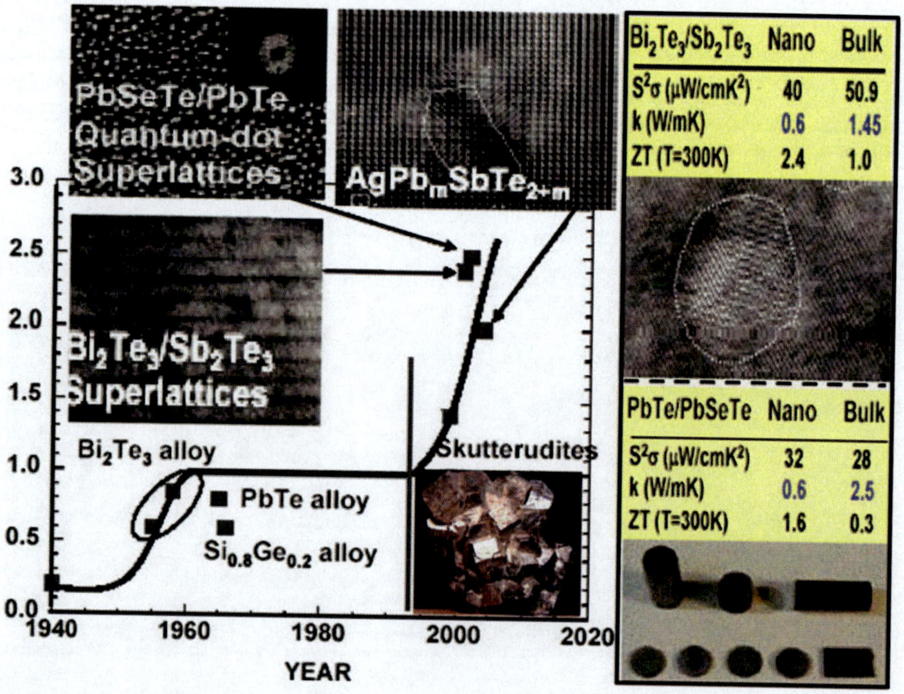


**Fig. 11.7** Advances in thermoelectric materials (Dresselhaus 2009)

with crystal structures that contain weakly bound atoms or molecules with large vibrational amplitudes (called ratters) at partially filled structural sites acting as effective phonon scatterers. Material systems, such as skutterudites (e.g., $CoSb_3$), clathrates (e.g., $Ba_8Ga_{16}Ge_{30}$) and Zintl phases belong to this category. The second approach is using low-dimensional materials (such as quantum well superlattices, quantum wires, and quantum dots) that would result in an enhancement of the $ZT$ by two mechanisms (Thiagarajan et al. 2009): (1) nanoscale features that introduce quantum confinement effects in the material lead to an enhancement of the power factor $S^2\sigma$, and (2) the use of the numerous interfaces in the nanostructures that scatter phonons more than the electrons, based on the difference in their respective scattering lengths, and thus reducing the thermal conductivity without adversely affecting the electrical conductivity as much.

However, the low-dimensional materials fabricated by atomic layer deposition techniques are not easily incorporated into commercial devices because they are slow and expensive to fabricate, and they cannot be fabricated in sufficient quantities. Modeling of phonon transport in superlattices indicated that the primary benefit from nanostructures, a reduced lattice thermal conductivity, does not require an atomically perfect interface or an exact geometry. All that is required is a material with a high density of interfaces, which can be present in any geometry. This discovery led to the idea of a nanocomposite, which retains the high density of interfaces but does not have a special geometry or structure, significantly simplifying the fabrication process and allowing the material to be produced in large quantities (Minnich et al. 2009).

A TE nanocomposite is a composite constructed by incorporating TE nanostructures in a matrix of a bulk TE material or compacting various TE nanostructures into bulk form. Several methods for the preparation of TE nanocomposites have been exercised. These methods to obtaining bulk samples with nanoscale features can be broadly classified into two categories (Thiagarajan et al. 2009): (1) compaction of nanoscale constituents (nanoparticles, nanowires, etc.) into bulk samples. The essence of all compaction techniques is to apply high pressure for densification, and often a rather high temperature to soften the material so that plastic deformation allows better filling and material flow by diffusion to remove the remaining porosity. The challenge is in achieving high density (and low porosity) without losing the nanoscale microstructure and keeping the material chemically pure. (2) In situ precipitation of nanoscale constituents by means of phase separation. The phase separation method of synthesizing nanostructures in situ in a bulk sample is inspired by precipitation process of a supersaturated solid solution, like hardening of aluminum. In this process, the TE materials will generally be heated up to the liquid-phase state, and then quenched to obtain a supersaturated solid solution or amorphous structure. The as-obtained metastable solid solution or amorphous structure will decompose into different dispersed phases after a nucleation and growth process by annealing at certain duration, and thus forms the nanoscale embedded precipitates in bulk matrix. The size of the precipitates increases as the duration and the temperature of the annealing process increase. Such spontaneous formation of nanoscale features is desirable because it minimizes the possibility of oxidation and the introduction of other forms of impurities, which would lead to degradation of electrical performance.

Therefore, either nanoparticles embedded in a host or a heterostructure geometry with nanoparticles of different materials adjacent to each other can be obtained. For the heterostructure geometry, when the two materials are the same the nanocomposite is essentially a material with nanometer-sized grains. Because these nanostructures have a size smaller than the phonon mean free path but greater than the electron or hole mean free path, phonons are more strongly scattered by the interfaces than are electrons or holes, resulting in a net increase in *ZT*. As nanocomposites are not formed by atomic layer deposition methods, they are significantly quicker and cheaper to fabricate than superlattices or other nanostructures, yet retain the same enhancements in TE properties. When the nanocomposite concept was still being developed, however, nanostructured materials which exhibited enhanced TE properties had only been created using atomic layer deposition techniques, and it was unclear whether any bulk process could create a thermodynamically stable, nanostructured material with an improved figure of merit. A still unresolved question is why certain nanostructured materials exhibit improved properties while others do not. Some fabrication methods, while capable of creating a nanostructured material, cannot produce a material with an improved *ZT* no matter how the fabrication conditions are adjusted (Minnich et al. 2009).

Despite these difficulties, several bulk processes have been developed which have successfully created stable nanocomposites with improved properties over those of their bulk counterparts. One method uses ball milling and hot pressing to create nanograined materials. In this technique, raw component elements, such as Bi and Te or Si and Ge, are ground into a nanoparticle dust using a ball milling machine. The resulting mixture is then hot pressed at an appropriate temperature and pressure. If the pressing temperature and duration are adjusted correctly, the nanoparticles will fuse together but leave the interface between each particle intact, creating a fully dense solid material with nanosize grains. This technique has been successfully implemented on n-type $Si_{80}Ge_{20}$, p-type $Si_{80}Ge_{20}$, and different types of $Bi_xSb_{2-x}Te_3$. The advantages of this method are that it is simple and inexpensive to fabricate the materials, and the technique can be applied to any material system. Another method to create a bulk nanocomposite is to use thermal processing techniques to induce the formation of nanoscale precipitates, an approach which has been successfully used in several material systems. By choosing the appropriate compounds and subjecting the material to a thermal processing procedure, a metastable solid solution of different elements can be made to undergo spinodal decomposition or nucleation and growth mechanisms to create nanoscale features. A related technique, known as matrix encapsulation, uses the fact that some materials are soluble in others in the liquid state but not in the solid state. By rapidly cooling the liquid mixture, the insoluble minority phase will precipitate, forming nanoparticles embedded in the host phase. These techniques have been applied with considerable success to several different solid solutions. High *ZT*s have been reported in n-type $AgPbTe_2/PbTe$, n-type $Pb_{1-x}Sn_xTe/PbS$, and p-type $Na_{1-x}Pb_mSb_yTe_{m+2}$ (Minnich et al. 2009).

Nanocomposites have been successful in increasing *ZT* by reducing the thermal conductivity below that of the bulk material while allowing for large scale, quick,

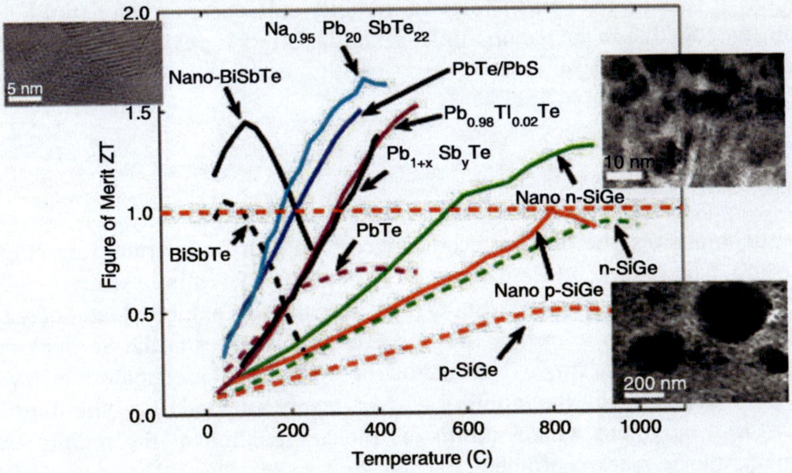


**Fig. 11.8** Figure of merit *ZT* of current state of the art thermoelectric materials *versus* temperature. The *dashed lines* show the maximum *ZT* values for bulk state of the art materials, and the *solid lines* show *ZT* values of nanocomposites (Minnich et al. 2009)

and inexpensive fabrication. Figure 11.8 summarizes the figure of merit *ZT* of the current state of the art TE materials versus temperature. The dashed lines show the maximum *ZT* values for bulk state of the art materials, and the solid lines show *ZT* values of some nanocomposites appropriated for different temperature ranges (Minnich et al. 2009). The first generation of nanocomposites has been developed, and efforts to use the materials in commercial devices are currently under way. However, developing the second generation of nanocomposites will require a more controlled design process where the TE properties of different nanostructures can be predicted and the desired structure created using a bulk procedure. The current design process of trial and error will not be sufficient. This goal can only be achieved with a fundamental understanding of transport in these complex materials and how their nanoscale structure affects bulk properties (Minnich et al. 2009).

While there is much to be understood about phonon and electronic transport in nanocomposites, this also means that there is still much room for innovation and improvement. Since the lattice thermal conductivity in nanocomposites has already been reduced, further improved materials will most likely feature some type of improvement in electrical properties combined with a reduction in lattice thermal conductivity. The flowing strategies to improve electrical properties might be used in the next generation of nanocomposite TE materials (Dresselhaus et al. 2007; Minnich et al. 2009).

### Increasing the Mobility

Nanostructured PbTe with encapsulated nanodots made of both Pb and Sb has an increased mobility over that of the bulk PbTe, resulting in a *ZT* a factor of two larger than that of bulk PbTe. This increase in mobility only occurs when both Pb and Sb are present together in the nanoparticle. A similar effect is possible in other material systems.

## Energy Filtering

In nanocomposites, the obvious scattering mechanism is electron grain boundary scattering, which does reduce the mobility but also preferentially scatters electrons with energy filtering. InGaAs/InGaAlAs superlattice exhibited an increase in power factor $S^2\sigma$ due to energy filtering effects. Increases in the Seebeck coefficient due to energy filtering in bulk nanostructured PbTe-based materials have also been reported, although the mobility was adversely affected. If a bulk fabrication process can be found which minimizes the degredation of the mobility while enhancing the Seebeck coefficient, the resulting material would have a net increase in power factor. This method is a good example of how nanostructuring can be used to improve electrical properties while still maintaining a low thermal conductivity.

## Resonant Levels to Enhance the Density of States

In most semiconductors, the addition of impurity atoms introduces states with energies in the band gap. If these states are very close to the conduction or valence band edges, as is the case for P or B in Si, the electrons (or holes) will be thermally activated into the conduction (or valence) bands and will be able to conduct. In some exceptional cases, however, the energy level created by the impurity can actually lie in the conduction or valence bands, creating a resonant level and a local maximum in the electronic density of states. For instance, it should be possible to create a material that uses both the resonant level and the nanocomposite concepts: one simply needs to add Tl to PbTe and generate the nanocomposite in the usual manner. The resulting material should have a lower thermal conductivity than the bulk value but significantly improved electrical properties due to the Tl doping. While Tl-doped PbTe did show a remarkable increase in $ZT$, Tl itself is highly toxic, making the commercial use of this material undesirable. Although resonant levels are not common, it is reasonable to believe that there are at least a few other materials in which impurities can introduce resonant levels near the band edges. These might be able to be identified using density functional theory or spectroscopic techniques to determine the impurity energy levels. These techniques together will hopefully help clarify the origin of the resonant level and allow researchers to fully realize the potential of this approach towards both enhancing the power factor and reducing the thermal conductivity.

## Reducing the Bipolar Effect

The major theme of nanocomposites has been a reduction of the lattice thermal conductivity. However, nanocomposites now have such a low lattice thermal conductivity that the other contributions to the thermal conductivity, the electronic thermal conductivity and the bipolar thermal conductivity, can be comparable in magnitude. The bipolar effect is a result of the fact that there are two types of carriers, electrons and holes. At high temperature many electrons are thermally excited from the valence band to the conduction band, leaving behind an equal number of holes. The presence of these two types of carriers allows heat conduction to take place even if the net electrical current is zero. Furthermore, because electrons and holes have opposite charges, the Seebeck coefficient is significantly suppressed if both electrons and holes are present. Thus, reducing the bipolar effect can improve the figure of merit in two ways. While many TE materials are usually so highly doped that this effect is not strong, for some materials at elevated temperature the bipolar thermal conductivity can be a significant component of the total thermal conductivity and the Seebeck coefficient can be markedly reduced.

## Reducing the Electronic Thermal Conductivity

The final strategy to enhance *ZT* is reducing the electronic thermal conductivity. At first, it might seem impossible to reduce the thermal conductivity from electrons or holes: after all, charge carriers must travel through the TE material, and when they do so they will inevitably carry heat. To implement this strategy it is necessary to search for materials with very strongly peaked density of states near the Fermi level. Another approach is to use quantum dots, as these 0-D structures inherently only have states at certain specific energies, giving a delta function density of states. The challenge here is to create quantum dots with the correct energy levels and in bulk quantities. Another way one could implement this strategy is to use the resonant level concept discussed earlier. The mechanism by which the Seebeck coefficient is enhanced is by a local maximum in the density of states contributed by the resonant level. If this effect could be enhanced, a simultaneous increase in Seebeck coefficient and reduction in electronic thermal conductivity can be obtained, leading to a dramatic increase in *ZT*.

Consequently, bulk nanocomposites have experimentally demonstrated a higher *ZT* than that of their bulk counterparts by reducing the thermal conductivity to values lower than were previously thought possible. More importantly for practical use, the materials can be fabricated inexpensively, quickly, and in a form that is compatible with existing TE device configurations. The first generation of bulk nanostructured materials is currently being commercialized, and it is promising that TE devices which use these materials will soon find more widespread applications. However, creating the second generation will require a fundamental understanding of carrier transport in these complex materials which is presently lacking. Continuous research in understanding carrier transport might lead to the next generation of bulk nanocomposites. Further research in this field will allow TEs to play a positive role in any future thermal management solution.

## Summary

TEC is a popular but inefficient way to remove heat from high-power-dissipating electrical components. The advantages of TEC are silent, compact, reliable, and durable. In addition, the cooling power can be modulated to maintain a fixed temperature. TEC uses Peltier effect to form a solid state heat pump, and works in conjunction with a heat sink or liquid cooling device to dissipate heat from an electronic packing system. A modern TEC is a semiconductor-based electronic component that functions as a small heat pump. By applying a low voltage DC power source to a TE module, heat will be moved through the module from one side to the other. One module face, therefore, will be cooled while the opposite face is heated simultaneously. A TE module may be used for both heating and cooling thereby making it highly suitable for precise temperature control applications.

The TE effect is the direct conversion of temperature differences to electric voltage and vice versa at the junctions of two different conductors. Traditionally, the TE effect encompasses three separately identified effects, the Seebeck effect, the Peltier effect, and the Thomson effect. These three effects, together with several other phenomena, form the basis of functional TE modules. A typical TEC component is an n- or p-type semiconductor (bismuth telluride) sandwiched between two conductors. The semiconductor is used because they can be optimized for pumping heat and because the type of charge carriers within them can be chosen. When an electronic field is applied to two ends of the n-type (doped with electrons) semiconductor, the electrons move towards the positive end to release heat, while the negative end absorbs heat. Therefore, the heat is transported in the direction of current flow. When a p-type semiconductor (doped with holes) is used instead, the holes move in a direction opposite the current flow. The heat is also transported in a direction opposite the current flow and in the direction of the holes. When the electrons fill a hole, they drop down to a lower energy level and release heat in the process. The most efficient configuration is where a p and n TE component is put electrically in series but thermally in parallel. When assembling it into an electronic package, one side is attached to a heat source and the other a heat sink that convects the heat away. In electronic packaging, TECs would be used to cool computer components to keep temperatures within design limits without the noise of a fan, or to maintain stable functioning when overclocking. A Peltier cooler with a heat sink or liquid cooling device can cool a chip to well below ambient temperature.

Peltier TEC performance is a function of ambient temperature, hot and cold side heat exchanger (heat sink) performance, thermal load, Peltier module (thermopile) geometry, and Peltier electrical parameters. TEC modules are considered to be highly reliable components due to their solid-state construction. For most applications they will provide long, trouble-free service. There have been many instances where TE modules have been used continuously for 20 or more years and the life of a module often exceeds the life of the associated equipment.

However, TECs in use today have a COP of only about 0.5. This is quite a low value compared with COPs of larger scale machines, such as air conditioners and refrigerators at levels of 3.0–5.0. For electronic component cooling, advanced emerging TE materials provide probability for improved efficiency of TECS. Two different approaches were taken for developing the next generation of new TE materials: one using new families of advanced bulk TE materials based on crystal structures that contain weakly bound atoms or molecules with large vibrational amplitudes at partially filled structural sites acting as effective phonon scatterers, such as skutterudites, clathrates and oxides; and the other using low-dimensional materials systems, such as quantum well superlattices, quantum wires, quantum dots, thin film, or band engineering structures. Low-dimensional materials systems are now being assembled as nanocomposites containing a coupled assembly of nanoclusters showing short-range low dimensionality embedded in a host material, thereby producing a bulk material with nanostructures and many interfaces that scatter phonons more effectively than electrons.

Bulk nanocomposites have experimentally demonstrated a higher *ZT* than that of their bulk counterparts by reducing the thermal conductivity to values lower than were previously thought possible. More importantly for practical use, the materials can be fabricated inexpensively, quickly, and in a form that is compatible with existing TE device configurations. The first generation of bulk nanostructured materials is currently being commercialized, and it is promising that TE devices which use these materials will soon find more widespread applications. However, creating the second generation will require a fundamental understanding of carrier transport in these complex materials which is presently lacking. Further research in this field will allow TEC to play a significant role in any future thermal management solution.

## References

Bar-Cohen A, Solbrekken GL, Yazawa K (2005) Thermoelectric powered convective cooling of microprocessors. IEEE Trans. Adv. Packag. **28(2)**: 231–239.

Bass JC, Allen DT, Ghamaty S, and Elsner NB (2004) New technology for thermoelectric cooling. 20th IEEE SEMI-THERM Symposium. http://www.hi-z.com/papers/IECEC 2004.pdf Accessed on 27 June 2010.

Boukai A et al (2008) Silicon nanowires as efficient thermoelectric materials. Nat. Lett. **451**: 168–171.

Buit RJ (1980) A simplified method for thermoelectric heat pump optimization. Third International Conference on Thermoelectric Energy Conversion, Arlington, Texas, May 12–14, 1980.

Bulusu A, Walker D (2008) Review of electronic transport models for thermoelectric materials. Superlattices Microstruct. **44**: 1–36.

Chakoumakos BC, Sales BC, Mandrus D, Keppens V (1999) Disparate atomic displacements in skutterudite-type LaFe3CoSb12, a model for thermoelectric behavior. Acta Crystallogr. B. **55**: 341–347.

Chein R, Huang G (2004) Thermoelectric cooler application in electronic cooling. Appl. Therm. Eng. **24(14–15)**: 2207–2217.

Cui Y (2009) Thermoelectric materials: ternary and higher oxides and tellurides. Ph.D thesis. University of Waterloo, Waterloo, Ontario, Canada. https://www.uwspace.uwaterloo.ca/bitstream/10012/4893/1/Cui_Yanjie.pdf. Accessed on 25 June 2010.

Da Silva LW et al (2003) Micro thermoelectric cooler fabrication: Growth and characterization of patterned Sb2Te3 and Bi2Te3 films. Proceeding of the 22nd International Conference on Thermoelectrics. La Grande-Motte, France, 2003, pp. 665–668.

Dresselhaus M (2009) Perspectives on recent advances in thermoelectric materials research. http://www1.eere.energy.gov/vehiclesandfuels/pdfs/thermoelectrics_app_2009/wednesday/dresselhauss.pdf. Accessed on 30 June 2010.

Dresselhaus MS et al (2005) New directions for nanoscale thermoelectric materials research. http://trs-new.jpl.nasa.gov/dspace/bitstream/2014/39339/1/05–3896.pdf. Accessed on 18 June 2010.

Dresselhaus MS et al (2007) New directions for low-dimensional thermoelectric materials. Adv. Mater. **19**: 1–12.

Ezzahri Y et al (2008) A comparison of thin film microrefrigerators based on Si/SiGe superlattice and bulk SiGe. Microelectronics J. **39**: 981–991.

Ferrotec (2010) Thermoelectric technical reference – Introduction to thermoelectric cooling. http://www.ferrotec.com/technology/thermoelectric/thermalRef01/. Accessed on 26 June 2010.

Goldsmid H (1986) Electronic refrigeration. Pion, London.

Harman T, Taylor P, Walsh M, LaForge B (2002) Quantum dot superlattice thermoelectric materials and devices. Science **297**: 2229–2232.

Hochbaum A et al (2008) Enhanced thermoelectric performance of rough silicon nanowires. Nat. Lett. **451**: 163–167.

Hornbostel MD et al (1997) Systematic study of new rare earth element–iron–antimony skutterudites synthesized using multilayer precursors. Inorg. Chem. **36**: 4270–4274.

Kim DS, Ferreira CAI (2008) Solar refrigeration options – A state-of-the-art review. Int. J. Refrig. **31**: 3–15.

Kimmel J (1999) Thermoelectric materials. http://physics.ucsd.edu/~phy152/ther.pdf. Accessed on 23 June 2010.

Luo X et al (2002) Electronic applications of flexible graphite. J. Electron. Mater. **31**: 534–544.

Madsen GKH, Singh DJ (2006) Boltz TraP: A code for calculating band-structure dependent quantities. Comput. Phys. Commun. **175**: 67–71.

Marlow Industries, Inc. (1998) Thermoelectric cooling systems design guide. http://www.lot-oriel.com/site/site_down/marlow_designguide_it01.pdf. Accessed on 28 June 2010.

Minnich AJ, Dresselhaus MS, Ren ZF, Chen G (2009) Bulk nanostructured thermoelectric materials: Current research and future prospects. Energy Environ. Sci. **2**: 466–479.

Nolas GS, Morelli DT, Tritt TM (1999) Skutterudites: A phonon-glass-electron crystal approach to advanced thermoelectric energy conversion applications. Annu. Rev. Mater. Sci. **29**: 89–116.

Nolas GS, Sharp J, Goldsmid HJ (2001a) Thermoelectrics: Basic principles and new materials developments. Springer-Verlag, Berlin

Nolas GS, Goldsmid H, Sharp J (2001b) Thermoelectrics: Basic principles and new materials developments. Springer, New York.

Ohtaki M (2002) Oxide thermoelectric materials: An overview with some historical and strategic perspectives. Oxide thermoelectrics 2002, pp. 159–180. Research signpost, Trivandrum, INDE (2002) (Monographie).

Ohta T, Kajikawa T, Uesuqi T, Tokiai T (1992) Thermoelectric material and process for production thereof. US Patent 5108515.

Roberts RB (1977) Absolute scale of thermoelectricity. Nature **265**: 226–227.

Rowe DM (1994) CRC handbook of thermoelectrics, CRC Press, Boca Raton.

Sales BC (2003) Filled skutterudites. In Handbook on the physics and chemistry of rare earths. Vol. 33. edited by K.A. Gschneidner Jr., J.C.G. Bilnzli and V.K. Pecharsky. Elsevier Science B.V., Amsterdam.

Sales BC, Mandrus D, Williams RK (1996) Filled skutterudite antimonides: A new class of thermoelectric materials. Science **272**: 1325–1328.

Sales BC, Mandrus D, Chakoumakos BC, Keppens V, Thompson JR (1997) Filled skutterudite antimonides: Electron crystals and phonon glasses. Phys. Rev. B. **56**: 15081–15089.

Sekine C, Uchiumi T, Shirotani I, Yagi T (1997) Metal-insulator transition in $PrRu_4P_{12}$ with skutterudite structure. Phys. Rev. Lett. **79**: 3218–3221.

Senthilkumar M, Vijayaraghavan R (2009) High-temperature resistivity and thermoelectric properties of coupled substituted $Ca_3Co_2O_6$. Sci. Technol. Adv. Mater. **10**: 1–5.

Shirotani I, Adachi T, Tachi K, Todo S, Nozawa K, Yagi T, Kinoshita M (1996) Electrical conductivity and superconductivity of metal phosphides with skutterudite-type structure prepared at high pressure. J. Phys. Chem. Solids **57**: 211.

Shirotani I et al (1997) Superconductivity of filled skutterudites $LaRu_4As_{12}$ and $PrRu_4As_{12}$. Phys. Rev. B. **56**: 7866–7869.

Simons RE, Chu RC (2000) Application of thermoelectric cooling to electronic equipment: A review and analysis. Annual IEEE Semiconductor Thermal Measurement and Management Symposium **19**: 1–9.

Tervo J, Manninen A, Ilola R, Hänninen H (2009) State-of-the-art of thermoelectric materials processing. http://www.vtt.fi/inf/pdf/workingpapers/2009/W124.pdf. Accessed on 28 June 2010.

Thiagarajan SJ, Wang W, Yang R (2009) Nanocomposites as high efficiency thermoelectric materials. http://spot.colorado.edu/~yangr/Publications/Yang_Thermoelectric_Nanocomposites_Annul_Review_of_Nanoresearch_2009_Final.pdf. Accessed on 29 June 2010.

Uchiumi T et al (1999) Superconductivity of $LaRu_4X_{12}$ ($X = P$, As and Sb) with skutterudite structure. J. Phys. Chem. Solids **60**: 689.

Venkatasubramanian R, Siivola E, Colpitts T, O'Quinn B (2001) Thin-film thermoelectric devices with high room-temperature figures of merit. Nature **413**: 597–602.

Watcharapasorn A et al (2002) Preparation and thermoelectric properties of $CeFe_4As_{12}$. J. Appl. Phys. **91**: 1344.

Yang R, Chen G (2005) Nanostructured thermoelectric materials: From superlattice to nanocomposites. http://spot.colorado.edu/~yangr/Publications/J8_MaterialIntegration.pdf. Accessed on 29 June 2010.

# Chapter 12
# Development and Application of Advanced Thermal Management Materials

**Abstract** As the demand for ease and access-speed increase, managing the thermal issues of the advanced electronic packaging would pose significant challenges. Each customer premise has its own special requirements that may complicate the required cooling scheme. The cooling constraints associated with customer premises impose stringent necessities on the component, board and system level power dissipation and cooling options. Together, these constraints create a set of opportunities and challenges that will dictate the thermal management requirements for the products for the next generation. In response to these needs, there have been revolutionary advances in thermal management materials and cooling approaches, such as low coefficient of thermal expansion (CTE), low-density materials with thermal conductivities up to 1,700 W/m K; increased reliability; reduced junction temperatures, reduced cost and weight; low CTE, thermally conductive printed circuit boards (PCBs), potentially eliminating the need for underfill; CTE matching allows direct attach with hard solders. There are a large and increasing number of microelectronic and optoelectronic applications with advanced thermal management technology, including: PCBs and PCB cold plates; heat sinks; microprocessor, RF (Radio Frequency) and power modules; heat spreaders and sinks; laser diode and light-emitting diode (LED) modules; thermoelectric coolers (TECs); plasma and liquid crystal displays (LCDs); detectors; and photovoltaics. This chapter covers the development roadmap of advanced thermal management materials, the growing array of applications, and future trends.

## Materials Development Routine and Methodology

Emerging electronic packaging technologies have an increasing demand for thermal management materials with new and dramatically improved properties. Achieving optimal performance with advanced thermal management materials may require cost-effective fabrication methods that incorporate novel chemicals, synthetic techniques, and metrology methods. The evaluation and successful integration of emerging materials within targeted applications will depend on optimal design in, and effective control for, specific sets of critical material and interface

X.C. Tong, *Advanced Materials for Thermal Management of Electronic Packaging*, Springer Series in Advanced Microelectronics 30,
DOI 10.1007/978-1-4419-7759-5_12, 

properties. Therefore, the development of a characterization and modeling infrastructure that supports the evolution of robust quantitative structure property correlations represents a strategic priority. These integrated tools would generate a more foundational understanding of novel thermal conduction mechanisms, contact resistance formation, and transport mechanisms and they would also drive new synthetic methods and cooling device concepts. Similarly, the fabrication of nanostructured materials, such as nanotubes or nanowires, for cooling devices or thermal interface interconnects would benefit from an improved understanding of the impact of process and structure on the resulting electronic, mechanical, thermal, and interface properties. In addition, modeling and simulation, aligned with validation experiments, will be needed to optimize the chemical and structural attributes that improve critical material properties and performance in targeted applications. As a result, the build-up of materials development routine and methodology would be necessary, which may cover establishing application target, materials requirements, and materials selection with optimal balance of cost and performance; design-in methodology of thermal managing components; development of lean processes; prototype fabrication and experimental validation; as well as production layout and quality assurance.

## *Establishing Application Target and Materials Requirements*

Establishing application target and materials requirements are primary to develop or optimize an efficient, cost-effective, and integratable thermal management solution, i.e. materials, techniques, mechanisms, or any combination thereof, for realizing significant improvement in heat removal from advanced high power and/or small form factor electronic devices. Because these devices mostly call for smaller and lighter, yet more powerful, electronic components, systems, and subsystems, one of the most significant and inherent challenges lies in ability of the components and systems to dissipate sufficient amounts of heat to enable the device and system to: (1) maintain thermal stability, (2) protect itself and neighboring components from thermal damage, and (3) adhere to the system's reliability requirements, which are impacted by its components' adherence to device-specific operating temperatures. As a result, broad electronic packaging industry requirements exist for new package component materials with thermal conductivity value generally stated as exceeding typical values for copper: >300 W/m K (through-plane). Package component materials that offer targeted CTE values and are capable of adequately dissipating heat loads from high heat flux semiconductor devices are required for a number of device types (Saums et al. 2005): (1) wide band gap (WBG) silicon carbide (SiC), gallium arsenide, and gallium nitride; (2) RF and MW (Microwave); (3) insulated-gate bipolar transistors (IGBTs) and other power semiconductor modules; (4) processors, ASICs (Application-Specific Integrated Circuits), and other integrated circuits (ICs). Some composites and hybrid structures have been developed to provide selective thermal conductivity increases with targeted CTE materials, for cost-effective high heat flux semiconductor module packaging. For example, a composite material structure utilizes both high-conductivity diamond pin inserts and very high in-plane conductivity graphite inserts within a cast Al-SiC metal matrix composite structure.

In order to develop or optimize the thermal management solution of a tactical radio system, for instance, it first needs to identify the performance and design requirements of the targeted joint tactical radio system (JTRS) component, subsystem, or line replaceable unit (LRU), and then utilize thermal modeling and simulation (M&S) to study the thermal characteristics of the identified JTRS system, subsystem, or LRU; model the identified communications components, systems, or subsystems with the proposed thermal management solution and simulate, characterize and determine the degree of improved thermal dissipation performance; develop and/or tailor the proposed thermal management solution to meet the requirements or changes identified in M&S; apply the proposed thermal management solution to a mockup JTRS system, subsystem, or LRU and retest; compare the predicted results from M&S with tested results from a mockup JTRS system, subsystem, or LRU. The thermal management solution shall provide one or more of the following benefits versus traditional heat sink/thermal management solutions (DOD 2010): A minimum 25% reduction in thermal management device/space volume, with comparable thermal performance; a minimum 20% increase in heat dissipation performance (e.g., 20% increase in power handling capability) while maintaining LRU form factor; and other quantitative benefits that demonstrate the clear advantages of the proposed solution. It is anticipated that the thermal management solution would be realized through the development and application of innovative thermal management materials, such as passive graphite foam, carbon nanotubes (CNTs), synthetic jets, and phase-change materials. However, the proposed solution may also, in conjunction with or in place of a material solution, pursue the development and application of thermal management techniques/mechanisms, such as heat pipes, evaporative cooling, and heat sinks. The proposed thermal management solution must be capable of meeting the environmental, safety, and health requirements outlined in industrial standards. Through delivery of the solution prototype and its specifications and test results, a closeout demonstration can be used to verify that the solution has been built to and satisfies thermal management objectives. The proposed and tested thermal management solution can then be integrated into actual target system or LRU for military application with improved thermal management within smaller form factor and/or more powerful tactical radios, for example.

Commercially, the market for improved thermal management exists in almost every sector of industry. Smaller and more powerful electronics have increasingly required innovative thermal management solutions through development of advanced thermal management materials, such as high through-plane thermal conductivity CTE-compatible materials capable of rapidly moving high heat loads from high heat flux hotspots. High in-plane thermal conductivity within such CTE-compatible materials is required to move high-heat flux loads in an *X*–*Y* orientation, for adequate spreading. Improvements in through-plane and in-plane thermal conductivity are needed in materials that have compatible and/or tailored CTE properties, to provide CTE compatibility with semiconductor package material sets. Purpose of in-plane spreading within a base plate, lid, or equivalent component is to maximize thermal performance of the heat sink solution employed, either liquid or air cooled. Maximizing in-plane spreading assists in remaining within an air cooling regime. A cost-effective, net-shape, high-volume capable manufacturing process is required to maintain lowest possible cost. High structural strength and stiffness and low density are important requirements. Material compatibility with soldering processes and other downstream assembly processes are also important requirements (Saums et al. 2005).

Semiconductor packaging materials which have CTE compatibility to other packaging components include Tungsten–copper alloys (W–Cu), Copper–molybdenum–copper laminates (CuMoCu), Kovar™, Al–SiC metal matrix composites, diamond–copper, diamond–silicon-carbide, SiC, and other SiC-based composites. Thermal conductivity enhancements for CTE-compatible semiconductor packaging materials have been introduced to manufacturing for structural components. These enhancements include the use of high thermal conductivity inserts, such as thermo-pyrolytic graphite (including annealed pyrolytic graphite or highly oriented pyrolytic graphite, HOPG), carbon fiber bundles, chemical vapor deposition (CVD) diamond, and diamond composites.

Semiconductor packaging materials with high thermal conductivity materials used in volume production today are limited to (1) copper, including OFHC (Oxygen-Free High Conductivity), and a variety of copper alloys; (2) Al–SiC metal matrix composites with TPG™ inserts; (3) SiC and diamond–silicon carbide; (4) CVD diamond submounts solder attached to ceramic substrates for optoelectronic and laser diode (LD) applications. Combining high thermal conductivity with specific CTE compatibility is limited to a small subset of semiconductor packaging materials, such as Al–SiC metal matrix composites with TPG and carbon fiber inserts, CVD diamond, and diamond composites. Market requirements for new copper replacement material have been driven by these industry applications segments (Saums et al. 2005): (1) Direct die solder attach for extremely high heat flux WBG semiconductors for military market. (2) High heat flux LED array submounts for solid state lighting. (3) Lids for traditional processor and IC packages: single chip modules and MCMs (Multi-Chip Modules). (4) Substrates for direct die solder attach for IC packaging designed for high heat fluxes. (5) RF and wireless module substrates. (6) High heat flux TEC submounts for optoelectronics modules. (7) IGBT power semiconductor modules for high reliability applications with direct bond copper on dielectrics. (8) Applications with extreme light weight requirements. (9) Active cooling system components for high heat flux direct die solder attach requirements. These market requirements include applications with both high and low volume production rates.

Establishing a base line package material with high conductivity core may be a promising option in advanced electronic packaging, as market requirements for new copper replacement material is driven by multiple packaging requirements

(Saums et al. 2005): (1) Achieve a reduction in package component material CTE to one or more values that are below the high relative value of copper. (2) Achieve a range of tailorable CTE values that will put alumina and aluminum nitride in slight compression. (3) Achieve thermal conductivity value greater than copper. (4) Maintain equivalent stiffness and strength to steel. (5) Achieve a reduction in density and component weight below that of copper and tungsten–copper. (6) Market requirements desire CTE compatibility combined with high thermal conductivity value: Generalized target conductivity value: >300 W/m K; Typical target CTE value ranges: 7.0–9.5 ppm/°C; 11.7–13.0 ppm/°C; Specific CTE targets for die matching: 2.0–5.5 ppm/°C; set associated target material values of critical importance across industry applications; baseline material manufacturing process must be cost-effective and high-volume capable; structural components require a variety of plating processes, surface finishes, feature sizes and capabilities, and attachment processes.

Al–SiC metal matrix composites can be one of the baseline package materials, which provide a combination of desirable material properties and high volume manufacturing processes with necessary attributes. Al–SiC is a metal matrix composite available commercially since 1990 (Saums et al. 2005). Selection of a suitable manufacturing process is the key to successful implementation in volume applications. Net-shape pressure infiltration casting process is proven as high volume capable and cost-effective. Metal matrix composites offer tailored CTE values appropriate for IC lids for attachment to package substrates manufactured from: Organic (FR-4) materials, HTCC (High Temperature Cofired Ceramics) ceramics (alumina, aluminum nitride), and LTCC ceramics (high-CTE value LTCC (Low Temperature Cofired Ceramics) products such as Kyocera GL550). Manufacturing process allows in situ casting of high conductivity core materials. Process is net shape, cost effective, volume capable, and compatible with a variety of plating and surface finishes and attachment processes. Cast components yield feature tolerances better than aluminum die casting, eliminating most machining cost.

Meanwhile, CVD diamond provides one of promising material enhancements with the following characteristics (Saums et al. 2005): Manufactured of 100% freestanding polycrystalline diamond; formed by CVD DC torch reactors; thermal conductivity: 1,400 W/m K (maximum), and 1,000–1,200 W/m K (typical); CTE: 1.5–1.8 ppm/°C; electrical resistivity: 1,015 Ω/cm (typical); density: 3.5 g/cm$^3$; cost in volume: approximately USD 1.00/mm$^3$; manufacturing specifications: typical sizes as insert pins in following examples:1.2 mm × 2.0 mm (L), 1.0 mm × 2.0 mm × 1.78 mm, and corner radius tolerance: <3 μm.

Emerging materials such as low-dimensional materials (nanotube, nanowire, graphene, and graphitic structures), macromolecules, self-assembly mechanisms and self-assembled materials, complex metal oxides, spin materials, interfaces, and heterointerfaces have exhibited potential in multiple application areas including emerging electronic devices and thermal management components. With the development and application of these materials, more and more innovative thermal management solution will come out with modified marketing targets and performance requirements.

## *Materials Selection with Optimal Balance of Cost and Performance*

Although advanced thermal management approaches are employing new materials and technologies, the constraints and design requirements have not changed greatly. Thermal management would not be the sole driving force behind new product designs. It must be subservient to other requirements and constraints. The major design goals for any thermal management solution are performance, cost, physical size, and reliability. An elegant technical approach will not see the light of day if it is not cost-effective and provide the appropriate levels of reliability and maintainability. As discussed in previous chapters, the use of technologies beyond forced-air cooling will be required for the next major generation of mainstream products. At the same time, the cost of cooling is beginning to be recognized as a major issue. The design selected for costing should represent the optimal balance (best value) between technical performance, reliability, acquisition cost, and operating cost. The cost of the best value design should be as low as is practicable, feasible, and reasonable, and consistent with performance and reliability (Garbincius 2006).

When choosing materials for device design, cost will always be a major factor. Cost estimates should be prepared in acknowledgment that there will be a continuous and aggressive program to drive down costs both in the design and in the fabrication phases of the project. While an estimator will endeavor to make his/her estimate clear and supportable, this should not be done at the expense of a higher cost. The detailed design and manufacture of components and systems may take place months or years into the future. In this interval requirements will be simplified, tolerances will be opened up as large as manageable, standard components will be used wherever possible, value engineering and cost trade-off studies will have been undertaken, and teams will have worked with potential suppliers to obtain the desired performance with the most cost-effective designs. Cost estimates should be prepared with the assumption that this process has occurred (Standford 2000).

Cost estimators should select a cost model that will result in the most cost-effective product or service for thermal management design. In the case where an industrial capability exists or could be developed, it is likely that an industrially obtained item will be more cost-competitive than the same item built or performed by a development department. Where industrial competence exists, much of the commercial products may also be more effectively performed by industry. In other cases, having a laboratory do the design and industry performing on a build-to-print contract would work the best. Cost estimates should be predicated on commonality and standardization to the maximum extent possible. Custom designs for every situation will be prohibitive financially. Using an available off-the-shelf part or condensing the number of custom parts or styles down to the absolute minimum will be necessary in order for the thermal management solution to be economically feasible.

The standardization of integrated thermal management and power electronics module design holds the key to large cost reductions and performance enhancements. It replaces complex power electronic circuits with a single module, reduces development and design costs for complex power circuits, and simplifies development and design of large electric power systems. Other key market-driving forces for pushing a standard module include better reliability, smaller footprints, and easier system integration.

## *Thermal Modeling and Design-In Methodology*

Modeling and simulation tools are used early in the design process to help understand the thermal performance of both passive and active cooling technologies and materials and the final reliability of the product. Thermal-mechanical-electrical modeling for interconnections and packaging is crucial as performance and reliability of integrated circuits is increasingly affected by interconnects and packaging. Electrical, thermal, and mechanical properties highly interact with each other and must therefore be simulated together. Reliability issues requiring modeling include electromigration, stress voiding, integrity and adhesion of thin films, surface roughness, package fracture, and corrosion. The capability to withstand the heat produced in the IC and to transport it off the chip is getting a top-level concern with further increasing densities. New materials are being introduced for thermal management design, and thermal modeling of the new materials in different levels of the electronic packaging is also required. Due to their variety and lack of knowledge of their properties, these new materials require large efforts on the development of models. Processing affects both material properties and the three-dimensional shape of interconnect. These nonidealities must be considered in the simulations (ITRS 2009).

Therefore, key drivers for using M&S early in product design are cost-effective, decreasing time to market, and reducing the physical complexities of thermal trends in electronic systems. The advantage of adopting modeling early in the design process is to allow potential thermal problems to be identified before building physical prototypes, which will improve product quality and reliability, and speed time to market (Young et al. 2006).

### Modeling Techniques

A variety of analysis and simulation methods have been used to predict thermal and thermomechanical behavior of materials and electronic-based products. Simulation using numerical modeling is inherently a computationally intensive process because there are many equations to be solved iteratively. Computer-aided design (CAD) systems are used to describe geometrical models, to manipulate data, and to display final results. Therefore, CAD systems with thermal analysis and computational fluid dynamics (CFD) modules become very powerful tools to investigate the thermal distribution within the electronics package. CFD technology has dramatically grown in the electronics thermal management. Particular products such as

FLOTHERM and Fluent (owned by ANSYS Inc.), are used to model the flow of air and liquids in electronic cooling systems and the resultant device temperatures. In addition, finite element analysis (FEA) tools used to predict mechanical behavior of electronic systems, and in particular, stresses due to temperature changes in the materials. Codes such as ANSYS and ABAQUS are particularly popular. These tools are used to predict the stress, strain, and resulting damage in the materials that make up an electronic system and these values are then used to assess the reliability of the overall product.

Although modeling tools are very powerful and provide extremely useful insights into thermal management technologies they need to be used with care. The materials data, boundary conditions, and failure models in particular as input to these codes need to be fully understood in terms of applicability. Furthermore, miniaturization, increased integration, and higher demands to quality have made it necessary to take thermal management and temperature-related failures seriously. Good thermal design is often required to achieve high reliability, low manufacturing cost (to avoid expensive cooling solutions), small size, and a predictable development time (to avoid redesigns and reduce time-to-market). With the availability of powerful computational tools, the design cycle time and the cost of designing new power electronics can be greatly reduced. In addition, the efforts of improving current computational tools as design tools have proved the future potential of computational analysis in various heat transfer analysis (Pan 2002).

### Thermal Modeling

CAD, CFD, and FEA are used extensively to predict physical phenomena such as fluid flow, heat transfer, electromagnetic interference, and mechanical stress. For thermal analysis of electronic systems, CFD tools will predict the governing heat transfer modes such as conduction, radiation, and convection. For each mode of heat transfer the governing partial differential equations are solved using numerical techniques programmed into software tools such as FLOTHERM and Fluent. Particularly, the surface temperature of the die and the junction temperature can be predicted with different level packages, and the heat removal paths and the thermal resistance for heat flow from the chip location to the ambient can also be calculated. The overall thermal resistance is made up of a combination of thermal resistances across each material in the whole package. Identifying how heat flows through the system and the locations of high thermal resistance throughout the electronic package would help to make design changes that will maximize heat removal and minimize junction temperature (Young et al. 2006).

Passive cooling technologies remove heat from the system primarily due to the high thermal conductivity of a particular material. Using modeling can help characterize the performance of the material both thermally and mechanically. The challenge with modeling these high thermal conductivity materials is the lack of accurate temperature-dependent thermal and thermomechanical data to allow meaningful modeling of anisotropic or isotropic behaviors.

Furthermore, the determination of the physical properties of thin film and bulk materials and the impact of these properties on the electrical, mechanical, and thermal properties of devices and ICs is becoming more significant across all aspects of semiconductor technology as new materials are being explored. The strong driving forces behind this are the physical limits of material systems used to date (Young et al. 2006).

Materials simulation tools that give insight to interrelationships between the physical properties of multilayer thin films and the electrical, thermal, and reliability aspects of the device or IC would allow the selection of options without the need for many and complex experimental characterizations. Both empirical and fundamental M&S materials are needed to aid in this understanding. Whereas at short time scales, insufficient availability of fundamental materials modeling capability will frequently require the use of phenomenological models, in the long-term first principle simulations will be indispensable. A special emphasis needs to be placed on the development of the material properties in this mesomaterial (between bulk and atomistic) range. M&S tools in the equipment, process, device, package, patterning, and interconnect topical areas are only as good as the input materials parameters. In many cases, these parameters are not known or only weakly approximated. Databases are needed that contain both experimental and, where not available, material parameters calculated from first principles. To calculate an increasingly large number of model parameters required in the databases demands an increasingly large computational and organizational effort. Therefore, the calculation process must be automated to a large degree, enabling an efficient workflow and a rapid reaction to changing technological requirements (ITRS 2009).

In many areas of materials modeling, a key problem is the approximate solution of Schrödinger's equation that leads to discrepancies between first principle simulations and experiment, possibly requiring readjustment and validation of the approximations. A second key problem is then to extract the desired physical parameters, using as few evaluations of the approximate solutions as possible. Often, both problems are solved within one monolithic simulation tool. A modularization of the simulations tools with respect to these two key problems would enable a faster and independent development and improvement of solution approaches for the two problems. With device active regions continuing to shrink to several tens of nanometers for the physical channel length and to the nanometer range for the effective oxide thickness of high-κ gate dielectric materials for instance, materials simulation and modeling tools that go from atomistic descriptions to continuum results will become more and critical. In the long term, the relevant material modeling approaches might be integrated into the modeling toolsets of the various topical areas. The materials modeling tools must be prepared for this integration. Specific materials modeling problems to be addressed for the different topical areas include the following (ITRS 2009): (1) Materials models are needed for improved (especially chemically amplified) resists, for advanced mask making and for multilayer mirrors to be used in EUV (Extreme-Ultraviolet) lithography. The impact of molecule sizes on the resist structure must be incorporated in the determination of line edge and line width roughness. (2) Interconnect performance and reliability will be strongly affected by the microstructure and the resultant change in conductance of copper, which must be taken into account in the simulation. (3) For processing, needs include codes with predetermined adjustable model parameters for ion implantation, dopant diffusion and activation and interdiffusion in thin films. A wide variety of dopants and codopants must be considered. (4) Most models used in device simulation can be considered as material models because they are based, for example, on the electronic structure of the semiconductor, i.e., dielectric properties, and channel transport properties, including quasiballistic transport. Major progress is needed here also due to shrinking dimensions, higher local electrical fields, and especially becasue of global and local strained channels including, strained substrates (sSi), SiGe, Ge, III-V, SOI, sSOI, GeOI, and other new materials.

Miniature heat pipes and high-performance heat sinks have also been investigated for thermal management as well as materials with high thermal conductivity other passive technologies such as thermosyphon loops. For thermosyphon loops and heat pipes, CFD technology can again be used although the challenge is to capture the physics relating to phase-change mechanics and the important effect of this on heat transfer rates. Using CFD to characterize heat sink performance is well established where it helps optimize fin thickness, spacing, and base plate shape and size (Young et al. 2006).

The most common form of active cooling through forced air convection using fans is well established, and codes such as FLOTHERM and Fluent have a number of libraries for different fan types from a modeling perspective. One challenge for CFD modeling in this area is the use of models for low Reynolds number flows which are typically found in electronics cooling, although new models to better predict this type of flow are developing. Modeling has been used here to help understand many novel active cooling processes such as ion wind, micropump design in terms of the effect of flow rates on back pressures and heat transfer coefficients, microchannel flow which is popular as it can disperse a lot of heat by using a small amount of fluid. Jet impingement and spray cooling are two other technologies that are being extensively researched and modeled. Jet impingement supplies a liquid coolant perpendicular to the cooling surface. It can accommodate critical heat fluxes up to 300 W/cm$^2$ and beyond, depending on the fluid used. Spray cooling is seen as a competitor to jet impingement cooling, and atomizes the fluid, forming coolant mist which contacts the high temperature items. Modeling the spray cooling process is particularly challenging because it involves modeling the atomization process, collision and coalescence of liquid particles, and vaporization on the cooling surface. TECs or Peltier devices act as heat pumps by force conducting heat from a hot to a cold plate. TECs are solid-state devices which generate electrical power with the Seebeck effect and converting electrical power into a temperature gradient with Peltier effect. Many CFD codes (such as FLOTHERM) have routines for modeling the effect of these devices (Young et al. 2006).

In addition, integrated modeling enables thermal, mechanical, electrical, and acoustic analysis to be undertaken at the same time. The trends in this area include (Young et al. 2006): (1) Cosimulation or multiphysics analysis tools, (2) integration with numerical optimization, (3) close integrattion with electronic computer-aided design and mechanical computer-aided design (MCAD), and (4) use of models with sensor data for prognostics and diagnostics.

## Reliability Modeling

Reliability is understood as the probability that a device will perform its required function under stated conditions for a specific period of time. A reliability prediction estimates the reliability of an electronics module as used in the field. This depends on the design of the module and the in-service environmental conditions it is subjected to during its lifetime. One relationship between junction temperature and semiconductor device reliability is the Arrhenius equation which can be used to estimate the failure rate, $\beta(T)$, as a function of processor temperature, $T$ (Young et al. 2006):

$$\beta(T) = A\mathrm{e}^{-\frac{E}{kT}}, \tag{12.1}$$

where $A$ is a constant, $k$ is the Boltzmann constant ($8.16173 \times 10^{-5}$), and $E$ is the activation energy. If the failure mode is considered to be in the copper interconnect metallization on the chip and the activation energy of the failure mechanism to be $E = 1.0$, then a rise in junction temperature from say $T_1 = 80°\mathrm{C}$ (353 K) to $T_2 = 100°\mathrm{C}$ (373 K) will result in acceleration in failure rate of (Young et al. 2006):

$$\frac{\beta(T_2)}{\beta(T_1)} = \mathrm{e}^{-\frac{E}{k}\left(\frac{1}{T_2}-\frac{1}{T_1}\right)}. \tag{12.2}$$

In this example when the juction temperature increases by 20°C, the failure rate will increase by a factor of 6. Of course, not all failures can be modeled using the above expression as they are much more complex and stress-related failure models are required.

The physics-of-failure approach to reliability combines mathematical modeling with accelerated life testing to predict the reliability of a product. This technique investigates the physics leading to a failure and identifies the root cause for that failure. In general, there are concerns about current accelerated life tests for reliability based on particular standards and the relationship between failures observed in these tests for current thermal management technologies and what would be seen in the field. The physics-of-failure approach with other techniques such as failure mode and effect analysis is seen as the correct way to predict and characterize reliability (Young et al. 2006).

### Design-In Methodology

Design-in technology enables the conception, implementation, and validation of the thermal management part of an integrated microelectronics design system. Elements of design-in technology include tools, libraries, manufacturing process characterizations, and methodologies. Design-in technology transforms conceived ideas and objectives of the electronic systems into manufacturable and testable representations. The role of design-in technology is to enable profits and growth of the semiconductor industry via cost-effective production of designs that fully exploit manufacturing capability. The process of designing and implementing an electronic packaging requires a large collection of techniques, or tools, and an effective methodology by which a designer's input predictably leads to a manufacturable product. While considerable attention is given to the tools needed, the equally important subject of design-in methodology is often neglected. Each technology generation requires designers to consider more issues; hence, new analysis methods and tools must be developed to evaluate new phenomena and aid the designer in making critical design decisions. An even greater challenge is determining the most effective sequence in which issues should be considered, and design decisions made, in order to minimize iterations (ITRS 2009a).

The cooling of electronics is often the least understood and most often ignored part of electronics design. All power electronics modules have a maximum junction temperature that cannot be exceeded. As for the integrated power electronics modules for example, the maximum junction temperature is set to be 125°C. This maximum junction temperature is directly related to the maximum power dissipation. Device longevity is, in many cases, attributed to keeping its junction temperature within these limits. Therefore, thermal design optimization of power electronics modules is important in increasing the device longevity. The most reliable and well-designed electronic device can malfunction or fail if it overheats. Good thermal design is often paramount in achieving high reliability, low manufacturing cost, small size and a predictable development time. The thermal design must be able to dissipate the maximum amount of power required to maintain operating junction temperature within limit. Thus, thermal design is an important aspect in the overall production design process to achieve the desired operational and reliability objectives for the electronic product.

## *Prototype Fabrication and Experimental Validation*

Nanometer process technology, increasing clock rate, mixed analog-digital-RF circuits, and highly integrated system on a chip (SOC) and slurry infiltration process present severe challenges to prototype fabrication and experimental validation. The experimental validation must cope with an enormous spectrum of problems ranging from high-level test synthesis for component-based design to noise/interference and power dissipation in extremely high-performance analog and RF subsystems. Many problems can be solved only if proper testability and experimental validation are considered and incorporated early in the design phase. Furthermore, the methodology precepts indicate a natural evolution of analyses into verifications into prototype tests, and the need for deeper unification of design with test. Effort and results (flows, vectors, fault models, sensitivities) from analyses in logical-circuit-physical implementation, and from design verification, must be reused during design and test. Across electronic packaging as varied as memory, digital signal processing, PE (Portable Executable), SOC, analog/mixed-signal/RF, memory protection unit, along with thermal management, high-level test challenges demand significant expansion of on-chip density functional theory, built-in self test, and testability features, and better integration to a preplanned manufacturing test equipment set during the chip planning phase of development (ITRS 2009a).

The overarching goal of functional verification is to ensure that a system's implementation fully matches its specification, that is, the planned intended behavior of the device. Unfortunately, due to the growing complexity of electronic packaging designs, functional verification is still an unresolved challenge. The current practice in industry is a partial verification process, developed under tight time and cost constraints, where only a few aspects of a design are addressed, checked, and verified. These aspects constitute only a vanishingly small fraction of the entire universe of possible design behaviors. The underlying reasons for this unmanageable complexity lie in the inability of verification to keep pace with highly integrated SOC designs and parallel chip multiprocessor systems, paired with highly interconnected communication protocols implementing distributed computation strategies. Because of the extremely high costs associated with flawed microchips, industrial development efforts devote a significant fraction of engineering time and resources to achieve functional correctness. Without major breakthroughs design verification will be a nonscalable, show-stopping barrier to further progress in the electronic packaging industry. There is hope for breakthroughs to emerge via a shift from ad hoc verification methods to more structured, formal processes. The mainstream methodology used in industry today attempts to verify the functionality of a system design by repeatedly building models, simulating them on a few selected vectors, and then patching any bugs that are exposed. To this end, logic simulation techniques predominate, because they can generate simulation vectors at a very high rate. However, the coverage of these tests is usually very low, and when a design error is found, the debugging phase entails manual analysis of very long and complex bug traces to narrow down the source of the problem. Additionally, these methods require substantial engineering effort to direct the verification activity toward specific design areas of critical quality, low coverage, etc. Traditional verification techniques are entrenched because the cost to transition to alternative methodologies is high. Hence, adoption of new and alternative formal and semiformal techniques, which have started to become available, is progressing slowly (ITRS 2009a).

More structured verification approaches organize the verification effort by generating a golden model of the system's intended behavior and comparing simulation outcomes with this model. Coverage metrics are collected to support some degree of confidence in the correctness of the design-under-verification. It is common to organize the verification activities hierarchically, by addressing first individual components, then the entire chip, and eventually the full system, so that bugs that are contained within a single unit can be addressed and solved earlier and more easily. Due to the complexity of full system simulation, prototype tests and hardware emulation are sometimes used instead of simulation, particularly for large-market designs, where the additional costs can be easily absorbed. Hardware emulation buys several orders of magnitude of performance improvement in simulation speed, and enables an early start to system integration and software development. At the same time, emulation represents only a constant-factor improvement. The most complex aspect of this methodology is the verification of the communication interface or interconnects between components; late or escaped bugs are often found in complex unverified interactions between units. Finally, there is a growing interest for formal and semiformal verification techniques: today, IC development teams are exploring new methodologies where mainstream validation is complemented by semiformal methods, using a handful of commercially available tools. This is a rapidly evolving arena: several new solutions, often integrating formal methods with simulation-based techniques in new and creative ways, appear each year both in academic environments and in new industrial developments (ITRS 2009a).

## *Production Layout and Quality Assurance*

Semiconductor factory extends across several manufacturing domains, including wafer manufacturing or fabrication, chip manufacturing that involves probe/e-test, backgrind, singulation, and product manufacturing where the final package is assembled and tested. Silicon substrate manufacturing and product distribution are outside the scope of factory integration. In order to clearly understand the integrated factory requirements and at the same time define measurable and actionable metrics, the factory integration is divided into five thrusts, or functional areas, that are required to perform semiconductor manufacturing. They are factory operations, production equipment, material handling, factory information and control systems, and facilities. Factory operations and its associated factory business model is a key driver of requirements and actions for the other four thrusts. Overall, these five thrusts are used to clarify how difficult challenges translate into technology requirements and potential solutions. In addition to these five thrust areas, the factory integration section also addresses the cross-cut issues and key focus areas that cut across all these five thrusts (ITRS 2009b).

Factory operations refers to the efficient and effective application of resources and integration of other facets of manufacturing such as information and control system, material handling, equipment, and facilities in order to maximize throughput, minimize cycle time, work-in-progress (WIP), and maintain lower operating cost. Cycle time now is gaining more importance both in high mix and low mix manufacturing. Cycle time shall have two different view axes from which it is to be evaluated; from utilization of the factory resources, especially production equipment utilization and from product viewpoint. Resource utilization viewpoint has been well studied and historically used to improve the throughput in order to maximize the profit. Product view axis is important for the delivery and cycle time management and control. Fabrication control methodology is also decisive to the cycle time. The performance characteristics of a factory that best defines its competitive posture will depend on a number of factors. For example, there is a well-established trade-off between cycle time and asset utilization, with higher asset utilization leading to higher cycle times. Thus, each factory must balance the value of lowering cycle time for the business segment(s) in which it participates versus the cost of the lower asset utilization. Similarly, a high-mix factory will have operational characteristics and will need decision support tools that will be different than those for a factory that is low mix and high volume. Asset utilization is captured in the form of equipment utilization, availability, and capacity degradation. Another metric, overall equipment effectiveness (OEE) is not explicitly mentioned in the operations table because the key elements of OEE, namely utilization and availability are captured in the technology tables. Once the fundamental performance characteristics have been established for a factory, there is a continuing need for increased productivity. This increased productivity is reflected either in lower cycle times or in increased throughput for the same capacity investment. It is also essential to address average number of wafers processed before a reticle change since the high-mix factory coupled with next wafer size (such as 450 mm) wafer would demand lesser number of wafers processed between reticle changes. These requirements are meant to provide guidance so that research can be better focused toward the innovations required to achieve these objectives. These innovations are envisioned to be in the form of new concepts, policies, models, algorithms, etc. These will be expressed in the form of software applications that would be developed and released to manufacturing. The software applications will be integrated into the overall factory information and control systems, either as decision support tools or as execution tools. The tools will help to drive factory productivity improvements to achieve factory operations objectives. Another aspect that is becoming critical to factory operations is the mask operations influence on factory operations. Mask operations can be considered as one of the key components of the semiconductor supply chain that produces the mask or reticle sets that are used by the fabrications. The complexity, shrinking and demand to keep die cost low influences mask manufacturing. Some of the key challenges in mask operations that need to be comprehended are: data explosions (time for object linking and embedding for process control calculations and data preparation for mask writer; time to send and load tape-out data into mask shop data system, etc.); rising cost of E-beam; supply chain synchronization with fabrications, and better process control needs. The factory operations potential solutions are classified into planning decision support (DS) tools at the strategic level and tools for running the factory at the tactical or execution level. The solution components for these two levels are quite different but are essential in order to manage high-mix factories effectively. The tactical tools need quick access to transactional data whereas the DS tools need large sets of data with several analysis/reporting options. The stringent engineering requirement is driving need for more data that would be resulting in data explosion. It is critical to not only collect necessary data but also to develop intelligent analytics and algorithms to identify and use the right signals to make data driven decisions, and reuse such intelligence as models in later occasions. The factory data shall be designed in accordance to these models for usages for high data utilization efficiencies. Demand information propagated over the factory network (Fab/sort/assembly/test) is neither accurate nor responsive, which results in poor factory and supply chain planning. Successful determination of where, when, and in what quantities the products are needed is essential for improving manufacturing productivity. The cost of capital equipment is significantly increasing and now constitutes more than 75% of wafer Fab capital cost and via depreciation a significant fraction of the fixed operating costs as well. Reducing the impact of these increasing costs on overall wafer costs requires improvements in equipment utilization, availability, and capacity loss due to set up (for high mix), tool dedication, etc. Effective factory scheduling also plays a key role in improving equipment utilization and it also leads to improved cycle time and on-time-delivery (OTD). In order to utilize the expensive production equipment effectively, it is imperative that effective scheduling and dispatching tools be utilized. Several factors complicate Fab scheduling. These include AMHS (Automated Material Handling System) that is not fully integrated with lot scheduling tools as well as scheduling policies that are not effectively integrated into lot scheduling tools. A real-time scheduling and dispatching tool integrated with AMHS and incorporating predictive maintenance, preventive maintenance scheduling, and resource scheduling policies are required to reduce WIP, improve OTD, and improve capacity utilization. Effective design and control of production equipment is central to controlling the cost of processing each unit area of silicon, and throughput improvements as measured in terms of processing time per wafer are an expected attribute of typical continuous improvement efforts. The industry's growth rate will not be sustainable in the future if increasing capitalization cost trends continue without significant improvement in productivity. There are several factors that impact productivity of the equipment. They include the following (ITRS 2009b):

1. Reducing intrinsic setup time that comes with high-mix factories through quick turn around setup options.
2. Improving equipment design to reduce losses from high product mixes (i.e., set-up time reduction or elimination), and reducing the percentage of nonvalue added time associated with equipment operation.
3. Finding breakthrough solutions for increasing equipment reliability, availability, and utilization.
4. Reducing variation within and between equipment and attaining chamber and tool matching.

5. Extending equipment lifetime to support multiple technology generations.
6. Achieving more effective use of utilities and consumables, including reduction of nonproduct wafers, while simultaneously reducing environmental impacts.
7. Reducing relative or normalized equipment capital cost (rate at which equipment cost increases vis-à-vis requirements for process capability) by speeding up the processing rate. This will drive reduced cost of ownership.

The production equipment potential solutions are prioritized toward attaining the improvements listed above. End-user understanding and analysis of equipment nonvalue added time, leading to targeted areas for improvement in intrinsic setup time reduction and other nonvalue added operations, can be realized through the development, standardization, and implementation of reliable and comprehensive equipment engineering data control systems. Improvements in the flexibility and agility of the equipment will also reduce cycle time and increase tool utilization. Reliability and utilization improvements can also be achieved by innovative solutions in the area of in situ monitoring, advance process control capability, design for manufacturing, smarter embedded controllers, self-diagnostics, remote diagnostic capability, and single-wafer level tracking and control. More energy-efficient equipment designs are achieved through the use of higher efficiency power distributions systems within the tool, more efficient tool-heat-load removal methods, and optimized recycling and reuse of water. An additional emerging area of opportunity for reducing utilities consumption is the concept of a "smart idle" or "sleep mode" operational model for equipment. Another high priority area is finding ways to extend the life of the equipment to support multiple technology generations. Other important areas are finding innovative solutions for ramp-up cycle time reduction and spares cost reduction. An additional emerging focus area requiring innovate solutions is the preventive control of airborne molecular contamination. Finally, efficient and cost-effective equipment development will be a critical milestone in the industry transition to the next wafer processing size. The principal goals of factory integration are maintaining cost per unit area of silicon, decreasing factory ramp time, and increasing factory flexibility to changing technology and business needs (ITRS 2009b).

The concept of quality certification and standards, however, breaks down when global competitiveness is at stake. International Organization for Standardization (ISO) 9000 certification has become a requirement for exports to Europe, and many countries have been forced to obtain ISO certification, not because it is a quality issue, but because it is a way of increasing market share. Quality is associated with the degree of conformance of the product to customer requirements, and thus, in a sense, with the degree of customer satisfaction. Implicit in quality products is an acceptable amount of reliability; that is, the product performs its intended function over its intended life under normal environmental and operating conditions. Reliability assessments are incorporated through simulation and qualification functions at the design and prototyping stages. With basic reliability designed in, quality control functions are then incorporated into the production line using in-line process controls and reduced human intervention through automation. Qualification includes activities to ensure that the nominal design and manufacturing specifications meet the reliability targets. In some cases, qualification of the manufacturing processes and of the pilot lots can be conducted together. Quality conformance for qualified products is accomplished through monitoring and control of critical parameters within the acceptable variations already established, perhaps during qualification. Quality conformance, therefore, helps to increase product yield and consequently to lower product cost (Satoh 1986).

Through endless improving next generation products, components, materials, equipment, or processes with dominating in materials, equipment, and manufacturing processes, it would be able to provide the highest quality and reliability for electronic packaging products with efficient thermal management solutions.

## Smart Composites and Multifunctional Materials for Thermal Management

Smart materials are materials that respond to environmental stimuli, such as temperature, moisture, pH, or electric and magnetic fields. For example, photochromic materials that change color in response to light; shape memory alloys (SMA) and polymers which change/recover their shape in response to heat and electro- and magnetorheological fluids that change viscosity in response to electric or magnetic stimuli Chen et al. 2002. Smart materials can be used directly to make smart thermal management systems or structures or embedded in conventional thermal management structures whose inherent properties can be changed to meet high value-added performance needs. Smart materials technology is relatively new to the thermal management of electronic packaging, but it has a strong innovative content that can truly create new thermal management solutions in chip level, component level, or system level, especially when combine with emerging nanomaterials (Quarshie 2005; Wax et al. 2003).

Many applications of smart materials have been explored and these applications would provoke exploration of smart materials and structures used for various thermal cooling solutions in electronic packaging. In the food industry, smart labels and tags could be used in the implementation of traceability protocols to improve food quality and safety, e.g. using thermochromic ink to monitor temperature history. In motorsport, the innovative use of smart materials, such as electrorheological fluids in active suspension systems, can help maintain this position and provide improvements which will eventually flow down into mainstream automotive applications. In construction, smart materials and systems could be used in smart buildings, for environmental control, security, and structural health monitoring, e.g. strainmeasurement in bridges using embedded fiber optic sensors. Magneto-rheological fluids have been used to damp cable-stayed bridges and reduce the effects of earthquakes. In aerospace, smart materials could find applications in smart wings, health and usage monitoring systems, and active vibration control in helicopter blades. In marine and rail transport, possibilities include strain monitoring using embedded fiber optic sensors. Smart textiles are also finding applications in sportswear that could be developed for everyday wear and for health and safety purposes. Any material format, such as speciality polymers, fibers and textiles, coatings, adhesives, composites, metals, and inorganic materials, can incorporate sensors or active functional materials including piezoelectrics, photochromics, thermochromics, electro and magneto rheological fluids, SMAs, aeroelastic-tailored and other auxetic materials (Matic 2003; Quarshie 2005).

For instance, microelectronic solder joints are typically exposed to aggressive thermomechanical cycling conditions during service. With thermomechanical cycling, strain localization occurs near solder bonding interfaces, where large, inelastic-shear strains accumulate, and eventually causing low-cycle fatigue failure of the joint. To mitigate the effects of strain localization within the joint the solder alloy is reinforced with a martensitic NiTi-based (up to 5 vol%), SMA. In this scheme, the SMA reinforcement deforms in shear concurrently with the solder during thermomechanical cycling and, subsequently, undergoes martensite-to-austenite ($M \rightarrow A$) transformation, placing the solder matrix next to the reinforcements in reverse shear. This is purported to reduce inelastic-strain localization within the solder and thus, enhance joint life (Dutta et al. 2004).

The ability to adapt a structure's performance at will is increasingly more attractive as system-operating scenarios become more space and logistics constrained. Currently multimission objectives are answered with multiple structures: a thin but smart materials skin that undergoes a radical but controlled change in mechanical strain, a coating that changes color on demand, and a shell that reconfigures shape to meet styling or mechanical performance criteria. These materials would enable the same system-level goals that are currently designed as subsystems; however they are more readily integrated into the larger engineering structures because they are lighter, smaller, less difficult to interface, and easier to maintain. Fueled by advances in biomaterials and nanotechnology, multifunctional materials are now emerging as a new interdisciplinary field that promises to provide a new level of functionality, adaptability, and tailoring to future engineered electronic packaging and thermal management systems (Christodolou and Venables 2003; Momoda 2005).

Similarly, multifunctional materials are typically a composite or hybrid of several distinct material phases in which each phase performs a different but necessary function such as structure, transport, logic, heat transfer and energy storage. Because each phase of the material is used to perform an essential function so there is little or no parasitic weight or volume, multifunctional materials promise to achieve performance flexibility but with higher weight and volume efficiency and potentially less maintenance than the traditional multicomponent, brassboard system solutions. In addition, as the integration scale in the material becomes finer and more distributed, reaction times may become faster and more autonomous. Multifunctionality within a material can be integrated on several dimensional scales with increasing interconnectivity between phases and engineering difficulty as the scale decreases. The drive for improved overall system performance determines the design of a multifunctional material, therefore its performance metrics are inherently different from its single component phases. In the component phases, the improvement of a single function, such as thermal conductivity, electrical conductivity, mechanical strain or force, energy density, etc. is maximized or minimized. In a multifunctional material however, a new materials design methodology is required in which the system-level performance is emphasized over the optimization of the individual functions. For example, materials integrating functions such as electromagnetics, thermal management, and optics into structures are under development. In addition, the increased knowledge of understanding of materials on the nanoscale level will increase the control and range of physical properties of materials while further decreasing the integration scale (Momoda 2005; Sigmund and Torquato 1999).

## Thermal Management Materials with Enhanced Electromagnetic Interference Shielding and Absorbing Performance

As more and more powerful electronic components are used in electronic devices with increased packaging densities, the combination of thermal management and electromagnetic (EMI) shielding and absorbing has become a development trend in electronic packaging and will provide an optimal option for producing more reliable electronic products. Electronic devices typically generate thermal emissions which should be dissipated effectively to preclude or minimize any undesirable effects. On the other hand, under normal operation, electronic equipment typically generates undesirable electromagnetic energy that can interfere with the operation of proximately located electronic equipment due to EMI transmission by radiation and conduction. The electromagnetic energy can exist over a wide range of wavelengths and frequencies. To minimize problems associated with EMI, sources of undesirable electromagnetic energy can be shielded and electrically grounded to reduce emissions into the surrounding environment. Alternatively, or additionally, susceptors of EMI can be similarly shielded and electrically grounded to protect them from EMI within the surrounding environment. Accordingly, shielding is designed to prevent both ingress and egress of electromagnetic energy relative to a barrier, a housing, or other enclosure in which the electronic equipment is disposed. Therefore, the optimal option would be to use thermal and EMI hybrid solutions, typically combining EMI shielding and absorbing materials with thermally conductive materials, such as those used for thermal management in association with electronic equipment with minimizing and suppressing the transmission of EMI.

### *Minimizing EMI from Heat Sinks*

Heat sinks play a vital role in enhancing circuit design reliability by keeping sensitive components within a particular operating-temperature range. Unfortunately, components and devices requiring the use of heat sinks are not well suited for protective treatment for EMI at its source, because such treatment would interfere with the operation of the heat sink. Moreover, the heat sinks can also add effects to overall circuit EMI by resonating the EMI that the temperature-sensitive components radiate (Tong 2009).

To reduce these kinds of effects, sources of EMI radiation should be placed away from the edges of the heat sink. If one dimension is longer than the others, then it is important to mount devices away from the ends of the longer dimension. It is even more important to avoid placing sources at the corners of the structure. Heat sinks with dimensions that are close to square or cubic yield the lowest levels of radiation enhancement. In the case of a nearly square structure, fin orientation should be configured so that the fins run longitudinally with the prime direction of excitement. Devices mounted in channels provide the lowest levels of EMI radiation, although a check should be carried out to make sure the channel will not form a resonant cavity. However, EMI generation methods in circuits are often complex, and multiple noise sources are often present. Therefore, the heat sink design should consider all factors to minimize effects of EMI for a particular electronic packaging system (Wang et al. 2002).

## *Combination of Board Level Shielding and Heat Dissipation*

A variety of hybrid solutions for electronic packaging design have developed to combine board level shielding and thermal cooling. For instance, Figure 12.1 shows a combination of the ventilating shield enclose with the PCB and heat dissipating system including thermal interface material (TIM) and heat sink (Tong 2009). This design can be used in board level shielding to increase heat transfer while maintaining shielding effectiveness, allowing heat sinks to be used inside the shielding

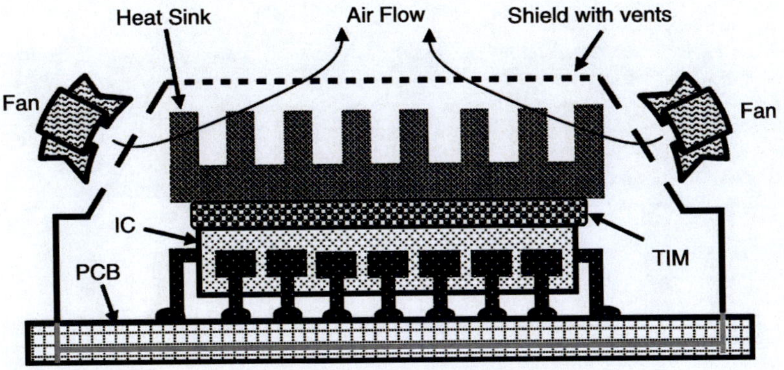


**Fig. 12.1** Illustration of the printed-circuit board and heat dissipating system enclosed with the ventilating electromagnetic interference shield (Tong 2009)

enclosure to eliminate EMI radiation from the heat sink and the electronic system. As shown in Figure 12.2, a unique combination of board level shielding and thermal dissipation was designed to utilize a heat sink as shielding lid by snapping into a board level shielding fence or frame with spring fingers (Tong 2009). Similar to the design shown in Figure 12.1, thermal path is made with the heat sink through the TIM, which conducts heat up through the heat sink. The four sides of the board level shielding frame is flexible in order to make the heat sink snap into the frame acting as a shield lid. Multiple frames can be made with a single heat sink lid to cool multiple components. The snap-in heat sink can be removable for component maintenance (Armstrong 2004). Another example is shown in Figure 12.3, by placing a TIM on the inside of a board level shield, the air gap between the shield and electronic power can be eliminated (Tong 2009). A heat spreader, heat sink, or heat pipe with additional TIM can be included to further enhance the thermal performance (English 2007).

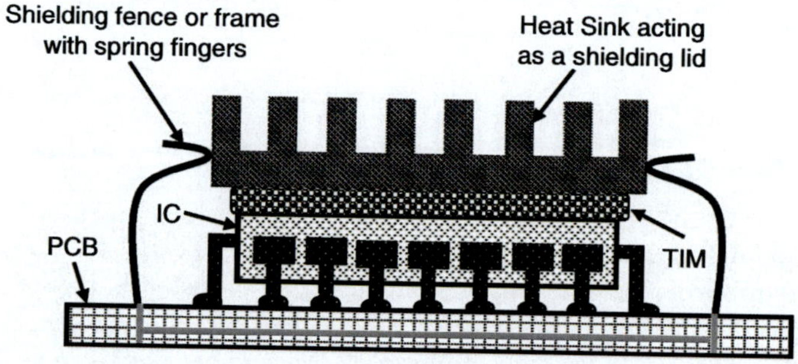


**Fig. 12.2** Illustration of that heat sink snaps into the shielding frame and acts as a shielding lid (Tong 2009)

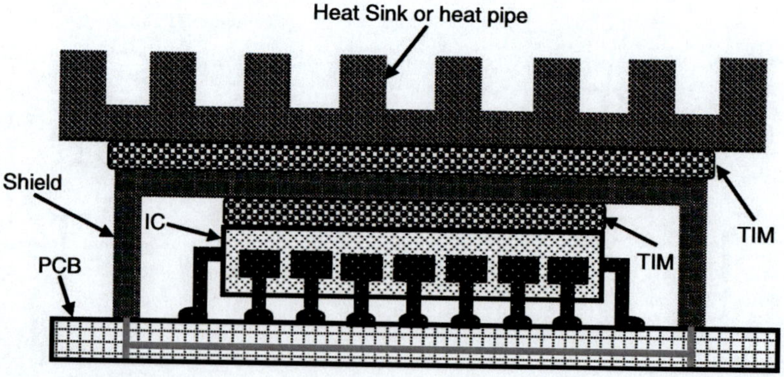


**Fig. 12.3** Combination of board level shield, heat sink or heat pipe, and thermal interface materials (Tong 2009)

## *Thermally Conductive EMI Absorbing Materials*

EMI absorbing materials are usually designed for attenuation and absorption of electromagnetic energy in frequencies ranging from the kHz to GHz of micrometer and millimeter waves. The thermal conductivity of magnetic absorbers can be achieved and enhanced by the addition of thermally conductive fillers. These fillers do not degrade the microwave absorbing property of the material while enhancing its thermal conductivity. The thermally conductive absorbers can be used as a thermal pad or any other forms of TIMs.

For instance, kind of thermally conductive composite absorbing materials have been developed. The thermally conductive EMI absorber is prepared by combining an EMI-absorbing material (40 vol% Fe ferrite particles) with a thermally conducting material (20 vol% AlN ceramic particles), each suspended within an elastomeric matrix (for example, silicone or isoprene). In application, a layer of thermally conductive EMI-absorbing material is applied between an electronic device and heat sink. Materials having electromagnetic-energy absorbing properties can be used to suppress the transmission of EMI over a broad range of frequencies. Such EMI-absorbing materials can provide substantial electromagnetic-shielding effectiveness, for example, up to about 25 dB or more at EMI frequencies occurring from about 2 GHz up to about 100 GHz. The EMI absorbers convert electromagnetic energy into another form of energy through a process commonly referred to as a loss. Electrical loss mechanisms include conductivity losses, dielectric losses, and magnetization losses. Conductivity losses refer to a reduction in EMI resulting from the conversion of electromagnetic energy into thermal energy. The electromagnetic energy induces currents that flow within an EMI absorber having a finite conductivity. The finite conductivity results in a portion of the induced current generating heat through a resistance. Dielectric losses refer to a reduction in EMI resulting from the conversion of electromagnetic energy into mechanical displacement of molecules within an absorber having a nonunitary relative dielectric constant. Magnetic losses refer to a reduction in EMI resulting from the conversion of electromagnetic energy into a realignment of magnetic moments within an EMI absorber. In general, the EMI absorber is selected from a group of materials such as electrically conductive material (metallic silver), carbonyl iron powder, SENDUST (an alloy containing 85% iron, 9.5% silicon, and 5.5% aluminum), ferrites, iron silicide, magnetic alloys, magnetic flakes, and combinations of them. The EMI absorber can be a magnetic material with a relative magnetic permeability greater than about 3.0 at approximately 1.0 GHz, and greater than about 1.5 at 10 GHz. The thermal conductor constituent is selected from materials, such as aluminum nitride, boron nitride, iron (Fe), metallic oxides and combinations of them. For example, the thermal conductor can include Fe-AlN (40 and 20% by volume, respectively) having a thermal conductivity value greater than about 1.5 W/m K. The composite matrix can be selected to have properties allowing it to conform to surface imperfections encountered in many heat-sink applications. Other desirable properties of the matrix material include ability for the material to accept and suspend a substantial volume of particles (for example, up to about 60% by volume) without compromising the other advantageous properties of the matrix material, such as conformability, compliance, and resilience. Generally, the matrix material is also substantially transparent to electromagnetic energy so that the matrix material does not impede the absorptive action of the EMI absorbers. For example, a matrix material exhibiting a relative dielectric constant of less than approximately 4 and a loss tangent of less than approximately 0.1 is sufficiently transparent to EMI. Values outside this range, however, are also contemplated. Usually the matrix material can be selected as a solid, a liquid, or a phase-change material, including thermoplastic materials and thermoset materials. Thermoplastic materials can be heated and formed, then reheated and re-formed repeatedly. The shape of thermoplastic polymer molecules is generally linear, or slightly branched, allowing them to flow under pressure when heated above the effective melting point. Thermoset materials can also be heated and formed; however, they cannot be reprocessed (that is, made to flow under pressure when reheated). Thermoset materials undergo a chemical as well as a phase change when they are heated. Their molecules form a three-dimensional cross-linked network. More specifically, the matrix material can be selected from elastomers, natural rubbers, synthetic rubbers, PDP, ethylene-propylene diene monomer (rubber, silicone, fluorosilicone, isoprene, nitrile, chlorosulfonated polyethylene, neoprene, fluoroelastomer, urethane, thermoplastics, such as thermoplastic elastomer (TPE), polyamide TPE and thermoplastic polyurethane, and combinations of them. The adhesive layer can be formulated using a pressure-sensitive, thermally conducting adhesive. A suitably selected phase-change material having properties of both a solid and a liquid can be chosen as a matrix. The matrix reflows allowing the EMI-absorbing particles and the thermally conductive particles of the composite material to flow into any gaps, such as those caused by surface imperfections. With an ability to flow into surface imperfections, a matrix formulated as a liquid, or phase-change material removes air gaps, thereby minimizing the thermal impedance between the device and an associated heat sink. The overall effect of removing air gaps reduces the thermal impedance between the electrical component and the heat sink, leading to improved heat transfer efficiency. The phase-change matrix material may be a mixture of a paraffin wax having a melting point of approximately 51°C and a 28% ethylene-vinyl acetate copolymer having a melting point of approximately 74°C. For example, a mixture of 95 parts by weight of the paraffin wax and 5 parts by weight of the ethylene-vinyl acetate copolymer may be used. Alternatively, a mixture of 25 parts by weight of the paraffin wax and 6 parts by weight of the ethylene-vinyl acetate copolymer may be used. Alternatively still, the matrix material may be a synthetic wax having a melting point of approximately 100°C and a molecular weight of approximately 1,000. The EMI absorber particles and thermally conducting particles are combined and suspended within either a solid matrix material, or a liquid matrix material. Once prepared, the composite thermal EMI shield can be applied between an electronic component and a heat sink or other thermal management device (Johnson 2009).

## *Thermally Conductive Metalized Plastic Housing for EMI Shielding*

Most high-frequency systems require some form of EMI shielding at the enclosure level. While a conductive metal housing is inherently an effective EMI barrier, plastic enclosure parts must be made conductive. This has typically been achieved using metallized paint, vacuum deposition, plating, or conductive composite process. Table 12.1 shows shielding effectiveness of some of these conductive coatings (Tong 2009). While each method offers intrinsic advantages, there is almost always a cost performance tradeoff necessary in the selection. The best of these systems accommodate the deepest recesses in a plastic housing part, where the smallest discontinuity provides a pathway for spurious EMI emissions.

On the other hand, managing heat can be crucial to maintaining the reliability and extending the life of the protected electronic components. A broad choice of cooling solutions is available for shunting away excess component heat, including heat sinks, heat pipes and flexible metal or ceramic heat spreaders. These typically require a thermally conductive interface material between the hot component and the heat-sinking device. Much of this hardware can potentially be replaced by conveying thermal energy through a transfer material and into the conductive conformal coating that is applied on the EMI shielding enclosures of portable electronics for instance. The all-metal coating is an effective heat conductor. The coating layer can spread and dissipate thermal energy along the housing surface. The metallic conformal coating, in layers as thin as 40 μm, will spread the heat generated by a semiconductor, and reduce the overall heat load that the plastic case must dissipate into the environment. This thermal spreading performance reduces the junction temperature of the device as well as reducing the hot spot on the plastic case. The 100 μm thick coating has a very significant effect in that it reduced the junction temperature by 15°C and make the case temperature not overreached 60°C. The case temperature reduction is important for portable electronic products because a case over 60°C can cause burns when touched. The metallic conformal coating process lends itself for use with a variety of metals and alloys that may have a greater initial thermal conductivity (Hannafin 2002).

**Table 12.1** Shielding effectiveness of typical conductive coatings

| Conductive coatings | Thickness | Shielding effectiveness (dB) | | |
|---|---|---|---|---|
| | | 1 GHz | 5 GHz | 10 GHz |
| All-over copper–nickel plating | 1.0–1.25 μm | 120 | 113 | 87 |
| Selective copper–nickel plating | 2.0–2.5 μm | 71 | 60 | 62 |
| Copper loaded paint | 0.025 mm | 70 | 81 | 63 |
| Copper–silver hybrid paint | 0.025 mm | 69 | 85 | 70 |
| Silver loaded paint | 0.0125 mm | 62 | 70 | 55 |
| Copper–stainless steel vapor deposition | 0.5 μm | 48 | 58 | 62 |
| Aluminum vapor deposition | 0.5 μm | 46 | 40 | 35 |

## Thermal Management Application in Computer Design

The demand for high-performance computers has resulted in an escalation of power dissipation as well as heat flux from component to system levels. At the same time, the desire for smaller form-factor computer system and lower operating temperatures is compounding the thermal challenge. Thermal design for a computer system can no longer be treated in isolation. Power and performance tradeoffs and smart circuit-design techniques are required to conserve power consumption. Materials and process improvements in packaging and thermal management technology are required to minimize thermal resistance and increase efficiency of heat dissipation. The concurrent development and packaging of all these elements is critical to ensure that from a cost and availability perspective a viable thermal design solution space exists.

The objective of thermal management is to ensure that the temperatures of all components in a computer system are maintained within their functional temperature range. Within this temperature range, a component is expected to meet its specified performance. Temperatures exceeding the maximum operating limit of a component may result in irreversible changes in the operating characteristics of this component. In a system environment, the processor temperature is a function of both system and component thermal characteristics. The system level thermal constraints consist of the local ambient air temperature and airflow over the processor as well as the physical constraints at and above the processor. The processor temperature depends in particular on the component power dissipation, the processor package thermal characteristics, and the processor thermal solution. All of these parameters are affected by the continued push of technology to increase processor performance levels (higher operating speeds) and packaging density. As operating frequencies increase and packaging size decreases, the power density increases while the thermal solution space and airflow typically become more constrained or remains the same within the system. The result is an increased importance on system design to ensure that thermal design requirements are met for each component, including the processor, in the system.

Careful management of the thermal design space from the chip to the system level is therefore critical to ensure a viable solution space for succeeding generations of processors in the computer system. The detail considerations includes (Viswanath et al. 2000): (1) packaging and system-level thermal budgets, such as power conservation features (clock gating, deep sleep, etc.) to optimize power dissipation and performance of the device; (2) optimal selection of TIMs, packaging integration, and modulation of critical-material properties for optimizing thermo-mechanical performance; (3) advanced thermal solutions at the system level, such as heat-sink technologies, chassis standards pertaining to air flow, board layout, heat-sink volumetrics, relating to performance, cost, and form factor in mobile, desktop, and server markets.

## *Design Baseline for Power Management and Performance Optimization*

The insatiable demand for higher performance processors has led to a steady escalation in power consumption across all computing market segments, including mobile and performance desktops as well as servers and workstations. Although improvements in process have been able to hold the power increase to reasonable levels, it is definitely tended higher. A similar trend is reflected in the average heat flux (power dissipated per unit die area) on the processor, indicating a linear increase over time. This is due to the fact that the power reduction obtained from architecture and process modifications is not commensurate with the scaling in die size, voltage, and frequency to support a cap in power consumption. In addition, the wider range of power and frequency offerings will enable performance and cost tradeoffs to be made between the various market segments (Viswanath et al. 2000).

The need for higher performance and an increased level of functional integration as well as die size optimization on the microprocessor leads to preferential clustering of higher power units on the processor. In turn, this leads to a higher heat-flux concentration in certain areas of the die and lower heat fluxes in other regions on the die, which manifest themselves as large temperature gradients on the die. This issue has become increasingly important for advanced microprocessor architectures, which requires that thermal designs have to meet stringent heat-flux requirements that are significant multiples of the average heat flux at the silicon-package interface. A similar trend has been seen in the mobile processor market segment. This market segment is typically constrained by battery life (~2–3 h). The additional constraint is that the form-factor must be small and light to allow for portability. The desktop market is more cost sensitive, and the mobile market is more space and weight sensitive. These sensitivities place bounds on the effective power removal capabilities of the chassis (Viswanath et al. 2000). The solutions to the problem has been explored such as (1) design and architecture of the microprocessor with optimal performance and power consumption; (2) development of novel cost-effective technologies including optimal thermal solutions in microprocessor and system packaging with emerging innovative materials such as ultrathin film and nanomaterials.

More specifically, efficient thermal management is critical for computing systems, which resulted from two major reasons: (1) The reliability of circuits (transistors) is exponentially dependent on the operating temperature of the junction. For example, small differences in operating temperature (order of 10–15°C) can result in approximately two times difference in the lifespan of the devices. (2) The speed of the microprocessor is a function of the operating temperature. At lower operating temperatures, due to reduced gate delay, microprocessors can operate at higher speeds. A secondary effect of lower temperatures is related to a reduction in idle power dissipation (also known as leakage power) of the devices, which manifests itself as reduction in overall power dissipation. Basically, the options improving a thermal design lie in management of power consumption by the processor, and management of the design and technology elements to meet the individual thermal-resistance budgets. Power consumption trends on the microprocessor are becoming an increasing area of concern due to the complexities in packaging technologies as well as system thermal design and cost. As a result, there is increasing emphasis on microprocessor architecture and design to contain and manage processor power against performance as well as die area (Viswanath et al. 2000).

### Thermal Design Targets

Power delivery designs are typically based on the theoretical maximum power that is drawn by the processor, which is based on a synthetic (power virus) code. The theoretical maximum power drawn is based on a synthetic code that is designed to use resident data from the on-chip caches. The pipelines and queues are maintained full to the best possible extent. Given the superscalar and superpipelined architectures of the microprocessor, such activity could conceivably occur over brief bursts of time, but would not likely be sustainable over long periods. Furthermore, if the thermal designs are done to a lower power target than the maximum power, the thermal capacity of the system may be able to support temporary bursts of power consumption over short durations, without violating the central processing unit (CPU) thermal specifications. For example, thermal designs targeting 75% of the maximum power may have little or no impact on system performance.

### Process Scaling

The most obvious power reduction is achieved through process technology optimization. Since $P \sim CV^2f$, significant reductions can be achieved at a given frequency by operating at a lower voltage as well as through a reduction in total capacitance (by reducing the die size).

### Circuit Design

There continues to be a significant emphasis to manage power through design techniques. Design techniques such as clock gating and functional unit block power downs are used to conserve active power consumption on maximum power applications. The key is to ensure that the power savings are achieved without adversely impacting the die size or performance of the processor. Speculative execution techniques can provide a significant performance advantage but at the cost of increased power consumption. Therefore, power/performance trade-off studies are carried out to ensure that the end result is a net gain.

### Smart Voltage Regulation

The output voltage of the voltage regulator is a function of the load on the die. When the current drawn from the power supply is high, the output voltage droops. At zero load, the voltage delivered to the processor is highest. Classical designs involved maintaining voltage regulator flat over the entire range of current load. New voltage regulator designs are able to respond with dynamic voltage outputs that guarantee safe microprocessor operation with power savings. For example, the use of this method has shown to provide a 10–12% reduction in CPU thermal design power in the mobile environment. In addition, the use of advanced virtual reality designs can result in idle power savings as well as improved battery life (Viswanath et al. 2000).

## *Packaging-Level Solutions*

The packaging technology for microprocessors of computing systems has primarily moved toward flip-chip attach for interconnecting the active side of silicon to an organic substrate. The substrate can be socketed in the case of pin grid array (PGA) and surface mounted in the case of ball grid array (BGA) packages. As is typical in flip-chip packaging, the primary mode of heat removal is from the back surface of the silicon. The thermal energy is removed ultimately by a heat sink or heat pipe to the surrounding ambient. The choice of packaging architecture used depends on the total power of the processor as well as the heat-flux density. Two basic architectures are identified: (1) FCxGA1 dealing with low-power processors, and (2) FCxGA2 dealing with medium- to high-power processors. The term xGA could stand for either PGA or BGA, and refers to the next level of interconnect. Figure 12.4 shows the basic implementation of these two architectures. FC-xGA1 deals with directly interfacing the die to the heat sink and therefore involves the design and development of TIM1 thermal-interface materials such as grease and phase-change films (PCFs) and associated retention mechanisms. FC-xGA2 architecture is designed to be scalable and meet the demands of medium- to high performance (power) processors. These involve the integration of a heat spreader to the back surface of

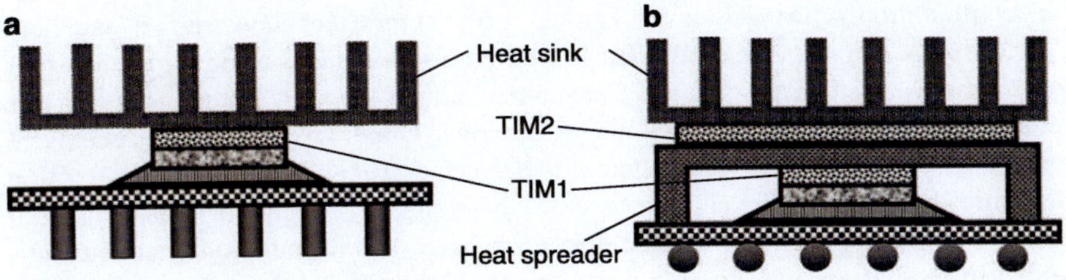


**Fig. 12.4** Schematics of **(a)** FC-xGA1 and **(b)** FC-xGA2 packaging architectures

silicon using TIM1 thermally conductive gels or epoxies, and the integration of the heat spreader to a heat sink using TIM2 thermal interface materials. The proper selection of the TIMs was discussed in Chapter 8. It is important to identify that the degree of voiding at the interface has a significant impact on thermal performance, since voids are air gaps that act as conduction heat transfer barriers. Microprocessor packages are designed to survive field-use conditions typically over 7–10 years. Packages are subject to a range of stress suites to ensure that the device meets performance specifications over the lifespan of operation (Viswanath et al. 2000).

FC-xGA1 packaging architecture encompasses packaging solutions for low- and medium-power microprocessors, predominantly encountered in the value and mobile processor market segments. Three TIMs are usually used for interfacing the heat sink to the die through a compliant interface material.

1. *Elastomeric Thermal Pads*. This class of materials (also known as gap-filler pads) is used to improve heat dissipation across large gaps, by establishing a conductive heat-transfer path between the mating surfaces. Thermal pads are typically 200–1,000 μm thick and are popular for cooling low-power devices, such as chip sets and mobile processors. The pad consists of a filled elastomer, with filler materials ranging from ceramic to boron nitride for varying thermal performance. Metal particles are seldom used due to the risk of dislodged particles resulting in electrical shorts. Another key requirement is that the pads need to be compliant: They should be capable of being compressed to within 25% of their total thickness. This is necessary due to the tolerance variation of large gap situations. The compressibility ensures that the pads can absorb the tolerance variation in assemblies. The tradeoff therefore is that the increase in filler materials (for lower thermal resistance) results in hardening of the pad and hence reduced compliance. Typical failure mechanisms are increased thermal resistance due to inadequate pressure or loss of contact at one or more surfaces. The thermal performance is also sensitive to the contact pressure at the mated surfaces. Some thermal pads have a thin layer of pressure-sensitive adhesive applied to promote adhesion at the interfaces. Nevertheless, the constraints discussed above significantly limit the thermal performance achievable with this class of materials (Viswanath et al. 2000).

2. *Thermal Greases*. Thermal greases offer several advantages over pads, including the ability to conform to the interfaces. They require no post-dispense processing (e.g., no cure) and they have higher effective thermal conductivity compared with other classes of materials. Greases have been used very successfully in combination with various packaging form-factors and have shown excellent performance. However, certain design and environmental considerations can preclude the use of thermal greases. Thermomechanical jeopardy to the processor was identified under certain environmental stresses. Under cyclic loading, extensive thermo-mechanical stresses exerted at the interface because of the relative motion (flexure) between the die and the base of a heat sink lead to loss of grease material from the interface. Under high-temperature bake, the formulation chemistry utilized in typical thermal greases can result in separation of the polymer and filler matrix due to the migration of the polymer component. The separation and loss of

polymeric material could result in poor wetability at the interfaces, resulting in an increase in thermal resistance, i.e., "dry-out." This phenomenon is strongly dependent on the temperature of the material with higher temperatures resulting in accelerated degradation. The failure mechanisms encountered are a strong function of the thermal grease operating temperature and the number of on/off cycles that the processor assembly has been subjected to. The rate of thermal degradation is also dependent on the surface finish of the mating surfaces (heat spreader surface vs. back side of silicon). The pump-out mechanism and phase separation mechanisms have an exponential dependence on temperature, with a twofold increase in degradation for every 10°C increase in average operating temperature of the interface material. For power cycling, the assembly between 0 and 100°C over 7,500 cycles results in a fourfold to sixfold increase in thermal resistance compared to a negligible increase in resistance for a 0–80°C exposure over 2,500 power cycles. Mechanical shock and vibration is related to retention of a heavy heat-sink mass interfaced to the die through a compliant material such as grease. Mechanical shock and vibration of heat-sink masses between 200 and 250 g indicate that the retention of the heat sink to the processor is critical to prevent die damage. During shock testing, relative motion between the heat sink and die could lead to mechanical damage to the die surface, with the corners of the die being highly susceptible to damage. This fail mechanism is strongly dependent on the design of the heat-sink retention feature as well as the mass of the heat sink used. Typically the heat-sink mass (or volume) is proportional to the power dissipation of the processor; a heavier heat sink is required to cool higher processor power. In the case of lighter die loading, die damage due to mechanical shock is not a concern. In summary, thermal grease-based materials are recommended for applications at lower operational temperatures (to alleviate phase separation), lower die loading (to alleviate mechanical damage), and lower power cycling requirements (to alleviate pump-out). However, the limitations in the use of thermal grease triggered the development of an alternate material (Viswanath et al. 2000).

3. *PCFs*. PCFs are a class of materials that undergoes a transition from a solid to semisolid phase with the application of heat; the material is in a liquid phase under die-operating conditions. This class of materials offers several advantages including the ability to conform to profiles of the mating surfaces, no post dispense processing (e.g., no cure), and ease of handling and processing due to its availability in a film format. However, from a formulation perspective, the polymers and filler combinations that can be utilized impose limitations on the thermal performance of these materials. PCFs are typically a polymer/carrier filled with thermally conductive filler, which change from a solid to a high-viscosity liquid (or semisolid) state at a certain transition temperature. The choice of materials is tailored such that the transition occurs below the operating temperature of the die. Key advantages of PCFs are related to their ability to conform to surfaces and their wetting properties, which significantly reduces the contact resistance at the different interfaces. These materials usually are reinforced with a fiberglass mesh, which acts as a core, providing mechanical rigidity. Due to this composite structure, PCF materials are able to withstand mechanical forces during shock and vibration, protecting the die

from mechanical damage. The semisolid state of these materials at elevated temperatures resolves issues related to “pump-out” under thermomechanical flexure. Typically, dispense processes required for thermal greases are throughput limiters. The manufacturing throughput of the assembly line is greatly improved because PCFs can be preattached to the base of a heat sink or heat spreader using a pick-and-place operation. The reliability of the PCF material assembled on the processor package was demonstrated through various stress suites. On average, it was observed that the thermal performance improves over the course of reliability stressing. Because the material is in a softened state during these stresses, the compressive loading at this interface resulted in a decrease in thickness as well as improved wettability (conformance) with the surface irregularities (Viswanath et al. 2000).

Different from the FC-xGA1, the FC-xGA2 architecture is designed to be scalable to meet the demands of medium- to high-performance (power) processors. The integration of the heat spreader on the die required the development and certification of a thermally conductive polymer, a heat spreader, and an adhesive sealant. The heat spreader helps in improving the diffusion of heat flux from the smaller die area to a much larger surface area. This in turn translates to improved thermal performance of the heat sink. For example, copper has a higher thermal conductivity than Al–SiC; it provides roughly 0.1°C/W lower thermal resistance due to improved heat spreading. The architecture precluded the use of thermal greases due to the thermomechanical failure mechanisms. Phase-change materials were precluded because they did not meet the stringent thermal-performance requirements. In addition, the need for a positive compressive load at the interface imposes limitations on the design of this packaging architecture. The highly conductive and commonly utilized metal-filled epoxy thermal polymers could not be used here because of several major obstacles. The high-modulus nature of these materials leads to delamination at the interface due to thermomechanical stresses. In addition, localized phase separation within the material resulted in high-contact resistance. In order to overcome these technological issues, novel chemistries were aggressively investigated. As a result, a thermally conductive, low-modulus gel was developed. The gel is typically a metal particle (Al) or ceramic particle-filled (aluminum oxide, zinc oxide, etc.) silicone polymer with low cross-link density. It combines the properties of both grease and cross-linked polymer; i.e., it is grease that can be slightly cured. Before cure, this material has properties similar to a grease: It has high-bulk thermal conductivities (2–5 W/m K) and conforms well to surface irregularities upon dispense and assembly. Post-cure, this material becomes a lightly cross-linked polymer with significantly lower modulus than epoxy systems. The cross linking reaction provides a high enough cohesive strength to the gel in order to circumvent the “pump-out” issues during temperature and power cycling. The modulus is maintained low enough (MPa range compared to GPa range of epoxies) so that the material can still absorb thermo-mechanical stresses to prevent interfacial delamination. The low-surface energies characteristic of silicones enable good wetting of the mated interfaces, which contribute to minimization of thermal resistance. The application of greases and elastomeric pads is restricted to low- and medium-power devices due to their inherently high-thermal resistance and limitations arising from reliability concerns. Several new material technologies including high-performance PCFs and metal particle-filled gels have been developed and integrated into the FC-xGA2 architecture to deal with high-power devices (Viswanath et al. 2000). Continued development in this area is necessary to satisfy the insatiable performance demands of the next generation of processors, as discussed in Chapter 8.

## *System-Level Solutions*

The primary goal of the system-level thermal solution is to extract the heat from the processor package and discharge it to the ambient air external to the chassis. Air cooling is the most widespread means of system-level cooling in desktop, workstation, and server segments. One or more system fans are employed to move the air within the chassis. Some of the parameters that affect the system thermal design are fan flow rate, acoustic limitations, ambient temperature, and the heat-sink volume. Air cooling employs convection of heat from the heat sink to the ambient air. The amount of heat that can be transferred by convection from the heat sink can be estimated using Newton's law of cooling (Viswanath et al. 2000):

$$\text{Power} = hA(T_s - T_a) \tag{12.3}$$

where $h$ is the heat transfer coefficient, $T_s$ is the temperature on the surface of the heat sink, $T_a$ is the ambient temperature, and $A$ is the total surface area of the heat sink. It is apparent from these equations that in order to increase the total heat transfer from the heat sink, one or more of the parameters $h$, $A$, or ($T_s$ - $T_a$) must be increased. An increase in the heat-sink surface area increases the amount of heat that the heat sink can release to the ambient air stream. The heat-sink surface area can be increased either by increasing the number of fins or by modifying the shape of the fins (i.e., using dimpled or wavy fins instead of rectangular fins). However, increasing the heat-sink surface area results in an increase in the pressure drop across the heat sink. This is because the viscous shear stress acts over a larger area creating a larger frictional force. The system fan must be capable of generating a large enough pressure head to overcome the frictional resistance to the flow of air across the heat sink. The temperature difference between the heat-sink surface and ambient air depends on the efficiency of heat spreading in the heat-sink base and fins. The amount of heat spreading depends primarily on the thermal conductivity of the heat sink material and on the heat sink geometry. For example, copper heat sinks have better spreading than aluminum heat sinks. Similarly, spreading is much better in a heat sink with a thicker base and thicker fins. Enhancing the heat spreading in the heat-sink base through the use of a vapor chamber adds to the total cost of the system cooling solution. The heat transfer coefficient on the heat-sink fins depends primarily on the air flow rate, the spacing between the fins, and the flow regime (i.e., laminar or turbulent) that exists on the heat-sink fins. A higher air flow results in higher heat transfer coefficients and a correspondingly higher pressure drop. The heat-transfer coefficient also depends on whether the flow is fully developed. Fin spacing and the air velocity can be used to determine if the flow is fully developed. In a fully developed flow, the boundary layers growing on adjacent fins merge within the heat sink. Closely spaced fins and lower air flow rates, which cause thicker boundary layers, will result in fully developed flow and lower heat-transfer coefficients. One way to break the growth of the boundary layer, thereby preventing the onset of a fully developed flow, is to use pin fins. Pin fins are usually formed by machining cross cuts across rectangular fins. The gap between the pin fins serves to break the boundary layer; a new boundary layer develops on each downstream fin resulting in a higher heat-transfer coefficient. This is usually accompanied by an increase in the pressure drop across the heat sink. Increasing the fin thickness would lead to a reduction in the number of fins and heat-transfer area. This implies that there would be an optimum fin thickness at which the increased heat spreading would offset the contribution from the decreased fin area to provide the maximum heat transfer from the heat sink. Any increase in finned surface area or change in heat-sink base material from aluminum to copper results in an increase in the heat-sink weight. Any enhancement to the heat transfer from the heat sink is usually accompanied by an increase in the pressure loss across the heat sink. In order to overcome the pressure loss, a larger fan may be required. Increasing the air flow rate to increase the heat transfer coefficient may also require a larger fan. Increased flow rates and larger fans typically result in increased fan noise. Noise attenuation schemes and larger system fans add to the total cost of the cooling solution (Viswanath et al. 2000).

In addition to selecting an optimum heat sink for cooling the processor, the system thermal designer must also consider the layout of the motherboard. In most desktop systems, one or two fans are used to provide air cooling for the processor and other auxiliary components such as memory chips, chipsets, graphics cards, etc. Consequently, the air gets preheated by these components before it reaches the processor heat sink. One simple technique to eliminate or reduce preheating effects is to supply outside ambient air to the processor by using a duct. Typically with this implementation, a second fan is required to provide air flow to cool the other components such as the chipset, memory, and graphics devices on the motherboard. The addition of a duct and a dedicated fan will increase the complexity and cost of the cooling solution. With the advent of multiprocessor systems, there is a greater need to control the ambient temperature local to the processor. In a dual-processor system, it is sometimes necessary (due to electrical layout requirements) to place the second processor downstream of the primary processor. The second processor is therefore in the shadow of the first processor, and the ambient temperature local to the downstream processor may be as much as 10–15°C higher than the upstream processor. One method to alleviate this problem is to lay out the processors in staggered rather than inline fashion on the motherboard (Viswanath et al. 2000).

The system thermal design in compact chassis for the mobile market segment (i.e., laptops and notebooks) poses a special set of challenges. There are severe space and weight constraints on the design of mobile products. Traditionally, mobile processors have been cooled via natural convection and radiation. With the increase in power dissipation, fans have been added to the notebook chassis to circulate the air and enhance the component cooling. In addition, heat pipes or heat spreaders have been used to transport the heat away from the processor to dissipate it at a remote location. One technique uses a copper plate to spread the heat over a large plate mounted just under the keyboard. Natural convection from the keyboard is used to dissipate the heat to the ambient air. A second technique utilizes a heat pipe to transport the heat from the processor through the hinges to the back surface of the display panel. A third concept utilizes a remote heat exchanger with a dedicated fan. A heat pipe is used to transport the heat from the processor module to the remote heat exchanger, which is typically located near the outer wall of the laptop chassis (Viswanath et al. 2000).

Therefore, the need for an integrated approach to deal with the thermal challenges posed by next-generation processors is clear. Improvements in one single area alone will not be able to satisfy the thermal budget requirements. Architecture and design techniques, process shrinks, and voltage scaling are critical to maintain a manageable power-frequency roadmap. Focus needs to continue on packaging materials and technologies to reduce interfacial resistance and improve heat spreading. Board layout designers need to pay attention to the layout of high- and medium-power devices in the vicinity of the microprocessor. System designers need to focus on optimizing air flow and preheating from other components in the chassis. The desired outcome would be to drive design and technology development concurrently at silicon, package, motherboard, and system-level packaging to ensure that thermal solutions can support the demand for increasing computing and communication needs (Viswanath et al. 2000).

## Thermal Management Application in Photonic LED Packaging

Similar to thermal management of electronic packaging, the photonic or optoelectronic industry has a similar daunting task: junction temperature management with demonstrated life test data, with improving performance, and increasing reliability. In fact, heat dissipation, thermal stresses, and cost are key packaging design issues for virtually all semiconductors, including photonic applications such as diode lasers, LEDs, solid state lighting, photovoltaics, displays, projectors, detectors, sensors, and laser weapons. This section will take the LED system as a representative, and give a brief review on its thermal characterization, and thermal solutions that are applied in existing products and the remaining challenges in different LED Systems.

The typical LED packaging configuration is flip chip soldering of the LED device to a silicon submount. The silicon submount provides the electrical interconnect to the die as well as electrostatic zener protection. The submount is secured within the package and wire bonded to the package lead frame. In order to maximize the heat extraction, the contact area from the submount to the die must be maximized. This requires an increase of the size and/or number of electrical interconnects in the LED packages. This requirement must be balanced with the fact that increasing the number of contacts increases die stress and has the potential to reduce packaging yields and reliability. Poor contacts to the die do not always manifest themselves in initial electrical or lumen testing. However, these poor contacts can result in localized heating on the LED and have a major impact on long-term reliability. The reliability of the LED, due to the local heating suffers both in the epitaxial layer and the epoxy encapsulation. The end result is accelerated loss of lumen output or ultimately premature device failure (Arik and Weaver 2004).

## *Thermal Characterization of LED Systems*

Since 1990s, improvements in the epilayer-growth and processing techniques have enabled the realization of mass-production of high-bright LEDs. Currently, high-power LEDs are replacing the cold cathode fluorescent lamp in the LCD display backlights and incandescent lamps for solid state lighting (Horng 1997). Traditionally, LEDs have been driven at a low power with a minimum power dissipation of 150 mW, which meant that most lighting applications required numerous low-power LEDs. The high-power visible LEDs have power dissipations ranging from 500 mW to as much as 10 W in a single package. With improving luminous efficacy, these high-power LED components can and will replace other lighting technologies in most applications. However, the demand for high power LEDs requires efficient thermal management since their performance decreases drastically with increasing chip (p–n junction) temperature. Thermal management of the LED packaging is crucial to the device performance and extended long term reliability, as the p–n junction temperature directly alters the performance and reliability of LEDs in the following ways (Saffa 2007):

1. *Reduced output power.* At constant operating current, the luminous efficiency decreases by about 5% for every 10°C rise in junction temperature.
2. *Reduced forward voltage.* At constant operating current, forward voltage decreases by about 20 mV for every 10°C rise in junction temperature.
3. *Shifted dominant wavelength.* Dominant wavelengths shift by about 2 nm for every 10°C change in junction temperature.
4. *Shifted color temperature.* White LEDs are more sensitive to changes in junction temperature because the color temperature changes significantly. LEDs emit white light by combining standard blue emission with a phosphor overcoat that absorbs the blue flux and reemits a wide range of wavelengths throughout the visible range. Reemission efficiency is highly dependent on the wavelength of the blue flux, which shifts as junction temperature changes. If the dominant wavelength of the blue LED shifts out of the efficient range of the phosphor, more blue flux escapes the package, which increases the color temperature.

5. *Reduced mean time to failure and accelerated degradation.* Catastrophic failure and LED degradation are mechanical and chemical processes which occur at rates described by the Arrhenius model. Their rates are inversely proportional to the exponent of the inverse of junction temperature.

p–n junction temperature depends on three factors: Power dissipation, thermal resistances of the substrate and assembly, and ambient conditions. Power dissipation determines how much heat is generated, while thermal resistances and ambient conditions dictate how efficiently heat is removed. All of the light and heat produced by an LED is generated at the p–n junction. Because the junction is very small, the heat generation rate per unit area is very large. A 1-W LED per $mm^2$ generates 100 $W/cm^2$. This rate is higher than many of today's high-power microprocessors. To maintain a low junction temperature, all methods, including conduction, convection, and radiation of removing heat from LEDs can be considered. LEDs are typically encapsulated in a light transmissive plastic, which is a very poor thermal conductor. Therefore, nearly all heat produced is conducted through the back side of the chip. Moreover, LEDs do not benefit from convection at the component level, because their surface area is too small. However, convective technologies such as fans, heat pipes, and liquid cooling can be used at the packaging or system level for thermal management of LEDs. In addition, heat sinks with large surface area are effective at radiating heat. All three types of heat transference become more efficient as temperature gradients increase. The LED junction temperature will rise at first until the rate of heat transference out of the system is equal to the rate of heat generation at the junction. As shown in Figure 12.5, the thermal characteristics of the LCD assembly are expressed by the following equations (Optek 2006; Luxeon 2006):

$$\Delta T_{J-A} = Q \cdot R\theta_{J-A} \tag{12.4}$$

$$\Delta T_{J-A} = Q \cdot [R\theta_{J-C} + R\theta_{C-A}] \tag{12.5}$$

$$\Delta T_{J-A} = Q \cdot [R\theta_{J-C} + R\theta_{C-S} + R\theta_{\text{TIM}} + R\theta_{H-A}] \tag{12.6}$$

where $Q = I_F \cdot V_F$, $I_F$ is the operating current and $V_F$ is the measured forward voltage of the LED. $\Delta T_{J\text{-}A}$ and $Q$ can be measured and $R\theta_{J\text{-}C}$ is LED designed data. $R\theta_{C\text{-}A}$ is the combined thermal resistance of the rest of the assembly. Equation (12.6) can be used to calculate $\Delta T_{J\text{-}A}$ if sufficient data can be obtained for the substrate, TIM, and heat sink. The multiple-component assembly's thermal characteristics are described by follow equations that are similar to those for single component assemblies (Optek 2006):

$$\Delta T_{J-An} = Q_n \cdot R\theta_{J-An} \tag{12.7}$$

$$\Delta T_{J-An} = Q_n \cdot R\theta_{J-Cn} + Q_{\text{Total}} \cdot R\theta_{C-A} \tag{12.8}$$

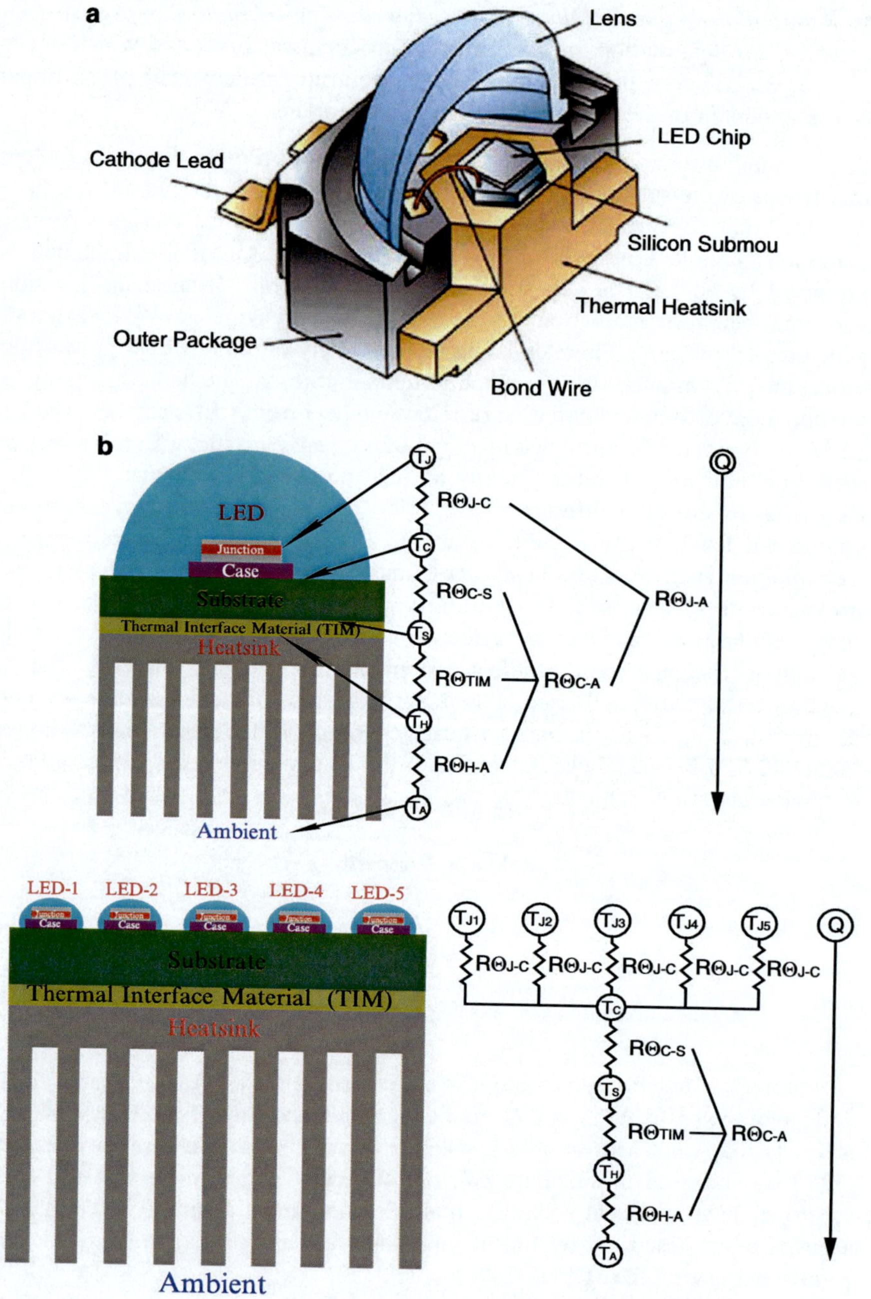


**Fig. 12.5** Schematics of light-emitting diode (LED) packaging: (**a**) LED device; (**b**) Thermal model for single-component assembly; (**c**) Thermal model for a multiple-component assembly

$$\Delta T_{J-An} = Q_n \cdot R\theta_{J-Cn} + Q_{\mathrm{Total}} \cdot [R\theta_{C-S} + R\theta_{\mathrm{TIM}} + R\theta_{H-A}] \tag{12.9}$$

$$R\theta_{C-A} = [\Delta T_{J-An}]/Q_n - R\theta_{J-Cn} \tag{12.10}$$

$T_C$ is the same for all components on the multiple-component assembly, and $R\theta_{C\text{-}A}$ can be derived based on one component's $\Delta T_{J\text{-}A}$, $Q$, and $R\theta_{J\text{-}C}$. For the same component and power dissipation, $\Delta T_{J\text{-}C}$ will be the same whether the LED is alone or is part of an array. In an array, however, the heat input of all LEDs must be transferred through the substrate, TIM, and heat sink. $\Delta T_{C\text{-}A}$ and $\Delta T_{J\text{-}A}$ increase considerably over single-component assemblies. When making a choice between 1 $p$-W component and $p$ 1-W components, the $p$-W component must have $p$ times lower thermal resistance than the 1-W component for the junction temperatures of both designs to be the same. The reality is that most package technologies for high-power LEDs have similar thermal resistances. Spreading the heat input to multiple components is recommended because less thermal management is required.

Therefore, recommendations for reducing junction temperature without compromising luminous flux can be summarized as (Optek 2006): (1) Use components with better luminous efficacy to reduce $I_F$ and $Q$, (2) increase the number of components at the same total power dissipation to reduce $R\theta_{J\text{-}C}$; (3) change to better packaged components to reduce $R\theta_{J\text{-}C}$; (4) use high thermal conductivity substrates to eliminate $R\theta_{C\text{-}S}$; (5) use high thermal conductivity heat sinks and TIMs to eliminate $R\theta_{C\text{-}S}$ and $R\theta_{\mathrm{TIM}}$, (6) increase the heat sink's surface area to reduce $R\theta_{H\text{-}A}$, and (7) add a fan, heat pipe or liquid cooling to reduce $R\theta_{H\text{-}A}$.

## *Design Guideline for LEDs with Efficient Thermal Dissipation*

As the LED heat escalates, several key characteristics may become apparent, which demonstrate the importance of proper LED thermal design. The forward voltage will begin to decrease. The decreasing voltage can impose an increased load on related LED driver components causing their temperature to increase as well. In resistor driven circuits, the forward current will increase. As the LED lights temperature continues to rise, the optical wavelength can shift. The increasing wavelength can cause orange LED lights to appear red or even white LED lights to appear bluish. This color shift typically intensifies with the AlInGaP technologies (red, orange, amber, and yellow). In addition, a thermally stressed LED lights will loose efficiency and light output will diminish. If the LED thermal management continues to race out of control, the LED junction may break down causing a state of complete thermal runaway. The result is typically catastrophic failure. Other affects of overstressed LEDs may include broken wire bonds, delaminating, internal solder joint detachment, damage to die-bond epoxy, and lens yellowing. Therefore, the thermal management design of an LED application is very important to ensure its reliability and optimum performance. The maximum junction temperature of the die inside the package is based on the allowable thermal stress of the package material which can not be exceeded to avoid a catastrophic failure of the device (Lunaraccents 2001).

### Efficiency of Thermal Transfer

The key to a successful design starts with the transfer of LED heat. Each custom LED lighting design involves the concept of efficiently transferring as much heat as possible away from LED p–n junction. The process begins within the LED lamp, where thermal energy released into an integrated slug can potentially exit the LED. Modern surface-mount LED lamps depend on the thermal efficiency of this slug. Traditional through-hole LEDs actually produce much less heat and can dissipate some into the actual wire leads. Other surface-mount LED lights rely on their power and ground pads to dissipate the heat. The slug found in modern LED lamps requires a secure bond with an underlying circuit board pad to provide an efficient means of heat transfer out of the LED lights. Corrupted solder joints or LED shifting caused during the assembly process can interrupt the flow of heat through the slug and into the pad. If the connection is reliable, the thermal energy transfer continues into the PCB copper surface area, or steel core in the case of a metal circuit board. From this point, the energy is typically transferred through a small copper plated hole in the printed circuit board, commonly referred to as the thermal via. The thermal via may lead to an extensive copper area on the opposite circuit board layer. Thermal energy is allowed to dissipate into this copper area and finally into the surrounding atmosphere. "Thermally enhanced PCB" is the common term for a circuit board equipped with such features utilized for LED thermal management.

### High-Efficiency Circuit Boards (PCBs)

A high-power LED lamp mounted on a standard PCB will dissipate thermal energy into the circuit board material as well as small thermal vias located under the LED itself. Although the thermal via is relatively small, it offers a path of high thermal conductivity, which is required to transfer heat away from the LED. Specialized PCBs offer a superior method of heat transfer. Rather than rely on a single thermal via, the entire circuit board provides a path of low thermal resistance, for LED thermal management. The metal circuit board offers high thermal conductivity because of its special composition. This allows heat to dissipate in all directions, as opposed to a single direction as found with the traditional FR4 circuit board. Reduced overall thermal resistance results in increased LED performance. Increasing LED drive currents as well as LED lifespan are now possibilities. An extended lifespan is now feasible, but would have been nearly impossible using a standard FR4 board. The specialized circuit board composition generally consists of ceramic-coated steel. Other materials may include steel, aluminum, aluminum nitride, and graphite core. Proper PCB design can reduce thermal resistance of a LED lamp assembly, and thus reduce the junction temperature of the LEDs. Conventional PCB design involves connecting various points on the board with traces of sufficient width to handle the current load. This process is usually visualized as adding traces to a blank PCB. For LED PCBs, this process should be reversed, visualized a removing metal only where needed to form the electrical circuit. Large metal pads surrounding the cathode leads of the LEDs are ideal. Comparably, as very little heat is conducted through the anode leads of the LED, additional metallization surrounding these leads does not help. The resistors should be located in remote portion of the PCB away from the LEDs, on a separate PCB, or in the wire harness if possible. If this is not possible, the resistors should be distributed evenly along the PCB to distribute the heat generated. In addition, the traces from resistors to metalized area surrounding cathode leads on the LEDs should be minimized to prevent resistors from heating adjacent LEDs. This can be accomplished by thinning down these traces, or by having metalized areas contacting the LEDs and resistors only contact the anode leads of the LED. Proper LED spacing should be controlled. Most of the electrical power in an LED is dissipated as heat. Tighter LED spacing provides a smaller area for heat dissipation, resulting in higher PCB temperatures and thus higher junction temperatures. The LEDs should be spaced as far apart as packaging and optical constrains will allow. Most applications use only a single row of LEDs at spacing greater than 15 mm which is ideal, as opposed to many amber turn signal applications which use a tightly spaced (less than 10 mm) $x$–$y$ array of LEDs.

## Thermal Dissipation from Enclosures with Proper Lamp Housing Design and Mounting of the LED Array

The most challenging custom LED lighting design is that which contains an airtight enclosure or housing. Some applications restrict the use of enclosure ventilation or fans for forced air convection. When heat generated by the LED lights cannot escape, the sealed enclosure will begin to develop an oven effect. As the temperature continues to increase, the LEDs may begin to suffer from thermal stress. The first solution may be to select an enclosure with a material composition rated for high thermal conductivity. If enough heat can escape through the enclosure walls, the LED lights and circuitry may function more normally. However, modern LEDs produce significant amounts of heat. Most fully enclosed configurations will always require additional thermal management. In some cases, forced air convection might be a possibility. An alternative solution features the hybrid housing, where the rear side of the housing incorporates an actual heat sink. For example, consider a 1-in diameter tube casing. The front half of this enclosure must consist of a transparent material such as polycarbonate. The polycarbonate front would mate with a somewhat similar component composed of aluminum. The rear side of the aluminum features a series of fins to aid in heat dissipation. The interior LEDs share a physical connection directly with the aluminum backing. If designed carefully, such an enclosure can provide an airtight environment for electronics, but will also allow for proper heat dissipation though the series of aluminum fins as a means of LED thermal management.

In general, LED lamp housings should be designed to provide a conductive path from the backside of the PCB to the lamp housing. This is typically accomplished by mounting the backside of the PCB directly to the lamp housing such that they are contacting one another across the entire length of the PCB. This mounting scheme can be improved by applying a thermally conductive pad between the PCB and the lamp housing. The thermally conductive pad conforms to the features on the backside of the PCB and provides a larger contact area for conduction. Often the PCB is mounted to the lamp housing on top of raised bosses. In this case, the area for conduction into the lamp housing is reduced to the contact area on the top side of the bosses, greatly reducing its effectiveness. Another common configuration mounts the PCB along its top and bottom edges to slots in the side of the lamp housing. Again, the area for conduction into the lamp housing is reduced to the contact areas of the slots, which reduces the effectiveness of conduction. If the PCB is mounted in such a way that conduction to the lamp housing is not effective (trapped air is a very poor conductor of heat), then allowances for convective cooling should be made. The most common techniques to take advantage of natural convection is to put holes in the top and bottom side of the lamp housing to allow for vertical air flow over the PCB. However, where the lamp housing must be sealed for environmental reasons, this type of approach is impractical.

## Circuit Design

Circuit design can help control the junction temperature of the LEDs through (1) minimizing fluctuations in the drive current (power input); and (2) dissipating a minimum amount of heat, or dissipating heat in such a way as to minimize its effect on LEDs. An ideal drive circuit will provide the same current to the LEDs even as ambient temperatures and battery voltages vary. Simple current control circuits can be designed to accomplish this task. Current control circuits are often too expensive and unnecessary for LED applications. The most common LED drive circuit consists of a current limiting resistor(s) and a silicon diode for reverse voltage protection in series with the LEDs. In this circuit design, the input current into the LEDs varies as the battery voltage changes. The current control characteristics of this type of circuit improve as larger resistors are used with fewer LEDs in series. However, circuits with fewer LEDs in series will have greater heat generation in the drive circuit.

If the LED drive circuit is in a remote location relative to the LEDs in the wire harness or on a separate PCB, the power dissipated by the drive circuit does not affect the junction temperature of the LEDs. Drive circuit heating is a concern when the drive circuit is on the same PCB as the LEDs. Drive circuit power dissipation, and thus heat generation is inversely proportional to the number of LEDs in series.

Circuits with fewer LEDs in series will have greater heat generation in the drive circuit. For most automotive applications in which the battery voltage is approximately 13 V, four LEDs in series is a good compromise between forward current control, heat generation, and minimum turn-on voltage for the LED array.

Current sources, which operate efficiently over a wide range on input voltages, can be designed using pulse-width modulation circuitry. Such circuits have the advantage of low heat dissipation, and large input voltage compliance. This type of power supply is traditionally used in applications where electrical efficiency and heat dissipation are of critical importance, such as a laptop computer. Due to their widespread adoption in other applications, the cost of components has decreased, and their availability has increased, making this an interesting alternative for driving LED arrays.

Drive circuitry can be designed which compensates for increasing ambient temperature by decreasing the forward current to the LED array. This allows the lamp designer to drive the LED array at a higher forward current by reducing the amount of current derating. Temperature compensation is achieved by incorporating temperature sensitive components into the drive circuitry, such as positive temperature coefficient resistors.

## External Heat Sinks and Forced Air Convection

When thermal energy generated exceeds the thermal energy dissipated, an additional means of cooling may be required to maintain LED performance. Some applications feature numerous LED lights within a refined space and cannot efficiently dissipate enough heat from circuit board copper alone. In these extreme but not rare cases, an external heat sink is always required. The external heat sink is an efficient and inexpensive method of expending the surface area necessary to dissipate heat generated by the LED array. Specific heat sink materials offer higher performance due to high thermal conductivity, and a lower thermal resistance. For example, copper is a coinage metal with a high thermal conductivity, and would perform excellent in a heat sink application for LED thermal management. Aluminum is also commonly utilized for heat sink manufacturing, although its thermal conductivity is nearly half that of the copper. Adding the feature of forced air convention can promote a cooler, more efficiently operating heat sink. A small electric fan forces air across the heat sink fins removing heat faster and more efficiently. This feature is often necessary when ambient airflow is limited or nonexistent.

Typical design tasks include deciding on the overall heat sink outline–the maximum space it may occupy; the fin structure and how it is made; cosmetic or radiant coatings on the heat sink; and the air side conditions. The air side conditions are most often natural convection for illumination applications, both because of reliability and ambient noise concerns. For projection applications, the diode light sources are in a unit along with other electronics, making a forced convection solution, either gas or liquid, acceptable.

Design goals for the analysis include not only accomplishing the mechanical design, but also determining suitable electrical operating conditions for the product. This might include the drive current level to use, as well as the temperature at which the device should operate. This temperature "limit" differs from the typical electrical component limits, in that the light output is maximized at a temperature lower than the allowable maximum temperature. An advantage of LED lighting over other types of illumination is supposed to be energy savings, but this savings would be negated if the device were to be operated at the maximum temperature. Other considerations for the temperature limit include reliability of thermally cycled connections and interfaces, and especially lumens maintenance (light output) over time. In fact, lifetime for a light source is often determined on the basis of relative brightness (Biber 2008).

### *Thermal Management Solutions and Challenges of LEDs*

Like thermal management of conventional electronic packaging, the thermal management of LEDs can also divided into passive and active thermal management systems. Passive thermal management systems have no moving parts or consumption of additional energy. They rely primarily on conduction and radiation to remove heat from the junction. The typical method is to attach LEDs to a thermally conductive substrate, such as a metal-core insulated metal substrate (IMS) or ceramic substrate, and then attach the substrate to a heat sink. Novel technologies such as Anotherm (anodized aluminum oxide thermal subtrate) make it possible to attach the LEDs directly to the heat sink. Heat is conducted to the heat sink and radiated from its surface. Thermal performance is enhanced by reducing the length and thermal resistances along the path to the heat sink and by increasing the surface area of the heat sink (Welwyn 2010).

Active thermal management systems involve convection by incorporating fans, heat pipes, and liquid cooling. These technologies enable significantly better thermal management and should be considered for ultrahot applications. In most cases, they are more complex and require better design to avoid decreasing the reliability of the system. These trade-offs are manageable if extreme thermal management is required. Additionally, low-attachment-stress wire bonding can be used for attaching wires directly on top of active device. This is good for edge emitting laser diodes and vertical-cavity surface-emitting LDs.

New materials for mounting and heat sinking have been developed for die attach (such as epoxies and solders) and heat sinking that absorb very little ultraviolet (UV) or visible light, and that do not degrade even at 175°C over 100,000 h. These materials must help minimize thermal expansion mismatch between chip and package. If they are used for heat sinking they must have very high thermal conductivities (such as diamond or diamond-like carbon). In addition, these materials must be environmentally friendly (e.g. solders must be lead free). Since indium-based lead free solders have low melting temperatures (~120°C), for instance, it may be necessary to use solder alloys like Sn–Ag (such as Sn3.9Ag0.6Cu) and Sn–Cu with the melting temperature of ~220°C, provided their processing temperatures are not too high. Moreover, underfills must also be developed that withstand 175°C and UV light (Tsao 2002).

Another approach is to use direct chip on board. The packaging approach has advantages and disadvantages that need to be explored. Advantages of this approach include reduced thermal impedance because the chip is directly placed on the board, removing the package level impedance. It also reduces overall cost by removing a portion of the packaging costs. As die efficiencies and thermal efficiencies improve this packaging approach allows for greater packaging density or lumens/cm$^2$. Disadvantages of this approach include that it, like any multichip module approach, requires known good bare die to achieve reasonable yields. Another disadvantage is the thermal-expansion mismatch between typical inexpensive epoxy resin based and metal core PCBs. This may drive the cost up due to the need for a lower CTE material based on ceramics (Tsao 2002).

## Thermal Management Application in Sustainable Energy Generation

This section focuses mainly on thermal management of batteries, fuel cells, and solar cell packaging.

### *Thermal Management of Batteries*

Battery performance, life, and cost directly affect the performance, life, and cost of the battery-powered electric devices, such as electric vehicles (EVs) and hybrid electric vehicles (HEVs). Battery temperature influences the availability of discharge power (for start up and acceleration), energy, and charge acceptance during energy recovery from regenerative braking. These affect vehicle drive-ability and fuel economy. Temperature also affects the life of the battery. Therefore, ideally, batteries should operate within a temperature range that is optimum for performance and life. The desired operating temperature range is different for different battery types with different electrochemistry. Usually, the optimum temperature range for the battery operation is much narrower than the specified operating range for the vehicle. For example, the desired operating temperature for a lead acid battery is 25–45°C, however the specified vehicle operating range could be −30 to 60°C. In addition to considering the temperature of a battery pack, uneven temperature distribution in a pack should also be considered. Temperature variation from module to module in a pack could lead to different charge/discharge behavior for each module. This, in turn, could lead to electrically unbalanced modules/packs, and reduced pack performance. For high temperature batteries such as ZEBRA and lithium metal polymer batteries, thermal management is considered an integral part of the battery pack and has been included in the design of the battery. The ambient temperature batteries such as valve regulated lead acid (VRLA), nickel metal hydride (NiMH), and lithium ion (Li-Ion) have also been designed recently with battery thermal management systems–some more elaborate than others (Pesaran 2001). A thermal management system could be designed with a range of methods; however, the basic performance of the management system is dictated by the thermal design of each cell or module. In order to find a high performance and cost effective cooling system, it is necessary to evaluate system thermal response and its sensitivity as a function of controllable system parameters (Pesaran et al. 1999).

The goal of a thermal management system design is to deliver a battery pack at an optimum average temperature dictated by life and performance trade-off with even temperature distribution or only small variations between the modules and within the pack. In addition, the pack thermal management system has to meet the requirements of specific applications, such as compact, lightweight, low cost, easily packaged, compatible, reliable, easily accessible for maintenance, use of low parasitic power, allow the pack to operate under a wide range of climate conditions, and proper ventilation if the battery generates potentially hazardous gases. A thermal management system may use air for heat/cooling/ventilation, liquid for cooling/heating, insulation, thermal storage such as phase-change materials, or a combination of these methods. The thermal management system may be passive (i.e. only the ambient environment is used) or active (i.e., a built-in source provides heating and/or cooling at cold or hot temperatures). The thermal management control strategy is done through the battery electronic control unit (Pesaran 2001).

Choice of heat transfer medium has a significant impact on the performance and cost of the battery thermal management system. The heat transfer medium could be air, liquid, phase-change material, or any combination. Heat transfer with air is achieved by directing/blowing the air across the modules. However, heat transfer with liquid could be achieved either through discrete tubing around each module; with a jacket around the module; submerging modules in a dielectric fluid for direct contact; or placing the modules on a liquid heated/cooled plate or heat sink. If the liquid is not in direct contact with modules, such as in tubes or jackets, the heat transfer medium could be water/glycol or even refrigerants, which are common automotive fluids. If modules are submerged in the heat transfer liquid, the liquid must be dielectric, such as silicon-based or mineral oils, to avoid any electrical shorts. Using the air as the heat transfer medium may be the simplest approach, but it may not be as effective as heat transfer by liquid. The rate of heat transfer between the walls of the module and the heat transfer fluid depends on the thermal conductivity, viscosity, density, and velocity of the fluid. For the same flow rate, the heat-transfer rate for most practical direct-contact liquids such as oil is much higher than with air because of the thinner boundary layer and higher fluid thermal conductivity. However, because of oil's higher viscosity and associated higher pumping power, a lower flow rate is usually used, making the oil heat transfer coefficient only 1.5–3 times higher than with air. Indirect-contact heat transfer liquids such as water or water/glycol solutions generally have lower viscosity and higher thermal conductivity than most oils, resulting in higher heat transfer coefficients. However, because the heat must be conducted through walls of the jacket/container or fins, indirect contact effectiveness decreases. Although liquid cooling/heating is more effective and takes up less volume, it has its drawbacks. It could have more mass, has a potential for leaks, need more components and could cost more. Maintenance and repair of a liquid cooled pack is more involved and costliness. Indirect liquid cooling, with either jackets or cold plate is easier to handle than direct liquid cooling (Pesaran 2001).

There are two methods for distributing air to a pack for cooling and/or heating. The first method is series cooling, where air enters from one end of the pack and leaves from the other, exposing the same amount of air to several modules. The second method is parallel cooling, where the same total airflow rate is split into equal portions, and each portion flows over a single module. Depending on the size and geometry of the modules, series-parallel combination could be configured. Parallel airflow usually provides a more even temperature distribution among the modules in the pack. In parallel flow design, however, distributing airflow uniformly to a large battery pack will require a careful design of the air manifold (Pesaran 2001).

The relative need of the thermal management of each of the batteries, such as VRLA, NiMH, Li-Ion, depends on the heat generation rate from each type of battery, its energy efficiency, and the sensitivity of performance to temperature. For example, NiMH batteries generate the most heat at high temperatures (>40°C) and are least efficient. At room temperature, NiMH generates less heat than VRLA and Li-Ion. The performance of a NiMH battery is more sensitive to temperature than VRLA and Li Ion batteries. Therefore, NiMH batteries need a more involved battery management control. This is also evident from various efforts to use the more effective liquid cooling for NiMH batteries. The concerns for Li Ion packs are safety and relatively poor performance at very cold temperatures. Since Li Ion batteries can deliver much more power and thus more heat for the same volume than either VRLA or NiMH, heat removal needs to be efficient. Thermal management also depends on the type of vehicle and where the pack will be located. For EV and series HEV, the pack is generally large and its thermal management system may need to be more elaborate, possibly incorporating liquid cooling, particularly for NiMH. However, for parallel HEVs, the pack is generally smaller and the thermal control could be achieved by a simpler air cooling/heating design, especially for Li Ion and VRLA (Pesaran 2001).

A well-designed thermal management system, for instance, is required to regulate EV and HEV battery pack temperatures evenly, keeping them within the desired operating range. Proper thermal design of a module has a positive impact on overall pack thermal management and its behavior. A thermal management system using air as the heat transfer medium is less complicated, though less effective, than a system using liquid cooling/heating. Generally, for parallel HEVs, an air thermal management system is adequate, whereas for EVs and series HEVs, liquid-based systems may be required for optimum thermal performance.

NiMH batteries require a more elaborate thermal management system than Li Ion and VRLA batteries. Li Ion batteries also need a good thermal management system because of safety and low temperature performance concerns. The location of the battery pack may also have a strong impact on the type of battery thermal management and whether the pack should be air cooled or liquid cooled.

## *Thermal Management of Fuel Cells*

The heart of a fuel cell is an electrochemical "cell" that combines a fuel and an oxidizing agent, and converts the chemical energy directly into electrical power, water and waste heat Burns et al. 2002. The typical fuel cells are acid based proton exchange membrane hydrogen–oxygen fuel cells. A hydrogen molecule reacts at the anode to create a pair of protons and electrons. The proton ion exchange membrane conducts the protons which were generated at the anode from the anode to the cathode. The electrons which were also generated at the anode are conducted through the electrical load that is connected to the fuel cell and also reach the cathode. The hydrogen protons and the electrons are reacted at the cathode with an oxygen atom to produce a molecule of water. A "stack" of cells connected electrically in series is usually employed in applications. Since all the energy that is not converted to electrical energy is dissipated as heat, the heat generated by a fuel cell can be calculated as (Burke 2008)

$$\begin{aligned} Q &= (\Delta H - W_a) m_{H_2O} (1{,}000\,\mathrm{W\,s/kJ}) \\ &= \left( -nFE_{\mathrm{therm}} + nF\left(\frac{I_a}{I_f}\right) E_a \right) m_{H_2O} \end{aligned} \tag{12.11}$$

where $Q$ is waste heat generated by the fuel cell, Watt; $m_{H_2O} = I_f/nF$, fuel consumption, moles of water produced per second. $I_a$ is actual fuel cell output current, Ampere. $I_f$ is theoretical fuel cell output current, Ampere. $n$ is number of electrons per mole of water formed. $F$ is Faraday's constant (96,487 A s/mol of electrons). $\Delta H$ is enthalpy of formation, kJ/mol of water formed. $W_a$ is actual electrical energy produced, kJ/mol. $E_{\mathrm{therm}}$ is thermal neutral cell voltage, Volt. $E_a$ is actual fuel cell voltage, Volt. Typically the actual fuel cell output current is approximately equal to the theoretical fuel cell output current. Assuming the two quantities are equal to each other, (12.11) can be further simplified to,

$$Q = I_f(-T_{\mathrm{therm}} + E_a) \tag{12.12}$$

Taking (12.12) and dividing by the active area of the fuel cell

$$\frac{Q}{A} = \frac{I_f}{A}(-E_{\mathrm{therm}} + E_a) = q = i_f(-E_{\mathrm{therm}} + E_a) \tag{12.13}$$

where $q$ fuel cell waste heat density, $W/cm^2$; $i_f$ theoretical fuel cell output current density, $mA/cm^2$; $A$ Total active area of the fuel cell stack, $m^2$.

Cooling plate can be used for fuel cell thermal management. Heat is absorbed from the fuel cells on either side of the cooling plate. Typically a liquid coolant is circulated within the plate and heat is removed convectively as the coolant passes through plate and out of the fuel cell stack. A passive cooling plate must conduct the heat within the plane of the plate out to one or more of the edges of the plate so that the heat can be transferred to a heat sink external to the fuel cell stack. The benefits of the passive approach are reductions in mass, system complexity, and parasitic power as well as improvements in system reliability. The key to making the passive approach workable is making the cooling plates light enough yet highly thermally conductive so that the heat can be effectively removed and also provide each cell in the fuel cell stack a thermally uniform heat sink. The total in-plane temperature differential across the cooling plate can be expressed as (Burke 2008):

$$\Delta T = \frac{qL^2}{2kt} \tag{12.14}$$

where $\Delta T$ total in-plane temperature differential across the cooling plate; $q$ heat density absorbed by the cold plate, $W/cm^2$; $L$ total length of the cooling plate, cm; $k$ thermal conductivity of the planar material, W/m K; $t$ thickness of the planar material, cm. For fuel cell applications it is ideal to have $\Delta T = 0$, but practically speaking, a $\Delta T \leq 3°C$ is acceptable. The reasons are twofold, first, the fuel cell chemical process kinetics are temperature sensitive. Nonuniformity in the temperature of the fuel cell electrodes means some areas of the electrodes are going to be more active than other areas on the fuel cell electrodes. The second reason is that the water movement within the fuel cell is affected by local variations in the water vapor pressure (which is a function of the water temperature). The fuel cell heat loads is typically $\leq$0.3 $W/cm^2$. The heat loads of fuel cells vary depending on their operating current density, but generally these heat loads are small when compared with microelectronics heat loads which can exceed 100 $W/cm^2$. For even small fuel cells, where the heat transmission distance is 5 cm or less, the conductivity requirements are high. While the thermal conductivity requirements are reduced for thicker plates, but thicker plates rapidly increase the mass of the fuel cell stack. Copper which has a thermal conductivity of 400 W/m K would be adequate only for heat transmission distances of 4 cm or less unless cooling plates $\geq$2 mm were used. Diamond, which has a thermal conductivity of 2,300 W/m K, would be adequate for heat transmission distances up to about 8 cm. Considering these material constraints, it is understandable why the development of passive thermal systems for fuel cells is limited (Burke 2008).

Aside from providing a uniform thermal environment for the fuel cell, it is also important that the mass of the cooling system be minimized. A metric used to evaluate the mass properties of a fuel cell system's cooling subsystem is to calculate the ratio of the waste heat handled by the cooling system to the mass of the cooling system components. For a typical liquid-cooled fuel cell system, the mass includes conductivity, and thickness of the cooling plate, the cooling cells within the cell stack as well as the cooling system components outside the cell stack (e.g., tubing, fittings, coolant, sensors, accumulator, etc.). This ratio should be calculated at the maximum heat dissipation since it is the maximum heat dissipation which sizes the cooling system. Metals such as copper (specific gravity of 8.89 and thermal conductivity of 400 W/m K) and aluminum (specific gravity of 2.7 and thermal conductivity of 180 W/m K) which are used in cooling applications in general, are usually unsuitable for fuel cell cooling except for very small fuel cells. The alternative materials are all carbon based: diamond, diamond composites, pyrolytic graphite, and graphite composites. HOPG has the best combination of high thermal conductivity and low specific gravity. In addition to these materials, heat pipes, which use phase changes to absorb and reject heat are capable of extremely high effective thermal conductivities, and also are candidates for passive cooling plates. Heat pipes are probably the only candidate capable of passively moving the heat for large fuel cells (transmission distances of $\geq$10 cm) (Burke 2008).

Conventional active cooling approaches, such as air blowing, air jetting, and gas–gas exchanging have been used for the fuel cell thermal management. Novel methods like active flow control are also under development.

## *Thermal Management of Solar Cell Packaging*

Solar cells, or photovoltaic (PV) cells, have the ability to convert sunlight directly into electricity. Current technology produces solar cells that are approximately 15% efficient in converting absorbed light into electricity. Concentrator photovoltaic cells, that is, PV cells coupled to sunlight concentrator optics, capture more of the electromagnetic spectrum and are more efficient, converting absorbed light into electricity at about 30% efficiency. Simultaneously, about 70% of the solar energy is converted to heat. Due to the small size of the PV cells and the high-energy absorption, the heat must be efficiently dissipated from the cell to the environment to prevent degradation of performance or damage of the cells. If properly collected, the dissipated heat may be beneficially used to heat structures or drive heat-enabled processes. This requires proper or optimal thermal management for solar cell packaging and power generation systems.

Typically solar cells employ passive cooling for thermal management; even something as simple as aluminum fins can lower the cell temperature by 15°C at 1sun, giving a 6% boost in output power. More often, one method of cooling the cell is to use a heat spreader to spread the heat generated in the cell, and then either passively or actively cool the cell by a heat sink or an exchanger, respectively. Active cooling with water cooled systems is another option, providing both increased electrical output and hot water for household use. Water cooling has several major disadvantages though, including increased structural and maintenance costs for pumps and plumbing plus the risk associated with water leakage and subsequent building damage. Additionally, while residential solar installations will likely benefit from hot water, businesses and utilities probably see this more as a liability than an advantage. Thermoelectric devices on the other hand have no moving parts, require essentially no maintenance, should last for the lifetime of the PV system and can run directly off the direct current (DC) power generated by the PV array. Thermoelectric devices can be used to harvest the waste heat from the PV module and convert it to electricity, or can also be used to cool the PV module to increase its power output. A typical thermoelectric device could increase the PV power output under 1,000 W/m$^2$ illumination by about 1.75%. The same thermoelectric device connected to actively cool a crystalline silicon PV module could increase the net power output by 10.8%. The thermoelectric cooling benefit is much greater for two times solar concentration, where a thermoelectric device could increase the power output by 29.3% (Najarian and Garnett 2006).

The thermal management for a combination of solar and fuel cells (CSF) is also very promising. For example, basic solar CSF power system consists of the solar power subsystem, the CSF energy storage subsystem, and the supporting subsystems. The solar power subsystem consists of PV arrays. The CSF subsystem consists of the Electrolyzer Subsystem and the Fuel Cell Subsystem with reactant storage tanks. In addition, in order to maintain operation, thermal management and electrical power management subsystems are required for the complete solar CSF power system. The basic system operates as follows (Equitech 2009): When the sun shines, the solar PV arrays not only put out enough DC power to satisfy the daytime loads, but also to supply power to the electrolyzer subsystem which dissociates water and stores the resulting hydrogen and oxygen gases in storage tanks. During the night, when the solar arrays are inactive, the fuel cell subsystem is turned on to supply the required nighttime power loads. Fuel cells consume the stored hydrogen and oxygen and produce electricity, water, and waste heat. The product water is stored and later separated back into its hydrogen and oxygen constituents by the solar-powered electrolyzer during the next daylight portion of the cycle. This solar CSF power system can accommodate a large variety of applications by designing variations to its basic system as well as through integration with other major systems. The thermal management of this system can borrow that of solutions from fuel cell systems as discussed previously.

## Perspective and Future Trends

Advanced thermal management materials and techniques have been applied to chip, PCB and system level designs to meet heat dissipation demands introduced by new generations of ICs that dissipate high current and high power. While chip makers are doing their part to improve thermal management for their devices, EMS providers are placing special attention on thermal management issues at the PCB design level. In these instances, remedies range from using applicable board materials to paying special attention to mounting holes. In between, there are more and more innovative as well as tried-and-proven materials and methods to improve thermal conductivity and heat dissipation (Khan 2008).

## *Electrothermal and Multiphysics Codesign and Software Solutions*

Electronic technology scaling continues in spite of tremendous technology development barriers, design challenges, and prohibitive costs. Miniaturization of electronic devices and circuits has led to the emergence of self-heating (Joule heating due to electron–phonon scattering) as a critical bottleneck to the performance and reliability of emerging microelectronic circuits and systems. Consequently, there is increasing emphasis on the electrothermal and multiphysics codesign of transistors, interconnects, and circuits. Therefore, it is crucial to identify theoretical, experimental, computational, and design bottlenecks for realizing the full power and potential of ICs through electrothermal and multiphysics codesign.

Serious challenges are being faced as technology scaling continues. Design of ICs is facing difficulties in sustaining supply and threshold voltage scaling to provide the required performance increase, limit energy consumption, control power dissipation, and maintain reliability. Increased leakage power and greater variation highlight the need for electrothermal codesign of electronic systems. Various process and design techniques are currently used to provide answers to the challenges faced in IC design, particularly the high power consumption. Clock gating, dual core design, sleep transistors, stack effect, and body bias are among the design techniques used to address power limitations. Carefully selected supply and threshold voltages are critical for operation in the desired design space. Consequently, distributed temperature and leakage sensors should be used to inform dynamic thermal management strategies. Clock throttling, register-file resizing, limiting issue width of processor, dynamic voltage scaling, clock gating, and decommissioning various sections of the chip are examples of the dynamic thermal solutions that are deployed. Careful placement of design units also helps with heat distribution and removal from the die. Microprocessors delivered MIPS (Million Instructions Per Second) and performance in the 1980s by pipelined architectures and in the 1990s by instruction-level parallelism. Thread-level parallelism (multithreading) and chip level multiprocessing (multicores) are the current approaches of choice. The modeling multiphysics, from the device to the circuit to the architecture level, of electrothermal transport in emerging microelectronics are important, which requires coupled electrothermal models incorporating both continuum Fourier and submicrometer effects. Heat transfer physics at the single device level, especially the occurrence of localized power density and temperature "hotspots," have been explore with Monte Carlo simulations of heat generation in quasiballistic devices, and device thermal resistances have been measured and reveal a power-law scaling of device thermal resistance with 1/L (device dimension such as source-to-drain channel length for a transistor) and indicate the possibility of electrothermal codesign from the bottom-up, i.e., starting at the device and materials level.

Electrothermal codesign in the development of a "hotspot" tool requires exploration of the design space for electrothermal optimization with the crucial aspect being modeling of both spatial and temporal variations in the electrical and thermal systems and their interactions. It is a future direction to do electrothermal codesign at the system level, including various active cooling and refrigeration systems to manage operating temperature of the system as well as cost and reliability issues. In addition, codesign in layout and power supply management is also needed. To do this the following critical issues should be addressed (Garimella et al. 2008):

1. *Dynamic thermal management strategies*. To deal with severe limitations posed by leakage and variation on circuit and system design, dynamic thermal solutions must be considered and deployed. Distributed temperature sensors are required for such dynamic approaches to thermal management. Temperature and leakage should be monitored in order to couple electrical with thermal parameters post Si design.
2. *Modeling electrical and thermal systems*. In order to perform electrothermal codesign, one needs to be able to simulate the electrical system, the thermal system, and their interactions. Modeling such systems is a crucial aspect of such codesign, where one needs to capture both desired spatial and temporal variations.
3. *Multiphysics modeling*. Strategies for controlling on-chip temperature, such as dynamic voltage and frequency scaling and migrating computation, require models for the prediction of temperature across a range of scales from the device to the circuit to the architecture level. At the larger scales, Fourier conduction models are adequate, while at the device and gate scales, coupled electrothermal models incorporating submicrometer effects (such as ballistic, boundary scattering, and confinement effects) are necessary. One representative solution is the development of compact models from detailed gate-level phonon transport models for successively coarser levels of abstraction that can be used to predict thermal runaway and develop low-power design strategies.
4. *Cost/performance issues*. The cost of cooling should be carefully analyzed as measured by $/W of cooling by considering various factors such as power density, scaling that provides lower material cost, efficiency that provides lower cooling operating cost, and reliability that lowers business operating cost.
5. *Lower component junction temperatures to conserve power and improve efficiency*. Devices operating at lower temperatures are more efficient. The cost of the power consumed can be greater than the cost of the equipment in examples such as servers. Overall system power consumption can be reduced by running devices cooler, but this comes at the cost of expending more energy on cooling. The balance point between device power and power for cooling needs to be determined.

Software tools for thermal design of electronics systems are indispensable for the development of cost-efficient, high-performance, and reliable cooling technologies for next-generation electronics devices, and particularly needed for the multiscale and multiphysics aspects of the electronics system design. A multidisciplinary evaluation of electronics systems, especially at the early design phase, is often required. In addition, multiscale modeling software is needed in electronics design ranging from circuit to package to system level. Current status and challenges to cope with thermal issues in the design of electronics systems include (Garimella et al. 2008):

1. The IC level thermal design can strongly interact with functionality, reliability, and leakage power of the electronics. As an example, functionality problems at the IC level were highlighted, in which case, the critical issues include both the large temperature peaks and the local temperature gradients. While most of the electronic heating is due to power dissipation at the silicon junction regions, increasing attention needs to be paid to self-heating effects of the billions of metal lines embedded in low-k dielectrics in order to control electromigration. As such, fine-grain, three-dimensional thermal analyses at the IC level and the reorganizations of the IC lay out can serve to enhance functionality and to avoid serious defects. Detailed IC level simulations can be accurately performed at present, and the interfaces to the IC design tools should be automated and efficient, in terms of IC lay-out geometry, material properties and power dissipation.
2. At the package, PCB, system, rack, and room levels, thermal design must be performed as an integral part of the mechanical and electrical design processes to develop cost-effective solutions that meet all physical, electrical and thermal design requirements. As such, a major challenge for present software tool development is to set up a tight, bidirectional integration between electronic design automation, MCAD, and thermal design environments, or even better, to fully embed analysis technology in the design environment. Thus, a fast and spontaneous thermal design validation throughout the electrical and mechanical design processes can be achieved, enabling the design to be driven by thermal requirements and vice versa. Further, inevitable overhead and potential for error associated with conventional data transfer methodologies is eliminated. To this end, intelligent automation for grid generation, solver settings, results interpretation and reliable compact model approaches are needed to shift the focus from analysis to design and to provide CFD for designers instead of requiring CFD specialists.
3. Electromagnetic compatibility solutions as well as thermomechanical modeling are expected to be increasingly used in the early design phase.
4. Computational cost is a constant concern for software developers, leading to automated model simplification on the one hand and faster computational procedures on the other.
5. Design optimization and sensitivity analyses should be considered for future development. Two modes of the use of CFD-based modeling tools are available. Modern day CFD tools can often be readily used to analyze experimentally observed phenomena. However, CFD does not always provide the answer in a straightforward manner due to the fact that input data are not always rigorously known and would in turn require other calculations, such as an

electrical simulation. Multiphysics simulation might bring a solution to this kind of uncertainties. Furthermore, inherent multiscale approaches needed to achieve simulation results within acceptable time frames might use subgrid scale models beyond their applicability. Progress in coupled analysis software development and in subgrid model development in general commercial CFD is therefore hampered by the fact that both coupling and modeling issues are very much physics-dependent. On the other hand, CFD tools can be used to formulate optimal design solutions. Much progress has been made and is still ongoing toward geometrical optimization: CAD-mesh generator interfaces, Cartesian meshes, hexahedral meshes, overlapping meshes, solution adapted meshes, etc. However, at present, grid generation still requires human intervention to examine overlaps, gaps, fineness, skewness, high aspect ratio, etc. With respect to physical parameter optimization, CFD tools need better postprocessor (post-analysis) tools such as batch processing, design of experiments, in-built multi-variable objective function optimization tools, and error estimators. Although implementation of these tools is easy to achieve, at present little or no automation is provided in CFD. The use of hierarchical models was highlighted that could be extended in order to fully exploit the presence of repetitive entities in package or system-level simulations. Furthermore, the optimization of new cooling strategies based on electroactuated flows, two-phase microchannels, spray cooling or nanofluids will require further developments in software tools with respect to multiscale and multiphysics approaches. Both at the package and system levels, challenges exist in the combined use of hierarchical models with multiscale approaches and to search for optimal modeling strategies in order minimize computational cost. Finally, systematic optimization tools based on gradient based optimization in combination with an adjoint approach as used in aerodynamic applications as well as the exploration of genetic algorithms for CFD seem to provide valuable perspectives for the development of software tools toward automatic design optimization.

Multiphysics modeling, i.e., the thermal-electrical-magnetic coupling for design would be needed in the future. The present procedures for compact models are really handshaking methods introducing minimal communication in between different levels of modeling accuracy. Furthermore, most software packages which have dedicated tools for electronics cooling include libraries for parameter characterization of contact resistance models and for power dissipation estimates. However, a fully integrated software tool is still missing and progress in the coming years is expected to happen incrementally. The perceived gaps and showstoppers toward progress in software tools and future directions include (Garimella et al. 2008):

1. Complex coupled physics with multiple length and time scales. A numerical method used for one type of physical problem may not be well suited for others. For example, finite volume methods work best for fluid flow applications where conservation equations are solved, while finite-element methods are more suited to the nonconservative structural mechanics equations. Thus, the complexity in the physics requires matching of various numerical paradigms, which is a challenge.

2. The length and time-scale spectra in many problems are still unsolvable using modern computing within realistic timeframes. Examples include direct simulation of turbulence and porous media. There is a greater need for so-called compact or subgrid scale models so that solution turnaround is faster.
3. Although the need for easy-to-use CFD tools for standard design problems has existed for several years, progress in this area has been rather incremental. However, improvements are expected in the near future due to merging of software vendors.
4. Model and parameter characterization, as well as sensitivity analysis, should be further incorporated into software tools to guide the designer toward optimal designs.
5. The modeling of innovative cooling strategies involving multiphysics and optimal numerical implementation for design optimization require further research. Furthermore, the accuracy of multiscale modeling and the optimal use of hierarchical models need more attention.
6. The use of multiscale and multiphysics modeling needs to be further elaborated for specific innovative cooling techniques.
7. Generic approaches toward optimization procedures in design tools with CFD should be further investigated for their applicability in electronics systems.
8. Further work toward an integrated software tool for electronics systems is needed. This should be further enriched by control and automated tools to support the nonexpert user and to reduce design time.
9. Both multiscale procedures and optimization tools should be further investigated. Systematic study is needed to determine the best solution strategy for specific applications. Systematic development of compact models for specific applications is needed.

Thermal modeling is used at the most basic levels of board layout, mapping, for investigating system airflow, heat sink design and other cooling mechanisms. For instance, with this tool, PCB designers can extend computer-aided design into prototyping and testing, thus saving considerable time and expense. Designers can also build a virtual prototype of the system and test the airflow and heat distribution at both the board and the system level. Equally importantly, thermal modeling gives the PCB designer the critical tool for conducting thermal fatigue failure analysis. In turn, these analyses can be modeled to provide failure prediction models. While board failures may not occur for a period of time, prediction models can forecast when certain PCB materials will incur thermal fatigue and cause field failures (Khan 2008).

## *Progress and Future Trends of High Heat Flux Thermal Management*

The future need is to create a data and technology base of flexible, controllable, and miniature hybrid active–passive and scalable cooling approaches that can apply across the entire range of future electronic products, and equally important, to overlay these approaches with pervasive sensing and control. This will create an energy-aware ecosystem; the management of energy as a key resource will be a requirement from an economic and sustainability standpoint for the high-tech ecosystem. This variability in heat generation must be utilized to enable balanced chip performance based on the most efficient provisioning of cooling resources. Flexibility must be devised at all levels, e.g., the cooling solution for low power devices must scale from passive to active in multiple steps if not in an analog manner. With these underlying flexibilities in heat generation and heat removal, one can overlay a low-cost sensing network and create a control system that can modulate the cooling resources and work in conjunction with a power scheduling mechanism to optimize energy usage. For example, flexibility can be implemented as separate policies that emphasize either sustainability or high performance. Invoking the sustainability policy will enable devices to work in a low-power mode with the cooling system operating at low power or in a passive manner. In contrast, invoking the performance policy will enable the cooling system to work at full power while providing high performance at the device level. The control policy will scale the cooling solution for modes in-between (Garimella et al. 2008).

The "right-hand turn" has been used in microprocessor power dissipation driven by the thermal bottleneck to Moore's Law, while calling for a holistic approach to optimization. For example, gaming systems have emerged as the new high heat flux battlefront because of cost and form factor considerations. Improvements in heat spreaders, especially two-phase spreaders, were cited as a very useful strategy. It is needed to effectively manage the localized microprocessor hotspots by mitigating the transient thermal gradients on a chip very quickly via rapid response thermal management strategies. Advancements are sought in reduction of spreading resistance using two-phase spreading, reduced interface resistance through the development of new TIMs, and improvements in heat sink resistance using single-phase, and even two-phase, microchannels. For instance, microchannel heat sinks that can interrupt the flow in a microchannel make maximum use of the high heat transfer coefficients in the entry region. Improvements in critical heat flux and microchannel condenser would be needed. Figure 12.6 illustrates futuristic cooling solution utilizing nano and micro technologies.

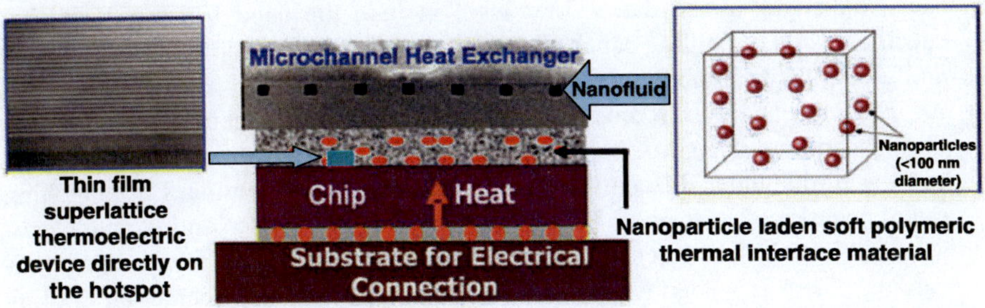


**Fig. 12.6** Illustration of futuristic cooling solution utilizing nano and micro technologies

In general, thermal management options for high heat flux applications along with the current challenges have been identified as follows (Garimella et al. 2008):

1. *Thermoelectric cooling*. As a promising technique for hot-spot mitigation, there is a need to improve *ZT* values from the current value of 1 to around 2.5. Improvements will most likely come from reductions in lattice thermal conductivity by scattering via melting/powder metallurgy techniques (filled skutterudites and clathrates), or by pursuing low-dimensional material such as quantum wells, superlattices, and quantum wires. Accurate property measurement is another ongoing challenge.
2. *Nanomaterials*. Improved TIMs using CNTs and improvements in nanofluids have shown promise. There is a need to improve property measurement/prediction capability. Agglomeration issues need to be addressed through functionalized nanoparticles coated with atomic-scale layers of hydrophobic materials.
3. *Microchannels*. As this technique is seeing greater industrial acceptance, there is a need for improved capability in both experimentation and analytical modeling. Specific issues include measurement of experimental heat loss, better understanding of inlet and exit plenum effects, better understanding of the conditions for the transition between macrochannel transport and microchannel transport, control of flow instabilities in phase-change schemes, improved prediction of critical heat flux, and development of better heat transfer fluids.
4. *Liquid cooling*. Liquid cooling systems of different forms, from spray cooling to heat pipe spreaders and thermosyphons, have been implemented in applications as varied as gaming systems and military electronics. Particular attention must be paid to containment and fluid compatibility issues especially in the context of overall cooling system cost, as well as to the importance of judicious fluid fill ratios in heat pipes and thermosyphons especially at high heat fluxes. Phase-change immersion cooling, while appropriate for prior-generation supercomputers, has not yet been embraced by system designers because of a lack of fully understanding of the boiling process, the role of enhanced surfaces, incipience hysteresis issues, and low critical heat flux values for dielectric fluids.
5. *Compression refrigeration cooling*. This staple of server thermal management systems is now being contemplated for mobile computing systems and desktop applications through the use of microrefrigeration systems. Some challenges include condensation issues, the need for extremely high reliability, and battery life. Challenges at the data center level include the need for improved site-specific design of cooling capability and the need for dynamic matching of local and global thermal parameters.
6. *Air cooling*. There is room for continued performance gains in air cooling through optimal design of heat sinks and air flows, and for a delineation of air cooling limits under different sets of constraints. Work continues in branching radial fins, skived heat sinks, ionic prime-movers, and active cooling with piezo-actuated jets. Jet impingement cooling continues to be viable for several applications. There is a need to pursue hybrid solutions that incorporate air cooling with heat pipes and thermoelectrics.

7. *Improvements in circuit physics and architecture*. All the techniques listed above go hand-in-hand with improvements in circuit physics and architecture such as strained silicon, multicore processors, stacked die leading to performance gains obtained by reducing RC (Resistive Capactive) delay, three-dimensional architecture, increased on-chip memory, and die-thinning.

As a result, microscale cooling systems are promising to meet the ever increasing cooling requirements due to advances in electronics and optical devices. Future research and development should focus on a good balance achieved between fundamental studies, analytical models, and verifiable experimentation for all components of the microscale system design. Multidisciplinary skills and collaboration are required for true integration of the cooling system into electronic packages and devices, and reducing the time between experimentation and implementation of cooling approaches. On the other hand, as chip makers move to the 65-nm node, several architectural innovations such as multicore processing are being applied to manage thermal issues and to positively impact power density by reducing the intensity of individual hotspots and spreading them over the chip area. However, with increased performance, chips will continue to be challenged with one or more hotspots, which have to be managed for both reliability and performance. Hotspot thermal management issues are expected to get potentially more severe when there is a move to the 45-nm technology node and beyond, due to increased transistor densities. Such thermal problems are also relevant to wireless, cellular, and mobile communication devices that rely on advanced algorithms for voice, video and error-free data operations. Component- and system-level heat flux densities will continue increasing as computers and other digital devices add increased functionality taking advantage of silicon process and architecture enhancements. Cooling strategies to manage these increasing fluxes will require a detailed understanding of the underlying physical limits of heat transfer mechanisms so that cooling capabilities can be maximized (Garimella et al. 2008).

Future solution strategies for thermal management of high heat fluxes encountered in microelectronics applications should focus on: methods of forecasting cooling demand, including the scaling of heat flux as a function of Moore's law, architecture and application environments and systems; identification of solutions and strategies to manage high heat flux. A number of cooling strategies, which employ enhanced air cooling, liquid cooling and solid state refrigeration, need to be cognizant of the technical, economic and environmental factors for approaches to be successful in large volume microelectronics applications. In addition, there are opportunities for radical innovation through nanotechnology-enabled concepts. Some of the significant advances made in development of different cooling solution strategies can be summarized as (Garimella et al. 2008):

1. Cooling solutions using microchannels have demonstrated the ability to handle W/mm at the package level; Higher critical heat flux values in microchannel flows might be attained by boosting the mass flux at the expense of increased pumping power, optimizing coolant pressure using a mixture of coolants, and optimizing the fin structure.

2. 6–9°C active hotspot cooling plus additional passive cooling have been demonstrated while pumping fluxes W/cm on the active side of the silicon chip in packaged devices using solid-state refrigeration; spot cooling accomplished with a vaporizing coolant has been developed.
3. Using CNT-based TIM materials, effective thermal conductivities in the range of 80 W/m K, i.e., at least one order of magnitude better than commercial grease and phase-change materials have been demonstrated.
4. Radical alternatives to air- and liquid-cooling approaches must consider whether phonons can be made coherent through focusing so that the heat built up is converted to photons and transmitted via an "optical chimney."

While considerable and impressive work has been done, significant opportunities exist for continued work on enhancing heat transfer through engineering of materials and interfaces. There are as-yet unrealized opportunities to increase heat transfer rates at solid–liquid interfaces. Novel nanoscale mechanisms need to be understood better and then engineered for efficient long-term solutions. There is considerable scope for improvement of frequency performance of high-performance electronics through site specific thermal management, i.e., through a general active–passive combination cooling strategy at the chip and package level. Such solutions are of interest to meet aggressive system-level energy-efficient cooling needs, as in large data centers. Performance gaps between the intrinsic capabilities and their in-system condition seen in a variety of technologies (e.g., TIM materials, thin-film thermoelectrics) must be understood and reduced if they are to be used for future thermal enhancements (Garimella et al. 2008).

## *Thermal Challenges for Next-Generation Military, Automotive, and Harsh-Environment Electronic Systems*

Significant electronics cooling issues exist for military, space, and automotive applications, which require high-reliability cooling solutions for periods of 10–20 years in severe environments. For example, future in situ NASA missions involve extended operation in extreme environments of −230 to 486°C temperature, 0–100 bar pressure, and greater than 5 Mrad radiation exposure. Without exception, NASA's environments are outside the catalog range of commercial and military electronic manufacturers. Currently, there is no design methodology for the semiconductor technologies of choice for these missions, such as complementary metal oxide semiconductors (CMOS), silicon on insulator (SOI) CMOS, and SiGe, as they will be exposed to a combination of radiation and cold temperatures. Cooling needs for military electronic systems require the electronic products to be able to survive temperature extremes of −55 to 125°C under vibration, shock, humidity, contamination, and acceleration conditions, and in any orientation. Self-contained cooling technologies are key to next-generation products, where common systems must be able to interface with a variety of application platforms that may use different overall system cooling schemes, such as fuel-cooling for high-altitude platforms, liquid cooling for mid-altitude platforms, and air cooling for low-altitude platforms. Inability to manage the heat dissipated by electronic components not only limits the potential power and performance of current military electronic systems, but constrains the packaging density and form factors for next-generation electronic systems. In addition, hermetic high-performance military electronic packaging is also required. The unique challenges in the thermal management of next-generation automotive electronic systems include: (1) higher semiconductor power density (up to 300 W/cm); (2) higher circuit board current density (300 A) and operating voltages of up to 700 V required by new high-power automotive applications (such as hybrid and electric vehicles); (3) frequency operation of 2.4 GHz and higher required by new wireless applications (such as satellite radio); (4) low-cost optical, IR (Infrared) or, microwave (up to 77 GHz) components and assemblies required by safety- and security-related applications; (5) high reliability required in the harsh automotive environment; and (6) very low-cost, high-volume, and small size and mass demanded by competition in the automotive electronics marketplace. Additionally, lower flow volume, high-performance liquid cooling technologies are called for in cooling applications in hybrid and electric vehicles in which the drive motor controllers switch hundreds of amps at hundreds of volts using power transistors and generate several kilowatts of waste heat. Adoption of novel cooling technologies and design redundancy to ensure reliability of military, space, and automotive electronics are affected by cost issues. Despite its energy efficiency, the reuse of waste heat from the engine to cool the electronics will add weight, size, and extra coolant loops and thereby raise reliability concerns in military and spacecraft applications. For automotive applications, cooling the power transistors and smart power devices, rather than the on-vehicle microprocessors, remains the major issue in the thermal design, since the performance of the latter is much lower than that of personal computers. In addition, for the automotive industry, pursuing thermal solutions is more economical than pushing high-temperature electronics (Garimella et al. 2008).

Therefore, future directions for next-generation military, automotive, and harsh-environment electronic systems may focus on following areas (Garimella et al. 2008):

1. Reliability issues. Fluid/vapor containment, dry out, fluid contamination over time, cooling loops for multiple thermal loads.
2. NASA's high-temperature (486°C) electronics needs for missions to Venus can only be supported using innovative solutions.
3. Low-cost, reliable, scaleable liquid cooling. This is primarily for system-level cooling rather than hotspots. Designs for miniature accumulator, miniature pump, and miniature condenser for cooling loop.
4. Improved TIMs and thermal gap fillers, and higher thermal conductivity/low CTE materials for direct die attach that are lower cost and density than currently used. New higher performance thermal interface and attach material is needed.

5. Adaptable/modular thermal management approaches to allow reuse of hardware across different platforms.
6. Low-impact transient thermal management, for example, phase-change cooling with wax provides W h/in.
7. Improved fan technology: high reliability, lower cost, lower noise, smaller size, higher pressure head.
8. Liquid cooling systems should address applications for military, aerospace, automotive, and other harsh environment applications. Technologies developed for commercial, room-temperature applications are not directly applicable to these products. Using water as the cooling liquid is not acceptable for many applications. Liquid cooling technology development needs to include wide-temperature-range dielectric fluids. A number of system-level cooling loop issues need additional investigation.
9. New correlations and concepts are needed for electronic packaging design. These include heat transfer coefficient relations for the two-phase regime to predict heat source temperature for a given heat source load, heat transfer coefficient relations for single-phase regime to predict heat source temperature for a given heat source load, heat transfer coefficient associated with loop's condenser, pressure drop determination for impingement jet/spray cooling module and pressure drop determination for loop's condenser, means of charging coolant into cooling loop, concepts for dampening or elimination of potential two-phase loop flow instabilities, and concepts for two-phase loop feedback flow control.

## Summary

Emerging electronic packaging technologies have increasing demand for thermal management materials with new and dramatically improved properties. Achieving optimal performance with advanced thermal management materials may require cost-effective fabrication methods that incorporate novel chemicals, synthetic techniques, and metrology methods. The evaluation and successful integration of emerging materials within targeted applications will depend on optimal design-in and effective control for specific sets of critical material and interface properties. Furthermore, the development of a characterization and modeling infrastructure would support the evolution of robust quantitative structure property correlations. These integrated tools can be used to generate a more foundational understanding of novel thermal conduction mechanisms, contact resistance formation, and transport mechanisms and drive new synthetic methods and cooling device concepts. Similarly, the fabrication of nanostructured materials, such as nanotubes or nanowires, for cooling devices or thermal interface interconnects would benefit from an improved understanding of the impact of process and structure on the resulting electronic, mechanical, thermal, and interface properties. In addition, modeling and simulation, aligned with validation experiments, will be needed to optimize the chemical and structural attributes that improve critical material properties and performance in targeted applications. As a result, the build-up of materials development routine and methodology would be necessary, which may cover establishing application target, materials requirements and materials selection with optimal balance of cost and performance; design-in methodology of thermal managing components; development of lean processes; prototype fabrication and experimental validation; as well as production layout and quality assurance.

One of the promising materials would be smart materials that respond to environmental stimuli, such as temperature, moisture, pH, or electric and magnetic fields. Smart materials can be used directly to make smart thermal management systems or structures or embedded in conventional thermal management structures whose inherent properties can be changed to meet high value-added performance needs. Smart materials technology is relatively new to the thermal management of electronic packaging, but it has a strong innovative content that can truly create new thermal management solutions in chip level, component level, or system level, especially when combine with emerging nanomaterials. As more and more powerful electronic components are used in electronic devices with increased packaging densities, the combination of thermal management and EMI shielding and absorbing has become a development trend in electronic packaging and will provide an optimal option for producing more reliable electronic products.

Thermal design for a computer system can no longer be treated in isolation. Power and performance tradeoffs and smart circuit-design techniques are required to conserve power consumption. Materials and process improvements in packaging and thermal management technology are required to minimize thermal resistance and increase efficiency of heat dissipation. Careful management of the thermal design space from the chip to the system level is therefore critical to ensure a viable solution space for succeeding generations of processors in the computer system.

Similar to thermal management of electronic packaging, the photonic or optoelectronic industry has a similar daunting task: junction temperature management with demonstrated life test data, with improving performance, and increasing reliability. Like thermal management of conventional electronic packaging, the thermal management of LEDs can also divided into passive and active thermal management systems. Passive thermal management systems have no moving parts or consumption of additional energy. They rely primarily on conduction and radiation to remove heat from the junction. The typical method is to attach LEDs to a thermally conductive substrate, such as a metal-core IMS substrate or ceramic substrate, and then attach the substrate to a heat sink. Novel technologies such as Anotherm (anodized aluminum oxide thermal substrate) make it possible to attach the LEDs directly to the heat sink. Heat is conducted to the heat sink and radiated from its surface. Thermal performance is enhanced by reducing the length and thermal resistances along the path to the heat sink and by increasing the surface area of the heat sink. Active thermal management systems involve convection by incorporating fans, heat pipes, and liquid cooling. These technologies enable significantly better thermal management and should be considered for ultrahot applications. In most cases, they are more complex and require better design to avoid decreasing the reliability of the system. These trade-offs are manageable if extreme thermal management is required.

Thermal management applications in sustainable energy generation, such as batteries, fuel cells, and solar cell packaging have been discussed. A thermal management system may use air for heat/cooling/ventilation, liquid for cooling/heating, insulation, thermal storage such as phase-change materials, or a combination of these methods. The thermal management system may be passive or active. Advanced thermal management materials and techniques have been applied to entire system designs to meet heat dissipation demands.

Electronic technology scaling continues in spite of tremendous technology development barriers, design challenges, and prohibitive costs. Miniaturization of electronic devices and circuits has led to the emergence of self heating as a critical bottleneck to the performance and reliability of emerging microelectronic circuits and systems. Consequently, there is increasing emphasis on the electrothermal and multiphysics codesign of transistors, interconnects, and circuits. Therefore, it is crucial to identify theoretical, experimental, computational, and design bottlenecks for realizing the full power and potential of ICs through electrothermal, and multiphysics codesign. The future needs is to create a data and technology base of flexible, controllable and miniature hybrid active–passive and scalable cooling approaches that can apply across the entire range of future electronic products, and equally important, to overlay these approaches with pervasive sensing and control. With these underlying flexibilities in heat generation and heat removal, one can overlay a low-cost sensing network and create a control system that can modulate the cooling resources and work in conjunction with a power scheduling mechanism to optimize energy usage.

Lower flow volume, high-performance liquid cooling technologies are called for in cooling applications in hybrid and electric vehicles in which the drive motor controllers switch hundreds of amps at hundreds of volts using power transistors and generate several kilowatts of waste heat. Adoption of novel cooling technologies and design redundancy to ensure reliability of military, space and automotive electronics are affected by cost issues. Despite its energy efficiency, the reuse of waste heat from the engine to cool the electronics will add weight, size, and extra coolant loops and thereby raise reliability concerns in military and spacecraft applications. For automotive applications, cooling the power transistors and smart power devices, rather than the on-vehicle microprocessors, remains the major issue in the thermal design, since the performance of the latter is much lower than that of personal computers. In addition, for the automotive industry, pursuing thermal solutions is more economical than pushing high-temperature electronics.

# References


Arik M, Weaver S (2004) Chip scale thermal management of high brightness LED packages. http://lib.semi.ac.cn:8080/tsh/dzzy/wsqk/SPIE/vol5530/5530–214.pdf. Accessed on 06 June 2010.

Armstrong MK (2004) Advanced PCB design and layout for EMC – Part 2: Segregation and interface suppression. EMC & Compliance Journal **3**: 32–42. http://www.compliance-club.com/keith_armstrong.asp. Accessed on 07 July 2010.

Biber C (2008) LED light emission as a function of thermal conditions. http://www.biberthermal.com/Reference_Links/Publications_List/LED_Light_Emission_protected.doc. Accessed on 11 July 2010.

Burke KA (2008) Advanced fuel cell system thermal management for NASA exploration missions. Sixth International Energy Conversion Engineering Conference and Exhibit (IECEC) sponsored by the American Institute of Aeronautics and Astronautics, Cleveland, Ohio, July 28–30, 2008.

Burns LD, McCormick JB, Barroni-Bird CE (2002) Vehicles of change. Scientific American 287(4): 64–73.

Chen X et al (2002) A thermally re-mendable cross-linked polymeric material. Science **295**: 1698–1702.

Christodolou L, Venables JD (2003) Multifunctional material systems: The first generation. JOM – The Member Journal of the Minerals, Metals & Materials Society 55(12): 39–45.

DOD (2010) Improved thermal management for high power and /or small form factor (SFF) tactical radios. http://www.dodsbir.net/Sitis/archives_display_topic.asp?Bookmark=28080. Accessed on 03 July 2010.

Dutta I et al (2004) Development of a novel adaptive lead-free solder containing reinforcements displaying the shape-memory effect. Journal of Electronic Materials **33**: 258–270.

English G (2007) Combined board level EMI shielding and thermal management. US Patent 7262369.

Equitech et al (2009) A comprehensive national energy policy based on integrated and advanced renewable energy systems and citizen cooperative land ownership. http://www.cesj.org/homestead/strategies/community/Equitech-ExecutiveSummary.pdf. Accessed on 12 July 2010.

Garbincius PH (2006) RDR cost estimating instructions & standards. http://www-ilcdcb.fnal.gov/RDR_Cost_Estimating_Instructions_23may06.pdf. Accessed on 05 July 2010.

Garimella SV et al (2008) Thermal challenges in next-generation electronic systems. IEEE Transactions on Components and Packaging Technologies 31(4): 801–815.

Hannafin J (2002) A novel approach to thermal management and EMI shielding via a metallic conformal coating on a plastic housing. Proceedings of the IMAPS, Telecom Hardware Solutions Conference, May 2002. http://vendor.parker.com/Groups/Seal/Divisions/Chomerics/Chomerics%20Product%20Library.nsf/05eb61a92fc220c78525695800744d42/5102fa16a41dd41d85256bc1004c61d2/$FILE/Article_ECOPLATE.pdf. Accessed on 08 July 2010.

Horng RH, Horng RH et al (1997) Thermal management design from chip to package for high power InGaN/sapphire LEDs applications. http://www.electrochem.org/meetings/scheduler/abstracts/214/2252.pdf. Accessed on 09 July 2010.

ITRS (2009) International technology roadmap for semiconductors – Modeling and simulation. http://public.itrs.net/Links/2009ITRS/2009Chapters_2009Tables/2009_Modeling.pdf. Accessed on 06 July 2010.

ITRS (2009a) International technology roadmap for semiconductors – Design. http://www.itrs.net/Links/2009ITRS/2009Chapters_2009Tables/2009_Design.pdf Accessed 06 July 2010.

ITRS (2009b) International technology roadmap for semiconductors – Factory integration. http://public.itrs.net/Links/2009ITRS/2009Chapters_2009Tables/2009_Factory.pdf. Accessed on 06 July 2010.

Johnson RN (2009) Thermally conductive EMI shield. US patent 7,608,326.

Khan Z (2008) New materials and techniques tackle PCB thermal management. http://rtcmagazine.com/magazine/articles/view/100944/pg: 1 Accessed on 03 July 2010.

Lunaraccents (2001) LED thermal management. http://www.lunaraccents.com/educational-LED-thermal-management.html. Accessed on 09 July 2010.

Luxeon (2006) Thermal design using LUXEON power light sources App Brief AB05 (6/06). http://www.philipslumileds.com/pdfs/AB05.pdf Accessed on 04 July 2010.

Matic P (2003) Overview of multifunctional materials. In Smart Structures and Materials 2003: Active Materials: Behaviors and Mechanics, D.C. Lagoudas, ed. Proceedings of SPIE Volume 5053, pp. 61–69.

Momoda LA (2005) The future of engineering materials: Multifunction for performance-tailored structures. http://www.nap.edu/openbook.php?record_id=11220&page=47 Accessed on 05 July 2010.

Najarian M, Garnett E (2006) Thermoelectrics and photovoltaics: Integration challenges and benefits. http://kammen.berkeley.edu/C226/5r.pdf Accessed on 06 July 2010.

Optek (2006) Thermal management of visible LEDs. http://www.optekinc.com/pdf/App%20Bulletin%20228.pdf. Accessed on 12 July 2010.

Pan Y-F (2002) Integrated thermal design and optimization study for active integrated power electronic modules (IPEMs). Master thesis. Virginia Polytechnic Institute and State University, Blacksburg.

Pesaran AA (2001) Battery thermal management in EVs and HEVs issues and solutions. Advanced Automotive Battery Conference, Las Vegas, Nevada, Feb. 6–8, 2001.

Pesaran AA, Burch S, Keyser M (1999) An approach for designing thermal management systems for electric and hybrid vehicle battery packs. Fourth Vehicle Thermal Management Systems Conference and Exhibition, London, UK, May 24–27, 1999.

Quarshie R (2005) Smart materials and related structures. http://www.berr.gov.uk/files/file18875.pdf. Accessed on 06 July 2010.

Saffa R (2007) Thermal management considerations for visible LEDs in general illumination and signage applications. http://www.ledjournal.com/images/Archived%20Articles/Thermal%20Management%20Considerations%20for%20Visible%20LEDs%20in%20General%20Illumination%20and%20Signage%20Applications.pdf. Accessed on 08 July 2010.

Saums D et al (2005) Implementation of diamond pins as thermal vias for high heat flux spreading in CTE-compatible lids and baseplates for semiconductor packaging. IMAPS Advanced Technology Workshop on Thermal Management 2005. http://www.sp3inc.com/pdf/dia_pins.pdf. Accessed on 01 July 2010.

Satoh Y (1986) Challenge to the Quality Revolution. NEC Corporation, Tokyo.

Sigmund O, Torquato S (1999) Design of smart composite materials using topology optimization. Smart Materials and Structures 8(9): 365–379.

Standford (2000) NLC general cost estimate guidance. http://www-project.slac.stanford.edu/lc/local/notes/dr/cost-contin.pdf. Accessed on 05 July 2010.

Tong XC (2009) Advanced Materials and Design for Electromagnetic Shielding. CRC Press, Boca Raton, USA.

Tsao JY (2002) Light emitting diodes (LEDs) for general illumination. http://apps1.eere.energy.gov/buildings/publications/pdfs/ssl/report_led_november_2002a_1.pdf. Accessed on 12 July 2010.

Viswanath R, Wakharkar V, Watwe A, Lebonheur V (2000) Thermal performance challenges from silicon to systems. Intel Technologies Journal **Q3**: 2000. http://www.intel.com/technology/itj/q32000/pdf/thermal_perf.pdf. Accessed on 03 June 2010.

Wang DG, Knighten JL, Muller PK (2002) An integrated vent, heatsink and EMI shield. IEEE 18th Annual Symposium on Semiconductor Thermal Measurement and Management, 12–14 March, 2002, pp. 125–131.

Wax SG, Fisher GM, Sands RR (2003) The past, present and future of DARPA's investment strategy in smart materials. JOM – The Member Journal of the Minerals, Metals & Materials Society 55(11): 17–23.

Welwyn (2010) The use of Anotherm in power LED lighting applications. http://www.hg-electronics.de/produktpartner/tt-electronics/downloads/Anotherm_HLfLLED.pdf. Accessed on 12 July 2010.

Young R et al (2006) Developments and trends in thermal management technologies – A mission to the USA. http://www.lboro.ac.uk/research/iemrc/documents/. . ./CB2007.pdf. Accessed on 26 February 2010.

# Appendix: Standards and Specifications for Evaluation of Thermal Management in Electronic Industry

## A. Standards for Thermal Analysis

| | |
|---|---|
| JEDEC Standard JESD51 | Methodology for the thermal measurement of component packages |
| JEDEC Standard JESD51-1 | Integrated circuit thermal measurement method–electrical test method (single semiconductor device) |
| JEDEC Standard JESD51-2 | Integrated circuits thermal test method environment conditions–natural convection (still air) |
| JEDEC Standard JESD51-3 | Low effective thermal conductivity test board for leaded surface mount packages |
| JEDEC Standard JESD51-4 | Thermal test chip guideline (wire bond type chip) contents |
| JEDEC Standard JESD51-5 | Extension of thermal test board standards for packages with direct thermal attachment mechanisms |
| JEDEC Standard JESD51-6 | Integrated circuit thermal test method environmental conditions–forced convection (moving air) |
| JEDEC Standard JESD51-7 | High effective thermal conductivity test board for leaded surface mount packages |
| JEDEC Standard JESD51-8 | Integrated circuit thermal test method environmental conditions–junction-to-board |
| JEDEC Standard JESD51-9 | Test boards for area array surface mount package thermal measurements |
| JEDEC Standard JESD51-10 | Test boards for through-hole perimeter leaded package thermal measurements |
| JEDEC Standard JESD51-11 | Test boards for through-hole area array leaded package thermal measurements |
| ISO 22007-1:2009 | Plastics – determination of thermal conductivity and thermal diffusivity–part 1: general principles |
| ISO 22007-2:2008 | Plastics–determination of thermal conductivity and thermal diffusivity – part 2: transient plane heat source (hot disc) method |

X.C. Tong, *Advanced Materials for Thermal Management of Electronic Packaging*, Springer Series in Advanced Microelectronics 30,
DOI 10.1007/978-1-4419-7759-5, © Springer Science+Business Media, LLC 2011

| | |
|---|---|
| ISO 22007-3:2008 | Plastics–determination of thermal conductivity and thermal diffusivity – part 3: temperature wave analysis method |
| ASTM C177-04 | Standard test method for steady-state heat flux measurements and thermal transmission properties by means of the guarded-hot-plate apparatus |
| ASTM C518-10 | Standard test method for steady-state thermal transmission properties by means of the heat flow meter apparatus |
| ASTM C1470-06 | Standard guide for testing the thermal properties of advanced ceramics |
| ASTM D5334-08 | Standard test method for determination of thermal conductivity of soil and soft rock by thermal needle probe procedure |
| ASTM D5470-06 | Standard test method for thermal transmission properties of thermally conductive electrical insulation materials |
| ASTM D5930-01 | Standard test method for thermal conductivity of plastics by means of a transient line-source technique |
| ASTM D2717-95 | Standard test method for thermal conductivity of liquids |
| ASTM E793-01 | Standard test method for enthalpies of fusion and crystallization by differential scanning calorimetry |
| ASTM E794-01 | Standard test method for melting and crystallization temperatures by thermal analysis |
| ASTM E831-00 | Standard test method for linear thermal expansion of solid materials by thermomechanical analysis |
| ASTM E928-01 | Standard test method for determination of purity by differential scanning calorimetry |
| ASTM E967-97 | Standard practice for temperature calibration of differential scanning calorimeters and differential thermal analyzers |
| ASTM E968-02 | Standard practice for heat flow calibration of differential scanning calorimeters |
| ASTM E1225-04 | Standard test method for thermal conductivity of solids by means of the guarded-comparative-longitudinal heat flow technique |
| ASTM E1131-98 | Standard test method for compositional analysis by thermogravimetry |
| ASTM E1269-01 | Standard test method for determining specific heat capacity by differential scanning calorimetry |
| ASTM E1356-98 | Standard test method for assignment of the glass transition temperatures by differential scanning calorimetry or differential thermal analysis |

| | |
|---|---|
| ASTM E1363-97e1 | Standard test method for temperature calibration of thermomechanical analyzers |
| ASTM E1545-00 | Standard test method for assignment of the glass transition temperature by thermomechanical analysis |
| ASTM E1582-00 | Standard practice for calibration of temperature scale for thermogravimetry |
| ASTM E1640-99 | Standard test method for assignment of the glass transition temperature by dynamic mechanical analysis |
| ASTM E1641-99 | Standard test method for decomposition kinetics by thermogravimetry |
| ASTM E1782-98 | Standard test method for determining vapor pressure by thermal analysis |
| ASTM E1824-02 | Standard test method for assignment of a glass transition temperature using thermomechanical analysis under tension |
| ASTM E1858-00 | Standard test method for determining oxidation induction time of hydrocarbons by differential scanning calorimetry |
| ASTM E1860-02 | Standard test method for elapsed time calibration thermal analyzers |
| ASTM E1867-01 | Standard test method for temperature calibration of dynamic mechanical analyzers |
| ASTM E1868-02 | Standard test method for loss-on-drying by thermogravimetry |
| ASTM E1877-00 | Standard practice for calculating thermal endurance of materials from thermogravimetric decomposition data |
| ASTM E1952-01 | Standard test method for thermal conductivity and thermal diffusivity by modulated temperature differential scanning calorimetry |
| ASTM E1953-02 | Standard practice for description of thermal analysis apparatus |
| ASTM E1970-01 | Standard practice for statistical treatment of thermoanalytical data |
| ASTM E2008-99 | Standard test method for volatility rate by thermogravimetry |
| ASTM E2009-99 | Standard test method for oxidation onset temperature of hydrocarbons by differential scanning calorimetry |
| ASTM E2038-99 | Standard test method for temperature calibration of dielectric analyzers |
| ASTM E2039-99 | Standard practice for determining and reporting dynamic dielectric properties |

| | |
|---|---|
| ASTM E2040-99 | Standard test method for mass scale calibration of thermogravimetric analyzers |
| ASTM E2041-01 | Standard method for estimating kinetic parameters by differential scanning calorimeter using the Borchardt and Daniels method |
| ASTM E2069-00 | Standard test method for temperature calibration on cooling of differential scanning calorimeters |
| ASTM E2070-00 | Standard test method for kinetic parameters by differential scanning calorimetry using isothermal methods |
| ASTM E2071-00 | Standard practice for calculating heat of vaporization or sublimation from vapor pressure data |
| ASTM E2092-00 | Standard test method for distortion temperature in three-point bending by thermomechanical analysis |
| ASTM E2113-00 | Standard test method for length change calibration of thermomechanical analyzers |
| ASTM E2160-01 | Standard test method for heat of reaction of thermally reactive materials by differential scanning calorimetry |
| ASTM E2206-02 | Standard method for force calibration of thermomechanical analyzers |
| ASTM E1952 | Standard test method for specific heat capacity and thermal conductivity by modulated temperature differential scanning calorimetry |
| IEEE Standard 98-2002 | Standard for the preparation of test procedures for the thermal evaluation of solid electrical insulating materials |

## B. Standards for Life and Reliability Evaluation

| | |
|---|---|
| JESD 47 | Stress-test–driven qualification of integrated circuits |
| JESD 22-A104-B | Temperature cycling |
| JESD 22A-101-B | Steady state temperature humidity bias life test |
| JESD 22-A100-B | Cycled temperature humidity bias life test |
| JESD 22-A110-B | Highly accelerated temperature and humidity stress testing |
| JESD 22-A102-C | Accelerated moisture resistance-unbiased autoclave |
| JESD 22-A103-B | High temperature storage life |
| JESD 22-A105-C | Power and temperature cycling |
| JESD 22-A106-A | Thermal shock |
| J-STD-020A | Moisture/reflow sensitivity classification for nonhermetic solid state surface mount devices |

| | |
|---|---|
| IEC 60300-1 (2003-06) | Dependability management–part 1 dependability program management |
| IEC 60300-2 (2004-03) | Dependability management–part 2 dependability program elements and tasks |
| IEC 60300-3-1 (2003-01) | Dependability management–part 3-1 application guide–analysis techniques for dependability–guide on methodology |
| IEC 60300-3-2 (2004-11-10) | Dependability management–part 3 application guide–section 2: collection of dependability data from the field |
| IEC 60300-3-3 (2005-08-29) | Dependability management–part 3 application guide–section 3: life cycle costing |
| IEC 60300-3-4 (2007-09) | Dependability management–part 3–4 guide to the specification of dependability requirements |
| IEC 60300-3-5 (2001-03) | Dependability management–part 3–5 application guide–reliability test conditions and statistical test principles |
| IEC 60300-3-6 (1997-11) | Dependability management–part 3 application guide–section 6: software aspects of dependability |
| IEC 60300-3-7 (1999-05) | Dependability management–part 3–7 application guide–reliability stress screening of electronic hardware |
| IEC 60300-3-9 (1995-12) | Dependability management–part 3 application guide–section 9: risk analysis of technological systems |
| IEC 60300-3-10 (2001-01) | Dependability management–part 3–10 application guide–maintainability |
| IEC 60300-3-11 (1999-03) | Dependability management–part 3–11 application guide–reliability centered maintenance |
| IEC 60300-3-12 (2001-12) | Dependability management–part 3–12 application guide–integrated logistic support |
| IEC 60300-3-3 (2005-08-29) | Dependability management–part 3: application guide–section 3: life cycle costing |
| IEC 60300-3-4 (2007-09) | Dependability management–part 3–4: guide to the specification of dependability requirements |
| IEC 60319 (1999-09) | Presentation and specification of reliability data for electronic components |
| IEC 61709 (1992-10) | Electronic components–reliability–reference conditions for failure rates and stress models for conversion |
| IEC 61160 (1992-09) | Formal design review |
| IEC 60812 (1985-07) | Analysis techniques for system reliability–procedure for failure mode and effects analysis |
| IEC 61025 (1990-10) | Fault tree analysis |
| IEC 61078 (1991-11) | Analysis techniques for dependability–reliability block diagram method |

| | |
|---|---|
| IEC 61165 (1995-01) | Application of Markov techniques |
| IEC 61882 (2001-05) | Hazard and operability studies–application guide |
| IEC 61649:2008 | Weibull analysis, edition 2.0 |
| IEC 61164 (2004) | Reliability growth–statistical test and estimation methods |
| IEC 61710 (2000-11) | Power law model–goodness-of-fit test and estimation methods |
| IEC 60605-2 (1994-10) | Equipment reliability testing–part 2: design of test cycles |
| IEC 60605-3-1 (1986-09) | Equipment reliability testing–part 3: preferred test conditions. Indoor portable equipment-low degree of simulation |
| IEC 60605-3-2 (1986-09) | Equipment reliability testing–part 3: preferred test conditions. Equipment for stationary use in weather protected locations–high degree of simulation |
| IEC 60605-3-3 (1992-11) | Equipment reliability testing–part 3: preferred test conditions–section 3: test cycle 3: equipment for stationary use in partially weather protected locations–low degree of simulation |
| IEC 60605-3-4 (1992-07) | Equipment reliability testing–part 3: preferred test condition–section 4: test cycle 4: equipment for portable and no-stationary use–low degree of simulation |
| IEC 60605-3-5 (1996-03) | Equipment reliability testing–part 3: preferred test condition–section 5: test cycle 4: ground mobile equipment–low degree of simulation |
| IEC 60605-3-6 (1996-08) | Equipment reliability testing–part 3: preferred test condition–section 6: test cycle 6: outdoor transportable equipment–low degree of simulation |
| IEC 60605-4 (2001-08) | Equipment reliability testing–part 4: statistical procedures for exponential distribution–point estimates, confidence intervals, prediction intervals and tolerance intervals |
| IEC 60605-6 (2007-05) | Equipment reliability testing–part 6: tests for the validity of the constant failure rate or constant failure intensity assumptions |
| IEC 61070 (1991-11) | Compliance test procedures for steady-state availability |
| IEC 61123 (1991-12) | Reliability testing–compliance test plans for success ratio |
| IEC 61124 (2006-03-15) | Reliability testing–compliance tests for constant failure rate and constant failure intensity |
| IEC 61649 (1997-05) | Goodness-of-fit tests, confidence intervals and lower confidence limits for Weibull distribution data |

| | |
|---|---|
| IEC 61650 (1997-08) | Reliability data analysis techniques–procedures for comparison of two constant failure rates and two constant failure (event) intensities |
| IEC 60410 (1973-01) | Sampling plans and procedures for inspection by attributes |
| IEC 61163-1 Ed. 2.0 (2006-06-26) | Reliability stress screening–part 1: repairable items manufactured in lots |
| IEC 61163-2 (1998-11) | Reliability stress screening–part 2: electronic components |
| IEC 61163-3 | Reliability stress screening–part 3: repairable single items |
| IEC 61713 (2000-06) | Software dependability through the software life-cycle process–application guide |
| IEC 61720 | Guide to techniques and tools for achieving confidence in software |
| IEC 60706-1 (1982-01) | Guide on maintainability of equipment–part 1: section one, two and three. Introduction, requirements and maintainability program |
| IEC 60706-2 (2006-03-20) | Maintainability of equipment–part 2: maintainability requirements and studies during the design and development phase |
| IEC 60706-3 (2006-04-27) | Maintainability of equipment–part 3–verification and collection, analysis and presentation of data |
| IEC 60706-4 (1992-09) | Guide on maintainability of equipment–part 4–section 8: maintenance and maintenance support planning |
| IEC 60706-5 (2007-09) | Guide on maintainability of equipment. Part 5–Testability and diagnostic testing |
| IEC 60706-6 (1994-12) | Guide on maintainability of equipment. Part 6–section 9: statistical methods in maintainability evaluation |
| IEC 62198 (2001-04) | Project risk management–application guidelines |
| IEC 60319 (1990-09) | Presentation and specification of reliability data for electronic components |
| IEC 60050-191 (1990-12) | International electrotechnical vocabulary, chapter 191: dependability and quality of service |
| IEC 61703 (2001-09) | Mathematical expressions for reliability, availability, maintainability and maintenance support terms |
| MIL-HDBK-189A | Reliability growth management |
| MIL-HDBK-217F | Reliability prediction of electronic equipment–notice F |
| MIL-HDBK-251 | Reliability/design thermal applications |
| MIL-HDBK-338B | Electronic reliability design handbook – revision B |
| MIL-HDBK-344A | Environmental stress screening of electronic equipment–revision A |

| | |
|---|---|
| MIL-HDBK-470A | Designing and developing maintainable products and systems – revision A |
| MIL-STD-471A | Maintainability verification/demonstration/evaluation–revision A |
| MIL-HDBK-472 | Maintainability prediction |
| MIL-STD-690D | Failure rate sampling plans and procedures–revision D |
| MIL-STD-721C | Definition of terms for reliability and maintainability–revision C |
| MIL-STD-756B | Reliability modeling and prediction–revision B |
| MIL-HDBK-781A | Handbook for reliability test methods, plans, and environments for engineering, development, qualification, and production–revision A |
| MIL-STD-781D | Reliability testing for engineering development, qualification and production–revision D |
| MIL-STD-785B | Reliability program for systems and equipment development and production–revision B |
| MIL-STD-790F | Established reliability and high reliability qualified products list systems for electrical, electronic, and fiber optic parts specifications–revision F |
| MIL-STD-882C | System safety program requirements |
| MIL-STD-882D | System safety |
| MIL-STD-1543B | Reliability program requirements for space and launch vehicles–revision B |
| MIL-STD-1629A | Procedures for performing a failure mode effects and criticality analysis–revision A |
| MIL-STD-2074 | Failure classification for reliability testing |
| MIL-STD-2155 | Failure reporting, analysis and corrective action systems |
| MIL-STD-2164 | Environmental stress screening process for electronic equipment |
| MIL-HDBK-2164A | Environmental stress screening process for electronic equipment–revision A |
| MIL-STD-2173 | Reliability centered maintenance requirements for naval aircraft, weapons systems and support equipment |
| MIL-P-24534A | Planned maintenance system: development of maintenance requirement cards, maintenance index pages, and associated documentation |
| MIL-HDBK-H 108 | Sampling procedures and tables for life and reliability testing (based on exponential distribution) |
| MIL-HDBK-189 | Reliability growth management |
| MIL-HDBK-263A | Electrostatic discharge control handbook for protection of electrical and electronic parts, assemblies and equipment (excluding electrically initiated explosive devices) |

| | |
|---|---|
| MIL-STD-690C | Failure rate sampling plans and procedures |
| MIL-STD-756B | Reliability modeling and prediction |
| MIL-HDBK-781 | Reliability test methods, plans and environments for engineering development, qualification and production |
| MIL-STD-781D | Reliability design qualification and production acceptance tests: exponential/distribution |
| MIL-STD-785B | Reliability program for systems and equipment, development and production |
| MIL-STD-790E | Reliability assurance program for electronic parts specifications |
| MIL-STD-2074 | Failure classification for reliability testing |
| MIL-STD-2155 | Failure reporting, analysis and corrective action system |

## C. Standards for Flammability and Toxicity Testing

| | |
|---|---|
| ASTM C542 | Flame propagation |
| ASTM C1166 | Flame propagation |
| ASTM D56 | Flash point by tag closed tester method |
| ASTM D93 | Close cup flash point |
| ASTM D229 | Method 1–burning rate, Method 2–flame resistance |
| ASTM D635 | Burning rate in vertical position |
| ASTM D1929 | Self and flash ignition temperatures |
| ASTM D2584 | Ignition loss of cured reinforced resins |
| ASTM D2843 | Smoke density from the burning or decomposition of plastics |
| ASTM D2859 | Flammability of finished textile floor covering materials (methenamine pill) |
| ASTM D2863 | Limiting oxygen index |
| ASTM D3014 | Flame height, burning time, loss of weight for plastics |
| ASTM D3065 | Flammability of aerosol products |
| ASTM D3675 | Surface flammability of flexible cellular materials |
| ASTM D3801 | Comparative extinguishing characteristics of solid plastics in a vertical position |
| ASTM D3874 | Ignition of materials by hot wire sources |
| ASTM D4804 | Flammability characteristics of non-rigid solid plastics (Methods A and B) |
| ASTM E84 | Surface flammability characteristics of building materials |
| ASTM E136 | Behavior of materials in a vertical tube furnace |
| ASTM E162 | Flame spread using a radiant panel |

| | |
|---|---|
| ASTM E603 | Full scale room burn test |
| ASTM E648 | Critical heat flux of floor coverings using a radiant panel |
| ASTM E659 | Auto ignition temperature of liquid chemicals |
| ASTM E662 | Optical smoke density |
| ASTM E681 | Flammability (explosion) limits of chemicals ASTM E800 HCN, hydrogen cyanide measurement using the colorimetric method |
| ASTM E906 | Heat and visible smoke release rates |
| ASTM E1317 | LIFT–lateral flame spread |
| ASTM E1321 | Material ignition and flame spread properties |
| ASTM E1354 | Cone calorimeter |
| ASTM E2058 (FM 4910) Screening Test | Screening test for clean room test |
| ASTM F814 | Smoke density for aerospace applications |
| AITM 2.0005 (7.1.6) | 60° flammability |
| AITM 2.0008 (7.3.3) | Optical smoke density |
| AITM 3.0005 (7.4) | Toxicity |
| ATS 1000.001 | Flammability, smoke, toxicity, and heat release |
| BELLCORE GR-63-CORE | Telecommunications equipment fire resistance |
| BIFMA X5.7 | Furniture flammability |
| BMS 13-48 | Smoke and toxicity |
| BSS 7230 | Flammability testing of aircraft materials to FAR Part 25, Appx F, Part 1 |
| BSS 7322 | Determination of heat release using OSU |
| BSS 7324 | 60° flammability, smoke, and toxicity |
| BSS 7238 | Optical smoke density |
| BSS 7239 | Toxic gas generation |
| CAL 106 | Resistance of a mattress or mattress pad to combustion from smoldering cigarettes |
| CAL 116 | Cigarette test of upholstered furniture |
| CAL 117 | Sect A, part I–flame retardance of resilient filling materials used in upholstered furniture |
| CAL 117 | Sect D, part II–cigarette smoldering screening test for bedding |
| CAL 121 | Flammability test procedure for mattresses in high risk occupancies |
| CAL 129 | Flammability test procedure for mattresses for use in public buildings |
| CAL 133 | Flammability test procedure for seating furniture for use in public occupancies |
| CFR vol. 16, 1610 | Std. for the flammability of clothing textiles (45° angle test) |
| CFR vol. 16, 1632.4 | Mattress flammability |

| | |
|---|---|
| CPAI 84 | Tent fabric flammability |
| DIN 4102, Part 1, Class A1 | Ignition (only) |
| DOT 173.338-18(b) (7) | Flammability using a red hot platinum wire |
| FAA/FAR 25.853 | Flammability tests |
| FAA/FAR 25.855 | Flammability tests |
| FED STD 191A, Method 5903.1 | Flammability of cloth in a vertical direction |
| FED STD 191A, Method 5903.2 | Flammability of cloth in a vertical direction |
| FMVSS302 | Automotive interior flammability |
| Halogen content | Determination of the halogen content level |
| IEC 331 | Fire characteristics of electrical cables |
| IEC 332-1 | Single wire/cable flame propagation |
| IEC 332-3 | Wire/cable bunch flame propagation |
| IEC 695-2-2 | Fire hazard testing; needle flame test |
| IEC 754-1 | Evolved combustion gases of wire/cable |
| IEC 754-2 | Acid gas |
| IEC 1034 (1,2) | Smoke density of wire/cable |
| IEEE 383 | Cable tray flame spread |
| IEEE 45 | Cable tray flame spread |
| IMO A.652(16) | Upholstered furniture flammability test |
| IMO A.653(16) | LIFT–flammability of marine surface finishes |
| ISO 4589-2 | Determination of burning behavior by oxygen index |
| ISO 5660 | Cone calorimeter |
| ISO 9705 | Room burn facility and modified (1/2 scale and 1/4 scale) room burn |
| MIL 2031 | Fire/toxicity tests for composites used in submarines |
| MIL C-24643 | Acid gas |
| Mobil 45 deg | Flammability test calculating weight loss for rigid urethane foams |
| MSC.41(64) | Smoke and toxicity products of combustion |
| NES 711 | Smoke |
| NES 713 | Toxicity |
| NES 715 | Temperature Index |
| NFPA 225 | Surface flammability characteristics of building materials |
| NFPA 258 | Optical smoke density |
| NFPA 264A | Cone calorimeter |
| NFPA 701 | Flame resistant textiles and fibers |
| NFPA 1971 | Flame resistance of cloth in a vertical direction (Fed. Std. 191A, Method 5903.1) |
| SMP 800-C | Combustion toxicity |

| | |
|---|---|
| UL 94 V Series | Vertical flammability |
| UL 94 HF Series | Horizontal flammability |
| UL 214 | Flame propagation of fabrics and films |
| UL 723 | Surface flammability characteristics of building materials |
| UBC 8-2 | Full scale room burn test |